Taylor's Manual of
PHYSICAL EVALUATION AND TREATMENT

*The Practitioner's Guide
to Joint, Nerve and
Soft Tissue Management*

Lyn Paul Taylor, MA, PT

Volume I

Slack Incorporated, 6900 Grove Road, Thorofare, NJ 08086

SLACK International Book Distributors

In Europe, the Middle East and Africa:
 John Wiley & Sons Limited
 Baffins Lane
 Chichester, West Sussex P019 1UD
 England

In Canada:
 McGraw-Hill Ryerson Limited
 330 Progress Avenue
 Scarborough, Ontario
 M1P 2Z5

In Australia and New Zealand:
 MacLennan & Petty Pty Limited
 P.O. Box 425
 Artarmon, N.S.W. 2064
 Australia

In Japan:
 Igaku-Shoin, Ltd.
 Tokyo International P.O. Box 5063
 1-28-36 Hongo, Bunkyo-Ku
 Tokyo 113
 Japan

In Asia and India:
 PG Publishing Pte Limited.
 36 West Coast Road, #02-02
 Singapore 0512

Foreign Translation Agent
 John Scott & Company
 International Publishers' Agency
 417-A Pickering Road
 Phoenixville, PA 19460
 Fax: 215-988-0185

Editor: Cheryl D. Willoughby
Managing Editor: Lynn C. Borders
Designer: Susan Hermansen
Production Manager: David Murphy
Publisher: Harry C. Benson

Copyright © 1990 by SLACK Incorporated

All rights reserved. No part of this book may be reproduced, stored in a retrieval system or transmitted in any form or by any means, electronic, mechanical, photocopying, recording or otherwise, without written permission from the publisher, except for brief quotations embodied in critical articles and reviews.

Printed in the United States of America

Library of Congress Catalog Card Number: 88-043547

ISBN 1-55642-114-1
Volume I Separately ISBN 1-55642-096-X
Volume II Separately ISBN 1-55642-113-3

Published by: SLACK Incorporated
 6900 Grove Road
 Thorofare, NJ 08086

Last digit is print number: 10 9 8 7 6 5 4 3 2 1

This work is dedicated to a group of professionals who have been instrumental in its production. As a group they have provided the basic knowledge, inspiration, emotional support and some of the motivational push necessary for its creation. This group includes:

Signe Brunnstrom

Margaret Bryce

Charlene Centorbi-vomDorp

Helen A. Desmond

Valerie V. Hunt

Margaret Knott

Leanore Krusell

Dee M. Lilly-Musada

Cynthia L. Loewenstein-Taylor

Roxey Morris

Margaret S. Rood

Claire W. Sherman

Amy Steinitz-Gross

Janet G. Travell

Helen C. Ziler

Contents

VOLUME I

List of: Syndromes	x
Treatable Causes	xi
Modalities	xii
Tables and Reference Documents	xv
Diagrams	xvi
Foreword	xxi
Preface	xxiii
Introduction	xxv
Achilles Tendonitis	1
Acne Syndrome	2
Adhesion Syndrome	3
Adverse Effects of Bed Rest	5
Ambulation With Assistive Devises	8
Antagonistic Muscles of Common Interest, Table of	12
Auricular Acupoint Treatment Protocols, Table of	15
Auricular Electroacustimulation	30
Pain Relief	34
Pathology Control	34
Auricular Acupoint Survey	40
Auricular Landmark & Acupoint Descriptions, Table of	42
Auxiliary Evaluation Techniques	57
Auscultatory Method of Measuring Blood Pressure	57
Spirometry	59
Auscultation	62
Bacterial Infection	65
Bell's Palsy Syndrome	70
Blood Pressure Control With Physiological Feedback Training	71
Blood Sugar Irregularities	77
Break-Through Bleeding Syndrome	80
Buerger's Disease Syndrome	81
Burn Syndrome	83
Bursitis Syndrome	85
Calcific Deposit	86
Calf Pain Syndrome	89
Capsulitis Syndrome	91
Carpal Tunnel Syndrome	94
Central Nervous System Dysfunction	96
Cerebral Palsy Syndrome	99
Cervical Dorsal Outlet Syndrome	101
Cervical (Neck) Pain Syndrome	105
Chest Pain Syndrome	107
Chondromalacia Syndrome	109
Chronic Bronchitis/Emphysema Syndrome	111
Colitis Syndrome	113
Cryotherapy	115
Suppression of Sympathetic Autonomic Hyperreactivity	121
Hypertonic Neuromuscular Management	125
Analgesia/Anesthesia	129
Soft Tissue Swelling Control	133
Soft Tissue Inflammation Control	136
Spasticity Control	140
Vasoconstriction	143
Evaluation of Capillary Response to Cold	147
Dental Pain	149
Developmental Reflexes, Table of	151
Diabetes Syndrome	152
Differential Skin Resistance Survey	154
Dupytren's Contracture Syndrome	157
Earache Syndrome	158
Elbow Pain Syndrome	160
Electrical Facilitation of Endorphin Production (Acusleep)	163
Electrical Stimulation (ES)	166
Muscle Lengthening	173
Muscle Reeducation/Muscle Toning	177
Increasing Blood Flow	181

contents

Reduction of Edema	185
Delaying the Atrophy of Denervated Muscle	188
Inhibition of Ionized Formation	191
Inflammation Control	195
Electrical Evaluation	198
Electrodermal Potential Monitry	201
Electroencephalometry	203
Mediation Facilitation	206
Hypertension Control	209
Electroencephalometric Evaluation	212
Electromyometric Feedback	217
Post Cerebral Accident Neuromuscular Reeducation	217
Post Peripheral Nerve Injury Neuromuscular Reeducation	222
Post Spinal Cord Injury Neuromuscular Reeducation	226
Neuromuscular Reeducation of Cerebral Palsy Conditions	231
Neuromuscular Reeducation of Neck and Shoulder Musculature	233
Neuromuscular Reeducation of Temporomandibular Musculature	238
Neuromuscular Reeducation of Low Back Musculature	244
Neuromuscular Reeducation of Torticollis	248
Post Immobilization Neuromuscular Reeducation	252
Breathing Pattern Correction	255
Muscle Toning	258
Electromyometric Evaluation	261
Determination of neuromuscular imbalance	262
Determination of neuromuscular hypertonicity	264
Determination of neuromuscular hypotonicity	266
Determination of neuromuscular dyscoordination	266
Determination of muscular innervation	270
Electrode Placements - Diagrams	271
Exercise	294
Muscle Strengthening	300
Improving Lung Function	306
Muscle Relengthening	310
Cardiovascular Conditioning	314
Gait Training	318
Muscle Stretching	322
Directed Functional Exercise	324
Guided Patterning	326
Temple Fay Approach	326
Brunnstrom Approach	328
Bobath Approach	330
Proprioceptive Neuromuscular Facilitation Approach	334
Rood Approach	335
Functional Testing	342
Evaluation of Joint Range of Motion	342
Evaluation of Muscle Strength	349
Evaluation of Functional Motor Skills	362
Evaluation of Ambulation Skills	366
Evaluation of Balance Skills	368
Evaluation of Developmental Reflexes	371
Pathological Neuromuscular Synergistic Pattern Evaluation	372
Exercises - Energy Consumption Rates, Table of	376
Exercises for Abdominal Muscle Toning, Table of	377
Exercises for Facial Muscles, Table of	378
Exercises for the Lower Extremities - Isometrics, Table of	379
Exercises for the Neck - Isometrics, Table of	381
Exercises for the Shoulder & Upper Extremities - Isometrics, Table of	382
Exercises - Isotonic (Weight Lifting Protocols), Tables of	383
Extrafusal Muscle Spasm	385
Facet Syndrome	388
Facial Muscle Function, Table of	390
Fasciitis Syndrome	391
Foot Pain Syndrome	392
Forearm Pain Syndrome	394
Frozen Shoulder Syndrome	396
Galvanic Skin Response Monitry	
General Relaxation	402
Desensitization	404
GSR assessment	407
Gel-foam Implant	409
Habitual Joint Positioning	410
Hand/Finger Pain Syndrome	414
Headache Pain Syndrome	417
Hiccup	419

contents

High Thoracic Back Pain Syndrome	421
Hydrotherapy	423
Wound (and Burn) Debridement	429
Circulation Enhancement	432
Muscle Tension Amelioration	436
Range of Motion Enhancement and Physical Reconditioning	439
Suppression of Sympathetic Hyperactivity	443
Hydrotherapeutic Tests	446
Gibbon-Landis Procedure	446
Cold Pressor	449
Test for Hypersensitivity to Cold	451
Hyperhidrosis	453
Hypermobile/Instable Joint	455
Hypertension	457
Hypertension Syndrome	460
Hypotension	462
Hysteria/Anxiety Reaction Syndrome	464
Insomnia	466
Insomnia Syndrome	468
Interspinous Ligamentous Strain	469
Iontophoresis	472
Pain Relief	476
Inflammation Amelioration	479
Edema Control	482
Ischemia Relief	485
Facilitation of Open Lesion Healing	487
Muscle Spasm Relief	489
Hypertonicity Relief	491
Gouty Trophi Dissolution	493
Sclerolytic (Scar Tissue) Formation Moderation	495
Calcific Deposit Absorption	498
Hyperhidrosis Control	500
Fungal InfectionEradication	502
Irregular Menses	504
Joint Ankylosis	506
Joint Approximation	509
Joint Mobilization	512
Joint Sprain Syndrome	516
Keloid Formation Syndrome	517
Kinesiology of Ambulation	518
Knee Pain Syndrome	520
Laser Stimulation	522
Ligamentous Strain	525
Localized Viral Infection	528
Low Back Pain Syndrome	531
Lower Abdominal Pain Syndrome	533
Lung Fluid Accumulation	535

VOLUME II

Massage	537
Muscle Fatigue (and/or Tension) Relief	541
Circulation Enhancement and Reduction of Edema	544
Acupressure	546
Reduction of Trigger Point Formations	549
Muscle Atrophy Prevention	552
Deep Soft Tissue Mobilization	554
Palpation Evaluation	558
Determination of trigger point formations	558
Determination of extrafusal muscle spasm or hypertonicity	560
Determination of soft tissue tenderness	561
Determination of fibroid tissue, scar tissue or calcific formation	564
Determination of soft tissue inflammation	566
Determination of soft tissue swelling (pitting edema)	568
Menorrhalgia	570
Menstrual Cramping Syndrome	572
Migraine (vascular headache)	574
Migraine (vascular) Headache Pain Syndrome	577
Muscle Spindle	580
Muscular Weakness	584
Myositis Ossificans Syndrome	586
Nausea/Vomiting	588
Necrotic Soft Tissue	590
Nerve Root Compression	592
Neurogenic Dermal Disorders (Psoriasis, Hives)	596
Neuroma Formation	598
Neuromuscular Paralysis Syndrome	600

contents

Neuromuscular Synergistic Patterns, Table of	602
Neuromuscular System	603
Neuromuscular Tonic Imbalance	609
Osteoarthritis Syndrome	613
Pathological Neuromuscular Dyscoordination	615
Pathological Neuromuscular Hypertonus	619
Pathological Neuromuscular Hypotonus	622
Peripheral Acupoint Treatment Protocols, Table of	625
Peripheral Acupoints and Their Locations, Table of	633
Peripheral Electrostimulation	645
Acupoint Electrical Stimulation Within an Inflamed Zone	649
Analgesia	651
Pathology Control	654
Peripheral Acupoint Survey	657
Peripheral Nerve Injury	660
Peripheral Nerve Root Dermatome Innervation, Table of	664
Phlebitis	666
Phlebitis Syndrome	668
Phobic Reaction Syndrome	69
Piriformis Syndrome	671
Pitting (Lymph) Edema Syndrome	674
Plantar Wart Syndrome	676
Pneumonia Syndrome	677
Positive Pressure	679
Edema Reduction	681
Phlebitis and/or Edema Control	684
Wound Healing Facilitation	686
Post Cerebral Vascular Accident (CVA) Syndrome	688
Post Immobilization Syndrome	691
Post Peripheral Nerve Injury Syndrome	683
Post Spinal Cord Injury Syndrome	695
Post Whiplash Syndrome	697
Postural Drainage	701
Psychoneurogenic Neuromuscular Hypertonus	703
Radiculitis Syndrome	706
Raynaud's Disease Syndrome	709
Reduced Lung Vital Capacity	710

Referred Pain Patterns of Interspinous Ligamentous Orgin, Table of	712
Referred Pain Patterns of Trigger Point Formation Origin, Table of	719
Referred Pain Patterns of Visceral Organ Origin, Table of	766
Referred Pain Syndrome	771
Reflex Zones on the Hands and Feet - Acupressure Massage, Table of	772
Restricted Joint Range of Motion	782
Scar Tissue Formation	786
Sciatica Syndrome	788
Sensory Stimulation	791
Neuromuscular Spasticity Control	794
Spastic Sphincter Control	796
Provoking Reflex Physiological Responses	797
Desensitization	799
Neurological Evaluation	800
Coordination, gait & equilibrium evaluation	800
Sensory evaluation	803
Reflex evaluation	806
Shingles (Herpes Zoster) Syndrome	808
Shoulder/Hand Syndrome	816
Shoulder Pain Syndrome	814
Sinus Pain Syndrome	817
Sinusitis	819
Soft Tissue Inflammation	821
Soft Tissue Swelling	825
Spastic Sphincter	828
Stretching Exercises for Muscle Lengthening, Table of	830
Superficial Thermal (Heat) Therapy	835
Circulation Enhancement	840
Lower Abdominal Pain Relief	844
Muscle Relaxation (Wet and Dry Sauna)	846
Excretory (Sweat) Facilitation	848
Sympathetic Hyperresponse	856
Tactile Hypersensitivity	852
Taping	854
Taping Techniques	854
Taping Procedures	855
Temperature Monitry	875
General Relaxation	879

Hypertension Control	882
Circulatory Disorder Control	885
Lower Intestinal Dysfunction (Colitis) Control	888
Vascular (Migraine) Headache Control	891
Dermal Temperature Evaluation	894
To determine capillary bed insufficiency	894
To determine the presence of colitis	895
To determine the presence of vascular headache potential	896
Temporomandibular Joint (TMJ) Pain Syndrome	897
Tendonitis Syndrome	900
Tennis Elbow (Lateral Epicondylitis) Syndrome	902
Thigh Pain Syndrome	904
Tic Douloureux Syndrome	906
Tinnitus	908
Tinnitus Syndrome	911
Toothache/Jaw Pain Syndrome	912
Torticollis (Wry Neck) Syndrome	914
Transcutaneous Nerve Stimulation	915
Pain Management	916
Prolonged Stimulation of Acupoints	921
Facilitation of Tissue Repair	922
Trigger Point Formation	927
Ultrahigh Frequency Sound (Ultrasound)	931
Phonophoresis	936
Desensitization	944
Scar Tissue Management	948
Calcium Deposit Management	952
Inflammation Control (Without Phonophoresis)	956
Bacterial Infection Eradication	957
Sinusitis Relief	961
Viral Infection Eradication	964
Open Wound Healing	967
Acupoint Stimulation	970
Gel-foam Dissolution	973
Trigger Point Determination	975
Autonomic Sympathetic Hyperactivity Suppression	978
Unhealed Dermal Lesion	981
Unhealed Dermal Lesion Syndrome	984
Vascular Insufficiency	985
Vertebral Traction	987
Cervical Traction	989
Lumbar Traction	991
Vibration	993
Neuromuscular Management	996
Spasticity Reduction	999
Sphincter Spasticity Control	1001
Postural Drainage Facilitation	1002
Densensitization of Soft Tissue	1003
Circulation Enhancement	1005
Vibration, Its Effect Upon Neuromuscular Tonicity	1006
Visceral Organ Dysfunction	1015
Wrist Pain Syndrome	1018
Bibliography	1020
Index	1035

List of Syndromes

Achilles Tendonitis	1	Knee Pain	520
Acne	2	Low Back Pain	531
Adhesion	3	Lower Abdominal Pain	533
Bacterial Infection	65	Menstrual Cramping	572
Bell's Palsy	70	Migraine (vascular) Headache Pain	577
Break Through Bleeding	80	Myositis Ossificans	586
Buerger's Disease	81	Neuromuscular Paralysis	600
Burn	83	Osteoarthritis	613
Bursitis	85	Phlebitis	668
Calf Pain	89	Phobic Reaction	69
Capsulitis	91	Piriformis	671
Carpal Tunnel	94	Pitting (Lymph) Edema	674
Cerebral Palsy	99	Plantar Wart	676
Cervical Dorsal Outlet	101	Pneumonia	677
Cervical (Neck) Pain	105	Post Cerebral Vascular Accident (CVA)	688
Chest Pain	107	Post Immobilization	691
Chondromalacia	109	Post Peripheral Nerve Injury	683
Chronic Bronchitis/Emphysema	111	Post Spinal Cord Injury	695
Colitis	113	Post Whiplash	697
Diabetes	152	Radiculitis	706
Dupytren's Contracture	157	Raynaud's Disease	709
Earache	158	Referred Pain	771
Elbow Pain	160	Sciatica	788
Facet	388	Shingles (Herpes Zoster)	808
Fasciitis	391	Shoulder/Hand	816
Foot Pain	392	Shoulder Pain	814
Forearm Pain	394	Sinus Pain	817
Frozen Shoulder	396	Temporomandibular Joint (TMJ) Pain	897
Hand/Finger Pain	414	Tendonitis	900
Headache Pain	417	Tennis Elbow (Lateral Epicondylitis)	902
High Thoracic Back Pain	421	Thigh Pain	904
Hypertension	457	Tic Douloureux	906
Hypertension	460	Tinnitus	911
Hysteria/Anxiety Reaction	464	Toothache/Jaw Pain	912
Insomnia	466	Torticollis (Wry Neck)	914
Insomnia	468	Unhealed Dermal Lesion	984
Joint Sprain	516	Wrist Pain	1018
Keloid Formation	517		

List of Treatable Causes

Bacterial Infection	65
Blood Sugar Irregularities	77
Calcific Deposit	86
Dental Pain	149
Extrafusal Muscle Spasm	385
Gel-foam Implant	409
Habitual Joint Positioning	410
Hiccup	419
Hyperhidrosis	453
Hypermobile/Instable Joint	455
Hypertension	457
Hypotension	462
Insomnia	466
Interspinous Ligamentous Strain	469
Irregular Menses	504
Joint Ankylosis	506
Joint Approximation	509
Ligamentous Strain	525
Localized Viral Infection	528
Lung Fluid Accumulation	535
Menorrhalgia	570
Migraine (Vascular Headache)	574
Muscular Weakness	584
Nausea/Vomiting	588
Necrotic Soft Tissue	590
Nerve Root Compression	592
Neurogenic Dermal Disorders (Psoriasis, Hives)	596
Neuroma Formation	598
Neuromuscular Tonic Imbalance	609
Pathological Neuromuscular Dyscoordination	615
Pathological Neuromuscular Hypertonus	619
Pathological Neuromuscular Hypotonus	622
Peripheral Nerve Injury	660
Phlebitis	666
Psychoneurogenic Neuromuscular Hypertonus	703
Reduced Lung Vital Capacity	710
Restricted Joint Range of Motion	872
Scar Tissue Formation	786
Sinusitis	819
Soft Tissue Inflammation	814
Soft Tissue Swelling	825
Spastic Sphincter	828
Sympathetic Hyperresponse	856
Tactile Hypersensitivity	852
Tinnitus	908
Trigger Point Formation	927
Unhealed Dermal Lesion	981
Vascular Insufficiency	985
Visceral Organ Dysfunction	1015

List of Modalities

Auricular Electroacustimulation	30
Auricular Acupoint Survey	40
Pain Relief	34
Pathology Control	37
Auxiliary Evaluation Techniques	57
Auscultation	62
Auscultatory Method of Measuring Blood Pressure	57
Spirometry	59
Cryotherapy	115
Analgesia/Anesthesia	129
Evaluation of Capillary Response to Cold	147
HypertonicNeuromuscular Management	125
Soft Tissue Inflammatio Control	136
Soft Tissue Swelling Control	133
Spasticity Control	140
Suppression of Sympathetic Autonomic Hyperreactivity	121
Vasoconstriction	143
Differential Skin Resistance Survey	154
Electrical Facilitation of Endorphin Production (Acusleep)	163
Electrical Stimulation (ES)	166
Delaying the Atrophy of Denervated Muscle	188
Electrical Evaluation	198
Increasing Blood Flow	181
Inflammation Control	195
Inhibition of Ionized Formation	191
Muscle Lengthening	173
Muscle Reeducation/Muscle Toning	177
Reduction of Edema	185
Electrodermal Potential Monitry	201
Electroencephalometry	203
Electroencephalometric Evaluation	212
Hypertension Control	209
Meditation Facilitation	206
Electromyometric Feedback	217
Breathing Pattern Correction	255
Electrode Placements—Diagrams	271
Electromyometric Evaluation	261
Determination of neuromuscular imbalance	262
Determination of neuromuscular hypertonicity	264
Determination of neuromuscular hypotonicity	266
Determination of neuromuscular dyscoordination	268
Determination of muscular innervation	270
Muscle Toning	258
Neuromuscular Reeducation of Cerebral PalsyConditions	231
Neuromuscular Reeducation of Low Back Musculature	244
Neuromuscular Reeducation of Neck and Shoulder Musculature	233
Neuromuscular Reeducation of Temporomandibular Musculature	238
Neuromuscular Reeducation Torticollis	248
Post Cerebral Accident Neuromuscular Reeducation	217
Post Immobilization Neuromuscular Reeducation	252
Post Peripheral Nerve Injury Neuromuscular Reeducation	222
Post Spinal Cord Injury Neuromuscular Reeducation	226
Exercise	294
Cardiovascular Conditioning	314
Directed Functional Exercise	324
Functional Testing	342
Evaluation of Ambulation Skills	366
Evaluation of Balance Skills	368
Evaluation of Developmental Reflexes	371
Evaluation of Functional Motor Skills	362
Evaluation of Joint Range of Motion	342
Evaluation of Muscle Strength	349
Pathological Neuromuscular Synergistic Pattern Evaluation	372
Gait Training	318
Guided Patterning	326
Bobath Approach	330
Brunnstrom Approach	328
Proprioceptive Neuromuscular Facilitation Approach	334
Rood Approach	335
Temple Fay Approach	326
Improving Lung Function	306
Muscle relengthening	310
Muscle strengthening	300
Muscle stretching	322

list of modalitites

Galvanic Skin Response Monitry	399
Desensitization	404
General relaxation	402
GSR assessment	407
Hydrotherapy	423
Circulation Enhancement	432
Hydrotherapeutic Tests	446
Cold Pressor	449
Gibbon-Landis Procedure	446
Test for Hypersensitivity to Cold	451
Muscle Tension Amelioration	436
Range of Motion Enhancement and Physical Reconditioning	439
Suppression of Sympathetic Hyperactivity	443
Wound (and Burn) Debridement	429
Iontophoresis	472
Calcific Deposit Absorption	498
Edema Control	482
Facilitation of Open Lesion Healing	487
Fungal Infection Eradication	502
Gouty Tophi Dissolution	493
Hyperhidrosis Control	500
Hypertonicity Relief	491
Inflammation Amelioration	479
Ischemia Relief	485
Muscle Spasm Relief	489
Pain Relief	476
Sclerolytic (Scar Tissue) Formation Moderation	495
Joint Mobilization	512
Laser Stimulation	522
Massage	537
Acupressure	546
Circulation Enhancement and Reduction of Edema	544
Deep Soft Tissue Mobilization	554
Muscle Atrophy Prevention	552
Muscle Fatigue(and/or Tension) Relief	541
Palpation Evaluation	558
Determination of trigger point formations	558
Determination of extrafusal muscle spasm or hypertonicity	560
Determination of soft tissue tenderness	561
Determination of fibroid tissue, scar tissue or calcific formation	564
Determination of soft tissue inflammation	566
Determination of soft tissue swelling (pitting edema)	568
Reduction of Trigger Point Formations	549

Peripheral Electroacustimulation	645
Acupoint Electrical Stimulation Within an Inflamed Zone	649
Analgesia	651
Pathology Control	654
Peripheral Acupoint Survey	657
Positive Pressure	679
Edema Reduction	681
Phlebitis and/or Edema Control	684
Wound Healing Facilitation	686
Postural Drainage	701
Sensory Stimulation	791
Desensitization	799
Neurological Evaluation	800
Coordination, gait and equilibrium evaluation	800
Sensory evaluation	803
Reflex evaluation	806
Neuromuscular Spasticity Control	794
Provoking Reflex Physiological Responses	797
Spastic Sphincter Control	796
Superficial Thermal (Heat) Therapy	835
Circulation Enhancement	840
Excretory (Sweat) Facilitation	848
Lower Abdominal Pain Relief	844
Muscle Relaxation (Wet and Dry Sauna)	846
Taping	844
Taping Procedures	855
Taping Techniques	854
Temperature Monitry	875
Circulatory Disorder Control	885
Dermal Temperature Evaluation	894
To determine capillary bed insufficency	894
To determine the presence of colitis	895
To determine the presence of vascular headache potential	896
General Relaxation	879
Hypertension Control	882
Lower Intestinal Dysfunction (Colitis) Control	888
Vascular (Migraine) Headache Control	891
Transcutaneous Nerve Stimulation	915
Facilitation of Tissue Repair	922
Pain Management	916
Prolonged Stimulation of Acupoints	921
Ultrahigh Frequency Sound (Ultrasound)	931
Acupoint Stimulation	970
Autonomic Sympathetic Hyperactivity Suppression	978

list of modalitites

Bacterial Infection Eradication	957	Vertebral Traction	987
Calcium Deposit Management	952	Cervical Traction	988
Desensitization	944	Lumbar Traction	989
Gel-foam Dissolution	973		
Increasing Circulation	940	Vibration	993
Inflammation Control (Without		Circulation Enhancement	1005
Phonophoresis)	956	Desensitization of Soft Tissue	1003
Open Wound Healing	967	Neuromuscular Management	996
Phonophoresis	936	Postural Drainage Facilitation	1002
Scar Tissue Management	948	Spasticity Reduction	999
Sinusitis Relief	961	Sphincter Spasticity Control	1000
Trigger Point Determination	975		
Viral Infection Eradication	964		

List of Tables

Antagonistic Muscles of Common Interest	12-14	Neuromuscular Synergistic Patterns	602
Auricular Acupoint Treatment Protocols	15-27	Peripheral Acupoint Treatment Protocols	625-632
Auricular Landmark & Auricular Acupoint Descriptions	42-56	Peripheral Acupoints and Their Locations	633-644
Developmental Reflexes	151	Peripheral Nerve Root Dermatome Innervation	664-665
Exercises - Energy Consumption Rates	376	Referred Pain Patterns of Interspinous Ligamentous Orgin	712-718
Exercises for Abdominal Muscle Toning	377	Referred Pain Patterns of Trigger Point Formation Orgin	719-788
Exercises for the Facial Muscles	378	Referred Pain Patterns of Visceral Organ Orgin	766-770
Exercises for the Lower Extremities - Isometrics	379-380	Reflex Zones on the Hands and Feet - Acupressure Massage	773-781
Exercises for the Neck - Isometrics	381	Stretching Exercises for Muscle Lengthening	830-834
Exercises for the Shoulder and Upper Extremities-Isometrics	382		
Exercises - Isotonic (Weight Lifting Protocols & Others)	383-384		
Facial Muscle Function	390		

List of Reference Documents

Adverse Effects of Bed Rest	5-7	Kinesiology of Ambulation	518-519
Ambulation with Assistive Devices	8-11	Muscle Spindle	580-583
Blood Pressure Control with Physiological Feedback Training	71-76	Neuromuscular System	603-608
Central Nervous System Dysfunction	96-98	Vibration, Its Effect Upon Neuromuscular Tonicity	1005-1013

List of Diagrams

Auricular Acupoints

Acupoints on Prominent Portions of the Auricle -First View	55
Acupoints on Prominent Portions of the Auricle - Second View	55
Acupoints on Concave Portions of the Auricle	56
Acupoints on the Posterior Portion of the Auricle	56

Auricular Anatomical Landmarks

Anatomical Landmarks of the Posterior Pinna	43
Anterior Landmarks of the Anterior Pinna	42

Blood Pressure

Representative Graph of the Hypernormal Hypertensive Response to Physiological Feedback Training	75
Representative Graph of the Normal Hypertensive Response to Physiological Feedback Training	74
Representative Graph of the Volatile Hypertensive Response to Physiological Feedback Training	74

Electrical Currents

Diagram of Common Electrical Current Wave Forms	167

Electrode Placements

Abdominal muscles	288
Anterior deltoid m.	284
Anterior tibialis m.	279
Back muscles	288
Biceps brachii m.	284
Biceps femoris m.	277
Brachioradialis m.	285
Corrugator supercilii m.	290
Depressor anguli oris & platysma muscles	292
Depressor labii inferior & mentalis muscles	292
Diaphragm m.	288
Finger extensor group (extensor digitorum communis, extensor indicis proprius, & extensor digiti quinti proprius)	284
Finger flexor group (flexor digitorum sublimis & flexor digitorum profundus)	286
Flexor digitorum brevis m.	281
Flexor pollicis brevis m.	287
Flexor pollicis longus m.	287
Frontalis m.	290
Gastrocnemius m.	278
Gluteus maximus m.	275
Gluteus medius m.	274
Gluteus minimus m.	274
Hamstring group (semitendinosus & semimembranosus)	278
Hip abductor group (abductor magnus, abductor brevis, abductor longus, pectineus, & gracilis)	275
Hip external rotator group (obturator externus, obturator internus, gemellus inferior, gemellus superior, & piriformis)	272
Hip flexor group (psoas major & iliacus)	272
Infraspinatus m.	283
Levator scapulae m.	282
Lower trapezius m.	282
Lumbrical m.	287
Masseter & pterygoid muscles	290
Masseter m.	288
Middle deltoid m.	284
Opponens group	287
Orbicularis oculi m.	291
Orbicularis oris m.	291
Pectoralis major m.	286
Peroneal muscles	280
Posterior deltoid m.	283
Posterior tibialis m.	280
Procerus m.	290
Pronator quadratus m.	286
Pterygoid group	289
Quadratus lumborum m.	273
Quadriceps group (vastus intermedius & rectus femoris)	277
Risorius & buccinator muscles	291
Sartorius m.	273
Scaleni muscles	289
Scapular abduction group (rhomboids & middle trapezius)	281
Serratus anterior m.	281
Shoulder extension group (latissimus dorsi & teres major)	282
Soleus m.	279
Sternocleidomastoideus m.	288
Supinator m.	285
Supraspinatus m.	283
Thumb extensor group (extensor pollicis longus & extensor pollicis brevis)	284
Toe extensor group (extensor digitorum longus & extensor hallucis longus)	280
Triceps m.	284
Upper trapezius & scaleni muscles	289

Upper trapezius m.	282
Vastus lateralis m.	276
Vastus medialis m.	276
Wrist extensor group (extensor carpi radialis longus, extensor carpi radialis brevis, & extensor carpi ulnaris)	285
Wrist flexor group (flexor carpi radialis & flexor carpi ulnaris)	285
Zygomaticus major m.	293
Zygomaticus minor & levator anguli oris muscles	291

Neuromuscular Systems

Action Pathways Between Components of the Neuromuscular Cybernetic System	608
Diagram of the Neuromuscular System	606-607
Disrupted Afferent Pathways Between the Cerebellum and the Cerebral Motor Cortex and Basal Ganglia Aggregate	94
Disrupted Muscle Spindle Afferent and Efferent Pathways	97
Disrupted Pathways Between Cerebellum, Cerebral Motor Cortex, and Basal Ganglia Aggregate	98
Muscle Spindle	583

Peripheral Acupoints

Acupoints on the Anterior Trunk	642
Acupoints on the Back - A	642
Acupoints on the Back - B	642
Acupoints on the Head - A	639
Acupoints on the Head - B	639
Acupoints on the Head -C	639
Acupoints on the Head -D	639
Acupoints on the Lower Extremity -A	643
Acupoints on the Lower Extremity -B	643
Acupoints on the Lower Extremity -C	644
Acupoints on the Lower Extremity -D	644
Acupoints on the Neck -A	640
Acupoints on the Neck -B	640
Acupoints on the Upper Extremity -A	640
Acupoints on the Upper Extremity -B	640
Acupoints on the Upper Extremity -C	641
Acupoints on the Upper Extremity -D	641

Peripheral Nerve Root Dermatome Innervation

Cutaneous Innervation of C4, C5, C6, C7, C8, T1 and T2	664
Cutaneous Innervation of Cranial Nerve V (Trigeminal Nerve), C2, and C3	664
Cutaneous Innervation of L2, L3, L4, L5, S1 and S2	665
Cutaneous Innervation of T3, T4, T5, T6, T7, T8, T9, T10, T11, T12, and L1	665

Positive Pressure

Positive Pressure Spiral Line Wrap	682

Referred Pain Patterns of Interspinous Ligamentous Origin

Referred Pain Patterns from Interspinous Ligaments of C5, C7, T3, T6	714
Referred Pain Patterns from Interspinous Ligaments of C6, C8, T5, T7	714
Referred Pain Patterns from Interspinous Ligaments of L3	717
Referred Pain Patterns from Interspinous Ligaments of L4	717
Referred Pain Patterns from Interspinous Ligaments of L5	718
Referred Pain Patterns from Interspinous Ligaments of S1 and S2	718
Referred Pain Patterns from Interspinous Ligaments of T11, L2	716
Referred Pain Patterns from Interspinous Ligaments of T10, T12	716
Referred Pain Patterns from Interspinous Ligaments of T2, T9, L1	715
Referred Pain Patterns from Interspinous Ligaments of T1, T4, T8	715

Referred Pain Patterns of Trigger Point Formation Origin

Abductor digiti quinti	754
Abductor hallucis	764
Abductor longus	761
Abductor pollicis	755
Anconeus	749
Anterior deltoid	746
Anterior digastric	731
Anterior tibialis	763
Biceps brachii (2)	750
Biceps femoris (hamstring)	762
Brachialis (superior & inferior)	750
Brachioradialis	753
Caudal rectus abdominus	760
Cervical multifidus	742
Coracobrachialis	738
Distal medial triceps	749
Dorsal interosseus	755
Dysmenorrhea	765
Extensor carpi radialis brevis	750
Extensor carpi radialis longus	765
Extensor carpi ulnaris	751
Extensor indicus proprius	753
Extensor of the fourth finger	752
External oblique [A]	759
External oblique [B]	759
First dorsal interosseus	755
Flexor carpi radialis	752
Flexor carpi ulnaris	752

Flexor digitorum sublimus-humeral head	754
Flexor digitorum sublimus-radial head	754
Flexor pollicis longus	754
Frontalis	731
Gastrocnemius	763
Gluteus medius	765
Gluteus minimus	761
Iliocostalis lumborum (L1)	757
Iliocostalis thoracis (T6)	758
Iliocostalis thoracis (T11)	758
Infraspinatus	736
Infraspinatus (abnormal)	736
Lateral pterygoid	731
Lateral teres major	737
Lateral triceps	749
Latissimus dorsi (abnormal)	744
Latissimus dorsi	743
Levator scapulae	735
Long toe extensors	763
Longhead triceps	749
Longissimus thoracis (L1)	756
Longissimus thoracis (T10-11)	757
Lower splenius cervicus	739
Lower trapezius [A]	741
Lower trapezius [B]	741
Masseter (deep)	728
Masseter (superficial A)	729
Masseter (superficial B)	729
Masseter (superficial C)	729
McBurney's point	759
Medial teres major	737
Medial triceps (deep)	748
Medial triceps (lateral)	748
Middle finger extensor	751
Middle trapezius [A]	740
Middle trapezius [B]	740
Middle trapezius [C]	740
Multifidus (L2-L3)	757
Multifidus (S1-S2)	757
Multifidus (S4)	756
Multifidus (T4-T5)	756
Orbicularis oris (orbital)	734
Occipitalis	732
Opponens pollicis	755
Pectoralis major (clavicular)	747
Pectoralis major (costal)	746
Pectoralis major (parasternal -2)	747
Pectoralis major (sternal)	747
Pectoralis minor	747
Peroneus longus	764
Platysma	735
Plenius capitus [B]	745
Posterior cervical group	733
Posterior deltoid	746
Posterior digastric	731
Pronator Teres	753
Pyramidalis	759
Rhomboids (2)	748
Scalenus (anterior, medius, and posterior)	735
Scalenus minimus	736
Semispinalis (capitus, cervicus)	732
Serratus anterior	745
Serratus posterior inferior	756
Serratus posterior superior (under the scapula)	744
Short toe extensors	764
Soleus	764
Splenius capitus [A]	734
Sternalis	748
Sternocleidomastoideus (deep)	734
Sternocleidomastoideus (superficial)	734
Subclavius	745
Suboccipital neck extensor (2)	732
Subscapularis	745
Supinator	750
Supraspinatus (muscle)	742
Supraspinatus (tendon)	743
Temporalis (posterior supra-auricular)	730
Temporalis (anterior)	729
Temporalis (middle A)	730
Temporalis (middle B)	730
Teres minor	738
Upper rectus abdominus	760
Upper trapezius [A]	733
Upper trapezius [B]	739
Vastus medialis	762
Zygomaticus (major)	735

Referred Pain Patterns of Visceral Organ Origin

A. Bladder Fundus, B. Uterine Fundus	768
A. Bladder Neck, B. Kidney Pelvis	769
A. Duodenum, B. Upper Colon, C. Testis	768
A. Gallbladder, B. Fallopian Tube, Ovary	769
Appendix	768
Ascending Aortic Arch (aneurysm)	770
Larynx	767
A. Lower Esophagus, Gastric Cardia, Jejunumm, Ileum, Duodenal Cap, B. Lower Colon	768
Myocarrdium	770
A. Prostate Gland, B. Pancreas	769
Seminal Vesicles, Uterine Cervis, or Upper Vagina	769
Upper Esophagus, Trachea, or Bronchi	767
Ureter	770

Reflex Zones on the Hands and Feet -Acupressure Massage

Reflex zones on the back of the hands	777
Reflex zones on the feet -A	781
Reflex zones on the feet -B	781
Reflex zones on the palm of the left hand	776
Reflex zones on the palm of the right hand	776

Spirometer

Spirometer Construction — 60

Taping

Taping Techniques
Taping Anchors — 862
Taping Basket Weave — 866
Taping Check Rein — 865
Taping Cover Up — 864
Taping Figure of Eight — 870
Taping Heel Locks — 863
Taping Spiral — 865
Taping Stirrups — 863

Taping Procedures

Anterior Thigh Muscle Stabilization Procedure — 869
Big Toe Stabilization Procedure — 866
Gastrocnemius Cramp Prevention Procedure — 867
Hamstring Origin Strain Relief Procedure — 870
Immobilization of the Lumbosacral Area Procedure — 871
Lateral Lower Abdominal Stabilization Procedure — 870
Prevention of Knee Hyperextension Procedure — 869
Rib Injury Support Procedure — 871
Shin Splint Support Procedure 1 — 867
Shin Splint Support Procedure 2 — 868
Single X Knee Stabilizing Procedure — 868
Upper Back Stabilization Procedure — 872
X Method of Shoulder Separation Reduction Procedure — 872

Ultrahigh Frequency Sound

Ultrasound Access to Frontal Sinus and Antrum — 961

Vertebral Traction

Cervical Horizontal Traction Apparatus — 989

Vibration Study

Representative Graphs of Myoelectric Response to Electrical Stimulation (Tests 5 and 6) — 1011
Representative Graphs of Myoelectric Response to Vibration -A (Tests 1 and 2) — 1009
Representative Graphs of Myoelectric Response to Vibration -B (Tests 3 and 4) — 1010

Foreword

As we enter a new electronic era and embrace a one world concept, all scientific fields merge as never before. With advanced ideas and improved communication we finally see that information overlaps rather than being exclusive to any discipline. Researchers of basic phenomena quickly extend their findings to everyday happenings. Philosophers integrate globally and applied scientists understand the interplay of electromagnetic happenings in the world and in living tissue. Together these concepts constitute a new model of the universe. Within this framework the contents of these books fit perfectly.

Specifically, the field of medicine is in massive transition. The era of hoped for rapid reversal for all maladies, with shots, pills and surgery is doomed. Stopgaps are no longer acceptable to many medical practitioners and most patients.

Patients today somehow behave differently. Many are intolerate of minor or chronic disturbances that affect their active lifestyles. They want to fully participate. They take more responsibility for their own health, requesting doctor's referrals for physical treatment. They want to know what they can do to remove the cause and assist the treatment. Average people seem to understand what we have always known, that the body usually heals itself if pointed in the right direction.

Many older people who have lived vital lives are emotionally intolerant of so-called aging problems. They want progress. This book was conceived from such patient needs.

We therapists are also entering a new period of growth when we question standard rote forms of treatment. Yet, because we have not learned to criticize, upgrade, or develop new modalities, we are at a loss to cope with individual variations in treatment responses or results.

For these reasons, there is a renaissance of holistic medicine—not the generalized treatment of the 1960's where all approaches were "shot-gunned" at the patient in hopes that something would connect. Now holistic treatment is energetic, sure, and geared to treatable causes for permanent solutions.

These books show a broad new vista of over 100 debilitating malfunctions, four times as many as we commonly accept, that are successfully treatable by physical therapy. But more importantly, their new problem-solving approach replaces the treatment of syndromes and gross symptoms of pain, stiffness, and weakness, with hot packs, ultrasound, massage, and stretching. These books move us into a deeper level of understanding the treatable causes. Surely physical therapists are not equipped to diagnose syndromes or determine etiologies, but we are qualified to diagnose all treatable causes and to select appropriate methods.

The author organizes treatable causes into 49 functional categories. He replaces the older anatomical classifications of bone, nerve, muscle, or connective tissue disturbances with the overriding causes of syndromes such as soft tissue inflammation, soft tissue swelling, and peripheral nerve degeneration. Although these concepts are not foreign to our way of thinking, the author's emphasis is. He points out that these are the primary causes of syndromes, not the end products, and therefore should be treated first.

Alphabetical indexing by syndromes of the text material practically follows physicians diagnosis and prescription. The author shows us that each syndrome is a generalization of aspects of many syndromes and that individuals have unique patterns of treatable causes.

The content of these volumes is comprehensive. There is an overall discussion of syndromes bringing together classic information from physical therapy pioneers. This discussion is clarified and integrated with newer electronic diagnostic and treatment techniques. These techniques are followed by a listing and discussion of all the treatable causes, including specific evaluative techniques to isolate the treatable causes.

Next, the author describes the most important modalities for efficient treatment of the causes, including an overview of techniques such as ultra sound and its deeper application with phonophoresis. Treatment notes contain special applications, like phonophoresis with ointment or creams or underwater.

Both the newer and the customary instruments are categorically discussed giving unique features of each, such as rotary verses linear vibrators. Instruments' characteristics are explained based on the energy sources of galvanic, DC, and AC currents and how these affect bodily tissues. New words from physics and electronics are defined to help therapists understand and evaluate new equipment. The author adds equipment details and protocols for successful treatment.

The special features section with information not available in current literature is noteworthy. Two examples are, instruments and techniques for evaluating high skin resistance, and a foolproof test of deep tissue inflammation. Because this test is so simple and accurate, it should replace the inefficient thermography.

In the past, when highly sensitive patients responded poorly or unpredictably, we downgraded the treatment to their tolerance. The author suggests using an antidote instead, to upgrade their

suggests using an antidote instead, to upgrade their pain tolerance by adding treatment, to stimulate endorphin production and raise the pain threshold. The author shows how biofeedback is not the parlance of specialists, it can be used profitably for a variety of treatable causes by regular practitioners. Computer, visual-auditory feedback activates the higher integrating mind to offset the alienated or lower level damaged brain.

I particularly applaud the emphasis on evaluative instruments to replace our and patients' subjective reports of treatment success. His stress on electromonitry is not aimed to replace classic joint range or muscle testing, but to give a finer objective measurement. There is a fine reference section containing acupoint charts and functional neuromuscular system mappings.

These volumes return us to the creative search firmly established by the historical physical therapy giants, reinstating the art with the science of healing. The author has skillfully woven hands-on knowingness of the past with the today's daringness of electrodynamics. I predict that these volumes will become classics, as important references to body therapy specialists as the PDR is for physicians. But additionally, these should stimulate the dynamism, even the glamour of the physical healing profession for new students, patients and doctors alike. The excellent contents should contribute immeasurably to the evolution of the profession.

Valerie V. Hunt
Professor Emeritus
Former Director Division of Physical Therapy
University of California, Los Angeles
Director, Bioenergy Fields Foundation
   Laboratory

Preface

Taylor's Manual of Physical Evaluation and Treatment begins with seven content sections that provide quick and efficient referencing. The first is a master Table of Contents listing all topics alphabetically as they appear in the two-volume set. The master Table of Contents is followed by the six principal topical subsets on Syndromes, Treatable Causes, Modalities, Reference Documents, Diagrams, and Tables.

The alphanumeric designations following each topic reference is included to help with system orientation.

The letter prefix preceeding each designation denotes a specific category of information.

Key: S = Syndrome
 C = Treatable Cause
 M = Modalities (General Treatment Techniques)
 R = Reference Document (generally a monograph devoted to the exploration of a particular topic of interest not otherwise discussed)
 T = Table or List

In the alphanumeric code following each prefix letter is a number relative to the listing of all the topics within a particular category.

Any additional letters and numbers appended to each alphanumeric code represent subtopic categories. For example, the first letter following the prefix letter and list number of the general treatment category represents a particular treatment technique. To illustrate, ultrahigh frequency sound (ultrasound) as a treatment technique category is represented by the designation M01, while the specific treatment technique of phonophoresis (driving chemicals into the soft tissues with ultrasound) is represented by the designation of M01A, and the definitive discussion of the phonophoresis subtopic is designated as M01A1 while its procedures of application are designated as M01A2. Further subcategorizing is denoted by additional subsequent letters and numbers.

Introduction

Over the last fifty years a large body of knowledge has accumulated relative to noninvasive techniques of evaluation and treatment which may be expeditiously utilized in the practice of soft tissue management. By in large, the professional fields engaged in this practice have failed to systematize this knowledge for themselves, relying instead upon authors from other fields (primarily medicine) to provide it with documents of reference. The end result has been that the bulk of the material important to soft tissue management is scattered throughout many books and journals developed (for the most part) in other fields and written from other points of view. As a result, the researching practitioner is often faced with surveying many, many documents before discovering an appropriate reference, which in the end may be fragmentary and fails to describe in practical terms how the technique or modality should be applied in the clinical setting. As a result, many valuable treatment techniques fail to be utilized or are misused (sometimes to the degree that they may be detrimental to the patient or are ineffective). This is especially true of evaluation and treatment techniques which have emerged out of the technical development and information explosion of the last 20 years in the electronics, medical, and other paramedical fields. Indeed, there often seems to be a rather impressive lack of reliable information on such techniques, even though they may have been reported, through one means or another, to be extremely useful in the clinical setting.

The document to follow was created and designed to supply the clinical practitioner with easy access to the details necessary for the clinical application of the majority of evaluation and treatment techniques. Techniques are divided into general categories (exercise, ultrasound, electrical stimulation, etc.) and first discussed in general terms relative to definition, procedures of application, conditions treated, precautions and contraindications, and any special notes which may be thought pertinent. Each general technique category is then divided into subcategories of specific treatment techniques (as phonophoresis is a subcategory of the general category of ultrasound) and discussed in the same (if more specific) terms as those used relative to the general category.

Some of the techniques discussed and described have not yet made their appearances in the literature commonly available to the clinical practitioner, but they all have been utilized successfully by various responsible practitioners in the various concerned fields. Some of the techniques described may prove to be controversial when reviewed by some, especially since some of them have appeared so recently (in the last 10 years), but successful application supplies its own defense. It is hoped that all the techniques described will be carefully considered and thoroughly explored by those practitioners who are searching for effective tools for application in the clinic.

Also included in this work are discussions of the most common syndromes which make up the bulk of the diagnoses which a clinical practitioner must generally contend. The attempt was made to describe each syndrome in terms of its classic definition and to draw attention to other syndromes which may be related to it by virtue of symptoms shared. Recognizing the fact that a syndrome is simply a collection of symptoms, an attempt was also made to list all of the *treatable causes* which could produce the symptoms responsible for the syndrome under discussion (a treatable cause is defined as the cause of a symptom which may be treatable through one or more of the noninvasive treatment techniques described herein). Also included in each syndrome discussion was a section in which any special notes relative to the treatment of the particular syndrome could be included. These notes might include, for instance, special amplitude levels, frequency of the treatment, or special reference to other sections in the text.

It is the view of the author that all pathological conditions should be viewed in terms of the treatable causes which are demonstrated (through evaluation) to be present. Consequently, in the following text, each treatable cause is separately discussed in terms of how it is defined, how its presence is commonly determined (evaluated), and its etiology. The discussion is terminated with a full listing of all the evaluation and treatment techniques which may be employed in addressing the particular treatable cause evaluation and treatment. This listing is further broken down into those techniques which are recommended for use if the treatable cause is acute or when chronically present. It should be noted that successful application (that which relieves a symptom or ameliorates a treatable cause) of a treatment technique will be directly dependent upon the correct determination of the presence of a particular treatable cause and that the tissues treated respond normally to the technique application (there are always exceptions).

Also in the following text are tables of various types (data listings, reference diagrams, etc.), which have been included because of their usefulness when easily accessible in the clinical setting, and various monographs which have been included to aid in special study in various areas.

This document was written in the hope that it might provide the clinical practitioner with an easy reference to the various tools and knowledge which may be applied in soft tissue management for the relief of the various pathological conditions which may affect our patients. It was also written with the intent of helping define the rather extensive limits of this field, and to help promote this discipline as the scientific field which many of us have determined it to be.

Volume I

ACHILLES TENDONITIS SYNDROME (S52)

Definition (S52A)

Achilles tendonitis (tenontitis) is defined as inflammation of that tendon which serves to connect the gastrocnemius muscle to the tuberosity of the calcaneus of the foot via the tendo calcaneus.

Achilles tendonitis has often been referred to as Achilles tenosinovitis. However, this is really a misnomer, since the Achilles tendon has no synovial sheath. In fact, a better name would be Achilles paratendonitis since the inflammation actually occurs in the loose connective tissue called paratenon surrounding the tendon.

Trauma or stress is usually the cause of inflammation around the Achilles tendon. The inflammation is characterized by palpation tenderness of the tendon and fine crepitus on motion. Swelling is frequently found in the region of the inflammation, and pain in the back of the heel is the outstanding symptom. Differential skin resistance (DSR) survey will characteristically demonstrate the skin resistance to be relatively high above the site of the deeper tendon inflammation. Running, jumping, dancing, or any activity which stretches the Achilles tendon will result in pain.

Achilles tendonitis has been shown to occur 5.3 times more frequently in females than in males.

Related Syndromes (S52B)

ADHESION (S56)
CALF PAIN (S50)
FASCIITIS (S20)
FOOT PAIN (S51)
JOINT SPRAIN (S30)
MYOSITIS OSSIFICANS (S67)
PITTING (LYMPH) EDEMA (S31)
POST IMMOBILIZATION (S36)
REFERRED PAIN (S01)
TENDONITIS (S28)

Treatable Causes Which May Contribute To The Syndrome (S52C)

Calcific deposit (C08)
Extrafusal muscle spasm (C41)
Habitual joint positioning (C48)
Hypermobile/instable joint (C10)
Ligamentous strain (C25)
Muscular weakness (C23)
Peripheral nerve injury (C07)
Restricted joint range of motion (C26)
Soft tissue inflammation (C05)
Soft tissue swelling (C06)
Trigger point formation (C01)

Treatment Notes (S52D)

Achilles tendonitis has responded well to eclectically applied ice packs (refer to Cryotherapy, Soft Tissue Inflammation Control [M10E]), phonophoresis of nonsteroidal antiinflammatories (refer to Ultrahigh Frequency Sound, Phonophoresis [M01A]), and immobilization of the ankle during ambulating (by splint, cast or taping).

Healing is promoted by preventing Achilles tendon stretching. Athletic taping of the heel to keep the foot in slight plantar flexion, utilizing double heel-locks, an athletic strap and a cover-up, has proven to be a sufficient ankle immobilizer (refer to Taping [M16]).

The following is a list of trigger point formations which may, singly or in combination, imitate or contribute to the pain accompanying Achilles tendonitis. It should be noted that it is possible for the given patient to experience pain throughout the entire stereotypical referred pain pattern or only in part or parts of it. Opposite to each listing is a description of the parts of the stereotypical referred pain pattern most commonly experienced by previous patients, relative to the given trigger point formation. Any descriptive reference which is underlined is a part of the pattern which has *not* been commonly experienced by previous patients. For complete description of each stereotypical referred pain pattern, refer to the Table of Referred Pain Patterns of Trigger Point Formation Origin [T005]. Please note the key to abbreviated words at the bottom of the list.

LOCATION (MUSCLE)	REFERRED PAIN ZONES (GENERAL)
Soleus	Achilles tendon & heel
Peroneus longus	L. d.1/4 calf behind l.malleolus

Key:
d.—distal
l.—lateral

Bibliography (S52E)

Scott, p. 401.
Shands, p. 407.

ACNE SYNDROME (S45)

Definition (S45A)

Acne is a common inflammatory disease of the skin which occurs in dermal areas where sebaceous glands are the largest, most numerous and most active (the face, back of the neck and the upper trunk).

The physiological process which creates acne begins with a lipokeratinous plug (comedo or blackhead), which partially or wholly obstructs a pilosebaceous orifice. As the process continues, the sebaceous duct ruptures and sebum is spilled into surrounding cells instead of coming to the surface. In reaction to the sebum irritation, the surrounding tissues isolate and contain the sebum in a superficial or deep noninflammatory cyst capped by a superficial pustule which blocks the pilosebaceous orifice. During further development, bacteria incubates within the cyst. The by-products of bacterial metabolism cause additional chemical irritation of the tissues, with attendant destruction and displacement of epidermal cells and the resultant formation of scar tissue. In the last stage of development, canalized inflamed lesions and infected sacs develop. Canalized lesions most frequently occur on the face, shoulders, upper trunk, and upper arms.

Contributory factors to the severity and incidence of acne vary.

Very humid and warm environments may produce "tropical acne" which is characterized by explosive, severe, and generalized eruptions.

Cold winter weather may produce exacerbations of acne, while improvements in the condition may occur during hot summer temperatures.

Drug consumption or diet may play a role in acne development. Bromides and iodides, for example, are notorious for precipitating acne-like eruptions, and chocolate, nuts, cola drinks and sometimes milk consumption (in excessive amounts) appear to aggravate acne already present.

The trauma attending clumsy attempts to extrude blackheads or superficial cysts, constant touching of lesions, or too much washing may increase the damage arising out of acne by forcing sebum into surrounding tissues, thereby speeding the development of acne and increasing the occurrence of scarring.

Acne-like eruptions are often associated with menses, with exacerbations of present acne occurring before, during or following flow.

Characteristically, onset of acne is usually just prior to or during puberty. Early lesions and a family history of acne may be precursors of severe acne development.

In its mildest form, acne usually persists for at least a year and often continues through college age, with severe disease remaining troublesome through middle age and beyond.

Related Syndromes (S45B)

ADHESION (S56)
BACTERIAL INFECTION (S63)
KELOID FORMATION (S38)
UNHEALED DERMAL LESION (S39)

Treatable Causes Which May Contribute To The Syndrome (S45C)

Bacterial infection (C09)
Necrotic soft tissue (C18)
Neurogenic dermal disorders (psoriasis, hives) (C49)
Scar tissue formation (C15)
Soft tissue inflammation (C05)
Soft tissue swelling (C06)
Unhealed dermal lesion (C12)

Treatment Notes (S45D)

The clinical treatment of acne, in the second and third stages, has been based upon the phonophoresis of vitamin E or other nonsteroidal anti-inflammatories (Ultrahigh Frequency Sound, Phonophoresis [M01A], Scar Tissue Management [M01D], Bacterial Infection Eradication [M01G], Open Wound Healing [M01J]). The thrust of such therapy is directed at killing bacteria which may be present and to facilitate formation of collagen which is closer to being normal tissue than the keloids which may form without intervention.

For phonophoresis of treatment of the acne condition, it is generally recommended that the ultrasound unit utilized should be preset to provide power at .6 watts/cm^2, in a pulsed wave form. The sound should be applied with a slow sweeping action of the sound head over the skin for up to four minutes in a 1 to 2 square inch area. To arrest an exacerbation of the acne condition, as few as three or four to more than a dozen sessions may be required.

Bibliography (S45E)

Ganong, pp. 355, 365.
Holvey, pp. 1390-1391.

ADHESION SYNDROME (S56)

Definition (S56A)

The adhesion syndrome is characterized by the abnormal sticking together of tissues which should slip or slide freely against one another. The adhesion itself is usually a fibrous tissue formed from a normal constituent of one of the involved tissues or organs; it is formed as a reparative or reactive process to soft tissue stress. Adhesions (fibrosis) may form as a result of disease, prolonged restricted motion of injured joints, surgical operations where surface skin becomes attached to underlying tendon(s) or muscle(s) and especially after surgery into the abdominal or pelvic cavities where pus has spilled from a ruptured abscess.

Adhesions may arise from injuries by direct violence such as fracture or direct external blow (without fracture) to muscle tissue or other soft tissue structures, injuries which cause soft tissue inflammation and considerable effusion or hemorrhage into muscle or other soft tissues. Fractures of the femur, for example, are occasionally associated with adhesion of the quadriceps to the newly formed callus; this causes knee extension lag and seriously limits joint function.

Additionally, adhesion development may result from local infection, spontaneous hemorrhage (as in hemophiliac conditions), and is often associated with joint capsulitis. The extensiveness of adhesion formation is to a large extent dependant upon the patient's propensity to excessive fibrotic development in response to trauma.

Tendon adhesions are most often seen in the wrist and hand, especially after multiple severe lesions of the tendons at the wrist where the tendons are inclined to adhere either to each other or to the skin after suture repair. Tendons may also become adherent in the palm or over the finger after suture repair or grafting.

Adhesive capsulitis (periarthritis) arises out of a low grade inflammatory process which results in edema and fibrosis. Elasticity of the periarticular tissues is lost as they become shortened and fibrotic, often fixing the joint and greatly reducing motion.

Related Syndromes (S56B)

ACHILLES TENDONITIS (S52)
ACNE (S45)
BACTERIAL INFECTION (S63)
BURNS (SECOND AND/OR THIRD DEGREE) (S72)
BURSITIS (S26)
CALF PAIN (S50)
CAPSULITIS (S27)
CARPAL TUNNEL (S35)
CERVICAL (NECK) PAIN (S73)
CHEST PAIN (S14)
COLITIS (S42)
DUPYTREN'S CONTRACTURE (S34)
ELBOW PAIN (S47)
FOOT PAIN (S51)
FOREARM PAIN (S54)
FROZEN SHOULDER (S64)
HAND/FINGER PAIN (S17)
HIGH THORACIC BACK PAIN (S48)
JOINT SPRAIN (S30)
KELOID FORMATION (S38)
KNEE PAIN (S46)
LOWER ABDOMINAL PAIN (S62)
LOW BACK PAIN (S03)
MYOSITIS OSSIFICANS (S67)
OSTEOARTHRITIS (S21)
PIRIFORMIS (S04)
PITTING (LYMPH) EDEMA (S31)
POST IMMOBILIZATION (S36)
POST PERIPHERAL NERVE INJURY (S23)
RADICULITIS (S29)
REFERRED PAIN (S01)
SCIATICA (S05)
SHOULDER/HAND (S57)
SHOULDER PAIN (S08)
TENDONITIS (S28)
TENNIS ELBOW (LATERAL EPICONDYLITIS) (S32)
THIGH PAIN (S49)
UNHEALED DERMAL LESION (S39)
WRIST PAIN (S16)

Treatable Causes Which May Contribute To The Syndrome (S56C)

Bacterial infection (C09)
Calcific deposit (C08)
Extrafusal muscle spasm (C41)
Habitual joint positioning (C48)
Interspinous ligamentous strain (C03)
Joint ankylosis (C42)
Joint approximation (C47)
Ligamentous strain (C25)
Localized viral infection (C27)
Muscular weakness (C23)
Nerve root compression (C04)
Neuroma formation (C14)
Peripheral nerve injury (C07)
Restricted joint range of motion (C26)
Scar tissue formation (C15)
Soft tissue inflammation (C05)
Soft tissue swelling (C06)

Trigger point formation (C01)
Unhealed dermal lesion (C12)
Vascular insufficiency (C11)
Visceral organ Dysfunction (C02)

Treatment Notes (S56D)

Adhesion between dermis and underlying tissue (fascia, muscle or tendon) may be treated successfully with the combination of phonophoresis of vitamin E (Ultrahigh Frequency Sound, Phonophoresis [M01A]) and soft tissue manipulation (Massage, Deep Soft Tissue Mobilization [M23F]). Pulsed ultrasound (one megahertz) should be applied over a vitamin E emulsion at .6 to 1.8 watts/cm^2 for a four to six-minute period over the adhesion site (the time increasing with the size of the area treated). Soft tissue manipulation should be performed at right angles (transversely) to the long axis of the fibers at the site of the adhesion.

The desired effect of the phonophoresis of vitamin E is to soften the collagen-like fibers at all levels of the adhesion to allow for a more immediate mobilization of the fascial, muscle, or tendon tissues under the site of dermal involvement by the deep transverse friction massage. The involved musculature should be relaxed during the massage.

For the treatment of adhesive capsulitis see Frozen Shoulder Syndrome [S64].

For the treatment of post immobilization adhesion see Post Immobilization Syndrome [S36].

Bibliography (S56E)

Basmajian, pp. 158, 273–274.
Bickley, pp. 52–53.
Salter, p. 244.
Shands, p. 426.
Travell, pp. 414–416.

ADVERSE EFFECTS OF BED REST (R008)

Possibly one of the dangerous aspects of serious illness or injury may be enforced long-term bed rest. Such inactivity may have detrimental effects upon the (1) musculoskeletal, (2) cardiovascular, (3) respiratory, (4) gastrointestinal, (5) genitourinary, (6) metabolic, (7) psychological, and (8) dermatological systems.

Musculoskeletal System (R008A)

The adverse effects of long-term bed rest upon the musculoskeletal system may include muscular atrophy, osteoporosis, joint stiffness or even joint contracture. To maintain muscular tone and the integrity of muscle tissue, the muscle must be regularly used. If unused, the muscle first weakens, then neuromuscular tone decreases and finally the muscle atrophies as muscle cells are lost and replaced by connective tissue. Muscle cells that are lost in this way are not replaced, even by reconditioning, so the system may be permanently weakened even if the patient is able to return to normal functional activities.

Osteoporosis may result from long-term bed rest if the bone is not subjected to longitudinal compressive stress. When the bone is formed, on-going osteoblastic (bone building) activity compensates for continuing osteoclastic (bone destroying) activity. The bone cells are laid down to withstand the force of gravity along the longitudinal line of the bone. In the normal active state, the continuation of these two activities functions two ways: to meet the skeletal needs for bone shape change should the direction of prolonged stresses upon the bone change, and to prevent the development of brittleness of the bone which is the normal consequence of bone aging.

Osteoclastic activity usually occurs over a three-week period as bone is destroyed by masses of osteoclasts (reportedly eating holes in the bone, cutting tunnels up to 1 mm. in diameter). When finished with their destructive activity, the osteoclasts become osteoblasts and reconstruction of the bone begins. The new bone is laid down in successive layers upon the inner surface of the cavitated area. Bone reconstruction continues for several months and lasts until the newly formed bone begins to encroach upon blood vessels which supply the area. The newly formed bone is deposited in proportion to the compressive load that the bone must endure. The structure of this bone is laid down to afford the best resistance to the compressive force (normally longitudinal in the long bones). Thus the external forces exerted against the bone determines its strength and even its shape. Long-term bed rest may be highly detrimental to the bone as the prone position fails to adequately stress the bone to promote or maintain strong bone formation, and any newly formed bone may not be laid down in the best formation to resist the compressive stresses which will be inflicted upon it when the patient assumes more stressful positions (sitting, standing, or even rolling over in bed). In the extreme, this lack of stress may result in seemingly spontaneous compression fractures of vertebrae, pelvis, hip, bones of the foot, leg bones, or even bones of the upper extremity.

Prolonged bed rest has been noted to cause joint stiffness seemingly unrelated to musculature atrophy or the process of bone destruction and formation. Such stiffness may be the result of the slowing of lymphatic circulation by virtue of muscle disuse. Muscle action is responsible for much of the fluid propulsion force within the lymphatic system which is responsible for the picking up of and recirculation of proteins which collect in synovial joints (shoulders, elbows, hands, hips and knees). These proteins, in excess, may have the effect of increasing pressure within the joint and thereby decreasing mobility, or (as some authorities have suggested) may decrease mobility by decreasing synovial fluid viscosity.

The above maladies may be prevented or their development slowed by the continued active use of the major muscle groups through active resistive exercise (isometric or isotonic) or other means. Electrical stimulation of these muscle groups has proven to be extremely valuable for the maintenance of muscle tone for bedridden patients, whether conscious or not, especially when intermittent tetanizing medium frequencies are utilized. Refer to Electrical Stimulation, Muscle Reeducation/Muscle Toning [M02B], Delaying Atrophy of Denervated Muscle [M02E] for details pertaining to application.

If the facilities to apply electrical stimulation are unavailable and the patient is unable to voluntarily exercise, the continued application of stress on the musculoskeletal system may be supplied through passive range of motion exercise accompanied by externally applied joint approximation (when safe to apply).

Cardiovascular System (R008B)

Adverse affects of prolonged bed rest upon the cardiovascular system may include the promotion of thrombus formation, edema of the extremities, orthostatic hypotension, and strain upon the heart (demand upon the heart may increase as much as thirty percent). These affects basically result from the disuse of the large muscle groups in the extremities. Both venous and lymphatic circulation are promoted by muscle action in the extremities. If the muscles are inactive, demand upon the heart increases as it is required to compensate for the loss of the positive pressure upon venous and lymphatic vessels contributed by action and tone of the the extremity muscles. This loss may result in lymph edema in the distal segments of both the upper and lower extremities and an increase in the possibility of cardiovascular congestion.

Venous congestion also seems to increase the probability of thrombus formation; the consequent venous circulation slowing apparently creates a situation in which spontaneous coagulating of platelets is promoted and a thrombus (and potentially an embolism) may be formed.

The problem of orthostatic hypotension which often accompanies prolonged bed rest stems from disuse of the carotid sinus in the neck, whose job it is to maintain blood pressure (especially that going to the brain) when postures are changed from reclining to sitting or standing. Apparently, if the carotid sinus is not regularly or frequently called upon to react to changes in posture, it loses its ability to respond quickly and the patient may experience dizziness as the blood pressure drops when changing the body posture from lying to sitting or standing. The dizziness will continue until the carotid sinus has compensated and adjusted the blood pressure up to the appropriate level.

Prevention of the above phenomena may be accomplished through the therapeutic use of electrical stimulation (refer to Electrical Stimulation, Increasing Blood Flow [M02C], Reduction of Edema [M02D]) to increase muscular activity, or through frequent changes in posture instituted as early as possible, when safe, by either helping the patient to voluntarily achieve sitting or standing postures or to involuntarily achieve them through the use of a mechanical bed, tilt table, or other mechanical device.

Respiratory System (R008C)

The respiratory system may be adversely affected by prolonged bed rest by virtue of the fact that the lying posture with accompanying slower and shallower than normal breathing is not conducive to the natural movement of mucal secretions out of the lungs and bronchial tree. Mucal secretion buildup in the respiratory structures may result in reduced vital capacity and a decrease in the amount of oxygen available for gaseous transfer in the lungs. If allowed to develop to extremes (which occurs far more frequently than most people realize) such mucal buildup may lead to hypostatic pneumonia, which is often a prelude to bacterial forms of pneumonia. If unrelieved, the patient may eventually drown (suffocate) to death.

To prevent such respiratory distress the patient should be discouraged from habitually semireclining (the semireclined position often associated with mechanical hospital bed), since this position promotes mucal secretion accumulation in the lower posterior lobes. The patient should be encouraged to lie flat and turn frequently or to sit in a full upright position, if possible, to promote natural postural drainage. The patient should also be encouraged to breathe deeply and to occasionally cough to help keep airways open. If mucal secretions begin to build up, postural drainage should be performed on the patient (refer to Postural Drainage [M14]) and positive pressure and other breathing exercises should be performed by the patient (refer to Exercise, Improving Lung Function [M12B]) to relieve and/or reverse the condition.

Gastro-intestinal System (R008D)

Normal gastro-intestinal function in human beings has been shown to depend to a given extent by the force of gravity; for the best gastro-intestinal function the human individual should spend at least part of the day standing or walking. If the patient is confined to bed or in a wheel chair for a prolonged period the end result may be trouble with persistent constipation, fecal impaction, and/or negative nitrogen balances in the blood and soft tissues.

These conditions may usually be prevented by elevating the patient, if safely possible, to an upright position (through the use a tilt table or other mechanical device, if necessary) and maintaining it for an hour a day, and by providing adequate fluids and an adequate diet. It is also reported that in-bed exercise may be helpful in preventing the gastro-intestinal difficulties.

Genito-urinary System (R008E)

Natural genito-urinary function depends to a given extent upon maintaining an upright position for at least part of the day. For the best, most effective passage of urine from the kidney and complete drainage of the ureter into the bladder the ureter should be maintained in a fairly vertical position for at least a part of the day.

Prolonged bed rest or wheelchair confinement may lead to urinary stasis which promotes bacterial infection development in the kidneys, ureters. Elevating the patient, if safely possible, to an upright position through the use a tilt table or other mechanical device and maintaining it for at least an hour a day, does much to promote genito-urinary function. Additionally, forced fluid intake of 300 cc's per day and frequent changes of body position, even when lying down or sitting, are said to promote more normal genito-urinary function and may be helpful in preventing infection.

Metabolic System (R008F)

Prolonged contact with bed and bed clothes inhibits the loss of body heat. The increased body temperature increases the dilation of surface blood vessels, increases sweating and loss of electrolytes, and increases in urinary excretion, thereby stressing the system. To compensate the patient should be provided with a high protein diet, adequate fluids (300 cc./day), and encouraged to sit up, away from the surface of the bed, as much as possible .

Psychological System (R008G)

The enveloping and protective nature of bed and bed clothes may effectively decrease sensory stimulation to the patient to the extent of promoting temporary abnormal psychological effects which include altered body perception, altered sensory perceptions, and hallucinations which may include imagined sounds, sights and bodily sensations. To prevent these psychotic effects the patient should be provided with additional external sensory stimulation which may include television, radio, social interactions (including games and conversation), small docile animals for petting and companionship, and exposure to the external activities of others.

Dermatological System (R008H)

If during prolonged bed rest the patient is not moved or does not move frequently to relieve points of pressure between the body and the bed surface, ischemia of the soft tissues over the point(s) of pressure may develop. Unrelieved, this ischemia may result in the breakdown of the soft tissues to form decubitus ulcers. The prevention of decubitus ulcers simply depends on the patient rolling or being rolled from one side of the body to another, changing from lying to sitting or standing positions.

Bibliography (R008I)

Guyton, pp. 260, 934-935.

AMBULATION WITH ASSISTIVE DEVICES (R005)

Various devices have been developed to assist ambulation when the ability of function of one or more of the extremities has been impaired. They include canes, crutches, walkers, and braces.

Canes (R005A)

Cane types (R005A01)

Straight canes Straight canes are constructed with a central straight shaft which is provided with a single (hemicircular) curved, flat top, or pistol grip handle on one end, and terminates on the other end in a single point, for contact with the floor or walking surface. The handle is generally designed to be grasped and manipulated with one hand and to provide a pressure platform for the palm or heel of the hand. Pistol grip handles are usually multiply-curved, designed to fit into the palm of the hand and to provide a prominent pressure platform for the heel of the supporting hand; it is generally considered to be the best designed to the handle types, providing the best support while taxing the soft tissues of the hand less when the pressure platform comes into play. The terminal point is usually fitted with a rubber tip for use on all surfaces, or a metal tip for use on softer or "hiking" surfaces. Straight canes are constructed from wood, metal or other strong material. Metal straight canes are generally adjustable, while wooden canes most usually must be cut to size.

Quad canes Quad canes are usually constructed with a flat top or pistol grip handle, and terminate in four rubber-tipped points of contact for maximum stability and can, in fact, actually stand alone. Quad canes are generally constructed from metallic substances and are usually made to be adjustable.

Fitting parameters (R005A02)

For most effective use, the cane should be just long enough to permit the patient to lean slightly over the cane when using it for support. As a general rule, the cane should be adjusted or cut so that when standing, with the supporting hand resting upon the cane handle, the patient's elbow will be in fifteen degrees of flexion. This cane length generally coincides with the distance between the floor and the head of the trochanter in the patient's hip.

Cane uses (R005A03)

When a cane is utilized to assist in ambulation, it is most commonly used to take some of the pressure off of a functionally impaired lower extremity. Generally, it is recommended that the cane be held in the hand on the side opposite that of the impaired lower extremity. Two gait patterns are most commonly utilized, the three point pattern and the two point pattern.

Cane gait patterns (R005A04)

Three point gait pattern

1) The patient should place the cane tip out in front of the feet, so that when swing phase of the impaired (involved) extremity is complete the arch of that foot is approximately the same directional distance from the foot completing the stance phase (the good leg) as the cane tip.
2) The involved extremity should then be taken into swing phase as the uninvolved extremity goes goes through the stance phase.
3) The patient should then lean on the cane, the weight of the body distributed between the involved extremity and the cane, as the uninvolved extremity begins, goes through and completes swing phase.
4) The cane tip should then be picked and placed again, a pace length in front of the feet, to begin a repeat of the process.

Two point gait pattern

1) The patient should swing the cane forward as the involved extremity goes through swing phase. The cane tip should come to rest on the walking surface just as the involved foot begins heel strike, the same directional length as the arch of the involved foot in front of the uninvolved foot.
2) The patient should then lean on the cane, the weight of the body distributed between the involved extremity and the cane, as the uninvolved extremity begins, goes through and completes swing phase.
3) The above process should then be repeated for the next pair of steps, as the cane and involved extremity once again begin swing phase.

Problems associated with cane ambulation (R005A05)

If a cane is used appropriately, it can do much to increase the stability and ambulatory range of the patient, by improving the balance and reducing the compressive force upon an involved extremity. As stability and additional support is gained, however, both speed and general mobility are lost and the rate of fatigue increases. The coordination of cane and leg motion is accomplished at a rate limited by the speed of cane placement; such coordination of arm and leg motion to create tripedal ambulation not only decreases the speed of ambulation but increases the amount of energy consumed in ambulation.

Tripedal ambulation with a cane poses the problem of increasing the space required to ambulate, especially if the cane enjoys the broad base of the quad cane; this increases the chances of collision with other individuals or objects with body parts or cane, as well as making ambulation on stairs increasing difficult.

Slippery surfaces are far more dangerous for a person utilizing a cane for ambulation because of the large amount of weight communicated intermittently to the cane; this is especially true for a single point cane, since a slip of the cane tip when weight is put on it makes a fall very nearly inevitable. Careless placement of the cane tip may also prove dangerous for the same reason. The main problem posed for a patient utilizing a cane lies in the false sense of security its use may impart to the patient. This is especially true for quad canes, with their innate stability, since a misplacement of the cane base may, as a result of leverage, have the effect of actually throwing the patient to the ground.

A cane may become extremely dangerous to the user if used after it is no longer essential. A person who uses a cane unnecessarily may forget its presence and inadvertently place its tip in a hole or on a slick surface, then falling, or maybe even tripping over it. Such cane use, especially over a long period of time, may add to the difficulty of reorienting the patient to a bipedal (unassisted) gait pattern, with its inherent feeling of instability.

Crutches (R005B)

Axillary crutches (R005B01)

Construction Axillary crutches, the most supportive of the crutch types, are usually constructed from wood or metal. An axillary crutch is equipped with a pressure plate which fits under the patient's axilla; like the cane, it has a handle (centrally placed) for hand control, and it usually terminates in a single contact point fitted with a rubber tip. The axillary crutch is designed to disperse the patient's compressive weight to the crutch handles through the patient's hands and to the axillary pressure plate positioned via its contact points with side of the chest wall and the upper arm. Crutches are correctly used in pairs.

Axillary crutch fitting Most axillary crutches are adjustable in length; the crutch length should be adjusted so that the axillary pressure plate is well below the axilla. The axillary crutch height should be determined by having the patient stand fully erect, with the crutch against the side of the body, and the pressure plate in position one or two inches below the axilla; this crutch length reduces the possibility of brachial plexus injury from direct pressure. The patient should be cautioned never to allow the axilla(s) to rest on the axilla pressure plate(s) because of the real danger of brachial plexus injury from the tremendous pressure that may be generated into the axilla if the body is so supported. The crutch handles are usually adjustable, as well, and should be adjusted so that they are at the level of the head of the trochanter, when the patient is standing straight up, with the hands on the handles, so that the elbows are in fifteen to thirty degrees of flexion.

Forearm crutches (Canadian canes) (R005B02)

Construction Forearm crutches are also most correctly used in pairs. They, like axillary crutches, generally terminate in a single shaft, fitted with a rubber tip. They are provided with cuffs (usually constructed of metal) adjusted to fit snugly around the proximal forearm. The cuffs are attached eight to ten inches above the handle, both positioned on the anterior aspect of the crutch. The portion of the staff occupied by the forearm cuff is angled rearward in a way which distributes much of the patient's body weight between the part of the crutch supporting the cuffs and the crutch handles and directs it down the rest of the shaft.

Fitting To fit the forearm crutch to the patient, the patient should be standing up straight with the crutches held up against the sides of the body. The cuff should be adjusted along the shaft so that it is one to two inches below the elbow crease. The handle should be adjusted to the level of the head of the trochanter so that when held in the hand of the fully erect patient the elbow is in fifteen to thirty degrees of flexion.

ambulation with assistive devices

Crutch uses (R005B03)

Axillary crutches Axillary crutches are generally used to provide the greatest amount of support, especially when one of the lower extremities is permitted only partial or no weight bearing. Axillary crutches may be used to "grade" the percentage of weight bearing afforded one of the lower extremities by simply using the appropriate gait pattern.

Forearm crutches Forearm crutches are designed to be less supportive then axillary crutches, requiring the patient's legs to support more of the patient's weight. This type of crutch is most commonly used in conjunction with adjustable fixed ankle joint short leg braces, not only to assist in ambulation but also to increase the demand upon leg muscles and to promote more normal gait patterns.

Crutch gait patterns (R005B04)

Two point gait The two point gait is performed by having the crutch on one side move forward as the leg on the contralateral side is moved forward to perform swing phase. They both should strike the surface of the floor simultaneously, and the crutch should take the pressure off the contralateral lower extremity as it goes through stance phase. As this is going on, the opposing crutch and lower extremity are performing the alternate function.

Three point gait The three point gait is performed by having both crutches move forward together while the patient's weight is supported by an uninvolved lower extremity. Both crutch tips should strike the floor simultaneously and bear the weight of stance phase as the uninvolved lower extremity is stepped through. The contralateral involved lower extremity doesn't touch the floor; it can be carried in front of the uninvolved lower extremity, or flexed at the hip and knee and carried below.

Four point gait The four point gait is performed by moving one crutch forward and following that by moving the contralateral forward. This is followed by bringing the other crutch forward and then the other foot. In this gait pattern the body weight is constantly distributed between three supports as the fourth is moved forward.

Swing-through gait To perform the swing-through gait, both crutches are moved forward, then the lower extremities are simultaneously swung forward to planted ahead to the crutch tips. The crutches are then brought forward to begin the process again.

Limitations associated with crutch ambulation (R005B05)

Ambulation utilizing forearm crutches is generally limited to the two and four point gait patterns. The three point and swing through gaits are considered dangerous because of the excessive amounts of pressure distributed to the patient's hands when the lower extremities and body are swung through.

It is permissible to use one forearm crutch as one would use a single point cane, but it is not acceptable to use only one axillary crutch. If the patient is stable enough to use a single axillary crutch a single point cane or forearm crutch should be used.

Walkers (R005C)

Walker types (R005C01)

Pick-up walker The pick-up walker is a stand alone contrivance which supports itself on four legs connected together with cross pieces to construct a three sided structure. The three closed sides of the structure provide stability for the patient while the open side allows the patient to walk into to the structure. In the central portions of the top two opposing sides, handles for grasping and to serve as pressure plates are generally provided; these handles allow the patient the means of lifting and maneuvering the walker.

Rolling walker The rolling walker is a stand alone contrivance which supports itself on four legs connected together with cross pieces to construct a three sided structure. The three closed sides of the structure provide stability for the patient while the open side allows the patient to walk into the structure. In the central portions of the top two opposing sides, handles for grasping and to serve as pressure plates are generally provided. The safest of the rolling walkers have a pair of wheels which are fitted into the ends of the two front legs. To move the walker the patient need only slide it forward.

Walker adjustment (R005C02)

Most walkers are adjustable. In general, the four legs of the walker should be the same length. They should be adjusted so that when the patient is standing erect within the structure, with the pronated hands on the walker handles, the elbows are flexed between fifteen and thirty degrees.

Walker gait patterns (R005C03)

The gait patterns utilized with either a pick-up or rolling walker are generally limited to two.

Walking pattern

(1) The walker should be either lifted and placed forward, or rolled forward and planted in place.

(2) The patient should lean upon the walker handles, and then walks up to and into the walker structure, taking several steps.

Care should be taken to avoid walking up too close to the front of the walker. If the patient's center of gravity is too far forward of the back legs, it may create instable forward leverage on the front legs which may throw both the walker and patient forward and down.

Swing pattern

(1) The walker should be either lifted and placed forward, or rolled forward and planted in place.

(2) The patient should lean upon the walker handles and then swing into the walker structure, planting the feet inside the structure before moving the walker and beginning the process again.

Care should be taken to avoid swinging up too close to the front of the walker. If the patient's center of gravity is carried too far forward of the back legs by momentum, it may create instable forward leverage on the front legs which may throw both the walker and patient forward and down.

Leg braces (R005D)

The only brace which is recommended for use here is the short leg brace which has been constructed of metal double upright bars, extending out of adjustable fixed ankle joints (Pope or Becker joints are often used). This type of brace is constructed as part of the shoe, with a metal shank which extends from the heel, past its juncture with the double uprights (which pass vertically past the malleoli), out to just behind the ball of the foot. The longer and broader this shank is the more stable the brace.

If constructed correctly, the short leg brace will provide stability to both the ankle and the knee by preventing any movement of the ankle joint and, by virtue of the mechanical advantage provided by the fixed right angle, by forcing the knee back into extension as the patient puts weight on it. Ambulation with this brace posses basically the same problems as ambulation without it, except ambulation is much slower without ankle flexibility, and the patient must learn to deal with the mechanical forces imparted to the lower extremity by the fixed ankle joint.

The greatest problem posed by short leg brace ambulation is the sense of security it offers to many of the patients who wear it. It may become such a symbol of security that the patient may be unwilling *not* to wear it, even after the patient's ability has made it unnecessary. This is regrettable since the lower extremity muscles which are compensated for may be caused to atrophy from disuse, the involved knee may begin to show signs of developing osteoarthritis from excessive hyperextension of the knee joint, and further development may be effectively blocked by both muscular weakness and pain and inherent feelings of insecurity.

Short leg braces equipped with spring joints are not recommended for use in any conditions since the flexibility of the joint fails to afford joint stability and may even foster the development of spasticity in the the gastrocnemius and soleus muscles which the action of the spring works against. Patients sometimes literally walk out of this type of brace, especially if suffering from extreme hypertonus of the lower extremity plantar flexors (a particular problem in some cerebral palsy victims).

Long leg braces are generally not recommended as an assistive device for ambulation. While they may serve to strengthen the rest of the body, no demand is made upon the musculature of the lower extremities, discouraging lower extremity muscular development. The energy cost of their use is generally so prohibitive that even patients who learn to use them efficiently finally give them up for general use. It is felt that if ambulation with braces is not possible utilizing short leg braces, it probably should not be indulged in except as an exercise to stimulate joint receptors to promote muscular activity.

TABLE OF ANTAGONISTIC MUSCLES OF COMMON INTEREST (T006)

The muscles below are generally grouped together according to the combined action they produce during normal contraction. The muscles which have their first letter capitalized are the prime movers of the action while those uncapitalized play an accessory role.

Homolateral rotation of the head
Sternocleidomastoideus, scaleni

Contralateral rotation of the head
Upper trapezius, splenius capitus, posterior cervical, levator scapulae

Anterior deviation of the jaw
Temporalis

Posterior deviation of the jaw
Masseter

Scapular abduction and upward rotation
Serratus anterior

Scapular adduction and downward rotation
Middle trapezius, Rhomboid major, Rhomboid minor, lower trapezius, upper trapezius

Scapular depression
Lower trapezius

Scapular elevation
Upper trapezius, Levator scapulae, rhomboid major, rhomboid minor

Shoulder flexion to 90 degrees
Anterior deltoid, Coracobrachialis, middle deltoid, biceps brachii, pectoralis major (clavicular fibers)

Shoulder extention
Latissimus dorsi, Teres major, Posterior deltoid, triceps (long head), teres minor

Shoulder abduction to 90 degrees
Middle deltoid, Supraspinatus, anterior deltoid, posterior deltoid, serratus anterior

Shoulder adduction
Pectoralis major, Latissimus dorsi

Shoulder horizontal abduction
Posterior deltoid, infraspinatus, teres major

Shoulder horizontal adduction
Pectoralis major, anterior deltoid

Shoulder external rotation
Infraspinatus, Teres minor, posterior deltoid

Shoulder internal rotation
Subscapularis, Pectoralis major, Latissimus dorsi, Teres major, anterior deltoid

Elbow flexion
Biceps brachii, Brachialis, brachioradialis, flexors of the wrist and fingers (arising from the medial epicondyle of the humerus)

Elbow extension
Triceps, extensors of wrist and fingers (those arising from the lateral concyle of the humerus)

Forearm supination
Biceps brachii, Supinator, brachioradialis,

Forearm pronation
Pronator teres, Pronator quadratus, flexor carpi radialis

Wrist flexion
Flexor carpi radialis, Flexor carpi ulnaris

Wrist extension
Extensor carpi radialis longus, Extensor carpi radialis brevis, Extensor carpi ulnaris

Flexion of the metacarpophalangeal joints of the fingers
Lumbricales, Dorsal interossei

Extension of the metacarpophalangeal joints of the fingers
Extensor digitorum communis, Extensor indicis proprius, Extensor digiti quinti proprius

Flexion of the proximal and distal interphalangeal joints of the fingers
Flexor digitorum sublimis, Flexor digitorum profundus

Finger abduction
Dorsal interossei, Abductor digiti quinti

Flexion of the metacarpophalangeal joint of the thumb
Flexor pollicis brevis

Flexion of interphalangeal joint of the thumb
Flexor pollicis longus

Thumb abduction
Abductor pollicis longus, Abductor pollicis brevis, palmaris longus

Opposition of thumb & fifth finger
Opponens pollicis, Opponens digiti quinti, abductor pollicis longus (to begin the action from the rest position), abductor pollicis brevis (to begin the action from the rest position)

Hip flexion
Psoas major, Iliacus, rectus femoris, sartorius, pectineus, tensor fasciae latae, adductor brevis, adductor longus, adductor magnus (oblique fibers)

Hip abduction
Gluteus medius, gluteus minimus, tensor fasciae latae, gluteus maximus (superior fibers)

Hip external rotation
Obturator externus, Obturator internus, Quadratus femoris, Gemellus inferior, Gemellus superior, Piriformis, sartorius, biceps femoris (long head), gluteus maximus

Knee flexion
Biceps femoris (long & short heads), Semitendinosus, Semimembranosus, popliteus, gracilis, sartorius, gastrocnemius

Ankle plantar flexion
Gastrocnemius, Soleus, Tibialis posterior, peroneus longus, peroneus brevis, flexor hallucis longus, flexor digitorum longus, plantaris

Foot inversion
Tibialis anterior, Tibialis posterior

Extention of the proximal and distal interphalangeal joints of the fingers
Extensor digitorum communis, Extensor indicis proprius, Extensor digiti quinti proprius

Finger adduction
Palmer interossei

Extension of the metacarpophalangeal joint of the thumb
Extensor pollicis brevis

Extension of the interphalangeal joint of the thumb
Extensor pollicis longus

Thumb adduction
Adductor pollicis obliquus, Adductor pollicis transversus

Anti-opposition of thumb and fifth finger
Abductor pollicis longus, Abductor pollicis brevis, Dorsal interossei, Abductor digiti quinti, palmaris longus

Hip extension
Gluteus maximus, Semitendinosus, Semimembranosus, Biceps femoris (long head)

Hip adduction
Adductor magnus, Adductor brevis, Adductor longus, Pectineus, Gracilis

Hip internal rotation
Gluteus minimus, Tensor fasciae latae, gluteus medius, semitendinosus, semimembranosus

Knee extension
Rectus femoris, Vastus intermedius, Vastus medialis, Vastus lateralis

Ankle dorsiflexion
Tibialis anterior, Extensor digitorum longus, Extensor hallucis longus, peroneus tertius

Foot eversion
Peroneus longus, Peroneus brevis

table of antagonistic muscles of common interest

Flexion of the toe metatarsophalangeal joints
Lumbricales, Flexor hallucis brevis, dorsal interossei, plantar interossei, flexor digiti quinti brevis, flexor digitorum longus, flexor digitorum brevis, flexor hallucis longus

Flexion of the toe interphalangeal joints
Flexor digitorum longus, Flexor digitorum brevis, Flexor hallucis longus

Toe abduction
Dorsal interossei, Abductor hallucis, Abductor digiti quinti

Extension of the toe metatarsophalangeal joints
Extensor digitorum longus, Extensor digitorum brevis, extensor hallucis longus

Extension of the toe interphalangeal joints
Extensor digitorum longus, Extensor hallucis longus

Toe adduction
Plantar interossei, Adductor hallucis

TABLE OF AURICULAR ACUPOINT TREATMENT PROTOCOLS (T004)

The acupoints listed below are listed according to general names and each is preceded by the point designation ascribed to it by Wexu or other authority.

Some controversy exists between authorities over exact acupoint sites. Acupoints are listed by the names assigned by the authority responsible for developing the protocol in question. A name in parenthesis is that assigned to that point by Wexu.

The treatment protocols listed below are based on the recommendations of acknowledged experts in the field of the therapeutic stimulation of acupoints. The authors of this document assume *no* responsibility for the clinical efficacy and/or dependability of any or all of the protocols listed.

Those protocols listed with a capital letter in parenthesis after the protocol name are generally used in combination with a peripheral acupoint protocol. In the Table of Peripheral Acupoint Treatment Protocols [T002], the protocols are listed under the same name, with the same, matching capital letter in parenthesis following it. Those protocols without a capital letter in parenthesis following it are generally used alone for treatment.

ABDOMEN, DISTENDED
M06 Small intestine, M08 Large intestine, M04 Stomach, H01 Sympathetic, F07 Abdomen, O05 San Jiao

ABDOMINAL PAIN FROM GASTROINTESTINAL CONDITIONS (A)
M04 Stomach, H01 Sympathetic, M06 Small intestine, M08 Large intestine, N06 Pancreas and gallbladder

ACNE
O02 Lung, D01 Internal secretion, E08 Testicle, A11 Cheek

ADDICTION, DRUG
O02 Lung, N09 Liver, N04 Kidney, I01 Shen Men, B08 Adrenal gland, E01 Brain axis, E10 Subcortex, E05 Occiput

ADDICTION, TOBACCO
A05 Lower jaw, A07 Upper jaw, A03 Tongue

ADNEXITIS
I05 Uterus, D02 Ovary, D01 Internal secretion, I01 (Shen Men) Divine gate, B08 Adrenal gland

AGOMPHIASIS
N04 Kidney, A05 Mandible, A07 Maxilla, E05 Occiput

ALOPECIA
N04 Kidney, O02 Lung, D01 Internal secretion, E05 Occiput

ALOPECIA AREATA
Corresponding regional points, N04 Kidney, O02 Lung, D01 Internal secretion

AMENORRHEA
I05 Uterus, D01 Internal secretion, D02 Ovary, B08 Adrenal gland

AMENORRHEA (B)
I05 Uterus, D01 Internal secretion, D02 Ovary, N04 Kidney, H01 Sympathetic

AMYOTROPHIC LATERAL SCLEROSIS
N04 Kidney, D01 Internal secretion, E01 Brain Axis, E05 Occiput, O05 San Jiao

ANAL PAIN (HEMORRHOIDS) (A)
K15 Lower rectum

ANAPHYLACTIC DERMATITIS
O02 Lung, D01 Internal secretion, E05 Occiput, B08 Adrenal gland, Corresponding regional points

ANAPHYLACTIC SHOCK (B)
I01 Shen Men, O01 Heart, B08 Adrenal gland, E10 Subcortex

ANAESTHESIA FOR CHOLECYSTECTOMY AND SPLENECTOMY (A)
N06 Gallbladder, N10 Spleen, F07 Abdomen, I01 Shen Men, O02 Lung, E10 Subcortex

ANAESTHESIA FOR CRANIAL SURGERY (A)
I01 Shen Men, E01 (brain axis) Brainstem, H01 Sympathetic, O02 Lung

ANAESTHESIA FOR OPEN HEART SURGERY (A)
F05 Neck, F06 (thorax) Chest, O02 Lung, N04 Kidney

table of auricular acupoint treatment protocols

ANAESTHESIA FOR THYROID SURGERY (A)
I01 Shen Men, E10 Subcortex, O02 Lung, F05 Neck

ANAESTHESIA FOR TOOTH EXTRACTION (A)
A01 (tooth extraction anesthesia point 1) Upper teeth, A08 (tooth extraction anesthesia point 2) Lower teeth

ANAESTHESIA FOR TOTAL LARYNGECTOMY (A)
O02 Lung, I01 Shen Men, B08 Adrenal gland, F05 Neck

ANAL FISSURE
K15 Lower segment of the rectum, I01 Shen Men, I08 (lower segment of the rectum) Constipation, N10 Spleen, M08 Large intestine, O02 Lung

ANEMIA (B)
N09 Liver, N10 Spleen, D01 Internal secretion, M04 Stomach, L01 Diaphragm, M06 Small intestine

ANEMIA, HYPOFERRIC
N09 Liver, N10 Spleen, D01 Internal secretion, L01 Diaphragm, M04 Stomach, M06 Small intestine

ANGINA PECTORIS (B)
I01 Shen Men, O01 Heart, O02 Lung, Corresponding regional points

ANKLE, ARTHRITIC PAIN (A)
G03 Ankle, I01 Shen Men, B08 Adrenal gland

ANKLE DISORDERS (B)
G03 Ankle joint, I01 Shen Men, E10 Subcortex

ANXIETY SYNDROME (B)
O02 Lung, I01 Shen Men, B08 Adrenal gland

APHASIA, HYSTERICAL
E04 Brain, E05 Occiput, O01 Heart, I01 Shen Men, N04 Kidney, E10 Subcortex

APPENDICITIS (B)
M07 Appendix, M08 Large intestine, H01 Sympathetic point, O02 Lung

APPENDICITIS (ACUTE and/or CHRONIC)
M07 Appendix, H01 Sympathetic, O02 Lung, M08 Large intestine, I01 (Shen Men) Divine gate

APPENDIX PAIN
M07 Appendix, I01 Shen Men, H01 Sympathetic, M04 Stomach, E10 Subcortex

APPENDIX PAIN (A)
M07 Appendix

ARM INTERMITTENT PAIN (NEURALGIA) (A)
I01 Shen Men, E10 Subcortex

ARTHRITIS OR BURSITIS OF THE ANKLE (B)
G03 Ankle joint, I01 Shen Men, N04 Kidney, E10 Subcortex, B08 Adrenal gland

ARTHRITIS OR BURSITIS OF THE ELBOW (B)
J05 Elbow, I01 Shen Men, N04 Kidney, E10 Subcortex, B08 Adrenal gland

ARTHRITIS OR BURSITIS OF THE FINGER (B)
J01 Fingers, I01 Shen Men, N04 Kidney, E10 Subcortex, B08 Adrenal gland

ARTHRITIS CONDITIONS OF THE FOOT (A)
G01 Toes, G03 Ankle, B08 Adrenal, I01 Shen Men, E01 (brain axis) Occiput, E10 Subcortex

ARTHRITIS OR BURSITIS OF THE HIP (B)
G05 Hip joint, I01 Shen Men, N04 Kidney, E10 Subcortex, B08 Adrenal gland

ARTHRITIS OR BURSITIS OF THE KNEE (B)
G04 Knee joint, G06 Knee, N04 Kidney, I01 Shen Men, E10 Subcortex, B08 Adrenal gland

ARTHRITIS OR BURSITIS OF THE LOWER BACK (B)
F03 Lumbosacral vertebrae, I01 Shen Men, N04 Kidney, E10 Subcortex, B08 Adrenal gland

ARTHRITIS OR BURSITIS OF THE SCAPULA (B)
J03 Shoulder joint, I01 Shen Men, N04 Kidney, E10 Subcortex, B08 Adrenal gland

ARTHRITIS OR BURSITIS OF THE SHOULDER (B)
J04 Shoulder, J02 Clavicle, I01 Shen Men, E10 Subcortex, N04 Kidney, B08 Adrenal gland

ARTHRITIS OR BURSITIS OF THE TOES (B)
G02 Toes, I01 Shen Men, E10 Subcortex, B08 Adrenal gland

ARTHRITIS OR BURSITIS OF THE WRIST (B)
J06 Wrist, I01 Shen Men, N04 Kidney, E10 Subcortex, B08 Adrenal Gland

ASTHMA
H01 Sympathetic, B08 Adrenal gland, E07 Ping-Chuan, B08 Adrenal gland, I01 Shen Men, O02 Lung, D01 Internal secretion, E05 Occiput

ASTHMA (B)
H01 Sympathetic, I06 Asthma, B08 Adrenal gland, O02 Lung, E01 (brain axis) Occiput, I01 Shen Men, N04 Kidney

table of auricular acupoint treatment protocols

ATELENCEPHALIA
  N04 Kidney, B05 Occiput, E01 Brain axis, I01 Shen Men, E10 Subcortex, D01 Internal secretion, E09 Forehead

BACKACHE, LOWER (B)
  F04 Sacral vertebrae, F03 Lumbar vertebrae, H04 Lumbago

BACKACHE, UPPER (B)
  F02 Thoracic vertebrae, I01 Shen Men, E10 Subcortex

BALDNESS (B)
  N04 Kidney, D01 Internal secretion, O02 Lung, E01 (brain axis) Occiput

BONE SPICULA POST ARTHRITIS OR PERIOSTITIS
  N04 Kidney, D01 Internal secretion, E05 Occiput, B08 Adrenal gland, Corresponding regional points

BRONCHITIS
  E07 Ping-Chuan, O02 Lung, B08 Adrenal gland, I01 Shen Men, O03 Bronchi, O04 Windpipe, H01 Sympathetic, E05 Occiput

BRONCHITIS (B)
  O03 Bronchi, I06 Asthma, B08 Adrenal gland, H01 Sympathetic, I01 Shen Men, E01 (brain axis) Occiput

BUERGER'S DISEASE
  H01 Sympathetic, N04 Kidney, O01 Heart, B08 Adrenal gland, N09 Liver, N10 Spleen, D01 Internal secretion, E10 Subcortex, E05 Occiput

CATARACT (B)
  A10 Eye, D03 Eye 1, D04 Eye 2, N09 Liver, N04 Kidney, O02 Lung, N10 Spleen

CELLULITIS
  Corresponding regional points, B08 Adrenal gland, I01 Shen Men, N10 Spleen

CEREBELLAR ATAXIA
  E01 Brain axis, F01 Cervical vertebrae, E05 Occiput, N04 Kidney, I01 Shen Men

CEREBRAL CONCUSSION
  E10 Subcortex, I01 (Shen Men) Divine gate, E01 Brain axis, O01 Heart, M04 Stomach, N04 Kidney

CEREBRAL MENINGITIS, SEQUELA
  N04 Kidney, E01 Brain axis, E05 Occiput, I01 Shen Men, O01 Heart, M04 Stomach, E10 Subcortex

CEREBRAL VASCULAR ACCIDENT SEQUELA
  Corresponding regional points, B08 Adrenal gland, E10 Subcortex, D01 Internal secretion, I01 Shen Men, E05 Occiput

CHALAZION
  A10 Eye, N09 Liver, N10 Spleen

CHEST DISCOMFORT
  O02 Lung, O01 Heart, H01 Sympathetic, F06 Thorax, I01 (Shen Men) Divine gate

CHEST PAIN
  I01 Shen Men, O01 Heart, H01 Sympathetic, F06 Thorax, Corresponding regional points, M04 Stomach, E10 Subcortex

CHEST PRESSURE
  H01 Sympathetic, F06 Thorax, O01 Heart, E05 Occiput, O02 Lung

CHICKENPOX OR MEASLES (B)
  E06 Parotid gland, D01 Internal secretion, B08 Adrenal gland, O02 Lung, E01 (brain axis) Occiput, I01 Shen Men

CHOLECYSTITIS
  H01 Sympathetic, I01 Shen Men, N06 Pancreas and gallbladder, N09 Liver, B08 Adrenal gland, D01 Internal secretion

CHOLELITHIASIS
  N06 Pancreas and gallbladder, H01 Sympathetic, I01 Shen Men, N09 Liver, M05 Duodenum

CHOLEPYRRHIN
  H01 Sympathetic, I01 Shen Men, N06 Gallbladder & pancreas, N09 Liver, B08 Adrenal gland, E10 Subcortex, O06 Muscle relax

CHONDROMALACIA
  Corresponding regional points, D01 Internal secretion, B08 Adrenal gland, E10 Subcortex, N04 Kidney, I01 Shen Men

CHORIONITIS
  O02 Lung, E05 Occiput, D01 Internal secretion, B08 Adrenal gland, N09 Liver, N10 Spleen, E04 Brain

COLDS
  B06 Internal nose, E09 Forehead, O02 Lung, B08 Adrenal gland, I01 (Shen Men) Divine gate, E05 Occiput, E10 Subcortex

table of auricular acupoint treatment protocols

COLIC (BILIARY) (A)
N06 Gallbladder, N09 Liver

COLIC, INTESTINAL
M06 Small intestine, H01 Sympathetic, I01 Shen Men, K12 Ear apex, P02 Upper abdomen, P01 Lower abdomen

COLITIS, ANAPHYLACTIC
M08 Large intestine, D01 Internal secretion, O02 Lung, H01 Sympathetic, I01 Shen Men, M06 Small intestine

COLITIS (B)
M04 Stomach, L02 (point of support) Ear center, F10 Abdomen, H01 Sympathetic, M06 Small intestine

COMA (B)
H01 Sympathetic, O01 Heart, B08 Adrenal gland, E10 Subcortex

COMMOTIO CEREBRI, SEQUELA
N04 Kidney, E01 Brain Axis, I01 Shen Men, E05 Occiput, O01 Heart, M04 Stomach, E10 Subcortex

CONJUNCTIVITIS (ACUTE OR FOLLICULAR)
A10 Eye, N09 Liver

CONJUNCTIVITIS, ANAPHYLACTIC
A10 Eye, N09 Liver, E05 Occiput, D01 Internal secretion

CONJUNCTIVITIS AND HERPES OPHTHALMIA (B)
A10 Eye, D03 Eye 1, D04 Eye 2, N09 Liver, N04 Kidney

CONSTIPATION
K15 Lower section of the rectum, I08 Lower section of the rectum, M08 Large intestine, E10 Subcortex, H01 Sympathetic, N10 Spleen

CONSTIPATION (B)
M08 Large intestine, K15 Lower section of the rectum, I08 (lower section of the rectum) Constipation, O02 Lung, H01 Sympathetic, E10 Subcortex, N10 Spleen

CORNEA ULCER
N04 Kidney, N09 Liver, A10 Eye, D04 Eye 2

CORONARY THROMBOSIS
O01 Heart, H01 Sympathetic, D01 Internal secretion, N04 Kidney, B08 Adrenal gland, M06 Small intestine, E10 Subcortex

COUGH (B)
I06 Asthma, B02 (visceral) Throat, B08 Adrenal gland, O02 Lung, E01 (brain axis) Occiput, I01 Shen Men

COUGHS (ACUTE and CHRONIC)
E07 Ping-Chuan, O02 Lung, B08 Adrenal gland, I01 Shen Men, O04 Windpipe, H01 Sympathetic, E05 Occiput

CUTANEOUS PRURITIS
I01 Shen Men, O02 Lung, E05 Occiput, D01 Internal secretion, B08 Adrenal gland

CYSTITIS
N01 Urinary bladder, N04 Kidney, B08 Adrenal gland, H01 Sympathetic, I01 Shen Men, E05 Occiput

DEAFNESS (B)
A12 Inner ear, E01 (brain axis) Occiput, C01 External ear, I01 Shen Men

DENTAL PAIN (B)
A07 (maxilla) Upper jowl, A05 (mandible) Lower jowl, I01 Shen Men, E02 Toothache pain, F01 (cervical vertebrae) Soft palate

DERMATITIS SEBORRHOICA
O02 Lung, D01 Internal secretion, N10 Spleen, E05 Occiput, B08 Adrenal gland

DERMATITIS SOLARIS
I01 Shen Men, O02 Lung, D01 Internal secretion, B08 Adrenal gland

DERMATOLOGICAL CONDITIONS (B)
I01 Shen Men, D01 Internal secretion, E01 (brain axis) Occiput, O02 Lung, B08 Adrenal gland, N04 Kidney, J10 Urticaria

DIABETES INSIPIDUS
E04 Brain, D01 Internal secretion, H01 Sympathetic, I01 Shen Men, N04 Kidney, N01 Urinary bladder

DIARRHEA (DYSENTERY)
M08 Large intestine, M06 Small intestine, H01 Sympathetic, I01 Shen Men, K15 Lower segment of the rectum, N10 Spleen

DIARRHEA (B)
M08 Large intestine, M06 Small intestine, O02 Lung, N10 Spleen, H01 Sympathetic

table of auricular acupoint treatment protocols

DIFFUSED LIGHT
N04 Kidney, N09 Liver, A10 Eye, D04 Eye 2, E05 Occiput

DIMINISHED KIDNEY FUNCTION
N04 Kidney, N01 Urinary bladder, H01 Sympathetic, I01 Shen Men, B08 Adrenal gland, E05 Occiput

DIPLOPIA
N04 Kidney, N09 Liver, D04 Eye 2, A10 Eye

DIZZINESS
E10 Subcortex, E09 Forehead, E05 Occiput, I01 Shen Men

DORSAL SPINE PAIN (A)
F02 Thoracic vertebrae, H01 Sympathetic

DOUBLE VISION
N04 Kidney, N09 Liver, A10 Eye

DRUG DEPENDENCE (B)
O01 Heart, O02 Lung(Each point should be stimulated for two minutes, at 80 hertz, every six to eight hours for six weeks to avoid withdrawal symptoms.)

DUODENAL ULCER
M05 Duodenum, H01 Sympathetic, I01 Shen Men, F10 Abdomen, E10 Subcortex, M04 Stomach, O02 Lung

DYSENTERY, BACTERIAL
M08 Large intestine, M06 Small intestine, K15 Lower segment of the rectum, I01 Shen Men, E05 Occiput, D01 Internal secretion, O02 Lung, H01 Sympathetic

ECZEMA
O02 Lung, D01 Internal secretion, E05 Occiput, M08 Large intestine

ECZEMA, INFANTILE
Corresponding regional points, O02 Lung, E05 Occiput, D01 Internal secretionEYE PAIN FROM CONJUNCTIVITIS and NEURALGIA (A)
A10 Eye

EDEMA
N04 Kidney, N01 Bladder, O01 Heart, N09 Liver, H01 Sympathetic, D01 Internal secretion

EDEMA (B)
N04 Kidney, N01 Bladder, H01 Sympathetic, N10 Spleen, N09 Liver, D01 Internal secretion, B08 Adrenal gland

EJACULATIO PRAECOX
I05 Uterus, K13 External genitalia, E08 Testicle, D01 Internal secretion, I01 Shen Men

ELBOW DISORDERS (B)
J05 Elbow, I01 Shen Men, E10 Subcortex

ELBOW PAIN (A)
J05 Elbow, I01 Shen Men

ELECTRIC OPHTHALMITIS
N04 Kidney, N09 Liver, A10 Eye, I01 Shen Men

EMPHYSEMA, ALVEOLAR
O02 Lung, O03 Bronchi, H01 Sympathetic, I01 Shen Men, E07 Ping Chuan, E05 Occiput, B08 Adrenal gland

ENDOMETRITIS
I05 Uterus, D01 Internal secretion, B08 Adrenal gland, D02 Ovary, I10 External genitalia

ENTERITIS
M08 Large intestine, K15 Lower segment of the rectum, H01 Sympathetic, I01 Shen Men, M06 Small intestine, N10 Spleen

ENTERITIS, ALLERGIC
M08 Large intestine, M06 Small intestine, H01 Sympathetic, D01 Internal secretion, N09 Liver, I01 (Shen Men) Divine gate

ENURESIS
N01 Urinary bladder, E10 Subcortex, L03 Branch, N04 Kidney

ENURESIS, NOCTURNAL (B)
N01 Bladder, N04 Kidney, K14 Urethra, I01 Shen Men, B08 Adrenal gland, H01 Sympathetic, K13 External genitalia

EPIDIDYMITIS
E08 Testicle, D01 Internal secretion, I01 Shen Men, B08 Adrenal gland, K13 External genitalia, I03 Ku-Kuan

EPILEPSY
I01 Shen Men, N04 Kidney, E05 Occiput, O01 Heart, M04 Stomach, E10 Subcortex

EPISTAXIS
B06 Internal nose, B08 Adrenal gland, E09 Forehead, I01 (Shen Men) Divine gate

ERUCTATION
O01 Shen Men, M04 Stomach, H01 Sympathetic, O04 Trachea

ERYSIPELAS
Corresponding regional points, O02 Lung, E05 Occiput, B08 Adrenal gland, D01 Internal secretion, I01 Shen Men

FACIAL PARALYSIS
A11 Cheek, E05 Occiput, A10 Eye, M01 Mouth, A07 Upper jaw, A05 Lower jaw, N09 Liver

FACIAL PARALYSIS (B)
A07 (maxilla) Upper jowl, A05 (mandible) lower jowl, A11 Cheek, I01 Shen Men, E01 (brain axis) Occiput, C01 External ear, J11 Minor occipital nerve

FACIAL SPASMS
A11 Cheek, I01 Shen Men, E10 Subcortex, E11 Tai Yang

FEVER (B)
O02 Lung, B08 Adrenal gland, D01 Internal secretion, I01 Shen Men

FEVER, LOW GRADE
K12 Apex of the auricle, B01 Apex of the tragus, B08 Adrenal gland, D01 Internal secretion, N09 Liver, N10 Spleen, I01 Shen Men

FINGER MOTOR DISORDER (FINGER RIGIDITY) (B)
J01 Fingers, I01 Shen Men, E10 Subcortex

FINGER SENSORY DISORDER (B)
J01 Finger, I01 Shen Men, E10 Subcortex

FOLLICULITIS
Corresponding regional points, E05 Occiput, O02 Lung, D01 Internal secretion, B08 Adrenal gland

FOOT DISORDERS (B)
G01 Toes, G02 Heel, G03 Ankle joint, I01 Shen Men

FOOT NEURALGIA (INTERMITTENT PAIN) (A)
I01 Shen Men, H01 Sympathetic, G01 Toes, G03 Ankle joint

FRACTURE, CONTUSION, SPRAIN, OR CRUSH INJURY
Corresponding regional points, I01 Shen Men, N04 Kidney, E10 Subcortex, B08 Adrenal gland

FRIGIDITY IN WOMEN (B)
K13 External genitalia, B08 Adrenal gland, D01 Internal secretion, O02 Lung, I01 Shen Men

FUNCTIONAL DISTURBANCES OF THE STOMACH AND/OR INTESTINE
M04 Stomach, M06 Small intestine, M08 Large intestine, H01 Sympathetic, I01 Shen Men, N10 Spleen, O05 San Jiao

FURUNCLE, CARBUNCLE OR PARONYCHIA
Corresponding regional points, I01 Shen Men, E05 Occiput, B08 Adrenal gland

FURUNCULOSIS OF THE EXTERNAL MEATUS
N04 Kidney, A12 Internal ear, D01 Internal secretion, C01 External ear

GALLBLADDER DISEASE (B)
N06 Gallbladder & pancreas, N09 Liver, H01 Sympathetic, M05 Duodenum

GALLBLADDER PARASITES
N06 Gallbladder & pancreas, N09 Liver, H01 Sympathetic, I01 Shen Men, E10 Subcortex

GALLBLADDER STONE
H01 Sympathetic, I01 (Shen Men) Divine gate, N06 Gallbladder & pancreas, N09 Liver, E10 Subcortex

GASTRALGIA (B)
M04 Stomach, H01 Sympathetic, I01 Shen Men, L02 (point of support) Ear Center, O02 Lung, M06 Small intestine, M08 Large intestine, E10 Subcortex, M02 Esophagus, P02 Upper abdomen, P01 Lower abdomen

GASTRIC ULCER
M04 Stomach, M06 Small intestine, H01 Sympathetic, E10 Subcortex, I01 Shen Men, N10 Spleen, O02 Lung, F10 Abdomen

GASTRITIS (ACUTE and/or CHRONIC)
M04 Stomach, M06 Small intestine, H01 Sympathetic, E10 Subcortex, I01 Shen Men, N10 Spleen, O02 Lung, F07 Abdomen

GASTRO-ENTERIC NEUROSIS
M04 Stomach, M09 Liver, H01 Sympathetic, I01 Shen Men, P02 Upper abdomen, P01 Lower abdomen, M05 Duodenum

GASTROPTOSIS
M04 Stomach, H01 Sympathetic, E10 Subcortex, I01 Shen Men, N09 Liver

GLAUCOMA
N04 Kidney, N09 Liver, D03 Eye 1, D04 Eye 2, A10 Eye

GLAUCOMA (B)
A10 Eye, D03 Eye 1, D04 Eye 2, N09 Liver, N04 Kidney, O02 Lung, N10 Spleen

GLOSSITIS
A03 Tongue, O01 Heart, I01 Shen Men, M01 Mouth, D01 Internal secretion

GYNECOMASTIA
D01 Internal secretion, E04 Brain, F08 Mammary gland

HALLUCINATIONS
N04 Kidney, N09 Liver, A10 Eye, E05 Occiput

HAND AND FINGER PAIN (A)
J01 Finger, J06 Wrist

HEADACHE
E10 Subcortex, E09 Forehead, E05 Occiput, I01 (Shen Men) Divine gate

HEADACHE (GENERAL) (A)
E05 Occiput, E09 Forehead, H01 Sympathetic, I01 Shen Men, E10 Subcortex

HEADACHE OF THE FRONTAL REGION (A)
E09 Forehead, H01 Sympathetic, I01 Shen Men, E10 Subcortex

HEADACHE OF THE OCCIPITAL REGION (A)
E05 Occiput, H01 Sympathetic, I01 Shen Men, E10 Subcortex

HEADACHE OF THE TEMPORAL REGION
H01 Sympathetic, I01 Shen Men

HEADACHE OF THE VERTEX (A)
E05 Occiput, E09 Forehead, H01 Sympathetic, I01 Shen Men, E10 Subcortex

HEART DISORDERS (B)
O01 Heart, H01 Sympathetic, I01 Shen Men, D01 Internal secretion

HEAT STROKE
E05 Occiput, O01 Heart, E10 Subcortex, B08 Adrenal gland

HEMATURIA
N04 Kidney, N01 Urinary bladder, N09 Liver, L01 Diaphragm, B08 Adrenal gland

HEMORRHOIDS (B)
K15 Lower segment of the rectum, M08 Large intestine, N10 Spleen, B08 Adrenal gland, E10 Subcortex

HEPATITIS
H01 Sympathetic, I01 (Shen Men) Divine gate, N09 Liver, D03 Eye 1, D04 Eye 2, N06 Gallbladder & pancreas

HEPATITIS (B)
N09 Liver, H01 Sympathetic, N10 Spleen, I07 Hepatitis point, K11 Liver Yang, N06 Gallbladder, D01 Internal secretion, N08 (pancreatitis point) Pancreas, M04 Stomach, E10 Subcortex

HEPATITIS, INFECTIOUS (ACUTE and CHRONIC)
N09 Liver, H01 Sympathetic, I01 Shen Men, N10 Spleen, K10 Liver Yang 1, K11 Liver Yang 2, N06 Gallbladder & pancreas, D01 Internal secretion, N04 Kidney

HERNIA, INCOMPLETE
P01 Lower abdomen, E10 Subcortex, D01 Internal secretion

HERPES PROGENITALIS (B)
O02 Lung, I01 Shen Men, K13 External genitalia, B08 Adrenal gland

HERPES SIMPLEX (B)
O02 Lung, I01 Shen Men, B08 Adrenal gland, D01 Internal secretion

HERPES ZOSTER (B)
O02 Lung, E01 (brain axis) Occiput, D01 Internal Secretion, B08 Adrenal gland

HICCUPS, PERSISTENT (B)
M06 Small intestine, M04 Stomach, N06 Gallbladder & pancreas, N08 (pancreatitis point) Pancreas, N10 Spleen

HIP DISORDERS (B)
G05 Hip joint, I01 Shen Men, N04 Kidney, E10 Subcortex, B08 Adrenal gland

HIP GENERAL (INCLUDING ARTHRITIC) PAIN (A)
H03 Buttock, G05 (hip joint) Femoral joint

HOARSENESS OF THE VOICE
B04 (pharynx and larynx) Throat, O01 Heart, O02 Lung, I01 Shen Men, D01 Internal secretion

HORDEOLUM
A10 Eye, N09 Liver, N10 Spleen

table of auricular acupoint treatment protocols

HYPERHIDROSIS
H01 Sympathetic, O02 Lung, D01 Internal secretion, E05 Occiput, B08 Adrenal gland

HYPERKINETIC CHILDREN (B)
I01 Shen Men, O01 Heart

HYPERTENSION
I04 Depressing point, Z01 Hypotensive groove, H01 Sympathetic, I01 Shen Men, O01 Heart, B08 Adrenal gland

HYPERTENSION (B)
B09 (high blood pressure) Lowering pressure point, O01 Heart, H01 Sympathetic, I01 Shen Men

HYPERTHYROIDISM
F09 Thyroid, D01 Internal secretion, E04 Brain, I01 Shen Men, F01 Cervical vertebrae

HYPERTROPHIC SPONDYLOPATHY
Corresponding regional points, D01 Internal secretion, B08 Adrenal gland, E10 Subcortex, N04 Kidney, I01 Shen Men

HYPOGLYCEMIA, FUNCTIONAL (B)
E04 (brain, hypophysis, pituitary body) Pituitary, B08 Adrenal, I01 Shen Men, D01 Internal secretion

HYPOPHYSIAL DWARFISM
N04 Kidney, D01 Internal secretion, E04 Brain, E08 Testicle (if male), D02 Ovary (if female)

HYPOTENSION (B)
H01 Sympathetic, O01 Heart, B08 Adrenal gland, E10 Subcortex

HYPOTHYROIDISM
F09 Thyroid, D01 Internal secretion, E04 Brain, I01 Shen Men

HYPOTHYROIDISM (B)
F02 (thoracic vertebrae) Thyroid, F09 (thyroid) Neck, D01 Internal secretion, I01 Shen Men

HYSTERIA
E10 Subcortex, E01 Brain axis, I01 (Shen Men) Divine gate, E05 Occiput, M04 Stomach, O01 Heart, N04 Kidney

HYSTERICAL MUTISM
E01 Brain axis, E05 Occiput, E09 Forehead, I01 (Shen Men) Divine gate

HYSTERICAL PARALYSIS
E10 Subcortex, I01 (Shen Men) Divine gate, E05 Occiput, O01 Heart, M04 Stomach, E01 Brain axis

IDIOTIC
E10 Subcortex, I01 (Shen Men) Divine gate, M04 Stomach, E01 Brain axis, O01 Heart, N04 Kidney

IMPAIRED HEARING
N04 Kidney, E05 Occiput, A12 Internal ear, C01 External ear

IMPOTENCE
K13 External genitalia, I10 External genitalia, E08 Testis, I05 Uterus, D01 Internal secretion, N04 Kidney, E10 Subcortex

IMPOTENCE (B)
I05 Uterus, K13 External genitalia, I10 External genitalia, E08 Testicle, N04 Kidney, D01 Internal secretion

INCONTINENCE OF URINE
N01 Urinary bladder, E04 Brain, L02 Point of support

INDIGESTION
M06 Small intestine, H01 Sympathetic, M04 Stomach, N06 Gallbladder & pancreas, N10 Spleen, M08 Large intestine, O05 San Jiao, I01 Shen Men

INDIGESTION (B)
M06 Small intestine, M04 Stomach, N06 Gallbladder & pancreas, N08 (pancreatitis point) Pancreas, O01 Heart, N10 Spleen

INFANTILE PARALYSIS, SEQUELA
Corresponding regional points, I01 Shen Men, B08 Adrenal gland, D01 Internal secretion, E10 Subcortex, E05 Occiput

INFECTION (B)
N09 Liver, H01 Sympathetic, I07 Hepatitis point, N10 Spleen, K11 Liver yang, B08 Adrenal, N04 Kidney, I01 Shen Men, D01 Internal Secretion, N06 Gallbladder & pancreas

INFERTILITY (B)
I05 Uterus, D01 Internal secretion, N04 Kidney, N09 Liver

INFLAMMATION OF THE ARTERIAL WALL
H01 Sympathetic, N04 Kidney, O01 Heart, B08 Adrenal gland, N09 Liver, N10 Spleen, E05 Occiput, D01 Internal secretion

INFLUENZA (AND OTHER VIRAL DISEASES) (B)
    B06 Internal nose, B08 Adrenal gland, E09 Forehead, O02 Lung

INSOMNIA (POOR QUALITY SLEEP)
    E10 Subcortex, O01 Heart, I01 Shen Men, E05 Occiput, N04 Kidney

INSOMNIA (B)
    I01 Shen Men, E10 Subcortex, N04 Kidney, E01 (brain axis) Occiput, O01 Heart

INSUFFICIENCY OF LACTATION (B)
    F08 Mammary gland

INTERCOSTAL NEURALGIA
    F06 Thorax, I01 Shen Men, E05 Occiput

INTERCOSTAL NEURALGIA (A)
    F06 (thorax) Chest

INTESTINAL OBSTRUCTION, PARALYTIC
    M08 Large intestine, M06 Small intestine, H01 Sympathetic, E10 Subcortex, F07 Abdomen

INTESTINAL TUBERCULOSIS
    M04 Large intestine, M06 Small intestine, H01 Sympathetic, I01 Shen Men, D01 Internal secretion, E05 Occiput, O05 San Jiao

IRREGULAR MENSES
    I05 Uterus, D02 Ovary, D01 Internal secretion, N04 Kidney, B08 Adrenal gland

ISCHIALGIA
    H02 Ischium, I01 Shen Men, N04 Kidney, E05 Occiput, B08 Adrenal gland

ITCHING
    O02 Lung, B08 Adrenal gland, N09 Liver, I01 (Shen Men) Divine gate, E10 Subcortex

JAW GENERAL PAIN (A)
    M01 Mouth

JOINT DISLOCATION, HABITUAL
    Corresponding regional points, D01 Internal secretion, B08 Adrenal gland, E10 Subcortex, N10 Spleen, N09 Liver

JOINT FRICTION
    Corresponding regional points, D01 Internal secretion, B08 Adrenal gland, E10 Subcortex, N04 Kidney, I01 Shen Men

JOINT TWISTING
    Corresponding joint, I01 (Shen Men) Divine gate, E10 Subcortex, B08 Adrenal gland

KNEE DISORDERS (B)
    G04 Knee joint, G06 Knee, I01 Shen Men, B08 Adrenal gland, N04 Kidney, E10 Subcortex

KNEE (INTERMITTENT) PAIN (NEURALGIA) (A)
    G06 Knee

LARYNGITIS (CHRONIC)
    B04 (pharynx and larynx) Throat, D01 Internal secretion, I01 Shen Men, O01 Heart, B08 Adrenal gland, O02 Lung, E03 Hou Ya

LEG GENERAL PAIN (A)
    G03 Ankle joint, I01 Shen Men, G06 Knee

LEUCORRHAGIA
    I05 Uterus, D02 Ovary, D01 Internal secretion

LEUCORRHEA
    I05 Uterus, D01 Internal secretion, I01 (Shen Men) Divine gate, D02 Ovary

LEUKOPENIA
    N09 Liver, N10 Spleen, N04 Kidney, D01 Internal secretion, O01 Heart, E05 Occiput, L01 Diaphragm, E10 Sympathetic

LOWER LEG CONSTANT PAIN (A)
    I01 Shen Men, B08 Adrenal gland

LOWER LEG INTERMITTENT PAIN (NEURALGIA) (A)
    I01 Shen Men, H01 Sympathetic

LUMBAR PAIN (A)
    F03 Lumbosacral, H03 Buttock, I01 Shen Men

LYMPHNODONCUS, MULTIPLE
    I03 Ku Kuan, E05 Occiput, D01 Internal secretion

MALARIA
    E10 Subcortex, D01 Internal secretion, N09 Liver, N10 Spleen, B08 Adrenal gland

MALARIA (B)
    E10 Subcortex, D01 Internal secretion, B08 Adrenal gland, N09 Liver, N10 Spleen

MAMMARY ABSCESS
    F08 Mammary gland, D01 Internal secretion, E05 Occiput, F06 Thorax, B08 Adrenal gland

MANIC DEPRESSIVE DISORDERS (B)
    N04 Kidney, E01 (brain axis) Occiput, O01 Heart, I01 Shen Men, M04 Stomach, F05 (neck) Brain stem, E10 Subcortex, J11 Minor occipital nerve

MASTITIS (B)
F08 Mammary gland, D01 Internal secretion, E01 (brain axis) Occiput, F06 Thorax, B08 Adrenal gland

MEMORY LOSS
E10 Subcortex, E01 Brain axis, I01 Shen Men, E05 Occiput, M04 Stomach, O01 Heart, N04 Kidney

MENIERE'S DISEASE
N04 Kidney, I01 Shen Men, E05 Occiput, A12 Internal ear, E10 Subcortex, M04 Stomach

MENIERE'S SYNDROME (B)
N04 Kidney, E01 (brain axis) Occiput, A12 Inner ear, I01 Shen Men, E10 Subcortex, M04 Stomach, J11 Minor occipital nerve

MENOPAUSE
I05 Uterus, D02 Ovary, D01 Internal secretion

MENORRHALGIA
I05 Uterus, H01 Sympathetic, I01 Shen Men, E10 Subcortex

MENSTRUAL PAIN (A)
I05 Uterus, E08 (testicle) Ovary, D01 (internal secretion) Endocrine

MENSTRUATION, EXCESSIVE (B)
I05 Uterus, D01 Internal secretion, N04 Kidney, H01 Sympathetic

METROPTOSIS
I05 Uterus, E10 Subcortex

MICTURATION, FREQUENT
N04 Kidney, N01 Urinary bladder, I01 Shen Men, I09 Urethra

MIGRAINE HEADACHE
E10 Subcortex, E11 Tai Yang, N04 Kidney, E05 Occiput, E09 Forehead

MIGRAINE HEADACHE (A)
E05 Occiput, E09 Forehead, H01 Sympathetic, I01 Shen Men, E10 Subcortex

MIGRAINE HEADACHE (B)
E11 (Tai Yang) Temple, J11 Minor occipital nerve, I01 Shen Men, N04 Kidney, E10 Subcortex

MILIARIA
O02 Lung, D01 Internal secretion, I01 Divine gate, B08 Adrenal gland, E05 Occiput, I01 Shen Men

MOTION SICKNESS
E05 Occiput, M04 Stomach, A12 Inner ear, I01 Shen Men

MOUTH ULCER
M01 Mouth, D01 Internal secretion, I01 Shen Men, A03 Tongue

MULTIPLE NEURITIS
Corresponding regional points, I01 Shen Men, B08 Adrenal gland, D01 Internal secretion

MULTIPLE SCLEROSIS OF THE LOWER EXTREMITIES (B)
H02 (Ischium) Sciatic, I01 Shen Men, N04 Kidney

MULTIPLE SCLEROSIS OF THE UPPER EXTREMITIES (B)
J04 Shoulder, I01 Shen Men, N04 Kidney

MUMPS (B)
E06 Parotid gland, D01 Internal secretion, A11 Cheek, E10 Subcortex

MUSCULAR ATROPHY OF THE LOWER EXTREMITIES (B)
H02 (Ischium) Sciatic, I01 Shen Men, N04 Kidney

MUSCULAR ATROPHY OF THE UPPER EXTREMITIES (B)
I01 Shen Men, N04 Kidney, J04 Shoulder, J05 Elbow

MUSCULOPHRENIC SPASM
L01 Diaphragm, I01 Shen Men, E10 Subcortex, K12 Ear apex

MUTISM (B)
A12 Inner ear, E01 (brain axis) Occiput, C01 Outer ear, I01 Shen Men

MYALGIA (B)
O06 Muscle relax, G07 Heat, I01 Shen Men, Corresponding regional points

MYOCARDITIS
O01 Heart, M06 Small intestine, H01 Sympathetic, I01 Shen Men, E05 Occiput

MYCOTIC STOMATITIS
M01 Mouth, D01 Internal secretion, B08 Adrenal gland, N10 Spleen, E05 Occiput

MYOPIA
N04 Kidney, N09 Liver, D04 Eye 2, A10 Eye

MYOPIA (B)
A10 Eye, D03 Eye 1, D04 Eye 2, N09 Liver, N04 Kidney, O02 Lung, N10 Spleen

NASAL AND SINUS DISORDERS (B)
B06 Internal nose, B08 Adrenal gland, E09 Forehead, D01 Internal secretion, O02 Lung

NASAL INFECTIONS (B)
B06 Internal nose, B08 Adrenal gland, E09 Forehead, D01 Internal secretion, O02 Lung

NAUSEA AND VOMITING (B)
O01 Heart, M04 Stomach, I01 Shen Men, E01 (brain axis) Occiput, H01 Sympathetic, M02 Esophagus, E10 Subcortex

NAUSEA and VOMITING
M04 Stomach, I01 Shen Men, E05 Occiput, H01 Sympathetic, E10 Subcortex, M02 Esophagus

NECK GENERAL PAIN (A)
F01 Cervical vertebrae, F05 Neck

NEPHRITIS
N01 Urinary bladder, N04 Kidney, B08 Adrenal gland, E10 Subcortex, I01 (Shen Men) Divine gate, H01 Sympathetic

NEPHRITIS (ACUTE)
N04 Kidney, N01 Urinary bladder, H01 Sympathetic, I01 Shen Men, N09 Liver, B08 Adrenal gland, D01 Internal secretion

NEPHROPATHIC SYNDROME
N04 Kidney, N01 Urinary bladder, H01 Sympathetic, I01 Shen Men, N07 Ascites, B08 Adrenal gland

NEURATHENIA
E10 Subcortex, I01 (Shen Men) Divine gate, M04 Stomach, O01 Heart, N04 Kidney

NEURODERMATITIS
Corresponding regional points, E05 Occiput, D01 Internal secretion, O02 Lung, B08 Adrenal gland, I01 (Shen Men) Divine gate

NEUROGENIC THIRST
D01 Internal secretion, E10 Subcortex, I01 (Shen Men) Divine gate, N04 Kidney

NIGHT BLINDNESS
N09 Liver, D04 Eye 2, A10 Eye

NOSE BLEEDING (B)
B06 Internal nose, B08 Adrenal gland, K12 Ear apex, O02 Lung, E09 Forehead

OPTIC ATROPHY
N04 Kidney, N09 Liver, A10 Eye

ORCHITIS
E08 Testicle, D01 Internal secretion, I01 Shen Men, B08 Adrenal gland, K13 External genitalia, I10 External genitalia, I03 Ku-Kuan

OTITIS MEDIA
N04 Kidney, A12 Internal ear, D01 Internal secretion, C01 External ear

OTITIS MEDIA (B)
N04 Kidney, A12 Inner ear, D01 Internal secretion, E01 (brain axis) Occiput, C01 External ear

PALPITATIONS (B)
O01 Heart, H01 Sympathetic, I01 Shen Men, M06 Small intestine, N10 Spleen

PANCREATITIS
N06 Gallbladder & pancreas, D01 Internal secretion, H01 Sympathetic, I01 Shen Men

PANCREATITIS (B)
N08 (pancreatitis point) Pancreas, D01 Internal secretion, H01 Sympathetic, I01 Shen Men

PAPILLITIS
N04 Kidney, N09 Liver, D03 Eye 1, D04 Eye 2, A10 Eye

PARALYSIS, HYSTERICAL
E10 Subcortex, I01 Shen Men, E05 Occiput, O01 Heart, Corresponding regional points, M04 Stomach, N04 Kidney

PARESIS OR PARALYSIS OF THE LOWER EXTREMITIES (B)
I01 Shen Men, I03 (Bu-Kuan, coxo-femoral joint) Thigh joint, G06 Knee, G04 Knee joint, H02 (ischium) Sciatic

PARESIS OR PARALYSIS OF THE UPPER EXTREMITIES (B)
J04 Shoulder, J03 Shoulder joint, I01 Shen Men, J02 Clavicle, B08 Adrenal gland

PAROTITIS
E06 Parotid, D01 Internal secretion, A11 Cheek, E10 Subcortex

PAROTITIS
E06 Parotid, D01 Internal secretion, I01 Shen Men

table of auricular acupoint treatment protocols

PELVIOPERITONITIS, CHRONIC
I05 Uterus, D02 Ovary, D01 Internal secretion, I02 Pelvic cavity

PERIARTHRITIS OF THE SHOULDER
J03 Shoulder joint, J04 Shoulder, I01 Shen Men, J02 Clavicle, B08 Adrenal gland

PERIODONTITIS
A07 Maxilla, A05 Mandible, M01 Mouth, B08 Adrenal gland, N04 Kidney

PERIPHERAL CIRCULATORY DISTURBANCES
Corresponding regions, D01 Internal secretion, B08 Adrenal gland

PERNIO (1 OR 2 DEGREES)
Corresponding regional points, I01 Shen Men, E05 Occiput, N10 Spleen, B08 Adrenal gland

PERTUSSIS
O02 Lung, O03 Bronchi, B08 Adrenal gland, I01 Shen Men, E07 Ping-Chuan, H01 Sympathetic, E05 Occiput

PHARYNGITIS (CHRONIC)
B04 (pharynx and larynx) Throat, I01 (Shen Men) Divine gate, B08 Adrenal gland, O01 Heart, D01 Internal secretion

PHTHISIS
O02 Lung, F06 Thorax, B08 Adrenal gland, D01 Internal secretion, E10 Subcortex, O05 San Jiao

PIRIFORMIS MUSCLE SPASM (B)
H02 (ischium) Sciatic nerve, H03 Buttocks, F04 Coccyx

PLEURAL ADHESIONS
F06 Thorax, B08 Adrenal gland, D01 Internal secretion, E10 Subcortex, I01 Shen Men

PLEURISY
O02 Lung, F06 Thorax, B08 Adrenal gland, D01 Internal secretion, E10 Subcortex, O05 San Jiao

PNEUMOEMPHYSEMA
H01 Sympathetic, E07 (Ping-Chuan) Asthma 2, O02 Lung, B08 Adrenal gland, E05 Occiput, F06 Thorax

PNEUMONIA, BRONCHIAL
O02 Lung, O03 Bronchi, H01 Sympathetic, I01 Shen Men, E07 Ping Chuan, B08 Adrenal gland, E05 Occiput, D01 Internal secretion

PNEUMONIA, LOBAR
O02 Lung, F06 Thorax, B08 Adrenal gland, D01 Internal secretion, I01 Shen Men, E10 Subcortex

POLIO OF THE LOWER EXTREMITIES (B)
H02 (ischium) Sciatic, I01 Shen Men, N04 Kidney

POLIO OF THE UPPER EXTREMITIES (B)
I01 Shen Men, N04 Kidney, J04 Shoulder, J05 Elbow

POLIO SEQUELA
Corresponding regional points, I01 Shen Men, B08 Adrenal gland, D01 Internal secretion, E10 Subcortex, E05 Occiput

POLLAKIURIA, PRECIPITANT URINATION
N01 Urinary bladder, N04 Kidney, I01 Shen Men

POST-MENSTRUAL PAIN (B)
I05 Uterus, D01 Internal secretion, N04 Kidney, H01 Sympathetic

POSTNATAL INVOLUTION PAIN
I05 Uterus, H01 Sympathetic, I01 Shen Men, E10 Subcortex

POST PARTUM UTERINE PAIN
I05 Uterus, H01 Sympathetic, I01 Shen Men, E10 Subcortex, N10 Spleen

PRE-MENSTRUAL CRAMPS (B)
I05 Uterus, D01 Internal secretion, N04 Kidney, H01 Sympathetic

PROLAPSE OF THE ANUS
K15 Lower segment of the rectum, M08 Large intestine, E10 Subcortex, N10 Spleen

PROSTATITIS
N02 Prostate gland, D01 Internal secretion, N04 Kidney, E10 Subcortex

PROSTATITIS
N02 Prostate, N01 Urinary bladder, D01 Internal secretion, N04 Kidney, E05 Occiput

PROSTATITIS (B)
D01 Internal secretion, B08 Adrenal gland, I02 Pelvis, I05 Uterus

PYLONEPHRITIS
N04 Kidney, N01 Urinary bladder, H01 Sympathetic, I01 Shen Men, N09 Liver, B08 Adrenal gland, D01 Internal secretion

table of auricular acupoint treatment protocols

RECTAL PROLAPSE
I08 Lower segment of the rectum, K15 Lower segment of the rectum, E10 Subcortex

RECTAL PROLAPSE (B)
K15 Lower segment of the rectum, M08 Large intestine, E10 Subcortex, N10 Spleen

RENAL CALCULUS
N04 Kidney, N03 Ureter, H01 Sympathetic, I01 Shen Men, E10 Subcortex

RENAL COLIC (A)
N04 Kidney

RENAL FAILURE, CHRONIC (B)
K15 Lower segment of the rectum, M04 Stomach, H01 Sympathetic, I01 Shen Men

RETENTION OF URINE
N04 Kidney, N01 Urinary bladder, H01 Sympathetic, K13 External genitalia, E10 Subcortex

RHEUMATOID ARTHRITIS
I01 Shen Men, N04 Kidney, D01 Internal secretion, E05 Occiput, Corresponding regional points, E10 Subcortex

RHEUMATOID CARDIOPATHY
O01 Heart, D01 Internal secretion, H01 Sympathetic, I01 Shen Men, M06 Small intestine, E10 Subcortex

RHINITIS, ALLERGIC
B06 Internal nose, D01 Internal secretion, O02 Lung, I01 (Shen Men) Divine gate

RHINITIS, ANAPHYLACTIC
B06 Internal nose, B08 Adrenal gland, E09 Forehead, O02 Lung, D01 Internal secretion

RHINITIS, SIMPLE OR ATROPHIC
B06 Internal nose, B08 Adrenal gland, E09 Forehead, O02 Lung

ROSACEA
B05 External nose, O02 Lung, D01 Internal secretion, B08 Adrenal gland

ROUND WORMS IN THE BILIARY TRACT
N06 Gallbladder & pancreas, H01 Sympathetic, I01 Shen Men, N09 Liver, M05 Duodenum

SCHIZOPHRENIA
E05 Occiput, I01 (Shen Men) Divine gate, M04 Stomach, E01 Brain axis, O01 Heart, N04 Kidney, E10 Subcortex,

SCIATICA (B)
H02 (ischium) Sciatic, I01 Shen Men, N04 Kidney

SCIATICA
H02 (ischium) Sciatic nerve, H03 Buttock, B08 Adrenal gland, I01 (Shen Men) Divine gate

SCIATIC PAIN (A)
H02 (ischium) Sciatic nerve

SECRETORY DISTURBANCES
D01 Internal Secretion, E04 Brain, E10 Subcortex, N10 Spleen, E08 Testicle (if male), D02 Ovary (if female)

SHEEHAN'S DISEASE
E04 Brain, N09 Liver, N10 Spleen, H01 Sympathetic, I05 Uterus, D01 Internal secretion

SHINGLES (HERPES ZOSTER)
Corresponding regional points, O02 Lung, E05 Occiput, D01 Internal secretion, B08 Adrenal gland, I01 (Shen Men) Divine gate

SHOCK
B08 Adrenal gland, E05 Occiput, O01 Heart, E04 Brain, E10 Subcortex

SHORTNESS OF BREATH
B06 Internal nose, B08 Adrenal gland, E09 Forehead, I01 Shen Men, O02 Lung

SHOULDER BURSITIS
J03 Shoulder joint, J04 Shoulder, I01 (Shen Men) Divine gate, B08 Adrenal gland

SHOULDER DISORDERS (B)
J04 Shoulder, J03 Shoulder joint, I01 Shen Men, J02 Clavicle, B08 Adrenal gland

SHOULDER PAIN (A)
J04 Shoulder, J03 Shoulder joint, J02 Clavicle

SINUSITIS
B06 Internal nose, B08 Adrenal gland, E09 Forehead

SNEEZING (RHINORRHEA) (B)
B06 Internal nose, B08 Adrenal gland, E09 Forehead, D01 Internal secretion, O02 Lung

SPRAIN, ANKLE (A)
G03 Ankle joint, I01 Shen Men, B08 Adrenal gland

table of auricular acupoint treatment protocols

SPRAIN, ANKLE (B)
G03 Ankle joint, I01 Shen Men, E10 Subcortex

SPRAIN, ELBOW (B)
J05 Elbow, I01 Shen Men, E10 Subcortex

SPRAIN, HIP (B)
G05 Hip joint, I01 Shen Men, E10 Subcortex

SPRAIN, KNEE (B)
G04 Knee joint, G06 Knee, I01 Shen Men, E10 Subcortex

SPRAIN, NECK (B)
F01 Cervical vertebrae, I01 Shen Men, E10 Subcortex

SPRAINS
Corresponding part, I01 (Shen Men) Divine gate, E10 Subcortex, B08 Adrenal gland

SPRAIN, WRIST (B)
J06 Wrist, I01 Shen Men, E10 Subcortex

STOMACH SPASM
M04 Stomach, M09 Liver, H01 Sympathetic, I01 Shen Men, P02 Upper abdomen, P01 Lower abdomen

SUNBURN
B08 Adrenal gland, D01 Internal secretion, O02 Lung, I01 Shen Men, M04 Stomach

SYNCOPE (B)
B08 Adrenal gland, E01 (brain axis) Occiput, O01 Heart, E04 Brain point, E10 Subcortex

TEETH, IMPACTED
A07 (maxilla) Upper jaw, A05 (mandible) Lower jaw

TEMPOROMANDIBULAR JOINT DISORDERS (B)
A05 Mandible, E11 (Tai Yang) Temple, I01 Shen Men

THROAT PAIN (A)
B02 (Visceral) Pharynx, A14 (tonsil 4) Tonsil

THROMBOPENIC PURPURA
N09 Liver, N10 Spleen, L01 Diaphragm, H01 Sympathetic, I01 Shen Men, D01 Internal secretion, E05 Occiput, O01 Heart

TINNITUS
A12 Inner ear, N04 Kidney

TINNITUS AURIUM
N04 Kidney, E05 Occiput, A12 Internal ear, C01 External ear

TOE SENSORY DISORDER (B)
G01 Toes, I01 Shen Men, E10 Subcortex

TONSILLITIS (B)
A14 Tonsil, B02 (visceral) Throat, K12 Ear Apex, I01 Shen Men

TONSILLITIS (ACUTE)
A14 Tonsil, A06 Tonsil 3, B04 Pharynx and larynx, B08 Adrenal gland, K01 Helix 1, K02 Helix 2, K03 Helix 3, K04 Helix 4, K05 Helix 5, K06 Helix 6

TOOTH, GROWTH-RETARDED
A07 Maxilla, A05 Mandible, I01 Shen Men, E02 Toothache, E03 Hou Ya

TOOTHACHE
A07 (maxilla) Upper jaw, A05 (mandible) Lower jaw, I01 Shen Men, E02 Toothache, E03 Hou Ya

TOOTHACHE IN THE UPPER JAW and LOWER JAW (A)
A01 (tooth extraction anesthesia point 1) Teeth (upper), A08 (tooth extraction anesthesia point 2) Teeth (lower)

TORTICOLLIS
F01 Cervical vertebrae, F05 Neck, I01 Shen Men

TORTICOLLIS (A)
F01 Cervical vertebrae, F05 Neck

TORTICOLLIS (B)
F01 Cervical vertebrae, I01 Shen Men, K13 External genitalia

TRACHITIS (ACUTE and/or CHRONIC)
I01 Shen Men, O04 Trachea, I06 Asthma 1, E07 Ping-Chuan, H01 Sympathetic, B08 Adrenal gland, E05 Occiput

TRIGEMINAL NEURALGIA
A11 Cheek, A07 Upper jaw, A05 Lower jaw, I01 Shen Men, E05 Occiput, E11 Tai Yang, C01 External ear

TRIGEMINAL NEURALGIA (A)
H01 Sympathetic, E10 Subcortex, I01 Shen Men, A14 (tonsil 4) Cheek, A08 (tooth extraction anesthesia point 2) Lower jaw, A01 (tooth extraction anesthesia point 1) Upper jaw

TRIGEMINAL NEURALGIA (B)
A11 Cheek, A07 (maxilla) Upper jowl, A05 (mandible) Lower jowl, C01 External ear, J11 Minor occipital nerve, I01 Shen Men, E01 (brain axis) Occiput

ULCEROUS VESTIBULUM NASI
B06 Internal nose, B08 Adrenal gland, E09 Forehead, O02 Lung

URESTERAL CALCULUS
N03 Ureter, N04 Kidney, H01 Sympathetic, I01 Shen Men

URETHRAL STONE
N01 Urinary bladder, N03 Ureter, H01 Sympathetic, N04 Kidney, I01 (Shen Men) Divine gate, E05 Occiput

URINARY INFECTIONS (B)
N01 Bladder, N04 Kidney, I09 Urethra, I01 Shen Men, B08 Adrenal gland, H01 Sympathetic, I10 External genitalia

URINARY LITHIASIS
N03 Ureter, N04 Kidney, H01 Sympathetic, E10 Subcortex, N01 Urinary bladder

URTICARIA
E05 Occiput, D01 Internal secretion, B08 Adrenal gland, I01 (Shen Men) Divine gate, N09 Liver, O02 Lung, E10 Subcortex

UTERINE HEMORRHAGE, FUNCTIONAL
I05 Uterus, E04 Brain, D01 Internal secretion, N09 Liver, N10 Spleen, N04 Kidney, B08 Adrenal gland

UTERUS, PROLAPSE
I05 Uterus, E10 Subcortex, I10 External genitalia

UVULAR OEDEMA
B04 Pharynx and larynx, I01 Shen Men, B08 Adrenal gland

VAGINAL DISCHARGE (B)
I05 Uterus, D01 Internal secretion, H01 Sympathetic, N04 Kidney

VARICELLA VULGARIS
Corresponding regional points, O02 Lung, D01 Internal secretion, B08 Adrenal gland, E05 Occiput, I01 Shen Men

VERRUCAPLANA
O02 Lung, D01 Internal secretion, E05 Occiput, M08 Large intestine

VERTIGO (B)
E01 (brain axis) Occiput, N09 Liver, J11 Minor occipital nerve, A12 Internal ear

VESICAL STONE
N01 Urinary bladder, N04 Kidney, H01 Sympathetic, I09 Urethra, I01 (Shen Men) Divine gate

VITILIGO
O02 Lung, D01 Internal secretion, E05 Occiput, B08 Adrenal gland, Corresponding regional points

VOMITING
M04 Stomach, H01 Sympathetic, I01 (Shen Men) Divine gate, E10 Subcortex

VULVAR PRURITIS
I10 External genitalia, E05 Occiput, B08 Adrenal gland, I01 Shen Men, O02 Lung, D01 Internal secretion

WHIPLASH INJURY, NECK (A)
F01 Cervical vertebrae, F05 Neck

WHIPLASH INJURY (TRAUMATIC CERVICAL SYNDROME) (B)
I01 Shen Men, B02 (visceral) Throat, A14 Tonsil, K12 Ear apex

WHOOPING COUGH
E07 (Ping-Chuan) Asthma 2, O02 Lung, B08 Adrenal gland, I01 (Shen Men) Divine gate, O03 Bronchi

WRIST DISORDERS (B)
J06 Wrist, I01 Shen Men, E10 Subcortex

WRIST PAIN (A)
J06 Wrist

AURICULAR ELECTROACUSTIMULATION (M06)

Definition (M0601)

Auricular electroacustimulation is a special branch of electro-acustimulation which utilizes acupoints found in, on, or closely associated with the pinnae of the ear.

Auricular electroacustimulation is relatively new to the field of acupoint stimulation. Historically, a few auricular acupoints were utilized for stimulation by the early Chinese acupuncturists, but it was not until the French acupuncturist and neurosurgeon Dr. Paul Nogier discovered an extensive and intricate acupoint system to be present in the ear that interest in it really began. Nogier's work prompted investigation of the auricular acupoint system by the modern Chinese. They were able to establish the presence of approximately two hundred points on each pinna.

Nogier and various subsequent investigators established, through experimentation and observation, definite nervous system correlations existing between various auricular acupoints and distinct somatic areas or organs. Subsequent experimentation and clinical experience demonstrated that stimulation of particular acupoints can affect specific physiological processes or body areas; a particular effect (reduction of pain, for example) will occur only if a particular point or set of points, correlated with the particular physiological response or area of the body involved, has been stimulated. The mechanisms responsible for these effects have not been established, only theory and speculation have been offered in explanation. It is clear from studies showing changes in endorphin levels and heart rates that the stimulation of auricular acupoints does not produce merely placebo effects.

Of major significance for those now investigating the system is the fact that the auricular acupoint (like its peripheral acupoint counterpart) has a distinctively lower skin resistance than the surrounding skin (usually between 40 and 130 kohm, as compared to the surrounding skin impedance of between 900 and 1100 kohm), and that points so discerned approximately correlate with the previous mapping conducted by Nogier and the Chinese investigators. Almost as important as the discovery of the electrical properties of the acupoint is the discovery that stimulation need not be limited to needle puncture or moxibustion stimulation techniques, but may be supplied by relatively brief electrical stimulation which, when appropriately applied, has predictably produced the same physiological changes reported when needle or moxibustion techniques are used, without the negative side affects associated with dermal invasion by needles (pain, possible infection, danger of dangerous penetration into vulnerable structures) or moxibustion (dermal irritation, burns, secondary infection, etc.).

Results from properly applied electrical stimulation of auricular acupoints are surprisingly rapid. Pain relief often occurs just seconds after commencement of stimulation. The relief afforded may last for only a few seconds if the originating cause has not been affected, or can be surprisingly long, sometimes affecting what appears to be permanent changes or relief.

Procedures of Application (M0602)

Establishing auricular/somatic correlates (M0602A)

Ear/somatic correlates may be demonstrated by applying the positive electrode probe from an electrical stimulator to an established peripheral acupoint, and then sweeping over the auricular skin with the negative probe. The stimulator should be capable of providing 1.5 to 3.0 volts and a frequency of 2 hertz (cycles per second), but adjusted to the subject's ear skin at the lowest possible voltage. When the auricular acupoint (approximately 0.1 mm. in diameter) which correlates with the somatic positive probe site is touched by the negative probe, the patient will feel the electrical stimulation, and a twitch of the muscle under the positive probe may occur. This is an exclusive response which will not occur when skin areas not correlated with the positive probe site are touched by the negative probe.

Technical clinical application (M0602B)

Precision of auricular acupoint location has proven to be absolutely essential for successful treatment through acupoint stimulation. As in the case of peripheral acupoints, auricular acupoints may be precisely located with an ohm-meter or skin resistance monitor designed to register the impedance difference between the acupoint and the surrounding dermis. The auricular acupoint may determined through a survey of skin resistance utilizing an ohm-meter sensitive enough to measure an impedance range of between 500 and 1500 kohm.

The acupoint is said to have an average impedance of 794 kohm while the surrounding skin has an impedance of 1,407 kohm. The instruments used to find acupoints are usually called point finders and often combine the ohm-meter function with a pulsed electrical current stimulator.

In preparation for the acupoint survey the skin surface of the ear should be cleansed of dirt, perspiration and oils (body or otherwise). A brief friction massage with a paper tissue dampened with alcohol should be sufficient. After this light debridement the area should be allowed to dry (to the touch) or gently patted dry.

To determine the precise acupoint location, the patient holds a "ground" electrode, typically a metal cylinder attached to the positive pole of the point finder by an electrode cable. An auricular acupoint survey is then conducted by sweeping over the pinnae skin with a hand-held, wand-like metal probe, insulated from the hand of the surveyor by a plastic handle, attached by an electrode cable to the negative pole of the point finder. When an acupoint is touched by the hand-held probe the point finder registers a sudden drop in impedance.

If a treatment protocol has been determined from the recommendation of an authoritative listing of appropriate acupoints, the head of the negative probe should be made to sweep the areas of the pinnae in which the recommended acupoints are indicated as located. The inexperienced therapist should refer directly to available diagrams, maps, or exact verbal descriptions of the ear acupoints to establish the approximate acupoint location (refer to the Table of Auricular Landmark & Auricular Acupoint Descriptions [T003] and associated diagrams).

Ideally, electrical stimulation of a located acupoint should commence immediately upon the precise determination of the acupoint location by pushing a stimulator activating button in the handle of the acupoint finder wand-like electrode, or by manually activating a switch on the point finder which converts it from a point finder to an electrical stimulator. If the point finder is separate from the stimulator, the acupoint should be precisely marked by surgical pen or other hypoallergenic, non-irritating inert ink and later electrically stimulated by a TNS unit or other electrical generator, usually through a hand-held electrode probe.

Prior to stimulation, the electrical generator should be preset at zero intensity and at the desired frequency with a pulse width (pulse duration) of between 100 to 200 milliseconds, or maximum for the unit used. When the acupoint has been determined and the stimulator activated, the intensity should be slowly increased until the limit of the patient's tolerance has been reached. The electrical generator should be able to supply (as a minimum) 0 to 50 microamperes with a driving force of 9 volts; most stimulation occurs at the 25 microampere level. Acupoints should be treated from 4 to 6 seconds, except in hyper-functional disorders (an element existent in most pain syndromes) in which case stimulation should continue for 16 seconds or more. In some cases stimulation should continue until analgesia has been effected or a major decrease in the pain level has been experienced by the patient. Longer continuous stimulation of some acupoints is possible through the use of earring-like "ear clip" electrode attachments which may actually clipped to the pinnae over the specific acupoint.

Needle application or electrical stimulation of needle electrodes is not within the scope of this discussion.

Frequency of acupoint stimulation relative to the part of the ear stimulated (M0602C)

Many authorities suggest that acupoints should be stimulated at different frequencies according to their relative positions on or around the pinnae of the ear:

Location	*Frequency Range*
Lobule	80–160 hertz
Helix	20–40 hertz
Cavum concha	5–10 hertz
Tragus	40–60 hertz
Triangular fossa	10–20 hertz
Antihelix	10–20 hertz

It should be noted, however, that some practitioners have found that a frequency setting of 2 to 4 hertz to be generally effective for the stimulation all auricular acupoints, regardless of location.

Partial List of Conditions Treated (M0603)

Conditions Treated in the Field (M0603A)

Aphasia
Aphonia
Cardiac arrhythmia
Conjunctivitis and herpes opthalmia
Constipation
Cough
Diarrhea
Dizziness (vertigo)
Drug dependence
Enuresis (nocturnal)
Fatigue
Fever
Gastralgia

Hyperkinesis
Hypoglycemia
Hypothyroidism
Impotence
Indigestion
Mastitis
Meniere's syndrome
Nose bleeding
Otitis media
Pancreatitis
Peptic ulcer
Post partum uterine pain
Sneezing
Sunburn
Syncope
Tachycardia
Tonsillitis

Treatable Causes   (M0603B)

Bacterial infection   (C09)
Blood sugar irregularities   (C31)
Dental pain   (C22)
Extrafusal muscle spasm   (C41)
Habitual joint positioning   (C48)
Hiccup   (C44)
Hypertension   (C39)
Hypotension   (C45)
Insomnia   (C46)
Irregular menses   (C19)
Menorrhalgia   (C20)
Migraine (vascular headache)   (C34)
Nausea/vomiting   (C36)
Neurogenic dermal disorders
   (psoriasis, hives)   (C49)
Sinusitis   (C16)
Soft tissue inflammation   (C05)
Soft tissue swelling   (C06)
Sympathetic hyperresponse   (C38)
Tinnitus   (C32)
Visceral organ Dysfunction   (C02)

Syndromes   (M0603C)

ACNE   (S45)
BACTERIAL INFECTION   (S63)
BREAK THROUGH BLEEDING   (S61)
BURNS (SECOND AND/OR THIRD
   DEGREE)   (S72)
CALF PAIN   (S50)
CAPSULITIS   (S27)
CARPAL TUNNEL   (S35)
CERVICAL DORSAL OUTLET (SCALENICUS
   ATTICUS/CERVICAL RIB)   (S09)
CERVICAL (NECK) PAIN   (S73)
CHEST PAIN   (S14)
CHONDROMALACIA   (S15)
CHRONIC BRONCHITIS/EMPHYSEMA
   (OBSTRUCTIVE PULMONARY DISEASE)   (S13)

COLITIS   (S42)
EARACHE   (S40)
ELBOW PAIN   (S47)
FOOT PAIN   (S51)
FOREARM PAIN   (S54)
HAND/FINGER PAIN   (S17)
HEADACHE PAIN   (S02)
HIGH THORACIC BACK PAIN   (S48)
HYPERTENSION   (S43)
HYSTERIA/ANXIETY REACTION   (S59)
INSOMNIA   (S71)
KNEE PAIN   (S46)
LOWER ABDOMINAL PAIN   (S62)
LOW BACK PAIN   (S03)
MENSTRUAL CRAMPING   (S60)
MIGRAINE (VASCULAR) HEADACHE   (S18)
MYOSITIS OSSIFICANS   (S67)
PHOBIC REACTION   (S44)
POST WHIPLASH   (S55)
REFERRED PAIN   (S01)
SHOULDER/HAND   (S57)
SHOULDER PAIN   (S08)
SINUS PAIN   (S33)
TEMPOROMANDIBULAR JOINT (TMJ)
   PAIN   (S06)
THIGH PAIN   (S49)
TIC DOULOUREUX   (S53)
TINNITUS   (S70)
TOOTHACHE/JAW PAIN   (S41)
WRIST PAIN   (S16)

Precautions and Contraindications   (M0604)

As a general precaution, auricular electroacustimulation should not be administered to pathologically weak, tired, or hungry people, or those which have not fully digested a meal, or those under the influence of alcohol; a noted exception is the administration of electroacustimulation for the relief of pain and/or improvement of physiological function to those patients who are terminally ill.

Auricular electroacustimulation should not be applied during the first five months of a woman's pregnancy, to preclude precipitating premature delivery or abortion. After this precautionary period, auricular electroacustimulation may be administered for the alleviation of pain but great care should be taken to avoid the stimulation of the uterus, ovary, internal secretion, pelvis and abdomen acupoints.

As with transcutaneous nerve stimulation (TNS or TENS) treatment, current flow from the electrical generator must not be allowed to cross a cardiac pacemaker or its cables as a precaution against inducing arrhythmias.

Patients suffering from anemia should be warned that auricular acustimulation may cause fainting, dizziness or nausea.

Patients with heart disease, anemia, diabetes, and hypotension should be watched carefully for signs of a developing autonomic response; symptoms may include sensations of fatigue or general muscular relaxation, drowsiness, and rapid decreases in both systolic and diastolic blood pressures.

Patients suffering from diabetes should be monitored for symptoms of developing diabetic coma, which may develop as metabolic insulin demand decreases as sympathetic slowing occurs. Symptoms may include dry and flushed skin, fever, dry mouth, intense thirst, absence of hunger, nausea and/or vomiting, abdominal pain (frequently seen), exaggerated air hunger, acetone breath odor, low blood pressure, weak and rapid pulse, lack of affect, and glassy stare. Such "diabetic shock" symptomology may be relieved by having the patient consume a small quantity of simple sugar, like a banana or substances high in fructose. Refer to Diabetes Syndrome [S10] or to Blood Sugar Irregularities [C31] for further details.

A thorough preliminary evaluation of both acute and chronic conditions before instituting treatment is very important. Because auricular electroacustimulation may eliminate pain, it can thereby mask symptoms of a disease process or pathology for which auricular electroacustimulation may not be indicated.

Notes Aside (M0605)

Reference is made in this work to acupoints which have been suggested by previous authority as beneficial (when stimulated electrically) in the treatment of specific maladies or to produce a desired affect upon the body; they are usually designated according to the body part with which they have reputedly been correlated. Certain points are designated differently, as a class of "general points" which apparently augment the effects produced by the stimulation of the other auricular acupoints, which include the Divine gate, Subcortex, Sympathetic and Internal secretion points.

For a description of auricular landmarks and a description of auricular acupoint locations refer to the Table of Auricular Landmark & Auricular Acupoint Descriptions [T003]. For auricular acupoint treatment protocols suggested by various specialized authorities in the field of auricular electroacustimulation, refer to the Table of Auricular Acupoint Treatment Protocols [T004].

Bibliography (M0606)

Cho M-H, pp. 113–121.

AURICULAR ELECTROACUSTIMULATION PAIN RELIEF (M06A)

Definition (M06A1)

Although auricular therapy has been used to good effect in the treatment of various "pathological" human maladies (as testified to by the literature and clinical experience) the most popular interest in it has revolved around its notable effect on acute and chronic pain perception.

Much of the apparent inhibitory effect upon the perception of pain afforded by auricular electroacustimulation has been laid at the feet of its ability to increase the production of endorphins (discussed in Peripheral Electroacustimulation [M05]). Endorphins have been shown to have an inhibitory effect upon the transmission of pain impulses along afferent nerves to the brain and within the brain itself. Whether this endorphin production is entirely responsible for such inhibition of pain impulses has yet to be proven. There are significant differences in patient responses to treatment and the variability in the longevity of the relief: one patient may receive complete relief for many hours from the pain accompanying a fractured bone, while another patient will report only a temporary reduction in the level of pain accompanying a sprained ankle. Despite these unknowns, auricular electroacustimulation has long been used effectively for the reduction and control of both acute and chronic pain; it is judged by many competent practitioners to be one of the most reliable modalities available for the noninvasive control and/or amelioration of pain, providing a dependable means (albeit often temporary) of relieving the patient's pain.

Auricular electroacustimulation is often most effectively utilized eclectically with other modalities, providing immediate relief of pain while other treatment techniques are afforded the opportunity to provide less immediate but more desirable long term "curative" effects upon primary and/or accompanying secondary cause(s) of the pain.

Auricular electroacustimulation is most remarkable for its ability to provide relief of pain in just seconds, and for the surprising longevity of its effects (often for days or even or longer). Refer to the Table of Auricular acupoint Treatment Protocols [T004] for recommended treatment protocols for the control of pain.

Procedures of Application (M06A2)

1) The skin surface of the pinna of the ear should be cleansed of dirt, perspiration and oils (body or otherwise). A brief friction massage with a paper tissue dampened with alcohol should be sufficient. After this light debridement the area should be allowed to dry (to the touch) or gently patted dry.

2) When a treatment protocol has been decided upon, the head of the negative probe should be made to sweep the part of the pinnae in which the recommended acupoints are located. The inexperienced therapist should refer directly to available diagrams, maps, or exact verbal descriptions of the ear acupoints to establish the approximate acupoint location (refer to the Table of Auricular Landmark & Auricular Acupoint Descriptions [T003] and associated diagrams, and Auricular Acupoint Treatment Protocols [T004] and associated diagrams).

3) Ideally, electrical stimulation of a located acupoint should commence immediately upon the precise determination of the acupoint location by pushing a stimulator activating button in the handle of the acupoint finder wand-like electrode, or by manually activating a switch on the point finder which converts it from a point finder to an electrical stimulator. If the point finder is separate from the stimulator, the acupoint should be precisely marked by surgical pen or other hypoallergenic, non-irritating inert ink and later electrically stimulated by a TNS unit or other electrical generator, usually through a hand-held electrode probe. Prior to stimulation, the electrical generator should be preset at zero intensity and at the desired frequency with a pulse width (pulse duration) of between 100 to 200 milliseconds, or maximum for the unit used. When the acupoint has been determined and the stimulator activated, the intensity should be slowly increased until the limit of the patient's tolerance has been reached.

4) To provide analgesia or anaesthesia, stimulation of each designated auricular acupoint should continue for at least 16 seconds or longer; alternatively, stimulation of the primary points (those correlated with the body part housing the source of

the pain) should continue until analgesia has been effected or a major decrease in the pain level has been experienced by the patient. If the correct protocol has been followed, i.e., the appropriate set of points to provide analgesia or anaesthesia to a given area have been accurately determined and adequately stimulated in the correct sequence, immediate pain reduction and/or relief may be experienced by the patient.

Partial List of Conditions Treated (M06A3)

Treatable Causes (M06A3A)

Calcific deposit (C08)
Dental pain (C22)
Extrafusal muscle spasm (C41)
Interspinous ligamentous strain (C03)
Joint approximation (C47)
Ligamentous strain (C25)
Nerve root compression (C04)
Phlebitis (C43)
Psychoneurogenic neuromuscular hypertonus (C29)
Restricted joint range of motion (C26)
Sinusitis (C16)
Soft tissue inflammation (C05)
Soft tissue swelling (C06)
Trigger point formation (C01)

Syndromes (M06A3B)

ACHILLES TENDONITIS (S52)
ADHESION (S56)
BURNS (SECOND AND/OR THIRD DEGREE) (S72)
BURSITIS (S26)
CALF PAIN (S50)
CAPSULITIS (S27)
CARPAL TUNNEL (S35)
CERVICAL DORSAL OUTLET (SCALENICUS ATTICUS/CERVICAL RIB) (S09)
CERVICAL (NECK) PAIN (S73)
CHEST PAIN (S14)
CHONDROMALACIA (S15)
COLITIS (S42)
EARACHE (S40)
ELBOW PAIN (S47)
FACET (S19)
FASCIITIS (S20)
FOOT PAIN (S51)
FOREARM PAIN (S54)
FROZEN SHOULDER (S64)
HAND/FINGER PAIN (S17)
HEADACHE PAIN (S02)
HIGH THORACIC BACK PAIN (S48)
JOINT SPRAIN (S30)
KNEE PAIN (S46)
LOWER ABDOMINAL PAIN (S62)
LOW BACK PAIN (S03)
MENSTRUAL CRAMPING (S60)
MIGRAINE (VASCULAR) HEADACHE (S18)
MYOSITIS OSSIFICANS (S67)
OSTEOARTHRITIS (S21)
PHLEBITIS (S68)
PIRIFORMIS (S04)
POST WHIPLASH (S55)
RADICULITIS (S29)
REFERRED PAIN (S01)
SCIATICA (S05)
SHOULDER/HAND (S57)
SHOULDER PAIN (S08)
SINUS PAIN (S33)
TEMPOROMANDIBULAR JOINT (TMJ) PAIN (S06)
TENDONITIS (S28)
TENNIS ELBOW (LATERAL EPICONDYLITIS) (S32)
THIGH PAIN (S49)
TOOTHACHE/JAW PAIN (S41)
WRIST PAIN (S16)

Precautions and Contraindications (M06A4)

As a general precaution, auricular electroacustimulation should not be administered to pathologically weak, tired, or hungry people, or those which have not fully digested a meal, or those under the influence of alcohol; a noted exception is the administration of electroacustimulation for the relief of pain and/or improvement of physiological function to those patients who are terminally ill.

Auricular electroacustimulation should not be applied during the first five months of a woman's pregnancy, to preclude precipitating premature delivery or abortion. After this precautionary period, auricular electroacustimulation may be administered for the alleviation of pain but great care should be taken to avoid the stimulation of the uterus, ovary, internal secretion, pelvis and abdomen acupoints.

As with transcutaneous nerve stimulation (TNS or TENS) treatment, current flow from the electrical generator must not be allowed to cross a cardiac pacemaker or its cables as a precaution against inducing arrhythmias.

Patients suffering from anemia should be warned that auricular acustimulation may cause fainting, dizziness or nausea.

Patients with heart disease, anemia, diabetes, and hypotension should be watched carefully for signs of a developing autonomic response; symptoms may include sensations of fatigue or general muscular relaxation, drowsiness, and rapid decreases in both systolic and diastolic blood pressures.

Patients suffering from diabetes should be monitored for symptoms of developing diabetic coma, which may develop as metabolic insulin demand decreases as sympathetic slowing occurs. Symptoms may include dry and flushed skin, fever, dry mouth, intense thirst, absence of hunger, nausea and/or vomiting, abdominal pain (frequently seen), exaggerated air hunger, acetone breath odor, low blood pressure, weak and rapid pulse, lack of affect, and glassy stare. Such "diabetic shock" symptomology may be relieved by having the patient consume a small quantity of simple sugar, like a banana or substances high in fructose. Refer to Diabetes Syndrome [S10] or to Blood Sugar Irregularities [C31] for further details.

A thorough preliminary evaluation of both acute and chronic conditions before instituting treatment is very important. Because auricular electroacustimulation may eliminate pain, it can thereby mask symptoms of a disease process or pathology for which auricular electroacustimulation may not be indicated.

Notes Aside   (M06A5)

Reference is made in this work to acupoints which have been suggested by previous authority as beneficial (when stimulated electrically) in the treatment of specific maladies or to produce a desired affect upon the body; they are usually designated according to the body part with which they have reputedly been correlated. Certain points are designated differently, as a class of "general points" which apparently augment the effects produced by the stimulation of the other auricular acupoints, which include the Divine gate, Subcortex, Sympathetic and Internal secretion points.

Electrical stimulation of the acupoints correlated with the source of the pain is essential if the patient's pain is to be relieved; electrical stimulation of uncorrelated acupoints will not affect the patient's perception of pain to any appreciable extent.

Bibliography   (M06A6)

Kitade, pp. 241–252.
Krause pp. 507–511.
Oliveri, pp. 12–16.
Paris, pp. 35–40.

AURICULAR ELECTROACUSTIMULATION PATHOLOGY CONTROL (M06B)

Definition (M06B1)

The ability to objectively demonstrate and precisely locate particular auricular acupoints through skin resistance survey and immediately apply effective electrical stimulation has done much to recommend auricular electroacustimulation to the clinical practitioner. It is increasingly being used to provide a means for quick relief of pain, and as regular treatment mode applied eclectically with other modalities, including continuous peripheral electroacustimulation, a treatment combination often recommended by acupoint stimulation specialists or "experts".

Slowly, auricular electroacustimulation is gaining a favorable reputation as an alternative to more traditional (western) treatment techniques for management of pathological conditions besides the symptom of pain; it often provides amazing beneficial effects upon abnormal physiological processes (apparently reestablishing normal "balances") without the risk of hazardous or detrimental side effects. A good many more "pathological" conditions or ailments may be successfully treated with auricular electroacustimulation then is generally realized or appreciated by western practitioners of all types, including many "traditional" acupuncturists.

In most western clinical settings utilizing auricular electroacustimulation, the treatment of "pathological" conditions for most practitioners is usually based upon the recommended acupoint protocols suggested by authoritative specialists in auricular acupoint stimulation; such treatment protocols were developed empirically, through trial-and-error by such experts during clinical practice. Such protocols may, in general, be relied upon to produce the desired effects if the correct protocol is selected for the particular condition to be treated (refer to Table of Auricular Acupoint Treatment Protocols [T004]). Of course, it is essential that the condition to be treated be correctly identified before a proper protocol may be selected (treatment disappointments may often be laid at the feet of a failure to correctly identify the condition to be treated). It should also be noted that often the order of auricular acupoint stimulation, frequency of the electrical impulse, the strength of the electrical stimulus, and the correct identification of acupoint location are important factors in providing successful results.

Procedure of Application (M06B2)

1) The skin surface of the pinna of the ear should be cleansed of dirt, perspiration and oils (body or otherwise). A brief friction massage with a paper tissue dampened with alcohol should be sufficient. After this light debridement the area should be allowed to dry (to the touch) or gently patted dry.

2) When a treatment protocol has been decided upon, the head of the negative probe should be made to sweep the part of the pinnae in which the recommended acupoints are located. The inexperienced therapist should refer directly to available diagrams, maps, or exact verbal descriptions of the ear acupoints to establish the approximate acupoint location (refer to the Table of Auricular Landmark & Auricular Acupoint Descriptions [T003] and associated diagrams).

3) Ideally, electrical stimulation of a located acupoint should commence immediately upon the precise determination of the acupoint location by pushing a stimulator activating button in the handle of the acupoint finder wand-like electrode, or by manually activating a switch on the point finder which converts it from a point finder to an electrical stimulator. If the point finder is separate from the stimulator, the acupoint should be precisely marked by surgical pen or other hypoallergenic, non-irritating inert ink and later electrically stimulated by a TNS unit or other electrical generator, usually through a hand-held electrode probe. Prior to stimulation, the electrical generator should be preset at zero intensity and at the desired frequency with a pulse width (pulse duration) of between 100 to 200 milliseconds, or maximum for the unit used. When the acupoint has been determined and the stimulator activated, the intensity should be slowly increased until the limit of the patient's tolerance has been reached.

4) To provide analgesia or anaesthesia, stimulation of each designated auricular acupoint should continue for at least 16 seconds or longer; alternatively, stimulation of the primary points (those correlated with the body part housing the source of the pain) should continue until analgesia has been effected or a major decrease in the pain level has been experienced by the patient. If the correct protocol has been followed, i.e., the appropriate set of points to provide analgesia or anaesthesia to a

given area have been accurately determined and adequately stimulated in the correct sequence, immediate pain reduction and/or relief may be experienced by the patient.

Partial List of Conditions Treated (M06B3)

Conditions Treated in the Field (M06B3A)

Aphasia
Aphonia
Cardiac arrhythmia
Conjunctivitis and herpes opthalmia
Constipation
Cough
Diarrhea
Dizziness (vertigo)
Drug dependence
Enuresis (nocturnal)
Fatigue
Fever
Gastralgia
Hyperkinesis
Hypoglycemia
Hypothyroidism
Impotence
Indigestion
Mastitis
Meniere's syndrome
Nose bleeding
Otitis media
Pancreatitis
Peptic ulcer
Post partum uterine pain
Sneezing
Sunburn
Syncope
Tachycardia
Tonsillitis

Treatable Causes (M06B3B)

Bacterial infection (C09)
Blood sugar irregularities (C31)
Dental pain (C22)
Extrafusal muscle spasm (C41)
Hiccup (C44)
Hypertension (C39)
Hypotension (C45)
Insomnia (C46)
Irregular menses (C19)
Menorrhalgia (C20)
Migraine (vascular headache) (C34)
Muscular weakness (C23)
Nausea/vomiting (C36)

Neurogenic dermal disorders
  (psoriasis, hives) (C49)
Sinusitis (C16)
Soft tissue inflammation (C05)
Soft tissue swelling (C06)
Sympathetic hyperresponse (C38)
Tinnitus (C32)
Visceral organ Dysfunction (C02)

Syndromes (M06B3C)

ACNE (S45)
BACTERIAL INFECTION (S63)
BREAK THROUGH BLEEDING (S61)
CALF PAIN (S50)
CAPSULITIS (S27)
CARPAL TUNNEL (S35)
CERVICAL DORSAL OUTLET (SCALENICUS
  ATTICUS/CERVICAL RIB) (S09)
CERVICAL (NECK) PAIN (S73)
CHEST PAIN (S14)
CHONDROMALACIA (S15)
CHRONIC BRONCHITIS/EMPHYSEMA
  (OBSTRUCTIVE PULMONARY DISEASE) (S13)
COLITIS (S42)
EARACHE (S40)
ELBOW PAIN (S47)
FOOT PAIN (S51)
FOREARM PAIN (S54)
HAND/FINGER PAIN (S17)
HEADACHE PAIN (S02)
HIGH THORACIC BACK PAIN (S48)
HYPERTENSION (S43)
HYSTERIA/ANXIETY REACTION (S59)
INSOMNIA (S71)
KNEE PAIN (S46)
LOWER ABDOMINAL PAIN (S62)
LOW BACK PAIN (S03)
MENSTRUAL CRAMPING (S60)
MIGRAINE (VASCULAR) HEADACHE (S18)
PHOBIC REACTION (S44)
POST WHIPLASH (S55)
REFERRED PAIN (S01)
SHOULDER/HAND (S57)
SHOULDER PAIN (S08)
SINUS PAIN (S33)
TEMPOROMANDIBULAR JOINT (TMJ)
  PAIN (S06)
THIGH PAIN (S49)
TIC DOULOUREUX (S53)
TINNITUS (S70)
TOOTHACHE/JAW PAIN (S41)
WRIST PAIN (S16)

Precautions and Contraindications (M06B4)

As a general precaution, auricular electroacustimulation should not be administered to pathologically weak, tired, or hungry people, or those which have not fully digested a meal, or those under the influence of alcohol; a noted exception is the administration of electroacustimulation for the relief of pain and/or improvement of physiological function to those patients who are terminally ill.

Auricular electroacustimulation should not be applied during the first five months of a woman's pregnancy, to preclude precipitating premature delivery or abortion. After this precautionary period, auricular electroacustimulation may be administered for the alleviation of pain but great care should be taken to avoid the stimulation of the uterus, ovary, internal secretion, pelvis and abdomen acupoints.

As with transcutaneous nerve stimulation (TNS or TENS) treatment, current flow from the electrical generator must not be allowed to cross a cardiac pacemaker or its cables as a precaution against inducing arrhythmias.

Patients suffering from anemia should be warned that auricular acustimulation may cause fainting, dizziness or nausea.

Patients with heart disease, anemia, diabetes, and hypotension should be watched carefully for signs of a developing autonomic response; symptoms may include sensations of fatigue or general muscular relaxation, drowsiness, and rapid decreases in both systolic and diastolic blood pressures.

Patients suffering from diabetes should be monitored for symptoms of developing diabetic coma, which may develop as metabolic insulin demand decreases as sympathetic slowing occurs. Symptoms may include dry and flushed skin, fever, dry mouth, intense thirst, absence of hunger, nausea and/or vomiting, abdominal pain (frequently seen), exaggerated air hunger, acetone breath odor, low blood pressure, weak and rapid pulse, lack of affect, and glassy stare. Such "diabetic shock" symptomology may be relieved by having the patient consume a small quantity of simple sugar, like a banana or substances high in fructose. Refer to Diabetes Syndrome [S10] or to Blood Sugar Irregularities [C31] for further details.

A thorough preliminary evaluation of both acute and chronic conditions before instituting treatment is very important. Because auricular electroacustimulation may eliminate pain, it can thereby mask symptoms of a disease process or pathology for which auricular electroacustimulation may not be indicated.

Notes Aside (M06B5)

Auricular acupoint protocols are often used eclectically or in concert with peripheral acupoint protocols. For a listing of auricular acupoint protocols refer to the Table of Auricular Acupoint Treatment Protocols [T004]. To coordinate auricular acupoint protocols with peripheral acupoint protocols refer to both the Table of Auricular Acupoint Treatment Protocols [T004] and the Table of Peripheral Acupoint Treatment Protocols [T002], matching the condition with the authority designation (A or B) for the auricular acupoint and peripheral acupoint protocols. Reference may also be made to other authoritative protocol listings found in various specialized publications.

Bibliography (M06B6)

Cho, pp. 113–121.
Vitou, pp. 275–280.

AURICULAR ELECTROACUSTIMULATION AURICULAR ACUPOINT SURVEY
(M06C)

Definition (M06C1)

It has been established that it is possible to determine the identity of a variety of pathological conditions by surveying the ear for the presence of active acupoints.

When pain is being produced in a body part, a visceral organ is suffering from a pathological condition, or a physiological process has been disrupted, the pertinent or corresponding auricular acupoint(s) will have lower impedance and heightened electrical conductivity when compared to the surrounding dermal areas of the ear. Such acupoints will additionally suffer heightened sensitivity to slight pressure and/or electrical stimulation. It is even purported by some that auricular loci may, upon close inspection, manifest morphological changes or local discoloration. To illustrate this point, a double blind study of this phenomena was performed, with the end result that physician evaluation of the auricular acupoints in the ears of forty patients produced a correct diagnosis 75.2% of the time. Obviously, this process may be of great aid in establishing the correct source of pain when confronted with a referred pain syndrome of less than obvious origin. It can also serve to supply an objective basis for the selection of an acupoint treatment protocol.

Procedures of Application (M06C2)

1) In preparation for the acupoint survey the skin surface of the ear should be cleansed of dirt, perspiration and oils (body or otherwise). A brief friction massage with a paper tissue dampened with alcohol should be sufficient. After this light debridement the area should be allowed to dry (to the touch) or gently patted dry.

2) To determine the precise acupoint location, the patient holds a "ground" electrode, typically a metal cylinder attached to the positive pole of the point finder by an electrode cable. An auricular acupoint survey is then conducted by sweeping over the pinnae skin with a hand-held, wand-like metal probe, insulated from the hand of the surveyor by a plastic handle, attached by an electrode cable to the negative pole of the point finder. When an acupoint is touched by the hand-held probe the point finder registers a sudden drop in impedance.

The twelve general auricular areas should be surveyed: the foot and toes, lower leg and ankle, upper leg and knee, hip and buttocks, lower back, upper back, neck, head, upper arm and shoulder, lower arm and elbow, wrist, and the hand and fingers. Active acupoints will be distinctive relative to their comparatively low impedance, high electrical conductivity, and heightened sensitivity to slight pressure or electrical stimulation. The acupoints so discovered should be marked (for reference or as prelude to treatment) and their presence recorded as evaluative findings.

3) If a treatment protocol has been determined from the recommendation of an authoritative listing of appropriate acupoints, the head of the negative probe should be made to sweep the areas of the pinnae in which the recommended acupoints are indicated as located. The inexperienced therapist should refer directly to available diagrams, maps, or exact verbal descriptions of the ear acupoints to establish the approximate acupoint location (refer to the Table of Auricular Landmark & Auricular Acupoint Descriptions [T003] and associated diagrams).

Partial List of Conditions Evaluated (M06C3)

Treatable Causes (M06C3A)

Habitual joint positioning (C48)
Hypermobile/instable joint (C10)
Interspinous ligamentous strain (C03)
Joint ankylosis (C42)
Joint approximation (C47)
Ligamentous strain (C25)
Migraine (vascular headache) (C34)
Phlebitis (C43)
Restricted joint range of motion (C26)
Sinusitis (C16)
Soft tissue inflammation (C05)
Tactile hypersensitivity (C24)
Trigger point formation (C01)
Visceral organ Dysfunction (C02)

Syndromes (M06C3B)

ACHILLES TENDONITIS (S52)
BURSITIS (S26)

CALF PAIN (S50)
CAPSULITIS (S27)
CARPAL TUNNEL (S35)
CERVICAL DORSAL OUTLET (SCALENICUS ATTICUS/CERVICAL RIB) (S09)
CERVICAL (NECK) PAIN (S73)
CHEST PAIN (S14)
CHONDROMALACIA (S15)
COLITIS (S42)
EARACHE (S40)
ELBOW PAIN (S47)
FASCIITIS (S20)
FOOT PAIN (S51)
FOREARM PAIN (S54)
HAND/FINGER PAIN (S17)
HEADACHE PAIN (S02)
HIGH THORACIC BACK PAIN (S48)
KNEE PAIN (S46)
LOWER ABDOMINAL PAIN (S62)
LOW BACK PAIN (S03)
MENSTRUAL CRAMPING (S60)
MIGRAINE (VASCULAR) HEADACHE (S18)
MYOSITIS OSSIFICANS (S67)
OSTEOARTHRITIS (S21)
PHLEBITIS (S68)
PIRIFORMIS (S04)
POST IMMOBILIZATION (S36)
POST WHIPLASH (S55)
RADICULITIS (S29)
REFERRED PAIN (S01)
SCIATICA (S05)
SHINGLES (HERPES ZOSTER) (S37)
SHOULDER/HAND (S57)
SHOULDER PAIN (S08)
SINUS PAIN (S33)
TEMPOROMANDIBULAR JOINT (TMJ) PAIN (S06)
TENDONITIS (S28)
TENNIS ELBOW (LATERAL EPICONDYLITIS) (S32)
THIGH PAIN (S49)
TOOTHACHE/JAW PAIN (S41)
WRIST PAIN (S16)

Precautions and Contraindications (M06C4)

A thorough preliminary evaluation of both acute and chronic conditions before instituting treatment is very important. Because auricular electroacustimulation may eliminate pain, it can thereby mask symptoms of a disease process or pathology for which auricular electroacustimulation may not be indicated.

Notes Aside (M06C5)

Determination of referred pain source (M06C5A)

To determine the source of referred pain, an auricular acupoint survey should be conducted over sites which correlate with body structures which could possibly be responsible for the production of the referred pain: points which correlate with trigger point formations, visceral organs, interspinous ligaments, or other structures which could be responsible for referring pain, as well as over the acupoints which correlate with site(s) perceived as being painful by the patient. If the pain is referred, the points over the correlate structure which is responsible will be demonstrated by the survey while the points over the site perceived as painful by the patient will not usually be demonstrated. Refer to Referred Pain Syndrome [S01] and/or the Table of Referred Pain Patterns of Trigger Point Formation Origin [T005] for descriptive details of possible referred pain sources.

Determination of pathological symptom source (M06C5B)

To establish the source of a pathological symptom, a survey should be made over the auricular acupoint sites which correlate with structures or physiological processes which might be associated with the given symptom. For a reference as to which auricular acupoints should be surveyed relative to a given pathological symptom, refer to the Table of Auricular Acupoint Treatment Protocols [T004]. A positive demonstration of the auricular acupoints indicates the source of the symptom.

Bibliography (M06C6)

Oleson, pp. 217–229.

TABLE OF AURICULAR LANDMARK & ACUPOINT DESCRIPTIONS (T003)

ANATOMICAL LANDMARKS OF THE PINNA (T003A)

Landmark Descriptions (T003A01)

Lobule : A soft, lobular appendage at the bottom of the pinna.

Tragus : A cartilagineous projection which protectively juts out over the auditory meatus.

Anterior incisure of the tragus : A notch above the tragus which separates the tragus from the crus of the helix above it.

Intertragic incisure : A notch below the tragus which separates it from the antitragus.

Antitragus : A cartilagineous projection above the lobule, opposite the tragus, and at the lower end of the antihelix

Antihelix : A semi-circular ridge formed about the auditory meatus.

Upper crura of the antihelix : Upper-lateral branch of the antihelix

Lower crura of the antihelix : Lower-medial branch of the antihelix

Triangular Fossa : The depression formed by the upper and lower crurae of the antihelix

Scaphoid Fossa : The depression between the helix and antihelix

Helix : Superior rim of the pinna, running from the crus of helix to the lobule.

Crus of helix : The root of the helix, taking form just above the auditory meatus.

Cymba conchae : The superior part of the depression formed between the crus of the helix and the lower crura.

Cavum conchae : The major depression of the pinna, lying below the crus of the helix and containing the auditory meatus.

External acoustic meatus : The external auditory meatus, situated in the cavum conchae.

Posterior auricular sulcus : The depression separating the antihelix from the antitragus.

Tuberculum supratragicum : A small superior projection of the tragus, just below the anterior incisure of tragus.

Auricular (Darwin's) tubercle : A small projection of the free edge of the posterior superior portion of the helix, facing the antihelix.

Posterior Auricle : The back of the pinna, irregularly convex.

For illustrative diagram, refer to Anatomical Landmarks of the Pinna [T003A02].

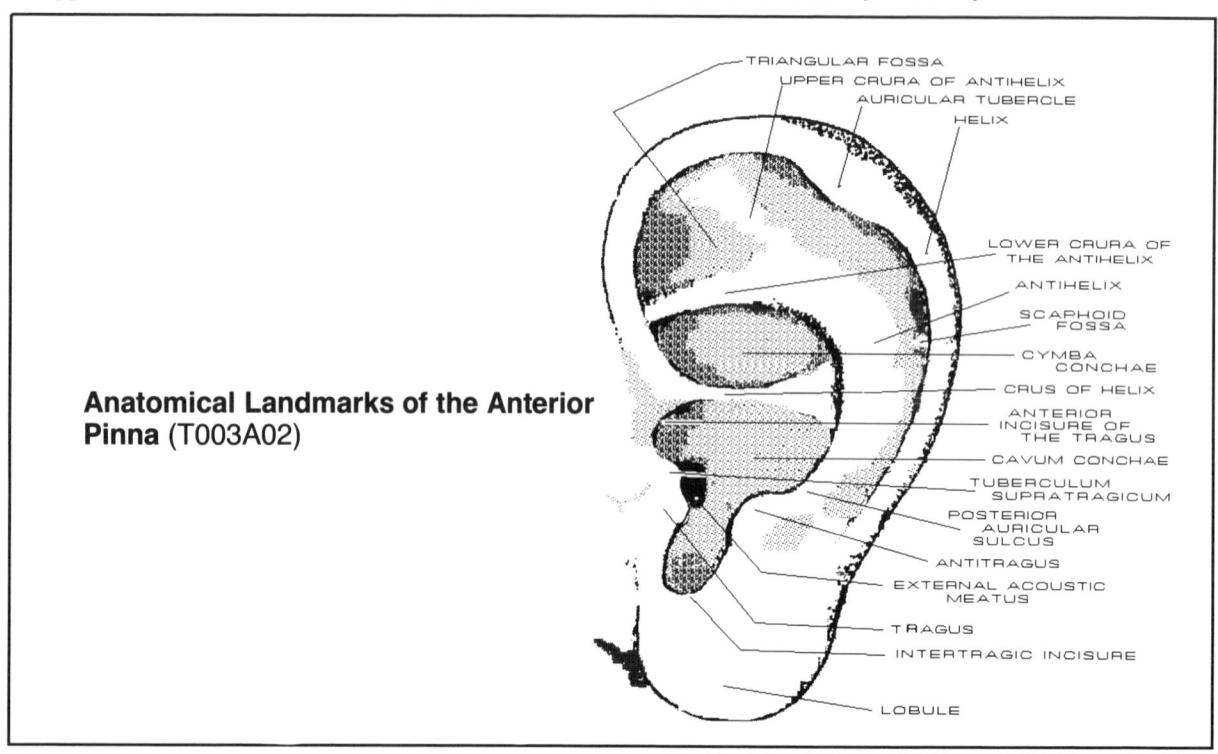

Anatomical Landmarks of the Anterior Pinna (T003A02)

table of auricular landmark & acupoint descriptions

Anatomical Landmarks of the Posterior Pinna (T003A03)

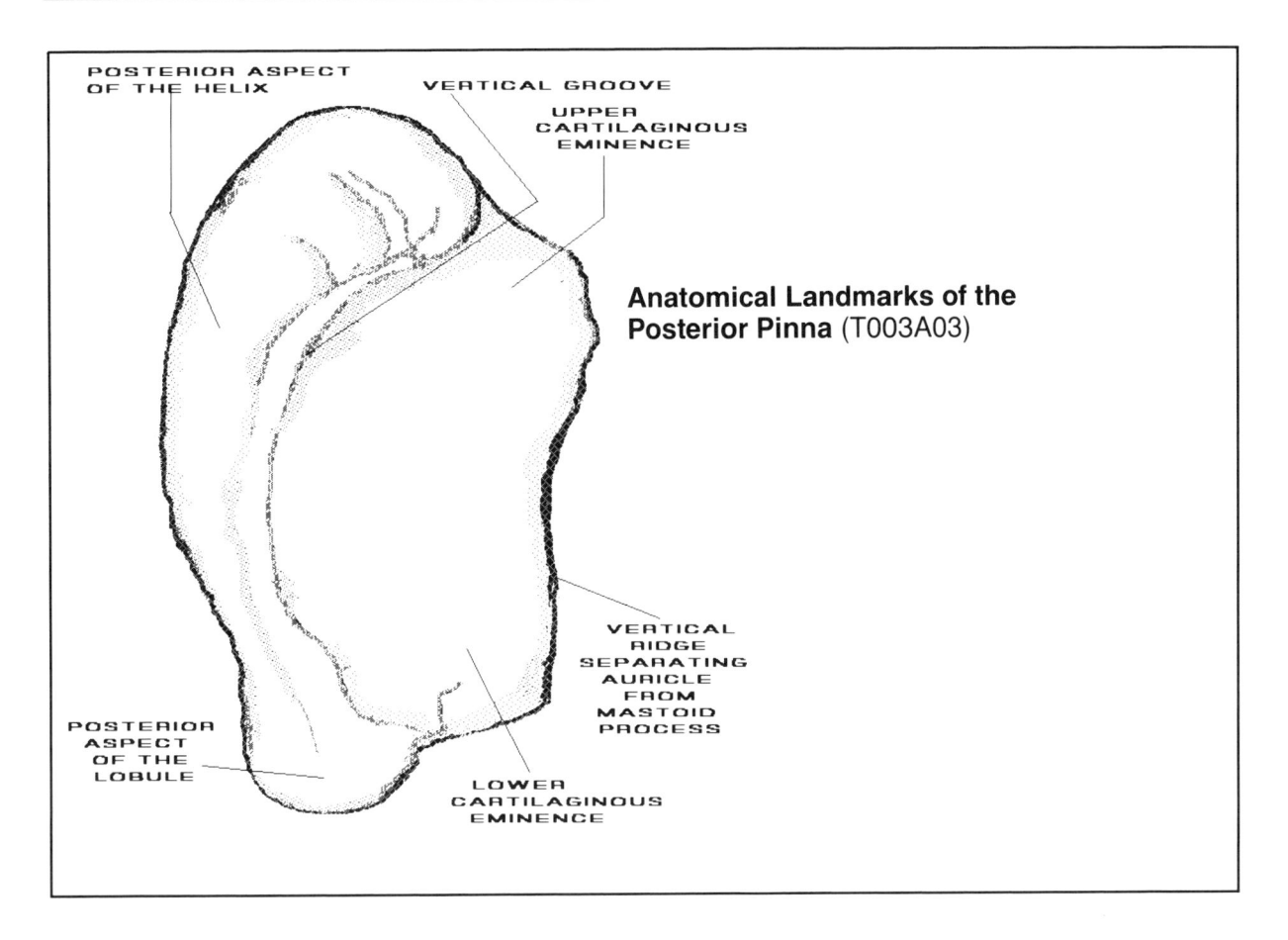

Labels: POSTERIOR ASPECT OF THE HELIX; VERTICAL GROOVE; UPPER CARTILAGINOUS EMINENCE; VERTICAL RIDGE SEPARATING AURICLE FROM MASTOID PROCESS; LOWER CARTILAGINOUS EMINENCE; POSTERIOR ASPECT OF THE LOBULE

AURICULAR ACUPOINTS AND THEIR LOCATION (T003B)

The auricular acupoint locations described here are approximations; exact location of such points depends upon an educated acupoint survey utilizing an ohm meter adjusted to register the relatively low impedance (high conductance) of the auricular acupoint. The point descriptions below and other graphic mapping may serve to establish which points have been located with the acupoint finder.

Points on the Lobule (T003B01)

*For descriptive orientation here, the anterior surface of the lobule is divided up and placed in reference to 9 rectangular shaped zones, each zone being occupied by varying amounts of lobule surface area. The *zone 1* is occupied by the most superior and anterior surface area of the lobule, *zone 2* the most superior middle area, *zone 3* the most superior and posterior area, *zone 4* the middle anterior area, *zone 5* the central area, *zone 6* the middle most posterior area, *zone 7* the anterior most inferior area, the *zone 8* middle most inferior area, and *zone 9* the most posterior most inferior surface area of the lobule. The most superior demarcation line of these zones is a horizontal line which is just contiguous with the most inferior point of the intertragic incisure; the most inferior horizontal demarcation line is just contiguous with the most inferior point of the lobule. The interior horizontal demarcation lines are equidistant from each other and to other related horizontal lines.

The first anterior vertical zone demarcation line lies just posterior to where the lobule joins the face, having its origin where it intersects with the most superior horizontal line and it terminates where it intersects with the most inferior horizontal line. Moving posteriorly, the second vertical line has its superior origin where the most inferior and posterior point of the intertragic incisure is contiguous with the superior horizontal line. The third vertical line has its origin on the superior horizontal line at a point which is directly inferior to where the antitragus and posterior auricular sulcus are contiguous. The fourth vertical line has its origin on the superior horizontal line at the point where the horizontal line terminates 1/3 to the distance across the anterior inferior section of the helix. Each vertical line is equidistant from that anterior to it.

table of auricular landmark & acupoint descriptions

Points on the Lobule   T003B01

Wexu Point	Common Point Name	Location
A01	Tooth extraction (upper teeth)	In the posterior inferior corner of zone 1.*
A02	Lower palate (lower jaw)	In the anterior superior corner of zone 2.*
A03	Tongue	At the center of zone 2.*
A04	Upper palate (upper jaw)	In the posterior inferior corner of zone 2.*
A05	Mandible	On the superior horizontal demarcation line, 3/4 of the distance from the third vertical line to the fourth.*
A06	Tonsil 3	On the superior horizontal demarcation line where it is contiguous with the middle point of the width of the helix.*
A07	Maxilla	Just posterior to the center of Zone 3.*
A08	Tooth extraction	Just posterior to the center of zone 4.*
A09	Nervousness point (neurathenia)	In the anterior inferior corner of zone 4 on the very edge of the lobule.*
A10	Eye	At the center of zone 5.*
A11	Cheek	On the third vertical demarcation line, at the midpoint between the second and third horizontal demarcation lines.*
A12	Inner ear	In the anterior superior corner of zone 6, 1/5 of the zone width from the anterior border and 1/4 of the height from relative demarcation lines.
A13	Helix 5	At the center of zone 6.*
A14	Tonsil 4	Just anterior of the midpoint between the second and third vertical demarcation lines and 1/3 of the zone height from the superior horizontal demarcation line of zone 8.*
A15	Helix 6	At the inferior tip of the lobule 1/3 of the distance from the second and third vertical demarcation line, just superior to the inferior demarcation of zone 8.*
A16	Cancer Line	A line of points running parallel and 1/4 of a zone width anterior to the the posterior inferior edge of the lobule, and found in zone 3, 5, 6 and 8.*

Points on the Tragus   T003B02

Wexu Point	Common Point Name	Location
B01	Apex of the Tragus	The superior most point on the upper projection of the tragus (if only only one projection is present), anterior to the anterior incisure of the tragus.
B02	Visceral Point	Just inferior to the apex of the tragus point (B01), just anterior to the free margin of the tragus.
B03	Thirst	Anterior to the tragus, at approximately the midpoint on a line connecting the external nose point and apex of the tragus point (B01 & B05).

Points on the Tragus   T003B02

Wexu Point	Common Point Name	Location
B04	Pharynx and Larynx	On the inner surface of the tuberculum supratragicum, or superior portion of the tragus if the tuberculum supratragicum is not present, opposite the external acoustic meatus.
B05	External Nose	Anterior to the tragus and at the center of the root of the tuberculum supratragicum and the tragus construction.
B06	Internal Nose	On the inner surface and at the center of the tragus just inferior to the pharynx and larynx point (B04).
B07	Hunger	Just inferior to the External nose point (B05) and at the inferior extreme of the root of the tragus.
B08	Adrenal gland	Just anterior to the brim of the most inferior extreme of the tragus formation, just superior to where the intertragic incisure begins.
B09	High blood pressure	Just inferior to the adrenal gland point (B08), approximately 1/2 the distance between it and the inferior extreme of the intertragic incisure.

Supra-tragic Incisure Points   T003B03

Wexu Point	Common Point Name	Location
C01	External Ear	Superior to the anterior extreme of the supratragic incisure, just inferior to the anterior extreme of the crus of helix.
C02	Heart point	Just posterior to the external ear point (C01).

Intertragic Incisure Points   T003B04

Wexu Point	Common Point Name	Location
D01	Internal secretion	At the inferior extreme of the intertragic incisure, on the interior aspect of its brim.
D02	Ovary	Lying just posterior to the internal secretion point (D01), on the interior aspect of the brim of the intertragic incisure.
D03	Eye 1	Just anterior to the internal secretion point on the external aspect of the brim of the intertragic incisure.
D04	Eye 2	Posterior and superior to the ovary point (D02), on the external aspect of the brim of the intertragic incisure.

Antitragus Points   T003B05

Wexu Point	Common Point Name	Location
E01	Brain Axis	At the juncture of the antihelix and the antitragus which forms the center of the posterior auricular sulcus.
E02	Toothache	Inside the ear, on the floor of the cavum conchae, superior and anterior to the brain axis point (E01).

table of auricular landmark & acupoint descriptions

Antitragus Points   T003B05

Wexu Point	Common Point Name	Location
E03	Hou-Ya (pharynx-tooth)	Just Anterior to the brain axis point, on the external aspect of the brim of the antitragus.
E04	Brain (pituitary)	Just inferior to the central portion of the antitragus ridge, on the interior aspect of the brim of the antitragus.
E05	Occiput	Anterior and inferior to the Hou-Ya point (E03) on the external aspect of the antitragus.
E06	Parotid (salivary)	On the external aspect of the antitragus brim, at the center of the antitragus ridge.
E07	Ping-Chuan (asthma)	Half way between the Hou-Ya and parotid glands points (E03 & E06), on the external aspect of the antitragus brim.
E08	Testicle	On the internal aspect of the brim of the antitragus, just inferior to the level of the parotid point (E06).
E09	Forehead	On the external aspect of the brim of the antitragus, just posterior to the inferior extreme of the antitragus.
E10	Subcortex	On the internal aspect of the brim of the antitragus, just inferior to the brain point (E04).
E11	Tai Yang (the sun)	Just anterior and inferior to the occiput point (E05).
E12	Vertex	Inferior to the Hou-Ya point (E03) and posterior to the Tai Yang point (E11).

Antihelix Points   T003B06

Wexu Point	Common Point Name	Location
F01	Cervical Vertebrae	Posterior and superior to the the brain axis point (E01) and the posterior auricular sulcus.
F02	Thoracic Vertebrae	At the center of the antihelix, approximately 1/4 of the distance superior to the inferior extreme of the antihelix.
F03	Lumbosacral Vertebrae	Posterior to the central portion of the antihelix 1/2 the distance from its distal extreme.
F04	Coccygeal Vertebrae	At the center of the antihelix, just posterior to the posterior extreme of the triangular fossa.
F05	Neck	On the anterior edge of the antihelix, approximately 1/5 of its distance superior to the posterior auricular sulcus.
F06	Thorax	On the interior border of the antihelix, on the same level as the crus of helix.
F07	Abdomen	On the interior border of the antihelix, 1/4 of the distance from its superior extreme.
F08	Mammary Gland	These are twin points, lying superior to the thoracic vertebrae point (F2), on the posterior and anterior aspect of the antihelix, forming an equilateral triangle with it.

Antihelix Points   T003B06

Wexu Point	Common Point Name	Location
F09	Thyroid Gland	On the posterior portion of the antihelix, approximately 1/5 of the distance superior to the posterior auricular sulcus.
F10	Outside Abdomen	On the posterior portion of the antihelix, posterior and superior to the level of the abdomen point (F07).

Points on the Upper Crura of the Antihelix   T003B07

Wexu Point	Common Point Name	Location
G01	Toes	At the posterior superior extreme of the upper crura of the antihelix.
G02	Heel	At the anterior superior extreme of the upper crura of the antihelix.
G03	Ankle joint	On the central portion of the upper crura of the antihelix, approximately 1/4 of the distance from its anterior extreme, posterior and inferior to the toes and heel points (G01 & G02).
G04	Knee joint	On the central portion of the upper crura of the antihelix, approximately 1/2 of the distance from its anterior extreme, posterior and inferior to the ankle joint point (G03).
G05	Hip joint	On the central portion of the upper crura of the antihelix, approximately 3/4 of the distance from its anterior extreme, posterior and inferior to the knee point (G04).
G06	Knee	On the posterior border of the upper crura, posterior and superior to the level of the hip joint point (G05).
G07	Heat	Just anterior to the central portion of the upper crura of the antihelix, just superior to its juncture with the antihelix.

Points on the Lower Crura of the Antihelix   T003B08

Wexu Point	Common Point Name	Location
H01	Sympathetic	At the anterior extreme of the lower crura of the antihelix.
H02	Ischium	On the superior border of the lower crura of the antihelix, approximately 1/2 of the distance from its anterior extreme.
H03	Buttock	On the superior border of the lower crura of the antihelix, approximately 1/4 of its distance from its posterior extreme, posterior and slightly superior to the ischium point (H02).
H04	Lumbago	On the superior border of the lower crura of the antihelix, just anterior of its posterior extreme, anterior and inferior to the coccygeal vertebrae point (F04)

Points in the Triangular Fossa   T003B09

Wexu Point	Common Point Name	Location
I01	Shen Men (Divine Gate)	On the floor of the triangular fossa, approximately midpoint on line drawn from the ankle point (G03) to the buttock point (H03).
I02	Pelvic Cavity	Posterior and inferior to the Shen Men point (I01), on the posterior ascending wall of the triangular fossa, just anterior of the juncture of the upper and lower crurae of the antihelix.
I03	Ku-Kuan	On the inferior ascending wall of the triangular fossa, just superior to the buttock point (H03).
I04	Hypotensive	On the superior ascending wall of the triangular fossa at its anterior superior extreme.
I05	Uterus	On the floor of the triangular fossa, approximately midway between the amterior extremes of the upper and lower crurae of the antihelix.
I06	Asthma (pants)	Posterior and inferior to the uterus point (I05).
I07	Hepatitis	Posterior and superior to the asthma point (I06).
I08	Lower Section of the Rectum	Just superior and posterior to the sympathetic point (H01).
I09	Urethra	Just superior to the lower section of the rectum point (I08) and anterior and inferior to the Asthma point (I06).
I10	External Genitalia	Just superior to the uterus point (I05).

Points in the Scaphoid Fossa (Scapha)   T003B10

Wexu Point	Common Point Name	Location
J01	Fingers	On the floor of the scaphoid fossa, under the superior extreme of the auricular tubercle.
J02	Clavicle	On the floor of the scaphoid fossa, at the level of the thoracic vertebrae point (F2).
J03	Shoulder Joint	On the floor of the scaphoid fossa, posterior to and at the level of the posterior mammary gland point (F08).
J04	Shoulder	On the floor of the scaphoid fossa, posterior to and at the level of the thorax point (F06).
J05	Elbow	On the floor of the scaphoid fossa, posterior to and at the level of the outside abdomen point (F10).
J06	Wrist	On the floor of the scaphoid fossa, under the inferior extreme of the auricular tubercle.
J07	Appendix 1	At the bottom of the scaphoid fossa, at its superior extreme.
J08	Appendix 2	On the floor of the scaphoid fossa, just superior to the shoulder point (J04).
J09	Appendix 3	At the bottom of the scaphoid fossa, just posterior to and slightly inferior to the thyroid gland point (F09).
J10	Urticaria Region	A zone of points found along the anterior ascending wall of the scaphoid fossa, beginning anterior and inferior to the fingers point (J01).

Points in the Scaphoid Fossa (Scapha)   T003B10

Wexu Point	Common Point Name	Location
J11	Minor Occipital Nerve	Posterior to the fingers point (J01) on the posterior ascending wall of the scaphoid fossa.

Points in the Helix   T003B11

Wexu Point	Common Point Name	Location
K01-K06	Helix 1-6	Helix 1 (K01) lies on the middle portion of the helix at the level of the midpoint of the auricular tubercle; subsequent helix points occur at equal intervals proceeding inferiorly with Helix 6 (K06) being on the A15 site and Helix 5 (K05) on A13.
K07	Tonsil 1	On the anterior superior central portion of the helix, on the level with and anterior to the toes point (G01).
K08	Tonsil 2	On the anterior central portion of the helix, on the level with and posterior to the posterior mammary point (F08).
K09	Tonsil 3	Same as A6.
K10	Liver Yang 1 (Kan-Yang 1)	Just superior to the Helix 1 point on the auricular tubercle.
K11	Liver Yang 2 (Kan-Yang 2)	On the central anterior portion of the helix at the inferior extreme of the auricular tubercle.
K12	Apex of the Auricle	On the anterior superior anterior portion of the helix, on the level with and anterior to the heel point (G02).
K13	External Genitalia	On the central anterior portion of the helix, just inferior to the level of the lower crura.
K14	Urethra	On the central anterior portion of the anterior helix, at the level of the large intestine point (M08) within the cymba conchae.
K15	Lower segment of the Rectum	On the central anterior portion of the anterior helix, at the level of the small intestine point (M06) within the cymba conchae.
K16	Anus	On the central anterior portion of the anterior helix, between and equidistant from the urethra and lower segment of the rectum points (K14 & K15).
K17	Blind Pile	On the central anterior portion of the anterior helix, on level with the juncture of the upper and lower crurae of the antihelix.

Points in the Crus of the Helix   T003B12

Wexu Point	Common Point Name	Location
L01	Diaphragm	On the central portion of the crus of the helix, just inferior to the small intestine point (M06) within the cymba conchae.
L02	Point of Support	On the central portion of the crus of the helix, inferior and posterior to the duodenum point (M05) within the cavum conchae.
L03	Branch	Just anterior and inferior to diaphragm point (L01), on the inferior portion of the crus of the helix.

table of auricular landmark & acupoint descriptions

Points around the Crus of the Helix T003B13

Wexu Point	Common Point Name	Location
M01	Mouth	Under the inferior brim of the crus of the helix, directly superior to the external acoustic meatus.
M02	Esophagus	Under the inferior brim of the crus of the helix, inferior to the small intestine point (M06).
M03	Cardia (cardiac orifice)	Under the inferior brim of the crus of the helix, inferior and anterior to the duodenum point (M05).
M04	Stomach	A zone of points which extends across the posterior neck of the crus of the helix, posterior to the cardia and duodenum points (M03 & M05).
M05	Duodenum	Superior to the crus of the helix, on the floor of the cymba conchae, just anterior to the origin of the crus of the helix.
M06	Small Intestine	Superior to the crus of the helix, on the floor of the cymba conchae, just anterior to the duodenum point (M05) and superior to the esophagus point (M02).
M07	Appendix	Superior to the crus of the helix, on the floor of the cymba conchae, anterior to the small intestine point (M06) and superior to the mouth point (M01).
M08	Large Intestine	Superior to the crus of the helix, on the floor of the cymba conchae at its anterior extreme.

Points in the Cymba Conchae T003B14

Wexu Point	Common Point Name	Location
N01	Bladder	Under the over hang of the lower crus of the helix, on the floor of the cymba conchae, inferior to the ischium point (H02).
N02	Prostate	Under the over hang of the lower crus of the helix, on the floor of the cymba conchae, inferior and posterior to the lower section of the rectum point (I08).
N03	Ureter	Under the over hang of the lower crus of the helix, on the floor of the cymba conchae, inferior to the buttock point (H03).
N04	Kidney	Under the over hang of the lower crus of the helix, on the floor of the cymba conchae, posterior to the ureter point (N03).
N05	Umbilical Region	On the floor of the cymba conchae, centrally located between the ureter and small intestine points (N03 & M06).
N06	Pancreas/Gall bladder	Under the over hang of the lower crus of the helix, on the floor of the cymba conchae, under the abdomen point (F07).
N07	Ascites	On the floor of the cymba conchae, centrally located between the kidney and duodenum points (N04 & M05).

Points in the Cymba Conchae   T003B14

Wexu Point	Common Point Name	Location
N08	Pancreatitis	On the floor of the cymba conchae, centrally located between the pancreas/gall bladder and stomach points (N06 & M04).
N09	Liver	Posterior and superior to the stomach point zone (M04).
N10	Spleen	Posterior and inferior to the stomach zone (M04), and directly inferior to the liver point (N09) and anterior to the antihelix.

Points in the Cavum Conchae   T003B15

Wexu Point	Common Point Name	Location
O01	Heart	At the center of the floor of the cavum conchae.
O02	Lung	Two points, one superior and the other inferior to the heart point (O01). The inferior point representing the lung on the ipsilateral side, while the superior point represents the contralateral lung.
O03	Bronchi	Just anterior and superior to the heart point (O01).
O04	Windpipe (trachea)	An irregularly shaped zone extending posteriorly from the edge of the external acoustic meatus toward the heart point (O01).
O05	San-Jiao (San-Chiao)	An irregularly shaped area inferior to the lung zone (O02) and the windpipe area, extending from the inferior edge of the external acoustic meatus meatus posteriorly toward a vertical line drawn through the lung and heart points (O02 & O01), assuming a half-crescent form.
O06	Muscle Relaxing	On the floor of the cavum conchae, just anterior to the neck point (F05).

Points around the External Acoustic Meatus   T003B16

Wexu Point	Common Point Name	Location
P01	Lower Abdomen	Within the external acoustic meatus, on the superior wall.
P02	Upper Abdomen	Within the external acoustic meatus, on the inferior wall.

Points on the Posterior Aspect of the Auricle   T003B17

Wexu Point	Common Point Name	Location
Z01	Bleeding & Depressing Groove	A series of points along the superior 1/3 of the curved vertical groove.
Z02	Upper Back	A series of points along a line just superior to the upper cartilaginous eminence.
Z03	Middle Back	A series of points along a line which runs along the middle of the posterior aspect of the auricle.
Z04	Lower Back	A series of points along a line just inferior to the lower cartilaginous eminence.

table of auricular landmark & acupoint descriptions

Points on the Posterior Aspect of the Auricle   T003B17

Wexu Point	Common Point Name	Location
Z05	Top of the Brain	On the posterior aspect of the helix, near the apex of the positive aspect of the auricle.
Z06	Spinal Cord 1	Inferior and medial to the brain point (Z05), superior to the curved vertical groove on the posterior aspect of the helix.
Z07	Antipyretics	On the posterior aspect of the helix, on the level of the spinal cord 1 point (Z06).
Z08	Tin Ying	On the posterior aspect of the helix, on the level of the series of upper back points (Z02).
Z09	Headache 1	Just inferior to the level of the upper back series of points (Z02) near the vertical ridge separating the ear from the mastoid process.
Z10	Headache 2	Medial and inferior to the headache 1 point (Z09) and just medial to the vertical ridge separating auricle from mastoid process.
Z11	Headache 3	Lateral and inferior to the headache 2 point (Z09) and just medial to the vertical groove behind the helix.
Z12	Neck	4 mm. inferior and lateral to the headache 2 point (Z10).
Z13	Sedation	3 mm. inferior and lateral to the neck point (Z12).
Z14	Brain Chi	Lateral to and at the level of the neck point (Z12), near the vertical groove behind the helix.
Z15	Groove 1	3 mm. inferior and medial to the brain chi point (Z14).
Z16	Groove 2	4 mm. inferior and lateral to the groove 1 point (Z15), just superior to the middle back point series line (Z03).
Z17	Epigastric Region	Inferior and lateral to the sedation point (Z13) and just superior to the middle back point series line (Z03).
Z18	Nerve Center	On the posterior aspect of the helix, 3 mm. directly inferior to the Tin Ying point (Z08).
Z19	Vertebral Column	On the posterior central aspect of the helix, 4 mm. inferior to the nerve center point (Z18).
Z20	Backache 1	4 mm. directly inferior and slightly medial to the vertebral column point (Z19).
Z21	Backache 2	On the posterior central aspect of the helix, just superior to the level of the middle back series line (Z03).
Z22	Fauces	Just lateral to the midpoint of the middle back series line (Z03) and slightly inferior to it.
Z23	Groove 3	Slightly more inferior to the middle back series line (Z03) then the preceding point (Z22), and just medial to the vertical groove behind the helix.
Z24	Ulcer	Approximately halfway between the previous 2 points but inferior to them; together they form an equilateral triangle.

Points on the Posterior Aspect of the Auricle T003B17

Wexu Point	Common Point Name	Location
Z25	Back	On the central portion of the posterior aspect of the helix, approximately 4 mm. inferior to the backache 2 point (Z21).
Z26	Lung	Approximately 4 mm. inferior to the back point (Z25) but medial to the central portion of the posterior aspect of the helix.
Z27	Lumbago	3 mm. from the lung point (Z26), but lateral and inferior to it; it combines to form an isosceles triangle with the lung and mesogastric region points (Z26 & Z28).
Z28	Mesogastric Region	Approximately 5mm. inferior to and in line with the lung point (Z26).
Z29	Cough & Asthma	Approximately 4 mm. inferior to the lumbago point (Z27).
Z30	Hypogastric Region	3 mm. inferior to and in direct line with the mesogastric region point (Z28).
Z31	Buttock	Approximately 3 mm. inferior to and in direct line with the cough & asthma point (Z29).
Z32	Gastro-Intestinal Tract	Midway between the middle back and lower back point series lines (Z03 & Z04) and approximately 2 mm. lateral to the vertical groove which divides the auricle from the skin over the mastoid process.
Z33	Epigastric Region	On the same level as the gastro-intestinal tract point (Z32) but lateral to it to the extent that it lies approximately 1 mm. medial to the vertical crease behind the helix.
Z34	Heart	Directly opposite the heart point (O01) on the anterior aspect of the auricle in cavum conchae and slightly inferior to the level of the epigastric region and gastro-intestinal points (Z33 & Z32).
Z35	Kidney Cell	Approximately 2 mm. inferior to and in direct line with the gastro-intestinal point (Z32).
Z36	Pai Ling 1	Approximately 2 mm. inferior to but slightly medial to the kidney cell point (Z35).
Z37	Yang Ho	Approximately 2 mm. inferior to and in direct line with the Pai Ling 1 point (Z36).
Z38	Appendix	Just superior to the level of the Pai Ling 1 point (Z36) and approximately 2 mm. medial from the vertical crease behind the helix.
Z39	Pai Ling 2	Approximately 4.5 mm. inferior to and slightly medial to the appendix point (Z38); just inferior to the lower back series line (Z04).
Z40	Lower Extremities	Inferior to the lower back point series line (Z04) and in direct line with and inferior to the kidney cell point (Z35).
Z41	Ear	Approximately 2.0 mm. inferior to but slightly medial to the lower extremities point (Z40).
Z42	Perineum 1	Inferior to but slightly medial to the Pai Ling 2 point (Z39).

Points on the Posterior Aspect of the Auricle T003B17

Wexu Point	Common Point Name	Location
Z43	Yi Shan	On the posterior aspect of the lobule, inferior to the perineum 2 and spinal cord 2 points (Z44 & Z45), forming an inverted isosceles triangle with them.
Z44	Perineum 2	On the posterior aspect of the lobule, lateral to and superior to the Yi Shan point (Z43); combined with the perineum 1 and Yi Shan points (Z42 & Z43) to form an inverted isosceles triangle.
Z45	Spinal Cord 2	Superior to and medial of the Yi Shan point (Z43), lying on the crease between the posterior aspect of the auricle and the lobule.
Z46	Yi Len	Inferior and medial to the Yi Shan point (Z43), near the inferior extent of the posterior aspect of the lobule.

Bibliography T003C

The Academy of Traditional Chinese Medicine, pp. 269–280.

An Explanatory Book of the Newest Illustrations of Acupunture Points, pp. 72–77.

Wexu, pp. 17–95.

table of auricular landmark & acupoint descriptions

Auricular Acupoint Locations (T003C)

Acupoints on Prominent Portions of the Auricle—First View (T003C01)

Acupoints on Prominent Portions of the Auricle—Second View (T003C02)

- ■ VISIBLE ACUPOINT
- □ ACUPOINT NOT VISIBLE FROM THE VIEW PROFFERED

- ■ VISIBLE ACUPOINT
- □ ACUPOINT NOT VISIBLE FROM THE VIEW PROFFERED

table of auricular landmark & acupoint descriptions

Acupoints on the Posterior Portion of the Auricle (T003C04)

Acupoints on the Concave Portions of the Auricle (T003C03)

- VISIBLE ACUPOINT
- ACUPOINT NOT VISIBLE FROM THE VIEW PROFFERED
- ACUPOINT ZONE

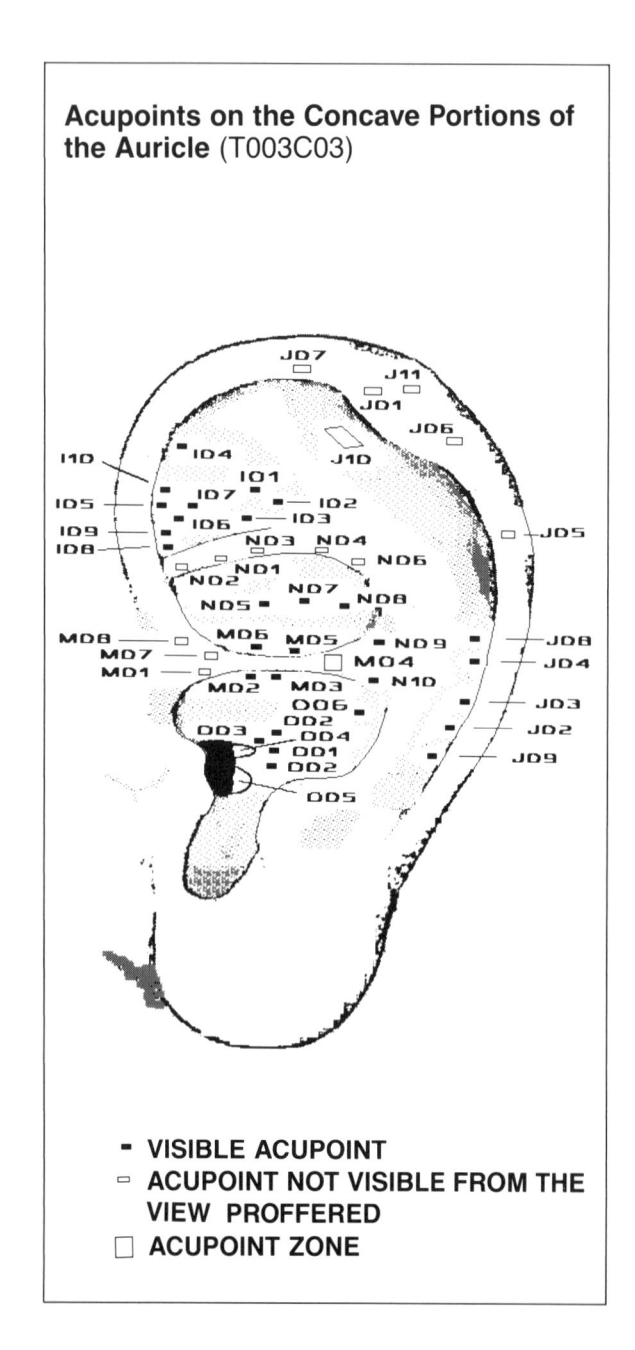

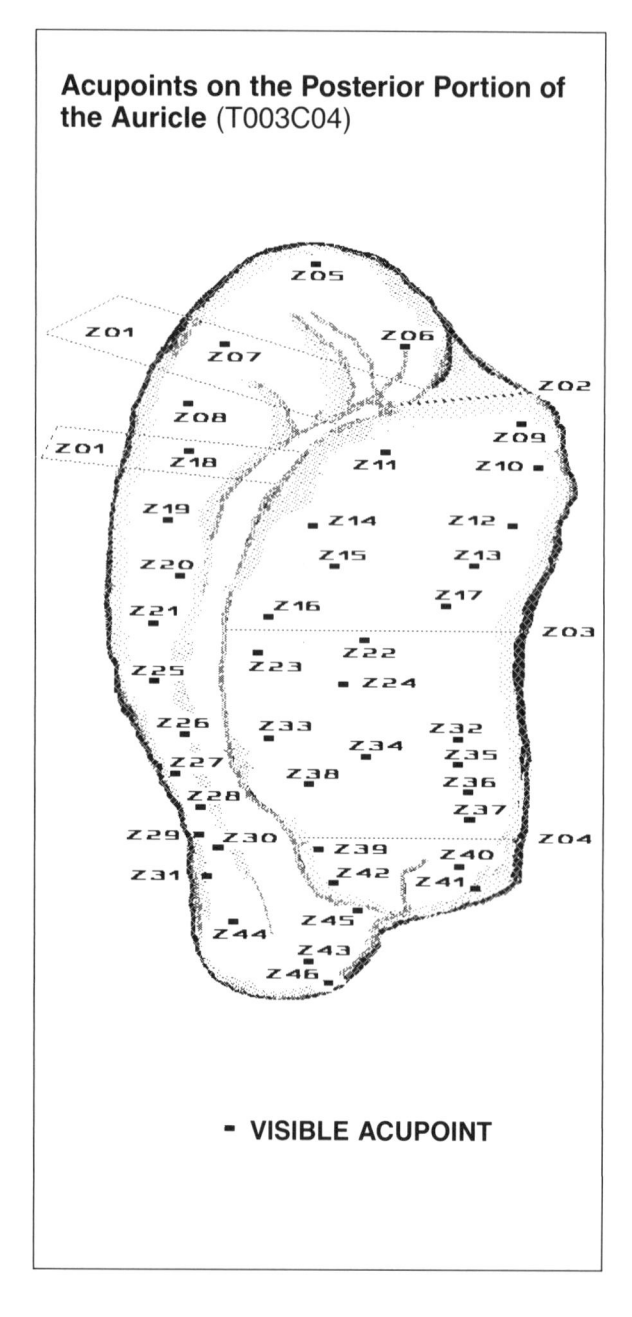

- VISIBLE ACUPOINT

AUXILIARY EVALUATION TECHNIQUES: AUSCULTATORY METHOD OF MEASURING BLOOD PRESSURE (M26A)

Definition (M26A1)

The auscultatory method for the determination of systolic and diastolic arterial blood pressures is conducted through the use of a stethoscope and a blood pressure cuff attached to a sphygmomanometer (mercury manometer).

Procedures of Application (M26A2)

Upper Extremity (M26A2A)

1) The patient should be in a comfortable sitting position. Pressures may be taken in semireclining or supine positions, but the blood pressure values will be affected by the differences in position and not directly comparable with the norms established for the sitting position.

2) The arm to be used for the blood pressure measurement procedure should be flexed and relaxed on the arm of the chair, in the patient's lap, or on the practitioner's supporting arm and/or hand.

3) For measurement of the blood pressure in the upper extremity, a blood pressure cuff should be wrapped around the humeral portion of the arm, just proximal to the elbow crease.

4) The head of the stethoscope should be placed over the artery in the antecubital space.

5) The blood pressure cuff should be inflated to well above the arterial systolic pressure, usually in excess of 140 Hg. or higher; i.e., the blood pressure cuff should be inflated to the point that the brachial artery is collapsed sufficiently that no blood flows into the lower artery during any part of the pressure cycle and no Korotkoff's sounds can be heard in the lower artery (the pressure registered by the pressure monitor attached to the blood pressure cuff may have to range from 130 to 240 mm. Hg. to be adequately high enough).

6) The pressure should be gradually decreased in the cuff by slowly letting out the air.

7) Just as the pressure in the cuff falls below the systolic pressure, blips (Korotkoff's sounds, created by the artery completely closing during diastole) will be picked up by the stethoscope and heard by the practitioner to occur in synchrony with the heart beat. The pressure level at which such sounds occur should be noted and recorded as the systolic pressure.

8) The pressure in the cuff should continue to be lowered slowly until the auditory blips disappear. At the point where the blips are no longer heard, when the artery no longer closes completely during diastole, the diastolic pressure becomes apparent; it should be noted and recorded as the diastolic pressure. This pressure will generally range from between 60 to 110 mm. Hg.

Normal systolic pressure in the upper extremity is commonly accepted by most practitioners to be below or equal to 160 mm. Hg., while normally the diastolic pressure is thought to be below or equal to 95 mm. Hg. for women of all ages (160/95). Normal for males over 45 is considered to be below or equal to 140/95, and 130/90 for those under 40. The ideal for all ages, whether male or female, is considered to be 120/80. It is thought by some that the blood pressure in the area of 110/70 is exceptionally good, and such pressure may be associated with exceptional longevity.

For further discussion of upper extremity blood pressure refer to Blood Pressure Control With Physiological Feedback Training [R007].

Lower Extremity (M26A2B)

1) The patient should be in a comfortable sitting position. The blood pressure may be taken in semireclining and lying positions, but the blood pressure values will be affected by each position and will not be directly comparable with the norms established for the sitting position.

2) The leg to be used for the blood pressure measurement procedure should be flexed and relaxed.

3) For measurement of the blood pressure in the lower extremity, a blood pressure cuff should be wrapped around the femoral portion of the leg, proximal to the knee.

4) The head of the stethoscope should be placed over the popliteal section of the femoral artery, just distal to the blood pressure cuff on the medial posterior aspect of the leg just proximal to the popliteal space.

5) The blood pressure cuff should be inflated to well above the arterial systolic pressure, to the point that the popliteal artery is collapsed so that no blood flows into the lower artery during any part of the pressure cycle and no Korotkoff's

sounds can be heard in the lower artery. The pressure registered by the pressure monitor attached to the blood pressure cuff may have to range from 130 to 240 mm. Hg. to be adequately high enough.

6) The pressure should be gradually decreased in the cuff by slowly letting out the air.

7) Just as the pressure in the cuff falls below the systolic pressure, blips known as Korotkoff's sounds, created by the artery completely closing during diastole), will be picked up by the stethoscope and heard by the practitioner to occur in synchrony with the heart beat. The pressure level at which this occurs should be noted and recorded as the systolic pressure.

8) The pressure in the cuff should continued to be lowered until the auditory blips disappear. At the point where the blips are no longer heard, when the artery no longer closes completely during diastole, the diastolic pressure becomes apparent; it should be noted and recorded as the diastolic pressure. This pressure will generally range from 60 to 110 mm. Hg.

Partial List of Conditions Evaluated (M26A3)

Treatable Causes (M26A3A)

Hypertension (C39)
Hypotension (C45)
Soft tissue swelling (C06)
Sympathetic hyperresponse (C38)
Vascular insufficiency (C11)
Visceral organ Dysfunction (C02)

Syndromes (M26A3B)

BURNS (SECOND AND/OR THIRD DEGREE) (S72)
HEADACHE PAIN (S02)
HYPERTENSION (S43)
HYSTERIA/ANXIETY REACTION (S59)
MIGRAINE (VASCULAR) HEADACHE (S18)
PHOBIC REACTION (S44)
POST SPINAL CORD INJURY (S24)

Notes Aside (M26A4)

Because each practitioner has a different hearing range, blood pressure measurements from a particular patient, taken periodically, over an extended period of time, should be made by the same person; unless this convention is observed, comparison of measurements with each other may be invalid, since absolute values may not be assumed if different practitioners have taken the different measurements.

Bibliography (M26A5)

Guyton, pp. 299–300.

AUXILIARY EVALUATION TECHNIQUES: SPIROMETRY (M26B)

Definition (M26B1)

Breathing functions which can be measured by spirometry (M26B1A)

Spirometric measurement is performed to determine tidal volume, inspiratory reserve volume, expiratory reserve volume, inspiratory capacity, and vital capacity.

Tidal volume (M26B1A01) *Tidal volume* is the volume of air inspired or expired with each normal breath. The normal volume is considered to be approximately 500 ml. for the young adult male (slightly less for the adultfemale).

Inspiratory reserve volume (M26B1A02) *Inspiratory reserve volume* is the extra volume of air which can be forcefully inspired beyond the normal tidal volume (a big breath). The normal volume is considered to be approximately 3000 ml. for the young adult male and less for the young adult female.

Expiratory reserve volume (M26B1A03) *Expiratory reserve volume* is the volume of air that can be forcefully expired beyond normal tidal expiration levels. The normal volume for a young adult male is approximately 1100 ml., and less for the young adult female.

Inspiratory capacity (M26B1A04) *Inspiratory capacity* is the volume of air that a person can breathe in, measured from the end of normal expiration to the point of full (maximal) distension of the lungs. It usually equals the tidal volume plus the inspiratory reserve volume or approximately 3500 ml. for the normal young adult female.

Vital capacity (M26B1A05) *Vital capacity* is the total (maximal) amount of air which can be expelled from the lungs; measurement begins after the subject has filled the lungs to maximum capacity to the end of maximum expiration. This should be approximately equal to the inspiratory reserve volume plus the tidal volume plus the expiratory reserve volume, or approximately 5800 ml., for the normal young adult male.

Spirometer construction (M26B1B)

The spirometer typically consists of a drum inverted in a chamber of water, its bottom above water line. A line attached to the bottom of the drum runs up to and across two pulleys and back down through a tubular stabilizing guide attached to the water chamber wall, to finally attach to a counter-balancing weight. A breathing tube, fitted with a mouthpiece, pierces the bottom of the water chamber. This tube then runs up through the water with its free end in the air space under the bottom of the drum. A scribing attachment is affixed to the counter-balance line, which is arranged to mark on a recorder strip (moving at a smooth, steady rate) any movement of the drum up or down.

Procedures of Application (M26B2)

General procedure (M26B2A)

1) The patient should be instructed to take the largest breath of free air as is possible and to blow all the accumulated air through the air tube. This volume of air, added to that already in the drum's air space, should provide sufficient air volume for testing measurement.
2) The air tube should then be pinched off to keep any air from escaping back out.
3) The resulting strip recording should be noted to be the vertical zero point.
4) The strip recorder should be turned on and kept running for each testing procedure.

The strip recording should be inverted for measurement and reading.

The number of breaths possible during a test is limited by the carbon dioxide build-up which occurs as normal breathing takes place.

Before other testing is made, the "old air" should be allowed to exhaust from the drum to beginning levels and be replaced by "new air" as in steps 1 and 2.

Breathing functions measured (M26B2B)

Tidal volume (M26B2B01) The patient should breathe normally in and out a few times through the air tube while the strip recorder runs; mea-

auxiliary evaluation techniques: spirometry

Spirometer Construction (M26B1C)

SPIROMETER

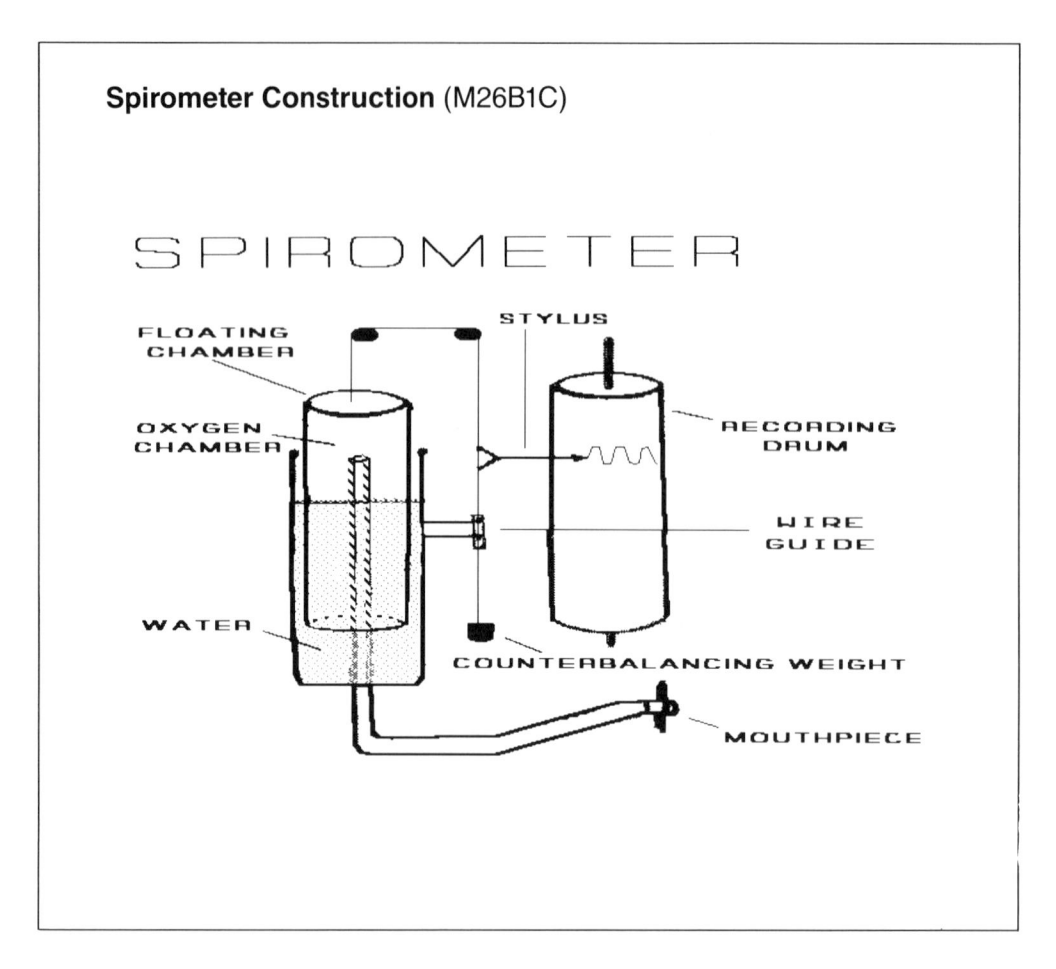

(Diagram labeled: FLOATING CHAMBER, OXYGEN CHAMBER, WATER, STYLUS, RECORDING DRUM, WIRE GUIDE, COUNTERBALANCING WEIGHT, MOUTHPIECE)

surement is made from the bottom of the normal expiration curve to the top of the normal inspiration curve.

Inspiratory reserve volume (M26B2B02) Following a couple of normal breaths, the patient should be instructed to take as full a breath as possible from the breathing tube. Following a normal exhalation, measurement is made from the top of the normal inspiration curve to the crest of the maximum inspiration curve.

Expiratory reserve volume (M26B2B03) Following a couple of normal breaths, the patient should be instructed to forcefully expire all possible air from the lungs. Measurement is made from the bottom of the normal expiration curve to the bottom of the maximum expiratory curve.

Inspiratory capacity (M26B2B04) Following a couple of normal breaths, the patient should be instructed to take as full an inspiration as is possible. Measurement is made from the bottom of the preceding normal expiration curve to the top maximum inspiration curve

Vital capacity (M26B2B05) Following a couple of normal breaths, the patient should be instructed to take as full an inspiration as is possible and then to forcefully exhale as much air as is possible. Measurement is made from the top of the maximum inhalation curve to the bottom of the maximum expiratory curve.

Partial List of Conditions Evaluated (M26B3)

Treatable Causes (M26B3A)

Bacterial infection   (C09)
Localized viral infection   (C27)
Lung fluid accumulation   (C13)
Pathological neuromuscular hypertonus   (C28)
Pathological neuromuscular hypotonus   (C33)
Reduced lung vital capacity   (C17)
Sympathetic hyperresponse   (C38)
Visceral organ Dysfunction   (C02)

Syndromes (M26B3B)

BACTERIAL INFECTION   (S63)
CEREBRAL PALSY   (S25)

CHEST PAIN (S14)
CHRONIC BRONCHITIS/EMPHYSEMA
 (OBSTRUCTIVE PULMONARY DISEASE) (S13)
HIGH THORACIC BACK PAIN (S48)
HYSTERIA/ANXIETY REACTION (S59)
NEUROMUSCULAR PARALYSIS (S22)
PNEUMONIA (S58)

POST CEREBRAL VASCULAR
 ACCIDENT (CVA) (S07)
POST SPINAL CORD INJURY (S24)

Bibliography (M26B5)

Guyton, pp. 460–462.

AUXILIARY EVALUATION TECHNIQUES: AUSCULTATION (M26C)

Definition (M26C1)

Auscultation is defined as listening to the sounds of the body; such listening is usually performed through the aid of a stethoscope. The most common uses of auscultation (not including blood pressure measurement) include the detection of normal or abnormal heart sounds, lung sounds, and joint and/or tendon sounds.

Procedures of Application (M26C2)

Heart Sounds (M26C2A)

For the evaluation of heart sounds, the stethoscope should be placed in strategic locations (vulvar locations) on the chest above the heart so that each valve may be heard; these strategic sites exist (1) between the second and third rib just lateral to the body of the sternum (aortic and pulmonic areas, respectively), (2) between the left fifth and seventh ribs just lateral to the distal aspect of the body of the sternum (tricuspid area), and (3) between the left fifth and sixth ribs just lateral to the joining of the ribs and their costal cartilage counterparts (mitral area). With the stethoscope placed over any one of the vulvar locations, the sounds from all the other valves can still be heard, though the particular valve monitored will be louder then the others; the stethoscope should be moved from one valve site to the next, noting the relative loudness of each valve and gradually picking out each component sound from the valve. Of particular interest are any of the following abnormal sounds:

Thrilling (murmur of aortic stenosis) (M26C2A01) Thrilling is an intensely harsh, vibratory sound produced by turbulent blood impinging against aortic walls. It is often so loud that it may be heard, unaided, several feet away from the patient.

Murmur of aortic regurgitation (M26C2A02) No sound is heard during systole, but during diastole a relatively high pitched "blowing" sound may be heard over the left ventricle as blood flows backward from the aorta into the left ventricle.

Murmur of mitral regurgitation (M26C2A03) A high frequency "blowing" sound is heard during systole as blood flows backward through the mitral valve.

Murmur of mitral stenosis (M26C2A04) This murmur is a low pitched, low frequency "blowing" sound, occurring usually during systole. It is often accompanied by an intense low frequency vibration which may felt, rather then heard, by placing a hand over the apex of the heart.

Lung Sounds (M26C2B)

The stethoscope should be placed on the chest wall over each of the lobes of the lung. The listening sites should include the superior, middle and inferior aspect of the right anterior and posterior rib cage, and the superior and inferior aspect of the left chest wall. Listening sites should also include paravertebral and parasternal areas which are coincidently over the bronchial tree.

The most common abnormal sounds associated with the lung are collectively called rales, and include:

Atelectatic (M26C2B01) A transitory light cracking sound which disappears after deep breathing or coughing.

Bubbling (M26C2B02) A moist sound like small bubbles bursting or moving; it is associated with pneumonia or small lung cavities.

Cavernous (M26C2B03) A hollow bubbling sound, caused by air entering a lung cavity partially filled with fluid.

Clicking (M26C2B04) A short sticking sound associated with the opening of small bronchi accompanying a deep breath; often heard in the early stages of pulmonary tuberculosis.

Consonating (M26C2B05) A resonant vibration-like sound; it is produced in a bronchial tube and is heard through consolidated lung tissue.

Crepitant (M26C2B06) A fine bubbling or crackling sound, associated with thin secretions present in the smaller bronchial tubes.

Dry (M26C2B07) A relatively loud rattling sound produced by a constriction in a bronchial tube or the presence of a viscid secretion narrowing the lumen.

Gurgling (M26C2B08) A coarse, burbling sound heard over large cavities or trachea nearly filled with fluid.

Guttural (M26C2B09) A deep, gurgling sound heard over the lung but resulting from upper airway obstruction.

Metallic (M26C2B10) A metallic clicking or rubbing caused by resonance in a large cavity.

Moist (M26C2B11) A bubbling sound heard over bronchial tubes containing mucal secretions.

Mucous (M26C2B12) A thick, bubbling sound heard over bronchial tubes containing mucus.

Palpable (M26C2B13) A low-pitched, sonorous sound which may be felt through the chest wall as well as heard.

Sibilant (M26C2B14) A whistling sound produced by air passing through lumen of a bronchus narrowed by viscus secretions.

Sonorous (M26C2B15) A cooing or snoring sound produced by the vibration of projecting masses of viscus secretions in large bronchi.

Joint and/or Tendon Sounds (M26C2C)

Sounds arising from the joints or muscle action are often sufficiently loud that they may be heard by both the patient and the practitioner, unaided. The specific site which generates the sound, however, may be best localized through the use of a stethoscope. The stethoscope should be placed over the body parts which may be responsible for the sound. The patient should be instructed to move the involved joints or contract the involved muscles while the stethoscope is being held in place. The sounds which should be listened for are tendon snapping sounds, tendon crepitus, and crepitus in the joints.

Partial List of Conditions Evaluated (M26C3)

Treatable Causes (M26C3A)

Bacterial infection (C09)
Calcific deposit (C08)
Habitual joint positioning (C48)
Hypermobile/instable joint (C10)
Joint ankylosis (C42)
Joint approximation (C47)
Ligamentous strain (C25)
Localized viral infection (C27)
Lung fluid accumulation (C13)
Reduced lung vital capacity (C17)
Restricted joint range of motion (C26)
Vascular insufficiency (C11)
Visceral organ dysfunction (C02)

Syndromes (M26C3B)

ACHILLES TENDONITIS (S52)
ADHESION (S56)
BACTERIAL INFECTION (S63)
BURSITIS (S26)
CALF PAIN (S50)
CAPSULITIS (S27)
CARPAL TUNNEL (S35)
CERVICAL (NECK) PAIN (S73)
CHRONIC BRONCHITIS/EMPHYSEMA
  (OBSTRUCTIVE PULMONARY DISEASE) (S13)
COLITIS (S42)
ELBOW PAIN (S47)
FOOT PAIN (S51)
FOREARM PAIN (S54)
HAND/FINGER PAIN (S17)
HEADACHE PAIN (S02)
HIGH THORACIC BACK PAIN (S48)
KNEE PAIN (S46)
OSTEOARTHRITIS (S21)
PNEUMONIA (S58)
POST IMMOBILIZATION (S36)
POST WHIPLASH (S55)
SHOULDER/HAND (S57)
SHOULDER PAIN (S08)
TEMPOROMANDIBULAR JOINT (TMJ)
  PAIN (S06)
TENDONITIS (S28)
TENNIS ELBOW (LATERAL
  EPICONDYLITIS) (S32)
THIGH PAIN (S49)
TOOTHACHE/JAW PAIN (S41)
WRIST PAIN (S16)

Notes Aside   (M26C4)

Auscultation may be used to good effect for the determination of decreased arterial capacity. The stethoscope should be placed over the pulse site and the pulse listened to; if the artery is affected by occlusion, it will produce a pulse sound which is muted compared with its counter part on the contralateral side. This procedure is most utilized for the evaluation of pedal and carotid artery pulses.

Auscultation is also useful for the determination of colitis; a stethoscope placed over the anterior lower abdominal quadrants may be utilized to detect the flow and presence of gas bubbles as a reflection of excessive lower intestinal motility.

Bibliography   (M26C5)

Guyton, pp. 350–352.
Salter, p. 57.

BACTERIAL INFECTION (C09)

Definition (C09A)

Bacterial infection is defined as the invasion by living pathological bacterial microorganisms of a bodily tissue when conditions are favorable to their growth.

Of interest here is bacterial infection of: (1) tissues near the surface of the body (dermal layers), (2) tissues directly concerned with gaseous exchange (nasal sinuses, bronchial passages, and lung) and (3) "deep" tissues of the body (colon, external layers of the bone, and/or joints).

Evaluation (C09B)

Bacterial infections of the skin may be primary or secondary to other dermatosis. They may be superficial or deep, or affect specialized structures like hair follicles, sebaceous, apocrine or eccrine glands. They may be characterized by calor, rubor, swelling and accumulation of pus within comedones, pustules, papules, inflamed nodules, infected cysts, and the canalizing inflamed and infected sacs. Infected sites are often pressure sensitive.

The symptoms of *acute* bacterial infection of the sinus tissues may include recurrent colds with copious post nasal discharge, complaints of frontal and temporal headaches with frequent complaints of infraorbital and supraorbital pain. In the acute stage these symptoms are increased by the patient leans forward. Mucous discharge may be seen to be present on the posterior pharyngeal wall, palpation tenderness of the tissues over the frontal sinuses and maxillary antra, pyrexia, anorexia, vertigo, anosmia, photophobia, transient toothache, generalized aches, periorbital edema and fever of usually less than 102° F. Higher fevers may be indicative of suppurative adenitis or bronchopneumonia.

The symptoms associated with *chronic* bacterial infection of the sinus tissues may include a fairly constant headache (generally less severe then the acute variety) in the forehead and supraorbital regions, usually arising from the frontal sinuses. Pain may seem to be present in the upper jaw teeth, cheek, frontal (forehead) region, occipital region, parietal region, behind the eyes, at the base of the nose, and/or in the neck. Emotional depression may commonly accompany other symptoms. A peculiarly "sour", slightly putrid odor associated with the head (but not the breath) may be present as a symptom of both acute and chronic syndromes.

Symptoms of bronchial and/or lung tissue bacterial infection may include, in the early stages, shaking chill, sharp chest pain, fever, headache, dry and hacking cough which may produce purulent sputum (painful paroxysms of cough are likely to be extreme. A pleural friction rub may often be heard. Gastrointestinal symptoms may include abdominal distention, jaundice, diarrhea, nausea and vomiting. Tenderness and rigidity of the right upper quadrant of the trunk may be present if the middle and/or lower lobes are involved. Herpes of the lips and face is often present. As infection progresses, the cough may be more productive of sputum which may be pinkish, blood-flecked or even rusty colored at the height of the disease. During the stage of resolution the sputum may to finally become yellow and mucopurulent. Dyspnea may be present with rapid respiration and a peculiar expiratory grunt. Rales and suppressed breath sounds may be heard over the involved area (refer to Auxiliary Evaluation Techniques, Lung Sounds [M26C2B]). The patient may sweat profusely, be cyanotic, have pulse rates of between 100 and 130 per minute, and be acutely ill with temperatures ranging between 101 and 105° F.

Symptoms of bacterial infection of the colon are those which may be seen in colitis: cramping pain in the abdomen, abdominal distension (bloating), excessive flatulence, diarrhea (sometimes bloody), anemia, perforation of the colon, generalized infection, arthritis and lesions of the skin.

Osteomyelitis is characterized by sudden pain in the affected bone and a sharp rise in temperature. Tenderness over the bone is present and movement may be painful and involuntarily restricted. Late in the process of the disease, swelling appears over the bone and often in the adjacent joint. Definitive diagnosis depends on laboratory evaluation of aspirated exudates.

Bacterial infection of the joint is marked by rubor, calor, swelling, restricted motion, and pain in the joint, with and/or without compressive pressure. Differentiating between arthritis caused by bacterial infection and other causes depends upon laboratory testing of aspirated joint fluids.

Etiology (C09C)

Bacterial infections most commonly result from invasion of the tissues by bacteria which are, or resemble, the pyogenic cocci. These bacteria characteristically cause the formation of pus (sometimes copious in amount) at or near the infected site. This group of bacterium are made of "specialists" which tend to invade specific body locations or tissue types. They are generally found to be present on the skin or other body surfaces of "normal" individuals and only invade when offered the opportunity of an open wound or lowered tissue resistance. Prominent examples of the pyogenic cocci include staphylococcus, streptococcus, and pneumococcus bacterium.

The spread of bacterial infection to adjoining tissues, after the initial invasion, generally results from the passive action of toxins (usually a pyogenic exotoxin or endotoxin) which have been formed from bacterial metabolism. If such toxins are allowed to come into contact with adjoining tissues, these tissues may become injured, thereby effectively lowering their resistance to subsequent invasion the by the multiplying bacteria.

Staphylococcus bacteria is found on the skin of the healthy individual and generally invades surface tissues when the skin has been broken by abrasion, tearing, incisions or punctures of various types (ragged or surgical), or when the tissues have deteriorated from pressure ulceration or follicular infection. Infection by staphylococcus bacteria may to produce a pustule, furuncle, or a carbuncle.

Streptococcus and pneumococcus bacteria are found in the oral cavity and pharynx of the healthy individual and, when resistance is low (when the autoimmune system is impaired or over taxed), attack sinus, bronchial and/or lung tissues to produce sinusitis, bronchitis and/or pneumonia. The staphlococcus bacteria have been known to attack respiratory tissue as well.

Occasionally, infection may occur in tissues below the skin from bacteria carried by the blood. Infection of organs not directly exposed to air may also occur; invasion by staphlococcus, for example, may cause gastroenteritis (pseudomembranous enteritis), osteomyelitis, or endocarditis.

Treatment Notes (C09D)

The bacteria which commonly invades dermal lesions has been clinically proven to be sensitive to exposure to ultrasound, which may prevent further development or completely eradicate the infection. Such bacteria have proven to be sensitive to one megahertz ultrasound applied to the infected site for short periods of time at fairly low ampli-tudes (.4 watts/cm^2 for four minutes to an area varying from 16 to 25 square centimeters). Phonophoresis of steroidal or nonsteroidal anti-inflammatories like vitamin E have been shown to increase effectiveness.

Bibliography (C09E)

Bickley, pp. 137–138.

TREATMENT OF BACTERIAL INFECTION (C09F)

Technique Options (C09F1)

Evaluation techniques (C09F1A)

Auxiliary evaluation techniques	**M26**
Spirometry	M26B (1)
Auscultation	M26C (1)
Lung sounds	M26C2C
Differential skin resistance survey	**M22**
Massage	**M23**
Palpation evaluation	M23
Determination of soft tissue tenderness	M23G3
Determination of soft tissue swelling	M23G6
Peripheral electroacustimulation	**M05**
Peripheral acupoint survey	M05D

Treatment techniques (C09F1B)

Auricular electroacustimulation	**M06**
Pain relief	M06A
Electrical facilitation of endorphin production (acusleep)	**M03**
Increasing pain tolerance	M0305A
Electrical stimulation (ES)	**M02**
Increasing blood flow	M02C
Facilitation of venous blood flow	M02C2A
Increasing capillary density	M02C2B
Hydrotherapy	**M15**
Wound (and burn) debridement	M15A
Circulation enhancement	M15B
Contrast baths to evaluate & treat migraine headache	M15B5A
Iontophoresis	**M07**
Pain relief	M07A
Ischemia relief	M07D
Facilitation of open lesion healing	M07E
Peripheral electroacustimulation	**M05**
Electrical stimulation of acupoints within an inflamed zone	M05A
Analgesia	M05B
Positive pressure	**M24**
Wound healing facilitation	M24C
Stasis ulcerations	M24C2A
Scar tissue formations	M24C2B
Postural drainage	**M14 (1)**
Transcutaneous nerve stimulation	**M04**
Prolonged stimulation of acupoints	M04B
Facilitation of tissue repair	M04C
Wound healing	M04C2A
Antibacterial effects of electrical stimulation	M04C5A

bacterial infection

Ultrahigh frequency sound (ultrasound) — M01
  Phonophoresis — M01A
    Phonmophoresis of vitamin E in wound healing — M01A5B
  Bacterial infection eradication — M01G
    Treatment of dermal and subdermal infections — M01G5A
      Sore throat treatment — M01G5B
  Sinusitis relief — M01H
  Open wound healing — M01J
  Acupoint stimulation — M01K
Vibration — M08
  Postural drainage facilitation — M08D (1)
(1) If lung infection

Techniques Suggested for the Acute Condition (C09F2)

Evaluation techniques (C09F2A)

Auxiliary evaluation techniques — M26
  Spirometry Auscultation — M26B (1)
  Auscultation — M26C (1)
    Lung sounds — M26C2C
Differential skin resistance survey — M22
Massage — M23
  Palpation evaluation — M23G
    Determination of soft tissue tenderness — M23G3
    Determination of soft tissue inflammation — M23G5
    Determination of soft tissue swelling — M23G6

Treatment techniques (C09F2B)

Auricular electroacustimulation — M06
  Pain relief — M06A
Electrical facilitation of endorphin production (acusleep) — M03
  Increasing pain tolerance — M0305A
Postural drainage — M14
Ultrahigh frequency sound (ultrasound) — M01
  Phonophoresis — M01A
    The phonophoresis of vitamin E in wound healing — M01A5B
  Bacterial infection eradication — M01G
    Treatment of dermal and subdermal infections — M01G5A
      Sore throat treatment — M01G5B
  Sinusitis relief — M01H
  Open wound healing — M01J
Vibration — M08
  Postural drainage facilitation — M08D (1)
(1) If lung infection

Techniques Suggested for the Chronic Condition (C09F3)

Evaluation techniques (C09F3A)

Auxiliary evaluation techniques — M26
  Spirometry — M26B (1)
  Auscultation — M26C (1)
    Lung sounds — M26C2C
Differential skin resistance survey — M22
Massage — M23
  Palpation evaluation M23G
    Determination of soft tissue tenderness — M23G3
    Determination of soft tissue inflammation — M23G5
    Determination of soft tissue swelling — M23G6
Peripheral electroacustimulation — M05
  Peripheral acupoint survey — M05D

Treatment techniques (C09F3B)

Electrical facilitation of endorphin production (acusleep) — M03
  Increasing pain tolerance — M0305A
Electrical stimulation (ES) — M02
  Increasing blood flow — M02C
    Facilitation of venous blood flow — M02C2A
    Increasing capillary density — M02C2B
Peripheral electroacustimulation — M05
  Electrical stimulation of acupoints within an inflamed zone — M05A
  Analgesia — M05B
Postural drainage — M14
Transcutaneous nerve stimulation — M04
  Prolonged stimulation of acupoints — M04B
  Facilitation of tissue repair — M04C
    Wound healing — M04C2A
    Antibacterial effects of electrical stimulation — M04C5A
Ultrahigh frequency sound (ultrasound) — M01
  Phonophoresis — M01A
    Phonophoresis of vitamin E in wound healing — M01A5B
  Bacterial infection eradication — M01G
    Treatment of dermal and subdermal infections — M01G5A
  Sinusitis relief — M01H
  Open wound healing — M01J
Vibration — M08
  Postural drainage facilitation — M08D (1)
(1) If lung infection

BACTERIAL INFECTION SYNDROME (S63)

Definition (S63A)

Bacterial infection syndromes are by definition multiple, since so many different types of tissues may be invaded by bacteria. These include tissues near the surface of the body (dermal layers), tissues directly concerned with gaseous exchange (nasal sinuses, bronchial passages, and lung), and "deep" tissues of the body (colon, external layers of the bone, and/or joints). The combination of symptoms which may make up each syndrome may vary a good deal depending on the type of tissue which has been invaded.

The symptoms which make up the syndrome associated with bacterial infection of dermal tissues may include calor, rubor, swelling and accumulation of pus within comedones, pustules, papules, inflamed nodules, infected cysts, the canalizing of inflamed and infected sacs, and palpation tenderness of infected sites.

The symptoms which make up the syndrome associated with *acute* bacterial infection of the sinus tissues may include recurrent colds with copious post nasal discharge, complaints of frontal and temporal headaches with frequent complaints of infraorbital and supraorbital pain. In the acute stage these symptoms are increased by the patient leans forward. Mucous discharge may be seen to be present on the posterior pharyngeal wall, palpation tenderness of the tissues over the frontal sinuses and maxillary antra, pyrexia, anorexia, vertigo, anosmia, photophobia, transient toothache, generalized aches, periorbital edema and fever of usually less than 102° F. Higher fevers may be indicative of suppurative adenitis or bronchopneumonia.

The symptoms which make up the syndrome associated with *chronic* bacterial infection of the sinus tissues may include a fairly constant headache (generally less severe then the acute variety) in the forehead and supraorbital regions, usually arising from the frontal sinuses. Pain may seem to be present in the upper jaw teeth, cheek, frontal (forehead) region, occipital region, parietal region, behind the eyes, at the base of the nose, and/or in the neck. Emotional depression may commonly accompany other symptoms. A peculiarly "sour", slightly putrid odor associated with the head (but not the breath) may be present as a symptom of both acute and chronic syndromes.

Symptoms which may combine to make up the syndrome associated with bacterial infection of the bronchial and/or lung tissues may include, in the early stages, shaking chill, sharp chest pain, fever, headache, dry and hacking cough which may produce purulent sputum (painful paroxysms of cough are likely to be extreme). A pleural friction rub may often be heard. Gastrointestinal symptoms may include abdominal distention, jaundice, diarrhea, nausea and vomiting. Tenderness and rigidity of the right upper quadrant of the trunk may be present if the middle and/or lower lobes are involved. Herpes of the lips and face is often present. As infection progresses, the cough may be more productive of sputum which may be pinkish, blood-flecked or even rusty colored at the height of the disease. During the stage of resolution the sputum may to finally become yellow and mucopurulent. Dyspnea may be present with rapid respiration and a peculiar expiratory grunt. Rales and suppressed breath sounds may be heard over the involved area (refer to Auxiliary Evaluation Techniques, Lung Sounds [M26C2B]). The patient may sweat profusely, be cyanotic, have pulse rates of between 100 and 130 per minute, and be acutely ill with temperatures ranging between 101 and 105° F.

Symptoms which may combine to make up the syndrome associated with bacterial infection of the tissues of the colon include cramping pain in the abdomen, abdominal distension (bloating), excessive flatulence, diarrhea (sometimes bloody), anemia, perforation of the colon, generalized infection, arthritis, and lesions of the skin.

The syndrome associated with bacterial infection of the surface layers of the bone may include the symptoms of sudden pain in the affected bone and a sharp rise in temperature, palpation tenderness over the infected site, involuntary pain restriction of movement involving the infected bone, the appearance of swelling over the infected bone and often within the adjacent joint.

The syndrome associated with bacterial infection of the joint is marked by rubor, calor, swelling, restricted motion, and pain in the joint, with and/or without compressive pressure.

Related Syndromes (S63B)

ACNE (S45)
ADHESION (S56)
BELL'S PALSY (S66)
BURNS (SECOND AND/OR THIRD
  DEGREE) (S72)
CALF PAIN (S50)
CERVICAL DORSAL OUTLET (SCALENICUS
  ATTICUS/CERVICAL RIB) (S09)
CERVICAL (NECK) PAIN (S73)

bacterial infection syndrome

CHEST PAIN (S14)
CHRONIC BRONCHITIS/EMPHYSEMA
  (OBSTRUCTIVE PULMONARY DISEASE) (S13)
COLITIS (S42)
EARACHE (S40)
ELBOW PAIN (S47)
FOOT PAIN (S51)
FOREARM PAIN (S54)
HAND/FINGER PAIN (S17)
HEADACHE PAIN (S02)
HIGH THORACIC BACK PAIN (S48)
KNEE PAIN (S46)
LOWER ABDOMINAL PAIN (S62)
LOW BACK PAIN (S03)
OSTEOARTHRITIS (S21)
PHLEBITIS (S68)
PITTING (LYMPH) EDEMA (S31)
PNEUMONIA (S58)
POST PERIPHERAL NERVE INJURY (S23)
REFERRED PAIN (S01)
SHOULDER/HAND (S57)
SHOULDER PAIN (S08)
SINUS PAIN (S33)
TEMPOROMANDIBULAR JOINT (TMJ)
  PAIN (S06)
THIGH PAIN (S49)
TINNITUS (S70)
TOOTHACHE/JAW PAIN (S41)
UNHEALED DERMAL LESION (S39)
WRIST PAIN (S16)

Treatable Causes Which May Contribute To The Syndrome (S63C)

Bacterial infection (C09)
Blood sugar irregularities (C31)
Dental pain (C22)

Irregular menses (C19)
Localized viral infection (C27)
Lung fluid accumulation (C13)
Menorrhalgia (C20)
Migraine (vascular headache) (C34)
Necrotic soft tissue (C18)
Nerve root compression (C04)
Neurogenic dermal disorders
  (psoriasis, hives) (C49)
Peripheral nerve injury (C07)
Phlebitis (C43)
Reduced lung vital capacity (C17)
Restricted joint range of motion (C26)
Scar tissue formation (C15)
Sinusitis (C16)
Soft tissue inflammation (C05)
Soft tissue swelling (C06)
Sympathetic hyperresponse (C38)
Tactile hypersensitivity (C24)
Tinnitus (C32)
Unhealed dermal lesion (C12)
Vascular insufficiency (C11)
Visceral organ Dysfunction (C02)

Treatment Notes (S63D)

Osteomyelitis (S63D1)

Clinical evidence has suggested that bacterial infection of the bone may be susceptible to treatment with ultrahigh frequency sound when the bone is close to the skin. Reference should be made to Ultrahigh Frequency Sound, Bacterial Infection Eradication [M01G].

Bibliography (S63E)

Bickley, pp. 137–138.

BELL'S PALSY SYNDROME (S66)

Definition (S66A)

Bell's palsy is caused by a neuritis and/or a compression of the facial nerve by soft tissue swelling, usually as it passes through the stylomastoid foramena in association with the mastoid process. Such soft tissue swelling may have resulted from chilling of the face, middle ear infection, tumors, fractures, meningitis, hemorrhage, infectious disease, and other less common causes.

Symptoms of Bell's palsy occur on the same side of the face as the lesion, and their number and severity depend upon the extent and the precise site of lesion along the facial nerve. If the lesion occurs outside the stylomastoid foramen, the corner of the mouth will droop and draw to the opposite side, deep facial sensation is lost, the patient may not be able to whistle, wink, close the involved eye (in which case tearing of the involved eye occurs), or wrinkle the forehead. If the lesion occurs in the facial canal and involves the chorda tympani nerve, the loss of the sense of taste in the anterior two-thirds of the tongue and a reduction of salivation on the affected side is added to the above symptoms. If the lesion is higher in the facial canal and involves the stapedius muscle, hyperacusis may be added. If the lesion on the facial nerve involves the geniculate ganglion, the onset is often acute, with pain behind and in the ear to accompany the paralysis. Lesion in the internal auditory meatus produces the paralysis with deafness from eighth nerve involvement. Lesions of the facial nerve from the pons results in the paralysis with additional various involvements of the V, VIII, VI, XI, and XII cranial nerves.

Most cases of Bell's palsy clear up in thirty days, but in some patients the paralysis may last for months, or indefinitely, with wide variations as to the extent of recovery and the permanency of neuromuscular deficits or defects. The speed and extent of recovery depends upon the speed of completeness of reinnervation.

Refer to Electromyometric Feedback, Electromyometric Evaluation [M18L] and Table of Facial Muscle Functions [T009].

Related Syndromes (S66B)

ADHESION (S56)
BACTERIAL INFECTION (S63)
EARACHE (S40)
HEADACHE PAIN (S02)
NEUROMUSCULAR PARALYSIS (S22)
POST PERIPHERAL NERVE INJURY (S23)
REFERRED PAIN (S01)
SINUS PAIN (S33)

Treatable Causes Which May Contribute To The Syndrome (S66C)

Bacterial infection (C09)
Calcific deposit (C08)
Extrafusal muscle spasm (C41)
Localized viral infection (C27)
Muscular weakness (C23)
Nerve root compression (C04)
Neuroma formation (C14)
Neuromuscular tonic imbalance (C30)
Pathological neuromuscular hypertonus (C28)
Pathological neuromuscular hypotonus (C33)
Pathological neuromuscular dyscoordination (C35)
Peripheral nerve injury (C07)
Scar tissue formation (C15)
Sinusitis (C16)
Soft tissue inflammation (C05)
Soft tissue swelling (C06)

Treatment Notes (S66D)

In acute or slowly developing cases of Bell's palsy evidence of soft tissue inflammation may be present in the area over or just anterior to the mastoid process of the skull; the area may be slightly swollen, tender to palpation, and may prove positive to differential skin resistance (DSR) survey (refer to Differential Skin Resistance Survey [M22]). The inflammation is treatable with phonophoresis of nonsteroidal antiinflammatories (refer to Ultrahigh Frequency Sound, Phonophoresis [M01A]), and multiple frequency pulsed galvanic electrical stimulation (refer to Electrical Stimulation, Reduction of Edema [M02D]). If applied early enough, the developing swelling process (if fluid) may be halted and reversed with a decrease of inflammation and swelling compression on the facial nerve.

Long-term facial nerve paralysis is often treatable with electromyometric feedback in a rehabilitation program of neuromuscular reeducation (refer to Electromyometric Feedback, Post Peripheral Nerve Injury Neuromuscular Reeducation [M18B] and the Table of Exercises for the Facial Muscles [T010]).

Bibliography (S66E)

Chusid, pp. 94–95.
Cooley, p. 267.

BLOOD PRESSURE CONTROL WITH PHYSIOLOGICAL FEEDBACK TRAINING (R007)

ABSTRACT (R007A)

This work has come out of empirical exploration of various types of feedback rather than combining feedback with other relaxation modalities as has been attempted by previous researchers. A treatment program to control vascular hypertension was developed using temperature training, alpha training and a home program. From the data gathered by taking pre-session and post-session blood pressures of a clinical sample of 65 patients, three types of patients were described and a fourth group that combines depression with hypertension was suggested.

INTRODUCTION (R007B)

Non-pharmacologic treatment of essential hypertension has not received wide spread acceptance in the medical community because the results, as reported in the literature, show that while blood pressure can be reduced more than 14 mm. Hg. in normal and hypertensive subjects, the reduction is inconsistent over time and is not effective for all subjects. The same effects have also been shown using biofeedback in conjunction with other relaxation methods. The type of feedback used in these studies was usually blood pressure feedback, although a few researchers have reported trying to find more efficient ways of achieving results with other forms of feedback.

By varying the type of feedback used in a clinical setting, we have found a treatment program that is effective for all of our hypertension patients. By observing their treatment progress, we have also been able to make some observations about different types of hypertensive patients and predict progress.

DEFINITION OF TERMS (R007C)

For the purposes of this study, we define blood pressure as the measure of peripheral vascular resistance by a sphygmomanometer. Regardless of age or medical history, blood pressure for women is usually less than or equal to 160 mm. Hg. systolic and 95 mm. Hg. diastolic (160/95). Men under age 45 have normal pressures at less than or equal to 130/90. For men over age 45, normal blood pressure is less than or equal to 140/95.

METHOD (R007D)

Instrumentation (R007D01)

The treatment of hypertension in our clinical setting employs instrumentation considered in the realm of biofeedback. Two units are commonly used: the temperature monitor and the electroencephalometer (EEM). Both of these units utilize auditory and visual feedback. Most of our units are produced by the Biofeedback Research Institute.

The visual temperature changes on the temperature monitor are registered in increments of one tenth of a degree Fahrenheit and displayed on a visual readout. The auditory feedback emits a changing tone as the temperature increases or decreases between degrees. For example, when the temperature being measured is between 85 and 86° F., the tone ascends as the temperature rises from 85.0 to 85.9° F. At 86.0° F., the tone returns to the low register and ascends again as it approaches the next whole degree. Our units also allow for minor tonal changes between each tenth of a degree which can help reduce boredom. The temperature unit is placed in contact with the skin through a thermistor electrode which can be attached to any skin surface. We employ the hand for temperature training for the control of vascular hypertension.

The EEM monitors brain wave frequencies theta (4-7 cycles per second or hz.), alpha-theta(4-13 hz.) and alpha (8-13 hz.). The amplitudes we monitor range from 0-100 microvolts (mv.). The machines can be set to monitor ranges of 0-3 mv., 0-10 mv., 0-30 mv. or 0-100 mv. Alternate units allow us to provide the patient feedback from 0-6 mv., 0-12 mv., 0-60 mv. or 0-120 mv. The amplitudes are reflected by a visual meter and by auditory feedback, usually supplied to the patient by earphones. We attach the meter to the patient via surface electrodes, generally over the occipital lobes. These units allow the patient to work at different sublevels which provides us with information on the patient's particular expertise.

Treatment Procedures (R007D02)

The blood pressure of each hypertensive patient is recorded at the beginning of each session, after each individual training procedure during that session, and at the end of the session (see *Record*

Keeping section below). We find that taking the patient's blood pressure three or four times each session is helpful not only in accumulating data on his or her progress, but in desensitizing the patient to the actual taking of the pressure. We take the blood pressure in the same position and by the same person each time. The position we use is with the patient sitting upright, arm relaxed, elbow flexed with the hand in lap. We may use the right side if the brachial pulse is hard to hear on the left. In some special cases which involve a left hemiplegic patient we use the right side.

After taking the first pressure reading, we ask the patient to assume a comfortable position. The monitor electrodes are attached to the third or fourth digit of the palmar surface of the left hand. We turn the unit on, but do not allow the patient to hear it or see the digital readout until it stabilizes at what we call the baseline temperature. We show the patient this reading and turn the sound on. Because we find a quicker response to training when auditory feedback alone is utilized, we turn the visual reading so the that patient cannot see it. The patient then listens to tone changes from a low to a high sound for each degree.

It is our experience that most patients have no difficulty in learning to "bring up the sound" at least a few degrees in the first session. Typically, temperature levels range from 80 to 86° F. on the first visit, and it is not uncommon for these patients to drive up hand temperature to approximately 96° F. at the end of the fourth and fifth training sessions.

The length of the temperature training session varies between individuals. Length is determined by patient's acceptance of training. In some cases the patient trains for fifteen minutes, takes a break, and then trains for another fifteen minutes. At other times, we allow a full half hour of training before breaking.

After temperature training, we take the blood pressure and go on to the EEM training. Once again, we encourage the patient to assume a comfortable position, and we attach the meter over the occipital lobes. We most commonly begin training at alpha levels at 0-10 mv., and then at alpha-theta at 0-10 mv. We instruct the patient to avoid unnecessary muscle activity (muscle activity is picked up by the unit as an alpha rhythm, masking the true response). The patient is also told to keep the eyes closed, because visual images, which are received in the occipital region (visual associative area), are received as beta waves or frequencies. The interference from these visual image impulses can prevent the patient from determining whether a meditative state has been achieved.

Having established the appropriate setting, we allow the patient to hear the EEM feedback tone. We instruct the patient to listen to the machine, and state that their goal is to achieve no or low sound, while high-pitched sounds need to be eliminated. We encourage the patient not to visualize or to attempt to control the machine, but to "allow it to happen."

It is important to note here that we do not put our patients in sound proof or dimly-lit rooms. Instead, we allow for normal lighting and, except for alpha training, we do not isolate them.

Record Keeping   (R007D03)

As was previously stated, an integral part of our clinical approach is to record each blood pressure reading before the session, after each procedure and at the end of the session. By being consistent, we have accumulated patient response data to the individual procedures, as well as to the overall effect on the blood pressure level as the sessions progress.

We graph the patient's progress after each session. We take the first and last blood pressures and record the systolic and diastolic levels on the graph. We connect the reading taken before the sessions with solid lines and those taken after the sessions with broken lines to compare pre- and post-session readings separately.

Home Program   (R007D04)

We urge the patient to engage in meditative techniques independently at home, even if not immediately successful. Our advice is to meditate twice during the day, mid-morning and mid-afternoon for ten minutes each session. We suggest that they use a timer to avoid focusing or worrying about the amount of time being spent, and that they create a situation in which they can avoid bombardment by phone calls or people coming into the environment.

To help with the home program, we usually supply the patient with a "mood" ring or temperature strip. These articles are specially treated to respond to changes in temperature by changes in color. This helps to verify success without a temperature unit.

As the patient becomes successful with "homework," we find a concomitant drop in pre-session blood pressure. Eventually, the patient is able to maintain a fairly low blood pressure throughout the day.

These homework periods actually serve the purpose of mental "coffee breaks." We also urge patients to take and record their own blood pressure at home, before and after each "coffee break" or three times a day. This record gives us a gauge of blood pressure responses to normal life stresses and enables us to judge the effectiveness of the total program.

DISCUSSION (R007E)

While some of the above treatment procedures are supported by data comparisons, and others by theoretical reasoning, all are supported by clinical observation.

We use the auditory feedback rather than the visual for training because we noted that the patient appears to externalize or rationalize what is seen, but tends to internalize what is heard. Early experience demonstrated that showing the visual readout tends to encourage the patient to *consciously* compete with the machine, thereby blocking progress. For example, a patient who insisted upon "seeing" the visual readout reached a definite plateau from which she made no progress for many minutes. She was able to raise the temperature once again when the visual feedback was taken from view and only auditory feedback was supplied. The same problem can arise when the patient begins consciously competing with the machine even with just auditory feedback. For example, if a patient is having trouble bringing the temperature up through auditory means and is working hard, but getting nowhere, and the therapist enters the room and engages the patient in conversation, the usual response to such an intervention is an immediate upward change in temperature as the patient's conscious attention is distracted from the meter. This reaction leads us to conclude that distracting the conscious mind facilitates rapid learning on an unconscious level.

We put the patient in as normal a setting as possible, because we have seen that by learning the meditative techniques in a more or less uninsulated environment, the patient will have a better chance of applying these techniques in a normal or non-clinical situation. Since most people are unable to find absolute isolation from sound or light, especially in a business setting, the best they can do for any homework we give them is to close the office door and have calls held. We feel that if we do not train in a comparable setting, the patient may never be able to lower blood pressure independently and will become dependent on the clinic's "therapeutic atmosphere." This would be, in our view, almost as objectionable as drug dependence.

We have observed that most of our hypertensive patients have a tendency to be overly sensitive to external stimuli, such as sound, light and color. This observation is supported by the research of Richter-Heinrich et al., who found that this was true even in the absence of any internal perception of sensitivity or psychological distress. Some of these patients complain about the color of the office walls (which are white). Our choice of instrumentation is partly based on this hypersensitivity. We usually begin with temperature training because it provides a peculiar sound stimuli linked with a goal. This usually provides enough motivation to induce the patient to tolerate the "noise." Once this tolerance is demonstrated, the patient is usually ready to tolerate the more irritating "noise" of alpha training.

During the first few sessions, alpha training may raise the blood pressure after the temperature training has lowered it; however, as the patient becomes desensitized to auditory feedback, this tendency decreases markedly. In fact, in the later sessions, the reverse is sometimes the case: the alpha training facilitates a drop in blood pressure, while temperature training causes it to rise. This switch occurs as the patient becomes accustomed to and adept at temperature training, and then starts to compete with the temperature unit to see how far the temperature will rise. This creates a psychologically hypertensive situation which causes blood pressure to go up automatically. This goal achievement is not present in alpha training, and it is not uncommon, in the later sessions, for us to eliminate the temperature training and simply use the alpha unit as the primary learning procedure. At this stage the patient is usually able to raise the hand temperature at will and has been doing so at home.

The reason for the home program, the mental "coffee break," is that it is apparent to us that many of those people we treat for hypertension are individuals who are keyed-up all day long. They are similar to the stereotypic "work-aholic" who works all day at a high energy level and then takes this energy home. Eventually, the internal unconscious tensions begin to be reflected as high blood pressure. These ten-minute relaxation periods break up their "rev-up cycle," and they become more normal in the way they handle ongoing work pressure. This change is a compromise between the needs of the person and their work situation that does not involve dramatic life style changes which might discourage many patients from dealing with the hypertension without medication.

If a patient follows through with the home program, independent blood pressure maintenance without therapeutic assistance, guidance or medications becomes possible. This independence is our therapeutic goal. This program has allowed us to help patients avoid getting on medication which at best would only artificially maintain lower blood pressure levels and to help some patients, with their physician's cooperation, reduce their levels of medication.

Before discussing the three types of hypertensive patients we have been able to identify, we will briefly mention one personality characteristic that is common to the greater population of our patients, and that is their reluctance to come in for treatment. We speculate that the defense mechanism of denial is responsible, that it is, indeed, a symptom of the syndrome itself. We have even considered that it may be at the root of the problem and considered naming it in the etiology. What we know for certain is that these patients are often "dragged" into or pressured into coming to our office by concerned spouses or friends, and that they often come in for their first appointment after having cancelled several appointments on previous dates. At this point, we take a very direct approach with them, suggesting that if they cannot find time to come in consistently (twice a week) and do their homework, that what they are doing in our office is not worth their time and money. We, in essence, fire them as patients if they will not comply with the program. Most patients respond well to this directness and settle down to work.

By comparing the graphs of 65 patients, we have discovered that there seem to be three distinct types of patients as defined by their responses to treatment.

First, we see that there is a parallel response in almost all patients' graphs, between systolic and diastolic pressures. That is, when the systolic pressure goes down during a session, so does the diastolic and vice-versa. There is, however, more similarity between pre-session pressures than post-session pressures when comparison is made between patients in each of the following groups.

Normal Hypertensives (R007E01)

The first group we have named Normal Hypertensives. These patients seem to follow a specific predictable pattern as illustrated by the first graph (see Representative Graph of the Normal Hypertensive Response to Physiological Feedback Training [R007E01A]). These patients experience an initial dip in their overall blood pressure during the first two or three sessions. Then a "crisis" occurs with a rise in blood pressure which may be even higher than the original reading. Following the "crisis," another dip occurs over several sessions which usually takes the patient to a new, lower level of overall blood pressure which exists both before and after treatment. Then there is a second "crisis" of higher blood pressure, after which there seems to be a general cooling-off period. The pressure drops consistently over the remaining sessions and plateaus at neither high nor low extremes, but at a normal level.

Volatile Hypertensives (R007E02)

The second group of patients are those we term Volatile Hypertensives. Behaviorally, they consistently express irritation, annoyance and frustration. This is just the tip of the iceberg, because their blood pressure reflects constant emotional

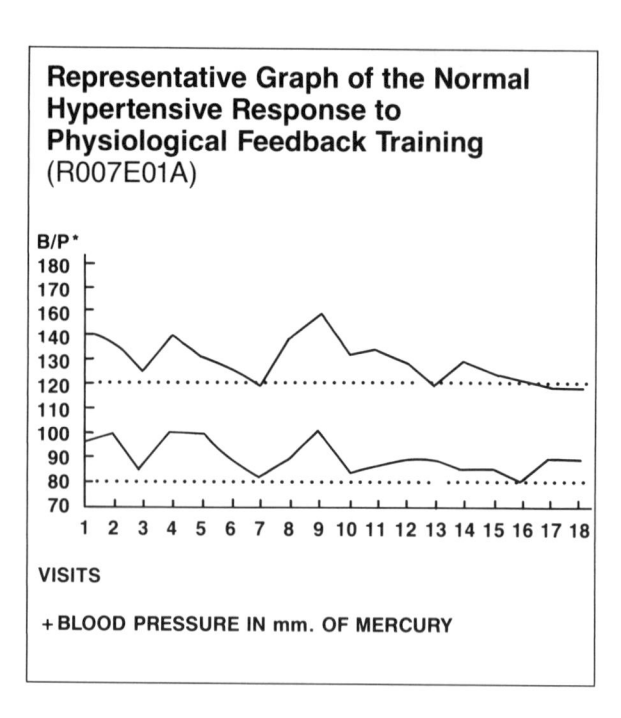

Representative Graph of the Normal Hypertensive Response to Physiological Feedback Training (R007E01A)

+ BLOOD PRESSURE IN mm. OF MERCURY

internalization with high systolic and diastolic coinciding with emotional outbursts, and lower readings with milder tempers. The second graph shows their response pattern, which is highly inconsistent, both pre- and post-session, as compared with the Normal Hypertensives (refer to Representative Graph of the Volatile Hypertensive Response to Physiological Feedback Training [R007E02A]). As was previously mentioned, there is a general parallel response between systolic and diastolic with all our patients, and the response to individual sessions also seems to have a parallel reflective qual-

Representative Graph of the Volatile Hypertensive Response to Physiological Feedback Training (R007E02A)

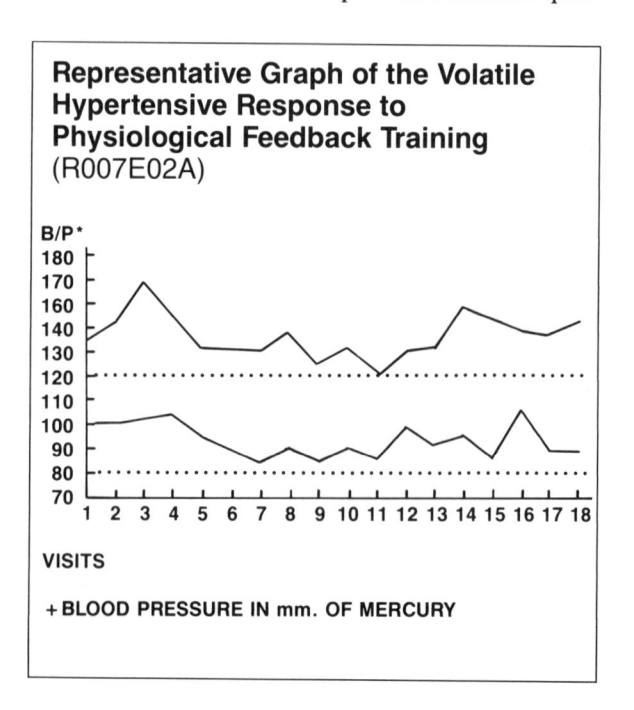

+ BLOOD PRESSURE IN mm. OF MERCURY

ity. In this group, aside from pre- and post-session parallels between systolic and diastolic pressures, patients do not compare well to each other. During their course of treatment they have erratic and unpredictable peaks, crises and remissions as reflected in their graphs. They also tend to be especially difficult people to handle in the therapeutic setting because their behavior tends to be erratic and unpredictable. We group these people, however, not by their behavior, but on the basis of their response to treatment.

Hypernormal Hypertensives (R007E03)

The third group of patients we have identified are those who have had apparent physiological damage. We call them Hypernormal Hypertensives. They respond much like the Normal Hypertensive group, with the same generally predictable pattern of two crises, but their minimal pressures, both systolic and diastolic, are still considered quite a bit above normal (see Representative Graph of the Hypernormal Hypertensive Response to Physiological Feedback Training [R007E03A]), thus implying vascular changes which have resulted in a decrease in vessel flexibility.

Depression and Hypertension (R007E04)

We noticed that there was a group of hypertensive individuals who did not respond positively to our treatment procedure and therefore did not fit into any of the three previously identified categories of patients. However, they often respond to anti-depression medication(s) with a drop in blood pressure.

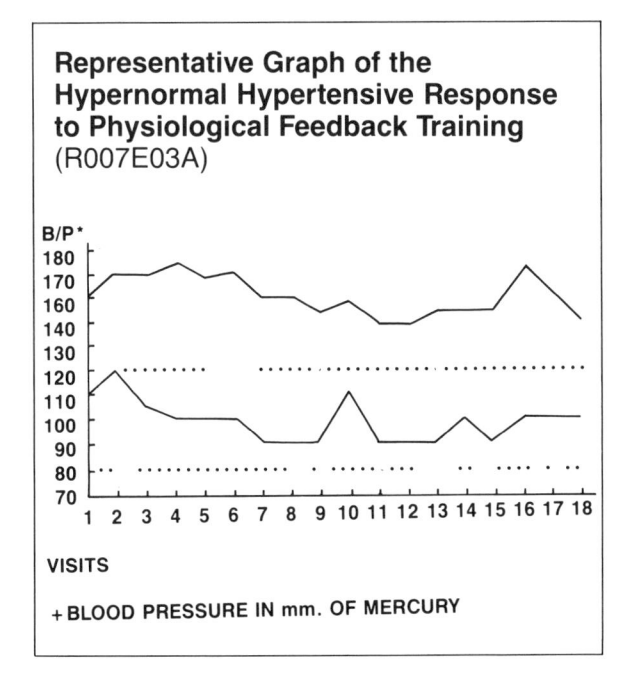

An evaluation with encephalometry reveals very high 8 to 12 hz. (alpha) spontaneously produced electrical brainwave activity (refer to Electroencephalometry, Electroencephalometry Evaluation [M21C]); such patients will demonstrate abnormally high activity levels within the alpha and theta (5 to 8 hz.) simply by closing their eyes. This observation is consistent with the study that showed that hypertensive people had an incidence of depression equal to that of the general population. Hypertension experienced by patients who suffer from endogenous depression, may be impossible to treat with the feedback techniques discussed here until the depression can be treated successfully through other means, and before the patient enters the feedback treatment program.

The question can be raised at this point as to whether we are viewing hypertension as a psychogenic or somatogenic disorder. We believe an answer cannot be given because the question is irrelevant; there is no difference. If a person's blood pressure is elevated, causing constriction of the small vessels in his left hand, and if we teach him to dilate those vessels and his pressure goes down, it does not matter why those vessels were constricted in the first place. If someone's brainwaves are of a certain frequency because of a learned depressive response to life, or if they are that way because of a physical response to certain genetic material, we can still help the person change brainwaves with feedback and thus change the affective experience of life. Another way to say this is that every emotion has a concomitant physical correlate and vice versa. If psychophysiological treatment can help a person change maladaptive responses through self regulation, it is rhetorical whether the etiology was physiological or psychological.

CONCLUSIONS (R007F)

From a clinical sample of 65 patients, we have developed a treatment program for essential hypertension that includes biofeedback with the temperature unit, the electroencephalometer and a home program. Examining graphs of the patients' systolic and diastolic blood pressures, both pre-session and post-session, we have identified three types of patients who follow three types of patterns of response during the lowering of their blood pressures. A fourth type of patient has been identified who has a treatable depression which must be attended to before being classified in one of the other three categories.

Patient response, illustrated by the graphing, not only allows the the therapist to review patient

Representative Graph of the Hypernormal Hypertensive Response to Physiological Feedback Training (R007E03A)

+ BLOOD PRESSURE IN mm. OF MERCURY

progress, but allows the patient to objectively observe results as well. In addition, new patients can be shown these graphs, thus encouraging an informed participation in their treatment.

There is, of course, a great diversity among individual responses. Each patient responds differently to each particular modality, no matter what the therapy. The remarkable part of our findings is that there is, indeed, an apparent pattern with Normal and Hypernormal Hypertensives. This finding, in itself, seems important, especially in demonstrating the efficacy of our particular treatment procedures.

REFERENCES (R007G)

Bali, pp. 637–646
Barr, pp. 339–342.
Blanchard, Haynes, pp. 445–451.
Blanchard, Youngs, Haynes, pp. 241–245.
Brener, pp. 1063–1064.
Elder, pp. 377–382.
Frankel, pp. 276–293.
Freedman, Bennet, pp. 134–141.
Freedman, Taub, pp. 335–347.
Frunkin, pp. 296–230.
Kristt, pp. 370–378.
Patel, pp. 1–41.
Richter–Heinrich, pp. 251–258.
Shapiro, Schwart, Turskey, pp. 296–304.
Shapiro, Turskey, et al., pp. 588–590.
Steptoe, pp. 417–424.
Sterman, pp. 207–222.
Sunwit, pp. 252–263.

BLOOD SUGAR IRREGULARITIES (C31)

Definition (C31A)

Of the chemical reactions within the cells, the greater proportion is concerned with the making of energy through the conversion of nutrients transported to the cells by the blood. The metabolism of carbohydrates is one of the most important of the energy-providing processes, and is almost entirely based upon the digestion of the monosaccharides glucose, fructose and galactose. To be used by the cells, the monosaccharides must be transported through the cell membrane into the cellular cytoplasm by a carrier which utilizes a mechanism called facilitated diffusion (which is not an active transport). The rate of transport is increased by insulin. The amounts of glucose that can diffuse in to most cells of the body without facilitation by insulin is far too little to supply the amount of glucose normally required for energy metabolism. In effect, the rate of carbohydrate utilization by the cells is controlled by the rate of insulin secreted by the pancreas. Insulin secretion by the pancreas is triggered by the presence of monosaccharides in the blood and the rate of production is normally proportional to the concentration of monosaccharides.

If the insulin production is insufficient blood glucose concentrations rise; prolonged high glucose concentrations may precipitate the diabetes mellitus syndrome.

If insulin production is overabundant and blood glucose concentration is low, the hypoglycemia syndrome may be produced with symptoms of insulin shock.

Evaluation (C31B)

Evaluation of blood sugar irregularities is dependant upon the observation of symptomatology and the patient's history.

Diabetes mellitus is characterized by polyuria, thirst, itching, hunger, weakness and weight loss. Long-term effects include atherosclerosis, microangiopathy, retinopathy and renal changes (diffuse and nodular sclerosis of the glomerulus leading eventually to renal failure), and, eventually, peripheral vascular claudication.

Hypoglycemia is characterized by sweating, flushing or pallor, numbness, chilliness, hunger, trembling, headache, dizziness, weakness, changes in pulse rate (usually accelerated), palpitations, elevated blood pressure, apprehensiveness and syncope. If the hypoglycemia goes unrelieved, signs of central nervous system involvement may become evident with restlessness, incoordination, thick speech, emotional instability, negativism, disorientation and diplopia, followed by tonic and clonic convulsions (often with marked salivation), and (in severe cases) coma and death.

Etiology (C31C)

If insulin production is inadequate to keep up with the monosaccharide levels in the blood, over a prolonged period of one to three weeks, the islets of the Langerhans organs in the pancreas responsible for insulin production, hypertrophy, and secretion of insulin increase correspondingly. If carbohydrate consumption increases still further, and is maintained at a high level, eventually beta cells of the islets of Langerhans may "burn out," causing them to lose their productive granules, resulting in complete or partial cessation of insulin production, precipitating diabetes mellitus.

Hypoglycemia may be induced by an exogenous overdose of insulin, from endogenous hyperinsulinism due to islet cell tumors or islet cell hyperplasia, hepatic insufficiency (causing glycogen depletion and impaired glucose production), adrenocortical deficiency (Addison's disease), hypopituitarism, hypothyroidism due to a decrease in glucose absorption, and hypothalamic lesions. Large sarcomas and other malignant tumors may be associated with hypoglycemia. "Functional" hypoglycemia may be due to over-secretion of insulin in response to glucose absorption, muscular exertion, pregnancy and/or lactation, or anorexia nervosa; it is reported that approximately seventy percent of all spontaneous hypoglycemia is caused by "functional" hyperinsulinism.

Treatment Notes (C31D)

Treatment of blood sugar irregularities is directed at helping the patient regulate metabolic rates through exercise and relaxation training. Refer to Exercise, Muscle Strengthening [M12A] and Table of Exercises - Energy Consumption Rates [T013] for further details.

Temperature training has proven to be markedly effective in decreasing the rate of metabolism, thereby decreasing the need for insulin. Refer to Temperature Monitry, General Relaxation (Sympathetic Hyperresponse Suppression)[M19A] for further details.

Bibliography (C31E)

Bickley, pp. 183–186.
Evans, pp. 241–254.
Guyton, pp. 786–788, 915–920.
Jette, pp. 339–342.
Holvey, pp. 325–341.

TREATMENT OF BLOOD SUGAR IRREGULARITIES (C31F)

Technique Options (C31F1)

Evaluation techniques (C31F1A)

Auricular electroacustimulation	**M06**
Auricular acupoint evaluative survey	M06C
Determination of pathological symptom source	M06C5B
Cryotherapy	**M10**
Evaluation of capillary response to cold	M10H
Determination of hyperreactivity to cold	M10H2A
Electroencephalometry	**M21**
Electroencephalometric evaluation	M21C
Galvanic skin response (GSR) monitry	**M20**
GSR assessment	M20C
Hydrotherapy	**M15**
Hydrotherapeutic tests	M15F
Gibbons-Landis procedure	M15F1
Test for hypersensitivity to cold	M15F3
Peripheral electroacustimulation	**M05**
Peripheral acupoint survey	M05D
Temperature monitry	**M19**
Dermal temperature evaluation	M19F
Determination of capillary bed insufficiency	M19F1

Treatment techniques (C31F1B)

Auricular electroacustimulation	**M06**
Pathology control	M06B
Cryotherapy	**M10**
Suppression of sympathetic autonomic hyperreactivity	M10A
Electrical facilitation of endorphin production (acusleep)	**M03**
Electroencephalometry	**M21**
Meditation facilitation	M21A
Galvanic skin response monitry	**M20**
General relaxation	M20A
Hydrotherapy	**M15**
Suppression of sympathetic hyperresponsiveness	M15E
Laser stimulation	**M27**
Massage	**M23**
Acupressure	M23C
Peripheral electroacustimulation	**M05**
Pathology control	M05C
Physiological dysfunction	M05C5A
Sensory stimulation	**M17**
Provoking reflex physiological responses	M17C
Prolonged heating for reflex physiological responses	M17C2B
Prolonged cooling for reflex physiological responses	M17C2B
Brief cooling for reflex physiological responses	M17C2C
Temperature monitry	**M19**
General relaxation (sympathetic hyperresponse suppression)	M19A
Circulatory disorder control	M19C
Transcutaneous nerve stimulation	**M04**
Prolonged stimulation of acupoints	M04B
Ultrahigh frequency sound (ultrasound)	**M01**
Acupoint stimulation	M01K
Autonomic sympathetic hyperactivity suppression	M01N
Vibration	**M08**
Circulation enhancement	M08G

Techniques Suggested for the Acute Condition (C31F2)

Evaluation techniques (C31F2A)

Auricular electroacustimulation	**M06**
Auricular acupoint evaluative survey	M06C
Determination of pathological symptom source	M06C5B
Cryotherapy	**M10**
Evaluation of capillary response to cold	M10H
Determination of hyperreactivity to cold	M10H2A
Galvanic skin response (GSR) monitry	**M20**
GSR assessment	M20C
Temperature monitry	**M19**
Dermal temperature evaluation	M19F
Determination of capillary bed insufficiency	M19F1

Treatment techniques (C31F2B)

Auricular electroacustimulation	**M06**
Pathology control	M06B
Cryotherapy	**M10**
Suppression of sympathetic autonomic hyperreactivity	M10A
Electrical facilitation of endorphin production (acusleep)	**M03**
Electroencephalometry	**M21**
Meditation facilitation	M21A
Galvanic skin response monitry	**M20**
General relaxation	M20A

Sensory stimulation   *M17*
   Provoking reflex physiological responses   M17C
      Prolonged heating for reflex
        physiological responses   M17C2A
      Prolonged cooling for reflex
        physiological responses   M17C2B
      Brief cooling for reflex physiological
        responses   M17C2C
Temperature monitry   *M19*
   General relaxation (sympathetic
      hyperresponse suppression)   M19A
   Circulatory disorder control   M19C
Ultrahigh frequency sound (ultrasound)   *M01*
   Autonomic sympathetic hyperactivity
      suppression   M01N
Vibration   *M08*
   Circulation enhancement   M08G

Techniques Suggested for the Chronic Condition   (C31F3)

Evaluation techniques   (C31F3A)

Electroencephalometry   *M21*
   Electroencephalometric evaluation   M21C

Galvanic skin response monitry   *M20*
   GSR assessment   M20C
Temperature monitry   *M19*
   Dermal temperature evaluation   M19F
      Determination of capillary bed
        insufficiency   M19F1

Treatment techniques   (C31F3B)

*Electrical facilitation of endorphin
production (acusleep)*   *M03*
Electroencephalometry   *M21*
   Meditation facilitation   M21A
Galvanic skin response monitry   *M20*
   General relaxation   M20A
Temperature monitry   *M19*
   Circulatory disorder control   M19C

BREAKTHROUGH BLEEDING SYNDROME (S61)

Definition (S61A)

Breakthrough bleeding (metrorrhagia) is abnormal bleeding from the uterus at times other than during normal menses. It may occur with ovulation, after exertion or coitus, from atypical hormonal balances, uterine malposition or subinvolution, cervical lacerations, cervicitis, endometritis, and pelvic inflammatory disease. Malignant tumors of the ovary or fallopian tube may be responsible for uterine bleeding. Benign fibroids or uterine polyps (cervical or intrauterine) may exist within the uterine wall, multiply or singly, and are the single most common cause of intermenstrual uterine bleeding during the childbearing years, and may even interfere with conception. Anemia may be a danger if bleeding is copious and/or prolonged. Episodes of breakthrough bleeding may sometimes be accompanied by hypermenorrhea, polymenorrhea, dysmenorrhea, leukorrhea and symptoms of pressure from adjacent structures. A history of spontaneous abortions is not uncommon.

Related Syndromes (S61B)

MENSTRUAL CRAMPING (S60)

Treatable Causes Which May Contribute To The Syndrome (S61C)

Bacterial infection (C09)
Blood sugar irregularities (C31)
Irregular menses (C19)
Localized viral infection (C27)
Visceral organ Dysfunction (C02)

Treatment Notes (S61D)

Breakthrough bleeding (metrorrhagia) has been treated successfully in a clinical setting with short-term auricular electroacu-stimulation. Refer to Auricular Electroacustimulation, Pathology Control [M06B], Table of Auricular Acupoint Treatment Protocols [T004] and Table of Auricular Landmark & Auricular Acupoint Descriptions [T003] for further details.

Bibliography (S61E)

Bickley, p. 283.
Holvey, pp. 674–675.

BUERGER'S DISEASE SYNDROME (S11)

Definition (S11A)

Thromboangiitis obliterans (Buerger's disease) is commonly associated with excessive smoking and typically occurs in young human males between the ages of 20 and 40, but it can occur in women (at a ratio of one female sufferer to 75 males) or nonsmokers of either sex. Fifty percent of reported cases occur in those of Jewish extraction. Precipitation of onset has been variously linked to excessive smoking, trauma, exposure to extreme temperatures, the use of ergot or epinephrine, and dermal infections.

Thromboangiitis obliterans is an obliterating, segmental, inflammatory disease affecting blood vessels of the extremities, which progresses from medium and small arteries. Structural changes associated with the perivascular reaction of round and giant cells in the arterial and venous walls may be seen with the intima to become roughened, and thrombosis may result. Fibrotic incorporation of the artery, vein and nerve may take place in the healing process. Reflex vasospasm involving normal arteries may further reduce blood supply. Lesions are episodic, producing complete and often permanent vascular obstruction leading in extreme cases to the need for amputation.

Onset may be gradual or sudden, with a precipitous development of gangrene. Most often the patient will complain of the gradual development of coldness, numbness, tingling or burning in the hands and/or feet before more objective evidence of the disease appears. Symptoms occur more rapidly with strenuous exercise. Persistent pain is experienced in the pregangrenous stage, which becomes severe in the ulcerated and gangrenous stages.

In the *Buerger's disease syndrome* rubor of the toes and foot in the dependent position and pallor in the same areas when the lower extremity is elevated above the level of the heart are signs of severe arterial insufficiency. Coldness of the involved extremities is typical with trophic changes in the nails and skin occurring early in the disease. As the disease progresses, arterial pulses become depressed in involved extremities. Long standing, untreated cases may result in skin and deeper tissue ulceration and gangrene. Refer to Hydrotherapy, Hydrotherapeutic Tests, Gibbons-Landis Procedure [M15F1], Cryotherapy, Evaluation of Capillary Response to Cold [M10H] and Temperature Monitry, Dermal Temperature Evaluation, Determination of Capillary Bed Insufficiency [M19F1] for descriptions of procedures for the evaluation of Buerger's disease.

Related Syndromes (S11B)

ADHESION (S56)
BACTERIAL INFECTION (S63)
CERVICAL DORSAL OUTLET (SCALENICUS ATTICUS/CERVICAL RIB) (S09)
DIABETES (S10)
FOOT PAIN (S51)
FOREARM PAIN (S54)
HAND/FINGER PAIN (S17)
MIGRAINE (VASCULAR) HEADACHE (S18)
RAYNAUD'S DISEASE (S12)
REFERRED PAIN (S01)
SHOULDER/HAND (S57)
UNHEALED DERMAL LESION (S39)
WRIST PAIN (S16)

Treatable Causes Which May Contribute To The Syndrome (S11C)

Bacterial infection (C09)
Blood sugar irregularities (C31)
Calcific deposit (C08)
Extrafusal muscle spasm (C41)
Habitual joint positioning (C48)
Joint ankylosis (C42)
Migraine (vascular headache) (C34)
Necrotic soft tissue (C18)
Neuroma formation (C14)
Peripheral nerve injury (C07)
Restricted joint range of motion (C26)
Scar tissue formation (C15)
Soft tissue inflammation (C05)
Soft tissue swelling (C06)
Sympathetic hyperresponse (C38)
Tactile hypersensitivity (C24)
Trigger point formation (C01)
Unhealed dermal lesion (C12)
Vascular insufficiency (C11)

Treatment Notes (S11D)

Temperature training has been demonstrated to be effective in ameliorating the Buerger's disease syndrome effect on dermal circulation of the involved extremity, if the training occurs early after onset. Temperature training involves training the patient to voluntarily dilate capillary vessels in the skin of the involved extremity; such training has been effective even to the point of reversing the trophic changes in nails and skin as well as improving the temperature of the involved extremity(s). Refer to Temperature Monitry, Circulatory Disorder Control [M19C] for further details.

Bibliography (S11E)

Bickley, p. 215.
Holvey, pp. 236–237.

BURNS (SECOND AND/OR THIRD DEGREE) SYNDROME (S72)

Definition (S72A)

Tissue injury which may be classified as a "burn" may be caused by thermal, electrical, radioactive, or chemical agent denaturation of protein resulting in cell injury and/or death. The degree of damage is dependant upon the type, duration and intensity of action by the agent.

The severity of the injury is gauged against the extent of the injury (percentage of total body area affected) and the depth of the injury. Burns are generally classified as three types: first, second and third degree.

The damage sustained from a first degree burn is limited to the outer layer of epidermis; symptoms include erythema, sensations of warmth or heat, sensitivity to touch, palpation tenderness, and overt pain.

Damage sustained from a second degree or partial thickness burn extends through the epidermis and involves the dermis, but leaves the sweat glands and hair follicles to provide the cells necessary for rapid regeneration of the epithelium. Second degree burn symptoms include intense pain and the formation of vesicles, blebs and/or bullae.

Third degree (full thickness) burns cause destruction of both the epidermis and dermis to the tissue level that sweat glands, hair follicles and nerve endings are destroyed. Symptoms of the third degree burn include charring of the skin, or (as in cases of scalding) coagulation of the skin (the skin is white and lifeless); the third degree burn site is initially painless and insensitive to touch, but intense pain and tactile hypersensitivity develops with the regrowth of exposed nerve endings.

Severe or extensive burns may initially be accompanied by varying degrees of low blood pressure, weak thready pulse, cold and clammy extremities, pale face, cold sweats, increased respiration, restlessness, confusion, anxiety, and oliguria. Cumulatively, these symptoms characterize the primary shock accompanying severe or extensive burns. The more life-threatening is the insidious or secondary shock which may develop as increased capillary permeability, resulting from blood vessel wall damage, allows copious amounts of fluid to exude into the wound area from the burned surfaces; the fluid thus lost is made up of water, plasma crystalloids and as much as two thirds of the plasma protein.

Of constant concern is the possibility of colonization of the burn site by opportunistic bacterium: pseudomonas aeruginosa, staphylococcus aureus, and streptococci are most likely to attack sites in the upper trunk and upper extremities, while coliform and clostridia bacteria may be a danger below the waist. The severe stress provokes autogenic high output of adrenocortical steroids, making the bacterial invasions are made more dangerous. The stress also suppresses the autoimmune defenses of the body, rendering the patient more vulnerable to infection and subsequent septicemia (the greatest threat to the patient's life). IgA and IgG immunoglobulins may fall to extremely low levels in forty-eight hours following the burn incident, to be gradually restored to normal levels over a seven to fourteen day period. Cellular immunity becomes severely depressed in the post-burn period and may remain depressed for up to two months.

There is also a remote possibility that the patient may develop nephrosis. The danger also exists that the patient may develop Curling's ulcers (ulcers of the stomach induced by stress) from the high steroid levels; such ulceration carries the threat of hemorrhage and/or perforation of the stomach wall.

The recovery of the second degree burn site, if there are no complications, follows the pattern of wound healing: (1) acute inflammation with inflammatory exudate to clear wound site of microorganisms and tissue debris, (2) contraction of wound edges, (3) ingrowth of capillary buds, (4) fibroblast proliferation, (5) synthesis and aggregation of tropocollagen, (6) reepithelialization, and (7) maturation and contraction of scar tissue. Third degree burn healing may depend upon skin grafting to provide the cells necessary for reepithelialization.

Related Syndromes (S72B)

ADHESION (S56)
BACTERIAL INFECTION (S63)
KELOID FORMATION (S38)
POST IMMOBILIZATION (S36)
UNHEALED DERMAL LESION (S39)

Treatable Causes Which May Contribute To The Syndrome (S72C)

Bacterial infection (C09)
Habitual joint positioning (C48)

Hyperhidrosis   (C37)
Hypotension   (C45)
Insomnia   (C46)
Joint ankylosis   (C42)
Lung fluid accumulation   (C13)
Muscular weakness   (C23)
Nausea/vomiting   (C36)
Necrotic soft tissue   (C18)
Peripheral nerve injury   (C07)
Reduced lung vital capacity   (C17)
Restricted joint range of motion   (C26)
Scar tissue formation   (C15)
Soft tissue inflammation   (C05)
Soft tissue swelling   (C06)
Sympathetic hyperresponse   (C38)
Tactile hypersensitivity   (C24)
Unhealed dermal lesion   (C12)
Vascular insufficiency   (C11)
Visceral organ Dysfunction   (C02)

Treatment Notes   (S72D)

Much of the physical treatment of the burn (second and/or third degree) in the recovery stage revolves around the careful debridement of necrotic tissue from the burn site without causing additional hemorrhage or damage to granulating tissue; debridement is performed to reduce the warm debris cover which may serve to enhance bacteria incubation and to facilitate the healing process. Refer to Hydrotherapy, Wound (and Burn) Debridement [M15A] for further details.

Bibliography   (S72E)

Bickley, pp. 48–49, 80–81.
Davis, pp. 1723–1724.
Holvey, pp. 1164–1171.
Wright, pp. 1217–1231.

BURSITIS SYNDROME (S26)

Definition (S26A)

Bursae are closed sacs associated with the joints, which are lined by specialized connective tissue and containing synovial fluid. Bursae are usually found where muscle or tendons move over the bony prominences and facilitates gliding movements of the muscle and/or tendons by diminishing friction.

Inflammation of the bursae may be caused by trauma (including unusual exercise), gout, acute or chronic infection (syphilis, tuberculosis, or rheumatoid arthritis). Common forms include olecranon, prepatellar, subdeltoid (subacromial), retrocalcaneal (Achilles), iliopectineal (iliopsoas), ischial, infrapatellar, trochanteric and bunion bursitis.

The *bursitis syndrome* is characterized by pain, effusion into the bursal sac (swelling), tenderness, and some limitation of motion in the joint associated with the inflamed bursa. In chronic bursitis the bursal wall becomes thickened and the lining endothelium degenerates. As the condition progresses the bursa may develop and contain adhesions, villi formation, tags and calcareous deposits, which may be accompanied by muscle atrophy and marked limitation of joint range of motion. Attacks of bursitis may last a few days or continue for months. During recovery from an episode of bursitis there is a tendency for reoccurrence in the same joint because of an increased sensitivity of the tissues involved.

Related Syndromes (S26B)

ADHESION (S56)
CAPSULITIS (S27)
CERVICAL DORSAL OUTLET (SCALENICUS ATTICUS/CERVICAL RIB) (S09)
CHONDROMALACIA (S15)
ELBOW PAIN (S47)
FASCIITIS (S20)
FROZEN SHOULDER (S64)
HAND/FINGER PAIN (S17)
KNEE PAIN (S46)
MYOSITIS OSSIFICANS (S67)
OSTEOARTHRITIS (S21)
POST IMMOBILIZATION (S36)
RADICULITIS (S29)
REFERRED PAIN (S01)
SHOULDER/HAND (S57)
SHOULDER PAIN (S08)
TEMPOROMANDIBULAR JOINT (TMJ) PAIN (S06)
TENDONITIS (S28)
TENNIS ELBOW (LATERAL EPICONDYLITIS) (S32)
TOOTHACHE/JAW PAIN (S41)
WRIST PAIN (S16)

Treatable Causes Which May Contribute To The Syndrome (S26C)

Calcific deposit (C08)
Extrafusal muscle spasm (C41)
Habitual joint positioning (C48)
Hypermobile/instable joint (C10)
Joint ankylosis (C42)
Joint approximation (C47)
Ligamentous strain (C25)
Muscular weakness (C23)
Nerve root compression (C04)
Neuromuscular tonic imbalance (C30)
Psychoneurogenic neuromuscular hypertonus (C29)
Restricted joint range of motion (C26)
Soft tissue inflammation (C05)
Soft tissue swelling (C06)
Tactile hypersensitivity (C24)
Trigger point formation (C01)

Treatment Notes (S26D)

Treatment of acute bursitis should be directed at reducing pain, inflammation and swelling, maintaining the range of motion and preventing adhesions in the associated joint(s).

Treatment of chronic bursitis involves reducing pain, inflammation and swelling, regaining lost range of motion of associated joint(s) by breaking adhesions, managing and reducing calcareous deposits within the bursa, and the determination and the elimination of the cause(s) of the chronicity (occupational factors or inadvisable sports).

Bibliography (S26E)

Cailliet, pp. 38–57.
Holvey, pp. 966–969.
Salter, pp. 249–250.
Scott, p. 326.
Shands, p. 7.

CALCIFIC DEPOSIT (C08)

Definition (C08A)

A calcific deposit is a calculus or concretion formed of inorganic matter (at least in part of calcium) in or on soft tissue (muscle, tendon or joint space) or upon bony structure (over regularly formed bone or cartilage). During the early stages of development, the calcium is reported to take the form of a paste like mass, which tends to expand as the calcium accumulates; such expansion irritates the under surface of involved bursa, tendons, and/or ligaments. Deposits which have been allowed to develop over a long period of time tend to become hardened, resulting in gritty sensations and/or sounds with movement of the tendon and/or joint.

Rapid deposition of calcium within the substance of a tendon, joint or muscle has been reported to cause excruciating, throbbing pain caused by increased local pressure.

Evaluation (C08B)

Establishing the existence of a calcific deposit is currently dependent upon the objective evidence of x-ray or other soft tissue scan, the patient's subjective evidence of condition history and pain site, and the relative subjective judgement based on the evidence of soft tissue palpation, auscultation of tendons and/or joints (refer to Auxiliary Evaluation Techniques, Auscultation, Joint and Tendon Sounds [M26C2C] and inflammatory measurement (refer to Differential Skin Resistance Survey [M22]). Acute episodes are accompanied by severe sharp pain, especially when movement of the involved joints or tendons are attempted, and the area over lying the calcium deposit is generally palpation tender; while the area over the calcium is still palpation tender, the chronic condition is marked by less severe pain, but that pain is constantly annoying during the day, seemingly more so at night when sleep is desired.

Etiology (C08C)

The body's serum and extracellular fluid are supersaturated with calcium. Excessive tissue stress seems to trigger healing processes which facilitate calcium precipitation from bodily fluid; calcium ions thus precipitated may agglutinate to form a "calcium deposit". Such calcium deposits occur as a part of the process produced by the organization and ossification of a hematoma or to callus formation following fractures or severe bruising of the bone. However, other tissue stresses have also been noted to facilitate calcium deposits, and it could be that bone spurs, free floating calcific bodies found in the joints, and calcification of tendon synovial sheathes generally result from an process analogous to that which produces the myositis ossificans syndrome in which proliferation of collagenoblasts or a metaplasic formation of local connective tissue cells into bone-forming cells seems to take place to aid or to begin the process of callus formation. In other words, the causal elements behind the development of calcium deposits in joints, soft tissues and bony prominences appear to be tissue trauma or stress and result from the body's attempt to provide a healing process, albeit misdirected.

Treatment Notes (C08D)

Treatment of the calcific deposit, be it bone spur, free floating deposit, or calcific tendon sheathe, should be directed at dissolving the deposit with ultrasound (or other means) or at desensitizing tissues which may be irritated by the deposit (refer to Ultrahigh Frequency Sound, Calcium Deposit Management [M01E], and/or Desensitization [M01C]).

Bibliography (C08E)

Bickley, pp. 53–56.
Salter, pp. 240–242.
Shands, pp. 442–444.

TREATMENT OF CALCIFIC DEPOSIT (C08F)

Technique Options (C08F1)

Evaluation techniques (C08F1A)

Auxiliary evaluation techniques	*M26*
Auscultation	M26C
Joint and/or tendon sounds	M26C2C
Cryotherapy	*M10*
Evaluation of capillary response to cold	M10H
Determination of hyperreactivity to cold	M10H2A
Differential skin resistance survey	*M22*
Exercise	*M12*
Functional testing	M12I
Evaluation of joint range of motion	M12I1
Evaluation of muscle strength	M12I2
Hydrotherapy	*M15*

calcific deposit

Hydrotherapeutic Tests	M15F
Test for hypersensitivity to cold	M15F3
Massage	*M23*
Palpation evaluation	M23G
Determination of fibroid, scar, or calcific formation	M23G4
Peripheral electroacustimulation	*M05*
Peripheral acupoint survey	M05D
Sensory stimulation	*M17*
Neurological evaluation	M17E
Sensory evaluation	M17E2

Treatment techniques (C08F1B)

Auricular electroacustimulation	*M06*
Pain relief	M06A
Cryotherapy	*M10*
Analgesia/anesthesia	M10C
Electrical facilitation of endorphin production (acusleep)	*M03*
Increasing pain tolerance	M0305A
Electrical stimulation (ES)	*M02*
Increasing blood flow	M02C
Facilitation of venous blood flow	M02C2A
Increasing capillary density	M02C2B
Inhibition of ionized formation (Ca++ or other ion deposit)	M02F
Exercise	*M12*
Muscle strengthening	M12A (1)
Hydrotherapy	*M15*
Range of motion enhancement and physical reconditioning	M15D (1)
Iontophoresis	*M07*
Pain relief	M07A
Gouty tophi dissolution	M07H
Calcific deposit absorption	M07J
Joint mobilization	*M13 (1)*
Laser stimulation	*M27*
Massage	*M23*
Acupressure	M23C
Deep soft tissue mobilization	M23F
Muscles	M23F2A
Ligaments	M23F2B
Tendon sheaths	M23F2C
Tendons	M23F2D
Peripheral electroacustimulation	*M05*
Analgesia	M05B
Transcutaneous nerve stimulation	*M04*
Prolonged stimulation of acupoints	M04B
Ultrahigh frequency sound (ultrasound)	*M01*
Calcium deposit management	M01E
Acupoint stimulation	M01K
(1) If arthritic in nature	

Techniques Suggested for the Acute Condition (C08F2)

Evaluation techniques (C08F2A)

Auxiliary evaluation techniques	*M26*
Auscultation	M26C
Joint and/or tendon sounds	M26C2C
Cryotherapy	*M10*
Evaluation of capillary response to cold	M10H
Determination of hyperreactivity to cold	M10H2A
Differential skin resistance survey	*M22*
Exercise	*M12*
Functional testing	M12I
Evaluation of joint range of motion	M12I1
Evaluation of muscle strength	M12I2
Massage	*M23*
Palpation evaluation	M23G
Determination of fibroid, scar, or calcific formation	M23G4
Sensory stimulation	*M17*
Neurological evaluation	M17E
Sensory evaluation	M17E2

Treatment techniques (C08F2B)

Auricular electroacustimulation	*M06*
Pain relief	M06A
Cryotherapy	*M10*
Analgesia / anesthesia	M10C
Electrical facilitation of endorphin production (acusleep)	*M03*
Increasing pain tolerance	M0305A
Electrical stimulation (ES)	*M02*
Inhibition of ionized formation (Ca++ or otherion deposit)	M02F
Joint mobilization	*M13 (1)*
Ultrahigh frequency sound (ultrasound)	*M01*
Calcium deposit management	M01E
(1) if arthritic in nature	

Techniques Suggested forthe Acute Condition (C08F3)

Evaluation techniques (C08F3A)

Auxiliary evaluation techniques	*M26*
Auscultation	M26C
Joint and/or tendon sounds	M26C2C

calcific deposit

Cryotherapy	***M10***
Evaluation of capillary response to cold	M10H
Determination of hyperreactivity to cold	M10H2A
Differential skin resistance survey	***M22***
Exercise	***M12***
Functional testing	M12I
Evaluation of joint range of motion	M12I1
Evaluation of muscle strength	M12I2
Massage	***M23***
Palpation evaluation	M23G
Determination of fibroid, scar, or calcific formation	M23G4
Peripheral electroacustimulation	***M05***
Peripheral acupoint survey	M05D
Sensory stimulation	***M17***
Neurological evaluation	M17E
Sensory evaluation	M17E2

Treatment techniques (C08F3B)

Auricular electroacustimulation	***M06***
Pain relief	M06A
Cryotherapy	***M10***
Analgesia / anesthesia	M10C
Electrical facilitation of endorphin production (acusleep)	***M03***
Increasing pain tolerance	M0305A
Electrical stimulation (ES)	***M02***
Inhibition of ionized formation (Ca^{++} or other ion deposit)	M02F
Exercise	***M12***
Muscle strengthening	M12A (1)
Joint mobilization	***M13(1)***
Massage	***M23***
Deep soft tissue mobilization	M23F
Muscles	M23F2A
Ligaments	M23F2B
Tendon sheaths	M23F2C
Tendons	M23F2D
Peripheral electroacustimulation	***M05***
Analgesia	M05B
Transcutaneous nerve stimulation	***M04***
Prolonged stimulation of acupoints	M04B
Ultrahigh frequency sound (ultrasound)	***M01***
Calcium deposit management	M01E

(1) If arthritic in nature

CALF PAIN SYNDROME (S50)

Definition (S50A)

Pain in the calf can be caused by trauma, nerve injury or impingement (sciatica), deep vein thrombosis, phlebitis, soft tissue inflammation and/or swelling, infection in soft tissues, bones or joints, muscle cramps or extrafusal spasm, intermittent claudication (insufficient blood reaching the muscles), and referred pain from trigger point formations.

Related Syndromes (S50B)

ACHILLES TENDONITIS (S52)
ADHESION (S56)
BACTERIAL INFECTION (S63)
DIABETES (S10)
FASCIITIS (S20)
MYOSITIS OSSIFICANS (S67)
PHLEBITIS (S68)
PIRIFORMIS (S04)
PITTING (LYMPH) EDEMA (S31)
POST CEREBRAL VASCULAR
   ACCIDENT (CVA) (S07)
POST IMMOBILIZATION (S36)
POST PERIPHERAL NERVE INJURY (S23)
POST SPINAL CORD INJURY (S24)
REFERRED PAIN (S01)
SCIATICA (S05)
SHINGLES (HERPES ZOSTER) (S37)
TENDONITIS (S28)
UNHEALED DERMAL LESION (S39)

Treatable Causes Which May Contribute To The Syndrome (S50C)

Bacterial infection (C09)
Blood sugar irregularities (C31)
Calcific deposit (C08)
Extrafusal muscle spasm (C41)
Habitual joint positioning (C48)
Hypermobile/instable joint (C10)
Interspinous ligamentous strain (C03)
Joint ankylosis (C42)
Joint approximation (C47)
Ligamentous strain (C25)
Localized viral infection (C27)
Muscular weakness (C23)
Nerve root compression (C04)
Neuroma formation (C14)
Neuromuscular tonic imbalance (C30)
Peripheral nerve injury (C07)
Phlebitis (C43)
Psychoneurogenic neuromuscular
   hypertonus (C29)
Restricted joint range of motion (C26)
Soft tissue inflammation (C05)
Soft tissue swelling (C06)
Tactile hypersensitivity (C24)
Trigger point formation (C01)
Unhealed dermal lesion (C12)
Vascular insufficiency (C11)

Treatment Notes (S50D)

Phlebitis without evident thrombosis responds remarkably well to multifrequency pulsed galvanic electrical stimulation; coupled with ongoing use of support hose, the patient can expect almost immediate relief of pain and increases in ambulatory ranges.

The following is a list of trigger point formations which may, singly or in combination, refer pain into the calf area. It should be noted that it is possible for the given patient to experience pain throughout the entire stereotypical referred pain pattern or only in part or parts of it. Opposite to each listing is a description of the parts of the stereotypical referred pain pattern most commonly experienced by previous patients, relative to the given trigger point formation. Any descriptive reference which is underlined is a part of the pattern which has *not* been commonly experienced by previous patients. For complete description of each stereotypical referred pain pattern, refer to the Table of Referred Pain Patterns of Trigger Point Formation Origin [T005]. Please note the key to abbreviated words at the bottom of the list.

LOCATION (MUSCLE)	REFERRED PAIN ZONES (GENERAL)
Gluteus miniumus	Gluteus maximus (excepting central area) & gluteus medius, p.l.leg, l.calf to p.of l. malleolus
Adductor longus	A. s.iliac crest into groin & down the a.med. thigh, central med. calf, med. malleolus & med. foot
Biceps femoris (hamstring)	P. d.central thigh, p.prox. calf
Vastus medialis	Vastus medialis, patella, knee joint
Gastrocnemius	P. med. calf, central thigh, behind med. malleolus, arch of the foot

calf pain syndrome

Anterior tibialis	A. calf just l.to tibia, *over metatarsal 1, big toe*	l.—lateral
		a.—anterior
		s.—superior
Long toe extensors	A. l.distal ⅔ of calf, over metatarsals & toes 2–5	&—and
		bord.—border
		i.—inferior
Soleus	Achilles tendon & heel	m.—middle
Peroneus longus	L. d.1/4 calf behind l.malleolus	d.—distal
		p.—posterior
		bil.—bilateral

Key:
maj.—major
prox.—proximal
med.—medial
para.—paraspinous

Bibliography (S50E)

Gardner, p. 157.

CAPSULITIS SYNDROME (S27)

Definition (S27A)

The articular capsules form complete envelopes for the moveable joints. Each capsule is made up of two strata, a white external fibrous tissue layer (stratum fibrosum) and an internal lining of synovial membrane (stratum synoviale). The synovial membrane is a loose structure of folds composed of connective tissue, fat and blood vessels which folds commonly surround the margin of articular cartilage, filling clefts and crevices created by the capsule. The synovial lining is filled with synovial fluid which acts as a lubricant to decrease friction of tendon motion through the joint. The synovial cavity normally contains only enough synovial fluid to moisten and lubricate the synovial surfaces. If injured, however, the synovial membrane may respond to the irritation of released inflammatories (histamine, bradykinin, and prostaglandins) by secreting excessive amounts of synovial fluid. As the resultant pressure mounts within the capsule, pain and discomfort develops which may be increased with motion of the involved joint and/or tendons associated with it.

The most commonly recognized capsulitis syndrome occurs to the shoulder's glenohumeral capsule. The presence of glenohumeral capsulitis may be established by the patient confirming pain when the involved shoulder is involuntarily or voluntarily abducted, while little discomfort is experienced in other shoulder ranges of motion. Voluntary shoulder abduction range of motion may be severely limited by pain. As the humerus approaches and exceeds 90 degrees, the patient will report increasing pain in the shoulder as the capsule is compressed between the humerus and acromial process (refer to Exercise, Functional Testing, Evaluation of Joint Range of Motion [M12I1]). Capsulitis may further be confirmed if relatively high skin resistance can be established over the capsule site through differential skin resistance survey (refer to Differential Skin Resistance Survey [M22]). There may accompanying modest soft tissue swelling visually apparent between the head of the humerus and the acromion. The space between the head of the humerus and the acromion will be palpation tender.

In some cases, the inflamed glenohumeral capsule may adhere to the humeral head and proceed to subsequently contract. As the process of capsule shrinkage slowly progresses (sometimes taking several months), the glenohumeral range of motion becomes increasingly limited, finally resulting in joint fixation or "freezing" in one position. At this point, the total ranges of motion of the shoulder are limited to the mobility that scapular motion, elevation and rotation afford the entire structure. As the condition progresses, the inflammation becomes subacute. Acute pain subsides to ultimately disappear when the shoulder is finally "frozen".

Related Syndromes (S27B)

ADHESION (S56)
CAPSULITIS (S27)
CERVICAL DORSAL OUTLET (SCALENICUS ATTICUS/CERVICAL RIB) (S09)
CHONDROMALACIA (S15)
ELBOW PAIN (S47)
FASCIITIS (S20)
FROZEN SHOULDER (S64)
HAND/FINGER PAIN (S17)
KNEE PAIN (S46)
MYOSITIS OSSIFICANS (S67)
OSTEOARTHRITIS (S21)
POST IMMOBILIZATION (S36)
POST PERIPHERAL NERVE INJURY (S23)
RADICULITIS (S29)
REFERRED PAIN (S01)
SHOULDER/HAND (S57)
SHOULDER PAIN (S08)
TEMPOROMANDIBULAR JOINT (TMJ) PAIN (S06)
TENDONITIS (S28)
TENNIS ELBOW (LATERAL EPICONDYLITIS) (S32)
TOOTHACHE/JAW PAIN (S41)
WRIST PAIN (S16)

Treatable Causes Which May Contribute To The Syndrome (S27C)

Calcific deposit (C08)
Extrafusal muscle spasm (C41)
Habitual joint positioning (C48)
Joint ankylosis (C42)
Joint approximation (C47)
Ligamentous strain (C25)
Muscular weakness (C23)
Nerve root compression (C04)
Neuromuscular tonic imbalance (C30)
Peripheral nerve injury (C07)
Psychoneurogenic neuromuscular hypertonus (C29)
Restricted joint range of motion (C26)
Soft tissue inflammation (C05)
Soft tissue swelling (C06)
Tactile hypersensitivity (C24)
Trigger point formation (C01)
Visceral organ Dysfunction (C02)

Treatment Notes (S27D)

Capsulitis, including early phases of glenohumeral capsulitis, has been clinically responsive to phonophoresis of nonsteroid antiinflammatories (refer to Ultrahigh Frequency Sound, Phonophoresis [M01A]), electrical stimulation of acupoints in the inflamed zone (refer to Peripheral Electroacustimulation, Peripheral Acupoint Survey [M05D] and Transcutaneous Nerve Stimulation, Prolonged Stimulation of Acupoints [M04B]) and multifrequency pulsed galvanic electrical stimulation of the area over the inflamed zone (refer to Electrical Stimulation, Inflammation Control [M02G]). Gentle range of motion exercise (Codman's exercise) has also been used to good effect in maintaining normal range of motion (refer to Table of Stretching Exercises for Muscle Lengthening [T016]).

The following is a list of trigger point formations which may, singly or in combination, imitate or contribute to the pain associated which shoulder capsulitis. It should be noted that it is possible for the given patient to experience pain throughout the entire stereotypical referred pain pattern or only in part or parts of it. Opposite to each listing is a description of the parts of the stereotypical referred pain pattern most commonly experienced by previous patients, relative to the given trigger point formation. Any descriptive reference which is underlined is a part of the pattern which has *not* been commonly experienced by previous patients. For complete description of each stereotypical referred pain pattern, refer to the Table of Referred Pain Patterns of Trigger Point Formation Origin [T005]. Please note the key to abbreviated words at the bottom of the list.

LOCATION (MUSCLE)	REFERRED PAIN ZONES (GENERAL)
Levator scapulae	(2) *Med. & l. bord. of scapula, joint, para.* C2–T2, upper trapezius
Scalenus	(4) *Shoulder*, med. & *l.arm*, p. thumb, *p.index & m.finger* trapezius D027
Infraspinatus	(2) *Subocciput*, deltoids, *l.arm, palm, p.hand*
Infraspinatus (abnormal)	Med. bord. & i.angle of scapula
Medial teres major	A/C joint, p.& m.deltoids, *p.forearm*
Lateral teres major	A/C joint, p.& m.deltoids, *p.forearm*
Teres minor	P. & m.deltoid, *med. & l.triceps*
Coracobrachialis	A. deltoid, shoulder joint,l. *triceps, p.arm & hand, p.m.finger*
Lower splenius cervicus	*Para. C1–C7*, upper trapezius
Upper trapezius [B]	*Subocciput, upper trapezius, l.p. neck*
Middle trapezius [A]	Para. C7–T4, *upper trapezius*
Middle trapezius [B]	S. scap.border (supraspinatus)
Lower trapezius [A]	Subocciput, *para. C1–C6*, upper *trapezius*, A/C joint
Lower trapezius [B]	M. scapula bord.
Cervical multifidus (C4–C5)	Subocciput, para. C1–C7, *upper trapezius*
Supraspinatus (muscle)	*Upper trapezius*, deltoids, *l.arm*
Supraspinatus (tendon)	Deltoids
Latissimus dorsi (abnormal)	A. delt., l.lower abdomen
Serratus posterior superior (under the scapula)	Scapula, p.deltoid, med. arm, wrist, med. hand, finger 5
Subclavius	Under clavicle, *a.deltoid*, biceps, brach., *l.hand, a.& p.f.s 1–3*
Subscapularis	*Scapula, axilla*, p.deltoid, *med. triceps, a.p.wrist*
Posterior deltoid	P. deltoid, *a.& m.deltoid, l. triceps*
Anterior deltoid	A. deltoid, *m.deltoid, l.upper arm*
Pectoralis major (costal)	(2) Breast, chest, *axilla*
Pectoralis major (clavicular)	(2) *Subclavical*, anterior deltoid
Pectoralis major (sternal)	(3) Pectoralis major, a.deltoid, *med. a. upper arm*, prox. a.forearm, *m.palm, a.f.s 3–5*
Pectoralis minor	*Pectoralis major*, a.deltoid, *med. a. arm, med. a.hand, a.f.s 3–5*
Sternalis	(3) Bil. parasternum, med. subclavicle, *a.deltoid, med. a.upper arm & elbow*
Rhomboids	(3) Med. bord. of scapula, *supraspinatus*
Longhead of the triceps	*Upper trapezius*, (2) p.deltoid, *p.upper arm*, l.p.forearm, *p.wrist*

Biceps brachii (2) *Upper trapezius, a.& m.deltoid, biceps,* a.elbow

Brachizlis (superior & inferior) (4) *A. deltoid, a.half m.deltoid, a.elbow,* a.& p.first metacarpal, l.hand, p.second metacarpal

Multifidus (T4–T5) Bil. para. T2–T7, l.over med. bord. of scapula

Iliocostalis thoracis (T6) *Para. T4–T10,* med. scapula, *a.lower med. chest*

Iliocostalis thoracis (T11) *Para. T11–S1,* l.bottem ribs, *l.bord. of the scapula,* a.lower abdominal quadrant

Key:
pect.—pectoralis
A/C—acromioclavicular
med.—medial
brach.—brachioradialis
para.—paraspinous

MP—metacarpal phalangeal
sterno.—sternocleidomastoideus
a.—anterior
s.—superior
&—and
maj.—major
bord.—border
f.—finger
l.—lateral
m.—middle
d.—distal
p.—posterior
prox.—proximal
bil.—bilateral
f.s.—fingers
i.—inferior

Bibliography (S27E)

Goss, p. 299.
Salter, p. 244.
Shands, pp. 426–427.

CARPAL TUNNEL SYNDROME (S35)

Definition (S35A)

The median nerve and the long flexor tendons traverse the wrist through an arch formed by the carpal bones, under the transverse carpal ligament which laterally crosses the arch. The structure thus formed is called the carpal canal or carpal tunnel. In this inelastic tunnel, compression of the median nerve or its blood supply may occur from synovitis, fibrosis of the flexor tendon sheaths, edema following Colle's fracture, Smith fracture, dislocated carpal bone, tumor, rheumatoid arthritis and/or tenosynovitis of the finger flexor tendons within the carpal tunnel. Should such compression occur, a common form of median neuropathy may result and produce symptomatology called the carpal tunnel syndrome.

Median nerve compression at the wrist causes motor weakness of the abductor pollicis brevis, opponens pollicis, first and second lumbrical muscles, and flexor pollicis brevis; it also causes numbness, burning, and tingling in the first three phalanges and often cyanosis of those three digits. Symptoms of the carpal tunnel syndrome are noted to have the unusual trait of "ascending" the arm and may thus imitate the cervical dorsal outlet syndrome or radiculitis. Symptoms most frequently occur at night or in the early morning and awaken the patient from sleep. Patients often complain of being "clumsy" with the involved hand and being prone to drop things.

Objective evidence of carpal tunnel syndrome includes insensitivity to pin prick in the index and middle finger, as well as loss of temperature, light touch and position senses. Paraesthesia and numbness of the index and middle fingers may be produced by sustained wrist flexion or extension, by manual compression of the radial and ulnar arteries, and/or prolongation of nerve conduction velocity (as demonstrated by electromyographic study). If muscle atrophy in the hand is apparent (denoting severe or prolonged involvement), it is usually of the short thumb abductor.

The carpal tunnel syndrome occurs most frequently to women and is usually a unilateral phenomena, but bilateral involvement is known to happen. The most common cause seems to be ischemia of the nerve arising from repeated wrist dorsiflexion and simultaneous contraction of the finger flexors (seen in handwork like sewing or knitting).

Related Syndromes (S35B)

ADHESION (S56)
FOREARM PAIN (S54)
HAND/FINGER PAIN (S17)
PITTING (LYMPH) EDEMA (S31)
POST IMMOBILIZATION (S36)
POST PERIPHERAL NERVE INJURY (S23)
REFERRED PAIN (S01)
SHOULDER/HAND (S57)
TENDONITIS (S28)
WRIST PAIN (S16)

Treatable Causes Which May Contribute To The Syndrome (S35C)

Calcific deposit (C08)
Extrafusal muscle spasm (C41)
Habitual joint positioning (C48)
Joint approximation (C47)
Ligamentous strain (C25)
Muscular weakness (C23)
Peripheral nerve injury (C07)
Restricted joint range of motion (C26)
Scar tissue formation (C15)
Soft tissue inflammation (C05)
Soft tissue swelling (C06)
Tactile hypersensitivity (C24)
Trigger point formation (C01)
Vascular insufficiency (C11)

Treatment Notes (S35D)

The carpal tunnel syndrome, if caused by daytime activities, requiring prolonged wrist dorsiflexion and simultaneous finger flexion (gripping) and not from space-occupying lesions or masses, is best treated with immobilization of the wrist in a neutral position by splinting on a day and night basis. In such a case, symptomatology may be reduced by phonophoresis of nonsteroidal antiinflammatories into the transverse carpal ligament and corresponding carpal tunnel. High differential skin resistance (DSR) in the area is indicative of soft tissue inflammation (refer to Differential Skin Resistance Survey [M22] and Ultrahigh Frequency Sound, Phonophoresis [M01A]) and prolonged electrical stimulation (low frequency TNS) of acupoints in the inflamed zone (refer to Peripheral Electroacustimulation, Peripheral Acupoint Survey [M05D] and Prolonged Stimulation of Acupoints [M04B]) or multifrequency pulsed galvanic electrical stimulation may help reduce pain, inflammation and interstitial swelling (refer to Electrical Stimulation, Inflammation Control [M02G] and Reduction of Edema [M02D]).

The following is a list of trigger point formations which may, singly or in combination, imitate or contribute to the pain associated with the carpal tunnel syndrome. It should be noted that it is possible for the given patient to experience pain throughout the entire stereo-typical referred pain pattern or only in part or parts of it. Opposite to each listing is a description of the parts of the stereotypical referred pain pattern most commonly experienced by previous patients, relative to the given trigger point formation. Any descriptive reference which is underlined is a part of the pattern which has *not* been commonly experienced by previous patients. For complete description of each stereotypical referred pain pattern, refer to the Table of Referred Pain Patterns of Trigger Point Formation Origin [T005]. Please note the key to abbreviated words at the bottom of the list.

LOCATION (MUSCLE)	REFERRED PAIN ZONES (GENERAL)
Scalenus	(4) Shoulder, med. & l.pect. maj., med. bord. of scapula, deltoids, *l.arm*, p.thumb, *p.index & m.finger*
Infraspinatus	(2) Subocciput, deltoids, *l.arm, palm, p.hand*
Upper latissimus dorsi	Latissimus dorsi, *med. arm, l.hand*, fingers 4 & 5
Serratus posterior superior (under the scapula)	Scapula, p.deltoid, med. arm, wrist, med. hand, finger 5
Serratus anterior	Lower a.l.chest, *med. a.arm & hand*, d.med. bord. of scapula, a.f.s 4 & 5
Subclavius	Under clavicle, *a.deltoid*, biceps, brach., *l.hand, a.& p.f.s 1–3*
Subscapularis	Scapula, axilla, p.deltoid, *med. triceps*, a.p.wrist
Pectoralis major (sternal)	Pectoralis major, (3) a.deltoid, *med. a. upper arm*, prox. a.forearm, *m.palm, a.f.s 3–5*
Pectoralis minor	Pectoralis major, a.deltoid, *med. a. arm, med. a.hand, a.f.s 3-5*
Medial triceps (deep fibers)	Med. a.elbow & *forearm, a.f.s 3-5*
Brachialis (superior & inferior)	A. deltoid, *a.half* (4) *m.deltoid, a.elbow*, a.& p.first metacarpal, l.hand, p.second metacarpal
Middle finger extensor	P. central forearm & middle finger
Palmaris longus	A. med. forearm, palm
Flexor carpi radialis	L. a.forearm, a.wrist, *palm*
Flexor carpi ulnaris	Med. a.d.forearm, a.med. wrist-hand
Brachioradialis	P. l.elbow, *l.forearm*, p.metacarpals 1 & 2
Pronator teres	A. *l.forearm* & wrist, a.metacarpal #1
Radial head of the flexor digitorum sublimis	A. l.wrist, central palm, a.m.finger
Humeral head of the flexor digitorum sublimis	A. of carpals & metacarpals & f.s. 4–5
Flexor pollicis longus	A. *thenar eminence* & thumb
Opponens pollicis	A. l.wrist, a.metacarpal & finger 1
Adductor pollicis	L. metacarpal 1, *a.& p.of metacarpals 1–2 & thumb*
First dorsal interosseus	*Palm*, f.2, p.metacarpals 2–5, l.p.f.5

Key:
TMJ—temporomandibular joint
A/C—acromioclavicular
med.—medial
para.—paraspinous
MP—metacarpal phalangeal
sterno.—sternocleidomastoideus
pect.—pectoralis
a.—anterior
s.—superior
&—and
maj.—major
bord.—border
f.—finger
l.—lateral
m.—middle
d.—distal
p.—posterior
prox.—proximal
bil.—bilateral
f.s.—fingers
i.—inferior

Bibliography (S50E)

Salter, pp. 274–275.
Shands, p. 227.

CENTRAL NERVOUS SYSTEM DYSFUNCTION (R006)

All components of the neuromuscular cybernetic system depend upon feedback loops. Each element in the system receives positive feedback from other elements and in turn responds, either directly or indirectly with negative feedback. This system is fully described in The Central Nervous System [R002]. Should any of these feedback loops be disrupted, the entire system may be rendered fully inoperable.

Dysfunction Resulting From Supraspinal Damage (R006A)

A disruption can occur if one of the supraspinal structures or one of the pathways between the structures is damaged or destroyed. In humans, non-fatal damage to the supraspinal neuromuscular system most commonly occurs to the cerebral motor cortex and to the afferent pathways from the cerebellum and the basal ganglia to the cerebral motor cortex as they transverse the internal capsule. The usual initial response of the cerebral motor cortex to the cessation of those afferent signals is a discontinuance of efferent motor signals to extrafusal muscle activity. The initial response of the basal ganglia is an interruption of facilitatory activity to intrafusal muscle. Damage of such magnitude usually results in temporary total flaccidity of the involved musculature. However, after a brief period of several days to six weeks or more, the cerebral motor cortex and the basal ganglia may adjust to the absence of incoming negative feedback from the cerebellum and will begin to send facilitatory motor messages to extrafusal and intrafusal muscle fibers, respectively. Without the coordinating and inhibitive influence of the cerebellum, hypertonus and reflex patterns of spasticity begin to appear. In the resulting altered cybernetic system (illustrated in Disrupted Afferent Pathways Between the Cerebellum and the Cerebral Motor Cortex and Basal Ganglia Aggregate [R006A01]), the supraspinal structures may make a motor decision and send the appropriate signals to the cerebral cortex, which then relays the decision to the cerebellum. The cerebral motor cortex then prepares to receive data either on a "learned" coordinating program or on the status of ongoing gross motor activity from the cerebellum. Because of the lost pathways from the basal ganglia and the cerebellum, that data never arrives and the cerebral motor cortex must make the choice of taking blind action or doing nothing at all.

If the cerebral motor cortex choose to take action, it simply facilitates extrafusal muscle activity

Disrupted Afferent Pathways Between the Cerebellum and the Cerebral Motor Cortex and Basal Ganglia Aggregate (R006A01)

* RED NUCLEUS, SUBSTANTIA NIGRA, OLIVE, SUBTHALAMUS, RETICULAR FORMATION

via the corticospinal tract and/ or activity in the basal ganglia, which, failing to receive selective inhibition programs from the cerebellum, can only act blindly itself, via the reticular formation, by facilitating intrafusal muscle contraction and thereby extrafusal muscle contraction through the tonic stretch reflex.

The clinical picture of this disruption varies from patient to patient according to the extent and site of the supraspinal damage and difference in individual response. If the cerebral motor cortex does not adjust to the absence of incoming information from the basal ganglia and the cerebellum, it may choose not to act or activate, and a state of functional flaccidity will result. As time passes, the systems may respond to proprioceptive stimulation delivered to the supraspinal structures via other pathways to the basal ganglia and begin to generate some activity, especially in the lower extremities, and hypertonus will develop.

Dysfunction Resulting From Damage to the Basal Ganglia (R006B)

The flaccid response provoked by severe damage to either the basal ganglia or to the supraspinal efferent motor pathways from the cerebral motor cortex (illustrated in Disrupted Pathways Between Cerebellum, Cerebral Motor Cortex, and Basal Ganglia Aggregate [R006B01]) will produce a condition which will not allow "muscle tone" to develop because of the consequent lack of basal ganglia facilitation of tonic gamma motor activity or structural flaccidity. Some individuals develop a modicum of voluntary phasic extrafusal muscle control, but they are neither able to maintain unconscious prolonged muscle contractions to any degree nor able to benefit from the motor programming provided to the tonic gamma system by the cerebellum. In such conditions, even small movements of the involved extremities require continuous, fully conscious attention and are therefore of small functional value.

Dysfunction as a Result of Peripheral Nerve Injury (R006C)

If the peripheral spinal nerve to the muscle is severed, both the motor and sensory nerves must regenerate to both the extrafusal and intrafusal muscle fibers from their respective neurons in the ventral "gray horn" of the spinal cord and the dorsal root ganglion. Clinical experience has led us to postulate that the motor nerve to the extrafusal muscle fibers is likely to reinnervate its end organs, because the reinnervation occurs primarily on the surface layers of the muscle, even if the deeper-lying motor and sensory end organs of the muscle spindle fail to be reinnervated (the result

Disrupted Pathways Between Cerebellum, Cerebral Motor Cortex, and Basal Ganglia Aggregate
(R006B01)

* RED NUCLEUS, SUBSTANTIA NIGRA, OLIVE, SUBTHALAMUS, RETICULAR FORMATION

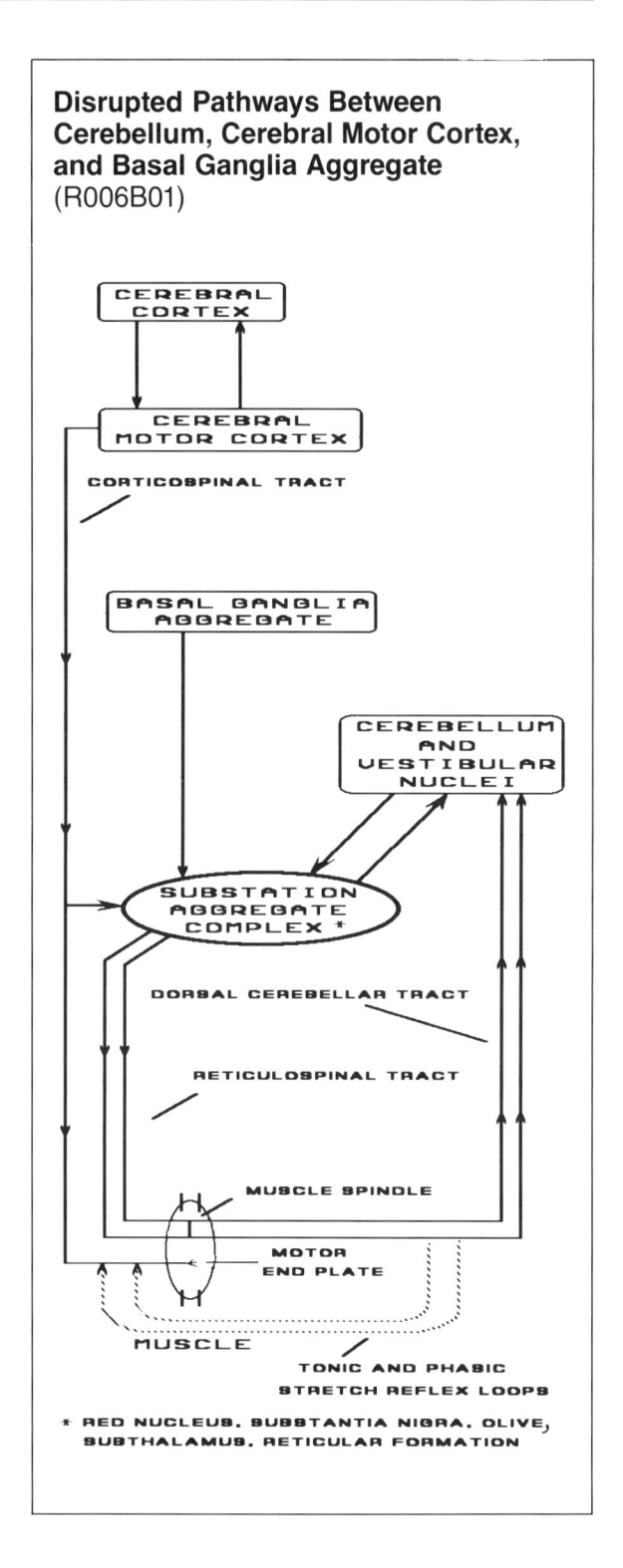

illustrated in Disrupted Muscle Spindle Afferent and Efferent Pathways [R006C01]). Theoretically, this condition leaves the neuromuscular system with only the option of contracting extrafusal muscle fibers, without the potential of maintaining tone or taking advantage of cerebellum programming provided through the tonic gamma system.

central nervous system dysfunction

This type of singular reinnervation is very rare. Most patients suffering from peripheral nerve injury syndromes lasting one to ten years develop muscle tone after neuromuscular reeducation with myometric feedback. It is likely that in such cases the tonic gamma and possibly the phasic gamma motor nerves and the tonic sensory nerves have reinnervated their end organs. However, the annulospiral sensory end organ may have failed to be reinnervated on the equatorial segment of the nuclear bag and nuclear chain fibers by virtue of its complex nature or the depth of tissue penetration required of the sensory neuron for end organ reinnervation. In any case, any muscle spindle sensory deficit will cause the cerebellum, without outside aid, to allow a state of functional flaccidity to exist because prior learned programming is dependent on constant phasic afferent input. However, the cerebellum can be "reprogrammed" through electromyometric feedback neuromuscular reeducation to do without phasic afferent input (refer to Toning of muscles paralyzed by peripheral nerve injury [M18K5A]).

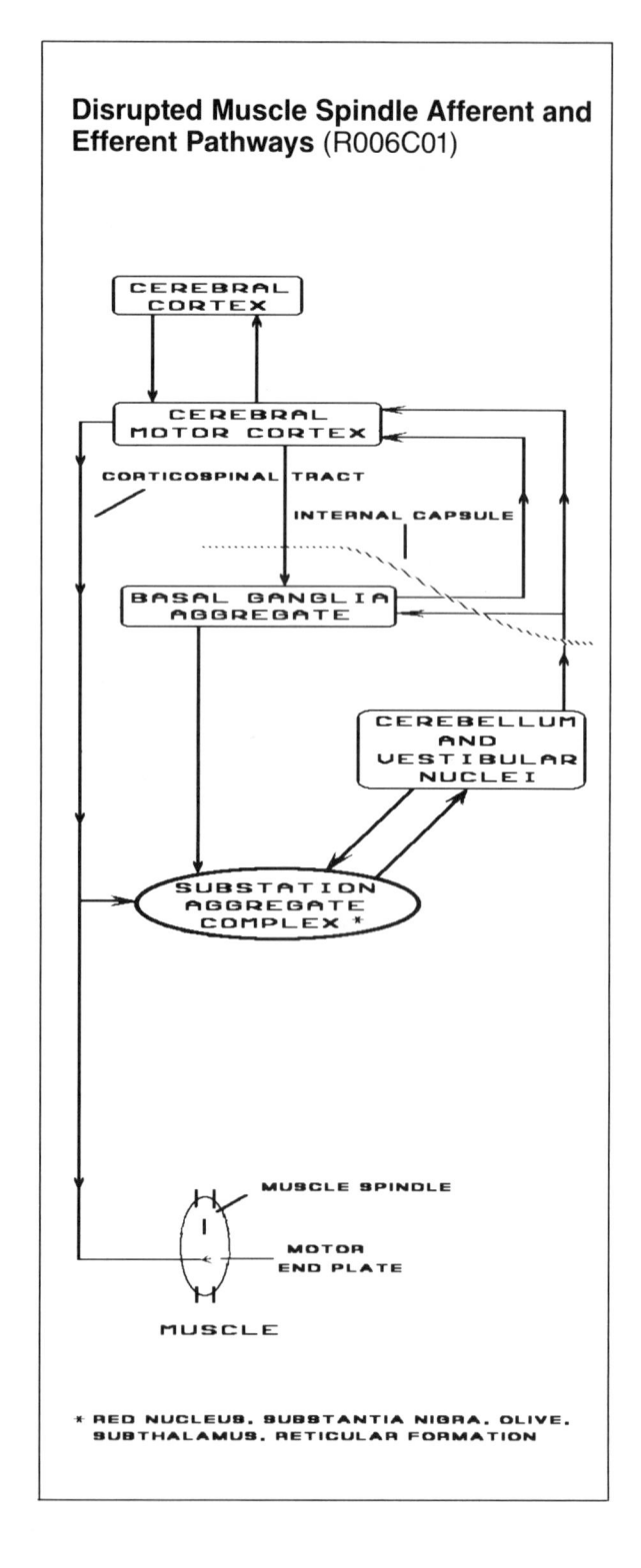

Disrupted Muscle Spindle Afferent and Efferent Pathways (R006C01)

```
              CEREBRAL
              CORTEX
                 ↕
              CEREBRAL
              MOTOR CORTEX  ←───────┐
                 ↓                   │
       CORTICOSPINAL TRACT           │
                      INTERNAL CAPSULE
                 ↓                   │
              BASAL GANGLIA  ←───────┤
              AGGREGATE              │
                 ↓                   │
                              CEREBELLUM
                                 AND
                              VESTIBULAR
                               NUCLEI
                 ↓              ↕
              SUBSTATION
              AGGREGATE
              COMPLEX *
                 ↓
                           MUSCLE SPINDLE
                            MOTOR
                            END PLATE
                 MUSCLE
```

* RED NUCLEUS, SUBSTANTIA NIGRA, OLIVE, SUBTHALAMUS, RETICULAR FORMATION

CEREBRAL PALSY SYNDROME
(S25)

Definition (S25A)

Cerebral palsy is defined as a paralysis or lack of muscle control as a result of injury to an immature nervous system, usually occurring before, during, or shortly following birth. This condition is sometimes accompanied by seizures, mental retardation, abnormal sensory perception, and/or impairment of sight, hearing and/or speech. One of the major causes of cerebral palsy conditions has historically been oxygen deprivation which can result from premature separation of the placenta and the uterine wall, strangulation of the umbilical cord or arterial blood supply to the brain, premature birth, or labor which is too short. Blood chemistry disorders may also damage immature neural tissues. These include chemical poisoning by alcohol, tobacco or other drugs, and incompatible blood factors: Rh factors or AB blood type mismatching between fetus and mother. Another major source of the cerebral palsy syndrome are the viral and bacterial diseases (rubella, for example) which may cause meningitis or meningitis-like conditions in the fetus or infant. Notable also for producing immature cerebrum damage are direct concussion or bruising of brain or spinal cord tissues; common sources include forceps-assisted deliveries, direct blows to the head after birth, or indirect trauma from unrelieved excessive intraventricular pressure (hydrocephalus).

Many widely divergent symptomatic profiles emerge from the cerebral palsy victim population by virtue of the variety and nature of the causes of the condition. Of especially complex symptomatology are the syndromes arising from neural damage caused by blood factor incompatibilities and diseases which affect the central nervous system. Victims often suffer sensory perceptual deficits along with various neuromuscular dysfunctions.

In terms of a neuromuscular reeducation, a large segment of the cerebral palsy victim population closely resembles the adult population of post CVA syndrome patients, especially when matched according to cause, extent of neural damage and the site of injury. Both groups could be classified together in three broad categories: (1) spastic patients who move involved joints stiffly and with difficulty because of neuromuscular hypertonus, (2) athetoid patients who experience involuntary and uncontrolled movements, and (3) ataxic patients who have problems with balance, difficulty making smooth coordinated movements of the joints, and poor depth perception.

Related Syndromes (S25B)

BACTERIAL INFECTION (S63)
NEUROMUSCULAR PARALYSIS (S22)
PNEUMONIA (S58)
POST CEREBRAL VASCULAR ACCIDENT
  (CVA) (S07)
POST IMMOBILIZATION (S36)
POST PERIPHERAL NERVE INJURY (S23)
POST SPINAL CORD INJURY (S24)
TORTICOLLIS (WRY NECK) (S65)
UNHEALED DERMAL LESION (S39)

Treatable Causes Which May Contribute To The Syndrome (S25C)

Extrafusal muscle spasm (C41)
Habitual joint positioning (C48)
Hyperhidrosis (C37)
Hypermobile/instable joint (C10)
Joint ankylosis (C42)
Joint approximation (C47)
Ligamentous strain (C25)
Muscular weakness (C23)
Nerve root compression (C04)
Neuroma formation (C14)
Neuromuscular tonic imbalance (C30)
Pathological neuromuscular hypertonus (C28)
Pathological neuromuscular hypotonus (C33)
Pathological neuromuscular
  dyscoordination (C35)
Peripheral nerve injury (C07)
Restricted joint range of motion (C26)
Spastic sphincter (C40)
Sympathetic hyperresponse (C38)
Tactile hypersensitivity (C24)
Vascular insufficiency (C11)
Visceral organ Dysfunction (C02)

Treatment Notes (S25D)

Rehabilitation of the neuromuscularly involved cerebral palsy victim normally entails electromyometric neuromuscular reeducation and resembles the program employed for the post cerebral vascular accident victims (refer to Electromyometric Feedback, Post Cerebral Accident Neuromuscular Reeducation [M18A]). Ideally, therapy should begin as soon as the child can appreciate audio and/or visual feedback and well before the onset of puberty. Experience has shown that prepuberty children respond much quicker than post puberty patients or their adult counterparts to therapy employing electromyometric feedback, and achieve functional results with surprising rapidity.

It is especially important to appreciate the role that may be played by dominating or undeveloped developmental reflexes in the cerebral palsy neuromuscular disability (refer to Exercise, Functional Testing, Evaluation of Developmental Reflexes [M12I6] and Table of Developmental Reflexes [T007]).

The general rule for dealing with the developmental reflexes in the cerebral palsy patient who has not gone through the normal developmental sequence is to facilitate developmental reflexes which have not yet developed with the use of electromyometric feedback, inhibit the developmental reflexes that are dominant, and develop functional abilities that are not reflex affected. The Brunnstrom principle is useful here: use a reflex, break the reflex, and work without the reflex. An attempt to draw the patient through the sequence of developmental reflexes should be made. General progress in the patient's functional abilities may well depend on it.

Of special note is the clinical evidence that in both adults and children cognition and the ability to speak are closely related to the ability to voluntarily inhibit to a given degree the wrist and finger flexors of the right hand. Work with aphasic patients indicates that a much greater appreciation of the role of neuromuscular development, as it relates to symbolic communication, must be realized before adequate objective intelligence quotient and intellectual competency testing of cerebral palsy or post CVA victims can be relied upon.

Bibliography (S25E)

Doman, pp. 257-262.

CERVICAL DORSAL OUTLET (SCALENICUS ATTICUS; CERVICAL RIB) SYNDROME (S09)

Definition (S09A)

The anterior scaleni muscle is attached to the first rib in front of the middle and inferior trunks of the brachial plexus and the subclavian artery. Should the anterior scaleni muscle become shortened and begin to compress the nerve trunks and the artery caught between it and the rib a brachial neuritis may occur. Weakness and atrophy of the intrinsic muscles of the hand and finally the muscles of the entire upper extremity may occur, occasionally leading to a claw-hand deformity. Paresthesia of the portion of the arm and forearm supplied by the ulnar nerve is sometimes present and often accompanied by radiating pain throughout the distribution of the nerve. Pallor, coldness and cyanosis may appear in the fingers of the involved hand. In very advanced cases, the mechanical pressure upon the artery may cause thrombosis and artery closure with resultant gangrene (which generally begins in the fingers). This syndrome is called the cervical dorsal outlet (CDO) syndrome.

The cervical dorsal outlet condition may be confirmed with a diminution of the radial pulse when the involved arm is elevated above the head (causing scapular adduction and depression) and the patient rotates the head toward the involved extremity and looks up at the involved hand. In more advanced cases complete cessation of palpable radial pulse will occur.

A congenital anomaly known as a "cervical rib" may occur which was formerly thought to be the major source of the above described syndrome. The cervical rib is a supernumerary unit of bony growth attached most often to the seventh cervical vertebra, but it may be alternately attached to the sixth. Cervical ribs usually occur bilaterally, with the cervical rib on one side being positioned higher on the spine than that on the contralateral side, and one is usually more highly developed than the other. When such extra ribs are a problem, the lower components of the brachial plexus are generally found to be in close approximation with them; the middle trunk of the seventh cervical nerve crossing over the transverse process of the seventh cervical vertebra, and the inferior trunk crossing over either the cervical rib or the fibrous band connecting it with the first rib. The mechanical factors of posture may cause the middle and lower portions of the brachial plexus to be stretched over the cervical rib, or the anterior scaleni muscle may act to compress those segments of the nerve against the cervical rib. Additionally, since the subclavian artery lies in front of the rib and behind the anterior scaleni muscle, it may also be affected by such forces. When compounded by the friction which may take place between the nerves, subclavian artery and the cervical rib when shoulder and arm are used, these actions may produce the previously described symptomatology associated with the CDO syndrome.

The CDO syndrome is most commonly associated with occupations which cause their members to advertently or inadvertently keep an arm or arms supported with the shoulder girdle raised. This posture may cause the anterior scalenus muscle(s) to tonically shorten, thereby precipitating a CDO syndrome when the shoulder girdle is allowed to drop and scaleni pressure comes to bear upon the nerve and/or subclavian artery. Potential victims of this disorder include teachers, students, truck drivers and accountants.

Related Syndromes (S09B)

ADHESION (S56)
BUERGER'S DISEASE (S11)
BURSITIS (S26)
CAPSULITIS (S27)
CARPAL TUNNEL (S35)
CERVICAL (NECK) PAIN (S73)
CHEST PAIN (S14)
CHRONIC BRONCHITIS/EMPHYSEMA
 (OBSTRUCTIVE PULMONARY DISEASE) (S13)
ELBOW PAIN (S47)
FASCIITIS (S20)
FOREARM PAIN (S54)
FROZEN SHOULDER (S64)
HAND/FINGER PAIN (S17)
HIGH THORACIC BACK PAIN (S48)
MYOSITIS OSSIFICANS (S67)
NEUROMUSCULAR PARALYSIS (S22)
PITTING (LYMPH) EDEMA (S31)
PNEUMONIA (S58)
POST IMMOBILIZATION (S36)
POST PERIPHERAL NERVE INJURY (S23)
POST WHIPLASH (S55)
RADICULITIS (S29)
RAYNAUD'S DISEASE (S12)
REFERRED PAIN (S01)
SHOULDER/HAND (S57)

SHOULDER PAIN (S08)
TENDONITIS (S28)
TENNIS ELBOW (LATERAL
  EPICONDYLITIS) (S32)
WRIST PAIN (S16)

Treatable Causes Which May Contribute To The Syndrome (S09C)

Calcific deposit (C08)
Extrafusal muscle spasm (C41)
Habitual joint positioning (C48)
Hyperhidrosis (C37)
Interspinous ligamentous strain (C03)
Joint ankylosis (C42)
Joint approximation (C47)
Lung fluid accumulation (C13)
Muscular weakness (C23)
Nerve root compression (C04)
Neuromuscular tonic imbalance (C30)
Peripheral nerve injury (C07)
Psychoneurogenic neuromuscular
  hypertonus (C29)
Reduced lung vital capacity (C17)
Restricted joint range of motion (C26)
Soft tissue inflammation (C05)
Soft tissue swelling (C06)
Tactile hypersensitivity (C24)
Trigger point formation (C01)
Vascular insufficiency (C11)
Visceral organ Dysfunction (C02)

Treatment Notes (S09D)

Treatment of the cervical dorsal outlet (CDO) syndrome centers around the relengthening of the scaleni muscles to reduce mechanical pressure upon the middle and inferior brachial plexus and subclavian artery trapped between the anterior scalenus muscle and the first rib. Prolonged and repeated stretch of the scaleni muscles as accomplished by the Hale's Regime exercise program will accomplish relengthening of the scaleni muscles (refer to Table of Stretching Exercises for Muscle Lengthening [T016]) but the immediate effect of this stretching is to immediately increase the discomfort experienced by the patient (a pulled muscle pulls back; a "tight" muscle or one in spasm will pull back even more). If diligently performed, the Hale's regime takes two weeks or more to accomplish full relief. Electrical stimulation coupled with vibration may be used to almost immediately tonically relengthen the scaleni muscles, giving almost complete and immediate relief. If performed correctly, only two treatment sessions may be required for full resolution (refer to Electrical Stimulation, Muscle Lengthening [M02A]).

The following is a list of trigger point formations which may, singly or in combination, imitate or contribute to the pain associated with the CDO syndrome. It should be noted that it is possible for the given patient to experience pain throughout the entire stereotypical referred pain pattern or only in part or parts of it. Opposite to each listing is a description of the parts of the stereotypical referred pain pattern most commonly experienced by previous patients, relative to the given trigger point formation. Any descriptive reference which is underlined is a part of the pattern which has *not* been commonly experienced by previous patients. For complete description of each stereotypical referred pain pattern, refer to the Table of Referred Pain Patterns of Trigger Point Formation Origin [T005]. Please note the key to abbreviated words at the bottom of the list.

LOCATION (MUSCLE)	REFERRED PAIN ZONES (GENERAL)
Posterior cervical group (Semispinalis capitus, semispinalis cervicis, C4 or C5 multifidus) (3)	Paraspinous area from nuchal line to mid thoracic area
Levator scapulae (2)	*Med. & l. bord. of scapula, joint, para.C2-T2, upper trapezius*
Scalenus (4)	*Shoulder, med. & l.pect. maj., med. bord. of scapula, deltoids, l.arm, p.thumb, p.index & m.finger*
Scalenus (minimus)	*L. triceps, p.forearm & hand & f.s*
Infraspinatus (2)	*Subocciput, deltoids, l.arm, palm, p.hand*
Medial teres major	*A/C joint, p.& m.deltoids, p.forearm*
Lateral teres major	*A/C joint, p.& m.deltoids, p.forearm*
Teres minor	*P. & m.deltoid, med. & l.triceps*
Coracobrachialis	*A. deltoid, shoulder joint.l.triceps, p.arm & hand, p.m.finger*
Lower splenius cervicus	*Para. C1-C7, upper trapezius*
Upper trapezius [B]	*Subocciput, upper trapezius, l.p.neck*
Middle trapezius [A]	*Para. C7-T4, upper trapezius*
Middle trapezius [B]	*S. scap.border (supraspinatus)*

Middle trapezius [C]	Goose flesh of l.triceps & brach.	Lateral triceps	L. triceps, *p.forearm, p.f.s 4-5*
Lower trapezius [A]	Subocciput, *para. C1-C6, upper trapezius,* A/C joint	Longhead of the triceps	*Upper trapezius,* (2) p.deltoid, *p.upper arm,* l.p.forearm, *p.wrist*
Cervical multifidus (C4-C5)	Subocciput, para. C1-C7, *upper trapezius*	Distal medial triceps	P. elbow
Supraspinatus (muscle)	*Upper trapezius,* (2) deltoids, *l.arm*	Anconeus	P. l.elbow
Supraspinatus (tendon)	Deltoids	Biceps brachii	*Upper trapezius,* a.& (2) *m.deltoid, biceps,* a.elbow
Upper latissimus dorsi	Latissimus dorsi, *med. arm, l.hand,* fingers 4 & 5	Brachialis (superior & inferior)	*A. deltoid, a.half m.deltoid, a.elbow,* a.& (4) p.first metacarpal, l.hand, p.second metacarpal
Latissimus dorsi (abnormal)	A. delt., l.lower abdomen	Supinator	L. elbow, *p.brach., p.metacarpals 1-2*
Serratus posterior superior (under the scapula)	Scapula, p.deltoid, med. arm, wrist, med. hand, finger 5	Extensor carpi radialis brevis	*P. d.2/3 forearm,* p.wrist & hand
Serratus anterior	Lower a.l.chest, *med. a.arm & hand,* d.med. bord. of scapula, *a.f.s 4 & 5*	Extensor carpi ulnaris	P. med. d.forearm & wrist & prox. hand
		Middle finger extensor	*P. central forearm* & middle finger
Subclavius	Under clavicle, *a.deltoid,* biceps, brach., *l.hand,* a.& *p.f.s 1–3*	Ring finger extensor	P. origin of brach., p.med. forearm, p.finger 4
Subscapularis	Scapula, axilla, p.deltoid, *med. triceps,* a. & *p.wrist*	Palmaris longus	*A. med. forearm,* palm
		Flexor carpi radialis	L. *a.forearm,* a.wrist, *palm*
Posterior deltoid	P. deltoid, a.& *m.deltoid, l.triceps*	Flexor carpi ulnaris	Med. a.d.forearm, a.med. wrist-hand
Anterior deltoid	A. deltoid, *m.deltoid, l.upper arm*	Brachioradialis	P. l.elbow, *l.forearm, p.metacarpals 1 & 2*
Pectoralis major (costal)	Breast, chest, *axilla* (2)	Pronator teres	A. *l.forearm* & wrist, *a.metacarpal #1*
Pectoralis major (clavicular)	*Subclavical,* anterior (2) deltoid	Extensor indicus	P. wrist, *p.metacarpals 2-4* & second MP joint
Pectoralis major (sternal)	Pectoralis major, (3) a.deltoid, *med. a. upper arm,* prox. a.forearm, *m.palm, a.f.s 3–5*	Radial head of the flexor digitorum sublimis	A. *l.wrist, central palm, a.m.finger*
Pectoralis minor	*Pectoralis major,* a.deltoid, *med. a. arm, med. a.hand, a.f.s 3-5*	Humeral head of the flexor digitorum sublimis	A. of *carpals & metacarpals & f.s. 4-5*
Sternalis	(3) Bil. parasternum, med. subclavicle, *a.deltoid, med. a.upper arm & elbow*	Flexor pollicis longus	A. *thenar eminence &* thumb
		Abductor digiti minimi	P. *l.metacarpal* & finger 5
Rhomboids	(3) Med. bord. of scapula, *supraspinatus*	Second dorsal interosseus	P. l.finger 3 (2)
Medial triceps (deep fibers)	Med. a.elbow & *forearm, a.f.s 3-5*	Opponens pollicis	A. l.wrist, a.metacarpal & finger 1
Medial triceps (l. fibers)	P. *l.elbow,* posterior brach.	Adductor pollicis	L. metacarpal l. *a.& p.of metacarpals 1-2 & thumb*

First dorsal interosseus	*Palm*, f.2, *p.metacarpals 2-5, l.p.f.5*	maj.—major bord.—border f.—finger
Extensor carpi radialis longus	P. l.elbow, p.l.forearm & wrist, p.carpals, p.metacarpals 1-4	l.—lateral m.—middle d.—distal

Key:
TMJ—temporomandibular joint
A/C—acromioclavicular
med.—medial
para.—paraspinous
MP—metacarpal phalangeal
sterno.—sternocleidomastoideus
pect.—pectoralis
a.—anterior
s.—superior
&—and

p.—posterior
prox.—proximal
bil.—bilateral
f.s.—fingers
i.—inferior

Bibliography   (S09E)

Cailliet: *Neck and Arm Pain*, pp. 92–98
Chusid, p. 294.
Salter, pp. 275-276.

CERVICAL (NECK) PAIN SYNDROME
(S73)

Definition (S73A)

Pain which is perceived to occur between the C7 vertebra and the base of the skull, either anteriorly or posteriorly, is defined as cervical (neck) pain.

Cervical (neck) pain may be referred from trigger point formations or visceral organs, created by compression of cervical nerve roots, caused by psychoneurogenic neuromuscular dysfunction (anxiety reaction, torticollis, etc.) or pathogenic neuromuscular dysfunction or may result from direct trauma to the soft and bony tissues in the neck. Trauma to the cervical tissues include bacterial infection, burns, localized viral infection and excessive mechanical stress.

Related Syndromes (S73B)

ACNE (S45)
ADHESION (S56)
BACTERIAL INFECTION (S63)
BELL'S PALSY (S66)
BURNS (SECOND AND/OR THIRD DEGREE) (S72)
CERVICAL DORSAL OUTLET (SCALENICUS ATTICUS/CERVICAL RIB) (S09)
CHEST PAIN (S14)
EARACHE (S40)
ELBOW PAIN (S47)
FASCIITIS (S20)
FOREARM PAIN (S54)
FROZEN SHOULDER (S64)
HAND/FINGER PAIN (S17)
HEADACHE PAIN (S02)
HIGH THORACIC BACK PAIN (S48)
HYSTERIA/ANXIETY REACTION (S59)
MIGRAINE (VASCULAR) HEADACHE (S18)
MYOSITIS OSSIFICANS (S67)
OSTEOARTHRITIS (S21)
PHOBIC REACTION (S44)
POST IMMOBILIZATION (S36)
POST PERIPHERAL NERVE INJURY (S23)
POST SPINAL CORD INJURY (S24)
POST WHIPLASH (S55)
RADICULITIS (S29)
REFERRED PAIN (S01)
SHINGLES (HERPES ZOSTER) (S37)
SHOULDER/HAND (S57)
SHOULDER PAIN (S08)
TEMPOROMANDIBULAR JOINT (TMJ) PAIN (S06)
TINNITUS (S70)
TORTICOLLIS (WRY NECK) (S65)
UNHEALED DERMAL LESION (S39)
WRIST PAIN (S16)

Treatable Causes Which May Contribute To The Syndrome (S73C)

Calcific deposit (C08)
Extrafusal muscle spasm (C41)
Habitual joint positioning (C48)
Hypermobile/instable joint (C10)
Interspinous ligamentous strain (C03)
Joint ankylosis (C42)
Joint approximation (C47)
Ligamentous strain (C25)
Localized viral infection (C27)
Muscular weakness (C23)
Nerve root compression (C04)
Neuroma formation (C14)
Neuromuscular tonic imbalance (C30)
Peripheral nerve injury (C07)
Psychoneurogenic neuromuscular hypertonus (C29)
Restricted joint range of motion (C26)
Soft tissue inflammation (C05)
Soft tissue swelling (C06)
Sympathetic hyperresponse (C38)
Tactile hypersensitivity (C24)
Tinnitus (C32)
Trigger point formation (C01)
Unhealed dermal lesion (C12)
Visceral organ Dysfunction (C02)

Treatment Notes (S73D)

Treatment of cervical (neck) pain must be directed at the particular treatable cause(s) determined to be responsible.

The following is a list of trigger point formations which may, singly or in combination, refer pain into the cervical area. It should be noted that it is possible for the given patient to experience pain throughout the entire stereotypical referred pain pattern or only in part or parts of it. Opposite to each listing is a description of the parts of the stereotypical referred pain pattern most commonly experienced by previous patients, relative to the given trigger point formation. Any descriptive reference which is underlined is a part of the pattern which has *not* been commonly experienced by previous patients. For complete description of each stereotypical referred pain pattern, refer to the Table of Referred Pain Patterns of Trigger Point Formation Origin [T005]. Please note the key to abbreviated words at the bottom of the list.

LOCATION (MUSCLE)		REFERRED PAIN ZONES (GENERAL)
Masseter (deep)		TMJ, lateral jaw, ear, *suboccipital*
Posterior digastric		Mastoid process, *jaw line, occiput*
Upper trapezius [A]	(2)	Posterior neck, occiput, temple, *eye*
Posterior cervical group (Semispinalis capitus, semispinalis cervicis, C4 or C5 multifidus)	(3)	Paraspinous area from nuchal line to mid thoracic area
Sternocleidomastoideus (superficial)	(3)	Sterno., occiput, across forehead & temple, *face, chin, superior lateral throat, top of head, behind & in eye*
Sternocleidomastoid (deep)	(3)	Sterno., ear, occiput, bil. forehead
Levator scapulae	(2)	*Med. & l. bord. of scapula, joint*, para.C2–T2, upper trapezius
Lower splenius cervicus		*Para. C1-C7,* upper trapezius
Upper trapezius [B]		Subocciput, *upper trapezius, l.p. neck*
Lower trapezius [A]		Subocciput, *para. C1–C6, upper trapezius,* A/C joint
Cervical multifidus (C4–C5)		Subocciput, para. C1–C7, *upper trapezius*

Key:
TMJ—temporomandibular joint
A/C—acromioclavicular
med.—medial
para.—paraspinous
brach.—brachioradialis
MP—metacarpal phalangeal
sterno.—sternocleidomastoideus
pect.—pectoralis
a.—anterior
s.—superior
&—and
maj.—major
bord.—border
f.—finger
l.—lateral
m.—middle
d.—distal
p.—posterior
prox.—proximal
bil.—bilateral
f.s.—fingers
i.—inferior

Bibliography (S73E)

Basmajian, pp. 85, 178–183.
Chusid, pp. 294, 333–334.
Salter, pp. 236–238.
Shands, pp. 412–421.

CHEST PAIN SYNDROME (S14)

Definition (S14A)

Chest pain can arise from the chest wall or from within the chest or abdomen. Pain may rise from the chest or associated structures. Its sources may include boils, abscesses, herpes zoster (shingles), pain in the female breast, Bornholm disease, neuralgia, injury or fracture of ribs, cervical dorsal outlet syndrome, any disease that can cause pressure upon the nerves innervating the chest wall (including cysts, tumors, or aneurysm of the aorta), various trigger point formations housed in muscles or other structures on the chest wall and/or in the neck and shoulder which refer pain to the chest area and from strain and/or extrafusal muscle spasm of the intercostal musculature or other musculature mounted on the chest wall. Pain arising from within the chest or abdomen can derive its source from pleurisy, emphysema, pneumothorax, pneumonia, pulmonary embolism, aneurism of the aorta, ischemic heart disease, pericarditis, obstruction of the esophagus by cancerous or benign tumor, by achalasia of the cardia, abscess in the liver or under the diaphragm, hernias through the diaphragm into the chest, diseases of the gallbladder and ulcers of the stomach (peptic or cancerous).

Infection and septic conditions such as cholecystitis or abscess of the liver or under the diaphragm will usually be accompanied by fever and general feelings of malaise.

Related Syndromes (S14B)

BACTERIAL INFECTION (S63)
CERVICAL DORSAL OUTLET (SCALENICUS ATTICUS/CERVICAL RIB) (S09)
CERVICAL (NECK) PAIN (S73)
CHRONIC BRONCHITIS/EMPHYSEMA (OBSTRUCTIVE PULMONARY DISEASE) (S13)
FASCIITIS (S20)
HIGH THORACIC BACK PAIN (S48)
HYSTERIA/ANXIETY REACTION (S59)
OSTEOARTHRITIS (S21)
PHOBIC REACTION (S44)
PNEUMONIA (S58)
POST PERIPHERAL NERVE INJURY (S23)
REFERRED PAIN (S01)
SHINGLES (HERPES ZOSTER) (S37)
UNHEALED DERMAL LESION (S39)

Treatable Causes Which May Contribute To The Syndrome (S14C)

Bacterial infection (C09)
Calcific deposit (C08)
Extrafusal muscle spasm (C41)
Habitual joint positioning (C48)
Interspinous ligamentous strain (C03)
Ligamentous strain (C25)
Localized viral infection (C27)
Lung fluid accumulation (C13)
Muscular weakness (C23)
Nerve root compression (C04)
Neuromuscular tonic imbalance (C30)
Peripheral nerve injury (C07)
Psychoneurogenic neuromuscular hypertonus (C29)
Reduced lung vital capacity (C17)
Restricted joint range of motion (C26)
Soft tissue inflammation (C05)
Soft tissue swelling (C06)
Sympathetic hyperresponse (C38)
Tactile hypersensitivity (C24)
Trigger point formation (C01)
Vascular insufficiency (C11)
Visceral organ Dysfunction (C02)

Treatment Notes (S14D)

Since so many sources of chest pain exist, careful evaluation must be made to rule out untreatable causes or treatment(s) which may be contraindicated for the existing pathology; a very careful screening must be made.

The following is a list of trigger point formations which may, singly or in combination, refer pain into the area of the chest. It should be noted that it is possible for the given patient to experience pain throughout the entire stereotypical referred pain pattern or only in part or parts of it. Opposite to each listing is a description of the parts of the stereotypical referred pain pattern most commonly experienced by previous patients, relative to the given trigger point formation. Any descriptive reference which is underlined is a part of the pattern which has *not* been commonly experienced by previous patients. For complete description of each stereotypical referred pain pattern, refer to the Ta-

chest pain syndrome

ble of Referred Pain Patterns of Trigger Point Formation Origin [T005]. Please note the key to abbreviated words at the bottom of the list.

LOCATION (MUSCLE)	REFERRED PAIN ZONES (GENERAL)
Scalenus	(4) *Shoulder*, med. & l.pect. maj., med. bord. of scapula, deltoids, *l.arm*, p.thumb, *p.index & m.finger*
Serratus posterior superior (under the scapula)	Scapula, p.deltoid, med. arm, wrist, med. hand, finger 5
Serratus anterior	Lower a.l.chest, *med. a.arm & hand*, d.med. bord. of scapula, *a.f.s 4 & 5*
Subclavius	Under clavicle, *a.deltoid*, biceps, brach., *l.hand, a.& p.f.s 1-3*
Pectoralis major (costal)	(2) Breast, chest, *axilla*
Pectoralis major (clavicular)	(2) *Subclavical*, anterior deltoid
Pectoralis major (parasternal)	(2) Parasternal, medial pectoralis major
Pectoralis major (sternal)	(3) Pectoralis major, a.deltoid, *med. a. upper arm*, prox. a.forearm, *m.palm, a.f.s 3-5*
Pectoralis minor	*Pectoralis major*, a.deltoid, *med. a. arm, med. a.hand, a.f.s 3-5*
Sternalis	(3) Bil. parasternum, med. subclavicle, *a.deltoid, med. a.upper arm & elbow*
Iliocostalis thoracis (T6)	*Para. T4-T10*, med. scapula, *a.lower med. chest*
External Oblique [A]	*Center of chest*, center of stomach s.to umbilicus, *opposite upper abdominal quadrant*

Key:
TMJ—temporomandibular joint
A/C—acromioclavicular
med.—medial
brach.—brachioradialis
para.—paraspinous
MP—metacarpal phalangeal
sterno.—sternocleidomastoideus
pect.—pectoralis
a.—anterior
s.—superior
&—and
maj.—major
bord.—border
f.—finger
l.—lateral
m.—middle
d.—distal
p.—posterior
prox.—proximal
bil.—bilateral
f.s.—fingers
i.—inferior

Bibliography (S14E)

Basmajian, pp. 85, 178–183.
Cailliet: *Neck and Arm Pain*, pp. 92–98.
Chusid, pp. 294, 333–334.
Salter, pp. 236–238, 275–276.
Scott, pp. 232–236.
Shands, pp. 412–421.

CHONDROMALACIA SYNDROME
(S15)

Definition (S15A)

Chondromalacia is a pathological degenerative process which occurs to the cartilage which composes the articular surface of the patella. If allowed to run its course, chondromalacia develops in four basic phases: (1) the patellar cartilage becomes swollen and begins to soften; (2) the soft tissues associated with the patella reportedly become fissured; (3) the cartilage surface begins to break down; and (4) the subchondral bone becomes exposed and eburnation and secondary degenerative changes begin to develop in the femoral trochlea and chondyles. Most commonly, the onset of chondromalacia is proceeded by a disabling trauma to the knee which seems to recover with a complete remission of symptoms. Several months or even years later, the patient may begin to display the symptoms of chondromalacia, including variable pain, "catch" in the knee, instability and/or locking of the knees during stance phase, general weakness of the knee, swelling of the knee, atrophy of the quadriceps muscles of the thigh and palpation tenderness of the patella. Also present may be subpatellar crepitation, synovial thickening and effusion.

Chondromalacia most often occurs to young adults, with a higher incidence reported among young women.

Related Syndromes (S15B)

BURSITIS (S26)
CAPSULITIS (S27)
KNEE PAIN (S46)
OSTEOARTHRITIS (S21)
PITTING (LYMPH) EDEMA (S31)
POST IMMOBILIZATION (S36)
REFERRED PAIN (S01)
SCIATICA (S05)
TENDONITIS (S28)

Treatable Causes Which May Contribute To The Syndrome (S15C)

Calcific deposit (C08)
Extrafusal muscle spasm (C41)
Hypermobile/instable joint (C10)
Joint approximation (C47)
Ligamentous strain (C25)
Muscular weakness (C23)
Restricted joint range of motion (C26)
Soft tissue inflammation (C05)
Soft tissue swelling (C06)
Trigger point formation (C01)

Treatment Notes (S15D)

Chondromalacia is traditionally treated with rest, phonophoresis (refer to Ultrahigh Frequency Sound, Phonophoresis [M01A]) and splinting or bracing which immobilizes the knee in full extension. Isometric exercise of the quadriceps muscles has proven helpful (refer to the Table of Exercises for the Back and Lower Extremities - Isometrics [T014]). Electrical stimulation may be used to good effect to reduce knee swelling and to speedily tone the quadriceps muscle group (refer to Electrical Stimulation, Reduction of Edema [M02D] and Muscle Reeducation/Muscle Toning [M02B]).

The following is a list of trigger point formations which may, singly or in combination, contribute to or imitate the pain associated with the chondromalacia syndrome. It should be noted that it is possible for the given patient to experience pain throughout the entire stereotypical referred pain pattern or only in part or parts of it. Opposite to each listing is a description of the parts of the stereotypical referred pain pattern most commonly experienced by previous patients, relative to the given trigger point formation. Any descriptive reference which is underlined is a part of the pattern which has *not* been commonly experienced by previous patients. For complete description of each stereotypical referred pain pattern, refer to the Table of Referred Pain Patterns of Trigger Point Formation Origin [T005]. Please note the key to abbreviated words at the bottom of the list.

LOCATION (MUSCLE)	REFERRED PAIN ZONES (GENERAL)
Gluteus minimus	Gluteus maximus (excepting central area) & gluteus medius, p.l.leg, l.calf to p.of l. malleolus
Adductor longus	A. s.iliac crest into groin & down the a.med. thigh, central med. calf, med. malleolus & med. foot
Biceps femoris (hamstring)	P. d.central thigh, p.prox. calf
Vastus medialis	Vastus medialis, patella, knee joint

Gastrocnemius	P. med. calf, central thigh, behind med. malleolus, arch of the foot	m.—middle d.—distal p.—posterior i.—inferior

Key:
maj—major
med.—medial
prox.—proximal
a.—anterior
s.—superior
&—and
l.—lateral

Bibliography (S15E)

Johnson, pp. 44–45.
Salter, pp. 214–216, 223.
Scott, pp. 327, 401.
Shands, p. 383.

CHRONIC BRONCHITIS/EMPHYSEMA (OBSTRUCTIVE PULMONARY DISEASE) SYNDROME (S13)

Definition (S13A)

Diffuse obstruction of the smaller bronchi and bronchioles by mucal formation or discharge is a prominent feature of bronchitis and/or pulmonary emphysema. Symptoms include cough, wheezing, shortness of breath and disturbances of gaseous exchange.

Bronchial obstruction reduces free exchange of air in and out of the lung. If unrelieved, such obstruction eventually causes hyperinflation of the lungs to occur. Persistent hyperinflation leads to stretched and narrowed alveolar capillaries which may lead to progressive trophic changes with the loss of elastic tissue and the dissolution of alveolar walls. The rupturing of the septa between alveoli causes the formation of air sacs of varying sizes, and a slow increase in lung size occurs. Consequently, the number of capillaries in the remaining alveolar walls becomes reduced and the pulmonary arterial vessels may become sclerotic, leading to a reduction in the area of alveolar membrane available for gaseous exchange.

Over time, if bronchial obstruction is not relieved, the thoracic cage tends to assume an inspiratory expanded position and the diaphragm is mechanically forced to flatten. As the diaphragm loses its relaxed dome shape it loses its ability to effect the wide swing of intrapleural positive and negative pressures which provide tidal air flow. As this occurs, the vital capacity of the lungs is reduced and the tidal volume decreases, while residual volume (dead air) increases. To compensate for the loss of diaphragm action, the accessory muscles are used to lift the rib cage to provide negative intrapleural pressure to provide for inspiration. Dyspnea is usually present and varies from mild distress on exertion to severe dyspnea and cyanosis at rest. Coughing is usually hard, spasmodic and tiring and initiated by any exertion (even talking). Sputum produced by the cough is usually produced in small amounts and is thick and viscid.

As bronchial obstruction continues and its effects upon the lung tissues increases, repeated episodes of bronchial infection become commonplace, which steps up mucal production and further disturbs gaseous exchange, eventually affecting cardiac function. Prognosis for life becomes increasing poorer as the maximum breathing capacity decreases with a commensurate decrease in arterial oxygen saturation, carbon dioxide retention and the increased possibility of congestive heart failure and/or polycythemia.

Related Syndromes (S13B)

CERVICAL DORSAL OUTLET (SCALENICUS ATTICUS/CERVICAL RIB) (S09)
CERVICAL (NECK) PAIN (S73)
CHEST PAIN (S14)
HIGH THORACIC BACK PAIN (S48)
HYSTERIA/ANXIETY REACTION (S59)
INSOMNIA (S71)
PNEUMONIA (S58)
SHOULDER PAIN (S08)

Treatable Causes Which May Contribute To The Syndrome (S13C)

Bacterial infection (C09)
Extrafusal muscle spasm (C41)
Localized viral infection (C27)
Lung fluid accumulation (C13)
Reduced lung vital capacity (C17)
Sympathetic hyperresponse (C38)
Visceral organ Dysfunction (C02)

Treatment Notes (S13D)

Physical therapy treatment of chronic bronchitis/emphysema should be directed at improving ventilation. Mucal secretion obstruction should reduced with postural drainage (active postural drainage by the therapist or passive postural drainage by the patient) and the insistence on deliberate coughing to raise mucal secretions and mucal plug expectoration (refer to Postural Drainage [M14]).

Diaphragm activity should be increased by teaching the patient abdominal breathing (refer to Electromyometric Feedback, Breathing Pattern Correction [M18J]). The patient should also be taught to promote lung elasticity by creating positive back pressure in the lungs through forced expiration against passage resistance (with blow bottle or garden hose). The patient should be taught how to exercise and reactivate the diaphragm utilizing manual compression of the abdomen during expiration (refer to Exercise, Improving Lung Function [M12B]). Elevating the foot of the bed to an angle of 12 to 22° will help during sleep by using the weight of the abdominal viscera to push the diaphragm up and forcing the diaphragm to contract to accomplish inspiration. A properly fitted abdominal belt may also be useful.

Bibliography (S13E)

Bickley, pp. 268–269.
Guyton, p. 512.
Holvey, pp. 1286–1292.

COLITIS SYNDROME (S42)

Definition (S42A)

Colitis is generally classified as an episodic chronic disorder of the colon. The cause of this colitis is generally idiopathic but in some cases a familial tendency may be demonstrated. Colitis has been variously called spastic colitis, irritable bowel syndrome, spastic colon, nervous diarrhea and irritable colon. The most common cause of colitis appears to be the somatization of emotional stress; i.e., the patient responds to emotionally stressful situations by unconsciously producing the symptoms of diarrhea, cramping pain in the abdomen, bloating, passing of gas and abdominal distention by increasing colon motility. However, food sensitivities and the sequelae of various bacterial and viral infections have also been noted to produce the same symptomatology.

Simple colitis is most often found in young women and rarely develops after the age of fifty. Men develop this condition much less frequently than do women.

Ulcerative colitis is an acute or chronic inflammatory condition of the colon and rectum and is characterized by bloody diarrhea. There are attendant ulcerations of variable size in the involved area. Accompanying complications such as anemia, perforation of the colon, generalized infection, arthritis and skin lesions may be present. As the process develops, the colon loses normal pliability and peristalsis, taking on the characteristics of the "lead pipe colon". The exact cause of ulcerative colitis is unknown but hypersensitivity and food allergies are suspect. Ulcerative colitis is sometimes regarded as precancerous.

Related Syndromes (S42B)

ADHESION (S56)
BACTERIAL INFECTION (S63)
HYSTERIA/ANXIETY REACTION (S59)
INSOMNIA (S71)
LOWER ABDOMINAL PAIN (S62)
MENSTRUAL CRAMPING (S60)
PHOBIC REACTION (S44)
POST SPINAL CORD INJURY (S24)
REFERRED PAIN (S01)

Treatable Causes Which May Contribute To The Syndrome (S42C)

Extrafusal muscle spasm (C41)
Insomnia (C46)
Menorrhalgia (C20)
Muscular weakness (C23)
Nausea/vomiting (C36)
Psychoneurogenic neuromuscular hypertonus (C29)
Sympathetic hyperresponse (C38)
Vascular insufficiency (C11)
Visceral organ Dysfunction (C02)

Treatment Notes (S42D)

Treatment of colitis should be directed at (1) symptom relief and (2) subversion of the patient's tendency to somatize emotional stress. Temperature training to increase lower abdominal temperature has been found to be useful symptom relief of colitis (refer to Temperature Monitry, Lower Intestinal Dysfunction (Colitis) Control [M19D]). Electrical stimulation of peripheral and/or auricular acupoints has also been found to be useful for the reduction of symptoms (refer to Peripheral Electroacustimulation, Pathology Control [M05C] and/or Auricular Electroacustimulation, Pathology Control [M06B]). Galvanized skin response *desensitization* aimed at decreasing patient somatic response to emotional stimuli has also proven valuable in the treatment of colitis (refer to Galvanic Skin Response Monitry, Desensitization [M20B]).

The following is a list of trigger point formations which may, singly or in combination, imitate or contribute to the pain associated with the colitis syndrome. It should be noted that it is possible for the given patient to experience pain throughout the entire stereotypical referred pain pattern or only in part or parts of it. Opposite to each listing is a description of the parts of the stereotypical referred pain pattern most commonly experienced by previous patients, relative to the given trigger point formation. Any descriptive reference which is underlined is a part of the pattern which has *not* been commonly experienced by previous patients. For complete description of each stereotypical referred pain pattern, refer to the Table of Referred Pain Patterns of Trigger Point Formation Origin [T005]. Please note the key to abbreviated words at the bottom of the list.

LOCATION (MUSCLE)	REFERRED PAIN ZONES (GENERAL)
Multifidi (L2)	Bil. para. T12-L4, *to p.iliac crest a.upper quadrant of the abdomen*

Multifidi (S1)	Bil. para. L4-S5, med. gluteus maximus, coccyx, *prox. p.hamstrings, a lower abdominal quadrant*	Key: maj—major prox.—proximal med.—medial bil.—bilateral
Iliocostalis thoracis (T11)	Para. T11-S1, l.bottom ribs, *l.bord. of the scapula,* a.lower abdominal quadrant	para.—paraspinous a.—anterior s.—superior &—and
External Oblique [A]	*Center of chest,* center of stomach s.to umbilicus, *opposite upper abdominal quadrant*	l.—lateral bord.—border m.—middle d.—distal p.—posterior
External Oblique [B]	*Bil. a.abdomen,* groin, a.med. thigh	i.—inferior
Pyramidalis	From umbilicus to groin	
McBurney's point	A. lower abdominal quadrant from midline to the l. extreme	
Dysmenorrhea	Rectus abdominus just distal to the umbilicus	

Bibliography (S42E)

Bickley, p. 311.
Gardner, p. 126.
Guyton, pp. 779–780.
Holvey, pp. 578–587.

CRYOTHERAPY (M10)

Definition (M1001)

Cryotherapy, simply stated, is the therapeutic use of locally applied cold to benefit the patient by affecting various physiological processes through the cooling of soft tissues.

Cooling occurs when heat is removed or lost from an object (or body soft tissues) through conduction of heat from one mass to another or by evaporation. Conduction occurs when heat is transferred from a warm body to a colder body by direct interaction of their molecules; the slow moving particles of the cold body derive energy from the faster moving particles of the warm body, thereby speeding up (becoming warmer), while conversely the faster particles of the warm body are made to slow down as they lose energy (become cooler). Conduction occurs when ice (as in an ice pack) or other cool mass is applied continuously to the skin, or when a body part is immersed in a cold or cool bath; as the body or body part cools the ice or cold water heats up as the process of conduction takes place.

Cooling may be more rapidly produced through the process of evaporation than through the process of conduction. Volatile molecular particles of a coolant spray sprayed on the skin, or a film of cold water (from a light massage with an ice cube) spread over the skin, absorbs energy (heat) from the skin and are so activated by the acquired heat that the liquid film becomes gaseous, taking the acquired heat (energy) with it as it leaves the body surface.

Cryotherapy has historically been used to provide pain relief, reduce fever, alleviate pain in and to slow damage to the soft tissues produced by thermal burns, to control of bleeding by producing vasoconstriction, and to prevent or reduce edema following soft tissue trauma. Cryotherapy has proven to be additionally useful in the clinical setting for the reduction of extrafusal and/or intrafusal muscle spasm, neuromuscular hypertonicity or spasticity from upper motor neuron lesion, to elevate the pain threshold for the relief of discomfort from muscle spasm, soft tissue inflammation and/or swelling, for the facilitation of muscle contraction during the process of neuromuscular reeducation, and for the retardation of the soft tissue inflammatory process or the destructive action of enzymes, especially in joint disease.

Cryotherapy is effective as a treatment modality because of the various physiological effects it has upon soft tissue. Obviously its effects are felt as a result of the cooling of those soft tissues. The cooling effect upon the soft tissue by an ice massage or ice pack will decrease as the depth of the tissues increases, and the cooling effect will vary according to the type of tissue being cooled. Fat, for example, is an insulating material and its presence will increase (relatively speaking) the amount of time required to cool underlying muscle tissue; conversely, the speed of cooling will increase as the amount of insulation provided by the fat content in or around the muscle tissue decreases.

Tissue temperatures in response to cooling (M1001A)

A 5 minute ice massage (or equivalent 10 minute ice pack) applied to the thigh:

Tissue Depth in Centimeters :	0.5	1	2	3	4
Tissue Temperature (Centigrade) :	12.4	8.8	4.1	1.1	0.4

A 5 minute ice massage (or equivalent 10 minute ice pack) applied to the calf:

Tissue Depth in Centimeters :	0.5	1	2	3	4
Tissue Temperature (Centigrade) :	13.2	6.2	1.1	0.2	0.0

A 10 minute ice massage (or equivalent 20 minute ice pack) applied to the thigh:

Tissue Depth in Centimeters :	0.5	1	2	3	4
Tissue Temperature (Centigrade) :	12.5	11.0	5.2	1.4	0.1

Effects of cooling upon nerve conduction (M1001B)

Muscle cooling has been shown to effect the afferents arising from muscle spindle flower-spray and annulospiral sensory endings; their rate of discharge decreases as the temperature of the muscle housing them decreases. As the temperature drops, the sensitivity of the muscle spindle initially increases (a minimal stretch was required for a spindle response at 30° C.), but as the temperature decreases further, the spindle's sensitivity decreases, requiring a bigger stretch to illicit a muscle spindle response. Some evidence has suggested that cooling may affect a decrease in the rate of muscle spindle discharge by causing a membrane hyperpolarization and lowering potassium concentration. Cooling may also affect the speed of impulse of the gamma efferents to the muscle spindle, decreasing gamma biasing of the muscle spindle intrafusal muscle fibers. Both these effects have a depressing effect upon membrane activity, and thereby a depressing effect upon the stretch reflex: a significant effective temperature drop occurs in the muscle after ten minutes of ice packing.

Cooling has been shown to increase the amplitude of the H response (essentially a phasic stretch reflex induced by electrical stimulation), which largely bypasses the muscle spindle, by increasing or facilitating alpha motor neuron excitability via the cold stimulation of the exteroceptors of the skin. If cooling is sufficient to drive the muscle tissue temperature at the neuromuscular junction to 5° C., blockage of the junction occurs; as the temperature drops toward 5° C., there are decreases in the amplitude and frequencies of the motor end-plate potentials with increases in their durations.

The effects of cooling on peripheral nerves are largely dependant upon myelination and fiber diameter. In general, all nerve fibers are affected by cooling. The small medullated fibers are affected first, then the large medullated, and finally the unmedullated. The smaller gamma efferent fibers are more sensitive to cold (frequency slowing more quickly and to a greater extent) than are the alpha efferents. Motor nerve conduction velocity decreases linearly at a rate of between 1.84 and 2.4 meters per second per degree (centigrade), as the temperature drops from 36 to 23° C. Proprioception sensory fibers (and others important for motor learning) are relatively insensitive to cooling since they are large diameter myelinated fibers.

Cooling has been found to be effective at reducing abnormal electrical activity in a muscle affected by post cerebral vascular accident spastic paralysis, slowing twitch response and decreasing neuromuscular tension (tonic setting), during and after application of sufficient cold. The effects of thirty minutes of ice packing or extremity immersion in an iced bath have been shown to last for three hours or more after cessation of the cold application. The reduction of abnormally high muscle tone by application of cold has also been shown to be possible without adversely affecting the training of motor tasks, allowing sufficient muscular function, strength and endurance, but at the price of increasing the time required to complete a neuromuscular task with the involved musculature.

Effects of cooling upon vasoconstriction (M1001C)

Vasoconstriction produced by cold application is thought to be caused by reflex vasoconstriction via the sympathetic fibers found in the tissues cooled, or by direct effect upon the blood vessels. Blood flow has been shown to decrease in the tissue to 18° C., but to markedly increase at 2° C. as a "Hunting reaction" designed to maintain a minimum temperature sufficient to prevent tissue damage occurs.

Effects of cooling upon capillary permeability (M1001D)

Cryotherapy has been shown to increase capillary permeability.

Procedures of Application (M1002)

Cryotherapy is most commonly applied through the utilization of (1) ice packs (including ice bags, frozen hydrocollators and commercial ice packs), (2) coolant sprays, (3) ice massage, and (4) cold (ice) or cool immersion baths.

Ice packing (M1002A)

The most effective ice pack is one made of towelling and crushed ice:

1) A towel should be spread out on a flat surface.
2) A one inch layer of crushed ice (enough to cover the treatment site) should be spread out in the middle of the towel.
3) The long edges of the towel should be folded to overlap each other to cover the ice.
4) The ends of the towel should then be rolled up toward each other to make handling easy.
5) Before being applied to the patient the surface of the ice pack to be in contact with the patient should be moistened with water.
6) The ice pack should then be unrolled in place over the treatment site.
7) In common use, the crushed ice pack should not be left in place longer than ten to twelve minutes, enough time to cause a vasodilation sufficient to create an erythemic (bright pink) reaction of the skin and/or local anesthesia.

Sensory sequence in response to prolonged cooling (M1002B)

The patient exposed to prolonged cooling will predictably experience (and exceptions will be found) a sequence of sensations before erythemia occurs; these include aching, burning, and finally numbness. The sensations occur as the sensory nerves in the skin, and to a degree in the tissues below, become overwhelmed by the cooling, finally culminating in the numbness of temporary sensory anesthesia.

Ice bag (M1002C)

An ice bag may be created by putting ice in a plastic bag or by freezing a moistened hydrocollator (commercial chemical-ice packs or bags are also available). If any of these is to be used, a damp towel should be wrapped around the ice bag or pack to prevent any part of it from touching the patient's unprotected skin. The towel-covered bag or commercial pack should then be placed over the treatment site. Ice bags or commercial packs are not as cold as the crushed ice pack (described above), and should remain in place for twenty minutes, or until an erythemic reaction and/or sensory anesthesia of the skin is produced.

Coolant sprays (M1002D)

Coolant sprays, capable of creating temperatures low enough to be considered therapeutically effective, including the ethyl chloride and Fluori-Methane [Trade name: 85% trichloromonofluoromethane and 15% dichlorodifluoromethane] are pharmaceutically available. Either spray exerts pressure in a bottle at room temperature, which provides the propulsive force to the thin directional stream of the particular spray which the specially designed bottles emit when triggered correctly. The coolant spray immediately begins evaporating when the coolant stream impacts with the body surface.

1) The coolant spray stream should be directed at the treatment site at a 30° angle, and the spray stream length should be about 18 inches, or 45 centimeters long.

2) The stream should be swept over the treatment site in regular, fairly rapid sweeps for several seconds.

Care should be taken to be sure that no area receives more then five seconds of the spray because freezing of the skin occurs if the spray is directed on one area for six seconds or more. For safety's sake, the instructions on the bottle containing the spray should be read and suggested precautions noted and respected.

Ice massage (M1002E)

A hand-held ice cube should be swept or rubbed over the treatment site, with a small towel or paper wrapping protecting the applicator's hand, until an erythemic reaction and/or a local, temporary, sensory anesthesia of the skin occurs; this usually takes three to seven minutes. The patient will usually experience the sequence of sensory sensations noted for the ice pack, but at an accelerated rate.

A handled ice massage applicator may be made by placing a tongue depressor (held in place upright by a long hair clip, clipped to the tongue depressor and placed across the mouth of the cup) in a small paper cup of water and freezing it. The cup should be removed when the water is frozen, before using it for treatment. Commercial home kitchen popsicle makers (sans sugar in the water) may also be used to produce a handled ice massage applicator.

Ice bath (M1002F)

An ice bath is created by filling a basin with water and enough ice to cool the water from 55 to 60° F. (12.8 to 15.6° C.); the basin should be large enough to contain the portion of the patient's body requiring treatment and the necessary of volume of water and ice. The body part to be treated should be immersed in the basin approximately ten minutes or until anesthesia of the body part is produced.

The patient can be expected to go through the sequence of sensory sensations previously described.

After the cryotherapeutic application chosen, the treated site and other dampened surfaces should be thoroughly dried and the treated area covered with dry garments. Additional therapeutic procedures should be performed at this point.

Partial List of Conditions Treated (M1003)

Conditions Treated in the Field (M1003A)

Acute muscle spasm
Acute muscle pain
Contusions
Hemorrhage
Sprains
Strains
Topical anesthesia prior to joint mobilization
Hypertonicity
Nervous agitation (to produce sedation)

Treatable Causes (M1003B)

Calcific deposit   (C08)
Extrafusal muscle spasm   (C41)
Habitual joint positioning   (C48)
Hypermobile/instable joint   (C10)

Hypertension   (C39)
Interspinous ligamentous strain   (C03)
Joint ankylosis   (C42)
Joint approximation   (C47)
Ligamentous strain   (C25)
Migraine (vascular headache)   (C34)
Muscular weakness   (C23)
Nausea/vomiting   (C36)
Nerve root compression   (C04)
Neuromuscular tonic imbalance   (C30)
Pathological neuromuscular hypertonus   (C28)
Pathological neuromuscular hypotonus   (C33)
Pathological neuromuscular dyscoordination   (C35)
Peripheral nerve injury   (C07)
Psychoneurogenic neuromuscular hypertonus   (C29)
Restricted joint range of motion   (C26)
Soft tissue inflammation   (C05)
Soft tissue swelling   (C06)
Sympathetic hyperresponse   (C38)
Tactile hypersensitivity   (C24)
Trigger point formation   (C01)
Vascular insufficiency   (C11)
Visceral organ Dysfunction   (C02)

Syndromes   (M1003C)

ACHILLES TENDONITIS   (S52)
BURNS (SECOND AND/OR THIRD DEGREE)   (S72)
BURSITIS   (S26)
CALF PAIN   (S50)
CAPSULITIS   (S27)
CARPAL TUNNEL   (S35)
CEREBRAL PALSY   (S25)
CERVICAL DORSAL OUTLET (SCALENICUS ATTICUS/CERVICAL RIB)   (S09)
CERVICAL (NECK) PAIN   (S73)
CHEST PAIN   (S14)
CHONDROMALACIA   (S15)
COLITIS   (S42)
DIABETES   (S10)
EARACHE   (S40)
ELBOW PAIN   (S47)
FACET   (S19)
FASCIITIS   (S20)
FOOT PAIN   (S51)
FOREARM PAIN   (S54)
FROZEN SHOULDER   (S64)
HAND/FINGER PAIN   (S17)
HEADACHE PAIN   (S02)
HIGH THORACIC BACK PAIN   (S48)
HYPERTENSION   (S43)
HYSTERIA/ANXIETY REACTION   (S59)
JOINT SPRAIN   (S30)
KNEE PAIN   (S46)
LOWER ABDOMINAL PAIN   (S62)
LOW BACK PAIN   (S03)

MENSTRUAL CRAMPING   (S60)
MIGRAINE (VASCULAR) HEADACHE   (S18)
MYOSITIS OSSIFICANS   (S67)
NEUROMUSCULAR PARALYSIS   (S22)
OSTEOARTHRITIS   (S21)
PIRIFORMIS   (S04)
PITTING (LYMPH) EDEMA   (S31)
POST CEREBRAL VASCULAR ACCIDENT (CVA)   (S07)
POST IMMOBILIZATION   (S36)
POST PERIPHERAL NERVE INJURY   (S23)
POST WHIPLASH   (S55)
RADICULITIS   (S29)
REFERRED PAIN   (S01)
SCIATICA   (S05)
SHOULDER/HAND   (S57)
SHOULDER PAIN   (S08)
SINUS PAIN   (S33)
TEMPOROMANDIBULAR JOINT (TMJ) PAIN   (S06)
TENDONITIS   (S28)
TENNIS ELBOW (LATERAL EPICONDYLITIS)   (S32)
THIGH PAIN   (S49)
TINNITUS   (S70)
TOOTHACHE/JAW PAIN   (S41)
WRIST PAIN   (S16)

Precautions and Contraindications   (M1004)

Adverse effects from cryotherapy are rare, but certain individuals have been shown to be systemically hypersensitive to cold applications (seemingly "allergic" to cold). These individuals may be divided up into three distinct groups, those whose tissues, upon cooling, release (1) histamine or histamine-like chemicals, (2) cold hemolysins and agglutinins, and (3) cryoglobulins.

The symptoms associated with histamine production triggered by cooling include cold urticaria, erythema, itching, sweating, facial flush, puffiness of the eyelids, laryngeal edema, and shortness of breath. More severe symptoms which may develop include syncope, hypotension, tachycardia, extra systoles, dysphagia, abdominal cramping, diarrhea, and/or vomiting. The resultant symptomatology may occur during therapy or several minutes or even hours after the technique has been applied. Such hypersensitivity to cold may be treatable through a program of desensitization to cold: the patient's hands should be immersed in water cooled to 50° F. for two minutes, twice daily for from 14 to 21 days.

Cold hemolysins and agglutinins produced in abnormal amounts in response to tissue cooling may produce significant anemia, malaise, chills, fever, paroxysmal cold hemoglobinuria of the kidney, and/or the dermal symptoms of cold urticaria, ulcerations, Raynaud's phenomenon, and acrocyanosis.

Abnormally high cryoglobulin production in response to cooling may produce chills and fever, and may adversely affect vision, causing blindness in the extreme, and hearing, causing deafness in the extreme. It may cause conjunctival hemorrhages and epistaxis, anemia with fibrinogenopenia, elevated sedimentation rates, and abnormal numbers of cryoglobulin inclusion cells. Other symptoms include erythema, itching, purpura, cold urticaria, Raynaud's phenomenon, ulcerations and necrosis, problems breathing (dyspnea), stomatitis, melena, and bleeding of the gums.

The adverse reactions to cryotherapy described above have been noted to occur in association with various systemic diseases including Raynaud's disease, Buerger's disease, lupus erythematosus, leucocyto-clastic vasculitis, rheumatoid disease, progressive systemic sclerosis, multiple myeloma, and pneumonia; consequently, cryotherapy involving immersion or the ice packing of the distal extremities of patient's suffering from these diseases may be contraindicated. Any cryotherapy applied to patients of this type must be supervised carefully, and should avoided if at all possible.

If arterial insufficiency is present (as is the case in some systemic diseases), cold applications would seem to be contraindicated because of their potent vasoconstricting effect which may potentially reduce already compromised blood flow through the extremity.

Some individuals, though not suffering from any particular disease process, respond to the natural vasopressor response to application of cold (capillary vasoconstriction) with failure to eventually compensate by capillary vasodilation, thus producing a Buerger's disease-like syndrome which may threaten the tissues with rapidly developing frostbite if exposure to the cold is not eliminated. This condition may cause the patient exquisite pain which may not be readily relieved. Such hyperreactivity to cooling may be readily tested for by observing the tissue response to ice packing. The normal dermal response to an applied crushed ice pack is to blanch (vasoconstrict) for the first three minutes after the ice pack is applied; the skin gradually develops an increasing erythemic reaction, becoming bright red after ten minutes of constant exposure. If the blanching phase does not end after the first three minutes of exposure to the ice pack, the patient may be considered hyperresponsive to exposure to cold (even without other symptoms) and the treatment should be immediately discontinued.

Though rare, frostbite is a potential danger inherent in cryotherapy, especially for techniques involving ice massage or the coolant sprays although reports suggest that such damage is rarely severe. Frostbite occurs as a consequence of over-exposure to cold which causes an increase in the capillary permeability. The consequent rapid loss of plasma strands red blood cells in the capillaries, which then clump together to form occlusive masses (clots) which finally undergo hyalinization. Treatment of frostbite involves the immediate immersion in warm water (109° F.). Soft tissue damage from frostbite is increased (1) as the length of time the tissues are frozen increases, (2) as the temperature of the tissues while in the frozen state decreases, and (3) as the duration of thawing increases.

Ice packing applied to the abdomen increases peristalsis in the stomach, small intestine and the colon and has been found to increase blood flow through the mucous membrane of the alimentary canal and to increase acid secretion in the stomach. Stomach cramps may, therefore, be aggravated and gastrointestinal upset be caused by ice packing of the anterior lower abdominal area; increased gastric acidity is undesirable in the presence of pathology of the stomach, particularly if peptic ulcers are present.

Extreme care must be exercised in the application of cryotherapy techniques to treatment sites which may suffer from disturbed skin sensations or sensory impairment. In such cases the patient's skin must be carefully tested for abnormal sensitivity to cold and normal reactions to its application (refer to Cryotherapy, Evaluation of Capillary Response to Cold [M10H]). Additionally, cryotherapy is considered by many to be contraindicated for treatment sites with open wounds. The judgement regarding cryotherapeutic applications to such sites should be based on the extent, nature and depth of the wound involved, and the chance of contaminating the wound. Advantages must be weighed against the risks. For example, there is little risk and a great deal of benefit to be gained by cryotherapeutic application to a freshly contused area, in spite of the presence of a shallow wound. It may be helpful in slowing or stopping bleeding if a dry sterilized ice bag is utilized or the treatment site protected from touch by the water in the ice bath (plastic bagging may be helpful); however, cryotherapeutic applications may do further damage to involved tissues if the wound is deep and/or the vasocirculation in the area is impaired.

Special note: Ethyl chloride spray is considered dangerous by some, by virtue of the fact that the extreme cold it produces may be too extreme. Constant diligence must be exercised to avoid frostbite. Ethyl chloride spray also poses a danger to both the patient and the practitioner because of its effectiveness as a general anesthetic; care must be taken that the gas created by evaporation is not inhaled by either party. Lastly, the gaseous form of ethyl chloride is extremely flammable and the danger of explosion exists if it is used in a poorly ventilated enclosure without precautions being taken to avoid sparks or open flames. In contrast, Fluori-Methane spray does not cool as fast or to the extent that ethyl chloride spray does, so frost bite is less likely to occur if the treatment site is not exposed to more than five seconds of spray time. Additionally, Fluori-Methane spray does not have a general anesthetic effect when inhaled, and it is nonflammable.

Notes Aside   (M1005)

Ice packing   (M1005A)

Ice packing has proven to be of value in the treatment of many ailments; it has, in the some clinical settings, supplanted the traditional use of heat or heat packing for the treatment of almost all musculoskeletal problems such as strains, spasms, trigger points and myositis. It most often facilitates return to normal function by reducing pain and the intensity of muscle spasm, without promoting the increased metabolism and metabolic by-product build up which is a normal consequence of the use of local heat applications like diathermy, heat lamps, or heat packing.

Cryotherapy has shown itself to be extremely valuable in the clinical setting when applied eclectically. Ice packing, for example, has shown itself to be of marked value when used just before the phonophoresis of antiinflammatories or other medications, since it increases soft tissue (capillary) permeability (refer to Ultrahigh Frequency Sound, Phonophoresis [M01A]). Ice packing is of inestimable value in the treatment of neuromuscular hypertonus and/or extrafusal and intrafusal muscle spasms accompanying acute muscle strain in conjunction with electrical stimulation (refer to Electrical Stimulation, Muscle Lengthening [M02A]), vibration (refer to Vibration, Neuromuscular Management [M08A]), and sometimes traction (refer to Vertebral Traction [M09]), especially for the treatment of postwhiplash injuries.

Contrast packs or baths   (M1005B)

Ice may be used together with heat in the form of *contrast packs* or *contrast baths* to increase capillary bed circulation and to reduce joint stiffness. If immersion baths are utilized, the hot bath should be 104° F. (40° C.) and the cold bath should be 59° F. (15° C.). The hot application by hot pack or immersion bath should last for ten minutes followed by one minute of cold exposure by ice pack or immersion bath. The heat should then be applied for four minutes followed by a one-minute cold exposure; this process should be repeated for a total treatment time of thirty minutes.

Cryotherapeutic effect upon cancerous growths   (M1005C)

Cancer tissue has been sited by some authorities to be less tolerant of cryotherapy then normal tissue.

Cryotherapeutic effect upon joint tissues   (M1005D)

Research has demonstrated that the cooling of a joint causes vasoconstriction and a decrease in interarticular temperature to occur. This has the effect of slowing enzyme action, thereby slowing destructive action. It has also been demonstrated that cryotherapy applied over the area of an overstretched ligament may increase the edematous swelling in the general area but decreased edema in the injured ligament.

Decreasing autonomic hyperactivity (sedation)   (M1005E)

Ice packing or ice massage of the cervical vertebrae (C7) paraspinous area has proven to be very effective for the arrest of anxiety attacks, hysteria, or the immediate temporary reduction of hypertension (refer to Cryotherapy, Suppression of Sympathetic Autonomic Hyperreactivity [M10A]).

Bibliography   (M1006)

Grant, pp. 233–238.
Haldovich, pp. 185–190.
Johnson, pp. 1238–1242.
Kowal, pp. 66–73.
Lehmann, pp. 563–602.
Michlovitz, pp. 73–97, 263–275, 277–294.
Murphy, pp. 112–115.
Newton, pp. 1034–1036.
Simpson, pp. 270–272.
Travell, pp. 65–70.

CRYOTHERAPY SUPPRESSION OF SYMPATHETIC AUTONOMIC HYPERACTIVITY
(M10A)

Definition (M10A1)

Sympathetic hyperactivity most often results from an autonomic sympathetic over response to stress or anxiety, usually associated with a direct, subjective, psychologically threatening experience. It is accompanied or preceded by patient subjective reports of feelings of apprehension, tension, fatigue (often sudden) and/or panic. Sympathetic hyperactivity is characterized by somatic symptomatology which may include sweating, palpitation, nausea, vomiting, diarrhea, urinary urgency, stomach cramps, sudden migraine headache, sudden pain in various body areas (from muscle imbalance or trigger point formation referred pain), acute asthmatic episodes, colitis, peptic ulceration, enteritis and/or heart disease (especially coronary artery disease).

The afferent autonomic nervous system communicates with the central nervous system via the dorsal root ganglia. Some of the most important of the dorsal root ganglia which serve as communication links between the two lie within the sixth cervical (C7) and third thoracic (T3) paravertebral areas. Symptomatology resulting from sympathetic hyperactivity may be affected (suppressed) by cryotherapy application above the paraspinous areas of C6 to T3 (bilaterally). Cooling the preganglionic autonomic nerve (B fibers) slows the autonomic nerve impulses conducted by small myelinated fibers from the sensory bodies of the various viscera, changing the central nervous system perception of them. The relationship between the efferent and afferent autonomic nerves is unclear, but the change of sensory perception afforded by afferent nerve cooling may reflexly (centrally or peripherally) effect efferent autonomic activity to the viscera, thereby affecting a visceral relaxation and/or dilation. Or these affects may come as a direct effect of cooling of both the afferent and efferent autonomic nerves; nausea, stomach cramps, migraine headache, acute asthmatic and colitis episodes may sometimes be almost immediately relieved by such cryotherapeutic application.

Procedures of Application (M10A2)

Cryotherapeutic devices (M10A2A)

Ice packing (M10A2A01) To create an ice pack, a towel should be spread out on a flat surface, and a one inch layer of crushed ice (enough to cover the treatment site) spread out in the middle of it. The long edges of the towel should be folded to overlap each other and to cover the ice. The ends may then be rolled up toward each other for easy handling. Before being applied to the patient the "bottom" of the ice pack should be moistened; the ice pack should then be unrolled in place over the treatment site. The patient will predictably experience (and exceptions will be found) a sequence of sensations before erythemia occurs, which include aching, burning and numbness. In practice, the ice pack is generally applied for a period of 10 to 15 minutes.

Ice bag (M10A2A02) Ice bags are not recommended for clinical use because of the extreme length of time it takes for them to be effective; they may take 30 to 60 minutes to be effective because they just simply are not cold enough.

Coolant sprays (M10A2A03) Coolant sprays are not used to suppress sympathetic autonomic hyperactivity because they cool only the skin and not the tissues beneath.

Ice massage (M10A2A04) An ice massage may be substituted for ice packing. The patient should be comfortably seated upright with the neck and shoulders bare. The skin above and around the C6 to T3 vertebrae should be lightly rubbed with a hand held ice cube (a small towel or paper wrapping of one end may be used to protect the practitioner's hand) or ice applicator. The treatment should continue until an erythemic reaction and a local temporary sensory anesthesia of the skin occurs; this usually takes several minutes (5 to 7) and the patient will usually experience the sequence of sensory experience noted for the ice pack, but at an accelerated rate.

A handled ice massage applicator may made by placing a tongue depressor (held in place upright by a long hair clip, clipped to the tongue depressor and placed across the mouth of the cup) in a small paper cup of water and freezing it. The cup should be removed when the water is frozen, before using it for treatment. Commercial home kitchen popsicle makers (sans sugar in the water) may also be used to produce a handled ice massage applicator.

Treatment protocol (M10A2B)

1) The patient should ideally be placed in a semireclined comfortable position (a reclining chair may be suitable).

2) The cryotherapeutic device should be applied over the spinous and bilateral paraspinous areas of C6 to T3. If ice packing is used, it should be a small pack measuring roughly 12 by 16 cm.

3) The cryotherapeutic device should be applied for an appropriate period. An ice pack should be left in place for at least 10 but no longer then 20 minutes (12 minutes is usually adequate).

4) Following application, the patient's skin should be thoroughly dried off.

5) The patient should leave the treatment area in warm dry clothing.

Partial List of Conditions Treated (M10A3)

Treatable Causes (M10A3A)

Blood sugar irregularities (C31)
Extrafusal muscle spasm (C41)
Hyperhidrosis (C37)
Hypertension (C39)
Insomnia (C46)
Migraine (vascular headache) (C34)
Muscular weakness (C23)
Nausea/vomiting (C36)
Neurogenic dermal disorders
 (psoriasis, hives) (C49)
Neuromuscular tonic imbalance (C30)
Psychoneurogenic neuromuscular
 hypertonus (C29)
Sympathetic hyperresponse (C38)
Trigger point formation (C01)
Visceral organ Dysfunction (C02)

Syndromes (M10A3B)

CALF PAIN (S50)
CERVICAL DORSAL OUTLET (SCALENICUS
 ATTICUS/CERVICAL RIB) (S09)
CERVICAL (NECK) PAIN (S73)
CHEST PAIN (S14)
COLITIS (S42)
EARACHE (S40)
ELBOW PAIN (S47)
FOOT PAIN (S51)
FOREARM PAIN (S54)
HAND/FINGER PAIN (S17)
HEADACHE PAIN (S02)
HIGH THORACIC BACK PAIN (S48)
HYPERTENSION (S43)
HYSTERIA/ANXIETY REACTION (S59)
INSOMNIA (S71)
KNEE PAIN (S46)
LOWER ABDOMINAL PAIN (S62)
LOW BACK PAIN (S03)
MIGRAINE (VASCULAR) HEADACHE (S18)
PHOBIC REACTION (S44)
POST WHIPLASH (S55)
RADICULITIS (S29)
REFERRED PAIN (S01)
SCIATICA (S05)
SHOULDER/HAND (S57)
SHOULDER PAIN (S08)
SINUS PAIN (S33)
TEMPOROMANDIBULAR JOINT (TMJ)
 PAIN (S06)
THIGH PAIN (S49)
TINNITUS (S70)
TOOTHACHE/JAW PAIN (S41)
WRIST PAIN (S16)

Precautions and Contraindications (M10A4)

Adverse effects from cryotherapy are rare, but certain individuals have been shown to be systemically hypersensitive to cold applications (seemingly "allergic" to cold). These individuals may be divided up into three distinct groups, those whose tissues, upon cooling, release (1) histamine or histamine-like chemicals, (2) cold hemolysins and agglutinins, and (3) cryoglobulins.

The symptoms associated with histamine production triggered by cooling include cold urticaria, erythema, itching, sweating, facial flush, puffiness of the eyelids, laryngeal edema, and shortness of breath. More severe symptoms which may develop include syncope, hypotension, tachycardia, extra systoles, dysphagia, abdominal cramping, diarrhea, and/or vomiting. The resultant symptomatology may occur during therapy or several minutes or even hours after the technique has been applied. Such hypersensitivity to cold may be treatable through a program of desensitization to cold: the patient's hands should be immersed in water cooled to 50° F. for two minutes, twice daily for from 14 to 21 days.

Cold hemolysins and agglutinins produced in abnormal amounts in response to tissue cooling may produce significant anemia, malaise, chills, fever, paroxysmal cold hemoglobinuria of the kidney, and/or the dermal symptoms of cold urticaria, ulcerations, Raynaud's phenomenon, and acrocyanosis.

Abnormally high cryoglobulin production in response to cooling may produce chills and fever, and may adversely affect vision, causing blindness in the extreme, and hearing, causing deafness in the extreme. It may cause conjunctival hemorrhages and epistaxis, anemia with fibrinogenopenia, elevated sedimentation rates, and abnormal numbers of cryoglobulin inclusion cells. Other symptoms include erythema, itching, purpura, cold urticaria, Raynaud's phenomenon, ulcerations and necrosis, problems breathing (dyspnea), stomatitis, melena, and bleeding of the gums.

The adverse reactions to cryotherapy described above have been noted to occur in association with various systemic diseases including Raynaud's disease, Buerger's disease, lupus erythematosus, leucocyto-clastic vasculitis, rheumatoid disease, progressive systemic sclerosis, multiple myeloma, and pneumonia; consequently, cryotherapy involving immersion or the ice packing of the distal extremities of patient's suffering from these diseases may be contraindicated. Any cryotherapy applied to patients of this type must be supervised carefully, and should avoided if at all possible.

If arterial insufficiency is present (as is the case in some systemic diseases), cold applications would seem to be contraindicated because of their potent vasoconstricting effect which may potentially reduce already compromised blood flow through the extremity.

Some individuals, though not suffering from any particular disease process, respond to the natural vasopressor response to application of cold (capillary vasoconstriction) with failure to eventually compensate by capillary vasodilation, thus producing a Buerger's disease-like syndrome which may threaten the tissues with rapidly developing frostbite if exposure to the cold is not eliminated. This condition may cause the patient exquisite pain which may not be readily relieved. Such hyperreactivity to cooling may be readily tested for by observing the tissue response to ice packing. The normal dermal response to an applied crushed ice pack is to blanch (vasoconstrict) for the first three minutes after the ice pack is applied; the skin gradually develops an increasing erythemic reaction, becoming bright red after ten minutes of constant exposure. If the blanching phase does not end after the first three minutes of exposure to the ice pack, the patient may be considered hyperresponsive to exposure to cold (even without other symptoms) and the treatment should be immediately discontinued.

Though rare, frostbite is a potential danger inherent in cryotherapy, especially for techniques involving ice massage or the coolant sprays although reports suggest that such damage is rarely severe. Frostbite occurs as a consequence of over-exposure to cold which causes an increase in the capillary permeability. The consequent rapid loss of plasma strands red blood cells in the capillaries, which then clump together to form occlusive masses (clots) which finally undergo hyalinization. Treatment of frostbite involves the immediate immersion in warm water (109° F.). Soft tissue damage from frostbite is increased (1) as the length of time the tissues are frozen increases, (2) as the temperature of the tissues while in the frozen state decreases, and (3) as the duration of thawing increases.

Ice packing applied to the abdomen increases peristalsis in the stomach, small intestine and the colon and has been found to increase blood flow through the mucous membrane of the alimentary canal and to increase acid secretion in the stomach. Stomach cramps may, therefore, be aggravated and gastrointestinal upset be caused by ice packing of the anterior lower abdominal area; increased gastric acidity is undesirable in the presence of pathology of the stomach, particularly if peptic ulcers are present.

Extreme care must be exercised in the application of cryotherapy techniques to treatmentsites which may suffer from disturbed skin sensations or sensory impairment. In such cases the patient's skin must be carefully tested for abnormal sensitivity to cold and normal reactions to itsapplication (refer to Cryotherapy, Evaluation of Capillary Response to Cold [M10H]). Additionally, cryotherapy is considered by many to be contraindicated for treatment sites with open wounds. The judgement regarding cryotherapeutic applications to such sites should be based on the extent, nature and depth of the wound involved, and the chance of contaminating the wound. Advantages must be weighed against the risks. For example, there is little risk and a great deal of benefit to be gained by cryotherapeutic application to a freshly contused area, in spite of the presence of a shallow wound. It may be helpful in slowing or stopping bleeding if a dry sterilized ice bag is utilized or the treatment site protected from touch by the water in the ice bath (plastic bagging may be helpful); however, cryotherapeutic applications may do further damage to involved tissues if the wound is deep and/or the vasocirculation in the area is impaired.

Treatment Notes (M10A5)

If an ice pack is used to suppress sympathetic autonomic hyperactivity, symptom relief usually begins at the end of the normal "blanching" period when erythema begins, after the ice pack has been in place for three minutes. At the end of twelve minutes the symptoms should be completely (albeit temporarily) relieved. Relief may last for several hours or be relatively permanent.

If ice massage is utilized, relief of the symptoms should begin to be felt within one or two minutes after commencement of application, and complete relief should be experienced after five minutes.

Bibliography (M10A6)

Brodal, pp. 704–713.
Chusid, pp. 100–101.
Goss, pp. 1009–1042.
Lehmann, pp. 571–573.
Johnson, pp. 1238–1242

CRYOTHERAPY HYPERTONIC NEUROMUSCULAR MANAGEMENT (M10B)

Definition (M10B1)

Hypertonic neuromuscular management is defined here as the reduction of neuromuscular hypertonicity resulting from extrafusal muscle spasm, intrafusal muscle spasm (the most common cause of the trigger point formation), and/or habitually "short" or tight muscles. Hypertonicity which results from upper motor neuron lesion or from peripheral nerve injury in part or all of the muscle is not included.

Cryotherapy may be used to reduce extrafusal or intrafusal muscle spasm and/or abnormally shortened muscles by cooling exteroceptors by inhibiting flower spray and annulospiral afferent activity in the skin. This has the effect of desensitizing the muscle spindle to stretch and effectively suppresses the phasic stretch reflex. Continued cooling with ice massage or ice packing further suppresses the phasic stretch reflex by decreasing the frequency and transmission speed of efferent gamma nerve impulse. Possibly the tonic stretch reflex is also inhibited by further cooling as continuing proprioception from the muscle spindle to the central nervous system is effected. After the stretch reflexes have been sufficiently suppressed, the cooled musculature becomes insensitive to passive stretch and the entire muscle, including the extrafusal and/or intrafusal muscle fiber bundles which may be in spasm, may be manually lengthened. This effectively provides for muscle tension and/or spasm relief by causing the involved muscle spindles to be reset at greater lengths and, as the direct effects of the cooling are lost with warming, the entire muscle may be allowed to remain lengthened. The direct effects of intense cooling of the muscle may last for several hours after cessation of the cold application, but the indirect effects may be more permanent if not over-ridden by central nervous system intervention.

Procedures of Application (M10B2)

Cryotherapeutic devices (M10B2A)

Ice packing (M10B2A01) To create an ice pack, a towel should be spread out on a flat surface, and a one inch layer of crushed ice (enough to cover the treatment site) spread out in the middle of it. The long edges of the towel should be folded to overlap each other and to cover the ice. The ends may then be rolled up toward each other for easy handling. Before being applied to the patient the "bottom" of the ice pack should be moistened; the ice pack should then be unrolled in place over the treatment site. The patient will predictably experience (and exceptions will be found) a sequence of sensations before erythemia occurs, which include aching, burning and numbness.

Ice bag (M10B2A02) Ice bags are not recommended for clinical use because of the extreme length of time it takes for them to be effective; they may take 30 to 60 minutes to be effective because they just simply are not cold enough.

Coolant sprays (M10B2A03) Coolant sprays may be used to relieve intrafusal muscle spasm (trigger point formation), via a sensory reflex relationship between the sensory organs in the skin and the involved muscle spindle. Such sprays have proven (for the most part) to be ineffective for the relief of extrafusal muscle spasms because they cool only the skin and not the tissues beneath.

Ice massage (M10B2A04) An ice massage may be substituted for ice packing. Generally, the ice massage may be applied via a hand held ice cube (a small towel or paper wrapping of one end may be used to protect the practitioner's hand) or ice applicator. The treatment should continue until an erythemic reaction and a local temporary sensory anesthesia of the skin occurs; this usually takes several minutes (5 to 7) and the patient will usually experience the sequence of sensory experience noted for the ice pack, but at an accelerated rate.

A handled ice massage applicator may made by placing a tongue depressor (held in place upright by a long hair clip, clipped to the tongue depressor and placed across the mouth of the cup) in a small paper cup of water and freezing it. The cup should be removed when the water is frozen, before using it for treatment. Commercial home kitchen popsicle makers (sans sugar in the water) may also be used to produce a handled ice massage applicator.

Treatment protocol (M10B2B)

1) The patient should ideally be placed in a semireclined comfortable position (a reclining chair may be suitable) which allows easy access to the dermal area covering the spasm.

2) The muscle housing the spasm (intrafusal or extrafusal) should be put on stretch, and the stretch maintained for the duration of the treatment.

3) The cryotherapeutic device selected should be applied to the dermis overlying the spasm for the appropriate amount of time (10 to 12 minutes is ice packing is used).

4) Following application, the patient's skin should be thoroughly dried off.

5) The patient should leave the treatment area in warm dry clothing.

Partial List of Conditions Treated (M10B3)

Treatable Causes (M10B3A)

Calcific deposit (C08)
Extrafusal muscle spasm (C41)
Habitual joint positioning (C48)
Interspinous ligamentous strain (C03)
Joint ankylosis (C42)
Joint approximation (C47)
Ligamentous strain (C25)
Nerve root compression (C04)
Neuromuscular tonic imbalance (C30)
Psychoneurogenic neuromuscular hypertonus (C29)
Restricted joint range of motion (C26)
Tinnitus (C32)
Trigger point formation (C01)

Syndromes (M10B3B)

ACHILLES TENDONITIS (S52)
CALF PAIN (S50)
CERVICAL DORSAL OUTLET (SCALENICUS ATTICUS/CERVICAL RIB) (S09)
CERVICAL (NECK) PAIN (S73)
CHEST PAIN (S14)
EARACHE (S40)
ELBOW PAIN (S47)
FACET (S19)
FASCIITIS (S20)
FOOT PAIN (S51)
FOREARM PAIN (S54)
FROZEN SHOULDER (S64)
HAND/FINGER PAIN (S17)
HEADACHE PAIN (S02)
HIGH THORACIC BACK PAIN (S48)
KNEE PAIN (S46)
LOWER ABDOMINAL PAIN (S62)
LOW BACK PAIN (S03)
MENSTRUAL CRAMPING (S60)
MIGRAINE (VASCULAR) HEADACHE (S18)
MYOSITIS OSSIFICANS (S67)
OSTEOARTHRITIS (S21)
PIRIFORMIS (S04)
POST IMMOBILIZATION (S36)
POST PERIPHERAL NERVE INJURY (S23)
POST WHIPLASH (S55)
RADICULITIS (S29)
REFERRED PAIN (S01)
SCIATICA (S05)
SHOULDER/HAND (S57)
SHOULDER PAIN (S08)
SINUS PAIN (S33)
TEMPOROMANDIBULAR JOINT (TMJ) PAIN (S06)
TENDONITIS (S28)
TENNIS ELBOW (LATERAL EPICONDYLITIS) (S32)
THIGH PAIN (S49)
TINNITUS (S70)
TOOTHACHE/JAW PAIN (S41)
TORTICOLLIS (WRY NECK) (S65)
WRIST PAIN (S16)

Precautions and Contraindications (M10B4)

Adverse effects from cryotherapy are rare, but certain individuals have been shown to be systemically hypersensitive to cold applications (seemingly "allergic" to cold). These individuals may be divided up into three distinct groups, those whose tissues, upon cooling, release (1) histamine or histamine-like chemicals, (2) cold hemolysins and agglutinins, and (3) cryoglobulins.

The symptoms associated with histamine production triggered by cooling include cold urticaria, erythema, itching, sweating, facial flush, puffiness of the eyelids, laryngeal edema, and shortness of breath. More severe symptoms which may develop include syncope, hypotension, tachycardia, extra systoles, dysphagia, abdominal cramping, diarrhea, and/or vomiting. The resultant symptomatology may occur during therapy or several minutes or even hours after the technique has been applied. Such hypersensitivity to cold may be treatable through a program of desensitization to cold: the patient's hands should be immersed in water cooled to 50° F. for two minutes, twice daily for from 14 to 21 days.

Cold hemolysins and agglutinins produced in abnormal amounts in response to tissue cooling may produce significant anemia, malaise, chills, fever, paroxysmal cold hemoglobinuria of the kidney, and/or the dermal symptoms of cold urticaria, ulcerations, Raynaud's phenomenon, and acrocyanosis.

Abnormally high cryoglobulin production in response to cooling may produce chills and fever, and may adversely affect vision, causing blindness in the extreme, and hearing, causing deafness in the extreme. It may cause conjunctival hemorrhages and epistaxis, anemia with fibrinogenopenia, elevated sedimentation rates, and abnormal numbers of cryoglobulin inclusion cells. Other symptoms include erythema, itching, purpura, cold urticaria, Raynaud's phenomenon, ulcerations and necrosis, problems breathing (dyspnea), stomatitis, melena, and bleeding of the gums.

The adverse reactions to cryotherapy described above have been noted to occur in association with various systemic diseases including Raynaud's disease, Buerger's disease, lupus erythematosus, leucocyto-clastic vasculitis, rheumatoid disease, progressive systemic sclerosis, multiple myeloma, and pneumonia; consequently, cryotherapy involving immersion or the ice packing of the distal extremities of patient's suffering from these diseases may be contraindicated. Any cryotherapy applied to patients of this type must be supervised carefully, and should avoided if at all possible.

If arterial insufficiency is present (as is the case in some systemic diseases), cold applications would seem to be contraindicated because of their potent vasoconstricting effect which may potentially reduce already compromised blood flow through the extremity.

Some individuals, though not suffering from any particular disease process, respond to the natural vasopressor response to application of cold (capillary vasoconstriction) with failure to eventually compensate by capillary vasodilation, thus producing a Buerger's disease-like syndrome which may threaten the tissues with rapidly developing frostbite if exposure to the cold is not eliminated. This condition may cause the patient exquisite pain which may not be readily relieved. Such hyperreactivity to cooling may be readily tested for by observing the tissue response to ice packing. The normal dermal response to an applied crushed ice pack is to blanch (vasoconstrict) for the first three minutes after the ice pack is applied; the skin gradually develops an increasing erythemic reaction, becoming bright red after ten minutes of constant exposure. If the blanching phase does not end after the first three minutes of exposure to the ice pack, the patient may be considered hyperresponsive to exposure to cold (even without other symptoms) and the treatment should be immediately discontinued.

Though rare, frostbite is a potential danger inherent in cryotherapy, especially for techniques involving ice massage or the coolant sprays although reports suggest that such damage is rarely severe. Frostbite occurs as a consequence of over-exposure to cold which causes an increase in the capillary permeability. The consequent rapid loss of plasma strands red blood cells in the capillaries, which then clump together to form occlusive masses (clots) which finally undergo hyalinization. Treatment of frostbite involves the immediate immersion in warm water (109° F.). Soft tissue damage from frostbite is increased (1) as the length of time the tissues are frozen increases, (2) as the temperature of the tissues while in the frozen state decreases, and (3) as the duration of thawing increases.

Ice packing applied to the abdomen increases peristalsis in the stomach, small intestine and the colon and has been found to increase blood flow through the mucous membrane of the alimentary canal and to increase acid secretion in the stomach. Stomach cramps may, therefore, be aggravated and gastrointestinal upset be caused by ice packing of the anterior lower abdominal area; increased gastric acidity is undesirable in the presence of pathology of the stomach, particularly if peptic ulcers are present.

Extreme care must be exercised in the application of cryotherapy techniques to treatment sites which may suffer from disturbed skin sensations or sensory impairment. In such cases the patient's skin must be carefully tested for abnormal sensitivity to cold and normal reactions to its application (refer to Cryotherapy, Evaluation of Capillary Response to Cold [M10H]). Additionally, cryotherapy is considered by many to be contraindicated for treatment sites with open wounds. The judgement regarding cryotherapeutic applications to such sites should be based on the extent, nature and depth of the wound involved, and the chance of contaminating the wound. Advantages must be weighed against the risks. For example, there is little risk and a great deal of benefit to be gained by cryotherapeutic application to a freshly contused area, in spite of the presence of a shallow wound. It may be helpful in slowing or stopping bleeding if a dry sterilized ice bag is utilized or the treatment site protected from touch by the water in the ice bath (plastic bagging may be helpful); however, cryotherapeutic applications may do further damage to involved tissues if the wound is deep and/or the vasocirculation in the area is impaired.

Special note: Ethyl chloride spray is considered dangerous by some, by virtue of the fact that the extreme cold it produces may be too extreme. Constant diligence must be exercised to avoid frostbite. Ethyl chloride spray also poses a danger to both the patient and the practitioner because of its effectiveness as a general anesthetic; care must be taken that the gas created by evaporation is not inhaled by either party. Lastly, the gaseous form of ethyl chloride is extremely flammable and the danger of explosion exists if it is used in a poorly ventilated enclosure without precautions being taken to avoid sparks or open flames. In contrast, Fluori-Methane spray does not cool as fast or to the extent that ethyl chloride spray does, so frost bite is less likely to occur if the treatment site is not exposed to more than five seconds of spray time. Additionally, Fluori-Methane spray does not have a general anesthetic effect when inhaled, and it is nonflammable.

Treatment Notes (M10B5)

If the instrument of cryotherapy is followed by agonist/ antagonist vibration, hypertonic neuromuscular management with cryotherapy is most effective for the treatment of extrafusal muscle spasm, intrafusal muscle spasm or trigger point formation or for muscle lengthening. Vibration adds to the central inhibition of the tonic elements of the muscle spindle, increasing effectiveness of treatment and the duration of treatment effects.

Bibliography (M10B6)

Grant, pp. 233–238.
Haldovich, pp. 185–190.
Johnson, pp. 1238–1242.
Lehmann, pp. 563–602.
Murphy, pp. 112–115.
Newton, pp. 1034–1036.
Travell, pp. 65–70.

CRYOTHERAPY ANALGESIA/ANESTHESIA (M10C)

Definition (M10C1)

Cryotherapy may be utilized to provide relief (analgesia) for localized pain arising from the soft tissues of the musculoskeletal system, as well as providing local anesthesia of dermal layers.

Pain relief or analgesia may result from cryotherapy through the elevation of the pain threshold as a result of cooling effects on the pain receptors and sensory fibers, the reduction of acute swelling and the inflammatory reaction resulting from trauma or other system insult, the reduction of painful spasticity and spasm through the relief of pain arising from vascular distention or circulatory ischemia and/or as a source of counter irritation.

Anesthesia arising from cryotherapy application results from the general super stimulation and overwhelming by unrelieved cooling of the sensory nerves and sensory nerve endings in the skin and the surface layers of the muscles below. Each class of receptors will fire in turn as they are cooled and their individual resistance to the cooling is exceeded; this produces a progressive sequence of sensations including aching, burning and (finally) numbness and/or local anesthesia generally reported by the patient.

Procedures of Application (M10C2)

Cryotherapeutic devices (M10C2A)

Anesthesia of the skin may be produced through coolant sprays, ice massage, ice packing, or ice bath; for particulars on the application of coolant sprays or ice bath refer to Cryotherapy [M10].

Ice packing (M10C2A01) The most commonly used and easily applied source of therapeutic cooling is the ice pack. The most effective ice pack is one made of towelling and crushed ice.

1) A towel should be spread out on a flat surface.
2) A one inch layer of crushed ice (enough to cover the treatment site) should be spread out in the middle of the towel.
3) The long edges of the towel should be folded to overlap each other to cover the ice.
4) The ends of the towel should then be rolled up toward each other to make handling easy.
5) Before being applied to the patient the "bottom" of the ice pack should be moistened with water.
6) The ice pack should then be unrolled in place over the treatment site.

Ice massage (M10C2A02) An ice massage may be substituted for ice packing. The patient should be comfortably seated upright with the neck and shoulders bare. The skin above and around the C6 to T3 vertebrae should be lightly rubbed with a hand held ice cube (a small towel or paper wrapping of one end may be used to protect the practitioner's hand) or ice applicator. The treatment should continue until an erythemic reaction and a local temporary sensory anesthesia of the skin occurs; this usually takes several minutes (5 to 7) and the patient will usually experience the sequence of sensory experience noted for the ice pack, but at an accelerated rate.

A handled ice massage applicator may made by placing a tongue depressor (held in place upright by a long hair clip, clipped to the tongue depressor and placed across the mouth of the cup) in a small paper cup of water and freezing it. The cup should be removed when the water is frozen, before using it for treatment. Commercial home kitchen popsicle makers (sans sugar in the water) may also be used to produce a handled ice massage applicator.

Treatment protocol (M10C2B)

1) The patient should ideally be placed in a semireclined comfortable position (a reclining chair may be suitable).
2) The area to be treated should be bared
3) The cryotherapeutic device should be applied to the area to in need of analgesia.
4) The cryotherapeutic device should be applied for an appropriate period. An ice pack should be left in place for at least 10 but no longer then 20 minutes (12 minutes is usually adequate).
5) Following application, the patient's skin should be thoroughly dried off.
6) The patient should leave the treatment area in warm dry clothing.

Partial List of Conditions Treated (M10C3)

Treatable Causes (M10C3A)

Calcific deposit (C08)
Extrafusal muscle spasm (C41)

Habitual joint positioning (C48)
Hypermobile/instable joint (C10)
Interspinous ligamentous strain (C03)
Joint ankylosis (C42)
Joint approximation (C47)
Ligamentous strain (C25)
Migraine (vascular headache) (C34)
Nerve root compression (C04)
Neuroma formation (C14)
Neuromuscular tonic imbalance (C30)
Peripheral nerve injury (C07)
Psychoneurogenic neuromuscular hypertonus (C29)
Restricted joint range of motion (C26)
Soft tissue inflammation (C05)
Soft tissue swelling (C06)
Tactile hypersensitivity (C24)
Trigger point formation (C01)

Syndromes (M10C3B)

ACHILLES TENDONITIS (S52)
ADHESION (S56)
BURNS (SECOND AND/OR THIRD DEGREE) (S72)
BURSITIS (S26)
CALF PAIN (S50)
CAPSULITIS (S27)
CARPAL TUNNEL (S35)
CERVICAL DORSAL OUTLET (SCALENICUS ATTICUS/CERVICAL RIB) (S09)
CERVICAL (NECK) PAIN (S73)
CHEST PAIN (S14)
CHONDROMALACIA (S15)
DIABETES (S10)
EARACHE (S40)
ELBOW PAIN (S47)
FACET (S19)
FASCIITIS (S20)
FOOT PAIN (S51)
FOREARM PAIN (S54)
FROZEN SHOULDER (S64)
HAND/FINGER PAIN (S17)
HEADACHE PAIN (S02)
HIGH THORACIC BACK PAIN (S48)
JOINT SPRAIN (S30)
KNEE PAIN (S46)
LOWER ABDOMINAL PAIN (S62)
LOW BACK PAIN (S03)
MENSTRUAL CRAMPING (S60)
MIGRAINE (VASCULAR) HEADACHE (S18)
MYOSITIS OSSIFICANS (S67)
OSTEOARTHRITIS (S21)
PIRIFORMIS (S04)
POST CEREBRAL VASCULAR ACCIDENT (CVA) (S07)
POST IMMOBILIZATION (S36)
POST PERIPHERAL NERVE INJURY (S23)
POST SPINAL CORD INJURY (S24)
POST WHIPLASH (S55)
RADICULITIS (S29)
REFERRED PAIN (S01)
SCIATICA (S05)
SHOULDER/HAND (S57)
SHOULDER PAIN (S08)
TEMPOROMANDIBULAR JOINT (TMJ) PAIN (S06)
TENDONITIS (S28)
TENNIS ELBOW (LATERAL EPICONDYLITIS) (S32)
THIGH PAIN (S49)
TINNITUS (S70)
TOOTHACHE/JAW PAIN (S41)
TORTICOLLIS (WRY NECK) (S65)
UNHEALED DERMAL LESION (S39)
WRIST PAIN (S16)

Precautions and Contraindications (M10C4)

Adverse effects from cryotherapy are rare, but certain individuals have been shown to be systemically hypersensitive to cold applications (seemingly "allergic" to cold). These individuals may be divided up into three distinct groups, those whose tissues, upon cooling, release (1) histamine or histamine-like chemicals, (2) cold hemolysins and agglutinins, and (3) cryoglobulins.

The symptoms associated with histamine production triggered by cooling include cold urticaria, erythema, itching, sweating, facial flush, puffiness of the eyelids, laryngeal edema, and shortness of breath. More severe symptoms which may develop include syncope, hypotension, tachycardia, extra systoles, dysphagia, abdominal cramping, diarrhea, and/or vomiting. The resultant symptomatology may occur during therapy or several minutes or even hours after the technique has been applied. Such hypersensitivity to cold may be treatable through a program of desensitization to cold: the patient's hands should be immersed in water cooled to 50° F. for two minutes, twice daily for from 14 to 21 days.

Cold hemolysins and agglutinins produced in abnormal amounts in response to tissue cooling may produce significant anemia, malaise, chills, fever, paroxysmal cold hemoglobinuria of the kidney, and/or the dermal symptoms of cold urticaria, ulcerations, Raynaud's phenomenon, and acrocyanosis.

Abnormally high cryoglobulin production in response to cooling may produce chills and fever, and may adversely affect vision, causing blindness in the extreme, and hearing, causing deafness in the extreme. It may cause conjunctival hemorrhages and epistaxis, anemia with fibrinogenopenia, elevated sedimentation rates, and abnormal numbers of cryoglobulin inclusion cells. Other symptoms include erythema, itching, purpura, cold urticaria, Raynaud's phenomenon, ulcerations and necrosis, problems breathing (dyspnea), stomatitis, melena, and bleeding of the gums.

The adverse reactions to cryotherapy described above have been noted to occur in association with various systemic diseases including Raynaud's disease, Buerger's disease, lupus erythematosus, leucocyto-clastic vasculitis, rheumatoid disease, progressive systemic sclerosis, multiple myeloma, and pneumonia; consequently, cryotherapy involving immersion or the ice packing of the distal extremities of patient's suffering from these diseases may be contraindicated. Any cryotherapy applied to patients of this type must be supervised carefully, and should avoided if at all possible.

If arterial insufficiency is present (as is the case in some systemic diseases), cold applications would seem to be contraindicated because of their potent vasoconstricting effect which may potentially reduce already compromised blood flow through the extremity.

Some individuals, though not suffering from any particular disease process, respond to the natural vasopressor response to application of cold (capillary vasoconstriction) with failure to eventually compensate by capillary vasodilation, thus producing a Buerger's disease-like syndrome which may threaten the tissues with rapidly developing frostbite if exposure to the cold is not eliminated. This condition may cause the patient exquisite pain which may not be readily relieved. Such hyperreactivity to cooling may be readily tested for by observing the tissue response to ice packing. The normal dermal response to an applied crushed ice pack is to blanch (vasoconstrict) for the first three minutes after the ice pack is applied; the skin gradually develops an increasing erythemic reaction, becoming bright red after ten minutes of constant exposure. If the blanching phase does not end after the first three minutes of exposure to the ice pack, the patient may be considered hyperresponsive to exposure to cold (even without other symptoms) and the treatment should be immediately discontinued.

Though rare, frostbite is a potential danger inherent in cryotherapy, especially for techniques involving ice massage or the coolant sprays although reports suggest that such damage is rarely severe. Frostbite occurs as a consequence of over-exposure to cold which causes an increase in the capillary permeability. The consequent rapid loss of plasma strands red blood cells in the capillaries, which then clump together to form occlusive masses (clots) which finally undergo hyalinization. Treatment of frostbite involves the immediate immersion in warm water (109° F.). Soft tissue damage from frostbite is increased (1) as the length of time the tissues are frozen increases, (2) as the temperature of the tissues while in the frozen state decreases, and (3) as the duration of thawing increases.

Ice packing applied to the abdomen increases peristalsis in the stomach, small intestine and the colon and has been found to increase blood flow through the mucous membrane of the alimentary canal and to increase acid secretion in the stomach. Stomach cramps may, therefore, be aggravated and gastrointestinal upset be caused by ice packing of the anterior lower abdominal area; increased gastric acidity is undesirable in the presence of pathology of the stomach, particularly if peptic ulcers are present.

Extreme care must be exercised in the application of cryotherapy techniques to treatment sites which may suffer from disturbed skin sensations or sensory impairment. In such cases the patient's skin must be carefully tested for abnormal sensitivity to cold and normal reactions to its application (refer to Cryotherapy, Evaluation of Capillary Response to Cold [M10H]). Additionally, cryotherapy is considered by many to be contraindicated for treatment sites with open wounds. The judgement regarding cryotherapeutic applications to such sites should be based on the extent, nature and depth of the wound involved, and the chance of contaminating the wound. Advantages must be weighed against the risks. For example, there is little risk and a great deal of benefit to be gained by cryotherapeutic application to a freshly contused area, in spite of the presence of a shallow wound. It may be helpful in slowing or stopping bleeding if a dry sterilized ice bag is utilized or the treatment site protected from touch by the water in the ice bath (plastic bagging may be helpful); however, cryotherapeutic applications may do further damage to involved tissues if the wound is deep and/or the vasocirculation in the area is impaired.

Special note: Ethyl chloride spray is considered dangerous by some, by virtue of the fact that the extreme cold it produces may be too extreme. Constant diligence must be exercised to avoid frostbite. Ethyl chloride spray also poses a danger to both the patient and the practitioner because of its effectiveness as a general anesthetic; care must be taken that the gas created by evaporation is not inhaled by either party. Lastly, the gaseous form of ethyl chloride is extremely flammable and the danger of explosion exists if it is used in a poorly ventilated enclosure without precautions being taken to avoid sparks or open flames. In contrast, Fluori-Methane spray does not cool as fast or to the extent that ethyl chloride spray does, so frost bite is less likely to occur if the treatment site is not exposed to more than five seconds of spray time. Additionally, Fluori-Methane spray does not have a general anesthetic effect when inhaled, and it is nonflammable.

Treatment Notes (M10C5)

The use of cryotherapy to produce analgesia and/or anesthesia is most remarkable when used in the acute phases following traumatic injury. Such use provides the patient with almost immediate relief from excruciating following whiplash, joint sprains and/or strains, muscle spasm, contusion and muscle strains (injuries often reported as sports related). The relief of pain, while comforting the patient, allows the practitioner the opportunity to treat the patient with techniques which might otherwise be too painful for the patient to endure (refer to Electrical Stimulation [M02], Vertebral Traction [M09], Taping [M16], and Massage, Deep Soft Tissue Mobilization [M23F]). Additionally, the application of cryotherapy may have the effect of slowing the pathological process of the injury. Its effects may include the retarding of swelling, the slowing or stopping bleeding into and/or out of the tissues or the prevention and/or slowing of the production of soft tissue inflammatories, their effects and their spread (refer to Soft Tissue Inflammation [C05] and Cryotherapy, Soft Tissue Inflammation Control [M10E]).

Bibliography (M10C6)

Grant, pp. 233–238.
Haldovich, pp. 185–190.
Lehmann, pp. 563–602.
Michlovitz, pp. 73–97, 263–275, 277–294.
Newton, pp. 1034–1036.
Travell, pp. 65–70.

CRYOTHERAPY SOFT TISSUE SWELLING CONTROL (M10D)

Definition (M10D1)

The primary therapeutic effect of cryotherapy upon soft tissue swelling is preventative by nature: soft tissue cooling produces a sympathetic nervous reflex which initially produces of vasoconstriction in the blood vessels (including the capillary beds) below the site of application, which initially prevents soft tissue swelling by suppressing fluid flow into the area. Authorities have suggested that intermittent cold applications (exclusively utilized) to induce vasoconstriction may have the beneficial effect of inhibiting swelling for as much as forty-eight hours after the initial trauma. Continuous cooling eventually causes vascular wall permeability to increase and will by actually promoting soft tissue swelling allow plasma to seep into adjacent tissues. Therefore, the most appropriate use of long duration cooling (ten to thirty minutes) is in the acute phase following trauma. After the first twenty-four hours, soft tissue cooling, exclusively utilized, becomes increasingly less beneficial, up to forty-eight hours after trauma, after which time it is of no benefit at all and may then be contraindicated.

Procedures of Application (M10D2)

The treatment of soft tissue swelling through soft tissue cooling is reported to be best accomplished through the use of cryotherapeutic devices which provide deep tissue cooling. These devices include ice packing, ice massage and ice baths. The most commonly applied and easiest to use is the ice pack.

Cryotherapeutic devices (M10D2A)

Ice packing (M10D2A01) The most effective ice pack is one made of towelling and crushed ice.

1) A towel should be spread out on a flat surface.
2) A one inch layer of crushed ice (enough to cover the treatment site) should be spread out in the middle of the towel.
3) The long edges of the towel should be folded to overlap each other to cover the ice.
4) The ends of the towel should then be rolled up toward each other to make handling easy.
5) Before being applied to the patient the "bottom" of the ice pack should be moistened with water.
6) The ice pack should then be unrolled in place over the treatment site.
7) In common use, the crushed ice pack should not be left in place longer then 10 to 12 minutes, enough time to cause a vasodilation sufficient to create an erythemic (bright pink) reaction of the skin and/or local anesthesia.

Ice massage (M10D2A02) An ice massage may be substituted for ice packing. The patient should be comfortably seated upright with the neck and shoulders bare. The skin above and around the C6 to T3 vertebrae should be lightly rubbed with a hand held ice cube (a small towel or paper wrapping of one end may be used to protect the practitioner's hand) or ice applicator. The treatment should continue until an erythemic reaction and a local temporary sensory anesthesia of the skin occurs; this usually takes several minutes (5 to 7) and the patient will usually experience the sequence of sensory experience noted for the ice pack, but at an accelerated rate.

A handled ice massage applicator may made by placing a tongue depressor (held in place upright by a long hair clip, clipped to the tongue depressor and placed across the mouth of the cup) in a small paper cup of water and freezing it. The cup should be removed when the water is frozen, before using it for treatment. Commercial home kitchen popsicle makers (sans sugar in the water) may also be used to produce a handled ice massage applicator.

Ice bath (M10D2A03) An ice bath is created by filling a basin with water and enough ice to cool the water from 55 to 60° F. (12.8 to 15.6° C.); the basin should be large enough to contain the portion of the patient's body requiring treatment and the necessary of volume of water and ice. The body part to be treated should be immersed in the basin approximately ten minutes or until anesthesia of the body part is produced.

The patient can be expected to go through the sequence of sensory sensations previously described.

After the cryotherapeutic application chosen, the treated site and other dampened surfaces should be thoroughly dried and the treated area covered with dry garments. Additional therapeutic procedures should be performed at this point.

Treatment protocol (M10D2B)

1) The patient should ideally be placed in a semireclined comfortable position (a reclining chair may be suitable).
2) The dermis to be treated should be bared.
3) The cryotherapeutic device should be applied to the injured body part.
4) The cryotherapeutic device should be applied for an appropriate period. An ice pack should be left in place for at least 10 minutes.
5) After the ice pack is removed, the patient's skin should be thoroughly dried off.
6) The patient should leave the treatment area in warm dry clothing.

Partial List of Conditions Treated (M10D3)

Treatable Causes (M10D3A)

Hypermobile/instable joint (C10)
Interspinous ligamentous strain (C03)
Joint approximation (C47)
Ligamentous strain (C25)
Restricted joint range of motion (C26)
Soft tissue swelling (C06)
Unhealed dermal lesion (C12)

Syndromes (M10D3B)

ACHILLES TENDONITIS (S52)
CALF PAIN (S50)
ELBOW PAIN (S47)
FACET (S19)
FOOT PAIN (S51)
FOREARM PAIN (S54)
HIGH THORACIC BACK PAIN (S48)
JOINT SPRAIN (S30)
KNEE PAIN (S46)
LOW BACK PAIN (S03)
OSTEOARTHRITIS (S21)
PITTING (LYMPH) EDEMA (S31)
POST IMMOBILIZATION (S36)
POST WHIPLASH (S55)
TENDONITIS (S28)
TENNIS ELBOW (LATERAL EPICONDYLITIS) (S32)
THIGH PAIN (S49)
TOOTHACHE/JAW PAIN (S41)
UNHEALED DERMAL LESION (S39)
WRIST PAIN (S16)

Precautions and Contraindications (M10D4)

Adverse effects from cryotherapy are rare, but certain individuals have been shown to be systemically hypersensitive to cold applications (seemingly "allergic" to cold). These individuals may be divided up into three distinct groups, those whose tissues, upon cooling, release (1) histamine or histamine-like chemicals, (2) cold hemolysins and agglutinins, and (3) cryoglobulins.

The symptoms associated with histamine production triggered by cooling include cold urticaria, erythema, itching, sweating, facial flush, puffiness of the eyelids, laryngeal edema, and shortness of breath. More severe symptoms which may develop include syncope, hypotension, tachycardia, extra systoles, dysphagia, abdominal cramping, diarrhea, and/or vomiting. The resultant symptomatology may occur during therapy or several minutes or even hours after the technique has been applied. Such hypersensitivity to cold may be treatable through a program of desensitization to cold: the patient's hands should be immersed in water cooled to 50° F. for two minutes, twice daily for from 14 to 21 days.

Cold hemolysins and agglutinins produced in abnormal amounts in response to tissue cooling may produce significant anemia, malaise, chills, fever, paroxysmal cold hemoglobinuria of the kidney, and/or the dermal symptoms of cold urticaria, ulcerations, Raynaud's phenomenon, and acrocyanosis.

Abnormally high cryoglobulin production in response to cooling may produce chills and fever, and may adversely affect vision, causing blindness in the extreme, and hearing, causing deafness in the extreme. It may cause conjunctival hemorrhages and epistaxis, anemia with fibrinogenopenia, elevated sedimentation rates, and abnormal numbers of cryoglobulin inclusion cells. Other symptoms include erythema, itching, purpura, cold urticaria, Raynaud's phenomenon, ulcerations and necrosis, problems breathing (dyspnea), stomatitis, melena, and bleeding of the gums.

The adverse reactions to cryotherapy described above have been noted to occur in association with various systemic diseases including Raynaud's disease, Buerger's disease, lupus erythematosus, leucocyto-clastic vasculitis, rheumatoid disease, progressive systemic sclerosis, multiple myeloma, and pneumonia; consequently, cryotherapy involving immersion or the ice packing of the distal extremities of patient's suffering from these diseases may be contraindicated. Any cryotherapy applied to patients of this type must be supervised carefully, and should avoided if at all possible.

If arterial insufficiency is present (as is the case in some systemic diseases), cold applications would seem to be contraindicated because of their potent vasoconstricting effect which may potentially reduce already compromised blood flow through the extremity.

Some individuals, though not suffering from any particular disease process, respond to the natural vasopressor response to application of cold (capillary vasoconstriction) with failure to eventually compensate by capillary vasodilation, thus producing a Buerger's disease-like syndrome which may threaten the tissues with rapidly developing frostbite if exposure to the cold is not eliminated. This condition may cause the patient exquisite pain which may not be readily relieved. Such hyperreactivity to cooling may be readily tested for by observing the tissue response to ice packing. The normal dermal response to an applied crushed ice pack is to blanch (vasoconstrict) for the first three minutes after the ice pack is applied; the skin gradually develops an increasing erythemic reaction, becoming bright red after ten minutes of constant exposure. If the blanching phase does not end after the first three minutes of exposure to the ice pack, the patient may be considered hyperresponsive to exposure to cold (even without other symptoms) and the treatment should be immediately discontinued.

Though rare, frostbite is a potential danger inherent in cryotherapy, especially for techniques involving ice massage or the coolant sprays although reports suggest that such damage is rarely severe. Frostbite occurs as a consequence of over-exposure to cold which causes an increase in the capillary permeability. The consequent rapid loss of plasma strands red blood cells in the capillaries, which then clump together to form occlusive masses (clots) which finally undergo hyalinization. Treatment of frostbite involves the immediate immersion in warm water (109° F.). Soft tissue damage from frostbite is increased (1) as the length of time the tissues are frozen increases, (2) as the temperature of the tissues while in the frozen state decreases, and (3) as the duration of thawing increases.

Ice packing applied to the abdomen increases peristalsis in the stomach, small intestine and the colon and has been found to increase blood flow through the mucous membrane of the alimentary canal and to increase acid secretion in the stomach. Stomach cramps may, therefore, be aggravated and gastrointestinal upset be caused by ice packing of the anterior lower abdominal area; increased gastric acidity is undesirable in the presence of pathology of the stomach, particularly if peptic ulcers are present.

Extreme care must be exercised in the application of cryotherapy techniques to treatment sites which may suffer from disturbed skin sensations or sensory impairment. In such cases the patient's skin must be carefully tested for abnormal sensitivity to cold and normal reactions to its application (refer to Cryotherapy, Evaluation of Capillary Response to Cold [M10H]). Additionally, cryotherapy is considered by many to be contraindicated for treatment sites with open wounds. The judgement regarding cryotherapeutic applications to such sites should be based on the extent, nature and depth of the wound involved, and the chance of contaminating the wound. Advantages must be weighed against the risks. For example, there is little risk and a great deal of benefit to be gained by cryotherapeutic application to a freshly contused area, in spite of the presence of a shallow wound. It may be helpful in slowing or stopping bleeding if a dry sterilized ice bag is utilized or the treatment site protected from touch by the water in the ice bath (plastic bagging may be helpful); however, cryotherapeutic applications may do further damage to involved tissues if the wound is deep and/or the vasocirculation in the area is impaired.

Treatment Notes (M10D5)

Cryotherapeutic applications to control swelling of soft tissues is most effectively applied to injured soft tissues shortly following injury, and is therefore becoming popular for the treatment of sports injuries. Facilities for providing ice packing or ice baths on the site of play are becoming increasingly prevalent, even in connection with amateur sports unrelated unrelated to school activities.

Bibliography (M10D6)

Guyton, pp. 241–251.
Kowal, pp. 66–73.
Lehmann, pp. 576–579.
McKean, pp. 82–92.
Michlovitz, pp. 79–81.
Murphy, pp. 112–115.
Shodell, pp. 78–82.

CRYOTHERAPY SOFT TISSUE INFLAMMATION CONTROL (M10E)

Definition (M10E1)

In humans, soft tissue inflammation is a physiological response to stress of soft issues. When traumatized (excessively stressed or irritated), the involved soft tissues produce inflammatories which are designed to promote the healing process. These inflammatories include histamine, bradykinin, and prostaglandins. The sources of the soft tissue inflammatory response include mechanical trauma (blows, puncture, tearing, burns, etc.), metabolite accumulation, bacterial toxins, viral toxins, enzyme irritations (as seen in the joints of rheumatoid victims) and others. Cryotherapy has proven to be of value in the treatment of soft tissue inflammation by: (1) slowing through tissue cooling the action of destructive enzymes and toxins, (2) decreasing the locally occurring metabolism, (3) desensitizing sensory nerves previously sensitized by histamine and bradykinin, and (4) retarding the production of prostaglandins by slowing the sensory nerve release of histamine.

Procedure of Application (M10E2)

The treatment of soft tissue inflammation by cooling is best performed by cryotherapeutic applications which provide deep tissue cooling, including ice packing, ice massage and ice baths. Of these devices, the most commonly applied and easiest to use is the ice pack.

Cryotherapeutic devices (M10E2A)

Ice packing (M10E2A01) The most effective ice pack is one made of towelling and crushed ice

1) A towel should be spread out on a flat surface.
2) A one inch layer of crushed ice (enough to cover the treatment site) should be spread out in the middle of the towel.
3) The long edges of the towel should be folded to overlap each other to cover the ice.
4) The ends of the towel should then be rolled up toward each other to make handling easy.
5) Before being applied to the patient the "bottom" of the ice pack should be moistened with water.
6) The ice pack should then be unrolled in place over the treatment site.
7) In common use, the crushed ice pack should not be left in place longer then 10 to 12 minutes, enough time to cause a vasodilation sufficient to create an erythemic (bright pink) reaction of the skin and/or local anesthesia.

Ice massage (M10E2A02) An ice massage may be substituted for ice packing. The patient should be comfortably seated upright with the neck and shoulders bare. The skin above and around the C6 to T3 vertebrae should be lightly rubbed with a hand held ice cube (a small towel or paper wrapping of one end may be used to protect the practitioner's hand) or ice applicator. The treatment should continue until an erythemic reaction and a local temporary sensory anesthesia of the skin occurs; this usually takes several minutes (5 to 7) and the patient will usually experience the sequence of sensory experience noted for the ice pack, but at an accelerated rate.

A handled ice massage applicator may made by placing a tongue depressor (held in place upright by a long hair clip, clipped to the tongue depressor and placed across the mouth of the cup) in a small paper cup of water and freezing it. The cup should be removed when the water is frozen, before using it for treatment. Commercial home kitchen popsicle makers (sans sugar in the water) may also be used to produce a handled ice massage applicator.

Treatment protocol (M10E2B)

1) The patient should ideally be placed in a semireclined comfortable position (a reclining chair may be suitable).
2) The dermis to be treated should be bared.
3) The cryotherapeutic device should be applied to the injured body part.
4) The cryotherapeutic device should be applied for an appropriate period. An ice pack should be left in place for at least 10 minutes
5) After the ice pack is removed, the patient's skin should be thoroughly dried off.
6) The patient should leave the treatment area in warm dry clothing.

Partial List of Conditions Treated (M10E3)

Treatable Causes (M10E3A)

Bacterial infection (C09)
Calcific deposit (C08)
Extrafusal muscle spasm (C41)
Habitual joint positioning (C48)
Hypermobile/instable joint (C10)
Interspinous ligamentous strain (C03)
Joint ankylosis (C42)
Joint approximation (C47)
Ligamentous strain (C25)
Localized viral infection (C27)
Nerve root compression (C04)
Neurogenic dermal disorders (psoriasis, hives) (C49)
Restricted joint range of motion (C26)
Soft tissue inflammation (C05)
Soft tissue swelling (C06)
Tactile hypersensitivity (C24)
Trigger point formation (C01)
Unhealed dermal lesion (C12)

Syndromes (M10E3B)

ACHILLES TENDONITIS (S52)
BACTERIAL INFECTION (S63)
BURNS (SECOND AND/OR THIRD DEGREE) (S72)
BURSITIS (S26)
CALF PAIN (S50)
CAPSULITIS (S27)
CARPAL TUNNEL (S35)
CERVICAL DORSAL OUTLET (SCALENICUS ATTICUS/CERVICAL RIB) (S09)
CERVICAL (NECK) PAIN (S73)
CHEST PAIN (S14)
CHONDROMALACIA (S15)
EARACHE (S40)
ELBOW PAIN (S47)
FACET (S19)
FASCIITIS (S20)
FOOT PAIN (S51)
FOREARM PAIN (S54)
FROZEN SHOULDER (S64)
HAND/FINGER PAIN (S17)
HEADACHE PAIN (S02)
HIGH THORACIC BACK PAIN (S48)
JOINT SPRAIN (S30)
KNEE PAIN (S46)
LOWER ABDOMINAL PAIN (S62)
LOW BACK PAIN (S03)
MIGRAINE (VASCULAR) HEADACHE (S18)
MYOSITIS OSSIFICANS (S67)
OSTEOARTHRITIS (S21)
PIRIFORMIS (S04)
POST IMMOBILIZATION (S36)
POST PERIPHERAL NERVE INJURY (S23)
POST WHIPLASH (S55)
RADICULITIS (S29)
REFERRED PAIN (S01)
SCIATICA (S05)
SHOULDER/HAND (S57)
SHOULDER PAIN (S08)
TEMPOROMANDIBULAR JOINT (TMJ) PAIN (S06)
TENDONITIS (S28)
TENNIS ELBOW (LATERAL EPICONDYLITIS) (S32)
THIGH PAIN (S49)
TINNITUS (S70)
TOOTHACHE/JAW PAIN (S41)
UNHEALED DERMAL LESION (S39)
WRIST PAIN (S16)

Precautions and Contraindications (M10E4)

Adverse effects from cryotherapy are rare, but certain individuals have been shown to be systemically hypersensitive to cold applications (seemingly "allergic" to cold). These individuals may be divided up into three distinct groups, those whose tissues, upon cooling, release (1) histamine or histamine-like chemicals, (2) cold hemolysins and agglutinins, and (3) cryoglobulins.

The symptoms associated with histamine production triggered by cooling include cold urticaria, erythema, itching, sweating, facial flush, puffiness of the eyelids, laryngeal edema, and shortness of breath. More severe symptoms which may develop include syncope, hypotension, tachycardia, extra systoles, dysphagia, abdominal cramping, diarrhea, and/or vomiting. The resultant symptomatology may occur during therapy or several minutes or even hours after the technique has been applied. Such hypersensitivity to cold may be treatable through a program of desensitization to cold: the patient's hands should be immersed in water cooled to 50° F. for two minutes, twice daily for from 14 to 21 days.

Cold hemolysins and agglutinins produced in abnormal amounts in response to tissue cooling may produce significant anemia, malaise, chills, fever, paroxysmal cold hemoglobinuria of the kidney, and/or the dermal symptoms of cold urticaria, ulcerations, Raynaud's phenomenon, and acrocyanosis.

Abnormally high cryoglobulin production in response to cooling may produce chills and fever, and may adversely affect vision, causing blindness in the extreme, and hearing, causing deafness in the extreme. It may cause conjunctival hemorrhages and epistaxis, anemia with fibrinogenopenia, elevated sedimentation rates, and abnormal numbers of cryoglobulin inclusion cells. Other symptoms include erythema, itching, purpura, cold urticaria, Raynaud's phenomenon, ulcerations and necrosis, problems breathing (dyspnea), stomatitis, melena, and bleeding of the gums.

The adverse reactions to cryotherapy described above have been noted to occur in association with various systemic diseases including Raynaud's disease, Buerger's disease, lupus erythematosus, leucocytoclastic vasculitis, rheumatoid disease, progressive systemic sclerosis, multiple myeloma, and pneumonia; consequently, cryotherapy involving immersion or the ice packing of the distal extremities of patient's suffering from these diseases may be contraindicated. Any cryotherapy applied to patients of this type must be supervised carefully, and should avoided if at all possible.

If arterial insufficiency is present (as is the case in some systemic diseases), cold applications would seem to be contraindicated because of their potent vasoconstricting effect which may potentially reduce already compromised blood flow through the extremity.

Some individuals, though not suffering from any particular disease process, respond to the natural vasopressor response to application of cold (capillary vasoconstriction) with failure to eventually compensate by capillary vasodilation, thus producing a Buerger's disease-like syndrome which may threaten the tissues with rapidly developing frostbite if exposure to the cold is not eliminated. This condition may cause the patient exquisite pain which may not be readily relieved. Such hyperreactivity to cooling may be readily tested for by observing the tissue response to ice packing. The normal dermal response to an applied crushed ice pack is to blanch (vasoconstrict) for the first three minutes after the ice pack is applied; the skin gradually develops an increasing erythemic reaction, becoming bright red after ten minutes of constant exposure. If the blanching phase does not end after the first three minutes of exposure to the ice pack, the patient may be considered hyperresponsive to exposure to cold (even without other symptoms) and the treatment should be immediately discontinued.

Though rare, frostbite is a potential danger inherent in cryotherapy, especially for techniques involving ice massage or the coolant sprays although reports suggest that such damage is rarely severe. Frostbite occurs as a consequence of over-exposure to cold which causes an increase in the capillary permeability. The consequent rapid loss of plasma strands red blood cells in the capillaries, which then clump together to form occlusive masses (clots) which finally undergo hyalinization. Treatment of frostbite involves the immediate immersion in warm water (109° F.). Soft tissue damage from frostbite is increased (1) as the length of time the tissues are frozen increases, (2) as the temperature of the tissues while in the frozen state decreases, and (3) as the duration of thawing increases.

Ice packing applied to the abdomen increases peristalsis in the stomach, small intestine and the colon and has been found to increase blood flow through the mucous membrane of the alimentary canal and to increase acid secretion in the stomach. Stomach cramps may, therefore, be aggravated and gastrointestinal upset be caused by ice packing of the anterior lower abdominal area; increased gastric acidity is undesirable in the presence of pathology of the stomach, particularly if peptic ulcers are present.

Extreme care must be exercised in the application of cryotherapy techniques to treatment sites which may suffer from disturbed skin sensations or sensory impairment. In such cases the patient's skin must be carefully tested for abnormal sensitivity to cold and normal reactions to its application (refer to Cryotherapy, Evaluation of Capillary Response to Cold [M10H]). Additionally, cryotherapy is considered by many to be contraindicated for treatment sites with open wounds. The judgement regarding cryotherapeutic applications to such sites should be based on the extent, nature and depth of the wound involved, and the chance of contaminating the wound. Advantages must be weighed against the risks. For example, there is little risk and a great deal of benefit to be gained by cryotherapeutic application to a freshly contused area, in spite of the presence of a shallow wound. It may be helpful in slowing or stopping bleeding if a dry sterilized ice bag is utilized or the treatment site protected from touch by the water in the ice bath (plastic bagging may be helpful); however, cryotherapeutic applications may do further damage to involved tissues if the wound is deep and/or the vasocirculation in the area is impaired.

Treatment Notes (M10E5)

Capillary permeability and phonophoresis (M10E5A)

The blood vessel permeability caused by prolonged cryotherapy application may be put to good use if phonophoresis of antiinflammatories into inflamed soft tissues is to be utilized. In the clinical setting, ice packing is most commonly used in connection with soft tissue inflammation control as a precursor to administration of antiinflammatories by phonophoresis (refer to Ultrahigh Frequency Sound, Phonophoresis [M01A]). An ice pack is applied for ten to twelve minutes before phonophoresis helps to increase capillary permeability and thereby insure a greater and deeper penetration of antiinflammatories.

Bibliography (M10E6)

Kowal, pp. 66–73.
Lehmann, pp. 576–579, 585–590.
McKean, pp. 82–92.
Michlovitz, pp. 73–97, 263–275, 277–294.
Murphy, pp. 112–115.
Shodell, pp. 78–82.

CRYOTHERAPY SPASTICITY CONTROL (M10F)

Definition (M10F1)

Clinical research has demonstrated that if a muscle is cooled, its passive resistance to stretch is reduced; in fact, the strength of antagonistic musculature is increased as spasticity (tonicity) of the agonist is reduced. Spasticity has been shown to be reduced in part for as long as ninety minutes after a twenty-minute cold application (ice packing or cold immersion bath) is terminated. It has also been demonstrated that repression of spasticity thus produced is not simply the result of skin cooling, with its consequent reflex inhibition of afferent muscle spindle activity (sensory fibers and gamma efferents), but may include indirect influences upon central nervous system structures and/or upon the contractile mechanism of the muscle itself via changes in muscle spindle proprioception input.

Regardless of the mechanism responsible for the inhibition of spasticity, spasticity has been shown to be temporarily reduced by cryotherapeutic applications; such inhibition may offer an opportunity to the practitioner to increase passive and/or (possibly) active range of motion in joints crossed by spastic musculature.

Procedures of Application (M10F2)

Spastic musculature may be effectively treated by cryotherapy to reduce hypertonicity (spasticity or clonus) with ice packing, or cold immersion (ice) bath.

Cryotherapeutic devices (M10F2A)

Ice packing (M10F2A01) The most effective ice pack is one made of towelling and crushed ice:

1) A towel should be spread out on a flat surface.
2) A one inch layer of crushed ice (enough to cover the treatment site) should be spread out in the middle of the towel.
3) The long edges of the towel should be folded to overlap each other to cover the ice.
4) The ends of the towel should then be rolled up toward each other to make handling easy.
5) Before being applied to the patient the "bottom" of the ice pack should be moistened with water.
6) The ice pack should then be unrolled in place over the treatment site.
7) In common use, the crushed ice pack should not be left in place longer then 10 to 12 minutes, enough time to cause a vasodilation sufficient to create an erythemic (bright pink) reaction of the skin and/or local anesthesia.

Ice massage (M10F2A02) An ice massage may be substituted for ice packing. The patient should be comfortably seated upright with the neck and shoulders bare. The skin above and around the C6 to T3 vertebrae should be lightly rubbed with a hand held ice cube (a small towel or paper wrapping of one end may be used to protect the practitioner's hand) or ice applicator. The treatment should continue until an erythemic reaction and a local temporary sensory anesthesia of the skin occurs; this usually takes several minutes (5 to 7) and the patient will usually experience the sequence of sensory experience noted for the ice pack, but at an accelerated rate.

A handled ice massage applicator may made by placing a tongue depressor (held in place upright by a long hair clip, clipped to the tongue depressor and placed across the mouth of the cup) in a small paper cup of water and freezing it. The cup should be removed when the water is frozen, before using it for treatment. Commercial home kitchen popsicle makers (sans sugar in the water) may also be used to produce a handled ice massage applicator.

Ice bath (M10F2A03) An ice bath is created by filling a basin with water and enough ice to cool the water from 55 to 60° F. (12.8 to 15.6° C); the basin should be large enough to contain the portion of the patient's body requiring treatment and the necessary of volume water and ice. The body part to be treated should be immersed in the basin generally until anesthesia of the body part is produced (approximately 10 minutes).

The patient can be expected to go through the sequence of sensory sensations in the previously described relative to ice packing.

Ice bag (M10F2A04) An ice bag is not used clinically to reduce spasticity because of its usual failure to provide sufficiently rapid cooling of the soft tissues. Likewise, it should be noted that coolant sprays have also been found to be inadequate for the task of reducing spasticity; it only cools the skin, and such surface cooling has been found to be insufficient affect the suppression of neuromuscular spasticity.

Treatment protocol (M10F2B)

1) The patient should ideally be placed in a semireclined comfortable position (a reclining chair may be suitable).

2) The dermis to be treated should be bared.

3) The spastic musculature should be put on stretch and maintained in that state for the duration of the treatment.

4) The cryotherapeutic device should be applied over the spastic musculature for an appropriate period. An ice pack should be left in place for at least 10 minutes.

5) Following application, the patient's skin should be thoroughly dried off.

6) The patient should leave the treatment area in warm dry clothing.

Partial List of Conditions Treated (M10F3)

Treatable Causes (M10F3A)

Extrafusal muscle spasm (C41)
Habitual joint positioning (C48)
Joint ankylosis (C42)
Pathological neuromuscular hypertonus (C28)
Pathological neuromuscular dyscoordination (C35)
Peripheral nerve injury (C07)
Restricted joint range of motion (C26)

Syndromes (M10F3B)

CALF PAIN (S50)
CEREBRAL PALSY (S25)
CERVICAL DORSAL OUTLET (SCALENICUS ATTICUS/CERVICAL RIB) (S09)
ELBOW PAIN (S47)
FOOT PAIN (S51)
FOREARM PAIN (S54)
FROZEN SHOULDER (S64)
HAND/FINGER PAIN (S17)
KNEE PAIN (S46)
NEUROMUSCULAR PARALYSIS (S22)
POST CEREBRAL VASCULAR ACCIDENT (CVA) (S07)
POST IMMOBILIZATION (S36)
POST PERIPHERAL NERVE INJURY (S23)
POST SPINAL CORD INJURY (S24)
SHOULDER/HAND (S57)
SHOULDER PAIN (S08)
THIGH PAIN (S49)
WRIST PAIN (S16)

Precautions and Contraindications (M10F4)

Adverse effects from cryotherapy are rare, but certain individuals have been shown to be systemically hypersensitive to cold applications (seemingly "allergic" to cold). These individuals may be divided up into three distinct groups, those whose tissues, upon cooling, release (1) histamine or histamine-like chemicals, (2) cold hemolysins and agglutinins, and (3) cryoglobulins.

The symptoms associated with histamine production triggered by cooling include cold urticaria, erythema, itching, sweating, facial flush, puffiness of the eyelids, laryngeal edema, and shortness of breath. More severe symptoms which may develop include syncope, hypotension, tachycardia, extra systoles, dysphagia, abdominal cramping, diarrhea, and/or vomiting. The resultant symptomatology may occur during therapy or several minutes or even hours after the technique has been applied. Such hypersensitivity to cold may be treatable through a program of desensitization to cold: the patient's hands should be immersed in water cooled to 50° F. for two minutes, twice daily for from 14 to 21 days.

Cold hemolysins and agglutinins produced in abnormal amounts in response to tissue cooling may produce significant anemia, malaise, chills, fever, paroxysmal cold hemoglobinuria of the kidney, and/or the dermal symptoms of cold urticaria, ulcerations, Raynaud's phenomenon, and acrocyanosis.

Abnormally high cryoglobulin production in response to cooling may produce chills and fever, and may adversely affect vision, causing blindness in the extreme, and hearing, causing deafness in the extreme. It may cause conjunctival hemorrhages and epistaxis, anemia with fibrinogenopenia, elevated sedimentation rates, and abnormal numbers of cryoglobulin inclusion cells. Other symptoms include erythema, itching, purpura, cold urticaria, Raynaud's phenomenon, ulcerations and necrosis, problems breathing (dyspnea), stomatitis, melena, and bleeding of the gums.

The adverse reactions to cryotherapy described above have been noted to occur in association with various systemic diseases including Raynaud's disease, Buerger's disease, lupus erythematosus, leucocyto-clastic vasculitis, rheumatoid disease, progressive systemic sclerosis, multiple myeloma, and pneumonia; consequently, cryotherapy involving immersion or the ice packing of the distal extremities of patient's suffering from these diseases may be contraindicated. Any cryotherapy applied to patients of this type must be supervised carefully, and should avoided if at all possible.

If arterial insufficiency is present (as is the case in some systemic diseases), cold applications would seem to be contraindicated because of their potent vasoconstricting effect which may potentially reduce already compromised blood flow through the extremity.

Some individuals, though not suffering from any particular disease process, respond to the natural vasopressor response to application of cold (capillary vasoconstriction) with failure to eventually compensate by capillary vasodilation, thus producing a Buerger's disease-like syndrome which may threaten the tissues with rapidly developing frostbite if exposure to the cold is not eliminated. This condition may cause the patient exquisite pain which may not be readily relieved. Such hyperreactivity to cooling may be readily tested for by observing the tissue response to ice packing. The normal dermal response to an applied crushed ice pack is to blanch (vasoconstrict) for the first three minutes after the ice pack is applied; the skin gradually develops an increasing erythemic reaction, becoming bright red after ten minutes of constant exposure. If the blanching phase does not end after the first three minutes of exposure to the ice pack, the patient may be considered hyperresponsive to exposure to cold (even without other symptoms) and the treatment should be immediately discontinued.

Though rare, frostbite is a potential danger inherent in cryotherapy, especially for techniques involving ice massage or the coolant sprays although reports suggest that such damage is rarely severe. Frostbite occurs as a consequence of over-exposure to cold which causes an increase in the capillary permeability. The consequent rapid loss of plasma strands red blood cells in the capillaries, which then clump together to form occlusive masses (clots) which finally undergo hyalinization. Treatment of frostbite involves the immediate immersion in warm water (109° F.). Soft tissue damage from frostbite is increased (1) as the length of time the tissues are frozen increases, (2) as the temperature of the tissues while in the frozen state decreases, and (3) as the duration of thawing increases.

Ice packing applied to the abdomen increases peristalsis in the stomach, small intestine and the colon and has been found to increase blood flow through the mucous membrane of the alimentary canal and to increase acid secretion in the stomach. Stomach cramps may, therefore, be aggravated and gastrointestinal upset be caused by ice packing of the anterior lower abdominal area; increased gastric acidity is undesirable in the presence of pathology of the stomach, particularly if peptic ulcers are present.

Extreme care must be exercised in the application of cryotherapy techniques to treatment sites which may suffer from disturbed skin sensations or sensory impairment. In such cases the patient's skin must be carefully tested for abnormal sensitivity to cold and normal reactions to its application (refer to Cryotherapy, Evaluation of Capillary Response to Cold [M10H]). Additionally, cryotherapy is considered by many to be contraindicated for treatment sites with open wounds. The judgement regarding cryotherapeutic applications to such sites should be based on the extent, nature and depth of the wound involved, and the chance of contaminating the wound. Advantages must be weighed against the risks. For example, there is little risk and a great deal of benefit to be gained by cryotherapeutic application to a freshly contused area, in spite of the presence of a shallow wound. It may be helpful in slowing or stopping bleeding if a dry sterilized ice bag is utilized or the treatment site protected from touch by the water in the ice bath (plastic bagging may be helpful); however, cryotherapeutic applications may do further damage to involved tissues if the wound is deep and/or the vasocirculation in the area is impaired.

Notes Aside   (M10F5)

Although cited by may other authorities as an aid in programs of rehabilitation for victims of upper motor neuron lesions of various types, it has proven to be a superfluous treatment technique in comparison to electromyometric feedback: inhibition externally inflicted is noteably inferior to inhibition internally exterted by the brain itself, if it can be accomplished. On the other hand, no effective substitute for the ice bath has been demonstrated for the the treatment of general spasticity accompanying multiple sclerosis or the catatonic-like states sometimes seen to result from cerebral concussion.

Bibliography   (M10F6)

Kowal, pp. 66–73.
Lehmann, pp. 563–602.
Michlovitz, pp. 73–97, 263–275, 277–294.
Murphy, pp. 112–115.

CRYOTHERAPY VASOCONSTRICTION (M10G)

Definition (M10G1)

Vasoconstriction may be affected by cold applications to the skin via affected sympathetic afferent and efferent nerve fibers innervating the skin and blood vessels (capillary beds included). This reflexive behavior has been used to good effect for the control of acute edema and for the control of bleeding (not discussed in this work), by the direct application of the cryotherapeutic instrument to the affected area.

Much ignored and as yet relatively unexplored is the fact that the sympathetic nervous vasoconstriction response produced by local application of a cryotherapy instrument also causes a vasodilation to occur in a distal area which is apparently in sympathetic vascular balance with the cooled area. In her lectures at the University of Southern California, Margaret Rood suggested that the sympathetic balancing of circulatory patterns could be used to good effect for the treatment and evaluation of the vascular or migraine headache. "Quick icing" of the pinnae of the ear causes vasoconstriction of the blood vessels (capillaries included) of the pinna, which reflexly causes the coinnervated temporal artery (and other important vessels) to simultaneously constrict as the reciprocally innervated vessels in the hand vasodilate. The vascular headache will immediately be temporarily relieved completely for at least several minutes. Sympathetic reflex balancing of circulatory patterns is tantalizing to the clinical practitioner and must be more fully explored for the obvious promise of treatment technique development.

Procedure of Application (M10G2)

Cryotherapeutic devices (M10G2A)

To control bleeding, ice packing, or ice massage are the most effective of the cryotherapeutic applications. Quick-ice is notably effective for causing reflex vasoconstriction.

Ice packing (M10G2A01) The most effective ice pack is one made of towelling and crushed ice:

1) A towel should be spread out on a flat surface.
2) A one inch layer of crushed ice (enough to cover the treatment site) should be spread out in the middle of the towel.
3) The long edges of the towel should be folded to overlap each other to cover the ice.
4) The ends of the towel should then be rolled up toward each other to make handling easy.
5) Before being applied to the patient the "bottom" of the ice pack should be moistened with water.
6) The ice pack should then be unrolled in place over the treatment site.
7) In common use, the crushed ice pack should not be left in place longer then 10 to 12 minutes, enough time to cause a vasodilation sufficient to create an erythemic (bright pink) reaction of the skin and/or local anesthesia.

Ice massage for Quick-ice (M10G2A02) A hand held ice cube should be quickly stroked over the treatment site several times and the skin blotted immediately after.

A handled ice massage applicator may made by placing a tongue depressor (held in place upright by a long hair clip, clipped to the tongue depressor and placed across the mouth of the cup) in a small paper cup of water and freezing it. The cup should be removed when the water is frozen, before using it for treatment. Commercial home kitchen popsicle makers (sans sugar in the water) may also be used to produce a handled ice massage applicator.

Treatment protocol (M10G2B)

Bleeding control (M10G2B01)

1) The patient should ideally be placed in a semireclined comfortable position (a reclining chair may be suitable).
2) The dermis to be treated should be bared.
3) The ice pack should be applied to the injured body part.
4) The ice pack should be applied for an appropriate period. An ice pack should be left in place for at least 10 minutes.
5) After the ice pack is removed, the patient's skin should be thoroughly dried off.
6) The patient should leave the treatment area in warm dry clothing.

Quick-ice treatment of the migraine headache (M10G2B02) The quick-ice technique, utilized to treat vascular headache may be performed with a hand held ice cube or an ice massage applicator. Ice should be applied to the skin of the each of the patient's pinnae, in three quick sweeps, both front and rear. The ice should then be applied briefly to one of the patient's temple, in a rapid circular motion, then drawn across the forehead to the other temple, which should in turn be massaged; this sequence should be repeated until the ice has been drawn across the forehead 3 times. A smooth flat surface of the ice should then be drawn once across each of the patient's closed eyelids. Relief of the vascular headache pain should be immediate and last several minutes or indefinitely.

Partial List of Conditions Treated   (M10G3)

Treatable Causes   (M10G3A)

Habitual joint positioning   (C48)
Hypermobile/instable joint   (C10)
Joint approximation   (C47)
Migraine (vascular headache)   (C34)
Soft tissue swelling   (C06)
Unhealed dermal lesion   (C12)
Vascular insufficiency   (C11)

Syndromes   (M10G3B)

ACHILLES TENDONITIS   (S52)
BUERGER'S DISEASE   (S11)
CHONDROMALACIA   (S15)
DIABETES   (S10)
ELBOW PAIN   (S47)
FOOT PAIN   (S51)
FOREARM PAIN   (S54)
HEADACHE PAIN   (S02)
JOINT SPRAIN   (S30)
KNEE PAIN   (S46)
LOW BACK PAIN   (S03)
MENSTRUAL CRAMPING   (S60)
MIGRAINE (VASCULAR) HEADACHE   (S18)
MYOSITIS OSSIFICANS   (S67)
OSTEOARTHRITIS   (S21)
PITTING (LYMPH) EDEMA   (S31)
POST IMMOBILIZATION   (S36)
RAYNAUD'S DISEASE   (S12)
SHOULDER/HAND   (S57)
SHOULDER PAIN   (S08)
TENDONITIS   (S28)
TENNIS ELBOW (LATERAL EPICONDYLITIS)   (S32)
THIGH PAIN   (S49)
TOOTHACHE/JAW PAIN   (S41)
UNHEALED DERMAL LESION   (S39)
WRIST PAIN   (S16)

Precautions and Contraindications   (M10G4)

Adverse effects from cryotherapy are rare, but certain individuals have been shown to be systemically hypersensitive to cold applications (seemingly "allergic" to cold). These individuals may be divided up into three distinct groups, those whose tissues, upon cooling, release (1) histamine or histamine-like chemicals, (2) cold hemolysins and agglutinins, and (3) cryoglobulins.

The symptoms associated with histamine production triggered by cooling include cold urticaria, erythema, itching, sweating, facial flush, puffiness of the eyelids, laryngeal edema, and shortness of breath. More severe symptoms which may develop include syncope, hypotension, tachycardia, extra systoles, dysphagia, abdominal cramping, diarrhea, and/or vomiting. The resultant symptomatology may occur during therapy or several minutes or even hours after the technique has been applied. Such hypersensitivity to cold may be treatable through a program of desensitization to cold: the patient's hands should be immersed in water cooled to 50° F. for two minutes, twice daily for from 14 to 21 days.

Cold hemolysins and agglutinins produced in abnormal amounts in response to tissue cooling may produce significant anemia, malaise, chills, fever, paroxysmal cold hemoglobinuria of the kidney, and/or the dermal symptoms of cold urticaria, ulcerations, Raynaud's phenomenon, and acrocyanosis.

Abnormally high cryoglobulin production in response to cooling may produce chills and fever, and may adversely affect vision, causing blindness in the extreme, and hearing, causing deafness in the extreme. It may cause conjunctival hemorrhages and epistaxis, anemia with fibrinogenopenia, elevated sedimentation rates, and abnormal numbers of cryoglobulin inclusion cells. Other symptoms include erythema, itching, purpura, cold urticaria, Raynaud's phenomenon, ulcerations and necrosis, problems breathing (dyspnea), stomatitis, melena, and bleeding of the gums.

The adverse reactions to cryotherapy described above have been noted to occur in association with various systemic diseases including Raynaud's disease, Buerger's disease, lupus erythematosus, leucocyto-clastic vasculitis, rheumatoid disease, progressive systemic sclerosis, multiple myeloma, and pneumonia; consequently, cryotherapy involving immersion or the ice packing of the distal extremities of patient's suffering from these diseases may be contraindicated. Any cryotherapy applied to patients of this type must be supervised carefully, and should avoided if at all possible.

If arterial insufficiency is present (as is the case in some systemic diseases), cold applications would seem to be contraindicated because of their potent vasoconstricting effect which may potentially reduce already compromised blood flow through the extremity.

Some individuals, though not suffering from any particular disease process, respond to the natural vasopressor response to application of cold (capillary vasoconstriction) with failure to eventually compensate by capillary vasodilation, thus producing a Buerger's disease-like syndrome which may threaten the tissues with rapidly developing frostbite if exposure to the cold is not eliminated. This condition may cause the patient exquisite pain which may not be readily relieved. Such hyperreactivity to cooling may be readily tested for by observing the tissue response to ice packing. The normal dermal response to an applied crushed ice pack is to blanch (vasoconstrict) for the first three minutes after the ice pack is applied; the skin gradually develops an increasing erythemic reaction, becoming bright red after ten minutes of constant exposure. If the blanching phase does not end after the first three minutes of exposure to the ice pack, the patient may be considered hyperresponsive to exposure to cold (even without other symptoms) and the treatment should be immediately discontinued.

Though rare, frostbite is a potential danger inherent in cryotherapy, especially for techniques involving ice massage or the coolant sprays although reports suggest that such damage is rarely severe. Frostbite occurs as a consequence of over-exposure to cold which causes an increase in the capillary permeability. The consequent rapid loss of plasma strands red blood cells in the capillaries, which then clump together to form occlusive masses (clots) which finally undergo hyalinization. Treatment of frostbite involves the immediate immersion in warm water (109° F.). Soft tissue damage from frostbite is increased (1) as the length of time the tissues are frozen increases, (2) as the temperature of the tissues while in the frozen state decreases, and (3) as the duration of thawing increases.

Ice packing applied to the abdomen increases peristalsis in the stomach, small intestine and the colon and has been found to increase blood flow through the mucous membrane of the alimentary canal and to increase acid secretion in the stomach. Stomach cramps may, therefore, be aggravated and gastrointestinal upset be caused by ice packing of the anterior lower abdominal area; increased gastric acidity is undesirable in the presence of pathology of the stomach, particularly if peptic ulcers are present.

Extreme care must be exercised in the application of cryotherapy techniques to treatment sites which may suffer from disturbed skin sensations or sensory impairment. In such cases the patient's skin must be carefully tested for abnormal sensitivity to cold and normal reactions to its application (refer to Cryotherapy, Evaluation of Capillary Response to Cold [M10H]). Additionally, cryotherapy is considered by many to be contraindicated for treatment sites with open wounds. The judgement regarding cryotherapeutic applications to such sites should be based on the extent, nature and depth of the wound involved, and the chance of contaminating the wound. Advantages must be weighed against the risks. For example, there is little risk and a great deal of benefit to be gained by cryotherapeutic application to a freshly contused area, in spite of the presence of a shallow wound. It may be helpful in slowing or stopping bleeding if a dry sterilized ice bag is utilized or the treatment site protected from touch by the water in the ice bath (plastic bagging may be helpful); however, cryotherapeutic applications may do further damage to involved tissues if the wound is deep and/or the vasocirculation in the area is impaired.

Treatment Notes (M10G5)

Cryotherapeutic applications to control swelling of soft tissues is most effectively applied to injured soft tissues shortly following injury. This is also true for the control of bleeding which has occurred as the result of soft tissue laceration or tearing, in which case its effectiveness is increased if it is applied with direct pressure to the wounds. Cryotherapy applications applied to an open wound should be as sterile as possible; cryotherapeutic applications may be inappropriate if the wound involved is too extensive or too deep.

The normal autonomic (sympathetic) nervous system response to local cryotherapy application of vasoconstriction in the application site normally causes a simultaneous vasodilation to occur in an area distal to the application site; this relationship of converse responses is apparently an expression of sympathetic vascular balance which exists as a normal function of the autonomic nervous system. As indicated previously, in her lectures at the University of Southern California, Margaret Rood suggested that the sympathetic balancing of circulatory patterns could be used to good effect for the (temporary) treatment of and differential evaluation of the vascular or migraine headache.

Quick-ice applied to the pinnae of the ear normally causes the vasoconstriction of the blood and capillaries of the pinna and the temporal artery. It also causes associated coinnervated capillary beds to simultaneously constrict as the vessels (including capillary beds) in the hand are made to reciprocally vasodilate. As a result, the vascular headache will be immediately (albeit temporarily) relieved, the pain disappearing completely for several minutes at least. A similar vascular relationship exists between the lower intestine and skin covering the anterior abdomen. Cooling of the abdominal skin causes an increase of peristalsis in the stomach, small intestine and the colon; it has been found to increase blood flow through the mucous membrane of the alimentary canal and to increase acid secretion in the stomach. Refer to

Superficial Thermal (Heat) Therapy, Circulation Enhancement, Reflex Effects of Superficial Heating [M11A1B] and Sensory Stimulation, Provoking Reflex Physiological Responses [M17C] for further examples of such vascular relationships and similar mechanisms.

Bibliography (M10G6)

Finnerty, pp. 135–137, 142–145.
Guyton, pp. 284–287.
Kowal, pp. 66–73.
Lehmann, pp. 404–441, 574–579.
Michlovitz, pp. 79–81, 99–117, 263–275.
Moor, pp. 15–17.
Murphy, pp. 112–115.
Shodell, pp. 78–82.

CRYOTHERAPY EVALUATION OF CAPILLARY RESPONSE TO COLD (M10H)

Definition (M10H1)

Hyperreactivity to cold (M10H1A)

Some individuals, though not suffering from any particular (or identifiable) disease process, respond with an inappropriate vasopressor response to prolonged cold application; i.e., their systems fail to eventually compensate with capillary vasodilation for the initial capillary vasoconstriction response to cooling. Thus, a Buerger's disease-like syndrome may be created which may threaten the tissues with rapidly developing frostbite if exposure to the cold is not eliminated and additionally cause the patient exquisite pain which may not be readily relieved. This hyperreactive response to cold may be readily tested for by observing the tissue response to ice packing.

Reflex responses to cold (M10H1B)

The normal autonomic (sympathetic) nervous system response to local cryotherapy application of vasoconstriction in the application site normally causes a simultaneous vasodilation to occur in an area distal to the application site. This relationship of converse responses is apparently an expression of sympathetic vascular balance which exists as a normal function of the autonomic nervous system. In her lectures at the University of Southern California, Margaret Rood suggested that the sympathetic balancing of circulatory patterns could be used to good effect for the (temporary) treatment of and differential evaluation of the vascular or migraine headache. "Quick icing" of the pinnae of the ear normally causes the vasoconstriction of the blood and capillaries of the pinna and the temporal artery. It also causes associated coinnervated capillary beds to simultaneously constrict as the vessels (including capillary beds) in the hand are made to reciprocally vasodilate. As a result, the vascular headache will be immediately (albeit temporarily) relieved, the pain disappearing completely for several minutes at least. A similar vascular relationship exists between the lower intestine and skin covering the anterior abdomen. Cooling of the abdominal skin causes an increase of peristalsis in the stomach, small intestine and the colon; it has been found to increase blood flow through the mucous membrane of the alimentary canal and to increase acid secretion in the stomach. Refer to Superficial Thermal (Heat) Therapy, Circulation Enhancement, Reflex Effects of Superficial Heating [M11A1B] and Sensory Stimulation, Provoking Reflex Physiological Responses [M17C] for further examples of such vascular relationships and similar mechanisms.

Procedures of Application (M10H2)

Determination of hyperreactivity to cold (M10H2A)

The normal dermal response to an applied crushed ice pack is to respond with vasoconstriction by blanching for the first three minutes after the ice pack is applied. The skin gradually develops an increasing erythemic reaction, becoming bright red after ten minutes of constant exposure. If the blanching phase does not end after the first three minutes of exposure to the ice pack the patient may be considered hyperresponsive to exposure to cold (even without other symptoms) and the treatment should be immediately discontinued.

Determination of vascular headache (M10H2B)

The quick-ice technique utilized to treat vascular headache may be performed with a hand-held ice cube or an ice massage applicator. Ice should be applied to the skin of the each of the patient's pinnae, in three quick sweeps, both front and rear. The ice should then be applied briefly to one of the patient's temples, in a rapid circular motion, then drawn across the forehead to the other temple, which should in turn be massaged. This sequence should be repeated until the ice has been drawn across the forehead three times. A smooth flat surface of the ice should then be drawn once across each of the patient's closed eyelids. Relief of the vascular headache pain should be immediate and last several minutes or indefinitely.

Determination of abdominal pain source (M10H2C)

Ice packing applied to the abdomen for from three to twelve minutes increases peristalsis in the stomach, small intestine and the colon, while also increasing blood flow through the mucous membrane of the alimentary canal and acid secretion in the stomach. Stomach cramps may therefore be aggravated and gastrointestinal upset be caused by ice packing of the anterior lower abdominal area, thus providing a way of confirming the source of chronic or acute abdominal pain, if the produced pain duplicates that previously experienced by the patient and all other evaluative tests are negative (refer to Table of Referred Pain Patterns of Trigger Point Formation Origin [T005], Table of Referred Pain Patterns of Visceral Organ Origin [T017], and Auricular Electroacustimulation, Auricular Acupoint Evaluative Survey [M06C]).

Partial List of Conditions Evaluated   (M10H3)

Treatable Causes   (M10H3A)

Migraine (vascular headache)   (C34)
Nausea/vomiting   (C36)
Peripheral nerve injury   (C07)
Sympathetic hyperresponse   (C38)
Tactile hypersensitivity   (C24)
Vascular insufficiency   (C11)
Visceral organ dysfunction   (C02)

Syndromes   (M10H3B)

BUERGER'S DISEASE   (S11)
COLITIS   (S42)
DIABETES   (S10)
FOOT PAIN   (S51)
HAND/FINGER PAIN   (S17)
HEADACHE PAIN   (S02)
HYPERTENSION   (S43)
LOWER ABDOMINAL PAIN   (S62)
MIGRAINE (VASCULAR) HEADACHE   (S18)
PITTING (LYMPH) EDEMA   (S31)
POST PERIPHERAL NERVE INJURY   (S23)
RADICULITIS   (S29)
RAYNAUD'S DISEASE   (S12)
REFERRED PAIN   (S01)
SHOULDER/HAND   (S57)
SINUS PAIN   (S33)
TEMPOROMANDIBULAR JOINT (TMJ) PAIN   (S06)
WRIST PAIN   (S16)

Notes Aside   (M10H4)

The utilization of quick-ice to establish the existence of a vascular (or migraine) headache should be a routine procedure in the evaluation of headache conditions.

Sympathetic reflex balancing of circulatory patterns is a tantalizing idea to the clinical practitioner and such mechanisms need to be more fully explored for the obvious promises it implies for additional treatment technique development and a better understanding of the neuroautonomic vascular system.

Bibliography   (M10H5)

Finnerty, pp. 135–137, 142–145.
Lehmann, pp. 404–441, 574.
Michlovitz, pp. 99–117, 263–275.
Moor, pp. 15–17.

DENTAL PAIN (C22)

Definition (C22A)

Dental pain is defined here as pain that has its origin in the mandible, maxillary, teeth, gingiva or palate.

Evaluation (C22B)

Evaluation of dental pain is dependent on patient reports of pain (as to a location and intensity) and the determination through physical examination that the pain is not the result of another treatable cause (trigger point formation referred pain, osteoarthritis of the temporomandibular joint, etc.).

Etiology (C22C)

The causes of dental pain include dental caries, pulpitis (inflammation of the dental pulp), dentoalveolar abscess, oral lichen planus (dermatosis producing hyperkeratotic changes in the mucosa), submaxillary cellulitis, tooth extraction, tooth fractures, fractures of the mandible and maxilla, and as the sequelae of surgical procedures performed on the mandible, maxilla, palate or gingiva.

Treatment Notes (C22D)

Auricular electroacustimulation has proven to be remarkably effective for temporarily relieving true dental pain (when used appropriately), especially for the temporary relief of toothache, post extraction and post palate surgery pain.

Bibliography (C22E)

Bickley, pp. 199–200.
Guyton, pp. 943–944.
Holvey, pp. 371–397.
Travell, pp. 219–289.

TREATMENT OF DENTAL PAIN (C22F)

Technique Options (C22F1)

Evaluation techniques (C22F1A)

Auricular electroacustimulation	*M06*
Auricular acupoint evaluative survey	M06C
Determination of pathological symptom source	M06C5B
Massage	*M23*
Palpation evaluation	M23G
Determination of trigger point formations	M23G1
Peripheral electroacustimulation	*M05*
Peripheral acupoint survey	M05D

Treatment techniques (C22F1B)

Auricular electroacustimulation	*M06*
Pain relief	M06A
Electrical facilitation of endorphin production (acusleep)	*M03*
Increasing pain tolerance	M0305A
Laser stimulation	*M27*
Massage	*M23*
Acupressure	M23C
Peripheral electroacustimulation	*M05*
Analgesia	M05B
Transcutaneous nerve stimulation	*M04*
Prolonged stimulation of acupoints	M04B
Ultrahigh frequency sound (ultrasound)	*M01*
Acupoint stimulation	M01K

Techniques Suggested for the Acute Condition (C22F2)

Evaluation techniques (C22F2A)

Auricular electroacustimulation	*M06*
Auricular acupoint evaluative survey	M06C
Determination of pathological symptom source	M06C5B
Massage	*M23*
Palpation evaluation	M23G
Determination of trigger point formations	M23G1

Treatment techniques (C22F2B)

Auricular electroacustimulation	*M06*
Pain relief	M06A
Auricular acupoint evaluative survey	M06C
Determination of pathological symptom source	M06C5B

Techniques Suggested for the Chronic Condition (C22F3)

Evaluation techniques (C22F3A)

Auricular electroacustimulation	*M06*
Auricular acupoint evaluative survey	M06C

    Determination of pathological symptom source   M06C5B
Peripheral electroacustimulation   ***M05***
    Peripheral acupoint survey   M05D

Treatment techniques (C22F3B)

Auricular electroacustimulation   ***M06***
    Pain relief   M06A

Electrical facilitation of endorphin production (acusleep)   ***M03***
    Increasing pain tolerance   M0305A
Peripheral electroacustimulation   ***M05***
    Analgesia   M05B
Transcutaneous nerve stimulation   ***M04***
    Prolonged stimulation of acupoints   M04B

TABLE OF DEVELOPMENTAL REFLEXES (T007)

REFLEX	AGE OF NORMAL APPEARANCE AND OF DISAPPEARANCE	APPROPRIATE DEVELOPMENTALLY-ASSOCIATED FUNCTIONS
[First Stage Developmental Reflexes]		
Flexor withdrawal	Birth to two months	1. Sucking
Extensor thrust	Birth to two months	2. Blowing
Crossed extension (1)	Birth to two months	3. Yelling
Crossed extension (2)	Birth to two months	4. Tongue thrust inhibition
		5. Tongue control exercise
[Second Stage Developmental Reflexes]		
Tonic labyrinthine supine	Birth to four months	1. Lift head from prone and maintain position
Tonic labyrinthine prone	Birth to four months	2. Self support on forearms
Positive supporting reaction	Birth to four months	3. Chewing
Negative supporting reaction	Birth to four months	
Asymmetrical tonic neck	One to four months	
[Third Stage Developmental Reflexes]		
Tonic lumbar	Four to six months	1. Rolling from supine to prone toward the involved side
[Fourth Stage Developmental Reflexes]		
Symmetrical tonic neck (1)	Six to eight months	1. Prone lying with hyperextension of the entire spine
Symmetrical tonic neck (2)	Six to eight months	2. Prone lying with the chin up
		3. Prone resting on elbows
		4. Crawling
		5. Self support on hands, elbows extended
[Fifth Stage Developmental Reflexes]		
Contralateral associative reactions		1. Hands and knees standing
Homolateral associative reactions		2. Creeping reciprocally
		3. Kneeling while maintaining balance
		4. Walking sideways with hand support
		5. Standing and maintaining balance against intermittent resistance

DIABETES SYNDROME (S10)

Definition (S10A)

Diabetes mellitus is a hereditary or developmental disease which expresses itself as a mismetabolism of carbohydrates due to an absolute or relative insufficiency of insulin. Diabetes can occur at any age, appearing as hyperglycemia, glycosuria, polyuria, polydipsia, polyphagia, or pruritus.

The exact cause of diabetes is unknown. Heredity is associated with fifty percent of the cases reported. Obesity, disorders of the endocrine glands other than the pancreas, infection, pancreatitis, pancreatic tumors, hemochromatosis and the use of some drugs (adrenocortical steroids and thiazide diuretics) have all been noted as possible precursors or precipitators of diabetes. It is a disease most commonly seen between fifty and sixty years of age with both sexes being affected equally.

Common symptoms of diabetes include polyuria, thirst, itching, hunger, weakness and weight loss. Large amounts of urine are passed daily, which results in excessive thirst and dryness of the skin.

The one of the primary symptoms contributing to the *diabetic syndrome* is premature arteriosclerosis. This brings attending danger of heart disease, the possibility of cerebral vascular accident (stroke), decreased arterial blood supply to the feet and small blood vessel involvement in the retina and kidney, which may produce diabetic retinopathy and albuminuria, respectively. As a result of the impaired blood circulation, infections may become common and healing time may become greatly protracted. Additionally, the depression of the arterial blood supply to the lower extremities often leads to intermittent claudication which produces cold feet, paresthesias, slow healing of injuries, local infection, ulceration and (if allowed to progress far enough) gangrene.

Ketosis is a common symptom of a diabetic episode. Symptoms of ketosis include lassitude, malaise, anorexia, thirst and polyuria. Should ketosis progress far enough, acidosis may develop, which increases the severity of these symptoms, and nausea, vomiting, dizziness, involuntary deep breathing and (if uncompensated for) coma may be added to the list of symptoms.

Diabetic neuritis may also develop. Early symptoms include tingling of the skin, paresthesias and numbness of the hands and feet, a constant, aching pain in the distal extremities (reported to be worse at night), diminished or lost vibration sense and decreased or absent lower extremity stretch reflexes. It has been noted that the autonomic and central nervous systems may also be involved, but is rarely seen in well-controlled diabetes.

Patients taking insulin may be subject to insulin reactions. Insulin shock (hypoglycemia) may occur if too much insulin has been taken or too little food eaten. Symptoms include blurred vision, weakness, dizziness, lack of affect, a glazed appearance of the eyes and a failure to respond to interrogation. Sugar in the form of fruit, such as oranges or bananas or candy should be administered immediately if these symptoms occur.

Related Syndromes (S10B)

BACTERIAL INFECTION (S63)
BUERGER'S DISEASE (S11)
FOOT PAIN (S51)
HAND/FINGER PAIN (S17)
HYPERTENSION (S43)
HYSTERIA/ANXIETY REACTION (S59)
PHOBIC REACTION (S44)
POST CEREBRAL VASCULAR ACCIDENT (CVA) (S07)
POST PERIPHERAL NERVE INJURY (S23)
RAYNAUD'S DISEASE (S12)
UNHEALED DERMAL LESION (S39)

Treatable Causes Which May Contribute To The Syndrome (S10C)

Bacterial infection (C09)
Blood sugar irregularities (C31)
Hypertension (C39)
Localized viral infection (C27)
Muscular weakness (C23)
Nausea/vomiting (C36)
Necrotic soft tissue (C18)
Neurogenic dermal disorders (psoriasis, hives) (C49)
Soft tissue inflammation (C05)
Soft tissue swelling (C06)
Sympathetic hyperresponse (C38)
Tactile hypersensitivity (C24)
Trigger point formation (C01)
Unhealed dermal lesion (C12)
Vascular insufficiency (C11)
Visceral organ dysfunction (C02)

Treatment Notes (S10D)

Physical therapy treatment of the diabetic syndrome can be preventative as well as symptom reductive.

Temperature training, teaching the patient to increase capillary dilation in the hands and feet, can be useful in preventing the progression of decreasing circulation. Such training has been shown to be effective in regaining normal skin temperatures in both hands and feet in diabetic patients who have not yet developed intermittent claudication sufficient to cause dermal breakdown or gangrene (refer to Temperature Monitry, Circulatory Disorder Control [M19C].

Exercise has been reported to be an effective way of controlling mild forms of diabetes; refer to Exercise [M12] for further reference.

Should intermittent claudication in the extremities become advanced enough to cause dermal breakdown, various treatment techniques may prove useful in restoring dermal circulation and precipitating healing. Such modalities include electrical stimulation (refer to Electrical Stimulation, Increasing Blood Flow [M02C]), contrast packs or baths (refer to Hydrotherapy, Circulation Enhancement [M15B]), vibration (refer to Vibration, Circulation Enhancement [M08F]), sensory stimulation (refer to Sensory Stimulation, Brushing [M1702A] and Provoking Reflex Physiological Responses [M17C]), auricular electroacustimulation (refer to Auricular Electroacustimulation, Pathology Control [M06B]), peripheral electroacustimulation (refer to Peripheral Electroacustimulation, Acupoint Electrical Stimulation within an Inflamed zone [M05A]), superficial heating (refer to Superficial Thermal (Heat) Therapy, Circulation Enhancement [M11A]), and massage (refer to Massage, Circulation Enhancement and Reduction of Edema [M23B]).

Bibliography (S10E)

Bickley, pp. 183–186.
Guyton, pp. 915–928.
Holvey, pp. 325–335.
Scott, pp. 417–418.

DIFFERENTIAL SKIN RESISTANCE SURVEY (M22)

Definition (M2201)

Differential skin resistance (DSR) survey is the measurement of skin resistance to establish the presence of zones of relatively high skin resistance, for the establishment and evaluation of soft tissue inflammatory conditions.

Zones of comparatively high skin resistance (relative to that of surrounding areas) have been clinically correlated with deep, subdermal inflammation of fascial, tendon, muscle, and joint tissues. When a deep tissue inflammation is present, zones of high skin resistance appear just over the extent of the distribution of the inflammation or inflammatory process.

The source of the correlation between high skin resistance and deep tissue inflammation is still unproven. It has been speculated that the effects on circulation within the skin and deeper tissues (including increased capillary and cellular permeability, arteriole dilation, and venule constriction) caused by the chemicals produced by irritated or stressed soft tissues (prostaglandins, histamine and bradykinin), may somehow affect an increase in the surface skin resistance right over the inflamed area through a mechanism that is the reverse of that triggered by the local application of heat (heat pack, hot air, etc.). That shunting mechanism takes blood up to the surface of the skin from the deeper tissues when the body attempts to cool the area directly under the heat source, thereby causing a decrease in the skin resistance at the site of application. If interpreted correctly, the effect of the heat produced by the inflamed deep tissues is to cause the shunting of blood away from the skin right over the inflammation into the deeper tissues, thereby affecting an increase in the skin resistance.

The DSR survey is performed through the use of an ohm meter especially designed to monitor the resistance to the passage of a small current of electricity through or over the surface of the skin. The instrument itself is a variation on the galvanic skin response (GSR) monitor instrumentation utilized for auricular or peripheral acupoint finding, and for GSR feedback monitry. In fact, with proper accessories and the proper setting of controllable variables, these instruments may be used to perform the DSR survey (with greater or lesser facility and convenience of application).

The GSR or ohm meter instrumentation utilized in the clinical setting for point finding or for DSR survey generally employ audio and visual feedback modes. Such instruments generally must be able to measure resistance levels ranging from 0 to 30 micro ohms (mohms). One GSR feedback monitor, for example, may be set to monitor 0 to 50 mohms or 10 mohm increments (0-10, 10-20, 20-30, 30-40, and 40-50 mohms) within the general range with the additional facility of being adjusted to monitor a 2 mohm range at the center of any of the 10 mohm incremental ranges; i.e., the instrument may be set so that feedback may be representative of resistance at different sensitivity levels.

Physically the GSR instrument communicates with the patient's skin via two insulated wire electrode cables. Ideally, the electrode cables connect with two identical electrode handles, fitted at the end with hollow brass or steel tips. These hollow tips are of a size to allow the insertion of a paper stick tipped with cotton. In use, these cotton tips are moistened and pressed against the skin to serve as a vector of a small electrical current, called a trickle current, which passes from one electrode to the other. The instrument registers how much of the current is passed between the two electrodes, which is a measure of the skin resistance. Comparison of skin resistance measurements taken from various areas of the skin comprise the process of DSR survey.

The DSR survey method has been shown to be an efficient, inexpensive method of verification or determination of the various soft tissue "itises" or soft tissue inflammation conditions, including capsulitis, bursitis, tendonitis, fasciitis, synovitis, myelitis, radiculitis, osteoarthritis and synovial tunnel syndromes. The DSR survey may be used not only as a preliminary evaluative tool, but also as a method of documenting condition improvement. The high skin resistance zones, correlated with soft tissue inflammatory processes, shrink in size and finally disappear as the inflammation is ameliorated with adequate treatment.

For detailed discussions on the other uses of GSR instrumentation, refer to Peripheral Electroacustimulation [M05], Auricular Electroacustimulation [M06] and Galvanic Skin Response Monitry [M20].

Procedures of Application (M2202)

1) The patient should be sitting or lying in a comfortable position.

2) The skin to be surveyed should be thoroughly cleansed (an alcohol rub may be sufficient).

differential skin resistance survey

3) The two hand-held electrode handles (with moistened cotton tipped ends) should be placed with unvarying equal light pressure against the patient' skin in the area to be surveyed, .25 to 2 inches apart.

4) The electrodes should be moved systematically in and around the area suspected of being inflamed. They should be moved in parallel .25 inches at a time, first in an inferior to superior direction and then in a medial to lateral direction to establish the encompassing limits of the high skin resistance. The skin over an inflamed area will be 1 or more mohms above that found over the surrounding skin. The limits should be marked by surgical or felt tipped pen if the area is to be subjected to subsequent treatment.

5) Once a zone of high skin resistance has been established, the degree of inflammatory entrenchment (how serious and roughly how old it is) may be tested by driving a small electrical current through the area for as little as four seconds. An area seriously inflamed and/or inflammation of long duration will not be affected by the current and the skin resistance will continue to remain high, while less involved areas will experience decreases in the resistance to varying degrees. The bigger the drop in the resistance and the more sustained the drop, the less entrenched or old the inflammation may be.

Partial List of Conditions Evaluated (M2203)

Treatable Causes (M2203A)

Bacterial infection (C09)
Calcific deposit (C08)
Habitual joint positioning (C48)
Hypermobile/instable joint (C10)
Interspinous ligamentous strain (C03)
Joint ankylosis (C42)
Joint approximation (C47)
Ligamentous strain (C25)
Localized viral infection (C27)
Nerve root compression (C04)
Peripheral nerve injury (C07)
Restricted joint range of motion (C26)
Soft tissue inflammation (C05)
Trigger point formation (C01)

Syndromes (M2203B)

ACHILLES TENDONITIS (S52)
ACNE (S45)
ADHESION (S56)
BACTERIAL INFECTION (S63)
BELL'S PALSY (S66)
BURSITIS (S26)
CALF PAIN (S50)
CAPSULITIS (S27)
CARPAL TUNNEL (S35)
CERVICAL DORSAL OUTLET (SCALENICUS ATTICUS/CERVICAL RIB) (S09)
CERVICAL (NECK) PAIN (S73)
CHEST PAIN (S14)
CHONDROMALACIA (S15)
DIABETES (S10)
EARACHE (S40)
ELBOW PAIN (S47)
FACET (S19)
FASCIITIS (S20)
FOOT PAIN (S51)
FOREARM PAIN (S54)
FROZEN SHOULDER (S64)
HAND/FINGER PAIN (S17)
HEADACHE PAIN (S02)
HIGH THORACIC BACK PAIN (S48)
JOINT SPRAIN (S30)
KNEE PAIN (S46)
LOWER ABDOMINAL PAIN (S62)
LOW BACK PAIN (S03)
MYOSITIS OSSIFICANS (S67)
OSTEOARTHRITIS (S21)
PHLEBITIS (S68)
PIRIFORMIS (S04)
POST IMMOBILIZATION (S36)
POST PERIPHERAL NERVE INJURY (S23)
POST WHIPLASH (S55)
RADICULITIS (S29)
REFERRED PAIN (S01)
SCIATICA (S05)
SHINGLES (HERPES ZOSTER) (S37)
SHOULDER/HAND (S57)
SHOULDER PAIN (S08)
SINUS PAIN (S33)
TEMPOROMANDIBULAR JOINT (TMJ) PAIN (S06)
TENDONITIS (S28)
TENNIS ELBOW (LATERAL EPICONDYLITIS) (S32)
THIGH PAIN (S49)
TINNITUS (S70)
TOOTHACHE/JAW PAIN (S41)
WRIST PAIN (S16)

Precautions and Contraindications (M2204)

There have been no precautions or contraindications sited for the DSR survey procedure. Practitioners should exercise caution when applying the DSR probe electrode ends so that the soft tissues do not suffer from the pressure of application. A respect for soft tissue sensitivity relative to the individual and the particular condition should be observed (adding irritation to already irritated tissues is not an especially good idea nor is it therapeutic).

Notes Aside (M2205)

The DSR survey procedure should be applauded as a breakthrough in soft tissue inflammation evaluation. It is relatively inexpensive, easily performed (with education and appropriate equipment), clean in application, and extremely objective. Its attributes are even more attractive in comparison with its nearest competitor, thermography, which is complicated, messy, and expensive.

Bibliography (M2206)

Rosenblatt, pp. 131–137.
McCarroll, pp. 131–137.

DUPYTREN'S CONTRACTURE SYNDROME (S34)

Definition (S34A)

Dupytren's contracture arises out of a chronic inflammation of the palmar fascia, leading to progressive fibrosis and contracture, principally of the pretendinous bands of the palmar aponeurosis to the fingers. Changes in the flexor tendons and/or digital joints occur only in extreme or very advanced cases. Dupytren's contracture occasionally (though rarely) develops in the plantar fascia of the foot.

Dupytren's contracture first appears as a small painless nodule or thickening in the palmar facia, over a flexor tendon in the metacarpal region. As the thickening progresses, a dimpling of the skin over the nodule begins to form and gradually develops into a thickened longitudinal band. As pressure begins to mount, a contracture of the tendon gradually develops which progressively forces the involved metacarpophalangeal and adjacent interphalangeal joints into flexion, while the distal interphalangeal joint retains a normal range of motion, until pulled into contact with the palm by an extreme development of the contracture. The fingers most often affected are the ring or little fingers, sometimes both. Characteristically, the inability to extend the proximal interphalangeal joint even when the wrist is flexed sets the Dupytren's contracture apart from contractures which occur as a secondary symptom of congenital malformation, spasticity, trauma or infection.

Most commonly seen to eventually afflict both hands, Dupytren's contracture develops asymmetrically, with one hand developing the condition much sooner than the other. Unilateral development occurs most frequently in the right hand. Dupytren's contracture condition afflicts two to three percent of the population. Most frequently seen in men between 55 and 75 years of age, it afflicts women less frequently but does so at an earlier age. The exact cause of Dupytren's contracture is unknown but it has been shown to have a familial association.

Related Syndromes (S34B)

ADHESION (S56)
KELOID FORMATION (S38)
POST IMMOBILIZATION (S36)
TENDONITIS (S28)

Treatable Causes Which May Contribute To The Syndrome (S34C)

Habitual joint positioning (C48)
Joint ankylosis (C42)
Muscular weakness (C23)
Restricted joint range of motion (C26)
Scar tissue formation (C15)

Treatment Notes (S34D)

Treatment of the Dupytren's contracture should be directed at inhibiting or ameliorating inflammation in the acute condition with phonophoresis of nonsteroidal antiinflammatories, or other means (refer to Ultrahigh Frequency Sound, Phonophoresis [M01A]). The chronic condition should be treated as a scar tissue, adhesion problem and treated with ultrasound and soft tissue massage to stretch or break adhesion (refer to Massage, Deep Soft Tissue Mobilization [M23F]).

Bibliography (S34E)

Basmajian, pp. 159–160.
Salter, p. 247.
Shands, pp. 453–454.
Travell, pp. 526–527.

EARACHE SYNDROME (S40)

Definition (S40A)

The causes of earache include soft tissue swelling and inflammation resulting from infection of the ear passage and/or middle ear, referred pain from trigger point formations housed in musculature associated with the temporomandibular joint, tongue and neck, and toothache or infections of the maxilla and mandible.

Related Syndromes (S40B)

BACTERIAL INFECTION (S63)
BELL'S PALSY (S66)
CERVICAL (NECK) PAIN (S73)
HEADACHE PAIN (S02)
HYSTERIA/ANXIETY REACTION (S59)
OSTEOARTHRITIS (S21)
PHOBIC REACTION (S44)
POST PERIPHERAL NERVE INJURY (S23)
REFERRED PAIN (S01)
SINUS PAIN (S33)
TEMPOROMANDIBULAR JOINT (TMJ) PAIN (S06)
TIC DOULOUREUX (S53)
TINNITUS (S70)
TOOTHACHE/JAW PAIN (S41)
TORTICOLLIS (WRY NECK) (S65)

Treatable Causes Which May Contribute To The Syndrome (S40C)

Bacterial infection (C09)
Calcific deposit (C08)
Dental pain (C22)
Extrafusal muscle spasm (C41)
Habitual joint positioning (C48)
Hypermobile/instable joint (C10)
Joint approximation (C47)
Ligamentous strain (C25)
Localized viral infection (C27)
Muscular weakness (C23)
Nausea/vomiting (C36)
Nerve root compression (C04)
Neuroma formation (C14)
Neuromuscular tonic imbalance (C30)
Peripheral nerve injury (C07)
Psychoneurogenic neuromuscular hypertonus (C29)
Restricted joint range of motion (C26)
Sinusitis (C16)
Soft tissue inflammation (C05)
Soft tissue swelling (C06)
Sympathetic hyperresponse (C38)
Tinnitus (C32)
Trigger point formation (C01)
Unhealed dermal lesion (C12)

Treatment Notes (S40D)

Many of the causes of the earache syndrome, including toothache, have proven to be amenable to noninvasive treatment techniques. Earache caused by toothache may be treatable as a symptom with auricular electroacustimulation. Stimulation should be directed at relieving the toothache pain which will coincidentally relieve the earache pain (refer to Auricular Electroacustimulation, Pain Relief [M06A] and Table of Auricular Acupoint Treatment Protocols [T004]).

Referred pain from the mandibular joint may have its source from osteoarthritis, bursitis or abnormal joint pressure arising from extrafusal muscle imbalance. Each of these possibilities should be explored during evaluation and existent components should be treated appropriately. Electromyometric (EMM) neuromuscular reeducation has been clinically notable for its effectiveness in the correction of mandibular joint musculature imbalance (refer to Electromyometric Feedback, Neuromuscular Reeducation of Temporomandibular Musculature [M18F]).

Referred pain from jaw and neck musculature originates from trigger point formations within that musculature. Trigger point formations should be treated as such; the ultimate effect of such treatment will be relief of the earache pain (refer to Electrical Stimulation, Muscle Reeducation/Muscle Toning [M02B] and Vibration, Neuromuscular Management [M08A]).

Earache which has resulted from infection of the ear passage has proven in some cases to be treatable with ultrahigh frequency sound directed at killing the infecting bacterial or viral colonies responsible for the build-up of pressure (refer to Ultrahigh Frequency Sound, Bacterial Infection Eradication [M01G] or Viral Infection Eradication [M01I]).

The following is a list of trigger point formations which may, singly or in combination, imitate or contribute to the pain associated with the earache syndrome. It should be noted that it is possible for the given patient to experience pain throughout the entire stereotypical referred pain pattern or only in part or parts of it. Opposite to each listing is a description of the parts of the stereotypical referred pain pattern most commonly experienced by previous patients, relative to the given trigger point formation. Any descriptive reference which is underlined is a part of the pattern which has *not* been commonly experienced by previous patients. For complete description of each stereotypical referred pain pattern, refer to the Table of Referred Pain Patterns of Trigger point Formation Origin [T005]. Please note the key to abbreviated words at the bottom of the list.

LOCATION (MUSCLE)	REFERRED PAIN ZONES (GENERAL)
Masseter (deep)	TMJ, lateral jaw, ear, *suboccipital*
Medial pterygoid	*Ear*, TMJ, *mandible, lateral throat*
Lateral pterygoid	Anterior maxilla, TMJ (2)
Posterior digastric	Mastoid process, *jaw line, occiput*
Splenius capitus [A]	*Lower hemicranium, ear,* behind eye
Sternocleido-mastoid (deep)	Sterno., ear, occiput, bil. forehead (3)

Key:
TMJ—temporomandibular joint
sterno.—sternocleidomastoideus
a.—anterior
bil.—bilateral
m.—middle

Bibliography (S40E)

Gardner, p. 89.

ELBOW PAIN SYNDROME (S47)

Definition (S47A)

The elbow is a hinge joint with a range of motion of 140 degrees of flexion. Pronation and supination at the elbow are made possible by rotation of the head of the radius on the capitellum of the humerus. Elbow stability is maintained by the fitting of the trochlea of the humerus deeply into the trochlear notch of the ulna, muscular and tendinous support, and by collateral ligaments. Stability of the radiohumeral joint is provided by the annular ligament which holds the proximal radial head in approximation with the radial notch of the ulna.

The wide range of motion in the elbow renders its structures liable to strains and sprains. The elbow is also in an exposed position, with a large dependency on tendon tissue, which make it liable to external trauma and attacks of bursitis. Because of the proximity of blood vessels and nerves on its anterior aspect, injuries of the elbow may occasionally lead to serious circulatory and neurological complications.

The sources of elbow pain are varied. They include joint strains and sprains, olecranon bursitis, radiohumeral bursitis (tennis elbow, or lateral epicondylitis), radiohumeral subluxation ("nursemaid's elbow" or pull syndrome, seen usually in children two to five years of age), posterior dislocation of the radius and ulna on the humerus, fractures of the bony structures in and around the elbow (olecranon process, supracondylar fractures of the humerus and proximal necks of the radius and ulna), traumatic myositis ossificans (ossification of a hematoma, extending into the muscle or other soft tissue with the effects of pain and range of motion limitations), any disease process that may cause inflammation and/or swelling of the joints and referred pain from structures outside of the elbow area.

Related Syndromes (S47B)

ADHESION (S56)
BACTERIAL INFECTION (S63)
BURSITIS (S26)
CAPSULITIS (S27)
CERVICAL DORSAL OUTLET (SCALENICUS ATTICUS/CERVICAL RIB) (S09)
CERVICAL (NECK) PAIN (S73)
FASCIITIS (S20)
FOREARM PAIN (S54)
FROZEN SHOULDER (S64)
HAND/FINGER PAIN (S17)
HYSTERIA/ANXIETY REACTION (S59)
JOINT SPRAIN (S30)
MYOSITIS OSSIFICANS (S67)
OSTEOARTHRITIS (S21)
PHOBIC REACTION (S44)
PITTING (LYMPH) EDEMA (S31)
POST CEREBRAL VASCULAR ACCIDENT (CVA) (S07)
POST IMMOBILIZATION (S36)
POST PERIPHERAL NERVE INJURY (S23)
POST WHIPLASH (S55)
RADICULITIS (S29)
REFERRED PAIN (S01)
SHINGLES (HERPES ZOSTER) (S37)
SHOULDER/HAND (S57)
SHOULDER PAIN (S08)
TENDONITIS (S28)
TENNIS ELBOW (LATERAL EPICONDYLITIS) (S32)
UNHEALED DERMAL LESION (S39)

Treatable Causes Which May Contribute To The Syndrome (S47C)

Bacterial infection (C09)
Calcific deposit (C08)
Extrafusal muscle spasm (C41)
Habitual joint positioning (C48)
Hypermobile/instable joint (C10)
Interspinous ligamentous strain (C03)
Joint ankylosis (C42)
Joint approximation (C47)
Ligamentous strain (C25)
Localized viral infection (C27)
Muscular weakness (C23)
Nerve root compression (C04)
Neuroma formation (C14)
Neuromuscular tonic imbalance (C30)
Peripheral nerve injury (C07)
Psychoneurogenic neuromuscular hypertonus (C29)
Restricted joint range of motion (C26)
Soft tissue inflammation (C05)
Soft tissue swelling (C06)
Sympathetic hyperresponse (C38)
Tactile hypersensitivity (C24)
Trigger point formation (C01)
Unhealed dermal lesion (C12)
Vascular insufficiency (C11)
Visceral organ dysfunction (C02)

Treatment Notes (S47D)

Elbow pain which has its origin from inflammation of associated tendons has proven to be treatable with tendon manipulation (refer to Massage,

elbow pain syndrome

Deep Soft Tissue Mobilization [M23F]), prolonged electrical stimulation of acupoints in the inflamed zones (refer to Peripheral Electroacustimulation, Acupoint Electrical Stimulation Within an Inflamed Zone [M05A]), phonophoresis of antiinflammatories (refer to Ultrahigh Frequency Sound, Phonophoresis [M01A]) and circulation enhancement with electrical stimulation (refer to Electrical Stimulation, Increasing Blood Flow [M02C]).

The following is a list of trigger point formations which may, singly or in combination, refer pain into the elbow area. It should be noted that it is possible for the given patient to experience pain throughout the entire stereotypical referred pain pattern or only in part or parts of it. Opposite to each listing is a description of the parts of the stereotypical referred pain pattern most commonly experienced by previous patients, relative to the given trigger point formation. Any descriptive reference which is underlined is a part of the pattern which has *not* been commonly experienced by previous patients. For complete description of each stereotypical referred pain pattern, refer to the Table of Referred Pain Patterns of Trigger point Formation Origin [T005]. Please note the key to abbreviated words at the bottom of the list.

LOCATION (MUSCLE)		REFERRED PAIN ZONES (GENERAL)
Scalenus	(4)	Shoulder, med. & l.pect. maj., med. bord. of scapula, deltoids, *l.arm*, p.thumb, *p.index & m.finger*
Scalenus (minimus)		L. triceps, *p.forearm & hand & f.s*
Infraspinatus	(2)	*Subocciput*, deltoids, *l.arm*, palm, *p.hand*
Coracobrachialis		A. deltoid, shoulder joint, *l. triceps, p.arm & hand, p.m.finger*
Supraspinatus (muscle)	(2)	*Upper trapezius*, deltoids, *l.arm*
Upper latissimus dorsi		Latissimus dorsi, *med. arm, l.hand*, fingers 4 & 5
Serratus posterior superior (under the scapula)		Scapula, p.deltoid, med. arm, wrist, med. hand, finger 5
Serratus anterior		Lower a.l.chest, *med. a.arm & hand*, d.med. bord. of scapula, *a.f.s 4 & 5*
Subclavius		Under clavicle, *a.deltoid*, biceps, brach., *l.hand, a.& p.f.s 1-3*
Subscapularis		*Scapula, axilla*, p.deltoid, *med. triceps, a.p.wrist*
Pectoralis major (sternal)	(3)	Pectoralis major, a.deltoid, *med. a. upper arm*, prox. a.forearm, *m.palm, a.f.s 3-5*
Pectoralis minor		*Pectoralis major*, a.deltoid, *med. a. arm, med. a.hand, a.f.s 3-5*
Sternalis	(3)	Bil. parasternum, med. subclavicle, *a.deltoid, med. a.upper arm & elbow*
Medial triceps (deep fibers)		Med. a.elbow & *forearm, a.f.s 3-5*
Medial triceps (l. fibers)		P. l.elbow, posterior brach.
Longhead of the triceps	(2)	*Upper trapezius*, p.deltoid, *p.upper arm*, l.p.forearm, *p.wrist*
Distal medial triceps		P. elbow
Anconeus		P. l.elbow
Biceps brachii	(2)	*Upper trapezius*, a.& m.deltoid, *biceps*, a.elbow
Brachialis (superior & inferior)	(4)	A. deltoid, *a.half m.deltoid, a.elbow*, a.& p.first metacarpal, l.hand, p.second metacarpal
Supinator		L. elbow, *p.brach.*, p.metacarpals 1-2
Ring finger extensor		P. origin of brach., p.med. forearm, p.finger 4
Brachioradialis		P. l.elbow, *l.forearm*, p.metacarpals 1 & 2
Pronator teres		A. *l.forearm* & wrist, a.metacarpal #1
Extensor carpi radialis longus		P. l.elbow, p.l.forearm & wrist, p.carpals, p.metacarpals 1-4

Key:
TMJ—temporomandibular joint
A/C—acromioclavicular
med.—medial
brach.—brachioradialis
para.—paraspinous
MP—metacarpal phalangeal
sterno.—sternocleidomastoideus
pect.—pectoralis
a.—anterior
s.—superior
&—and
maj.—major
bord.—border

f.—finger
l.—lateral
m.—middle
d.—distal
p.—posterior
prox.—proximal
bil.—bilateral
f.s—fingers
i.—inferior

Bibliography (S47E)

Basmajian, pp. 144, 273–279.
Salter, pp. 245–246.
Scott, pp. 141–154.
Shands, pp. 436–444.

ELECTROACUSLEEP (M03)

Definition (M0301)

Electroacusleep is a composite name for a technique which has been used clinically to electrically induce sleep, provide a state of light sedation, or to induce an autonomic sympathetic reaction. It has been known as electronarcosis, electrosleep, electrosleep therapy, transcerebral electrotherapy, cerebral electrotherapy, cerebral electro-therapy and acusleep. Various current forms and variations in the patient "hook-up" have been utilized to produce essentially the same effect. Generally these techniques employ a very weak, intermittent direct current, applied through electrodes attached to various sites on the head. Some experimentation and treatment protocols have involved electrode placements on both the head and on a distal extremity, with attention paid to the effects of the direction of continuous d/c current flow. Variations in pulse frequencies (1 to 135 hertz), pulse widths (durations, .1 to 2 milliseconds), amplitudes (.5 to 15 microamperes) and session durations (10 to 30 minutes) have been utilized with varying degrees of effectiveness. Patients who have been successfully treated are said to report relief from feelings of fatigue and general irritability, improved dispositions, increased energy, feelings of euphoria, cheerfulness and a sense of general relaxation. They also generally report an improved ability to sleep.

Some objective physiological changes have been reported to be affected by the process of electroacusleep. These changes include measurable increases in serum thyroxine, catecholamines and alterations in the levels of seventeen of the ketosteroids. Also noted have been increases in pain tolerance (short term for high frequencies and long term for low frequencies) which have been associated by some researchers with the production of endorphins, apparently produced as a result of direct or indirect electrical (especially low frequency) stimulation of periaqueductal or paraventricular gray matter in the brain and the dorsal horn cells in the spinal cord.

Endorphins are a group of opiate-like chemicals which activate the descending dorsal horn cell pathways (and perhaps other inhibitory pathways as well) to produce serotonin, which has the effect of preventing afferent pain impulses from reaching the brain (center of perception). Endorphin production is the body's number one pain inhibiting mechanism. Possibly as a consequence of this precipitation of endorphin production, electroacusleep has proven to be effective not only for the treatment of insomnia and emotional agitation or anxiety but for the treatment (at least in part) of painful conditions as well.

To produce the effects of electroacusleep, earlier investigators assumed that the currents used actually entered the cranium via the surface electrodes (as much as forty-five percent of the current was "calculated" to enter the brain), but this concept was later challenged on the basis that the currents utilized were actually too weak to pass from the skin through the skull and other tissues to effect changes in the brain. Regardless of the failure of this current to enter the brain, it has been shown that it can precipitate electrical changes within the brain by virtue of the rhythmic nature of the peripheral stimulation used. This "electromassage" is apparently what induces the central nervous system to affect changes in its hormone levels, including the endorphin production.

Procedure of Application (M0302)

1) The patient should be placed in a position which will be comfortable for the duration of the treatment. The treatment room need not be dimly lighted and/or quiet for the procedure to be effective.

2) The electrical stimulation unit should be capable of providing a current level of between 25 to 500 microamperes, at a pulse rate of from 1 to 4 hertz. It should be preset at the desired frequency level and at the 25 microampere intensity level.

3) Surface electrodes should be attached to the ear lobes with earclips, over the mastoid processes behind the ears, or over each side of the forehead.

4) The stimulation unit should be turned on and the intensity level slowly raised until the patient is able to perceive the electrical pulse, or until the patient feels dizzy. Should the patient become dizzy the intensity level should be lowered until the dizziness disappears.

5) Stimulation should last from ten to twenty minutes. Electroacusleep has a sedative-like effect upon the patient which becomes increasingly more pronounced as the stimulation time increases. If the treatment is successful, at the end of the session the patient should report a general feeling of relaxation or a slight sense of euphoria which may last up to twenty-four hours.

Partial List of Conditions Treated (M0303)

Conditions treated in the field (M0303A)

Psychosis
Encephalitis
Toxemia of pregnancy
Enuresis
Hypertension
Anxiety
Depression (non-endogenous)
Insomnia
Nightmares
Irritability
Migraine
Neurasthenia
Alcoholism
Substance abuse
Drug addiction withdrawal

Treatable Causes (M0303B)

Calcific deposit (C08)
Dental pain (C22)
Extrafusal muscle spasm (C41)
Habitual joint positioning (C48)
Hypertension (C39)
Insomnia (C46)
Interspinous ligamentous strain (C03)
Joint ankylosis (C42)
Joint approximation (C47)
Ligamentous strain (C25)
Localized viral infection (C27)
Menorrhalgia (C20)
Migraine (vascular headache) (C34)
Nerve root compression (C04)
Neurogenic dermal disorders (psoriasis, hives) (C49)
Neuromuscular tonic imbalance (C30)
Psychoneurogenic neuromuscular hypertonus (C29)
Sinusitis (C16)
Soft tissue inflammation (C05)
Sympathetic hyperresponse (C38)
Trigger point formation (C01)
Visceral organ dysfunction (C02)

Syndromes (M0303C)

BURNS (SECOND AND/OR THIRD DEGREE) (S72)
BURSITIS (S26)
CALF PAIN (S50)
CAPSULITIS (S27)
CARPAL TUNNEL (S35)
CERVICAL (NECK) PAIN (S73)
CHEST PAIN (S14)
COLITIS (S42)
EARACHE (S40)
ELBOW PAIN (S47)
FACET (S19)
FASCIITIS (S20)
FOOT PAIN (S51)
FOREARM PAIN (S54)
FROZEN SHOULDER (S64)
HAND/FINGER PAIN (S17)
HEADACHE PAIN (S02)
HIGH THORACIC BACK PAIN (S48)
HYPERTENSION (S43)
HYSTERIA/ANXIETY REACTION (S59)
INSOMNIA (S71)
JOINT SPRAIN (S30)
KNEE PAIN (S46)
LOWER ABDOMINAL PAIN (S62)
LOW BACK PAIN (S03)
MENSTRUAL CRAMPING (S60)
MIGRAINE (VASCULAR) HEADACHE (S18)
MYOSITIS OSSIFICANS (S67)
OSTEOARTHRITIS (S21)
PHOBIC REACTION (S44)
PIRIFORMIS (S04)
POST IMMOBILIZATION (S36)
POST WHIPLASH (S55)
RADICULITIS (S29)
REFERRED PAIN (S01)
SCIATICA (S05)
SHINGLES (HERPES ZOSTER) (S37)
SHOULDER/HAND (S57)
SHOULDER PAIN (S08)
SINUS PAIN (S33)
TEMPOROMANDIBULAR JOINT (TMJ) PAIN (S06)
TENDONITIS (S28)
TENNIS ELBOW (LATERAL EPICONDYLITIS) (S32)
THIGH PAIN (S49)
TINNITUS (S70)
TOOTHACHE/JAW PAIN (S41)
WRIST PAIN (S16)

Precautions and Contraindications (M0304)

Electroacusleep is contraindicated for patients suffering from severe endogenous or involutional depressions, epilepsy, brain tumors, cerebrovascular disorders and heart disease.

Transient dizziness during and sometimes shortly after treatment has been noted to occur to some patients. Dizziness is most often associated with a current intensity between 200 to 500 microamperes, and it usually lasts only until the intensity is reduced.

Transient blurred vision has been reported to occur to some patients shortly after treatment, by investigators who have tried applying electrodes over patients' eyes, but it is said to be a very temporary experience. The literature does not specify how long it can last.

Electroacusleep has proven to be ineffective as a treatment for the "normal" sleep problems of the elderly.

Electroacusleep should generally not be applied for longer than twenty minutes at a time if the patient is to perform tasks other then sleeping afterward. It has been demonstrated to provide very effective sedation for most subjects.

Notes Aside (M0305)

Increasing pain tolerance (M0305A)

Electroacusleep has proven to be effective in elevating the pain tolerance of patients suffering from both chronic and acute pain syndromes, as well as serving as a mild sedative.

Reducing the effects of drug withdrawal (M0305B)

Electroacusleep has shown itself to be effective in reducing the symptoms of codeine and valium withdrawal. It has also provided an aid to compulsive joggers who may be provided with the endorphin surge that they are addicted to without the stressful physical activity of prolonged running, thus allowing them to submit to treatment which may be precluding such behavior temporarily.

Relieving the insomnia syndrome (M0305C)

Electroacusleep has been very effective in the treatment of acute insomnia (unassociated with aging), often relieving the condition in one or two sessions.

Bibliography (M0306)

Akil, pp. 961–962.
Andersson, pp. 569–578.
Brown, pp. 402–410.
Muftic, pp. 40–46.
Nias, pp. 766–773.
Rivlin, pp. 46–49.
Sjolund, pp. 382–384.
Tannenbaum, pp. 131–136.
Von Richthofen, Mellor, 1979: pp. 1264–1271.
Von Richthofen, Mellor, 1980: pp. 212–218.
Weinberg, pp. 35–39

ELECTRICAL STIMULATION OF MUSCLE (OR DEEP) TISSUE (M02)

Definition (M0201)

Electrical stimulation of human tissues is an old procedure dating from the attempts of the early Greeks to utilize electrical eels for therapeutic purposes. The modern rediscovery of electricity and its uses in physical medicine date from the early eighteenth century, when, once again, electrical stimulation generated by electric eels ("live black torpedoes") was applied therapeutically to relieve headaches and to effect neuromuscular paralysis.

Therapeutically applied electrical stimulation has had a checkered history, enjoying long periods of popularity and respectable use interspersed with periods of widespread misuse by "medical" charlatans who treated everything from psychiatric conditions to cancerous tumors. One must suppose that the simplicity of electricity-producing equipment, especially static or direct current, coupled with seemingly "magical" effects that electrical currents can have on human tissues, has unavoidably led to its exploitation by unscrupulous "practitioners" at the expense of the unsophisticated and gullible.

Often billed as a near panacea for the "cure" of most human physical ills, over time the therapeutic use of electrical stimulation has gradually, in the minds of many, become associated with quackery, and therefore beneath the use of scrupulous and sophisticated practitioners. This state of affairs is unfortunate, since there are many physical ills that in fact may be improved or corrected by appropriately applied electrical stimulation. It is our purpose here to explore the various forms of electrical stimulation and their applications relative to muscle or deep tissue.

Terms related to electrical stimulation (M0201A)

Modern instrumentation that is used to apply electricity to the body is designed (by and large) to be used without detailed knowledge of the instrument's internal circuitry or the physics responsible for the production of electricity. However, some knowledge of the basic principles of electrical stimulation is required for intelligent application and an understanding of the variable results that come from the necessary trial and error that is a regular feature of aggressive clinical therapy. A few terms are defined below to help the reader confronted by literature in this area.

Coulomb (M0201A01) A coulomb is a basic unit of charge theoretically produced by 6.28×10^{18} electrons. Most therapeutic electrical stimulators have a low pulse charge, expressed in microcoulombs (10^{-6} coulombs).

Joule (M0201A02) A Joule is a basic expression for work of electricity, i.e., work is the force required to move a charge.

$$\text{Joule} = \text{coulomb} \times \text{voltage}$$

Volt (M0201A03) A volt is a unit of measure which indicates the amount of potential energy (joule) each unit of charge (coulomb) contains.

$$\text{Voltage} = \text{Joule} / \text{coulomb}$$

Ampere (M0201A04) An ampere usually abbreviate to amp, is the unit amount of charge (current) that is flowing. Most therapeutic electrical stimulators have a low average current of less than 1.5 milliamps and relatively high peak currents of 60 to 100 milliamps.

$$\text{Amps} = \text{coulomb} / \text{second}$$

Ohm (M0201A05) An ohm is a unit expressing the amount of resistance offered by a current conductor.

Ohm's Law = The current (amps) is directly proportional to potential (volts) and inversely proportional to resistance (ohms).

Pulse duration (M0201A06) The pulse duration is the amount of time the current flows in one direction. Pulse duration is measured when the current level is at fifty percent of its peak, usually expressed in microseconds.

Pulse frequency (M0201A07) The pulse frequency is the expression of the number of pulses produced per second (hertz or cycles/second).

Burst frequency (M0201A08) The burst frequency is the number of trains of impulses produced per second; it is dependent upon the stimulation on and stimulation off duty cycle selected.

electrical stimulation of muscle (or deep) tissue

Current density (M0201A09) The current density is the amount of current per unit area. The smaller the electrode, the greater the current density, which makes the stimulus perceptually stronger to the patient.

Resistance to current (M0201A10) The human body is made up of tissues and fluids that vary in their electrical conductivity and, conversely, their resistance to the passage of electricity through them. Tissue conductivity is proportionally related to the tissue's water content: the higher the water content the greater the conductivity and the lower the tissue's resistance. The water content of muscle is 72 to 75%; of the brain, 68%; of fat, 14 to 15%; and of the peripheral nerve, skin, and bone, 5 to 16%. Resistance varies in direct proportion to the distance between electrodes; the further the electrical stimulus must travel the greater the resistance.

Monopolar technique of electrode placement (M0201A11) One electrode is placed on the treatment area and the other is placed in a remote location on the patient's body. This arrangement provides a rather general stimulation pattern because of the multiple parallel pathways the current may take from one electrode to the other. One may also expect variations in the responses produced by the electrical stimulation because of the number of nerves and other structures possibly encountered by the current as it passes.

Bipolar technique of electrode placement (M0201A12) Both cathode and anode electrodes are placed on the treatment area, in relative proximity, one to another. This arrangement provides for rather specific stimulation of structures with few variations in responses.

Electrotherapeutic current forms (M0201B)

Electrotherapeutic currents are generally derived from the commercial lighting circuit (alternating current or a/c) or from batteries (direct current or d/c). These basic currents are modified by transformers, other electromagnetic or thermionic devices, and/or complex circuitry beyond our scope here to produce various therapeutic current forms. The therapeutic current forms include galvanic (square wave), interrupted galvanic, surged interrupted galvanic, sinusoidal, alternating, surged alternating, faradic, surged faradic, and other "hybrid" wave forms (generally produced by combining 2 or more wave forms). The variables of electrical wave form production include voltage, amperage, mode flow (direction), pulse frequency, and pulse duration (width).

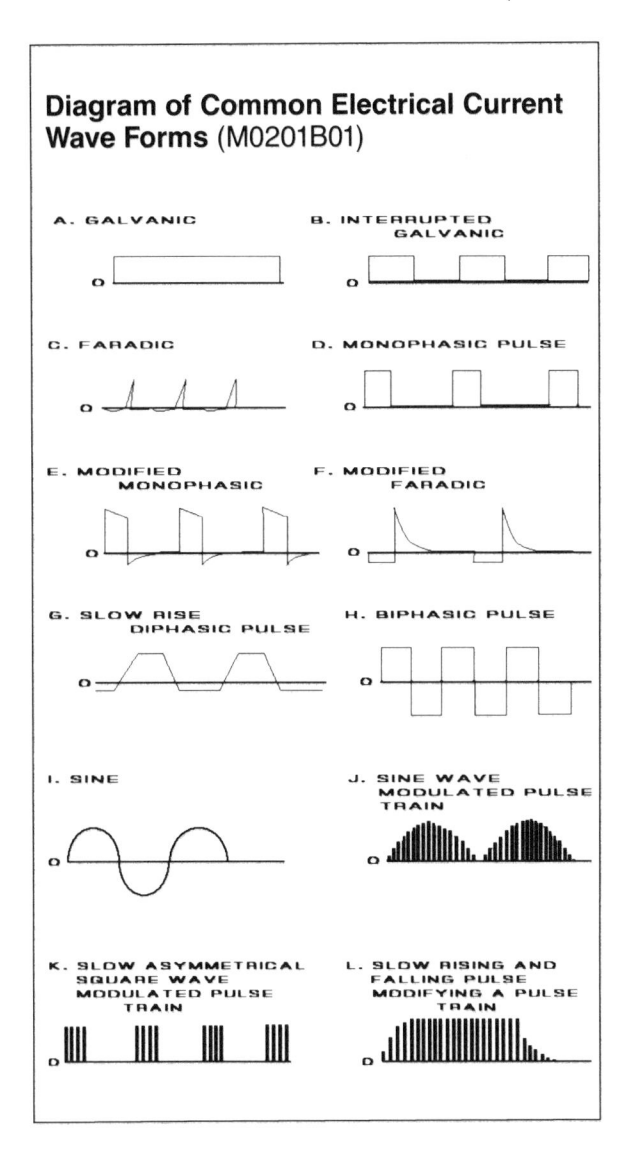

Diagram of Common Electrical Current Wave Forms (M0201B01)

A. GALVANIC
B. INTERRUPTED GALVANIC
C. FARADIC
D. MONOPHASIC PULSE
E. MODIFIED MONOPHASIC
F. MODIFIED FARADIC
G. SLOW RISE DIPHASIC PULSE
H. BIPHASIC PULSE
I. SINE
J. SINE WAVE MODULATED PULSE TRAIN
K. SLOW ASYMMETRICAL SQUARE WAVE MODULATED PULSE TRAIN
L. SLOW RISING AND FALLING PULSE MODIFYING A PULSE TRAIN

Action of electrical current upon soft tissues (M0201C)

Electrical stimulation is applied to the patient through the use of a pair of electrodes placed on the patient's body. The electricity is passed from the cathode or negative pole electrode through the tissues to the other, anode or positive pole, or dispersive, electrode, thus completing an electrical circuit.

electrical stimulation of muscle (or deep) tissue

Electrical currents passed through muscle or nervous tissue from an external source (electrical stimulator) will be partially depolarized in the region of the cathode and hyperpolarized in the region of the anode. If the current is sufficiently strong, the degree of depolarization will reach or exceed a critical level which will produce a muscle contraction or firing of the nerve. At the anode, as the circuit is completed, the electrical changes induced by the current are overcompensated for, so that some degree of irritability is present at the anode. If sufficiently great, the irritability will cause a muscle contraction or nerve firing to occur under the anode. If the current level is adequate to elicit single neuron impulse or single muscle fiber contraction it is said to be a *minimal stimulus*. If a stronger stimulus is required to excite all of a group of nerve fibers or denervated muscle fibers, it is called a *maximal stimulus*. Any stimulus greater than the maximal stimulus is called a *supramaximal stimulus*.

The factors which determine the adequacy of a stimulus to either elicit muscle contraction or provoke firing of nervous tissue are the pulse frequency, the duration of the stimulus (pulse duration), and the magnitude or amplitude of the current (voltage and/or amperage levels). The minimal duration of an effective electrical stimulus, sufficient to provoke contraction or firing, is .03 microseconds for a nerve fiber and 1.0 microseconds for an innervated muscle fiber. The strength of electrically-induced muscle contraction is related to the intensity and pulse duration of the stimulus: the greater the intensity and pulse duration the greater the strength of contraction.

Equipment used in electrical stimulation (M0201D)

The electrical stimulation units currently utilized for the stimulation of muscle or deep tissues can be generally classified into six categories:

High voltage units (M0201D01) These units have a high peak current of 500 milliamperes or greater with a low average current of less than 1 milliampere. They are constant voltage generators with a pulse charge of approximately 4 microcoulombs and their pulse durations usually range from 5 to 8 microseconds. They generally provide a variety of duty cycles and pulse frequencies from which to choose.

TENS units (M0201D02) These units are not generally considered muscle stimulation units, but if the voltage amplitudes are taken high enough both muscle contraction and motor nerve firing can be provoked. The transcutaneous nerve simulator (TENS or TNS) unit is usually a constant current generator with a pulse duration of 200 microseconds or less. Most units have a 50 to 60-milliampere peak current with a 50 to 60-volt driving force. The average current is usually 1.5 milliamps or less. The pulse charge is usually 12 to 15 microcoulombs but may some times approach 25 microcoulombs (AAMI standards set the ceiling for these units at 25 microcoulombs). Many TENS units provide a variety of pulse durations, pulse frequencies, and burst frequencies or duty cycles from which to choose.

Portable neuromuscular stimulation units (M0201D03) These units generally employ a constant current, which generally has a peak current of 100 milliamps with a driving force of between 50 and 100 volts. They generally provide a choice of duty cycles, pulse frequencies, peak currents, and ramps (the time it takes for the current level to rise from zero to its peak).

High Frequency (medium frequency) units (M0201D04) These units by definition generate more then 1000 cycles per second with popular models producing 2500 cycles per second. The 2500-cycle per second units employ a usual duty cycle of 0 milliseconds on and 10 milliseconds off. In this case, the 2500-cycle current is interrupted at 1/100 of a second on and 1/100 of a second off with a fifty percent duty cycle producing 50 bursts per second with 25 cycles per burst. The 2500-cycle per second unit generally has a peak current of 130 milliamperes with an average current level of between 80 and 100 milliamperes RMS (root mean square). These units provide a variety of duty cycles, ramps, and peak currents from which to choose; they can create a muscle contraction of sixty percent or greater of maximal isometric contraction potential.

Interferential units (M0201D05) These units are constant current generators which create a pulse frequency of 4000 to 5000 cycles per second. The interferential units generally employ two electrical sine wave circuits, one of which has a fixed frequency while the other varies its frequency; when the two wave forms intersect an interferential frequency results. The interferrential unit usually has a peak current of 60 milliamperes.

Low voltage electrical stimulators (M0201D06) These units have low peak currents, low voltage driving forces that can be alternate or direct current, and their pulse durations are usually large (measured in milliseconds or seconds). If utilizing a direct current, they have the ability to produce thermal and chemical effects and can be used in iontophoresis.

The electrical energy from the electrical stimulators is conveyed to the patient via conducting cables. The cables are plastic- or rubber-insulated flexible copper or silver wires. The thickness of the cable depends upon the amount of current to be carried by the conductor: the greater the current, the thicker the cable needs to be. These cables may be a uniform color or color-coded according to function; if color-coded, the wire to the negative cathode electrode is conventionally black, and to the positive anode electrode, red. (The electric current carried along the cable length eventually leads to some crystallization of the conducting wire which may lead to breaks in the wire at the sites where the most bending or movement (stress) of the cable occurs; this is usually close to the electrode connections and occasionally near the connection with the electrical stimulator.)

An electrode is a medium which intervenes between the cable from the electrical stimulator and the patient's body (only surface electrodes will be discussed). It generally consists of a good conducting material whose shape and form can be adapted to conform the body contour. Electrode mediums include water (a patient's body part immersed in the same bath of water with the end of one of the conducting cables, not both), metal foil (usually made from an alloys of lead, tin, and zinc), moist-pad electrodes (usually water bearing sponges between a metal foil electrode and the patient's skin), flexible carbon electrodes, and self adhesive conductive (goopy) pads.

Electrodes are usually employed in pairs, often of equal size. Between two electrodes of equal size the current density beneath each is equal; if one is twice as large as the other, the current density under the smaller one will be twice as great as that under the larger. As the current spreads between two electrodes across the body its density must gradually decrease so that midway between them the density is the least. The nearer the electrodes are to one another the greater the density of the current between them. The higher the current density the greater the effect on the tissues effected.

Procedures of Application (M0202)

1) Before the patient is introduced into the treatment environment, the equipment should be checked for proper connection with its power source, between the electrode cables and the stimulator, and between the electrode cables and the electrodes. All equipment should be checked before use to confirm good working order, and all controls should be off or set in the zero position.

2) The patient should be placed in a position which will remain comfortable for the duration of the treatment session. A reclining, over-stuffed arm chair is ideal in most cases because the neck, arms, legs, anterior abdomen, and the back, with some skillful maneuvering, are all accessible to the therapist and the patient is provided with supported comfort.

3) Before applying electrodes, the skin surface to be treated should be inspected for recent scars, evidence of peripheral injury, dermal lesions, or other causes of disturbed sensation. Such tissue may be hypersensitive to electrical stimulation and care should be taken to build up electrical amplitudes gradually to prevent undue discomfort should the skin resistance suddenly drop, magnifying the effects of the stimulation on those areas.

4) If low frequency sponge pad electrodes are being used, they should be moistened with a saline solution or water and placed over the chosen treatment sites, and care should be taken that they are in good even contact with the skin. In general the negative cathode electrode should be placed on the muscle's motor point, where the motor nerve is most superficial as it innervates the muscle, and when stimulated, provokes the greatest muscle contraction. If metal or flexible carbon electrodes are used, a proper electrode paste should be used to insure adequate contact. Once the best site(s) for electrode placement have been determined, elastic strapping, weighting the electrodes, or strategic placement of the electrode(s) under the patient's body or body parts may be necessary to insure good electrode contact. If the body part is to be immersed for treatment, it should be placed in the bath with the negative cathode electrode, several inches apart.

5) A watch or timer should be set for the length of treatment. The electrical stimulator should be turned on, and the amplitude increased until a visible contraction takes place, always staying within the patient's range of tolerance.

6) The patient should be allowed to become comfortable with the current, and then the intensity should be slowly increased as the patient's tolerance allows. This process should be continued until the desired degree of contraction is reached. The patient should be closely monitored for excessive muscle cramping, joint compression, pain, and autonomic sympathetic hyperresponse to the stimulation. The patient should *never* be left out of sight or sound once the treatment has been started unless the patient can turn off the instrument should the need arise.

7) At the end of the session, if not automatically shut off, the intensity of the stimulator should be gradually decreased until it is switched off. Re-

move all electrodes from the patient and inspect each site of stimulation carefully, record any abnormal tissue response and return all controls to zero. The patient should rest for several minutes before being allowed to leave.

Partial List of Conditions Treated (M0203)

Conditions treated in the field (M0203A)

Acrocyanosis
Alienation of motor control (hysteria)
Burns
Contusions
Deep vein thrombosis
Dupytren's contracture
Endarteritis obliterans
Extremity weakness due to upper motor neuron damage
Fecal incontinence (not discussed here)
Hysteria
Migraine
Muscle weakness or dysfunction from partial or total denervation
Myalgia
Myotonia congenita
Periarthritis humeroscapularis
Relaxation of spasticity (upper motor neuron lesions)
Scoliosis
Spasticity
Sudeck's atrophy
Urinary incontinence (not discussed here)
Varicosis
Ventilator dependency (phrenic nerve stimulation, not discussed here)
Vesical weakness and loss of ejaculation (not discussed here)

Treatable Causes (M0203B)

Extrafusal muscle spasm (C41)
Habitual joint positioning (C48)
Hypermobile/instable joint (C10)
Hypotension (C45)
Interspinous ligamentous strain (C03)
Joint ankylosis (C42)
Joint approximation (C47)
Muscular weakness (C23)
Nerve root compression (C04)
Neuromuscular tonic imbalance (C30)
Pathological neuromuscular hypertonus (C28)
Pathological neuromuscular hypotonus (C33)
Pathological neuromuscular dyscoordination (C35)
Peripheral nerve injury (C07)
Phlebitis (C43)
Psychoneurogenic neuromuscular hypertonus (C29)
Restricted joint range of motion (C26)
Soft tissue inflammation (C05)
Soft tissue swelling (C06)
Tactile hypersensitivity (C24)
Tinnitus (C32)
Trigger point formation (C01)
Vascular insufficiency (C11)

Syndromes (M0203C)

ADHESION (S56)
BELL'S PALSY (S66)
CALF PAIN (S50)
CEREBRAL PALSY (S25)
CERVICAL DORSAL OUTLET (SCALENICUS ATTICUS/CERVICAL RIB) (S09)
CERVICAL (NECK) PAIN (S73)
CHEST PAIN (S14)
CHONDROMALACIA (S15)
DIABETES (S10)
EARACHE (S40)
ELBOW PAIN (S47)
FACET (S19)
FOOT PAIN (S51)
FOREARM PAIN (S54)
FROZEN SHOULDER (S64)
HAND/FINGER PAIN (S17)
HEADACHE PAIN (S02)
HIGH THORACIC BACK PAIN (S48)
KNEE PAIN (S46)
LOWER ABDOMINAL PAIN (S62)
LOW BACK PAIN (S03)
MENSTRUAL CRAMPING (S60)
MIGRAINE (VASCULAR) HEADACHE (S18)
NEUROMUSCULAR PARALYSIS (S22)
OSTEOARTHRITIS (S21)
PHLEBITIS (S68)
PIRIFORMIS (S04)
PITTING (LYMPH) EDEMA (S31)
POST CEREBRAL VASCULAR ACCIDENT (CVA) (S07)
POST IMMOBILIZATION (S36)
POST PERIPHERAL NERVE INJURY (S23)
POST SPINAL CORD INJURY (S24)
POST WHIPLASH (S55)
RADICULITIS (S29)
REFERRED PAIN (S01)
SCIATICA (S05)
SHOULDER/HAND (S57)
SHOULDER PAIN (S08)
SINUS PAIN (S33)
TEMPOROMANDIBULAR JOINT (TMJ) PAIN (S06)
TENDONITIS (S28)
TENNIS ELBOW (LATERAL EPICONDYLITIS) (S32)

THIGH PAIN (S49)
TINNITUS (S70)
TORTICOLLIS (WRY NECK) (S65)
WRIST PAIN (S16)

Precautions and Contraindications (M0204)

Care should be taken to avoid over-fatigue of the muscle stimulated. Stimulation should stop when the muscle begins to respond with less vigor.

Electrical muscle stimulation is generally contraindicated for victims of cancer or other malignant disease if the site to be stimulated is physically associated with the cancerous growth. It is also contraindicated for those suffering from sensory deficits, not including those suffering from hysterical sensory abnormalities.

Care should be taken not to put electrodes too close together. The closer electrodes are to one another, the greater the current density becomes between their edges and an *edge-effect* may be produced which causes unpleasant burning sensations and superficial or deep burns may ultimately occur.

Electrical burns may occur when an excess of electricity (current density) is applied to the skin or mucous membrane. Electrical burns may appear in the clinical setting as a passing erythema, blistering of superficial skin layers, or deep tissue coagulation. The tissue damage produced by the deep electrical burn occurs in a roughly conical area, extending from the apex of the cone on the skin surface, where the original electrical contact occurred, fanning out into the deeper layers. Just following the injury, the burn site appears rather small and inconsequential but becomes more alarming as the damaged tissues subsequently slough off and ulceration occurs. These burns are slow to heal, prone to infection, and if sufficiently deep, may be followed by extensive unsightly scarring.

Electric shock may be caused if the patient, while being electrically stimulated, makes contact with a grounded object such as a water pipe, radiator, or electric circuit. This is especially serious if a large area of the patient's body is subjected to the shock (as in a bath or treatment tub) and may pose the danger of cardiac arrest. Electric shock may also occur if the electrical stimulator suffers transformer breakdown, in which case the hightension, low frequency current may "jump" to the patient and produce an electrical burn as well as shock.

Great care must be taken in the application of all electrical appliances to patients with cardiac pacemakers. Electrodes should not be placed in the vicinity of a pacemaker nor should the current flow from the cathode to the anode electrode be allowed to cross the pacemaker or implanted wires.

In general for high frequency stimulation, treatment should be precluded where joint repairs, rheumatoid and collagen diseases are present, or other conditions in which tendons, joint capsules or the joint may be threatened or weakened by stimulation. In general, electrodes should not be placed over scar tissue, skin irritations or open skin lesions. High frequency stimulation should not be applied over the thorax, abdomen, or back extensor musculature; such stimulation is contraindicated for patients with cardiac demand pacemakers, cardiac arrhythmias, ST segment depression or conduction block. If the patient suffers increased sweating, salivation and/or has complaints of nausea when being stimulated with high frequency stimulation, the stimulation should be discontinued.

Low frequency currents are contraindicated for treatment of myopathies (muscular dystrophy, myotonia congenita, polymyositis, muscle-involved connective tissue disease, cortisone myopathies, myopathy due to metastatic cancer, and myopathy from hyperthyroidism), Parkinson's syndrome, amyotrophic lateral sclerosis, syringomyelia, and peripheral neuritis and neuronitis.

Electrical currents should not be allowed to flow across a pregnant uterus.

It is recommended that the therapist test the apparatus on him or herself shortly before it is used to test the patient. This avoids unwarranted negative findings and provides a safeguard against injuring the patient through faulty equipment. During pre-testing, the current levels which will be used on the patient should be used on the tester; this approach takes courage because such currents by necessity are often high and uncomfortable.

Notes Aside (M0205)

When applying electrodes, care should be taken not to overlap cathode and anode electrodes or to let any conductive material (electrode cream or gel) form a conductive bridge between the two. Either situation will cause a completion of the circuit without involving or affecting the patient's tissues (the basic reason why inserting cathode and anode electrodes into a basin or treatment tub with the patient affects the patient so little).

When applying electrical stimulation, a gradual increase of intensity is preferred because of the tendency of skin resistance to break down suddenly after having been exposed to electrical current for several minutes. If the apparent lack of tissue response (resulting from high skin resistance) persuades the therapist to raise the intensity to a relatively high level before adjustment takes place and the skin resistance breaks down suddenly, the patient may pay for the therapist's lack of patience by experiencing additional pain and future treatment may be put in jeopardy because of the patient's acquired fear.

Bibliography (M0206)

Alon, pp. 890–895.
Baker, pp. 1967–1974.
Currier, pp. 915–921.
Delitto, McKowen, McCarthy, Shively, Rose, pp. 45–50.
Delitto, Rose, pp. 1704–1707.
Eckerson, pp. 483–490.
Mohr, Carlson, Sulentic, Landry, pp. 607–609.
Newton, pp. 1593–1596.
Nitz, pp. 219–222.
Owens, pp. 162–168.
Packman-Braun, pp. 51–56.
Selkowitz, pp. 186–196.
Shriber, pp. 110–123, 139–147.
Stillwell, pp. 1–64, 65–108, 124–173.
Wong, pp. 1209–1214.

ELECTRICAL STIMULATION OF MUSCLE (OR DEEP) TISSUE MUSCLE LENGTHENING (M02A)

Definition (M02A1)

Electrical stimulation to cause muscle lengthening has proven to be valuable in the treatment of conditions in which extrafusal or intrafusal muscle spasm is present, from habitual shortening of musculature or enforced joint immobilization or when musculature is hypertonically shortened (imbalanced) relative to antagonistic hypotonically lengthened musculature as a result of post upper and lower motor neuron lesion syndromes.

A muscle may be therapeutically lengthened with any of the electrical stimulators which produce a visible muscle contraction approaching tetany. The best results, with the fewest adverse side effects such as subsequent and additional soreness, are generally provided by those instruments which have a 5 to 10-second stimulation on, followed by a 5 to 10-second off duty cycle. The muscles so stimulated are allowed to alternately contract and relax, preventing rapid muscle fatigue and metabolite build-up. Other stimulus forms have also been shown to be effective for muscle lengthening, including continuous 4 to 8-hertz electrical stimulation which is strong enough to produce a vigorous contraction of the muscles stimulated. When this form is used, stimulation is usually continued for ten minutes.

The basic principle behind the process of muscle lengthening with electrical stimulation is based upon the neurogenic tendency to tonically balance apposing musculature through reciprocal inhibition; i.e., as a muscle contracts its antagonist reciprocally lengthens.

Antagonist stimulation to affect agonist lengthening (M02A1A)

To take advantage of the neurological relationship requiring neuromuscular reciprocal inhibition, the hypertonic or short muscle should be passively stretched by external force or by body part positioning while electrical stimulation is applied to the muscle's antagonist, compelling it to maintain a visible or tetanic contraction for five to ten seconds; this causes it to contract in its shortened range while the previous short muscle is inhibited and made to "relax" in its stretched or lengthened range.

Agonist stimulation to affect agonist lengthening (M02A1B)

Conversely, a short muscle may be lengthened by putting it on stretch, in a fixed position, and electrically stimulating it to cause a visible tetanic contraction for five to ten seconds and then alternately allowing it to rest for the same period. The involved joint(s) should not be allowed to move during entire length of treatment, which will run ten to fifteen minutes. The effect of this approach is to cause the stimulated muscle's antagonist to be inhibited in a shortened range while the stimulated muscle assumes a reciprocally lengthened range by virtue of the servo system employed by the muscle's spindles.

Alternate stimulation to affect agonist lengthening (M02A1C)

The above approaches may be put together by putting the agonist (short) muscle on stretch, and then first electrically stimulating the agonist for five to ten seconds and then alternately electrically stimulating the antagonist for the same lenth of time. This procedure may only be applied if the available electrical stimulator provides a duty cycle which alternates between independent electrode pairs or sets. The joint(s) involved should be held in a fixed position for the entire ten to fifteen-minute treatment session.

If any of the above approaches is used to lengthen a muscle, the involved joint(s) should remain in the fixed position for fifteen minutes following cessation of the electrical stimulation, before the patient is allowed to to assume another position. An unpublished pilot study has demonstrated that an electrically stimulated hypertonic muscle increases its myoelectric activity for eight minutes after cessation of electrical stimulation, but its myoelectric activity falls off for at least another six minutes following that.

Muscle lengthening will occur by virtue of the servo system employed by each opposing muscle's spindles and their relationship based on reciprocal inhibition.

Procedures of Application (M02A2)

General application considerations (M02A2A)

1) Before the patient is introduced into the treatment environment, the equipment should be checked for proper connection with its power source, between the electrode cables and the stimulator, and between the electrode cables and the electrodes. All equipment should be checked before use to confirm good working order, and all controls should be off or set in the zero position.

2) The patient should be placed in a position which will remain comfortable for the duration of the treatment session. A reclining, over-stuffed arm chair is ideal in most cases because the neck, arms, legs, anterior abdomen, and the back, with some skillful maneuvering, are all accessible to the therapist and the patient is provided with supported comfort.

3) Before applying electrodes, the skin surface to be treated should be inspected for recent scars, evidence of peripheral injury, dermal lesions, or other causes of disturbed sensation. Such tissue may be hypersensitive to electrical stimulation and care should be taken to build up electrical amplitudes gradually to prevent undue discomfort should the skin resistance suddenly drop, magnifying the effects of the stimulation on those areas.

4) If low frequency sponge pad electrodes are being used, they should be moistened with a saline solution or water and placed over the chosen treatment sites, and care should be taken that they are in good even contact with the skin. In general the negative cathode electrode should be placed on the muscle's motor point, where the motor nerve is most superficial as it innervates the muscle, and when stimulated, provokes the greatest muscle contraction. If metal or flexible carbon electrodes are used, a proper electrode paste should be used to insure adequate contact. Once the best site(s) for electrode placement have been determined, elastic strapping, weighting the electrodes, or strategic placement of the electrode(s) under the patient's body or body parts may be necessary to insure good electrode contact. If the body part is to be immersed for treatment, it should be placed in the bath with the negative cathode electrode, several inches apart.

5) A watch or timer should be set for the length of treatment. The electrical stimulator should be turned on, and the amplitude increased until a visible contraction takes place, always staying within the patient's range of tolerance.

6) The patient should be allowed to become comfortable with the current, and then the intensity should be slowly increased as the patient's tolerance allows. This process should be continued until the desired degree of contraction is reached. The patient should be closely monitored for excessive muscle cramping, joint compression, pain, and autonomic sympathetic hyperresponse to the stimulation. The patient should *never* be left out of sight or sound once the treatment has been started unless the patient can turn off the instrument should the need arise.

7) At the end of the session, if not automatically shut off, the intensity of the stimulator should be gradually decreased until it is switched off. Remove all electrodes from the patient and inspect each site of stimulation carefully, record any abnormal tissue response and return all controls to zero. The patient should rest for several minutes before being allowed to leave.

Specific procedures applied for muscle lengthening (M02A2B)

The muscle to be lengthened should be put on stretch and the involved joint(s) fixed, usually by judicious body and/or body part positioning, but sometimes strapping may be necessary. The electrodes should then be secured over the muscle or muscle group to be stimulated. If only one muscle or muscle group is to be stimulated, for best results, the electrodes should be placed in a bipolar fashion with both cathode and anode electrodes on the same muscle or muscle group. The cathode electrode should be placed over the muscle or the muscle group's dominant muscle motor point.

As the patient becomes comfortable with the current the intensity should be increased as the patient's comfort allows until tetany or near tetany is produced.

Electrical stimulation should last five to ten seconds, followed by a rest period of at least that long (some stimulators are designed to provide a fifty-second rest between stimulation periods) to avoid excessive fatigue and to reduce the tendency to spasm. Typically, treatment sessions vary from ten to fifteen minutes.

If the technique of alternation of electrical stimulation of first one antagonistic and then the other is used, electrodes should be applied in a monopolar fashion, in which independent cathode electrodes are placed over each antagonist with the anode dispersive electrode over an area distal but roughly equidistant from the cathode electrodes. The patient should be positioned as described above with the short muscle put on stretch and the joint position fixed. After turning it on, the electrical stimulator will first drive current from one cathode electrode and then from the other, usually in a 5 to 10-second cycle. Treatment length usually varies from ten to fifteen minutes.

Partial List of Conditions Treated (M02A3)

Treatable Causes (M02A3A)

Extrafusal muscle spasm (C41)
Habitual joint positioning (C48)
Hypermobile/instable joint (C10)
Joint approximation (C47)
Nerve root compression (C04)
Neuromuscular tonic imbalance (C30)
Pathological neuromuscular hypertonus (C28)
Pathological neuromuscular hypotonus (C33)
Peripheral nerve injury (C07)
Psychoneurogenic neuromuscular hypertonus (C29)
Restricted joint range of motion (C26)
Trigger point formation (C01)

Syndromes (M02A3B)

ACHILLES TENDONITIS (S52)
BELL'S PALSY (S66)
CALF PAIN (S50)
CEREBRAL PALSY (S25)
CERVICAL DORSAL OUTLET (SCALENICUS ATTICUS/CERVICAL RIB) (S09)
CERVICAL (NECK) PAIN (S73)
CHEST PAIN (S14)
EARACHE (S40)
ELBOW PAIN (S47)
FOOT PAIN (S51)
FOREARM PAIN (S54)
FROZEN SHOULDER (S64)
HAND/FINGER PAIN (S17)
HEADACHE PAIN (S02)
HIGH THORACIC BACK PAIN (S48)
KNEE PAIN (S46)
LOWER ABDOMINAL PAIN (S62)
LOW BACK PAIN (S03)
MENSTRUAL CRAMPING (S60)
MIGRAINE (VASCULAR) HEADACHE (S18)
NEUROMUSCULAR PARALYSIS (S22)
OSTEOARTHRITIS (S21)
PIRIFORMIS (S04)
POST CEREBRAL VASCULAR ACCIDENT (CVA) (S07)
POST IMMOBILIZATION (S36)
POST PERIPHERAL NERVE INJURY (S23)
POST SPINAL CORD INJURY (S24)
POST WHIPLASH (S55)
RADICULITIS (S29)
REFERRED PAIN (S01)
SCIATICA (S05)
SHOULDER/HAND (S57)
SHOULDER PAIN (S08)
SINUS PAIN (S33)
TEMPOROMANDIBULAR JOINT (TMJ) PAIN (S06)
TENDONITIS (S28)
TENNIS ELBOW (LATERAL EPICONDYLITIS) (S32)
THIGH PAIN (S49)
TINNITUS (S70)
TOOTHACHE/JAW PAIN (S41)
TORTICOLLIS (WRY NECK) (S65)
WRIST PAIN (S16)

Precautions and Contraindications (M02A4)

Care should be taken to avoid over-fatigue of the muscle stimulated. Stimulation should stop when the muscle begins to respond with less vigor.

Electrical muscle stimulation is generally contraindicated for victims of cancer or other malignant disease if the site to be stimulated is physically associated with the cancerous growth. It is also contraindicated for those suffering from sensory deficits, not including those suffering from hysterical sensory abnormalities.

Care should be taken not to put electrodes too close together. The closer electrodes are to one another, the greater the current density becomes between their edges and an *edge-effect* may be produced which causes unpleasant burning sensations and superficial or deep burns may ultimately occur.

Electrical burns may occur when an excess of electricity (current density) is applied to the skin or mucous membrane. Electrical burns may appear in the clinical setting as a passing erythema, blistering of superficial skin layers, or deep tissue coagulation. The tissue damage produced by the deep electrical burn occurs in a roughly conical area, extending from the apex of the cone on the skin surface, where the original electrical contact occurred, fanning out into the deeper layers. Just following the injury, the burn site appears rather small and inconsequential but becomes more alarming as the damaged tissues subsequently slough off and ulceration occurs. These burns are slow to heal, prone to infection, and if sufficiently deep, may be followed by extensive unsightly scarring.

Electric shock may be caused if the patient, while being electrically stimulated, makes contact with a grounded object such as a water pipe, radiator, or electric circuit. This is especially serious if a large area of the patient's body is subjected to the shock (as in a bath or treatment tub) and may pose the danger of cardiac arrest. Electric shock may also occur if the electrical stimulator suffers transformer breakdown, in which case the high-tension, low frequency current may "jump" to the patient and produce an electrical burn as well as shock.

Great care must be taken in the application of all electrical appliances to patients with cardiac pacemakers. Electrodes should not be placed in the vicinity of a pacemaker nor should the current flow from the cathode to the anode electrode be allowed to cross the pacemaker or implanted wires.

In general for high frequency stimulation, treatment should be precluded where joint repairs, rheumatoid and collagen diseases are present, or other conditions in which tendons, joint capsules or the joint may be threatened or weakened by stimulation. In general, electrodes should not be placed over scar tissue, skin irritations or open skin lesions. High frequency stimulation should not be applied over the thorax, abdomen, or back extensor musculature; such stimulation is contraindicated for patients with cardiac demand pacemakers, cardiac arrhythmias, ST segment depression or conduction block. If the patient suffers increased sweating, salivation and/or has complaints of nausea when being stimulated with high frequency stimulation, the stimulation should be discontinued.

Low frequency currents are contraindicated for treatment of myopathies (muscular dystrophy, myotonia congenita, polymyositis, muscle-involved connective tissue disease, cortisone myopathies, myopathy due to metastatic cancer, and myopathy from hyperthyroidism), Parkinson's syndrome, amyotrophic lateral sclerosis, syringomyelia, and peripheral neuritis and neuronitis.

Electrical currents should not be allowed to flow across a pregnant uterus.

It is recommended that the therapist test the apparatus on him or herself shortly before it is used to test the patient. This avoids unwarranted negative findings and provides a safeguard against injuring the patient through faulty equipment. During pre-testing, the current levels which will be used on the patient should be used on the tester; this approach takes courage because such currents by necessity are often high and uncomfortable.

Notes Aside (M02A5)

Scaleni relengthening (M02A5A)

One technique involves having the patient lying supine or in a semi-reclined position with the involved extremity stretched straight out from the shoulder and fully supported from beneath (an end table with a pillow on it may suffice). The patient's head should be rotated toward the involved hand throughout the treatment. The electrical stimulation should be applied directly to the involved scaleni (the negative electrode over the scaleni and the positive electrode usually on the low back) at an amplitude strong enough to cause a visible contraction from the anterior scaleni or the muscles innervated by the compressed portion of the brachial plexus. Stimulation should continue for ten minutes. Ideally, the polarity of the electrical current should be then be reversed for a five-minute period. Following this procedure, linear vibration should be applied to the origin, insertion or tendon of each of the involved scaleni muscle's accessible antagonists, including the opposing scaleni. The vibrator should deliver approximately 1.5 to 2.0 pounds per square inch at a frequency of 30 hertz. Each site should be vibrated for one minute. The patient should be instructed to rest for between 6 to 12 minutes before changing arm, shoulder and head position. If successful, the involved scaleni should be tonically reset and the patient may experience immediate, dramatic (albeit sometimes temporary), complete relief of the pain provoked by the CDO syndrome. The above described treatment program may have to be repeated several times (at daily intervals) to effect complete relief without return.

Muscle lengthening with E.S., vibration reinforced (M02A5B)

In general, for best results, muscle lengthening with electrical stimulation should be followed immediately, during the patient's fifteen-minute post stimulation period, by linear vibration of the "short" muscle and then the antagonistic muscle(s). The involved joint(s) should remain fixed during the linear vibration in the position in which electrical stimulation was applied; each muscle site (origin, insertion or tendon) should be vibrated for one minute. The linear vibration has the effect of reinforcing the reciprocal inhibition set up by the electrical stimulation; linear vibration of the appropriate muscle site facilitates an involuntary contraction (myoelectric activity) of the muscle vibrated and reciprocally inhibits contraction (myoelectric activity) in antagonistic musculature (refer to Vibration, Neuromuscular Management [M08A]).

Bibliography (M02A)

Alon, pp. 890–895.
Guyton, pp. 651–655.
Owens, pp. 162–168.
Packman-Braun, pp. 51–56.
Shriber, pp. 110–123, 139–147.
Stillwell, pp. 1–64, 65–108, 124–173.

ELECTRICAL STIMULATION OF MUSCLE (OR DEEP) TISSUE MUSCLE REEDUCATION/MUSCLE TONING (M02B)

Definition (M02B1)

Research has confirmed that electrical stimulation, if appropriately applied, may be used effectively to increase tone, strengthen, improve endurance and increase the size of innervated muscle. It has *not* been shown, regardless of electrical stimulation unit type, to be superior to more traditional forms of voluntary exercise for building tone and strength, including isometric exercise, but several studies have shown electrical stimulation to be nearly as effective. In fact, some research has demonstrated that high voltage pulsed electrical stimulation of at least at 30 hertz may be used to cause involuntary isometric muscular contraction (tetany against resistance) without stressing the cardiovascular system by increasing heart rate or blood pressure.

Review of literature suggests that more study is needed to establish which types of electrical stimulators and which electrical forms are most effective for the increasing of muscle tone and strength, which methods of application are most efficient, and which muscle types will respond best to electrical stimulation.

Muscle toning (as reflected by myoelectric activity) with electrical stimulation seems to be most effectively accomplished by those electrical stimulation units capable of producing currents strong enough to produce tetany (or near tetany) while being fairly comfortable for the patient during application. Such stimulation units should provide a duty cycle of a 10 to 15-second stimulation period, followed by a 10 to 50-second off period. Such stimulation is usually best provided by a high voltage unit, portable neuromuscular stimulation unit, or high frequency ("medium frequency") unit.

Procedures of Application (M02B2)

General application considerations (M02B2A)

1) Before the patient is introduced into the treatment environment, the equipment should be checked for proper connection with its power source, between the electrode cables and the stimulator, and between the electrode cables and the electrodes. All equipment should be checked before use to confirm good working order, and all controls should be off or set in the zero position.

2) The patient should be placed in a position which will remain comfortable for the duration of the treatment session. A reclining, over-stuffed arm chair is ideal in most cases because the neck, arms, legs, anterior abdomen, and the back, with some skillful maneuvering, are all accessible to the therapist and the patient is provided with supported comfort.

3) Before applying electrodes, the skin surface to be treated should be inspected for recent scars, evidence of peripheral injury, dermal lesions, or other causes of disturbed sensation. Such tissue may be hypersensitive to electrical stimulation and care should be taken to build up electrical amplitudes gradually to prevent undue discomfort should the skin resistance suddenly drop, magnifying the effects of the stimulation on those areas.

4) If low frequency sponge pad electrodes are being used, they should be moistened with a saline solution or water and placed over the chosen treatment sites, and care should be taken that they are in good even contact with the skin. In general the negative cathode electrode should be placed on the muscle's motor point, where the motor nerve is most superficial as it innervates the muscle, and when stimulated, provokes the greatest muscle contraction. If metal or flexible carbon electrodes are used, a proper electrode paste should be used to insure adequate contact. Once the best site(s) for electrode placement have been determined, elastic strapping, weighting the electrodes, or strategic placement of the electrode(s) under the patient's body or body parts may be necessary to insure good electrode contact. If the body part is to be immersed for treatment, it should be placed in the bath with the negative cathode electrode, several inches apart.

5) A watch or timer should be set for the length of treatment. The electrical stimulator should be turned on, and the amplitude increased until a visible contraction takes place, always staying within the patient's range of tolerance.

6) The patient should be allowed to become comfortable with the current, and then the intensity should be slowly increased as the patient's tolerance allows. This process should be continued un-

til the desired degree of contraction is reached. The patient should be closely monitored for excessive muscle cramping, joint compression, pain, and autonomic sympathetic hyperresponse to the stimulation. The patient should *never* be left out of sight or sound once the treatment has been started unless the patient can turn off the instrument should the need arise.

7) At the end of the session, if not automatically shut off, the intensity of the stimulator should be gradually decreased until it is switched off. Remove all electrodes from the patient and inspect each site of stimulation carefully, record any abnormal tissue response and return all controls to zero. The patient should rest for several minutes before being allowed to leave.

Specific procedures for the increase of muscle tone (M02B2B)

To tone muscle with electrical stimulation the best results seem to come from putting the muscle or muscle group on stretch (relative to the patient's comfort) and fixing the involved joint(s) to prohibit them from moving in response to provoked contractions (i.e., isometric contraction). The electrodes should be placed over the muscle or muscle group to be stimulated in a bipolar fashion with both cathode and anode electrodes on the same muscle or muscle group. The cathode electrode should be placed over the muscle or the muscle group's dominant muscle motor point.

The electrodes should be securely attached over the chosen muscle sites. The electrical stimulator should be turned on and the intensity slowly increased until a visible contraction develops. As the patient gets used to the stimulation the current should be gradually increased again until tetany or near tetany is produced. Stimulation should continue until between ten and twenty contractions have taken place. If the options are available, a duty cycle of 10 to 15 seconds of stimulation followed by 50 seconds off should be used for maximum toning and a minimum of muscle fatigue.

Treatment may occur daily but suitable results have come from treatments occurring every other day or twice a week.

Partial List of Conditions Treated (M02B3)

Treatable Causes (M02B3A)

Habitual joint positioning (C48)
Hypermobile/instable joint (C10)
Joint ankylosis (C42)
Ligamentous strain (C25)
Muscular weakness (C23)
Neuromuscular tonic imbalance (C30)
Pathological neuromuscular hypotonus (C33)
Psychoneurogenic neuromuscular hypertonus (C29)
Restricted joint range of motion (C26)
Trigger point formation (C01)

Syndromes (M02B3B)

CALF PAIN (S50)
CEREBRAL PALSY (S25)
CERVICAL (NECK) PAIN (S73)
CHEST PAIN (S14)
COLITIS (S42)
EARACHE (S40)
ELBOW PAIN (S47)
FOOT PAIN (S51)
FOREARM PAIN (S54)
FROZEN SHOULDER (S64)
HAND/FINGER PAIN (S17)
HEADACHE PAIN (S02)
HIGH THORACIC BACK PAIN (S48)
HYSTERIA/ANXIETY REACTION (S59)
KNEE PAIN (S46)
LOWER ABDOMINAL PAIN (S62)
LOW BACK PAIN (S03)
MENSTRUAL CRAMPING (S60)
MIGRAINE (VASCULAR) HEADACHE (S18)
NEUROMUSCULAR PARALYSIS (S22)
OSTEOARTHRITIS (S21)
POST CEREBRAL VASCULAR ACCIDENT (CVA) (S07)
POST IMMOBILIZATION (S36)
POST PERIPHERAL NERVE INJURY (S23)
POST WHIPLASH (S55)
RADICULITIS (S29)
SCIATICA (S05)
SHOULDER/HAND (S57)
SHOULDER PAIN (S08)
SINUS PAIN (S33)
TEMPOROMANDIBULAR JOINT (TMJ) PAIN (S06)
THIGH PAIN (S49)
WRIST PAIN (S16)

Precautions and Contraindications (M02B4)

Care should be taken to avoid over-fatigue of the muscle stimulated. Stimulation should stop when the muscle begins to respond with less vigor.

Electrical muscle stimulation is generally contraindicated for victims of cancer or other malignant disease if the site to be stimulated is physically associated with the cancerous growth. It is also contraindicated for those suffering from sensory deficits, not including those suffering from hysterical sensory abnormalities.

Care should be taken not to put electrodes too close together. The closer electrodes are to one another, the greater the current density becomes between their edges and an *edge-effect* may be produced which causes unpleasant burning sensations and superficial or deep burns may ultimately occur.

Electrical burns may occur when an excess of electricity (current density) is applied to the skin or mucous membrane. Electrical burns may appear in the clinical setting as a passing erythema, blistering of superficial skin layers, or deep tissue coagulation. The tissue damage produced by the deep electrical burn occurs in a roughly conical area, extending from the apex of the cone on the skin surface, where the original electrical contact occurred, fanning out into the deeper layers. Just following the injury, the burn site appears rather small and inconsequential but becomes more alarming as the damaged tissues subsequently slough off and ulceration occurs. These burns are slow to heal, prone to infection, and if sufficiently deep, may be followed by extensive unsightly scarring.

Electric shock may be caused if the patient, while being electrically stimulated, makes contact with a grounded object such as a water pipe, radiator, or electric circuit. This is especially serious if a large area of the patient's body is subjected to the shock (as in a bath or treatment tub) and may pose the danger of cardiac arrest. Electric shock may also occur if the electrical stimulator suffers transformer breakdown, in which case the high-tension, low frequency current may "jump" to the patient and produce an electrical burn as well as shock.

Great care must be taken in the application of all electrical appliances to patients with cardiac pacemakers. Electrodes should not be placed in the vicinity of a pacemaker nor should the current flow from the cathode to the anode electrode be allowed to cross the pacemaker or implanted wires.

In general for high frequency stimulation, treatment should be precluded where joint repairs, rheumatoid and collagen diseases are present, or other conditions in which tendons, joint capsules or the joint may be threatened or weakened by stimulation. In general, electrodes should not be placed over scar tissue, skin irritations or open skin lesions. High frequency stimulation should not be applied over the thorax, abdomen, or back extensor musculature; such stimulation is contraindicated for patients with cardiac demand pacemakers, cardiac arrhythmias, ST segment depression or conduction block. If the patient suffers increased sweating, salivation and/or has complaints of nausea when being stimulated with high frequency stimulation, the stimulation should be discontinued.

Low frequency currents are contraindicated for treatment of myopathies (muscular dystrophy, myotonia congenita, polymyositis, muscle-involved connective tissue disease, cortisone myopathies, myopathy due to metastatic cancer, and myopathy from hyperthyroidism), Parkinson's syndrome, amyotrophic lateral sclerosis, syringomyelia, and peripheral neuritis and neuronitis.

Electrical currents should not be allowed to flow across a pregnant uterus.

It is recommended that the therapist test the apparatus on him or herself shortly before it is used to test the patient. This avoids unwarranted negative findings and provides a safeguard against injuring the patient through faulty equipment. During pre-testing, the current levels which will be used on the patient should be used on the tester; this approach takes courage because such currents by necessity are often high and uncomfortable.

Notes Aside (M02B5)

Muscle toning with electrical stimulation has proven to be remarkably effective for retrieving muscle tone in conditions which has has been lost as a secondary effect of long term inflammatory conditions (including chronic post whiplash, low back pain, and bursitic hip syndromes) or disuse from prolonged bed rest (refer to Adverse Effects of Bed Rest [R008]). If the electrical stimulation is applied correctly and appropriately, muscle strength can be improved without risk of reinflammation of the previously involved soft tissue through strain which is often a consequence of voluntary exercise.

Bibliography (M02B6)

Alon, pp. 890–895.
Baker, Parker, Sanderson, pp. 1967–1974.
Baker, Yeh, Wilson, Waters, pp. 1495–1506.
Currier, Lehman, Lightfoot, pp. 1508–1512.
Currier, Mann: Muscular Strength Development by Electrical Stimulation in Healthy Individuals. pp. 915–921.
Currier, Mann: Pain Complaint. pp. 318–323.
Delitto, McKowen, McCarthy, Shively, Rose, pp. 45–50.
Delitto, Rose, pp. 1704–1707.
Eckerson, Axelgaard, pp. 483–490.
Kramer, pp. 31–38.
Kramer, Lindsay, Magee, Mendryk, Wall, pp. 324–331.
Laughman, Youdas, Garrett, Chao, pp. 494–499.
Mohr, Carlson, Sulentic, Landry, pp. 606–609.

electrical stimulation of muscle (or deep) tissue muscle reeducation/muscle toning

Nitz, pp. 219–222. Owens, pp. 162–168.
Packman-Braun, pp. 51–56.
Selkowitz, pp. 186–195.
Soo, pp. 333–337.

Urabe, pp. 283.
Walker, Currier, Threlkeld, pp. 481–485.
Walmsey, Letts, Vooys, pp. 10–17.
Wong, pp. 1209–1214.

ELECTRICAL STIMULATION OF MUSCLE (OR DEEP) TISSUE INCREASING CIRCULATION (M02C)

Definition (M02C1)

It has long been accepted that electrical stimulation of muscle tissue may be used therapeutically to improve circulation by utilizing provoked muscle contractions to rhythmically *squeeze* associated blood vessels. This process induces the muscles to provide artificially the "pumping" action required by nature to facilitate venous blood flow and to add impetus to lymphatic circulation. Consequently, electrical stimulation may be used to good effect when the patient is bed ridden or has lost the ability to voluntarily contract the muscles necessary for normal blood and lymph circulation. This process also provides an efficient means of preventing post-surgical thrombophlebitis and reducing lymph edema which is often a consequence of post cerebral vascular accident or other syndromes in which the muscles are prevented from contracting or are unable to do so voluntarily.

Less known or appreciated is the effect that electrical stimulation may have upon capillary development. Research undertaken in the late 1970's has demonstrated that chronic low frequency electrical stimulation can increase capillary-muscle fiber ratio and the number of capillaries present per cross-sectional area. The formation of new capillaries may be accompanied by an increase in the total capillary surface area and the "sprouting" of large capillaries. Increases in capillary density (20% after 4 days of stimulation, 50% after 14, and 100% after 28 days) are apparently not a consequence of the action of the electrical current itself, but in response to the muscular need for additional blood supply to support the demand of increased muscular activity. Similar changes have been demonstrated to occur in connection with endurance or aerobic exercise- (repetitive isotonic contraction) induced hypoxia in humans and animals (non-primates), and it should be noted that both high frequency electrical stimulation (30 hertz and above) and isometric exercise has failed to demonstrate the ability to increase capillary density.

As a consequence of the ability to increase capillary density, low frequency electrical stimulation may be used to good effect in the treatment of conditions stemming from impaired or decreased circulation as seen in cases of fully developed shoulder/hand syndromes or in the late stages of diabetes mellitus when claudication of peripheral vascular beds has begun.

Procedures of Application (M02C2)

General application considerations (M02C2A)

1) Before the patient is introduced into the treatment environment, the equipment should be checked for proper connection with its power source, between the electrode cables and the stimulator, and between the electrode cables and the electrodes. All equipment should be checked before use to confirm good working order, and all controls should be off or set in the zero position.

2) The patient should be placed in a position which will remain comfortable for the duration of the treatment session. A reclining, over-stuffed arm chair is ideal in most cases because the neck, arms, legs, anterior abdomen, and the back, with some skillful maneuvering, are all accessible to the therapist and the patient is provided with supported comfort.

3) Before applying electrodes, the skin surface to be treated should be inspected for recent scars, evidence of peripheral injury, dermal lesions, or other causes of disturbed sensation. Such tissue may be hypersensitive to electrical stimulation and care should be taken to build up electrical amplitudes gradually to prevent undue discomfort should the skin resistance suddenly drop, magnifying the effects of the stimulation on those areas.

4) If low frequency sponge pad electrodes are being used, they should be moistened with a saline solution or water and placed over the chosen treatment sites, and care should be taken that they are in good even contact with the skin. In general the negative cathode electrode should be placed on the muscle's motor point, where the motor nerve is most superficial as it innervates the muscle, and when stimulated, provokes the greatest muscle contraction. If metal or flexible carbon electrodes are used, a proper electrode paste should be used to insure adequate contact. Once the best site(s) for electrode placement have been determined, elastic strapping, weighting the electrodes, or strategic placement of the electrode(s) under the patient's body or body parts may be necessary to insure good electrode contact. If the body part is to be immersed for treatment, it should be placed in the bath with the negative cathode electrode, several inches apart.

5) A watch or timer should be set for the length of treatment. The electrical stimulator should be turned on, and the amplitude increased until a visible contraction takes place, always staying within the patient's range of tolerance.

6) The patient should be allowed to become comfortable with the current, and then the intensity should be slowly increased as the patient's tolerance allows. This process should be continued until the desired degree of contraction is reached. The patient should be closely monitored for excessive muscle cramping, joint compression, pain, and autonomic sympathetic hyperresponse to the stimulation. The patient should *never* be left out of sight or sound once the treatment has been started unless the patient can turn off the instrument should the need arise.

7) At the end of the session, if not automatically shut off, the intensity of the stimulator should be gradually decreased until it is switched off. Remove all electrodes from the patient and inspect each site of stimulation carefully, record any abnormal tissue response and return all controls to zero. The patient should rest for several minutes before being allowed to leave.

Facilitation of venous blood flow   (M02C2B)

To facilitate venous blood flow (and prevent thrombophlebitis) the electrodes should be placed over the muscle (or muscle group) in a bipolar fashion with both cathode and anode electrodes on the same muscle group and the cathode electrode over the muscle group's dominant muscle motor point. The electrodes should be securely attached over the chosen muscle sites.

The electrical stimulator should be preset to deliver pulsed square wave or faradic current flow at a pulse frequency of 4 hertz for a twenty-minute period. The stimulator should be turned on and the intensity slowly increased until a visible contraction develops. As the patient gets used to the sensation of electrical stimulation the current should be gradually increased again until the contractions are quite brisk.

Treatment should occur daily, but suitable results have come from treatments occurring every other day and two or three times a week.

Increasing capillary density   (M02C2C)

To increase capillary density, the electrodes may be placed as suggested above or a bath may be used for a more generalized effect. If a bath is utilized, the patient's distal extremity (foot or hand or hand and forearm) should be placed in the bath along with the cathode electrode placed several inches away, and the anode secured in place on a dry area over the motor point of a muscle in the same synergistic pattern as the major muscle group to be stimulated.

The electrical stimulator should be preset to deliver a pulsed square wave or faradic current flow at a pulse frequency of between 4 and 14 hertz (10 hertz would be ideal), for a twenty-minute period. The stimulator should be turned on, and the intensity slowly increased until a visible contraction of the muscle or muscle group develops. As the patient gets used to the sensation of electrical stimulation, the current should be gradually increased again until the contractions are quite brisk.

For best results treatments should occur daily.

Partial List of Conditions Treated   (M02C3)

Treatable Causes   (M02C3A)

Habitual joint positioning   (C48)
Hypotension   (C45)
Phlebitis   (C43)
Soft tissue inflammation   (C05)
Soft tissue swelling   (C06)
Vascular insufficiency   (C11)

Syndromes   (M02C3B)

BUERGER'S DISEASE   (S11)
CALF PAIN   (S50)
CERVICAL DORSAL OUTLET (SCALENICUS ATTICUS/CERVICAL RIB)   (S09)
DIABETES   (S10)
ELBOW PAIN   (S47)
FOOT PAIN   (S51)
FOREARM PAIN   (S54)
HAND/FINGER PAIN   (S17)
KNEE PAIN   (S46)
PHLEBITIS   (S68)
PITTING (LYMPH) EDEMA   (S31)
POST CEREBRAL VASCULAR ACCIDENT (CVA)   (S07)
POST IMMOBILIZATION   (S36)
POST PERIPHERAL NERVE INJURY   (S23)
POST SPINAL CORD INJURY   (S24)
RAYNAUD'S DISEASE   (S12)
SHOULDER/HAND   (S57)
TENNIS ELBOW (LATERAL EPICONDYLITIS)   (S32)
THIGH PAIN   (S49)
WRIST PAIN   (S16)

Precautions and Contraindications (M02C4)

Care should be taken to avoid over-fatigue of the muscle stimulated. Stimulation should stop when the muscle begins to respond with less vigor.

Electrical muscle stimulation is generally contraindicated for victims of cancer or other malignant disease if the site to be stimulated is physically associated with the cancerous growth. It is also contraindicated for those suffering from sensory deficits, not including those suffering from hysterical sensory abnormalities.

Care should be taken not to put electrodes too close together. The closer electrodes are to one another, the greater the current density becomes between their edges and an edge-effect may be produced which causes unpleasant burning sensations and superficial or deep burns may ultimately occur.

Electrical burns may occur when an excess of electricity (current density) is applied to the skin or mucous membrane. Electrical burns may appear in the clinical setting as a passing erythema, blistering of superficial skin layers, or deep tissue coagulation. The tissue damage produced by the deep electrical burn occurs in a roughly conical area, extending from the apex of the cone on the skin surface, where the original electrical contact occurred, fanning out into the deeper layers. Just following the injury, the burn site appears rather small and inconsequential but becomes more alarming as the damaged tissues subsequently slough off and ulceration occurs. These burns are slow to heal, prone to infection, and if sufficiently deep, may be followed by extensive unsightly scarring.

Electric shock may be caused if the patient, while being electrically stimulated, makes contact with a grounded object such as a water pipe, radiator, or electric circuit. This is especially serious if a large area of the patient's body is subjected to the shock (as in a bath or treatment tub) and may pose the danger of cardiac arrest. Electric shock may also occur if the electrical stimulator suffers transformer breakdown, in which case the high-tension, low frequency current may "jump" to the patient and produce an electrical burn as well as shock.

Great care must be taken in the application of all electrical appliances to patients with cardiac pacemakers. Electrodes should not be placed in the vicinity of a pacemaker nor should the current flow from the cathode to the anode electrode be allowed to cross the pacemaker or implanted wires.

In general for high frequency stimulation, treatment should be precluded where joint repairs, rheumatoid and collagen diseases are present, or other conditions in which tendons, joint capsules or the joint may be threatened or weakened by stimulation. In general, electrodes should not be placed over scar tissue, skin irritations or open skin lesions. High frequency stimulation should not be applied over the thorax, abdomen, or back extensor musculature; such stimulation is contraindicated for patients with cardiac demand pacemakers, cardiac arrhythmias, ST segment depression or conduction block. If the patient suffers increased sweating, salivation and/or has complaints of nausea when being stimulated with high frequency stimulation, the stimulation should be discontinued.

Low frequency currents are contraindicated for treatment of myopathies (muscular dystrophy, myotonia congenita, polymyositis, muscle-involved connective tissue disease, cortisone myopathies, myopathy due to metastatic cancer, and myopathy from hyperthyroidism), Parkinson's syndrome, amyotrophic lateral sclerosis, syringomyelia, and peripheral neuritis and neuronitis.

Electrical currents should not be allowed to flow across a pregnant uterus.

It is recommended that the therapist test the apparatus on him or herself shortly before it is used to test the patient. This avoids unwarranted negative findings and provides a safeguard against injuring the patient through faulty equipment. During pre-testing, the current levels which will be used on the patient should be used on the tester; this approach takes courage because such currents by necessity are often high and uncomfortable.

Notes Aside (M02C5)

Increasing circulation with electrical stimulation of muscle tissue by ultimately increasing capillary density in the distal extremity may be a major breakthrough for the treatment of entrenched conditions of peripheral circulatory insufficiency, as seen in diabetes and in the later stages of the shoulder/hand syndrome. In the latter case, it may be one of the last hopes of restoring capillary circulation and preserving the upper extremity as a functional limb.

As noted earlier, when increased capillary bed density is desired, the best results are affected when treatment occurs daily for at least thirty days. If performed clinically, such treatment may become extremely expensive in terms of time and money. In such cases, it may be advisable for the practitioner to arrange for the patient to acquire an adequate electrical stimulator, which may be utilized in the patient's home, if the patient or other associated individual can be adequately trained to correctly administer the treatment.

Bibliography (M02C6)

Brown, p. 241.
Brown, Gogia, pp. 662–667.
Currier, Petrilli, Threlkeld, pp. 937–943.
Hudlicka, p. 141.
Liu, pp. 340–350.
Mohr, Akers, Wessman, pp. 526–533.
Myrhage, pp. 73–90.
Salmons, pp. 94–105.
Stillwell, pp. 130–131.
Walker, pp. 481–485.

ELECTRICAL STIMULATION OF MUSCLE (OR DEEP) TISSUE REDUCTION OF EDEMA (M02D)

Definition (M02D1)

It has long been accepted that electrical stimulation of muscle tissue may be used therapeutically to improve circulation by utilizing provoked muscle contractions to rhythmically *squeeze* associated blood vessels, which induces the muscles to artificially provide the "pumping" action required by nature to facilitate venous blood flow and to add impetus to lymphatic circulation. Consequently, electrical stimulation may be used to good effect when the patient has lost the ability to voluntarily contract the muscles necessary. This provides a means of reducing lymph edema which is often a consequence of post cerebral vascular accident or other syndromes in which the muscles are prevented from contracting or are unable to do so.

Aside from the mechanical pumping action that electrically induced muscle contraction can provide, electrical current flow of the pulsed direct current variety has the apparent ability to carry or drive fluid out of edematous tissue, possibly by virtue of electrical ion transfer (refer to Electrical Stimulation, Inhibition of Ionized Formation [M02F]).

Procedures of Application (M02D2)

General application considerations (M02D2A)

1) Before the patient is introduced into the treatment environment, the equipment should be checked for proper connection with its power source, between the electrode cables and the stimulator, and between the electrode cables and the electrodes. All equipment should be checked before use to confirm good working order, and all controls should be off or set in the zero position.

2) The patient should be placed in a position which will remain comfortable for the duration of the treatment session. A reclining, over-stuffed arm chair is ideal in most cases because the neck, arms, legs, anterior abdomen, and the back, with some skillful maneuvering, are all accessible to the therapist and the patient is provided with supported comfort.

3) Before applying electrodes, the skin surface to be treated should be inspected for recent scars, evidence of peripheral injury, dermal lesions, or other causes of disturbed sensation. Such tissue may be hypersensitive to electrical stimulation and care should be taken to build up electrical amplitudes gradually to prevent undue discomfort should the skin resistance suddenly drop, magnifying the effects of the stimulation on those areas.

4) If low frequency sponge pad electrodes are being used, they should be moistened with a saline solution or water and placed over the chosen treatment sites, and care should be taken that they are in good even contact with the skin. In general the negative cathode electrode should be placed on the muscle's motor point, where the motor nerve is most superficial as it innervates the muscle, and when stimulated, provokes the greatest muscle contraction. If metal or flexible carbon electrodes are used, a proper electrode paste should be used to insure adequate contact. Once the best site(s) for electrode placement have been determined, elastic strapping, weighting the electrodes, or strategic placement of the electrode(s) under the patient's body or body parts may be necessary to insure good electrode contact. If the body part is to be immersed for treatment, it should be placed in the bath with the negative cathode electrode, several inches apart.

5) A watch or timer should be set for the length of treatment. The electrical stimulator should be turned on, and the amplitude increased until a visible contraction takes place, always staying within the patient's range of tolerance.

6) The patient should be allowed to become comfortable with the current, and then the intensity should be slowly increased as the patient's tolerance allows. This process should be continued until the desired degree of contraction is reached. The patient should be closely monitored for excessive muscle cramping, joint compression, pain, and autonomic sympathetic hyperresponse to the stimulation. The patient should *never* be left out of sight or sound once the treatment has been started unless the patient can turn off the instrument should the need arise.

7) At the end of the session, if not automatically shut off, the intensity of the stimulator should be gradually decreased until it is switched off. Remove all electrodes from the patient and inspect each site of stimulation carefully, record any abnormal tissue response and return all controls to zero. The patient should rest for several minutes before being allowed to leave.

Specific procedures to reduce edema in distal sites (M02D2B)

To facilitate lymph circulation or decrease edema, for best effect a bath may be used for a generalized effect. The patient's swollen distal extremity (foot or hand or hand and forearm) should be placed in the bath along with the cathode electrode, placed several inches apart. If a single joint is swollen, a wet cloth (wet wrap) should be placed over the swollen area or around the joint with the cathode placed over the most swollen area. If a dual cathode is available, each electrode should be placed on either side of the joint. The anode, a large dispersive pad, should be placed and secured over the patient's low back.

The electrical stimulator should be preset to deliver a pulsed square wave current flow at a pulse frequency of 28 hertz for a twenty-minute period. The stimulator should be turned on and the intensity slowly increased until a visible contraction develops. As the patient gets used to the sensation of electrical stimulation the current should be gradually increased until a mild tetany occurs.

Treatment should occur daily, but suitable results have been reported to result from treatments occurring every other day, or two or three times a week.

Specific procedures to reduce edema in proximal sites (M02D2C)

The electrical stimulator should be able to supply a pulsed square wave current form in a frequency from 4 to 28 hertz. The electrodes should be of the same size and placed in a monopolar fashion with the anode or positive pole over the swollen area. Karya type pads have proven to be remarkably effective for joint space placement. The cathode or negative electrode should be placed in a remote area. The electrodes should both be secured in place.

The electrical stimulation unit should be preset to deliver square wave pulses, continuously, at the desired frequency of between 4 and 28 hertz for twenty minutes. The electrical stimulation unit should then be turned on and the intensity increased turned up until the patient begins to feel the stimulus. When the patient has gotten used to the sensation of the electrical stimulation, the intensity should gradually be increased until the patient's tolerance level is reached. As the patient accommodates to the current level the intensity may once again be increased to tolerance.

In combination with other treatment techniques (principally ultrasound), daily treatment sessions, although no more than fifteen, have proved to be the most successful at reducing or inhibiting calcium deposits (ionic formations); however, sessions occurring two or three times a week for several weeks have also proven effective.

Partial List of Conditions Treated (M02D3)

Treatable Causes (M02D3A)

Hypermobile/instable joint (C10)
Hypotension (C45)
Ligamentous strain (C25)
Pathological neuromuscular hypertonus (C28)
Pathological neuromuscular hypotonus (C33)
Peripheral nerve injury (C07)
Soft tissue swelling (C06)

Syndromes (M02D3B)

CARPAL TUNNEL (S35)
JOINT SPRAIN (S30)
KNEE PAIN (S46)
OSTEOARTHRITIS (S21)
PITTING (LYMPH) EDEMA (S31)
POST CEREBRAL VASCULAR ACCIDENT (CVA) (S07)
POST IMMOBILIZATION (S36)
POST PERIPHERAL NERVE INJURY (S23)
SHOULDER/HAND (S57)

Precautions and Contraindications (M02D4)

Care should be taken to avoid over-fatigue of the muscle stimulated. Stimulation should stop when the muscle begins to respond with less vigor.

Electrical muscle stimulation is generally contraindicated for victims of cancer or other malignant disease if the site to be stimulated is physically associated with the cancerous growth. It is also contraindicated for those suffering from sensory deficits, not including those suffering from hysterical sensory abnormalities.

Care should be taken not to put electrodes too close together. The closer electrodes are to one another, the greater the current density becomes between their edges and an *edge-effect* may be produced which causes unpleasant burning sensations and superficial or deep burns may ultimately occur.

Electrical burns may occur when an excess of electricity (current density) is applied to the skin or mucous membrane. Electrical burns may appear in the clinical setting as a passing erythema, blistering of superficial skin layers, or deep tissue coagulation. The tissue damage produced by the deep electrical burn occurs in a roughly conical area, extending from the apex of the cone on the skin surface, where the original electrical contact occurred, fanning out into the deeper layers. Just following the injury, the burn site appears rather small and inconsequential but becomes more alarming as the damaged tissues subsequently slough off and ulceration occurs. These burns are slow to heal, prone to infection, and if sufficiently deep, may be followed by extensive unsightly scarring.

Electric shock may be caused if the patient, while being electrically stimulated, makes contact with a grounded object such as a water pipe, radiator, or electric circuit. This is especially serious if a large area of the patient's body is subjected to the shock (as in a bath or treatment tub) and may pose the danger of cardiac arrest. Electric shock may also occur if the electrical stimulator suffers transformer breakdown, in which case the high-tension, low frequency current may "jump" to the patient and produce an electrical burn as well as shock.

Great care must be taken in the application of all electrical appliances to patients with cardiac pacemakers. Electrodes should not be placed in the vicinity of a pacemaker nor should the current flow from the cathode to the anode electrode be allowed to cross the pacemaker or implanted wires.

In general for high frequency stimulation, treatment should be precluded where joint repairs, rheumatoid and collagen diseases are present, or other conditions in which tendons, joint capsules or the joint may be threatened or weakened by stimulation. In general, electrodes should not be placed over scar tissue, skin irritations or open skin lesions. High frequency stimulation should not be applied over the thorax, abdomen, or back extensor musculature; such stimulation is contraindicated for patients with cardiac demand pacemakers, cardiac arrhythmias, ST segment depression or conduction block. If the patient suffers increased sweating, salivation and/or has complaints of nausea when being stimulated with high frequency stimulation, the stimulation should be discontinued.

Low frequency currents are contraindicated for treatment of myopathies (muscular dystrophy, myotonia congenita, polymyositis, muscle-involved connective tissue disease, cortisone myopathies, myopathy due to metastatic cancer, and myopathy from hyperthyroidism), Parkinson's syndrome, amyotrophic lateral sclerosis, syringomyelia, and peripheral neuritis and neuronitis.

Electrical currents should not be allowed to flow across a pregnant uterus.

It is recommended that the therapist test the apparatus on him or herself shortly before it is used to test the patient. This avoids unwarranted negative findings and provides a safeguard against injuring the patient through faulty equipment. During pre-testing, the current levels which will be used on the patient should be used on the tester; this approach takes courage because such currents by necessity are often high and uncomfortable.

Notes Aside (M02D5)

The electrical stimulation for the reduction of edema technique is remarkable for its ability to immediately reduce the swelling associated with strained, sprained, or immobilized joints. When edema occurs in distal joints (knees, elbows, wrists, or ankles), electrical stimulation utilized to reduce it should generally be followed by taping of the involved joint (refer to Taping [M16]) or the fitting of positive pressure garments (refer to Positive Pressure, Phlebitis and/or Edema Control [M24B]) to prevent swelling redevelopment and/or to stabilize the joint to prevent further joint trauma and further consequent swelling.

If edema has resulted from the effects of soft tissue inflammation, it may be advisable to first treat the inflammation with peripheral electroacustimulation (refer to Peripheral Electroacustimulation within an Inflamed Zone), ice packing (refer to Cryotherapy, Soft Tissue Inflammation Control [M10E]) and phonophoresis (refer to Ultrahigh Frequency Sound, Phonophoresis [M01A]).

Bibliography (M02D6)

Alon, pp. 890–895.
Guyton, pp. 242–248.
Owens, pp. 162–168.
Shriber, pp. 110–123, 139–147.
Stillwell, p. 151.
Wong, pp. 1209–1214.

ELECTRICAL STIMULATION OF MUSCLE (OR DEEP) TISSUE DELAYING THE ATROPHY OF DENERVATED MUSCLE (M02E)

Definition (M02E1)

Electrical stimulation may be used to delay the atrophy of denervated muscle, however, the type of current, the intensity, the frequency (pulses per second), electrode placement, and the length of treatment session must all be ideal for the treatment to be successful. Current forms which may be used include faradic (in the first two weeks after injury), sine wave, continuous direct current, and pulsed square wave. Of these, pulsed square wave is the most comfortable to the patient and consequently the most effective.

For the activation of denervated muscles, the ideal electrical stimulator must provide pulsed square waves of at least 50 millisecond duration and at a frequency of from 4 to 20 pulses per second. The intensity utilized should be high enough to stimulate the deep muscle fibers for a tetanic contraction which includes the entire muscle (not just surface fibers), but within the tolerance levels of the patient. For best results, the electrodes should be placed at the two ends of the muscle belly (if to be singly stimulated) or across the muscles just inside the tendon attachments at each end of the group of muscles to be stimulated. The treatment session length should be limited to the endurance level of the muscles stimulated. Two or three contractions may be enough to begin with but the number may be increased gradually to thirty or forty as the muscle's endurance improves. When the intensity must be raised to maintain the strength of contraction, the endurance level of the muscle has been reached and stimulation should be ceased in order to prevent over-fatigue.

Nerve regeneration may take up to three years (for brachial or lumbosacral plexus nerve injury) before reinnervation is complete. Strictly speaking, treatment with appropriate electrical stimulation to prevent complete atrophy of the involved musculature may be justified for the full term until reinnervation is complete and normal function is restored. The literature suggests that electrical stimulation of denervated muscle is only really effective in slowing atrophy if it is performed five days a week.

Procedures of Application (M02E2)

1) Before the patient is introduced into the treatment environment, the equipment should be checked for proper connection with its power source, between the electrode cables and the stimulator, and between the electrode cables and the electrodes. All equipment should be checked before use to confirm good working order, and all controls should be off or set in the zero position.

2) The patient should be placed in a position which will remain comfortable for the duration of the treatment session. A reclining, over-stuffed arm chair is ideal in most cases because the neck, arms, legs, anterior abdomen, and the back, with some skillful maneuvering, are all accessible to the therapist and the patient is provided with supported comfort.

3) Before applying electrodes, the skin surface to be treated should be inspected for recent scars, evidence of peripheral injury, dermal lesions, or other causes of disturbed sensation. Such tissue may be hypersensitive to electrical stimulation and care should be taken to build up electrical amplitudes gradually to prevent undue discomfort should the skin resistance suddenly drop, magnifying the effects of the stimulation on those areas.

4) If low frequency sponge pad electrodes are being used, they should be moistened with a saline solution or water and placed over the chosen treatment sites, and care should be taken that they are in good even contact with the skin. In general the negative cathode electrode should be placed on the muscle's motor point, where the motor nerve is most superficial as it innervates the muscle, and when stimulated, provokes the greatest muscle contraction. If metal or flexible carbon electrodes are used, a proper electrode paste should be used to insure adequate contact. Once the best site(s) for electrode placement have been determined, elastic strapping, weighting the electrodes, or strategic placement of the electrode(s) under the patient's body or body parts may be necessary to insure good electrode contact. If a hand-held probe is to be used, the motor point should be marked on the skin with an indelible marker to avoid straying with the probe.

5) A watch or timer should be set for the length of treatment. The electrical stimulator should be turned on, and the amplitude increased until a visible contraction takes place, always staying within the patient's range of tolerance.

6) The patient should be allowed to become comfortable with the current, and then the intensity should be slowly increased as the patient's tolerance allows. This process should be continued until the desired degree of contraction is reached. The patient should be closely monitored for excessive muscle cramping, joint compression, pain, and autonomic sympathetic hyperresponse to the stimulation. The patient should *never* be left out of sight or sound once the treatment has been started unless the patient can turn off the instrument should the need arise.

7) At the end of the session, if not automatically shut off, the intensity of the stimulator should be gradually decreased until it is switched off. Remove all electrodes from the patient and inspect each site of stimulation carefully, record any abnormal tissue response and return all controls to zero. The patient should rest for several minutes before being allowed to leave.

Partial List of Conditions Treated (M02E3)

Treatable Causes (M02E3A)

Muscular weakness (C23)
Nerve root compression (C04)
Pathological neuromuscular hypotonus (C33)
Peripheral nerve injury (C07)

Syndromes (M02E3B)

BELL'S PALSY (S66)
CERVICAL DORSAL OUTLET (SCALENICUS ATTICUS/CERVICAL RIB) (S09)
NEUROMUSCULAR PARALYSIS (S22)
POST PERIPHERAL NERVE INJURY (S23)
SCIATICA (S05)

Precautions and Contraindications (M02E4)

Care should be taken to avoid over-fatigue of the muscle stimulated. Stimulation should stop when the muscle begins to respond with less vigor.

Electrical muscle stimulation is generally contraindicated for victims of cancer or other malignant disease if the site to be stimulated is physically associated with the cancerous growth. It is also contraindicated for those suffering from sensory deficits, not including those suffering from hysterical sensory abnormalities.

Care should be taken not to put electrodes too close together. The closer electrodes are to one another, the greater the current density becomes between their edges and an edge-effect may be produced which causes unpleasant burning sensations and superficial or deep burns may ultimately occur.

Electrical burns may occur when an excess of electricity (current density) is applied to the skin or mucous membrane. Electrical burns may appear in the clinical setting as a passing erythema, blistering of superficial skin layers, or deep tissue co-agulation. The tissue damage produced by the deep electrical burn occurs in a roughly conical area, extending from the apex of the cone on the skin surface, where the original electrical contact occurred, fanning out into the deeper layers. Just following the injury, the burn site appears rather small and inconsequential but becomes more alarming as the damaged tissues subsequently slough off and ulceration occurs. These burns are slow to heal, prone to infection, and if sufficiently deep, may be followed by extensive unsightly scarring.

Electric shock may be caused if the patient, while being electrically stimulated, makes contact with a grounded object such as a water pipe, radiator, or electric circuit. This is especially serious if a large area of the patient's body is subjected to the shock (as in a bath or treatment tub) and may pose the danger of cardiac arrest. Electric shock may also occur if the electrical stimulator suffers transformer breakdown, in which case the hightension, low frequency current may "jump" to the patient and produce an electrical burn as well as shock.

Great care must be taken in the application of all electrical appliances to patients with cardiac pacemakers. Electrodes should not be placed in the vicinity of a pacemaker nor should the current flow from the cathode to the anode electrode be allowed to cross the pacemaker or implanted wires.

In general for high frequency stimulation, treatment should be precluded where joint repairs, rheumatoid and collagen diseases are present, or other conditions in which tendons, joint capsules or the joint may be threatened or weakened by stimulation. In general, electrodes should not be placed over scar tissue, skin irritations or open skin lesions. High frequency stimulation should not be applied over the thorax, abdomen, or back extensor musculature; such stimulation is contraindicated for patients with cardiac demand pacemakers, cardiac arrhythmias, ST segment depression or conduction block. If the patient suffers increased sweating, salivation and/or has complaints of nausea when being stimulated with high frequency stimulation, the stimulation should be discontinued.

Low frequency currents are contraindicated for treatment of myopathies (muscular dystrophy, myotonia congenita, polymyositis, muscle-involved connective tissue disease, cortisone myopathies, myopathy due to metastatic cancer, and myopathy from hyperthyroidism), Parkinson's syndrome, amyotrophic lateral sclerosis, syringomyelia, and peripheral neuritis and neuronitis.

Electrical currents should not be allowed to flow across a pregnant uterus.

It is recommended that the therapist test the apparatus on him or herself shortly before it is used to test the patient. This avoids unwarranted negative findings and provides a safeguard against injuring the patient through faulty equipment. During pre-testing, the current levels which will be used on the patient should be used on the tester; this approach takes courage because such currents by necessity are often high and uncomfortable.

Notes Aside (M02E5)

A great deal of controversy in the literature revolves around the question of whether or not electrical stimulation is helpful in delaying or slowing atrophy of denervated muscle. It would seem from the sources reviewed that successful treatment involves performing the technique exactly right, five days a week, until reinnervation occurs. The failures that have raised the question would seem to have derived from a divergence from technique and inadequate number of treatment sessions. Everything can't be fixed on a two- or three-times-a-week basis, and sloppy uninformed technique is never effective. Careful patient training in self-stimulation, using a home-based electrical stimulation unit may serve to decrease the cost of treatment in terms of time and dollars ordinarily eaten up by daily clinical treatment.

Bibliography (M02E6)

Shriber, pp. 184–189.
Stillwell, pp. 154–165.

ELECTRICAL STIMULATION OF MUSCLE (OR DEEP) TISSUE INHIBITION OF IONIZED FORMATION (M02F)

Definition (M02F1)

Of all the therapeutic uses of electrical stimulation, the most historically notorious and detrimental to the continued use of electrical stimulation in medical, paramedical, or physical therapy practice is the use of electrical stimulation to influence ionic distribution. What is surprising is that some of that some of the treatment techniques based on the electrical manipulation of ionized chemicals were remarkably successful; that very success may have been what led it into medical disrepute.

Historically, any treatment technique which may be successfully utilized to treat a variety of conditions may become the victim of the over-enthusiastic and all too early take on the label of a "panacea" for the treatment of almost all ills. When it becomes obvious that the treatment technique has its limitations and does not live up to earlier exaggerated expectations, enthusiasm may not only wane but the technique may be denigrated as useless or dangerous by former practitioner hopefuls, regardless of its specific valuable applications. This process occurred to electrical ionic manipulation, therapeutic electrical stimulation of muscle or deep tissue, ultrahigh frequency sound, and (most recently) biofeedback. For a treatment technique to survive this process, a long period of utilization of the treatment techniques for its predictable beneficial effects by a few practical clinicians is necessary to reestablish its reputation and to bring it back into responsible general use. If, however, the modality falls into the hands of the lay public, with the assistance of exploitative charlatans and quacks, the inevitable failures through misapplication or inappropriate use may have the effect of almost permanently discrediting it. This happened to electrical manipulation of ions (without the addition of chemicals) early in this century, when it was discovered that tumors could be shrunk by the simple application of direct current electrical stimulation when correctly applied. The necessary equipment was readily available to the lay public. Charlatans and quacks took advantage of the situation by not only selling apparatus to an unsophisticated public but also by treating patients as "electrical practitioners." The eventual result was that laws were drafted to help prevent such abuse, which was good, but meanwhile the technique was rejected by the medical field as "obviously" disreputable and probably of no value. This condemnation was so adamant and widespread that it is now difficult to find any reference to it in the literature, beyond general allusions to it in the historical development of electrical stimulation as a modality. Consequently, the use of electrical stimulation to reduce tumors or to inhibit the development of ionic formations (calcium deposits) is generally no longer seriously discussed or considered as a subject for clinical research.

Iontophoresis of therapeutic chemicals managed to survive in a weakened state until recently; it is now enjoying a period of clinical rediscovery and enthusiasm (refer to Iontophoresis [M07]). Surprisingly, many physical therapy practitioners have recently rediscovered the effectiveness of treatment of soft tissue swelling (refer to Electrical Stimulation, Reduction of Edema [M02D]), calcific deposits inherent in calcific tendonitis (tenosynovitis), bone spurs, and osteoarthritis by low frequency pulsed square wave electrical stimulation.

Procedures of Application (M02F2)

General application considerations (M02F2A)

1) Before the patient is introduced into the treatment environment, the equipment should be checked for proper connection with its power source, between the electrode cables and the stimulator, and between the electrode cables and the electrodes. All equipment should be checked before use to confirm good working order, and all controls should be off or set in the zero position.

2) The patient should be placed in a position which will remain comfortable for the duration of the treatment session. A reclining, over-stuffed arm chair is ideal in most cases because the neck, arms, legs, anterior abdomen, and the back, with some skillful maneuvering, are all accessible to the therapist and the patient is provided with supported comfort.

3) Before applying electrodes, the skin surface to be treated should be inspected for recent scars, evidence of peripheral injury, dermal lesions, or

other causes of disturbed sensation. Such tissue may be hypersensitive to electrical stimulation and care should be taken to build up electrical amplitudes gradually to prevent undue discomfort should the skin resistance suddenly drop, magnifying the effects of the stimulation on those areas.

4) If low frequency sponge pad electrodes are being used, they should be moistened with a saline solution or water and placed over the chosen treatment sites, and care should be taken that they are in good even contact with the skin. In general the negative cathode electrode should be placed on the muscle's motor point, where the motor nerve is most superficial as it innervates the muscle, and when stimulated, provokes the greatest muscle contraction. If metal or flexible carbon electrodes are used, a proper electrode paste should be used to insure adequate contact. Once the best site(s) for electrode placement have been determined, elastic strapping, weighting the electrodes, or strategic placement of the electrode(s) under the patient's body or body parts may be necessary to insure good electrode contact. If the body part is to be immersed for treatment, it should be placed in the bath with the negative cathode electrode, several inches apart.

5) A watch or timer should be set for the length of treatment. The electrical stimulator should be turned on, and the amplitude increased until a visible contraction takes place, always staying within the patient's range of tolerance.

6) The patient should be allowed to become comfortable with the current, and then the intensity should be slowly increased as the patient's tolerance allows. This process should be continued until the desired degree of contraction is reached. The patient should be closely monitored for excessive muscle cramping, joint compression, pain, and autonomic sympathetic hyperresponse to the stimulation. The patient should *never* be left out of sight or sound once the treatment has been started unless the patient can turn off the instrument should the need arise.

7) At the end of the session, if not automatically shut off, the intensity of the stimulator should be gradually decreased until it is switched off. Remove all electrodes from the patient and inspect each site of stimulation carefully, record any abnormal tissue response and return all controls to zero. The patient should rest for several minutes before being allowed to leave.

Specific procedures for reduction of calcium deposit (M02F2B)

The electrical stimulator should be able to supply a pulsed square wave current form in a frequency from 4 to 28 hertz. The electrodes should be of the same size and placed in a monopolar fashion with the anode or positive pole over the suspected calcium (ionic formation) deposit. Karya type pads have proven to be remarkable effective for joint space placement. The cathode or negative electrode should be placed in a remote area. The electrodes should both be secured in place.

The electrical stimulation unit should be preset to deliver square wave pulses, continuously, at the desired frequency of 28 hertz for twenty minutes. The electrical stimulation unit should then be turned on and the intensity turned up until the patient begins to feel it. When the patient has gotten used to the sensation of the electrical stimulation the intensity should be gradually increased until the patient's tolerance level is reached. As the patient accommodates to the current level the intensity may once again be increased to tolerance.

Specific procedures to reduce soft tissue swelling (M02F2C)

Because of the ionic nature of fluid associated with soft tissue inflammation, iontophoresis may also be used for treatment of soft tissue swelling. The electrodes should be of the same size and placed in a monopolar fashion with the anode or positive pole over the swollen area. Karya type pads have proven to be remarkably effective for joint space placement. The cathode or negative electrode should be placed in a remote area. The electrodes should both be secured in place.

The electrical stimulation unit should be preset to deliver square wave pulses, continuously, at the desired frequency of between 4 and 28 hertz for twenty minutes. The electrical stimulation unit should then be turned on and the intensity increased until the patient begins to feel the stimulus. When the patient has gotten used to the sensation of the electrical stimulation, the intensity should gradually be increased until the patient's tolerance level is reached. As the patient accommodates to the current level the intensity may once again be increased to tolerance.

In combination with other treatment techniques (principally ultrasound), daily treatment sessions, although no more than fifteen, have proven to be the most successful for the reduction of calcium deposits (ionic formations) or soft tissue inflammation; however, sessions occurring two or three times a week for several weeks have also proven effective.

Partial List of Conditions Treated (M02F3)

Treatable Causes (M02F3A)

Calcific deposit (C08)
Habitual joint positioning (C48)

Joint ankylosis (C42)
Nerve root compression (C04)
Restricted joint range of motion (C26)
Soft tissue inflammation (C05)
Soft tissue swelling (C06)

Syndromes (M02F3B)

ACHILLES TENDONITIS (S52)
CERVICAL (NECK) PAIN (S73)
ELBOW PAIN (S47)
FACET (S19)
FOOT PAIN (S51)
FROZEN SHOULDER (S64)
HAND/FINGER PAIN (S17)
HEADACHE PAIN (S02)
HIGH THORACIC BACK PAIN (S48)
KNEE PAIN (S46)
LOW BACK PAIN (S03)
MYOSITIS OSSIFICANS (S67)
OSTEOARTHRITIS (S21)
POST IMMOBILIZATION (S36)
RADICULITIS (S29)
SCIATICA (S05)
SHOULDER/HAND (S57)
SHOULDER PAIN (S08)
TEMPOROMANDIBULAR JOINT (TMJ)
   PAIN (S06)
TENDONITIS (S28)
TENNIS ELBOW (LATERAL
   EPICONDYLITIS) (S32)
WRIST PAIN (S16)

Precautions and Contraindications (M02F4)

Care should be taken to avoid over-fatigue of the muscle stimulated. Stimulation should stop when the muscle begins to respond with less vigor.

Electrical muscle stimulation is generally contraindicated for victims of cancer or other malignant disease if the site to be stimulated is physically associated with the cancerous growth. It is also contraindicated for those suffering from sensory deficits, not including those suffering from hysterical sensory abnormalities.

Care should be taken not to put electrodes too close together. The closer electrodes are to one another, the greater the current density becomes between their edges and an *edge-effect* may be produced which causes unpleasant burning sensations and superficial or deep burns may ultimately occur.

Electrical burns may occur when an excess of electricity (current density) is applied to the skin or mucous membrane. Electrical burns may appear in the clinical setting as a passing erythema, blistering of superficial skin layers, or deep tissue coagulation. The tissue damage produced by the deep electrical burn occurs in a roughly conical area, extending from the apex of the cone on the skin surface, where the original electrical contact occurred, fanning out into the deeper layers. Just following the injury, the burn site appears rather small and inconsequential but becomes more alarming as the damaged tissues subsequently slough off and ulceration occurs. These burns are slow to heal, prone to infection, and if sufficiently deep, may be followed by extensive unsightly scarring.

Electric shock may be caused if the patient, while being electrically stimulated, makes contact with a grounded object such as a water pipe, radiator, or electric circuit. This is especially serious if a large area of the patient's body is subjected to the shock (as in a bath or treatment tub) and may pose the danger of cardiac arrest. Electric shock may also occur if the electrical stimulator suffers transformer breakdown, in which case the high-tension, low frequency current may "jump" to the patient and produce an electrical burn as well as shock.

Great care must be taken in the application of all electrical appliances to patients with cardiac pacemakers. Electrodes should not be placed in the vicinity of a pacemaker nor should the current flow from the cathode to the anode electrode be allowed to cross the pacemaker or implanted wires.

In general for high frequency stimulation, treatment should be precluded where joint repairs, rheumatoid and collagen diseases are present, or other conditions in which tendons, joint capsules or the joint may be threatened or weakened by stimulation. In general, electrodes should not be placed over scar tissue, skin irritations or open skin lesions. High frequency stimulation should not be applied over the thorax, abdomen, or back extensor musculature; such stimulation is contraindicated for patients with cardiac demand pacemakers, cardiac arrhythmias, ST segment depression or conduction block. If the patient suffers increased sweating, salivation and/or has complaints of nausea when being stimulated with high frequency stimulation, the stimulation should be discontinued.

Low frequency currents are contraindicated for treatment of myopathies (muscular dystrophy, myotonia congenita, polymyositis, muscle-involved connective tissue disease, cortisone myopathies, myopathy due to metastatic cancer, and myopathy from hyperthyroidism), Parkinson's syndrome, amyotrophic lateral sclerosis, syringomyelia, and peripheral neuritis and neuronitis.

Electrical currents should not be allowed to flow across a pregnant uterus.

It is recommended that the therapist test the apparatus on him or herself shortly before it is used to test the patient. This avoids unwarranted negative findings and provides a safeguard against injuring the patient through faulty equipment. During pre-testing, the current levels which will be used on the patient should be used on the tester; this approach takes courage because such currents by necessity are often high and uncomfortable.

Notes Aside   (M02F5)

Electrical stimulation applied to prevent or reduce calcific deposit is often best applied after the application of ultrahigh frequency sound applied for the same purpose (refer to Ultrahigh Frequency Sound, Calcium Deposit Management [M01E]). It is an integral part of the eclectic treatment of bone spurs, calcium deposits in the joint(s), and calcium deposits associated with the tendons or ligaments.

Bibliography   (M02F6)

Shriber, pp. 110–123, 139–147.
Stillwell, pp. 36–39.
Wong, pp. 1209–1214.

ELECTRICAL STIMULATION OF MUSCLE (OR DEEP) TISSUE INFLAMMATION CONTROL (M02G)

Definition (M02G1)

Control or reduction of soft tissue inflammation by electrical stimulation of muscle or deep tissues (as opposed to iontophoresis of superficial tissues) appears from a review of the available literature to have not been well explored. However, clinical empirical experience has shown that sequential multifrequency pulsed square wave electrical stimulation may be used effectively as an aid in the control or reduction of deep soft tissue inflammation.

Procedures of Application (M02G2)

General application considerations (M02G2A)

1) Before the patient is introduced into the treatment environment, the equipment should be checked for proper connection with its power source, between the electrode cables and the stimulator, and between the electrode cables and the electrodes. All equipment should be checked before use to confirm good working order, and all controls should be off or set in the zero position.

2) The patient should be placed in a position which will remain comfortable for the duration of the treatment session. A reclining, over-stuffed arm chair is ideal in most cases because the neck, arms, legs, anterior abdomen, and the back, with some skillful maneuvering, are all accessible to the therapist and the patient is provided with supported comfort.

3) Before applying electrodes, the skin surface to be treated should be inspected for recent scars, evidence of peripheral injury, dermal lesions, or other causes of disturbed sensation. Such tissue may be hypersensitive to electrical stimulation and care should be taken to build up electrical amplitudes gradually to prevent undue discomfort should the skin resistance suddenly drop, magnifying the effects of the stimulation on those areas.

4) If low frequency sponge pad electrodes are being used, they should be moistened with a saline solution or water and placed over the chosen treatment sites, and care should be taken that they are in good even contact with the skin. In general the negative cathode electrode should be placed on the muscle's motor point, where the motor nerve is most superficial as it innervates the muscle, and when stimulated, provokes the greatest muscle contraction. If metal or flexible carbon electrodes are used, a proper electrode paste should be used to insure adequate contact. Once the best site(s) for electrode placement have been determined, elastic strapping, weighting the electrodes, or strategic placement of the electrode(s) under the patient's body or body parts may be necessary to insure good electrode contact. If the body part is to be immersed for treatment, it should be placed in the bath with the negative cathode electrode, several inches apart.

5) A watch or timer should be set for the length of treatment. The electrical stimulator should be turned on, and the amplitude increased until a visible contraction takes place, always staying within the patient's range of tolerance.

6) The patient should be allowed to become comfortable with the current, and then the intensity should be slowly increased as the patient's tolerance allows. This process should be continued until the desired degree of contraction is reached. The patient should be closely monitored for excessive muscle cramping, joint compression, pain, and autonomic sympathetic hyperresponse to the stimulation. The patient should *never* be left out of sight or sound once the treatment has been started unless the patient can turn off the instrument should the need arise.

7) At the end of the session, if not automatically shut off, the intensity of the stimulator should be gradually decreased until it is switched off. Remove all electrodes from the patient and inspect each site of stimulation carefully, record any abnormal tissue response and return all controls to zero. The patient should rest for several minutes before being allowed to leave.

Specific procedures for inflammation control (M02G2B)

The electrical stimulator should be able to supply a pulsed square wave current form in a frequency from 4 to 80 hertz. The electrodes should be placed in a monopolar fashion with the cathode or negative pole over the inflamed zone. If possible the electrode should be large enough to cover the area; karya type pads have proven to be remarkably effective for joint space placement. The anode or positive electrode, of equal size or larger, should be placed in a remote area. The electrodes should both be secured in place.

The electrical stimulation unit should be preset to deliver square wave pulses, continuously, at 4 hertz. The timer (if present) should be set for at least fifteen minutes. The electrical stimulation unit should then be turned on and the intensity slowly increased until a mild visible contraction occurs. After one minute the frequency should be increased to 5 hertz; after each succeeding minute the frequency should be progressed to 8, 14, 20, 28, 35, 45, 60, and 80 hertz. After one minute of stimulation at 80 hertz, the pattern should be performed in reverse order from 60 to 4 hertz at thirty-second intervals. If the patient's affected muscle(s) develop tetany as the frequency is increased, the intensity may be decreased if the patient's comfort so demands, and again increased as the frequency is reduced in line with the patient's comfort.

Daily treatment sessions for the treatment of soft tissue inflammation utilizing multiple frequency electrical stimulation eclectically with other modalities have been proven to be the most successful at reducing inflammation of soft tissues, but sessions two or three times a week for several weeks have also proven to be effective.

Partial List of Conditions Treated   (M02G3)

Treatable Causes   (M02G3A)

Interspinous ligamentous strain   (C03)
Joint ankylosis   (C42)
Joint approximation   (C47)
Ligamentous strain   (C25)
Nerve root compression   (C04)
Restricted joint range of motion   (C26)
Soft tissue inflammation   (C05)
Tactile hypersensitivity   (C24)
Trigger point formation   (C01)

Syndromes   (M02G3B)

ACHILLES TENDONITIS   (S52)
BURSITIS   (S26)
CALF PAIN   (S50)
CAPSULITIS   (S27)
CARPAL TUNNEL   (S35)
CERVICAL DORSAL OUTLET (SCALENICUS ATTICUS/CERVICAL RIB)   (S09)
CERVICAL (NECK) PAIN   (S73)
CHEST PAIN   (S14)
CHONDROMALACIA   (S15)
EARACHE   (S40)
ELBOW PAIN   (S47)
FACET   (S19)
FASCIITIS   (S20)
FOOT PAIN   (S51)
FOREARM PAIN   (S54)
FROZEN SHOULDER   (S64)
HAND/FINGER PAIN   (S17)
HEADACHE PAIN   (S02)
HIGH THORACIC BACK PAIN   (S48)
JOINT SPRAIN   (S30)
KNEE PAIN   (S46)
LOWER ABDOMINAL PAIN   (S62)
LOW BACK PAIN   (S03)
MENSTRUAL CRAMPING   (S60)
MIGRAINE (VASCULAR) HEADACHE   (S18)
MYOSITIS OSSIFICANS   (S67)
OSTEOARTHRITIS   (S21)
PIRIFORMIS   (S04)
POST IMMOBILIZATION   (S36)
POST WHIPLASH   (S55)
RADICULITIS   (S29)
REFERRED PAIN   (S01)
SCIATICA   (S05)
SHOULDER/HAND   (S57)
SHOULDER PAIN   (S08)
SINUS PAIN   (S33)
TEMPOROMANDIBULAR JOINT (TMJ) PAIN   (S06)
TENDONITIS   (S28)
TENNIS ELBOW (LATERAL EPICONDYLITIS)   (S32)
THIGH PAIN   (S49)
WRIST PAIN   (S16)

Precautions and Contraindications   (M02G4)

Care should be taken to avoid over-fatigue of the muscle stimulated. Stimulation should stop when the muscle begins to respond with less vigor.

Electrical muscle stimulation is generally contraindicated for victims of cancer or other malignant disease if the site to be stimulated is physically associated with the cancerous growth. It is also contraindicated for those suffering from sensory deficits, not including those suffering from hysterical sensory abnormalities.

Care should be taken not to put electrodes too close together. The closer electrodes are to one another, the greater the current density becomes between their edges and an *edge-effect* may be produced which causes unpleasant burning sensations and superficial or deep burns may ultimately occur.

Electrical burns may occur when an excess of electricity (current density) is applied to the skin or mucous membrane. Electrical burns may appear in the clinical setting as a passing erythema, blistering of superficial skin layers, or deep tissue coagulation. The tissue damage produced by the deep electrical burn occurs in a roughly conical area, extending from the apex of the cone on the skin surface, where the original electrical contact occurred, fanning out into the deeper layers. Just following the injury, the burn site appears rather small and inconsequential but becomes more alarming as the damaged tissues subsequently slough off and ulceration occurs. These burns are slow to heal, prone to infection, and if sufficiently deep, may be followed by extensive unsightly scarring.

Electric shock may be caused if the patient, while being electrically stimulated, makes contact with a grounded object such as a water pipe, radiator, or electric circuit. This is especially serious if a large area of the patient's body is subjected to the shock (as in a bath or treatment tub) and may pose the danger of cardiac arrest. Electric shock may also occur if the electrical stimulator suffers transformer breakdown, in which case the high-tension, low frequency current may "jump" to the patient and produce an electrical burn as well as shock.

Great care must be taken in the application of all electrical appliances to patients with cardiac pacemakers. Electrodes should not be placed in the vicinity of a pacemaker nor should the current flow from the cathode to the anode electrode be allowed to cross the pacemaker or implanted wires.

In general for high frequency stimulation, treatment should be precluded where joint repairs, rheumatoid and collagen diseases are present, or other conditions in which tendons, joint capsules or the joint may be threatened or weakened by stimulation. In general, electrodes should not be placed over scar tissue, skin irritations or open skin lesions. High frequency stimulation should not be applied over the thorax, abdomen, or back extensor musculature; such stimulation is contraindicated for patients with cardiac demand pacemakers, cardiac arrhythmias, ST segment depression or conduction block. If the patient suffers increased sweating, salivation and/or has complaints of nausea when being stimulated with high frequency stimulation, the stimulation should be discontinued.

Low frequency currents are contraindicated for treatment of myopathies (muscular dystrophy, myotonia congenita, polymyositis, muscle-involved connective tissue disease, cortisone myopathies, myopathy due to metastatic cancer, and myopathy from hyperthyroidism), Parkinson's syndrome, amyotrophic lateral sclerosis, syringomyelia, and peripheral neuritis and neuronitis.

Electrical currents should not be allowed to flow across a pregnant uterus.

It is recommended that the therapist test the apparatus on him or herself shortly before it is used to test the patient. This avoids unwarranted negative findings and provides a safeguard against injuring the patient through faulty equipment. During pre-testing, the current levels which will be used on the patient should be used on the tester; this approach takes courage because such currents by necessity are often high and uncomfortable.

Notes Aside (M02G5)

Decreasing soft tissue inflammation (M02G5A)

Multiple frequency electrical stimulation has proven to be effective for the control and/or reduction of soft tissue inflammation, especially when utilized eclectically with other treatment techniques including phonophoresis (refer to Ultrahigh Frequency Sound, Phonophoresis [M01A]), soft tissue manipulation (refer to Massage, Deep Soft Tissue Mobilization [M23F]), peripheral electroacustimulation (refer to Peripheral Electroacustimulation, Acupoint Electrical Stimulation within an Inflamed Zone [M05A]), auricular electroacustimulation (refer to Auricular Electroacustimulation, Pathology Control [M06B]), and/or others.

Decreasing phlebitis (M02G5B)

Soft tissue inflammation associated with phlebitis which has not produced thrombosis has proven to be effectively treatable with multiple frequency electrical stimulation alone. Considering the nature of phlebitis, the establishment of an active treatment modality for its reduction is remarkable. In general, positive pressure hosiery should be utilized to maintain treatment gains and/or to prevent further development of the phlebitis between sessions (refer to Positive Pressure, Phlebitis and/or Edema Control [M24B]).

ELECTRICAL STIMULATION OF MUSCLE (OR DEEP) TISSUE ELECTRICAL EVALUATION (M02H)

Definition (M02H1)

Electrical evaluation deals with the reaction of muscle and/or motor nerves to electrical stimuli. Many pathological conditions of the central and peripheral nervous systems are accompanied by typical reactions to electrical stimulation.

Peripheral nerve injury (M02H1A)

When a motor nerve is severed, and both sensory and motor connections (afferent and efferent elements, respectively) with the muscle(s) are interrupted, both the damaged nerve and denervated muscle(s) degenerate. A quick electrical stimulation of the denervated muscle will fail to elicit a response from the muscle(s) or from either element of the involved nerve. A longer and more intense stimulus may bring about a responsive contraction in the muscle if contractile tissue still exists in the muscle; failure to respond to such a prolonged and intense electrical stimulus may be sufficient to prove that neither a conductive nerve nor contractile muscle tissue is present, indicating a very late stage of unrecovered peripheral damage or progressive muscular atrophy.

Faradic and galvanic tests of innervation (M02H1A01) A good test of muscle innervation, or lack thereof, is to contrast the effects of faradic and galvanic current stimulation on nerve and/or muscle. Faradic current stimulation at adequate intensity levels will produce a continuous tetanic contraction of an innervated muscle for the duration of the current flow. Faradic current stimulation, because of its brief pulse duration, will fail to elicit a contraction from a denervated muscle if ten days of degeneration have occurred. In contrast, an adequately intense galvanic current will produce a brisk, unsustained contraction when the stimulus is suddenly "made" or started, or broken, but not while the current is continuously flowing. The response of denervated muscle to adequately intense galvanic stimulation is a sluggish, wormlike contraction when the current is suddenly made or broken.

Reaction of degeneration (M02H1A02) The lack of response to faradic current stimulation and the sluggish response to galvanic current stimulation are collectively called the *reaction of degeneration* and are characteristic of denervated musculature. The *reaction of degeneration* is seen in lesions of the bulbar nucleus, compression of intracranial nerve trunks, lesions of the anterior horns, lesions of the spinal cord (including the anterior horns), compression of peripheral nerve trunks within the spinal canal (cauda equina injury), traumatic lesion of the peripheral nerve trunks outside of the spinal column, toxic neuritis (alcohol or lead poisoning, diabetes, focal infections, etc.), infectious neuritis (typhoid, syphilis, influenza, tuberculosis, polyneuritis), and viral neuritis (Bell's palsy, etc.).

Hyperresponse to electrical stimulation (M02H1B)

Neuromuscular hyperresponsiveness to electrical stimulation, regardless of wave form, is a characteristic symptom of an upper motor neuron insult, which results in cerebral motor cortex damage (post cerebral vascular accident or cerebral concussion), and a peripheral nerve injury, shortly after onset. It is also symptomatic of the early developmental stages of tetanus, spasmophilia, chorea, and various spinal cord diseases.

Reduced neuromuscular response to tetanic currents (M02H1C)

Reduced neuromuscular responsiveness to electrical stimulation otherwise sufficiently intense to be tetanizing (often to the point of failing to be visually and/or palpably discernible), is found in advanced stages of the post cerebral vascular syndrome with hemiplegia, tabes dorsalis, paralysis agitans, in later stages of spinal cord diseases, and with the muscular atrophy resulting from disuse associated with various types of arthritis, myositis and other chronic disabling diseases.

Rare reactions to electrical stimulation (M02H1D)

Rarer reactions to electrical stimulation include the *myotonic* and *myasthenic* reactions.

Myotonic reaction (M02H1D01) The *myotonic* reaction is evident when muscles remain in tetany for up to twenty seconds after the electrical stimulus has ceased. The myotonic response is a typical response characteristic of myotonia congenita (Thomsen's disease).

Myasthenic reaction (M02H1D02) A *myasthenic* reaction is evident when, after an initial normal response to a tetanic current, the force of muscular contraction progressively and rapidly diminishes. Usually after a sufficient rest of several minutes the process may be repeated with the same pattern of response reoccurring. The myasthenic reaction is characteristic of, and gets its name from, myasthenia gravis.

Tests not discussed (M02H1E)

Strength duration, chronaxie, rheobase, nerve conduction velocity and electromyographic measurements are not discussed here.

Procedures of Application (M02H2)

General procedural considerations (M02H2A)

1) The room in which the testing takes place should be well lighted so that any muscle twitch may be visible.
2) The patient should be in a comfortable position to avoid fatigue.
3) The tester should be able to reach all parts of the apparatus as well as the muscles to be tested without changing position.
4) The area over the muscle to be tested should be warmed (a few minutes under a thermaphore sufficiently warm to cause a gentle hyperemia) to decrease the skin resistance and to minimize the discomfort of the testing.
5) The electrodes to be used in the testing should be well moistened with a saline solution if sponge electrode surfaces are used. Saline solution should remain close at hand to allow the test electrode to be remoistened frequently during a long testing session.
6) The anode should be a relatively large dispersive electrode placed in a remote, muscle-poor area such as the chest, the small of the back, or the sacrum, away from the muscle to be tested.
7) The cathode test electrode usually consists of a small flat metal disk or rectangle (2 to 16 square centimeters), covered with several layers of gauze or chamois or a thin layer of sponge. This testing surface is mounted on a handle with a built-in make or break switch that is usually thumb controlled.

Galvanic testing (M02H2B)

For *galvanic testing*, the test electrode should be placed over the muscle's suspected motor point (usually located near the origin of the muscle belly). Advance the rheostat (intensity control) slowly, from zero upward. Press the make or break switch at regular intervals as the intensity is increased, and watch for the first visible contraction (the *threshold of excitation*, the minimum current level necessary for a response); with repeated tests the muscle response will increase in force as the skin resistance breaks down in response to current flow.

Faradic testing (M02H2C)

For *faradic testing*, the test electrode should be placed over the muscle's suspected motor point. The intensity control should be advanced up from zero slowly and the make or break switch thumbed on with each small advance until a visible contraction occurs. When the current is sufficient a tetanic contraction should occur that lasts as long as the current flows. To establish the motor point more concisely, turn the intensity level down a little and move the electrode end over the surface area, thumbing the switch with each move, to establish the area of greatest sensitivity to stimulation; the motor point will produce the strongest contraction.

Partial List of Conditions Evaluated (M02H3)

Treatable Causes (M02H3A)

Muscular weakness (C23)
Nerve root compression (C04)
Neuromuscular tonic imbalance (C30)
Pathological neuromuscular hypotonus (C33)
Peripheral nerve injury (C07)

Syndromes (M02H3B)

BELL'S PALSY (S66)
CERVICAL DORSAL OUTLET (SCALENICUS ATTICUS/CERVICAL RIB) (S09)
HYSTERIA/ANXIETY REACTION (S59)
NEUROMUSCULAR PARALYSIS (S22)
POST CEREBRAL VASCULAR ACCIDENT (CVA) (S07)
POST IMMOBILIZATION (S36)
POST PERIPHERAL NERVE INJURY (S23)
POST SPINAL CORD INJURY (S24)
SCIATICA (S05)
SHOULDER/HAND (S57)

Precautions and Contraindications (M02H4)

Care should be taken to avoid over-fatigue of the muscle stimulated. Stimulation should stop when the muscle begins to respond with less vigor.

Electrical muscle stimulation is generally contraindicated for victims of cancer or other malignant disease if the site to be stimulated is physically associated with the cancerous growth. It is also contraindicated for those suffering from sensory deficits, not including those suffering from hysterical sensory abnormalities.

Care should be taken not to put electrodes too close together. The closer electrodes are to one another, the greater the current density becomes between their edges and an edge-effect may be produced which causes unpleasant burning sensations and superficial or deep burns may ultimately occur.

Electrical burns may occur when an excess of electricity (current density) is applied to the skin or mucous membrane. Electrical burns may appear in the clinical setting as a passing erythema, blistering of superficial skin layers, or deep tissue coagulation. The tissue damage produced by the deep electrical burn occurs in a roughly conical area, extending from the apex of the cone on the skin surface, where the original electrical contact occurred, fanning out into the deeper layers. Just following the injury, the burn site appears rather small and inconsequential but becomes more alarming as the damaged tissues subsequently slough off and ulceration occurs. These burns are slow to heal, prone to infection, and if sufficiently deep, may be followed by extensive unsightly scarring.

Electric shock may be caused if the patient, while being electrically stimulated, makes contact with a grounded object such as a water pipe, radiator, or electric circuit. This is especially serious if a large area of the patient's body is subjected to the shock (as in a bath or treatment tub) and may pose the danger of cardiac arrest. Electric shock may also occur if the electrical stimulator suffers transformer breakdown, in which case the high-tension, low frequency current may "jump" to the patient and produce an electrical burn as well as shock.

Great care must be taken in the application of all electrical appliances to patients with cardiac pacemakers. Electrodes should not be placed in the vicinity of a pacemaker nor should the current flow from the cathode to the anode electrode be allowed to cross the pacemaker or implanted wires.

In general for high frequency stimulation, treatment should be precluded where joint repairs, rheumatoid and collagen diseases are present, or other conditions in which tendons, joint capsules or the joint may be threatened or weakened by stimulation. In general, electrodes should not be placed over scar tissue, skin irritations or open skin lesions. High frequency stimulation should not be applied over the thorax, abdomen, or back extensor musculature; such stimulation is contraindicated for patients with cardiac demand pacemakers, cardiac arrhythmias, ST segment depression or conduction block. If the patient suffers increased sweating, salivation and/or has complaints of nausea when being stimulated with high frequency stimulation, the stimulation should be discontinued.

Low frequency currents are contraindicated for treatment of myopathies (muscular dystrophy, myotonia congenita, polymyositis, muscle-involved connective tissue disease, cortisone myopathies, myopathy due to metastatic cancer, and myopathy from hyperthyroidism), Parkinson's syndrome, amyotrophic lateral sclerosis, syringomyelia, and peripheral neuritis and neuronitis.

Electrical currents should not be allowed to flow across a pregnant uterus.

It is recommended that the therapist test the apparatus on him or herself shortly before it is used to test the patient. This avoids unwarranted negative findings and provides a safeguard against injuring the patient through faulty equipment. During pre-testing, the current levels which will be used on the patient should be used on the tester; this approach takes courage because such currents by necessity are often high and uncomfortable.

Notes aside (M02H5)

Electrical evaluation, as presented above, has proven quite valuable for differentiating between cauda equina nerve root injuries and spinal cord injuries, or establishing elements of both in the same patient.

Bibliography (M02H6)

Baker, Parker, Sanderson, pp. 1967–1974.
Packman-Braun, pp. 51–56.
Shriber, pp. 148–162.
Stillwell, pp. 1–64, 65–108, 124–173.

ELECTRODERMAL POTENTIAL MONITRY (M25)

Definition (M2501)

Electrodermal potential (EDP) monitry is theoretically the measurement of the electrical potential existing across the cell membrane; it is an area of study as yet to be well-explored relative to its importance and/or its role in the physiology of humans or any other species.

It has been demonstrated that if the patient is provided with audio and/or visual feedback directly representative of EDP activity the patient can learn to voluntarily change it. The importance of this potential ability is still unestablished, with the exception of individuals suffering from various neurogenic dermal disorders (psoriasis, hives, etc.) who have been able to control some symptoms associated with the disorder (at least temporarily) by voluntarily changing the EDP over a site manifesting the disorder. Why this is effective and what such changes imply have still to be established. Activity norms have yet to be established for EDP and treatment protocols are as yet rather empirical and unsystematized. Emotional and/or autonomic correlates with EDP activity levels have not as yet been established.

EDP is measured with an instrument similar to the galvanic skin response (GSR) monitor. It communicates with the patient via three electrode cables: two active and one ground, which are attached to electrodes coupled to the skin by electrode paste and affixed to the skin by tape or strapping. The EDP instrument is designed to register microvolt potential of either positive or negative polarity. If feedback modes are available they usually include audio feedback, which reflects the frequency of EDP activity (by clicks which may increase or decrease) and amplitude (by rising or falling pitch), and representative visual feedback in the form of a meter or light bar.

Procedures of Application (M2502)

1) The patient should be sitting or semireclining in a position which will remain comfortable for the duration of the session.

2) The electrodes should be affixed to the treatment site with tape or strapping, if they are not self adhering. One active electrode is placed on one side of the treatment site, the ground electrode on the opposite side, and the remaining active electrode in a remote area, well away from the treatment site.

3) The EDP instrument should be turned on, but the feedback modes should remain unavailable to the patient.

4) The instrumentation should be allowed to adjust to the patient's EDP activity level and then the feedback modes adjusted to suitable levels so that any changes (increases or decreases) will be perceptibly appreciated by the patient and/or the practitioner; the EDP level should be noted and recorded.

5) If feedback training is to be employed, the feedback modes should be presented to the patient (the sound turned on and/or the visual feedback mode made visible).

6) The patient should then be instructed to increase or decrease the level of EDP activity by increasing or decreasing the apparent feedback levels. For psoriasis or neurodermatitis the EDP activity should be increased in favor of becoming more positive.

7) The patient should then be left alone with the instrument and only interrupted if audio levels indicate to the practitioner that the level is either too high or too low and the patient is no longer able to perceive feedback change. The practitioner then should reenter the patient's environment to adjust the feedback levels to more satisfactory levels.

8) The session should last from twenty to fifty minutes.

9) At the end of the session the EDP level should be noted and recorded.

10) The electrodes should be removed from the patient and the skin cleansed of electrode paste.

Treatment sessions are usually conducted once or twice a week, with at least one day between each session.

Partial List of Conditions Treated (M2503)

Conditions treated in the field (M2503A)

Neurodermatitis
Psoriasis
Sclerodermatitis
Hives

Treatable Causes (M2503B)

Neurogenic dermal disorders (psoriasis, hives) (C49)
Soft tissue inflammation (C05)

Syndromes (M2503C)

ACNE (S45)

Precautions and Contraindications (M2504)

There are no precautions or contraindications directly associated with EDP training in the literature reviewed. However, caution should be observed when temperature training is performed with patients suffering from diabetes.

The general relaxation effect of the EDP training may decrease metabolism and reduce demand for insulin. An insulin reaction may occur in a diabetic patient who has taken an insulin supplement just prior to temperature training, failed to previously consume adequate amounts of food or failed to digest what has been eaten, or taken too large a dose of insulin before beginning training. The first symptoms of an insulin reaction include feelings of extreme hunger, nervousness (general agitation), heavy perspiration, palpitation of the heart, and occasional double vision. These symptoms may be relieved by consumption of eight ounces of orange juice which has had two teaspoonfuls of regular sugar added to it. If it is unavailable, any sugar or carbohydrate-containing food, like bread or crackers, will serve. If the first symptoms are unrelieved, more serious symptoms may appear which include delirium, convulsions and/or unconsciousness (coma). To relieve these symptoms, intravenous injection of glucose may be necessary.

Notes Aside (M2505)

Since EDP feedback training is relatively unexplored, suggestions of how to help the patient learn such control are limited to suggesting that the patient must intensely relate to the audio feedback and keep their conscious mind out of the way of the unconscious, subliminal learning.

Bibliography (M2506)

Green, pp. 1–26.

ELECTROENCEPHALOMETRY
(M21)

Definition (M2101)

The electroencephalograph (EEG) has long been an instrument primarily dedicated to research and the clinical evaluation of supraspinal neurological function. Electroencephalographic instruments are designed to graph the electrical (wave) patterns produced by the various functioning elements of the brain. Such instrumentation is expensive and highly sophisticated and as a consequence has historically remained in the hands of relatively few, and this has kept experimentation with the EEG at a minimum. Some experimentation did take place, however, and it was established that brain wave patterns and activity could be voluntarily modified if appropriate feedback, representative of such activity, was presented to the patient on an immediate and qualitative basis, and the patient was allowed time to learn to change it.

The value of being able to voluntarily change brain wave patterns has become apparent from the exploration conducted by various important researchers. Their work demonstrated that such modification could be used to induce meditative states which could be used to reduce blood pressure, slow the metabolic rate, decrease pulse rate, lability, and intestinal motility, all symptoms of sympathetic autonomic hyperactivity. Additional benefits have said to include increased powers of concentration and ability to sleep, a general feeling of well-being and heightened awareness. Such attributes have always been associated with different forms of meditation, but with EEG feedback training, such states appeared to be more easily and consistently induced.

The ability to indulge in or conduct EEG feedback training generally remained in the hands of a few formal researchers and their subjects until technological advances produced the electroencephalometer (EEM), a crude, fairly reliable, relatively inexpensive instrument which could be used to monitor general brain wave frequency and amplitude ranges, and provide the patient with representative feedback in the form of audio and visual modes which replaced the graphic functions of the EEG. It was demonstrated, through research and clinical observation with the EEM, that if appropriately applied, many of the benefits previously claimed to be produced by EEG feedback training could be produced through EEM feedback training.

The EEM instrument most commonly seen in the clinical setting is a type of brain wave monitor designed to provide appropriate feedback to certain brain wave electrical activity. The feedback provided generally takes the form of audio feedback which is modulated from a high to low pitch when a given frequency of brain wave activity is produced in sufficient amounts. Low range pitches generally designate successful satisfaction of the criteria of brain wave electric activity produced at a given frequency which meets or exceeds a preset amplitude (a threshold usually associated with amplitudes of 3, 5, 6, 10, 30, 50, 60, or 120 microvolts). Most EEM instruments are geared to foster the production of brain wave activity in the frequency ranges of 5 to 7.5 (theta), 7.5 to 8 (alpha-theta), and/or 8 to 12 hertz (alpha); some instruments provide the option of selecting which of these frequencies is to be fostered and a full range of amplitude thresholds, while others are fixed to monitor only one frequency (usually the 8 to 12 hertz frequency range) within a given amplitude range (3 to 10 microvolts, for example).

If the EEM instrument being utilized provides a choice of monitoring any of the three basic frequency levels and can be set to appreciate various amplitude ranges, the general aim of EEM training is for the patient to progressively learn to decrease the general frequency of the electrical brain wave activity, one frequency range at a time (from highest to lowest), while progressively learning to also increase the amplitude of the electrical activity being produced (from a level of 3 to levels exceeding 30 microvolts); i.e., the patient learns to produce brainwave electrical activity at a given frequency and at (or beyond) a given maximal amplitude level before proceeding to the learn to selectively produce activity at the next lower frequency range at any given amplitude level.

The EEM instrument communicates with the patient via electrode cables (usually three: two live pickup and one ground) which terminate in surface electrodes which are coupled by electrode paste to the patient's scalp and affixed in place with supportive (usually elastic) strapping. Such instruments generally provide audio feedback (for benefit of the patient) which varies in pitch, dropping from high to low pitches as the brain wave frequency is reduced and produced in sufficient amplitudes. Visual feedback in the form of a meter or light bar is provided only for the convenience of the practitioner's observation and to make record keeping possible; the visual feedback is generally synchronous with the auditory feedback, with high meter readings indicating successfully achieved frequency production at acceptable amplitudes, and low readings denoting failure to achieve the desired frequency range and/or adequate amplitude levels. During training the patient is generally required to have the eyes closed and the neck muscles relaxed since visual activity, eye and eyelid movement, and neck muscle contraction may interfere with training. The electrical activity produced by such functions may be picked up by the EEM and effectively mask the brain wave electrical activity being produced.

Procedures of Application (M2102)

1) The patient should be in a sitting or semi-reclining position with head and/or neck supported, which will remain comfortable for the duration of the training session.

2) The training process should be clearly explained and the feedback levels denoting success and failure illustrated.

3) The EEM surface electrodes should be coupled with electrode paste to the patient's scalp and affixed in place over the occipital lobes, superior to the nuchal line with strapping or other device.

4) If the treatment environment has a level of background noise significant enough to be distracting, the patient should be provided with earphones which will provide the audio feedback from the EEM and screen out other noise.

5) The patient should be instructed to close the eyes for the duration of the training.

6) The treatment room light should be dimmed, but not so low that meter reads can not be discerned by the practitioner.

7) The EEM instrument, preset to foster a given frequency within a particular amplitude range, should be turned turned on and the audio feedback volume adjusted to patient comfort but at levels that will permit the patient to hear the feedback easily, without straining.

8) The initial range of EEM activity should be noted and recorded.

9) The patient should be left alone and undisturbed for ten to fifteen minutes.

10) The instrument should be turned off and the patient allowed to rest for five minutes before continuing.

11) After the five-minute rest, the instrument should be turned on again and the patient left alone, undisturbed for another ten to fifteen minute period.

12) The final levels of EEM activity should be noted and recorded.

13) The instrument should be turned off, the electrodes removed from the patient, and the electrode paste thoroughly cleansed from the patient's skin.

14) The patient should be allowed to rest for several minutes before rising and leaving.

Partial List of Conditions Treated (M2103)

Treatable Causes (M2103A)

Blood sugar irregularities (C31)
Extrafusal muscle spasm (C41)
Hypertension (C39)
Insomnia (C46)
Migraine (vascular headache) (C34)
Nausea/vomiting (C36)
Neurogenic dermal disorders (psoriasis, hives) (C49)
Neuromuscular tonic imbalance (C30)
Psychoneurogenic neuromuscular hypertonus (C29)
Restricted joint range of motion (C26)
Sympathetic hyperresponse (C38)
Tinnitus (C32)
Trigger point formation (C01)
Visceral organ dysfunction (C02)

Syndromes (M2103B)

CALF PAIN (S50)
CERVICAL (NECK) PAIN (S73)
CHEST PAIN (S14)
COLITIS (S42)
DIABETES (S10)
EARACHE (S40)
ELBOW PAIN (S47)
FOOT PAIN (S51)
FOREARM PAIN (S54)
HAND/FINGER PAIN (S17)
HEADACHE PAIN (S02)
HIGH THORACIC BACK PAIN (S48)
HYPERTENSION (S43)
HYSTERIA/ANXIETY REACTION (S59)
INSOMNIA (S71)
KNEE PAIN (S46)
LOWER ABDOMINAL PAIN (S62)
LOW BACK PAIN (S03)
MIGRAINE (VASCULAR) HEADACHE (S18)
PHOBIC REACTION (S44)
POST WHIPLASH (S55)
RADICULITIS (S29)
REFERRED PAIN (S01)
SCIATICA (S05)
SHOULDER/HAND (S57)
SHOULDER PAIN (S08)
SINUS PAIN (S33)
TEMPOROMANDIBULAR JOINT (TMJ) PAIN (S06)
THIGH PAIN (S49)
TINNITUS (S70)

TOOTHACHE/JAW PAIN (S41)
WRIST PAIN (S16)

Precautions and Contraindications (M2104)

Epilepsy (M2104A)

EEM feedback training at the 5 to 12 hertz frequency range should not be employed by patients who suffer from epileptic seizures. Such training may trigger an epileptic seizure, which on an EEG appears as an exaggerated, sustained train of 8 to 12 hertz high amplitude spikes.

Endogenous depression (M2104B)

EEM feedback training at the 5 to 12 hertz frequency range should not be employed by patients who suffer from primary or endogenous depression or hypoglycemia; such training may precipitate an attack of depression (melancholia).

Diabetes (M2104C)

Caution should be observed when EEM training (general relaxation or desensitization) is performed with patients suffering from diabetes. The depression of sympathetic activity may decrease metabolism and reduce demand for insulin thereby increasing the possibility of an insulin reaction occurring, especially if the diabetic patient has (1) taken an insulin supplement just prior to EEM training, (2) failed to previously consume adequate amounts of food or failed to digest what has been eaten, or (3) taken too large a dose of insulin before beginning training. The first symptoms of an insulin reaction include feelings of extreme hunger, nervousness (general agitation), heavy perspiration, palpitation of the heart, and occasional double vision. These symptoms may be relieved by consumption of eight ounces of orange juice which has had two teaspoonfuls of regular sugar added to it. If it is unavailable, any sugar or carbohydrate-containing food, like bread or crackers, will serve. If the first symptoms are unrelieved, more serious symptoms may appear which include delirium, convulsions and/or unconsciousness (coma). To relieve these symptoms, intravenous injection of glucose may be necessary.

Notes Aside (M2105)

EEM training has proven to be remarkably successful as a technique for the treatment of colitis, especially when eclectically applied in conjunction with auricular electroacustimulation (refer to Auricular Electroacustimulation, Pathology [M06B]), and peripheral electroacustimulation (refer to Peripheral Electroacustimulation, Pathology Control [M05C]). The induced meditative state which results from successful training seems to have an inhibitory effect upon sympathetic hyperactivity, while apparently fostering parasympathetic activity to the point of balancing the autonomic system, reestablishing homeostasis.

Bibliography (M2106)

Engstrom, pp. 1261–1262.
Green, pp. 1–26.
Sterman, Friar, pp. 89–95.

ELECTROENCEPHALOMETRY MEDITATION FACILITATION
(M21A)

Definition (M21A1)

Electroencephalometry (EEM) training has been demonstrated to be useful in helping patients achieve a meditative state; such states have been shown to depress sympathetic autonomic activity to the extent of producing a reduction in blood pressure, a decrease in pulse rate, a slowing of metabolic processes, and decreased intestinal motility. Less measurable benefits are said to include a decrease in emotional lability, increased powers of concentration, increased ability to sleep, a general feeling of well being, heightened awareness, and a increased susceptibility to hypnosis. Meditation states induced by EEM training are efficiently and reliably induced; they require little practice and no great powers of concentration or self-discipline on the part of the patient.

The EEM instrument most commonly seen in the clinical setting is a type of brain wave monitor designed to provide appropriate feedback to certain brain wave electrical activity. The feedback provided generally takes the form of audio feedback which is modulated from a high to low pitch when a given frequency of brain wave activity is produced in sufficient amounts. Low range pitches generally designate successful satisfaction of the criteria of brain wave electric activity produced at a given frequency which meets or exceeds a preset amplitude (a threshold usually associated with amplitudes of 3, 5, 6, 10, 30, 50, 60, or 120 microvolts). Most EEM instruments are geared to foster the production of brain activity in the frequency ranges of 5 to 7.5 (theta), 7.5 to 8 (alpha-theta), and/or 8 to 12 hertz (alpha); some instruments provide the option of selecting which of these frequencies is to be fostered and a full range of amplitude thresholds, while others are fixed to monitor only one frequency (usually the 8 to 12 hertz frequency range) within a given amplitude range (3 to 10 microvolts, for example).

If the EEM instrument being utilized provides a choice of monitoring any of the three basic frequency levels and can be set to appreciate various amplitude ranges, the general aim of EEM training is for the patient to progressively learn to decrease the general frequency of the electrical brain wave activity, one frequency range at a time (from highest to lowest), while progressively learning to also increase the amplitude of the electrical activity being produced (from a level of 3 to levels exceeding 30 microvolts); i.e., the patient learns to produce brainwave electrical activity at a given frequency and at (or beyond) a given maximal amplitude level before proceeding to the learn to selectively produce activity at the next lower frequency range at any given amplitude level.

The EEM instrument communicates with the patient via electrode cables (usually three: two live pickup and one ground) which terminate in surface electrodes which are coupled by electrode paste to the patient's scalp and affixed in place with supportive (usually elastic) strapping. Such instruments generally provide audio feedback (for benefit of the patient) which varies in pitch, dropping from high to low pitches as the brain wave frequency is reduced and produced in sufficient amplitudes. Visual feedback in the form of a meter or light bar is provided only for the convenience of the practitioner's observation and to make record keeping possible; the visual feedback is generally synchronous with the auditory feedback, with high meter readings indicating successfully achieved frequency production at acceptable amplitudes, and low readings denoting failure to achieve the desired frequency range and/or adequate amplitude levels. During training the patient is generally required to have the eyes closed and the neck muscles relaxed since visual activity, eye and eyelid movement, and neck muscle contraction may interfere with training. The electrical activity produced by such functions may be picked up by the EEM and effectively mask the brainwave electrical activity being produced.

Procedures of Application (M21A2)

1) The patient should be in a sitting or semi-reclining position, with head and/or neck supported, which will remain comfortable for the duration of the training session.

2) The training process should be clearly explained and the feedback levels denoting success and failure illustrated.

3) The EEM surface electrodes should be coupled with electrode paste to the patient's scalp and affixed in place over the occipital lobes, superior to the nuchal line with strapping or other device.

4) If the treatment environment has a level of background noise significant enough to be distracting, the patient should be provided with earphones which will provide the audio feedback from the EEM and screen out other noise.

5) The patient should be instructed to close the eyes for the duration of the training.

6) The treatment room light should be dimmed, but not so low that meter reads can not be discerned by the practitioner.

7) The EEM instrument, preset to foster a given frequency within a particular amplitude range, should be turned turned on and the audio feedback volume adjusted to patient comfort but at levels that will permit the patient to hear the feedback easily, without straining.

8) The initial range of EEM activity should be noted and recorded.

9) The patient should be left alone and undisturbed for ten to fifteen minutes.

10) The instrument should be turned off and the patient allowed to rest for five minutes before continuing.

11) After the five-minute rest, the instrument should be turned on again and the patient left alone, undisturbed for another ten to fifteen-minute period.

12) The final levels of EEM activity should be noted and recorded.

13) The instrument should be turned off, the electrodes removed from the patient, and the electrode paste thoroughly cleansed from the patient's skin.

14) The patient should be allowed to rest for several minutes before rising and leaving.

Partial List of Conditions Treated (M21A3)

Treatable Causes (M21A3A)

Blood sugar irregularities (C31)
Extrafusal muscle spasm (C41)
Insomnia (C46)
Migraine (vascular headache) (C34)
Nausea/vomiting (C36)
Nerve root compression (C04)
Neurogenic dermal disorders (psoriasis, hives) (C49)
Neuromuscular tonic imbalance (C30)
Psychoneurogenic neuromuscular hypertonus (C29)
Restricted joint range of motion (C26)
Sinusitis (C16)
Sympathetic hyperresponse (C38)
Tinnitus (C32)
Trigger point formation (C01)
Visceral organ dysfunction (C02)

Syndromes (M21A3B)

ACNE (S45)
CALF PAIN (S50)
CERVICAL (NECK) PAIN (S73)
CHEST PAIN (S14)
COLITIS (S42)
DIABETES (S10)
EARACHE (S40)
ELBOW PAIN (S47)
FOOT PAIN (S51)
FOREARM PAIN (S54)
HAND/FINGER PAIN (S17)
HEADACHE PAIN (S02)
HIGH THORACIC BACK PAIN (S48)
HYSTERIA/ANXIETY REACTION (S59)
INSOMNIA (S71)
KNEE PAIN (S46)
LOWER ABDOMINAL PAIN (S62)
LOW BACK PAIN (S03)
MIGRAINE (VASCULAR) HEADACHE (S18)
PHOBIC REACTION (S44)
POST WHIPLASH (S55)
RADICULITIS (S29)
REFERRED PAIN (S01)
SCIATICA (S05)
SHOULDER/HAND (S57)
SHOULDER PAIN (S08)
SINUS PAIN (S33)
TEMPOROMANDIBULAR JOINT (TMJ) PAIN (S06)
THIGH PAIN (S49)
TINNITUS (S70)
TOOTHACHE/JAW PAIN (S41)
WRIST PAIN (S16)

Precautions and Contraindications (M21A4)

Epilepsy (M21A4A)

EEM feedback training at the 5 to 12 hertz frequency range should not be employed by patients who suffer from epileptic seizures. Such training may trigger an epileptic seizure, which on an EEG appears as an exaggerated, sustained train of 8 to 12 hertz high amplitude spikes.

Endogenous depression (M21A4B)

EEM feedback training at the 5 to 12 hertz frequency range should not be employed by patients who suffer from primary or endogenous depression or hypoglycemia; such training may precipitate an attack of depression (melancholia).

Diabetes (M21A4C)

Caution should be observed when EEM training (general relaxation or desensitization) is performed with patients suffering from diabetes. The depression of sympathetic activity may decrease metabolism and reduce demand for insulin thereby increasing the possibility of an insulin reaction occurring, especially if the diabetic patient has (1) taken an insulin supplement just prior to EEM training, (2) failed to previously consume adequate amounts of food or failed to digest what has been eaten, or (3) taken too large a dose of insulin before beginning training. The first symptoms of an insulin reaction include feelings of extreme hunger, nervousness (general agitation), heavy perspiration, palpitation of the heart, and occasional double vision. These symptoms may be relieved by consumption of eight ounces of orange juice which has had two teaspoonfuls of regular sugar added to it. If it is unavailable, any sugar or carbohydrate-containing food, like bread or crackers, will serve. If the first symptoms are unrelieved, more serious symptoms may appear which include delirium, convulsions and/or unconsciousness (coma). To relieve these symptoms, intravenous injection of glucose may be necessary.

Notes Aside (M21A5)

EEM facilitation of meditation states has been markedly effective when used in the eclectic treatment of colitis, especially when augmented by temperature monitry training (refer to Temperature Monitry, Lower Intestinal Dysfunction (Colitis) Control [M19D]), electrical stimulation of auricular acupoints (refer to Auricular Electroacustimulation, Pathology Control [M06B]), and continuous electrical stimulation of peripheral acupoints (refer to Peripheral Electroacustimulation, Pathology Control [M05C]).

Bibliography (M21A6)

Engstrom, pp. 1261–1262.
Green, pp. 1–26.
Patel, pp. 1–41.

ELECTROENCEPHALOMETRY HYPERTENSION CONTROL (M21B)

Definition (M21B1)

Electroencephalometric (EEM) feedback training has been shown to be effective for the depression of sympathetic autonomic activity to the extent of producing a reduction in blood pressure, a decrease in pulse rate, a slowing of metabolic processes, and decreased intestinal motility. Less measurable benefits are said to include a decrease in emotional lability, increased powers of concentration, increased ability to sleep, and a general feeling of well-being. Producing such effects, EEM training lends itself well to a program dedicated to the control of hypertension.

The EEM instrument most commonly seen in the clinical setting is a type of brain wave monitor designed to provide appropriate feedback to certain brain wave electrical activity. The feedback provided generally takes the form of audio feedback which is modulated from a high to low pitch when a given frequency of brain wave activity is produced in sufficient amounts. Low range pitches generally designate successful satisfaction of the criteria of brain wave electric activity produced at a given frequency which meets or exceeds a preset amplitude (a threshold usually associated with amplitudes of 3, 5, 6, 10, 30, 50, 60, or 120 microvolts). Most EEM instruments are geared to foster the production of brain wave activity in the frequency ranges of 5 to 7.5 (theta), 7.5 to 8 (alpha-theta), and/or 8 to 12 hertz (alpha); some instruments provide the option of selecting which of these frequencies is to be fostered and a full range of amplitude thresholds, while others are fixed to monitor only one frequency (usually the 8 to 12 hertz frequency range) within a given amplitude range (3 to 10 microvolts, for example).

If the EEM instrument being utilized provides a choice of monitoring any of the three basic frequency levels and can be set to appreciate various amplitude ranges, the general aim of EEM training is for the patient to progressively learn to decrease the general frequency of the electrical brain wave activity, one frequency range at a time (from highest to lowest), while progressively learning to also increase the amplitude of the electrical activity being produced (from a level of 3 to levels exceeding 30 microvolts); i.e., the patient learns to produce brainwave electrical activity at a given frequency and at (or beyond) a given maximal amplitude level before proceeding to the learn to selectively produce activity at the next lower frequency range at any given amplitude level.

The EEM instrument communicates with the patient via electrode cables (usually three: two live pickup and one ground) which terminate in surface electrodes which are coupled by electrode paste to the patient's scalp and affixed in place with supportive (usually elastic) strapping. Such instruments generally provide audio feedback (for benefit of the patient) which varies in pitch, dropping from high to low pitches as the brain wave frequency is reduced and produced in sufficient amplitudes. Visual feedback in the form of a meter or light bar is provided only for the convenience of the practitioner's observation and to make record keeping possible; the visual feedback is generally synchronous with the auditory feedback, with high meter readings indicating successfully achieved frequency production at acceptable amplitudes, and low readings denoting failure to achieve the desired frequency range and/or adequate amplitude levels. During training the patient is generally required to have the eyes closed and the neck muscles relaxed since visual activity, eye and eyelid movement, and neck muscle contraction may interfere with training. The electrical activity produced by such functions may be picked up by the EEM and effectively mask the brain wave electrical activity being produced.

For a detailed description of hypertension and its treatment with EEM feedback training the clinical setting, in concert with other treatment techniques, refer to Blood Pressure Control with Physiological Feedback Training [R007].

Procedures of Application (M21B2)

1) The patient should be in a sitting or semi-reclining position, with head and/or neck supported, which will remain comfortable for the duration of the training session.

2) The patient's blood pressure should be taken and recorded.

3) The training process should be clearly explained and the feedback levels denoting success and failure illustrated.

4) The EEM surface electrodes should be coupled with electrode paste to the patient's scalp and affixed in place over the occipital lobes, superior to the nuchal line with strapping or other device.

5) If the treatment environment has a level of background noise significant enough to be distracting, the patient should be provided with ear-

phones which will provide the audio feedback from the EEM and screen out other noise.

6) The patient should be instructed to close the eyes for the duration of the training.

7) The treatment room light should be dimmed, but not so low that meter reads can not be discerned by the practitioner.

8) The EEM instrument, preset to foster a given frequency within a particular amplitude range, should be turned turned on and the audio feedback volume adjusted to patient comfort but at levels that will permit the patient to hear the feedback easily, without straining.

9) The initial range of EEM activity should be noted and recorded.

10) The patient should be left alone and undisturbed for ten to fifteen minutes.

11) The instrument should be turned off. The patient's blood pressure should be taken again and recorded. The patient should be allowed to rest for five minutes before continuing.

12) After the five-minute rest, the instrument should be turned on again and the patient left alone, undisturbed for another ten to fifteen-minute period.

13) The final levels of EEM activity should be noted and recorded.

14) The instrument should be turned off. The patient's blood pressure should be taken again and recorded.

15) The electrodes should be removed from the patient, and the electrode paste thoroughly cleansed from the patient's skin.

16) The patient should be allowed to rest for several minutes before rising and leaving.

Partial List of Conditions Treated (M21B3)

Treatable Causes (M21B3A)

Blood sugar irregularities (C31)
Hypertension (C39)
Insomnia (C46)
Sympathetic hyperresponse (C38)
Visceral organ dysfunction (C02)

Syndromes (M21B3B)

CHEST PAIN (S14)
DIABETES (S10)
HEADACHE PAIN (S02)
HYPERTENSION (S43)
HYSTERIA/ANXIETY REACTION (S59)
INSOMNIA (S71)
PHOBIC REACTION (S44)
TINNITUS (S70)

Precautions and Contraindications (M21B4)

Epilepsy (M21B4A)

EEM feedback training at the 5 to 12 hertz frequency range should not be employed by patients who suffer from epileptic seizures. Such training may trigger an epileptic seizure, which on an EEG appears as an exaggerated, sustained train of 8 to 12 hertz high amplitude spikes.

Endogenous depression (M21B4B)

EEM feedback training at the 5 to 12 hertz frequency range should not be employed by patients who suffer from primary or endogenous depression or hypoglycemia; such training may precipitate an attack of depression (melancholia).

Diabetes (M21B4C)

Caution should be observed when EEM training (general relaxation or desensitization) is performed with patients suffering from diabetes. The depression of sympathetic activity may decrease metabolism and reduce demand for insulin thereby increasing the possibility of an insulin reaction occurring, especially if the diabetic patient has (1) taken an insulin supplement just prior to EEM training, (2) failed to previously consume adequate amounts of food or failed to digest what has been eaten, or (3) taken too large a dose of insulin before beginning training. The first symptoms of an insulin reaction include feelings of extreme hunger, nervousness (general agitation), heavy perspiration, palpitation of the heart, and occasional double vision. These symptoms may be relieved by consumption of eight ounces of orange juice which has had two teaspoonfuls of regular sugar added to it. If it is unavailable, any sugar or carbohydrate-containing food, like bread or crackers, will serve. If the first symptoms are unrelieved, more serious symptoms may appear which include delirium, convulsions and/or unconsciousness (coma). To relieve these symptoms, intravenous injection of glucose may be necessary.

Notes Aside (M21B5)

EEM feedback training may be used to reduce or control various forms of chronic hypertension. For the immediate reduction of acute and/or extreme hypertension EEM feedback may be inappropriate, since such training may initially increase the patient's neuroautonomic tension thereby increasing the blood pressure. Acute and/or extensive hypertension may best be treated with modalities which will immediately reduce the blood pressure to more acceptable levels. These modalities include ice packing (refer to Cryotherapy, Suppression of Sympathetic Autonomic Hyperreactivity [M10A]), auricular acustimulation (refer to Auricular Electroacustimulation, Pathology Control [M06B]), and acusleep (refer to Electrical Facilitation of Endorphin Production [M03]).

Bibliography (M21B6)

Blanchard, pp. 241–245.
Brener, pp. 1063–1064.
Erbeck, pp. 63–72.
Frankel, pp. 276–293.
Frunkin, pp. 296–230.
Green, pp. 1–26.
Hatch, pp. 119–138.
Kristt, pp. 370–378.
Shapiro, Schwartz, Turskey, pp. 296–304.
Shapiro, Tursky, et al., pp. 588–590.
Steptoe, pp. 252–263.

ELECTROENCEPHALOMETRY ELECTROENCEPHALOMETRIC EVALUATION (M21C)

Definition (M21C1)

Electroencephalometry (EEM) may be used as a gross evaluation tool for the determination of the possible presence of primary endogenous depression and/or the presence of epileptic seizure disorders. Both of these conditions are remarkable for the presence of very high amplitude, 5 to 12 hertz electroencephalometer activity. During evaluation the patient suffering from either of these conditions will demonstrate EEM activity which resembles an exaggerated, high amplitude, alpha, 8 to 12 hertz state of meditation, without any feedback being provided to the patient.

The most effective EEM feedback instruments used for evaluation may be preset to register the production of brain activity in the frequency ranges of 8 to 12 (alpha), 7.5 to 8 (alpha-theta), or 5 to 7.5 hertz (theta) and for amplitude ranges of 6, 12, 30, 60, or 120 microvolts. They provide visual feedback to the practitioner in the form of strip recordings, video graphics, and a meter or light board.

The EEM instrument communicates with the patient via electrode cables (usually three: two live pickup and one ground) which terminate in surface electrodes which are coupled by electrode paste to the patient's scalp and affixed in place with supportive (usually elastic) strapping. High meter readings or graphic responses indicate high amplitude activity at whichever frequency level the instrument is set to monitor.

Procedures of Application (M21C2)

1) The patient should be in a sitting or semi-reclined position, with head and/or neck supported, which will remain comfortable for the duration evaluative session.

2) The EEM surface electrodes should be coupled with electrode paste to the patient's scalp and affixed in place over the occipital lobes, superior to the nuchal line, with strapping or other device.

3) The patient should be instructed to keep the eyes closed for the duration of the treatment.

4) The light in the room should be dimmed, but not so low that meter reads or graphic responses cannot be discerned by the practitioner.

5) The EEG instrument, preset to foster a given frequency within a particular amplitude range, usually the 8 to 12 hertz range, should be turned on.

6) The activity ranges reflected by the EEM should be noted and recorded.

7) Following evaluation, the instrument should be turned off, the electrodes removed from the patient, and the electrode paste thoroughly cleansed from the patient's skin.

Partial List of Conditions Evaluated (M21C3)

Treatable Causes (M21C3A)

Blood sugar irregularities (C31)
Insomnia (C46)
Migraine (vascular headache) (C34)
Nausea/vomiting (C36)
Neurogenic dermal disorders (psoriasis, hives) (C49)
Visceral organ dysfunction (C02)

Syndromes (M21C3B)

ACNE (S45)
COLITIS (S42)
HEADACHE PAIN (S02)
HYSTERIA/ANXIETY REACTION (S59)
INSOMNIA (S71)
MIGRAINE (VASCULAR) HEADACHE (S18)
PHOBIC REACTION (S44)

Notes Aside (M21C4)

If the EEM evaluation for the establishment of endogenous primary depression and/or epileptic disorders is suspected to be positive, EEM feedback training should be precluded for that particular patient, and the patient should be guided to seek more differential evaluation in a more formal setting utilizing electroencephalography (EEG).

Bibliography (M21C5)

Green, pp. 1–26.
Sterman, pp. 89–95.

ELECTROMYOMETRIC FEEDBACK (M18)

Definition (M1801)

Myoelectric activity (M1801A)

When a muscle contracts it produces an electric current; an electromyometer is an instrument designed to pick up some of that myogenic current and to register its amplitude and/or frequency. In most cases this qualitative feedback is reinterpreted into a mode or modes of feedback which may be used by the practitioner either to help evaluate various neuromuscular conditions or to provide the patient with direct information regarding the myoelectric activity of the particular muscle being monitored.

Myoelectric feedback (M1801B)

The various feedback modes provided by the electromyometer, reflective of electronic input, may include: (1) visual feedback, provided by meter readings using a needle meter and/or light bars, paper graphing, computer graphics, and/or robotic displays of trains, robots, etc.; (2) audio feedback, generally provided by electronic speakers which turn the electronic input into variable pitch-summated tones (a functional melding of amplitude and frequency measurements), interrupted monotonal auditory pulses (clicks, usually representative of frequency), or continuous summated tones which simultaneously change pitch and volume level with changes in frequency and amplitude; and/or (3) tactile feedback which allows the patient to "feel" myoelectric changes in the muscle being monitored via tactile and/or proprioceptive nerve stimulation provided by mechanical vibration, transcutaneous nerve stimulation (TNS or TENS), and/or other contraction-building electrical stimulation. Feedback modes are generally set to reflect myoelectric change when the base myoelectric activity has exceeded a given amplitude level, ruling out background electrical noise. An additional option may be available, however, which allows a feedback mode (or modes) to be arbitrarily set by the practitioner to provide or extinguish feedback when a particular level of frequency and/or amplitude above the base level has been attained or adequately suppressed. This level is generally called the "threshold".

Electromyometric (EMM) feedback may be utilized therapeutically to teach or show the patient how to modify otherwise unconscious (subliminal) neuromuscular activity by providing an additional cybernetic loop, albeit external, between the patient's neuromuscular control mechanisms (cerebral cortex, cerebellum, reticular formation, etc.) and the musculature. EMM feedback has been demonstrated to effectively provide for interplay between the conscious and unconscious (subliminal) levels of the "mind", allowing each to learn from the other. Electromyometry is unique in this capacity, contrary to the current popular belief systems regarding the effects of hypnosis and conscious "imagery." Clinical research and experience has demonstrated that this facility provided by electromyometric feedback often gives the patient an opportunity to regain lost neuromuscular control resulting from pathological interruption of internal cybernetic feedback loops, or to over ride or unlearn emotionally induced neuromuscular response patterns stemming from the need for psychogenic defense mechanisms or other psychoneurogenic sources.

Subliminal learning pattern (M1801C)

Clinical evidence and experience has demonstrated that electromyometric feedback training must follow a particular pattern for meaningful, efficient subliminal learning to take place. This pattern involves the *facilitation* of myoelectric activity from musculature which is weak or inactive, *inhibition* of myoelectric activity from musculature which is overactive (hypertonic or spastic), and *balance* of the myoelectric activity from facilitated and inhibited muscles, increasing the myoelectric activity from one while suppressing myoelectric activity from the other. These newly gained neuromuscular skills are finally used together as part of an integrated neuromuscular *function:* a neuromuscular activity or activity performed without monitry or other external proprioceptive aid or cluing mechanism. The final goal is to use the activity enough to retain the consequently learned neuromuscular skill; i.e., use it or lose it.

The balance phase of this process involves the supply of simultaneous, mutually exclusive electromyometric feedback from opposing musculature to the patient in such as way that the differences between the two are easily perceived as being separately derived. The opposing musculature may be across a joint (biceps vs. triceps), members of antagonistic synergistic muscle groups (finger extensors vs. the clavicular portion of the pectoralis major), or as muscle members of the same synergy where differentiation between muscle members is desired (finger flexion vs. thumb flexion, or diaphragm vs. breathing accessory muscles).

Procedures of Application (M1802)

Electromyometric feedback instrumentation (M1802A)

Electromyometric feedback is generally supplied by electromyometer instrumentation connected to the patient via electrode cables. The electrode cables terminate as they connect to surface electrodes: pre-molded, plastic-collared surface electrodes, surface electrode "buttons", or disposable self-adhesive types. Most types of surface electrodes are composed of silver plated copper, and often treated with silver nitrate. The surface electrodes are generally coupled to the patient's skin with a conductive gel or paste, and are usually affixed to the patient's skin with some form of tape or by various kinds of strapping. Double or single-sided adhesive collars are commonly used.

EMM instrumentation employed for neuromuscular reeducation should be sensitive enough, when translated into a feedback mode, to reflect minute changes in myoelectric activity to one tenth of a microvolt. The electromyometers should be efficient enough to provide feedback which reflects almost simultaneously the changes in myoelectric activity as they occur. The quicker the translation of myoelectric activity into feedback the better and more immediate the patient's learning will be. The meters or light bars should be large enough for the patient to see small changes in activity and the audio feedback should be loud enough to clearly hear changes in pitch or pulse frequency. The same principles apply if computer enhancement of auditory and/or visual feedback is employed: feedback should immediately reflect myoelectric changes at a sensitivity level which adequately represents myoelectric change. It must demonstrate myoelectric change that the patient can immediately perceive and appreciate as dynamic change.

Dual electromyometry and feedback modes (M1802B)

In the balance phase of the learning process, if computer enhancement is not available, two electromyometers will be needed to simultaneously reflect myoelectric activity from the antagonistic musculature. The meters from each will need to be simultaneously visible to the patient, and the audio feedback from each perceivable by the patient as being distinctly different, one from another, enough to make differentiation between the two myoelectric activities clearly apparent. If the auditory feedback from the two machines is not distinctly different to the patient, the auditory feedback mode from one of the machines will have to be turned off. The remaining sound should be allowed to come from the machine monitoring the hardest muscle to control, which is usually the machine monitoring the muscle to be inhibited.

It has been clinically demonstrated that the pulsed or broken, clicking auditory feedback mode seems to be most effectively employed to reflect myoelectric inhibition, while the continuous variable pitch tone seems to be most effectively employed to reflect myoelectric facilitation; however, as with most variables, this should always depend upon empirical trial and error relative to the individual patient.

Tactile feedback mode (M1802C)

Some EMM instrumentation provides for tactile and/or proprioceptive feedback modes. These may include mechanical vibration (rotary or linear), transcutaneous electrical nerve stimulation (TNS or TENS), and/or electrical stimulation strong enough to cause muscle contraction. Most commonly, these modes are used to proportionally reflect the myoelectric activity of only one muscle, even when used during the balance phase of training, but if multiple instrumentation is available, tactile or proprioceptive feedback modes, if cleverly arranged, could conceivably be used to reflect mutually exclusive myoelectric activity from two monitored muscles.

Electromyometric tactile and/or proprioceptive feedback is produced by vibratory or electrical stimulation units which have circuits electrically independent of the electromyometric circuitry but controlled by it. This is to prevent the electrical activity which provides the tactile feedback from feeding back to the equipment and inadvertently masking myoelectric activity as it is picked up.

The process which allows control of a tactile stimulator is one of transduction. Typically, electromyometers which provide tactile and/or proprioceptive feedback usually have an auxiliary analogue output which communicates with a transducing mechanism acting as a prompter for a response from the tactile stimulator which is parallel, if not equal, to electromyometer analogue output. As the myoelectric activity increases or decreases the reflected analogue EMM output ,through transduction which doesn't allow any mixing of circuit electromyometer and stimulator electrical currents, correspondingly causes the stimulator amplitude to proportionally increase or decrease. Such a machine should provide the option of increasing or decreasing its level of stimulation as the electromyometer analogue output increases.

In use, the site designated for tactile and/or proprioceptive feedback stimulation should be selected carefully, in view of the fact that the sources of tactile and/or proprioceptive stimulation discussed here are generally facilitatory of muscle tension and tend to increase myoelectric activity below the site of stimulation, and are consequently reciprocally inhibitory of antagonistic muscular activity. If a muscle is to be inhibited through the use of electromyometry, the site selected to be tactily stimulated should not be over the monitored muscle or a synergistic partner, but over a direct antagonist or a member of an antagonistic synergistic group. Conversely, if facilitation is desired, the tactile feedback stimulation site should not be over a muscle antagonistic to a muscle to be facilitated but over the muscle to be facilitated or over one of its synergistic partners.

If an electrical stimulator is used to provide tactile and/or proprioceptive electromyometric feedback, the placement site(s) of the electrical stimulator electrodes must be far enough away from the electromyometer pickup electrodes to prevent pickup of the electrical stimulator currents by the electromyometer electrodes. Electromyometric pickup of the electrical current produced by a TNS unit is less likely to occur than is pickup from a muscle contraction-inducing stimulator, because TNS stimulators provide a comparatively low voltage, low amperage electrical stimulation and are less likely to be picked up by electromyometry than the relatively high voltage high amperage muscle contraction-inducing electrical currents.

The electromyometer analogue output which may be used to drive a tactile and/or proprioceptive feedback units may also be used to provide auxiliary visual feedback in the form of electric trains and other electrically controlled robotics. The transduction of the analogue output may be used to prompt the increase or decrease of visually appreciated electrically driven machine movements which are proportionally parallel to the electromyometer analogue output.

Computer enhancement of electromyometry (M1802D)

Computer enhancement of electromyometric feedback has proven to be very effective, providing graphical representations of myoelectric activity which, if properly designed, are easily appreciated by the patient as visual "games" which help provide motivation and more concrete feelings of accomplishment by the patient than seems to be provided by unenhanced meter readings and/or auditory feedback. Computer enhancement may also in many cases, if so designed, further include further supplementations or variations of auditory feedback modes, and additionally may be used to coordinate all feedback mode presentations (visual, auditory, tactile and/or proprioceptive) and any other auxiliary stimulation modes (faradic, pulsed galvanic, inferential, or medium frequency electrical stimulation) not used as feedback modes.

EMM-triggered electrically induced muscular contraction (M1802E)

Electrical stimulation of muscle contraction has been used in conjunction with electromyometry, where the electromyometer is made to act as the trigger of electrical stimulation. To accomplish this, an electrical stimulator is engineered to begin stimulation when a given EMM activity level or threshold is reached. When electrical stimulation begins, and while the electrically induced neuromuscular contraction is occurring, electromyometry pickup is automatically halted to prevent damage from occurring to the circuitry. Effectiveness of this procedure seems limited to the facilitatory phase of neuromuscular training, and this technical procedure should not be classed as a tactile or proprioceptive feedback procedure since the amplitude of stimulation is not tied to the amplitude of myoelectric activity.

Partial List of Conditions Treated (M1803)

Treatable Causes (M1803A)

Extrafusal muscle spasm (C41)
Habitual joint positioning (C48)
Hypermobile/instable joint (C10)
Hypertension (C39)
Interspinous ligamentous strain (C03)
Joint approximation (C47)
Muscular weakness (C23)
Nerve root compression (C04)
Neuromuscular tonic imbalance (C30)
Pathological neuromuscular hypertonus (C28)
Pathological neuromuscular hypotonus (C33)
Pathological neuromuscular dyscoordination (C35)
Peripheral nerve injury (C07)
Psychoneurogenic neuromuscular hypertonus (C29)
Restricted joint range of motion (C26)
Spastic sphincter (C40)
Sympathetic hyperresponse (C38)
Tinnitus (C32)
Trigger point formation (C01)

Syndromes (M1803B)

BELL'S PALSY (S66)
CALF PAIN (S50)
CEREBRAL PALSY (S25)
CERVICAL DORSAL OUTLET (SCALENICUS ATTICUS/CERVICAL RIB) (S09)
CERVICAL (NECK) PAIN (S73)
CHEST PAIN (S14)
EARACHE (S40)
ELBOW PAIN (S47)
FACET (S19)
FOOT PAIN (S51)
FOREARM PAIN (S54)
FROZEN SHOULDER (S64)
HAND/FINGER PAIN (S17)
HEADACHE PAIN (S02)
HIGH THORACIC BACK PAIN (S48)
HYPERTENSION (S43)
HYSTERIA/ANXIETY REACTION (S59)
KNEE PAIN (S46)
LOWER ABDOMINAL PAIN (S62)
LOW BACK PAIN (S03)
MENSTRUAL CRAMPING (S60)
MIGRAINE (VASCULAR) HEADACHE (S18)
MYOSITIS OSSIFICANS (S67)
NEUROMUSCULAR PARALYSIS (S22)
PHOBIC REACTION (S44)
POST CEREBRAL VASCULAR ACCIDENT (CVA) (S07)
POST IMMOBILIZATION (S36)
POST PERIPHERAL NERVE INJURY (S23)
POST SPINAL CORD INJURY (S24)
POST WHIPLASH (S55)
RADICULITIS (S29)
REFERRED PAIN (S01)
SCIATICA (S05)
SHOULDER/HAND (S57)
SHOULDER PAIN (S08)
SINUS PAIN (S33)
TEMPOROMANDIBULAR JOINT (TMJ) PAIN (S06)
TENDONITIS (S28)
TENNIS ELBOW (LATERAL EPICONDYLITIS) (S32)
THIGH PAIN (S49)
TINNITUS (S70)
TOOTHACHE/JAW PAIN (S41)
TORTICOLLIS (WRY NECK) (S65)
WRIST PAIN (S16)

Precautions and Contraindications (M1804)

No precautions or contraindications have been listed or detailed for electromyometric feedback, ex-cept in connection with the stimulation provided by tactile and proprioceptive feedback modes. For precautions and contraindications associated with these feedback modes refer to the Electrical Stimulation, *Precautions and Contraindications* [M0204], Transcutaneous Nerve Stimulation, *Precautions and Contraindications* [M0404], and/or Vibration, *Precautions and Contraindications* [M0804] sections.

Notes Aside (M1805)

Hierarchy of subliminal attention getting (M1805A)

Research evidence has demonstrated the existence of a hierarchy of subliminal attention getting by respective feedback modes. Tactile and/or proprioceptive feedback modes arouse subliminal attention to a greater degree than do auditory auditory modes, and auditory modes arouse subliminal attention to a greater degree than do visual feedback modes. The subliminal attention-getting facility seems to be related to the level at which the primary sensory input of the feedback mode takes place. Tactile and/or proprioceptive sensory input makes its primary impression on mechanisms related to the cerebellum (very deep in the brain structure), while auditory feedback is primarily received by mid-brain structures and visual sensory input is primarily directed to the visual cerebral cortex (very high up). With this hierarchy in mind, it would seem most obvious that EMM instrumentation which utilizes tactile, auditory and visual feedback modes is superior to instrumentation which employs fewer modes or is limited to visual and/or auditory feedback. It would also seem that those modes of feedback which arouse the greatest degree of subliminal attention should be employed to help the patient deal with the most difficult tasks; for instance, musculature which is generally the hardest to inhibit should be monitored by an electromyometer which provides audio and tactile feedback modes while visual feedback may be all that is necessary for the patient to gain mastery over a muscle which should be facilitated.

Bibliography (M1806)

Braud, pp. 67–89.
Brudny, pp. 925–932.
Hampstead, pp. 113–125.
Johnson, Lee, pp. 826–831.
Marinacci, pp. 57–71.
Swaan, pp. 9–11.

ELECTROMYOMETRIC FEEDBACK POST CEREBRAL ACCIDENT NEUROMUSCULAR REEDUCATION
(M18A)

Definition (M18A1)

The cerebral accident (M18A1A)

A cerebral accident is defined here as a traumatic injury of cerebral tissue or related structures which has resulted in the disruption of communication loops or pathways between various brain structures including damage to structures themselves since they are an integral part of any communication loop they participate in. Such injuries include cerebral vascular occlusion by thrombus, bruising or laceration of cerebral tissue or related structures resulting from concussion or other externally induced trauma of the cranium and/or brain (contra-coup concussion, and damage resulting from surgical procedures upon the brain are included), and other sources of cerebral tissue and related structure damage including anoxia, excessive anesthesia, sequelae of encephalitic disease or poisoning, and excessive intracranial pressure (unrelieved hydrocephalus, subdural or other meningeal layer hematoma, or cerebral edema). Because of the large number of causes, the population of cerebral accident victims is extremely large. Regardless of the source of damage, if cerebral tissue or related structure(s) have been damaged beyond the body's ability to repair, a post cerebral vascular accident (CVA) syndrome may result. Of interest to us here is the post CVA syndrome whose symptomatology includes impairment of neuromotor function.

Post CVA syndromes involving motor dysfunction are classified in groups according to the extent of impairment. Such classifications generally include hemiplegia (half body paralysis), quadriplegia (full body paralysis), hemiparesis (half body weakness or dyscoordination), and quadriparesis (full body weakness or dyscoordination). By and large, paralysis or paresis fall into two categories (1) spastic or hypertonic paralysis or paresis, and (2) flaccid or hypotonic paralysis. In the latter category, (a) functional flaccidity may occur, which is eventually replaced by spasticity (especially if encouraged), or (b) structural flaccidity may occur, in which case tone will never develop in the affected musculature (apparently due to damage sustained not only to the pathways between the cerebellum and the motor cortex, but also to the pathways between the motor cortex and the basal ganglia). (Refer to Central Nervous System Dysfunction [R006]).

EMM neuromuscular reeducation of post CVA paralysis (M18A1B)

Electromyometric feedback has been found to be useful in the neuromuscular reeducation of post CVA victims suffering from neuromuscular dysfunction. Clinical evidence and experience has demonstrated that electromyometric feedback training must follow a particular pattern for meaningful, efficient subliminal learning to take place. This pattern involves the *facilitation* of myoelectric activity from musculature which is weak or inactive, *inhibition* of myoelectric activity from musculature which is overactive (hypertonic or spastic), and *balance* of the myoelectric activity from facilitated and inhibited muscles, increasing the myoelectric activity from one while suppressing myoelectric activity from the other. These newly gained neuromuscular skills are finally used together as part of an integrated neuromuscular *function*: a neuromuscular activity or activity performed without monitry or other external proprioceptive aid or cluing mechanism. The final goal is to use the activity enough to retain the consequently learned neuromuscular skill; i.e., use it or lose it.

The balance phase of this process involves the supply of simultaneous, mutually exclusive electromyometric feedback from opposing musculature to the patient in such a way that the differences between the two are easily perceived as being separately derived. The opposing musculature may be across a joint (biceps vs. triceps), members of antagonistic synergistic muscle groups (finger extensors vs. the clavicular portion of the pectoralis major), or as muscle members of the same synergy where differentiation between muscle members is desired (finger flexion vs. thumb flexion, or diaphragm vs. breathing accessory muscles).

Spastic hemiplegic EMM neuromuscular reeducation (M18A1C)

For the rehabilitation of spastic hemiplegia or paresis, the practitioner is faced with the added complication of having to help the patient overcome the dominance of pathological developmental and/or synergistic reflex patterns apparent on the involved side (refer to the Table of Developmental Reflexes [T007] and Table of Neuromuscular Synergistic Patterns [T008]). Simply stated, the general method for doing this is to utilize electromyometric feedback techniques to inhibit the spastic muscle components of an identified pathological developmental reflex or synergistic pattern while facilitating the muscle components of a developmental reflex at the next higher level and/or a normal synergistic pattern.

Flaccid hemiplegic EMM neuromuscular reeducation (M18A1D)

For the rehabilitation of flaccid hemiplegia, the first objective is to help the patient generate, through electromyometric feedback, activity in all of the muscles involved, with the goal of developing normal developmental reflexes if possible and repeating the process described above for treatment of spastic hemiplegia. Failure to develop tone in the involved musculature, whether spastic or not, confirms structural flaccidity. Such flaccidity can be overcome to the extent that the patient can be taught to voluntarily activate involved musculature, even in coordinated functional synergies, but the benefit may remain limited since resting tone will never develop and the patient will be forced to "start over" each time movement of an involved extremity is desired.

Bilateral hemiplegic EMM neuromuscular reeducation (M18A1E)

For the rehabilitation of spastic or flaccid quadriplegia, the above processes described are valid but the process is very complex. Generally, the two sides of the patient's body are asymmetrical, with each side displaying different pathological developmental reflexes and synergistic reflex patterns. Involvement on one side will usually be greater than on the other. In practice, rehabilitation of a patient suffering from bilateral hemiplegia is much like treating two patients with hemiplegia of opposing sides at the same time.

Procedures of Application (M18A2)

Electromyometric instrumentation (M18A2A)

Instrumentation required for neuromuscular reeducation with electromyometric feedback of musculature affected by a post CVA syndrome includes one or two electromyometers with the most sophisticated feedback modes available, which can be preset to register different sensitivity levels. Common sensitivity settings include the 0 to 10, 0 to 50, 0 to 100, and 0 to 500 microvolt ranges. The basic electromyometer has both audio and visual (meter or light bar) feedback modes. These modes can be enhanced by built-in feedback mode variables (different sounds and multiple visual displays) or by external computer enhancement of both modes. Some types of EMM instrumentation also offer tactile feedback modes and/or robotic-mechanical displays.

Electrode placement (M18A2B)

The electromyometers are attached to the patient via electrode cables which terminate by attaching to surface electrodes affixed to the skin over the muscles monitored.

Evaluation of the post cerebral vascular accident (M18A2C)

The patient should be placed in a comfortable sitting, semi-reclining or supine position which allows easy access to the muscles suspected of being effected by the post cerebral vascular accident.

During the evaluation phase, the patient should remain unexposed to the feedback modes, unable to see, hear or feel changes in myoelectric production as reflected by EMM feedback.

Each suspect muscle should be individually monitored in three phases: (1) at rest, (2) during an attempted voluntary sustained contraction of at least six seconds, and (3) during a subsequent relaxation of effort.

A muscle which is hypotonic may be unable to produce any increase in myoelectric activity which is discernible above the resting level, or may produce an increase in myoelectric activity to the order of 1 or 2 microvolts above resting levels.

A hypertonic muscle will demonstrate a fairly high level of resting myoelectric activity, generally in excess of six microvolts, and may display on the meter or video screen the rhythmic deflection pattern associated with spasticity. Increases in myoelectric activity levels produced by hypertonic muscles tend to be lower then normal. Generally the patient may only be able to generate between 1 and 150 microvolts above resting levels, generating such activity slowly, taking several seconds to achieve maximum levels. A return to resting levels may likewise be very slow, sometimes taking minutes.

Neuromuscular reeducation utilizing EMM feedback (M18A2D)

Simply stated, the process of electromyometric feedback neuromuscular reeducation follows the subliminal learning pattern of facilitation, inhibition, balance and function.

Facilitation phase (M18A2D01) If spasticity and/or rigidity is present in the muscles monitored, the facilitation phase may be as important as the obviously important inhibition phase. If this is the case, the patient should be asked to attempt to increase the myoelectric activity produced by the spastic muscle for five or six seconds and then to relax (stop trying). In some cases the patient will be able to increase the myoelectric activity above resting levels, anywhere from 1 to 150 microvolts, but in other cases the the activity level will not rise sufficiently high enough to be differentiated from resting level activity; however, when the patient gives up the attempt to facilitate myoelectric activity (to relax), a drop in myoelectric activity below previous resting levels may be discernible. Having recognized this change in feedback levels, the patient may have been provided with enough of a gauge of inhibition to begin controlling the muscle to the extent of being able to contract it voluntarily. If immediate ability to voluntarily contract the muscle is not forthcoming, this exercise at least gives the patient an introduction to how the supraspinal structures may begin to relate on a subliminal level to muscles who fail to provide proprioceptive input. The patient may be able to come back to the facilitation phase when the inhibition phase has provided adequate suppression of the spastic activity to make small increases in myoelectric activity discernible.

If the musculature is hypotonic or flaccid in the facilitation phase, the hypotonic muscle or synergistic muscle group should be connected by surface electrodes and electrode cables to an electromyometer. The patient should then be asked to increase myoelectric activity produced by the muscle or muscle group as reflected by the provided feedback. During the facilitation phase the electromyometer should be set a sensitivity level which best demonstrates positive or increasing myoelectric change. The patient must be able to see, hear and/or feel changes in myoelectric activity during the entire procedure to subliminally learn from them. If a muscle is flaccid, the EMM will initially have to be set at the level of greatest sensitivity (0 to 6 or 0 to 10 microvolt ranges) so that any increase in myoelectric activity will be substantially demonstrated by all the feedback modes.

If the muscle is more active, the EMM should be set at a less sensitive level (0 to 30 or 0 to 50 microvolt ranges, for example) to provide a gauge of success as well as providing the patient with an opportunity to strive to produce more activity at observable levels.

The patient should be asked to increase the myoelectric activity generated by the muscle or synergistic muscle pair, as reflected by the available feedback, up to an activity level determined to be attainable by the practitioner. Parameters in the form of limits visually displayed on the video screen, or demonstrated by audio feedback changes or thresholds should be set that will adequately demonstrate failure or success. The parameter which is designated to denote a successful attempt should be potentially attainable, but enough out of reach that the patient must exert a concentrated effort to attain it. If electromyometric feedback is being used to increase the strength of muscle contraction or the general tone, the patient should be asked to hold the maximum level of activity for five or six seconds or more, with at least a five- or six-second rest period between each attempt.

Inhibition phase (M18A2D02) In the inhibition phase, a muscle or synergistic muscle group (usually spastic, hypertonic or comparatively hyperactive) antagonistic to the muscle(s) formerly facilitated, should be monitored by electromyometry and the patient should be asked to actively inhibit the myoelectric activity, as reflected by the representative feedback. Usually a spastic or hypertonic muscle (or synergistic muscle pair) antagonistic to the muscle(s) focused on during the inhibitory stage should be monitored. The patient should be asked to actively inhibit or decrease the myoelectric activity produced by the muscle monitored. The patient should be asked to increase activity from the muscle(s) by a small amount, less than ten microvolts and then to actively suppress the activity level below the the previous resting level. The attempt to increase myoelectric activity from the muscle(s) should last no longer than six seconds, but the suppression stage may last until adequate inhibition is attained (from six seconds to several minutes) or until failure must be admitted and another attempt made. The electromyometer should be set at a sensitivity level which will adequately demonstrate successful inhibition (optimally at the 0 to 10 microvolt sensitivity level).

Balance phase (M18A2D03) In the balance phase, two EMM setups are required to allow the patient the opportunity of attempting to facilitate the initially facilitated muscle or synergistic muscle pair while simultaneously inhibiting the initially inhibited muscle or synergistic muscle pair.

The electromyometer monitoring the muscle(s) to be facilitated should be set at the least sensitive level that best magnifies and reflects increases in myoelectric activity without exceeding a maximum setting level (over 50 microvolts if the sensitivity level is set at a 0 to 50 microvolt range). The EMM monitoring the muscle(s) to be inhibited should be set at the most sensitive level that best magnifies and reflects decreases in myoelectric activity (shows the largest range of feedback, ideally at the 0 to 10 microvolt sensitivity level). The choice of feedback modes relative to the muscle and neuromuscular problem is of paramount importance. When in the balance phase of neuromuscular reeducation, the feedback modes most attention-arousing should be used in conjunction with the EMM monitoring the hardest muscle to control, which is usually the muscle to be inhibited if spasticity is present.

Functional phase (M18A2D04) In the functional phase is designed to provide the patient with an opportunity to use the antagonistic muscles or synergistic pairs in a functional activity that requires the previously facilitated muscle to contract while the previously inhibited muscle or synergistic pair is required to remain inactive or perform an eccentric contraction (lengthen as its antagonist contracts). The functional exercises used in this last stage must usually be improvised by the practitioner, often requiring a great deal of thought, since the exercise should employ the antagonistic muscles in a correctly balanced manner while performing a somewhat practical (usable) maneuver.

Computer enhanced feedback (M18A2E)

Computer enhancement of electromyometric feedback has been shown to greatly add to patient appreciation and participation in the neuromuscular reeducation program by making feedback more graphic and easier to understand in terms of visual and auditory feedback modes for both patient and practitioner. Some of the graphic designs currently in use for single EMM activity enhancement have shown themselves to be remarkably suited for increasing patient motivation and enthusiastic participation. One of the more effective graphic designs currently available requires the patient to keep the EMM activity line showing on the computer's video screen between parallel lines which cross the screen in various patterns or shapes. After the activity line has transversed the screen, a numerical readout in one corner of the screen tells the patient what percentage of the time the activity line was above, within, or below the parallel line pattern. The time it takes to cross the screen can be preset by the operator.

During the balance phase, computer enhancement is especially effective if the computer instrumentation allows representative activity from both EMM's to appear simultaneously on the video monitor as distinctly different lines. Effectiveness increases if the patient can clearly relate the two activity lines to a graphic screen design which helps demonstrate their relative activity levels, as with, for example, two fixed parallel lines laid out across the screen, each related to one of the action lines.

Session termination (M18A2F)

At the end of the treatment session the electromyometer and associated feedback mode instrumentation should be turned off.

The surface electrodes should be carefully removed and the electrode paste carefully cleansed from the patient's skin before the patient is allowed to leave the treatment area. A surface electrode count should be conducted after removal from the skin to prevent them from leaving when the patient does.

Partial List of Conditions Treated (M18A3)

Treatable Causes (M18A3A)

Hypermobile/instable joint   (C10)
Ligamentous strain   (C25)
Muscular weakness   (C23)
Neuromuscular tonic imbalance   (C30)
Pathological neuromuscular hypertonus   (C28)
Pathological neuromuscular hypotonus   (C33)
Pathological neuromuscular
   dyscoordination   (C35)
Restricted joint range of motion   (C26)
Soft tissue swelling   (C06)

Syndromes (M18A3B)

CEREBRAL PALSY   (S25)
NEUROMUSCULAR PARALYSIS   (S22)
PITTING (LYMPH) EDEMA   (S31)
POST CEREBRAL VASCULAR ACCIDENT
   (CVA)   (S07)
SHOULDER PAIN   (S08)

Precautions and Contraindications (M18A4)

No precautions or contraindications have been listed or detailed for electromyometric feedback, except in connection with the stimulation provided by tactile and proprioceptive feedback modes. For precautions and contraindications associated with these feedback modes refer to the Electrical Stimulation, *Precautions and Contraindications* [M0204], Transcutaneous Nerve Stimulation, *Precautions and Contraindications* [M0404], and/or Vibration, *Precautions and Contraindications* [M0804] sections.

It should be noted that neuromuscular reeducation utilizing electromyometric feedback may be of little value if the patient suffers from structural flaccidity. Such rehabilitative measures may prove to be esoteric from a functional point of view since without resting tone the patient must "start over" from flaccid joint positions when initiating motion and must always exert extraordinary attention when engaged in coordinated muscle activity with involved extremities.

Neuromuscular reeducation utilizing electromyometric feedback has proven to be effective for helping regain lost neuromuscular function for spastic and functionally flaccid post CVA patients, up to five years post injury. After five years, habitual neuromuscular behavior patterns seem to extinguish any muscular control gains made with electromyometric feedback, rendering the EMM neuromuscular reeducation program esoteric. The most impressive results in regaining functional control of involved musculature, through the use of electromyometric feedback, come two years post onset. At that point, what nature has been able to return to the patient has been returned (usually in spite of the therapeutic measures taken during those two years); any further functional gain must be considered impressive.

The above point of view notwithstanding, the earlier the initiation of neuromuscular reeducation utilizing electromyometric feedback the better the prognosis for the patient, if it is competently performed. Much of the symptomatology traditionally expected to be a part of the post CVA syndrome might be avoided with early initiation of electromyometric feedback, including: (1) subluxation and/or dislocation of the involved shoulder (or early reduction of same if electromyometric feedback is not begun soon enough to avoid it), (2) hyperextension of the involved knee during ambulation, and (3) much of the hypertonic spasticity which may potentially develop in flexion synergists in the upper extremity and extension synergists in the lower extremity.

The practitioner should be prepared to handle any cerebral motor seizures which may inadvertently occur as a result of motor cortex damage. Seizure occurrence during electromyometric feedback training is exceedingly rare; while none have been reported to date, they remain possible whenever a brain-injured patient is being treated.

If a seizure should occur, the patient should be placed in a side-lying position and an object placed between the teeth to prohibit the patient from biting the tongue. The object placed between the teeth should not be hard enough to cause tooth damage when bitten (no metal, porcelain, or glass) nor small enough to swallow; a wooden tongue depressor, rubber bite bar, thick wooden pencil, or a relatively soft plastic pen may serve. If the patient is aware of seizure development and reports it, ice packing of the posterior base of the head, the neck and paraspinous C7 vertebral areas bilaterally has been shown to be effective in preventing the onset of the some types of seizures. Following the seizure, the patient may be disoriented and should be prevented from leaving the protective environment without adequate careful guidance by clinical personnel or informed family members.

Notes Aside (M18A5)

EMM feedback and the aphasic post CVA victim (M18A5A)

It has been demonstrated through work with aphasic post CVA victims that the ability to speak and cognate is intrinsically related to the neuromuscular ability to voluntarily inhibit myoelectric activity of the wrist and finger flexors of the right hand. Until such inhibition occurs, speech therapy may be an exercise in futility. Electromyometric feedback has proven to be remarkably effective for providing the means of regaining the voluntary inhibition of right wrist and finger flexor myoelectric hyperactivity, which is apparently essential for regaining cognition and reducing aphasia in post CVA victims up to ten years post onset.

Bibliography (M18A6)

Andrews, pp. 530–532.
Basmajian, pp. 231–236.
Binder, pp. 886–892.
Honer, pp. 299–303.
Johnson, Garton, pp. 320–325.
Marinacci, pp. 57–71.
Mills, pp. 190–193.
Review of EMG Biofeedback Findings on Applications to Stroke Patients' Upper Extremities & Lower Extremities, pp. 1456–1459.
Taylor, pp. 29–80.
Winchester, pp. 1096–1103.
Wolf, pp. 1449–1455.
Wolf, Binder-Macleod, pp. 1405–1413.

ELECTROMYOMETRIC FEEDBACK POST PERIPHERAL NERVE INJURY NEUROMUSCULAR REEDUCATION (M18B)

Definition (M18B1)

Peripheral nerve injury (M18B1A)

The type peripheral nerve injury of interest to us here is that which occurs to the proprioceptive (afferent) and motor (efferent) innervation of the muscle. Nerves may be damaged by severance, crushing, deprivation of blood, or from the action of toxic substances.

If a peripheral nerve injury is severe enough, the damaged axons "die" and degenerate back to their cell bodies (Wallerian degeneration), and then begin to regrow or regenerate back to their end organ (sensory end organs in the muscle spindle or to extrafusal and intrafusal motor end plates). Degeneration proceeds back to the first Node of Ranvier, the beginning point of regeneration. In a best case scenario, regeneration will occur down the pathway provided by the neurilemma shell, which has remained after Wallerian degeneration to serve as a guide for regrowth to the end organ.

If both the afferent and efferent divisions of the motor nerve have been severed, function in the involved muscle will be completely lost until regeneration of the efferent nerve supply to the motor end plates on the extrafusal and/or intrafusal muscle fibers. If afferent nerve fibers are able to reinnervate the sensory nerve end organs of the involved muscle spindles, full neuromuscular function will return to the involved muscles and only toning or the rebuilding of muscle strength and coordination may be required.

Reinnervation (M18B1B)

If reinnervation of sensory nerve end organs on the muscle spindle fails to occur but reinnervation of the extrafusal and intrafusal motor end plates, or the extrafusal motor end plates by themselves, occurs, neuromuscular reeducation utilizing electromyometric feedback will be essential if voluntary neuromuscular control is to be restored to the involved muscle(s). If sensory input (proprioception) from the muscle is not available, the brain can not "feel" the muscle and without auxiliary (electromyometric) feedback will not know that the muscle(s) can be turned on or how to do it.

How much of a muscle or muscle group is affected by denervation depends upon how much of the peripheral nerve has been damaged (number and types of neurons affected). If all of the nerve fibers have not been damaged in a particular peripheral nerve and partial innervation of the muscle exists, weakened or impaired sensory and/or motor function may still remain. If only the sensory nerve fibers have been permanently damaged, motor function may remain potentially available, if the brain can be shown through electromyometric feedback how to use the muscle without feeling it.

It is interesting to note that if reinnervation of the muscle spindle efferents occurs without the proprioceptive elements, the effected musculature may become hypertonic. If the afferent and efferent innervation of the muscle spindle fails to occur, the muscle will tend to be hypotonic and appear flaccid even if voluntary extrafusal control has developed through electromyometric feedback neuromuscular reeducation.

If a whole muscle group (all of the facial nerves, for example) has been affected by denervation, and proprioceptive reinnervation is only partially complete (a few muscles being reinnervated) the whole group of muscles will tend to respond to demand as a group (sometimes to the extent of imitating the neuromuscular effects of collateral reinnervation). Electromyometric feedback may be essential in to a program of neuromuscular reeducation which may be able to teach the patient how to consciously and subliminally differentiate between the individual muscles so effected, during normal function.

Procedures of Application (M18B2)

Electromyometric instrumentation (M18B2A)

Instrumentation required for neuromuscular reeducation with electromyometric feedback of musculature affected by a post peripheral nerve injury includes one or two electromyometers with the most sophisticated feedback modes available, which can be preset to register different sensitivity levels. Common sensitivity settings include the 0 to 10, 0 to 50, 0 to 100, and 0 to 500 microvolt ranges. The basic electromyometer has both audio and visual (meter or light bar) feedback modes.

These modes can be enhanced by built-in feedback mode variables (different sounds and multiple visual displays) or by external computer enhancement of both modes. Some types of EMM instrumentation also offer tactile feedback modes and/or robotic-mechanical displays.

Electrode placement (M18B2B)

The muscles to be monitored should be connected to EMM instrumentation via electrode cables which terminate as they attach to surface electrodes affixed to the skin over the muscles which are thought to be effected by denervation.

Evaluation of the post peripheral nerve injury (M18B2C)

The patient should be placed in a comfortable sitting, semi-reclining or supine position which allows easy access to the muscles suspected of being effected by denervation. The muscles which are normal counterparts of the denervated ones should also be monitored so that the degree of impairment may be determined.

During the evaluation phase, the patient should remain unexposed to the feedback modes, unable to see, hear or feel changes in myoelectric production as reflected by EMM feedback.

Each individual suspect muscle should be monitored in two phases, at rest and during an attempted voluntary contraction, sustained for at least six seconds.

A muscle which is hypotonic as a result of peripheral nerve injury may be unable to produce any increase in myoelectric activity which is discernible above the resting level, or may produce an increase in myoelectric activity which is just a little less than that produced by a normal contralateral counterpart.

A hypertonic muscle is one which demonstrates a myoelectric level of activity at rest which is above that produced at rest by its normal counterpart. In the contraction phase the muscle may produce activity which matches or exceeds that produced by a normal counterpart. Contraction may develop slowly and subsequent decreases in myoelectric activity to resting levels may take longer then normal.

Neuromuscular reeducation utilizing EMM feedback (M18B2D)

Simply stated, the process of electromyometric feedback neuromuscular reeducation of neuromuscular function effected by peripheral nerve injury follows the pattern of facilitation, inhibition, balance and function.

Facilitation phase (M18B2D01) In the facilitation phase, the patient should be asked to increase myoelectric activity produced by the muscle or muscle pair, as reflected by the available feedback modes. The electromyometer should be set at a level of sensitivity which best demonstrates positive or increased myoelectric change. The patient must be able to see, hear and/or feel changes in myoelectric activity during the entire procedure to subliminally learn from them.

If a muscle is flaccid the unit will most commonly have to be set at the level of greatest sensitivity (0 to 6 or 0 to 10 microvolt ranges) so that any increase in myoelectric activity will be substantially demonstrated by all the feedback modes. If the muscle is more active the EMM should be set at a less sensitive level (0 to 30 or 0 to 50 microvolt ranges, for example) to provide a gauge of success as well as an opportunity to strive to produce more activity at observable levels.

The patient should be asked to increase the myoelectric activity generated by the muscle or synergistic muscle pair, as reflected by the available feedback, up to a parameter level of activity set by the practitioner. This can be mechanically set if a computer is being used, or as suggested arbitrary levels of activity measured by threshold responses, meter measurement marks or frequency rates. Parameters should be set or suggested that will adequately demonstrate failure or success. The parameter which is designated to denote a successful attempt should be potentially attainable, but enough out of reach that the patient must exert a concentrated effort to attain it. If electromyometric feedback is being used to increase the strength of muscle contraction or the general tone, the patient should be asked to hold the maximum level of activity for six seconds or more, with at least a six-second rest period between each attempt.

Inhibition phase (M18B2D02) In the inhibition phase, the muscle or synergistic muscle group which has been formerly facilitated and may have become hypertonic or comparatively hyperactive should be monitored by the electromyometer and the patient asked to increase activity from the muscle(s) by a small amount (less than 10 microvolts) and then to actively suppress the activity level below the the previous resting level. The attempt to increase myoelectric activity from the muscle(s) should last no longer than six seconds, but the suppression stage should last from six seconds to several minutes, until adequate inhibition is attained or until failure must be admitted and another attempt made. The electromyometer should be set at a sensitivity level which will adequately demonstrate patient inhibitory success (optimally at the 0 to 10 microvolt sensitivity level).

In the latter stages of neuromuscular reeducation of musculature affected by peripheral nerve injury, the muscle(s) selected to be inhibited is often in the same synergy as the muscle selected to be facilitated. Such a selection is made to provide the patient with the opportunity of differentiating between synergistic muscles.

Balance phase (M18B2D03) In the balance phase, two EMM setups are required to allow the patient the opportunity of attempting to facilitate the initially facilitated muscle or synergistic muscle pair while simultaneously inhibiting an antagonistic muscle or muscle group, or a member of the synergistic muscle group requiring differentiation. Such attempts should last for five or six seconds followed by a relaxation phase which lasts at least five or six seconds or until the inhibited muscle's myoelectric activity returns to resting levels.

The electromyometer monitoring the muscle(s) to be facilitated should be set at the least sensitive level that best magnifies and reflects increases in myoelectric activity without exceeding the maximum setting level (over 50 microvolts if the sensitivity level is set at 0 to 50 microvolts). The EMM monitoring the muscle(s) to be inhibited should be set at the most sensitive level that best magnifies and reflects decreases in myoelectric activity (shows the largest range of feedback, ideally at the 0 to 10 microvolt sensitivity level).

The choice of feedback modes relative to the muscle and neuromuscular problem is of paramount importance. When in the balance phase of neuromuscular reeducation, the feedback modes which are the most attention-getting should be used in conjunction with the electromyometer monitoring the hardest muscle to control, which is usually the muscle to be inhibited if spasticity is present.

Functional phase (M18B2D04) During the functional phase, the patient should be asked to perform a functional activity which requires coordination of the previously facilitated and inhibited muscles. This activity should be performed first while the muscles involved are monitored and the EMM activity fed back to the patient, and then without EMM instrumentation. The functional exercises used must usually must be improvised by the practitioner. These often require a great deal of thought since the exercise should employ the antagonistic muscles in the correctly balanced manner while performing a somewhat practical, useful maneuver.

Computer enhanced feedback (M18B2E)

Computer enhancement of electromyometric feedback has been shown to greatly add to patient appreciation and participation in the neuromuscular reeducation program by making feedback more graphic and easier to understand in terms of visual and auditory feedback modes for both patient and practitioner. Some of the graphic designs currently in use for single EMM activity enhancement have shown themselves to be remarkably suited for increasing patient motivation and enthusiastic participation. One of the more effective graphic designs currently available requires the patient to keep the EMM activity line showing on the computer's video screen between parallel lines which cross the screen in various patterns or shapes. After the activity line has transversed the screen, a numerical readout in one corner of the screen tells the patient what percentage of the time the activity line was above, within, or below the parallel line pattern. The time it takes to cross the screen can be preset by the operator.

During the balance phase, computer enhancement is especially effective if the computer instrumentation allows representative activity from both EMM's to appear simultaneously on the video monitor as distinctly different lines. Effectiveness increases if the patient can clearly relate the two activity lines to a graphic screen design which helps demonstrate their relative activity levels, as with, for example, two fixed parallel lines laid out across the screen, each related to one of the action lines.

Session termination (M18B2F)

At the end of the treatment session the electromyometer and associated feedback mode instrumentation should be turned off.

The surface electrodes should be carefully removed and the electrode paste carefully cleansed from the patient's skin before the patient is allowed to leave the treatment area. A surface electrode count should be conducted after removal from the skin to prevent them from leaving when the patient does.

Partial List of Conditions Treated (M18B3)

Treatable Causes (M18B3A)

Extrafusal muscle spasm   (C41)
Habitual joint positioning   (C48)
Hypermobile/instable joint   (C10)
Joint ankylosis   (C42)
Ligamentous strain   (C25)
Muscular weakness   (C23)
Nerve root compression   (C04)
Neuromuscular tonic imbalance   (C30)

Pathological neuromuscular hypertonus (C28)
Pathological neuromuscular hypotonus (C33)
Pathological neuromuscular dyscoordination (C35)
Peripheral nerve injury (C07)
Restricted joint range of motion (C26)
Spastic sphincter (C40)

Syndromes (M18B3B)

BELL'S PALSY (S66)
CEREBRAL PALSY (S25)
CERVICAL DORSAL OUTLET (SCALENICUS ATTICUS/CERVICAL RIB) (S09)
NEUROMUSCULAR PARALYSIS (S22)
POST IMMOBILIZATION (S36)
POST PERIPHERAL NERVE INJURY (S23)
POST SPINAL CORD INJURY (S24)
SCIATICA (S05)

Precautions and Contraindications (M18B4)

No precautions or contraindications have been listed or detailed for electromyometric feedback, except in connection with the stimulation provided by tactile and proprioceptive feedback modes. For precautions and contraindications associated with these feedback modes refer to the Electrical Stimulation *Precautions and Contraindications* [M0204], Transcutaneous Nerve Stimulation *Precautions and Contraindications* [M0404], and Vibration *Precautions and Contraindications* [M0804] sections.

Notes Aside (M18B5)

Differentiation between synergists (M18B5A)

One of the most difficult phases of a program of neuromuscular reeducation of muscles effected by peripheral nerve injury is the process of differentiation of one muscle in a synergistic muscle group from the other members. It is a problem which is created early in the facilitation phase when much time and effort is spent upon facilitating the synergists: the subliminal learning is so successful relative to turning the synergists on that selectively turning them off becomes very difficult. If differentiation between synergists is possible, a great deal more treatment time may be spent upon achieving it than is spent upon any other aspect of the neuromuscular reeducation process.

Synergists effected by collateral reinnervation (M18B5B)

Differentiation between synergists, even with EMM feedback instrumentation, may in fact be impossible if collateral reinnervation has occurred amongst synergists and they are forced to act together by virtue of sharing innervation.

Bibliography (M18B6)

Marinacci, pp. 57–71.
Swaan, pp. 9–11.
Taylor, pp. 33–44, 81–91.

ELECTROMYOMETRIC FEEDBACK POST SPINAL CORD INJURY NEUROMUSCULAR REEDUCATION
(M18C)

Definition (M18C1)

Spinal cord injury (M18C1A)

A spinal cord injury may result from the cord being crushed, cut, bruised, over-stretched, chemically corrupted, or its nervous pathways (columns) obstructed by abnormal formations or deposits. All such damage may result in the destruction of spinal cord-based axons and/or nerve cell bodies and/or the interruption of afferent and/or efferent spinal cord nervous pathways or columns, thereby interfering with communication between the supraspinal structures and the muscles. One of the most serious of spinal cord injuries, not discussed here, is the interruption and/or damage of the vascular system around a spinal cord lesion site. This type of injury precludes functional regeneration of the spinal cord nerves and promotes the dissolution of all supportive tissue in the area, causing an "isolation dystrophy" to occur; this points to a hypersensitivity of neural tissue to nutritional deprivation. Neurons cease to function normally within a few seconds following alteration in circulation.

It is has been historically accepted that the adult human is incapable of replacing neurons lost to, or of causing neuronal division within, the central nervous system. No hard evidence has been brought forward to date to refute this view.

Spinal cord regeneration (M18C1B)

Without the ability to replace damaged neurons or to cause mitosis of remaining neurons, the adult human must depend upon axonal nerve sprouting and collateral sprouting for reinnervation following injury to nervous tissues. Unlike the peripheral nerve, however, the spinal cord nerve which has been damaged and attempts to regenerate is not provided with Schwann cells (neurolemma shells) to provide a pathway for the regeneration, and without preexisting intact guiding pathways, undirected axonal regeneration may occur which is easily misdirected and prevented from reaching appropriate end organs. Inappropriate reinnervation of end organ structures which are in close proximity to the lesion site may also effectively defeat any additional axonal growth. It has been shown that while the presence of denervated tissue (by virtue of Wallerian degeneration) is a stimulus for spinal cord nerve regeneration, it also promotes collateral nerve sprouting from undamaged neurons. These collateral sprouts may reinnervate denervated end organs prior to the arrival of the appropriate regenerating axons and may deny these sites to them since the regenerated axons are not equipped to displace the collateral nerve sprouts (unlike their peripheral nerve counterparts).

Another deterrent to functional spinal cord nerve regeneration is the formation of scar tissue. Scar tissue can potentially obliterate access across the lesion site, stemming the advance of regenerating axons. This mechanical barrier is the most serious impediment to spinal cord nerve regeneration. Scar tissue may even be a threat to those who have had minimal immediate disruption of function from spinal cord injury, because of the tendency for scar tissue to develop over a long period of time, often years, at and around the site of lesion. As it develops, it gradually constricts the spinal cord, cutting off nerve pathways and stricturing the vascular supply.

Regardless of the stumbling blocks, evidence has emerged which supports the possibility of spinal cord recovery, at least in part. Regenerating axonal fibers have been shown to be capable of carrying across spinal cord lesion gaps and of even penetrating scar tissue formation. Missing from the literature reviewed were studies making it clear that regeneration was equally prevalent among ascending as well as descending neurons. Indeed, evidence suggests that what regeneration has been demonstrated to occur has been limited to the ventral areas of the spinal cord and that unmyelinated fibers might be responsible. Such reinnervation provides the patient with the potential for interneuronal communication, providing a channel for positive feedback required to produce the efferent element of the tonic stretch reflex without providing for negative feedback from the muscles affected. This would (and apparently does) have the effect of producing intense hypertonicity (rigidity and/or spasticity) of the involved musculature.

EMM reeducation of musculature effected by spinal injury (M18C1C)

The problem of neuromuscular reeducation of the post spinal cord injured patient is similar to that faced by the post peripheral nerve injury patient: the patient's supraspinal centers may be faced with having the possibility of sending messages to the muscles, but may be unable to receive messages back. Electromyometric feedback has proven to be useful, in some cases, for reducing the spasticity (uninhibited phasic stretch reflex) and rigidity (uninhibited tonic stretch reflex), by providing the patient's supraspinal structures with a measure of myoelectric activity from the muscles. The patient may be able to learn to use such feedback to effectively stem or reduce facilitatory efferent communication to the musculature. If the patient is able to benefit from this kind of training and can learn the trick of selectively turning the facilitatory efferent activity to some of the affected muscles off while allowing or increasing facilitatory efferent communication to other effected muscles, thereby increasing tone and causing them to contract, coordinated motion may be the result.

Procedures of Application (M18C2)

Electromyometric instrumentation (M18C2A)

Instrumentation required for neuromuscular reeducation with electromyometric feedback of musculature affected by a post spinal cord injury includes one or two electromyometers with the most sophisticated feedback modes available, which can be preset to register different sensitivity levels. Common sensitivity settings include the 0 to 10, 0 to 50, 0 to 100, and 0 to 500 microvolt ranges. The basic electromyometer has both audio and visual (meter or light bar) feedback modes. These modes can be enhanced by built-in feedback mode variables (different sounds and multiple visual displays) or by external computer enhancement of both modes. Some types of EMM instrumentation also offer tactile feedback modes and/or robotic-mechanical displays.

Electrode placement (M18C2B)

The muscles to be monitored should be connected to EMM instrumentation via electrode cables which terminate as they attach to surface electrodes affixed to the skin over the muscles which are thought to be effected by denervation.

Evaluation of the post spinal cord injury (M18C2C)

The patient should be placed in a comfortable sitting, semi-reclining or supine position which allows easy access to the muscles suspected of being effected by the post cerebral vascular accident.

During the evaluation phase, the patient should remain unexposed to the feedback modes, unable to see, hear or feel changes in myoelectric production as reflected by EMM feedback.

Each suspect muscle should be individually monitored in three phases: (1) at rest, (2) during an attempted voluntary sustained contraction of at least six seconds, and (3) during a subsequent relaxation of effort.

A muscle which is hypotonic may be unable to produce any increase in myoelectric activity which is discernible above the resting level, or may produce an increase in myoelectric activity to the order of 1 or 2 microvolts above resting levels.

A hypertonic muscle will demonstrate a fairly high level of resting myoelectric activity, generally in excess of six microvolts, and may display on the meter or video screen the rhythmic deflection pattern associated with spasticity. Increases in myoelectric activity levels produced by hypertonic muscles tend to be lower then normal. Generally the patient may only be able to generate between 1 and 150 microvolts above resting levels, generating such activity slowly, taking several seconds to achieve maximum levels. A return to resting levels may likewise be very slow, sometimes taking minutes.

Neuromuscular reeducation utilizing EMM feedback (M18C2D)

Simply stated, the process of electromyometric feedback neuromuscular reeducation of neuromuscular function effected by spinal cord injury follows the pattern of facilitation, inhibition, balance and function.

Facilitation phase (M18C2D01) If spasticity and/or rigidity is present in the muscles monitored, the facilitation phase may be as important as the obviously important inhibition phase. If this is the case, the patient should be asked to attempt to increase the myoelectric activity produced by the spastic muscle for five or six seconds and then to relax (stop trying). In some cases the patient will be able to increase the myoelectric activity above resting levels, anywhere from 1 to 150 microvolts, but in other cases the the activity level will not rise sufficiently high enough to be differentiated from resting level activity; however, when the patient gives up the attempt to facilitate myoelectric activity (to relax), a drop in myoelectric activity below previous resting levels may be discernible. Having recognized this change in feedback levels, the patient may have been provided with enough of a gauge of inhibition to begin controlling the muscle to the extent of being able to contract it voluntarily. If immediate ability to voluntarily contract the muscle is not forthcoming, this exercise at least gives the patient an introduction to how the supraspinal structures may begin to relate on a subliminal level to muscles who fail to provide proprioceptive input. The patient may be able to come back to the facilitation phase when the inhibition phase has provided adequate suppression of the spastic activity to make small increases in myoelectric activity discernible.

If the musculature is initially hypotonic, the patient should, in the facilitation phase, be asked to increase myoelectric activity produced by the muscle or muscle group, as reflected by the available feedback modes. The electromyometer should be set at a level of sensitivity which best demonstrates positive or increased myoelectric change. The patient must be able to see, hear and/or feel changes in myoelectric activity during the entire procedure to subliminally learn from them.

If a muscle is hypotonic the electromyometer will most commonly have to be set at the level of greatest sensitivity (0 to 6 or 0 to 10 microvolt ranges) so that any increase in myoelectric activity will be substantially demonstrated by all the feedback modes. If the muscle is more active the unit should be set at a less sensitive level (0 to 30 or 0 to 50 microvolt ranges, for example) to provide a gauge of success as well as providing the patient with an opportunity to strive to produce more activity at observable levels.

The patient should be asked to increase the myoelectric activity generated by the muscle or synergistic muscle pair, as reflected by the available feedback, up to a parameter level of activity set by the practitioner. This can be mechanically set if a computer is being used, or as suggested arbitrary levels of activity measured by threshold responses, meter measurement marks or frequency rates. Parameters should be set or suggested that will adequately demonstrate failure or success. The parameter which is designated to denote a successful attempt should be potentially attainable, but enough out of reach that the patient must exert a concentrated effort to attain it. If electromyometric feedback is being used to increase the strength of muscle contraction or the general tone, the patient should be asked to hold the maximum level of activity for six seconds or more, with at least a six-second rest period between each attempt.

Inhibition phase (M18C2D02) In the inhibition phase, the muscle or synergistic muscle group which has been formerly facilitated and may have become hypertonic or comparatively hyperactive should be monitored by the electromyometer and the patient asked to increase activity from the muscle(s) by a small amount (less than 10 microvolts) and then to actively suppress the activity level below the the previous resting level. The attempt to increase myoelectric activity from the muscle(s) should last no longer than six seconds, but the suppression stage should last from six seconds to several minutes, until adequate inhibition is attained or until failure must be admitted and another attempt made. The electromyometer should be set at a sensitivity level which will adequately demonstrate patient inhibitory success (optimally at the 0 to 10 microvolt sensitivity level).

In the latter stages of neuromuscular reeducation of musculature affected by spinal cord injury, the muscle(s) selected to be inhibited is often in the same synergy as the muscle selected to be facilitated. Such a selection is made to provide the patient with the opportunity of differentiating between synergistic muscles.

Balance phase (M18C2D03) In the balance phase, two EMM setups are required to allow the patient the opportunity of attempting to facilitate the initially facilitated muscle or synergistic muscle pair while simultaneously inhibiting an antagonistic muscle or muscle group, or a member of the synergistic muscle group requiring differentiation. Such attempts should last for five or six seconds followed by a relaxation phase which lasts at least five or six seconds or until the inhibited muscle's myoelectric activity returns to resting levels.

The electromyometer monitoring the muscle(s) to be facilitated should be set at the least sensitive level that best magnifies and reflects increases in myoelectric activity without exceeding the maximum setting level (over 50 microvolts if the sensitivity level is set at 0 to 50 microvolts). The EMM monitoring the muscle(s) to be inhibited should be set at the most sensitive level that best magnifies and reflects decreases in myoelectric activity (shows the largest range of feedback, ideally at the 0 to 10 microvolt sensitivity level).

The choice of feedback modes relative to the muscle and neuromuscular problem is of paramount importance. When in the balance phase of neuromuscular reeducation, the feedback modes which are the most attention-getting should be used in conjunction with the electromyometer monitoring the hardest muscle to control, which is usually the muscle to be inhibited if spasticity is present.

Functional phase (M18C2D04) During the functional phase, the patient should be asked to perform a functional activity which requires coordination of the previously facilitated and inhibited muscles. This activity should be performed first while the muscles involved are monitored and the EMM activity fed back to the patient, and then without EMM instrumentation. The functional exercises used must usually must be improvised by the practitioner. These often require a great deal of thought since the exercise should employ the antagonistic muscles in the correctly balanced manner while performing a somewhat practical, useful maneuver.

Computer enhanced feedback (M18C2E)

Computer enhancement of electromyometric feedback has been shown to greatly add to patient appreciation and participation in the neuromuscular reeducation program by making feedback more graphic and easier to understand in terms of visual and auditory feedback modes for both patient and practitioner. Some of the graphic designs currently in use for single EMM activity enhancement have shown themselves to be remarkably suited for increasing patient motivation and enthusiastic participation. One of the more effective graphic designs currently available requires the patient to keep the EMM activity line showing on the computer's video screen between parallel lines which cross the screen in various patterns or shapes. After the activity line has transversed the screen, a numerical readout in one corner of the screen tells the patient what percentage of the time the activity line was above, within, or below the parallel line pattern. The time it takes to cross the screen can be preset by the operator.

During the balance phase, computer enhancement is especially effective if the computer instrumentation allows representative activity from both EMM's to appear simultaneously on the video monitor as distinctly different lines. Effectiveness increases if the patient can clearly relate the two activity lines to a graphic screen design which helps demonstrate their relative activity levels, as with, for example, two fixed parallel lines laid out across the screen, each related to one of the action lines.

Session termination (M18C2F)

At the end of the treatment session the electromyometer and associated feedback mode instrumentation should be turned off.

The surface electrodes should be carefully removed and the electrode paste carefully cleansed from the patient's skin before the patient is allowed to leave the treatment area. A surface electrode count should be conducted after removal from the skin to prevent them from leaving when the patient does.

Partial List of Conditions Treated (M18C3)

Treatable Causes (M18C3A)

Extrafusal muscle spasm (C41)
Habitual joint positioning (C48)
Hypermobile/instable joint (C10)
Hypotension (C45)
Ligamentous strain (C25)
Muscular weakness (C23)
Neuromuscular tonic imbalance (C30)
Pathological neuromuscular hypertonus (C28)
Pathological neuromuscular hypotonus (C33)
Pathological neuromuscular dyscoordination (C35)
Peripheral nerve injury (C07)
Restricted joint range of motion (C26)
Spastic sphincter (C40)

Syndromes (M18C3B)

CEREBRAL PALSY (S25)
NEUROMUSCULAR PARALYSIS (S22)
POST IMMOBILIZATION (S36)
POST PERIPHERAL NERVE INJURY (S23)
POST SPINAL CORD INJURY (S24)

Precautions and Contraindications (M18C4)

No precautions or contraindications have been listed or detailed for electromyometric feedback, except in connection with the stimulation provided by tactile and proprioceptive feedback modes. For precautions and contraindications associated with these feedback modes refer to the Electrical Stimulation *Precautions and Contraindications* [M0204], Transcutaneous Nerve Stimulation *Precautions and Contraindications* [M0404], and Vibration *Precautions and Contraindications* [M0804] sections.

Notes Aside (M18C5)

Cauda equina injury (M18C5A)

Flaccidity has been said to be encountered in the post spinal cord injured patient; however, it is now thought that such cases are generally the result of intraspinal column peripheral nerve (cauda equina) injury and not the result of injury to true spinal cord tissue at all. If reinnervation is not inhibited by scar tissue formation or chemical retardation (valium therapy is noted to inhibit reinnervation of nervous end organs), voluntary muscle return may be expected to occur spontaneously, at least to some degree, two or three years following injury.

Time limits (M18C5B)

Of all the patients evaluated suffering from a post spinal cord injury suffered less than five years previously, all were unable to voluntarily decrease myoelectric associated with spasticity using EMM feedback as a guide. All were unable to voluntarily increase myoelectric activity from involved musculature, except for individuals who had experienced a bruising of the spinal cord (even without severance) or had an injury which was questionably or definitely a cauda equina injury less than three years old. Patients who are more than three to five years post spinal cord injury (disregarding spinal cord bruise and caudal equina injury) may not be able to benefit from neuromuscular reeducation utilizing EMM feedback if the involved muscles have atrophied to the point that muscle fibers have been replaced by connective tissue.

Paraplegic ambulation (M18C5C)

If a paraplegic patient has hip and low back back musculature sufficiently strong enough to permit hyperextension of the back, and enough quadratus lumborum strength to the voluntarily elevate the pelvis while standing, the patient may be a candidate for ambulation with short leg braces. The short leg braces which may make this possible are constructed of parallel double upright bars and adjustable fixed ankle joints attached to a boot or shoe. The lower shank of the brace should be long enough to provide a good lever arm, terminating just behind the ball of the foot. Walking may be accomplished through the progressive reliance upon pickup walkers, forearm crutches (Canadian canes), and regular canes.

Bibliography (M18C6)

Brudny, pp. 925–932.
Davidoff, pp. 9–11.
Guth, pp. 1–43.
Marinacci, pp. 57–71.
Swaan, pp. 9–11.
Taylor, pp. 94–108.

ELECTROMYOMETRIC FEEDBACK NEUROMUSCULAR REEDUCATION OF CEREBRAL PALSY (M18D)

Definition (M18D1)

Cerebral palsy (CP), as a condition, was first called infantile paralysis (probably a more definitive name). It is a term which actually denotes any paralysis or lack of neuromuscular control which has resulted from injury to an immature nervous system usually occurring before, during or shortly after birth. Symptoms of the cerebral palsy condition may include the symptoms of seizures, mental retardation, abnormal sensory perceptions, impaired cognition, impaired sight, hearing and/or speech. Essentially, cerebral palsy is a blanket term for a great number of conditions of varying causes, severity, and etiology. In general, CP population may be divided up between those suffering from supraspinal structural damage (caused by disease, cerebral vascular accident, or external trauma of the cerebrum), peripheral nerve injury, and spinal cord injury.

The CP group suffering from supraspinal structural damage may have general characteristics similar to those most commonly associated with adult counterparts such as *spasticity, flaccidity, rigidity,* and/or *ataxia* (problems with balance and/or depth perception), but may additionally be subjected to various forms of athetosis and other dyscoordination movement patterns. Such stereotypic movement patterns have long been thought to be characteristic of the CP condition, but in fact only a relatively small part of the CP population suffer from this symptom.

Those CP patients who have been the victims of spinal cord injury (including impaired development) or peripheral nerve injury will suffer from much the same symptomatology as that associated with their adult counterparts.

Electromyometric feedback neuromuscular reeducation of the individual CP patient should follow the process suggested for the particular post nervous system insult which has created the syndrome (post CVA, post peripheral nerve injury, or spinal cord injury). If athetosis is an element in the patientr's symptomatology, especially if it is present in all four extremities, the neuromuscular reeducation program may be more complex, involving modalities affecting proprioception through tactile and motor sensory input. The rules for neuromuscular reeducation are basically the same for the patient suffering CP condition as it would be for an adult counterpart condition.

Procedures of Application (M18D2)

For recommendations as which procedures of application may profitably be employed in a program of neuromuscular reeducation utilizing electromyometry, refer to the treatment technique appropriate for the treatment of the individual syndrome. If the condition stems from supraspinal structure damage, refer to the treatment technique section Electromyometric Feedback, Post Cerebral Vascular Accident Neuromuscular Reeducation [M18A]. If it stems from a peripheral nerve injury refer to Electromyometric Feedback, Post Peripheral Nerve Injury Neuromuscular Reeducation [M18B]. If the CP condition stems from a spinal cord injury, refer to the Electromyometric Feedback, Post Spinal Cord Injury Neuromuscular Reeducation [M18C] section.

Partial List of Conditions Treated (M18D3)

Treatable Causes (M18D3A)

Extrafusal muscle spasm (C41)
Habitual joint positioning (C48)
Hypermobile/instable joint (C10)
Joint ankylosis (C42)
Ligamentous strain (C25)
Muscular weakness (C23)
Neuromuscular tonic imbalance (C30)
Pathological neuromuscular hypertonus (C28)
Pathological neuromuscular hypotonus (C33)
Pathological neuromuscular
   dyscoordination (C35)
Peripheral nerve injury (C07)
Psychoneurogenic neuromuscular
   hypertonus (C29)
Restricted joint range of motion (C26)
Spastic sphincter (C40)

Syndromes (M18D3B)

BELL'S PALSY (S66)
CEREBRAL PALSY (S25)
CERVICAL DORSAL OUTLET (SCALENICUS
   ATTICUS/CERVICAL RIB) (S09)
NEUROMUSCULAR PARALYSIS (S22)

POST CEREBRAL VASCULAR ACCIDENT
  (CVA) (S07)
POST IMMOBILIZATION (S36)
POST PERIPHERAL NERVE INJURY (S23)
POST SPINAL CORD INJURY (S24)
SHOULDER/HAND (S57)

Precautions and Contraindications (M18D4)

No precautions or contraindications have been listed or detailed for electromyometric feedback, except in connection with the stimulation provided by tactile and proprioceptive feedback modes. For precautions and contraindications associated with these feedback modes refer to the Electrical Stimulation, *Precautions and Contraindications* [M0204], Transcutaneous Nerve Stimulation, *Precautions and Contraindications* [M0404], and/or Vibration, *Precautions and Contraindications* [M0804] sections.

Effect of puberty upon CP rehabilitation (M18D4A)

The remarkable benefits which may be bestowed by a program of neuromuscular reeducation of neuromuscular reeducation utilizing EMM feedback upon the CP patient have proven to be limited, by in large, to the period preceding the onset of puberty. After puberty the central nervous system seems to lose most of its prepuberty plastic adaptability. Changes in the patient's neuromuscular control get relatively harder to affect as the age of puberty passes and the child matures toward adulthood.

Athetosis (M18D4B)

Athetosis is exaggerated or heightened by emotional response, good or bad. A calm work-like environment is essential in the preliminary stages of electromyometric feedback neuromuscular reeducation of the athetoid CP patient since emotionally evoked somatic over-response may interfere with the step-by-step process which seems necessary for an organized program of neuromuscular reeducation.

Seizure activity (M18D4C)

The practitioner should be prepared to handle any cerebral motor seizures which may inadvertently occur as a result of motor cortex damage. Seizure occurrence during electromyometric feedback training is exceedingly rare; while none have been reported to date, they remain possible whenever a brain-injured patient is being treated.

If a seizure should occur, the patient should be placed in a side-lying position and an object placed between the teeth to prohibit the patient from biting the tongue. The object placed between the teeth should not be hard enough to cause tooth damage when bitten (no metal, porcelain, or glass) nor small enough to swallow; a wooden tongue depressor, rubber bite bar, thick wooden pencil, or a relatively soft plastic pen may serve. If the patient is aware of seizure development and reports it, ice packing of the posterior base of the head, the neck and paraspinous C7 vertebral areas bilaterally has been shown to be effective in preventing the onset of the some types of seizures. Following the seizure, the patient may be disoriented and should be prevented from leaving the protective environment without adequate careful guidance by clinical personnel or informed family members.

Notes Aside (M18D5)

Various treatment techniques may be useful in helping reduce spasticity or athetoid dyscoodinated motion. These techniques include cryotherapy (refer to Cryotherapy, Hypertonic Neuromuscular Management [M10B]), vibration (refer to Vibration, Spasticity Reduction [M18B]), and brushing (refer to Sensory Stimulation, Neuromuscular Spasticity Control [M17A]).

The practice of operant conditioning has proven effective when dealing with the immature CP patient; i.e., rewarding correct behavior and not rewarding incorrect behavior. Operant conditioning is especially valuable for the CP patient who is skilled at interpersonal manipulation; many of the "games" that the patient tries to set up to foster conflict and demonstrate personal control may be avoided by setting up the relational roles of judge and performer early in the program.

Bibliography (M18D6)

Asato, pp. 1447–1451.
Brudny, pp. 925–932.
Doman, pp. 257–262.
Marinacci, pp. 57–71.
Swaan, pp. 9–11.

ELECTROMYOMETRIC FEEDBACK NEUROMUSCULAR REEDUCATION OF NECK AND SHOULDER MUSCULATURE (M18E)

Definition (M18E1)

Neck and shoulder muscular hypertonus (M18E1A)

It has been clinically demonstrated that if the musculature supporting the structures of shoulders and neck become relatively hypertonic (producing more myoelectric activity then normal) pain will result. The causes of the pain produced may include unequal pressure upon the involved vertebral discs, intrafusal muscle spasm (trigger point formations) accompanying the hypertonicity, extrafusal muscle spasm existing as an exaggeration of the hypertonicity, strain placed upon ligaments, excessive muscular pressure upon involved sensory nerve roots, or any combination of these causative agents. Referred pain patterns may be produced from sensory nerve root pressure or from trigger point formation which may produce headaches, pain in the upper anterior and/or posterior upper trunk, pain down the arms, and/or in the hands. This may be accompanied by some degree of numbness of either all or part of the involved upper extremity.

Acute neuromuscular hypertonus of the neck and/or shoulder musculature generally occurs in response to muscle strain and/or other trauma to the muscles or associated structures (ligaments, facets, etc.). Such trauma may trigger an automatic protective mechanism which expresses itself as an unrelieved phasic stretch reflex (prolonged extrafusal spasm), or as a balanced or imbalanced splinting of the musculature (general muscle tetanus of the involved muscles to prevent further injury). Such acute responses may be temporarily or permanently relieved by modes of treatment which usually do not include electromyometry (refer to Auricular Electroacustimulation, Pain Relief [M06A], Cryotherapy, Hypertonic Neuromuscular Management [M10B], Electrical Facilitation of Endorphin Production (Acusleep) [M03A], Electrical Stimulation, Muscle Lengthening [M02A], Vertebral Traction, Cervical Traction [M09A], and/or Vibration, Neuromuscular management [M08A]).

Chronic hypertonus of the neck and shoulder musculature generally either occurs as supraspinal structures fail to eventually correct neuromuscular hypertonus created during the acute phase of injury or is provoked by the supraspinal structures themselves. Neuromuscular tonicity of the neck and shoulder musculature is generally controlled through supraspinal structure influence over the muscle spindle, or tonic stretch reflex (refer to Neuromuscular System [R002] and/or Muscle Spindle [R001]).

If chronic neuromuscular hypertonicity occurs in the neck and shoulder musculature, it is generally as an unconscious, subliminal, supraspinal response to a psychological need for a physical defense from an emotional stress. If this occurs, supraspinal structures unconsciously induce the muscle spindle system to selectively maintain hypertonicity of the neck and shoulder musculature to create pain (via the causes listed above) or a sensation of extreme tension.

Many who suffer from psychoneurogenic neck and/or shoulder musculature hypertonus, as an expression of somatized stress, report the onset of the condition to follow an emotionally traumatizing conflict occurring before, during or shortly following a physically traumatic injury of the neck and/or shoulder musculature or related structures. It has been postulated that the supraspinal structures learn on a subliminal level that it is easier to deal with physical pain than with emotional pain. Evidently, because of the mechanisms involved in the system attempts to compensate for or correct the damage done by the injury, the supraspinal structures learn how to self-induce pain by producing selective muscular hypertonus. Clinical research has verified that the gamma motor system (which is responsible for maintaining muscle tone) is so attuned to supraspinal structure influence that general muscle tone is directly affected by mood changes, and clinical observation seems to confirm that the degree of neuromuscular response is "learned" in association with particular environmental circumstances or stimuli, much as any other fine motor skill is "programmed". When emotional stress is somatized, it is simply the natural neuromuscular response to emotional tension, exaggerated and specified to a given muscle or muscle group.

Electromyometric feedback has proven to be effective in the treatment of hypertonus of the shoulder and neck musculature which has resulted from the somatization of emotional stress, good or bad.

Neck and shoulder muscles (M18E1B)

The muscles most commonly involved in psychoneurogenic shoulder and neck hypertonus are the upper trapezius, sternocleidomastoideus, and scaleni muscles. The treatment of choice for shoulder and neck neuromuscular hypertonus involving these muscles is electromyometric feedback. The process of neuromuscular reeducation utilizing electromyometric feedback is the same as that used for neuromuscular reeducation of pathological neuromuscular dysfunction (post CVA, post peripheral nerve injury and/or post spinal cord injury), and follows the basic learning phase pattern of facilitation, inhibition, balance and function.

Procedures of Application (M18E2)

EMM instrumentation (M18E2A)

Instrumentation required for neuromuscular reeducation with electromyometric feedback of psychoneurogenic hypertonic shoulder and neck musculature includes one or two electromyometers with the most sophisticated feedback modes available, which can be preset to register different sensitivity levels. Common sensitivity settings include the 0 to 10, 0 to 50, 0 to 100, and 0 to 500 microvolt ranges. The basic electromyometer has both audio and visual (meter or light bar) feedback modes. These modes can be enhanced by built-in feedback mode variables (different sounds and multiple visual displays) or by external computer enhancement of both modes. Some types of EMM instrumentation also offer tactile feedback modes and/or robotic-mechanical displays.

Electrode placement (M18E2B)

The patient should be sitting in an upright position which will be comfortable for the duration of the session.

During the evaluation phase, the patient should remain unexposed to the feedback modes, unable to see, hear or feel changes in myoelectric production as reflected by EMM feedback.

The muscles to be monitored should be connected to the electromyometer via electrode cables which terminate as they connect to surface electrodes affixed to the skin over the muscles to be monitored. These muscles usually include the upper trapezius, scaleni, and diaphragm muscles, unilaterally or bilaterally monitored. Electrode placement on the skin over the upper trapezius muscle(s) should be above the site of the upper trapezius trigger point formation. For the scaleni muscle group monitory electrodes should be placed on the skin above the scaleni muscle at the approximate midpoint of an imaginary line drawn along the base of the neck from the origin of the sternocleidomastoideus muscle to the anterior edge of the upper trapezius muscle. If the diaphragm is to be monitored, the two active electrodes should be symmetrically placed on either side of the anterior body midline, approximately three inches apart, just distal to edge of the last rib structure.

If unilateral monitoring is utilized, the muscle on one side is monitored exclusively with both active electrodes. If bilateral monitry is used, contralateral opposing muscles (both upper trapezius or scaleni groupings) are simultaneously monitored by sharing an electrode cable, from which one of the two active pickup leads is attached to an electrode placed over each muscle or muscle group.

Evaluation (M18E2C)

Each of the muscles or muscle groups in question should be attached to an electromyometer and monitored during attempted muscular relaxation, breathing and talking. Hypertonus will be marked by relatively high myoelectric production from the muscle monitored, usually in excess of 9 to 10 microvolts at rest (sometimes as high as 30 to 50 microvolts at rest or during breathing inspiration) and between 30 and 100 microvolts during speech. Relative hypotonus of one of the muscles or muscle groups of interest here is marked by myoelectric activity below 3 microvolts at rest (sometimes seen from the diaphragm during inspiration) with little change during speech. The normal myoelectric activity range for these muscles or muscle groups at rest is generally considered to be between 6 to 8 microvolts and during speech, below 12 microvolts.

Subliminal learning pattern (M18E2D)

The neuromuscular reeducation electromyometric feedback learning pattern of facilitation, inhibition, balance and function is utilized.

Facilitation phase (M18E2D01) Typically, in the facilitation phase the diaphragm muscle is monitored. The choice of this muscle as an antagonist of upper trapezius and/or scaleni muscle activity was determined through trial and error empirical experimentation. The patient is asked to voluntarily increase myoelectric activity, as reflected by the available feedback modes, from the diaphragm muscle.

During the facilitation stage the electromyometers should be set at a sensitivity level which best demonstrates positive or increasing myoelectric change. The patient must be able to see, hear and/or feel changes in myoelectric activity during the entire procedure to subliminally learn from them. If the diaphragm is severely under active, the unit will most commonly have to be set initially at the level of greatest sensitivity (0 to 6 or 0 to 10 microvolt ranges) so that any increase in myoelectric activity will be substantially demonstrated by all the feedback modes. If the muscle is more active, the unit should be set at a less sensitive level (0 to 30 or 0 to 50 microvolt ranges, for example) to provide a gauge of success as well as providing the patient with an opportunity to strive to produce more activity at observable levels.

Inhibition phase (M18E2D02) In the inhibition phase, either the upper trapezius or the scaleni muscle(s) is monitored. The least hypertonic of the two is usually the first to be monitored at this stage. The patient should be asked to voluntarily facilitate the myoelectric activity from the monitored muscle, reflected as electromyometric feedback, and then to voluntarily inhibit the myoelectric activity below the previous resting level.

The patient should be asked to increase activity from the muscle(s) by a small amount (less than 10 microvolts) and then to actively suppress the activity level below the the previous resting level. The attempt to increase myoelectric activity from the muscle(s) should last no longer than six seconds, but the inhibition stage should last from six seconds to several minutes, until adequate inhibition is attained or until failure must be admitted and another attempt made. The electromyometer should be set at a sensitivity level which will adequately demonstrate patient inhibitory success (optimally at the 0 to 10 microvolt sensitivity level, but less sensitive levels may initially be used if resting levels exceed the 0 to 10 microvolt range).

Balance phase (M18E2D03) In the balance phase, two EMM setups are required to allow the patient the opportunity of attempting to facilitate the diaphragm while inhibiting either the upper trapezius or scaleni muscle(s). Once the ability to inhibit the hypertonic muscle (usually between 6 to 8 microvolts), while simultaneously facilitating activity from the diaphragm (above 10 microvolts), has been demonstrated, the muscle found initially to be the most hypertonic (either the upper trapezius or scaleni muscles(s)) is then monitored and the learning process repeated to this stage.

During the balance phase the electromyometer monitoring the muscle(s) to be facilitated should be set at the least sensitive level that best magnifies and reflects increases in myoelectric activity, without exceeding the maximum setting level (over 50 microvolts if the sensitivity level is set at 0 to 50 microvolts). The electromyometer monitoring the muscle(s) to be inhibited should be set at the most sensitive level that best magnifies and reflects decreases in myoelectric activity (shows the largest range of feedback, ideally at the 0 to 10 microvolt sensitivity level).

The choice of feedback modes relative to the muscle and neuromuscular problem is of paramount importance. When in the balance phase of neuromuscular reeducation, the feedback modes which are most attention-getting should be used in conjunction with the electromyometer monitoring the hardest muscle to control, which is usually the hypertonic muscle.

Functional phase (M18E2D04) In the functional stage only the most hypertonic of the muscles is monitored. The patient is asked to talk (an emotion-packed exercise) while keeping the muscle's myoelectric activity, as reflected by electromyometric feedback, below an arbitrarily determined level. This level should be lowered as the patient's control increases. During the talking phase the patient should (1) slowly count out loud (whispering is initially permissible) from one to ten (the patient should be cautioned not to say more then a few numbers on a single breath because as the air runs out the shoulder and neck muscular tension increase), (2) say out loud the names of people involved in the patient's life, and (3) talk conversationally with someone in the same room. Each step should be attempted while striving to keep the feedback levels below the designated level. These are successive steps and the patient should be directed to proceed from one step to the next only as control has been demonstrated.

Computer enhancement of electromyometry (M18E2E)

Computer enhancement of electromyometric feedback has been shown to greatly add to patient appreciation and participation in the neuromuscular reeducation program by making feedback more graphic and easier to understand (for the patient as well as the practitioner) in terms of visual and auditory feedback modes. Some of the graphic designs currently in use for single electromyometer activity enhancement have shown themselves to be remarkably suited for increasing patient motivation and enthusiastic participation. One of the more effective graphic designs currently available requires the patient to keep the EMM activity line showing on the computer's video screen between parallel lines which cross the screen in various patterns or shapes. A multiple tooth spike shape is used for diaphragm facilitation to avoid hyperventilation. After the activity line has transversed the screen, a numerical readout in one corner of the screen tells the patient what percentage of the time the activity line was above, within, or below the parallel line pattern. The time it takes to cross the screen can be preset by the operator.

During the balance phase, computer enhancement is especially effective if the computer instrumentation allows representative activity from both electromyometers to appear on the video monitor simultaneously as distinctly different lines. Effectiveness increases if the patient can clearly relate the two to a graphic screen design which helps demonstrate their relative activity levels. Two fixed parallel lines across the screen placed relative to the action lines may be adequate.

During the functional phase, if computer enhancement is used, a single activity line from the monitored hypertonic muscle is compared to a preset line which represents the arbitrary level of control. If the activity level from the electromyometer exceeds or goes above the line during one of the talking exercises, the patient is said to be out of control and should be instructed to discontinue talking until the activity level is once again below the control line.

Session termination (M18E2F)

At the end of the session the electrodes should be carefully removed from the patient's skin and the electrode paste cleaned off. The patient should be instructed to practice the "diaphragm breathing" procedure, on their own, for a few minutes each day (as described above) especially if the patient becomes aware of tension developing in shoulder and/or neck musculature. Such practice should continue until shoulder, neck, and/or headache has disappeared.

Partial List of Conditions Treated (M18E3)

Treatable Causes (M18E3A)

Extrafusal muscle spasm   (C41)
Habitual joint positioning   (C48)
Interspinous ligamentous strain   (C03)
Psychoneurogenic neuromuscular hypertonus   (C29)
Restricted joint range of motion   (C26)
Sympathetic hyperresponse   (C38)
Trigger point formation   (C01)

Syndromes (M18E3B)

CERVICAL DORSAL OUTLET (SCALENICUS ATTICUS/CERVICAL RIB)   (S09)
CERVICAL (NECK) PAIN   (S73)
CHEST PAIN   (S14)
EARACHE   (S40)
HEADACHE PAIN   (S02)
HIGH THORACIC BACK PAIN   (S48)
HYPERTENSION   (S43)
HYSTERIA/ANXIETY REACTION   (S59)
MIGRAINE (VASCULAR) HEADACHE   (S18)
OSTEOARTHRITIS   (S21)
PHOBIC REACTION   (S44)
POST IMMOBILIZATION   (S36)
POST WHIPLASH   (S55)
RADICULITIS   (S29)
REFERRED PAIN   (S01)
SHOULDER/HAND   (S57)
SHOULDER PAIN   (S08)
SINUS PAIN   (S33)
TEMPOROMANDIBULAR JOINT (TMJ) PAIN   (S06)
TINNITUS   (S70)

Precautions and Contraindications (M18E4)

No precautions or contraindications have been listed or detailed for electromyometric feedback, except in connection with the stimulation provided by tactile and proprioceptive feedback modes. For precautions and contraindications associated with these feedback modes refer to the Electrical Stimulation *Precautions and Contraindications* [M0204], Transcutaneous Nerve Stimulation *Precautions and Contraindications* [M0404], and Vibration *Precautions and Contraindications* [M0804] sections.

Notes Aside (M18E5)

EMM treatment of tension headache (M18E5A)

This technique is the primary source of permanent relief for the tension headache. Successful, long-term relief of the tension headache syndrome, stemming from hypertonus of the shoulder and/or cervical musculature, usually occurs within three to five sessions, if coupled with an exercise program which tones offending musculature and if the patient is able and willing (even on a subliminal level) to be deprived of this popular symptom of a defense mechanism. If the patient really needs the tension headache syndrome, only temporary relief may be possible, with the syndrome reoccurring when the patient most needs it.

EMM desensitization (M18E5B)

Additionally, this technique may be used as a desensitization process via the functional stage to help reduce psychoneurogenic induced hypertension or anxiety reactions. This is especially important if the patient is unable to utilize galvanic skin response (GSR) monitry.

Bibliography (M18E6)

Pritchard, pp. 165–175.
Swaan, pp. 9–11.

ELECTROMYOMETRIC FEEDBACK NEUROMUSCULAR REEDUCATION OF TEMPOROMANDIBULAR MUSCULATURE (M18F)

Definition (M18F1)

The TMJ syndrome (M18F1A)

The temporomandibular joint (TMJ) pain syndrome is characterized by limited TMJ range of motion with severe pain in and around the joint. Faulty occlusion of the teeth has been considered a causative or contributing factor by most authorities, with little attention being paid to the influence that the joint's combined musculature may have upon TMJ function and the TMJ pain syndrome.

The TMJ is, for the most part, controlled by the combined and coordinated action of the pterygoid externus, pterygoid internus, digastric, suprahyoid (mylohyoid, geniohyoid, and stylohyoid), temporalis and masseter muscles. The pterygoid externus and internus muscles generally work together to move the mandible laterally and/or forward, to the contralateral side, while the digastric and suprahyoid muscles depresses the mandible, and the largest and strongest of the jaw muscles, the temporalis and masseter muscles, together elevate the mandible and close the jaw.

If hypertonus develops in any one of the muscles or muscle groups associated with the TMJ, relative to the others, jaw function will be effected to the degree of the imbalance produced. This varies with the number and nature of the muscles which are hypertonic. For example, if the pterygoids on one side of the jaw are hypertonic, the mandible may be forced forward so that the anterior surface of the lower incisors on the homolateral side are made to abut up against the posterior surface of the upper incisors on the homolateral side to the extent (in extreme cases) that stability of those upper incisors within the maxilla may be threatened. Likewise, if the masseter and temporalis muscles on one side are hypertonic, relative to the muscles on the contralateral side, prolonged approximation of the homolateral TMJ with increased pressure between the upper and lower molars and premolars on that side may occur which may eventually result in abnormal wearing of joint cartilage surfaces and/or enamel surfaces of the involved teeth, effectively creating faulty occlusion of the teeth.

One of the most consistent symptoms accompanying hypertonus of the jaw musculature is pain. The causes of the resulting pain may include unequal pressure upon the articular cartilage surfaces of the TMJ, intrafusal muscle spasm (trigger point formations) accompanying the hypertonicity, extrafusal muscle spasm existing as an exaggeration of the hypertonicity, strain placed upon ligaments and/or tendinous tissues of origin or insertion, pressure communicated to the teeth and/or bone of the mandible or maxilla, or any combination of these. Referred pain patterns may be produced from joint pressure or trigger point formations which may produce headaches, jaw or tooth pain, earache, throat pain, facial pain, suboccipital or occipital pain, and/or posterior cervical neck pain.

Acute hypertonus of TMJ musculature (M18F1A01) Acute neuromuscular hypertonus of the TMJ musculature generally occurs in response to muscle strain and/or other trauma to the muscles or associated structures (ligaments, articular surfaces, etc.). Such trauma may trigger an automatic protective mechanism which expresses itself as an unrelieved phasic stretch reflex (prolonged extrafusal and/or intrafusal spasm), or as a balanced or imbalanced splinting of the musculature (general muscle tetanus of the involved muscles to prevent further injury). Such acute symptoms may be temporarily or permanently relieved by modes of treatment which usually do not include electromyometry (refer to Auricular Electroacustimulation, Pain Relief [M06A], Cryotherapy, Hypertonic Neuromuscular Management [M10B], Electrical Facilitation of Endorphin Production (Acusleep) [M03A], Electrical Stimulation, Muscle Lengthening [M02A], and/or Vibration, Neuromuscular management [M08A]).

Chronic hypertonus of the TMJ musculature (M18F1A02) Chronic hypertonus of the TMJ musculature generally either occurs as supraspinal structures fail to correct neuromuscular hypertonus created during an acute phase of injury, or is provoked by the supraspinal structures themselves. Neuromuscular tonicity of the TMJ musculature is generally controlled through supraspinal structure influence over the muscle spindle, or tonic stretch reflex (refer to Neuromuscular System [R002] and/or Muscle Spindle [R001]).

If chronic neuromuscular hypertonicity occurs in TMJ musculature, it is generally as an unconscious, subliminal, supraspinal response to a psychological need for a physical defense from an emotional stress. If this occurs, supraspinal structures unconsciously induce the muscle spindle system to selectively maintain hypertonicity of a muscle or muscle group associated with the TMJ to create pain (via the causes listed above) or extreme tension to restrict TMJ range of motion and to promote the sensation of extreme tension.

Many who suffer from hypertonus of musculature associated with the TMJ, as an expression of somatized stress, report the onset of the condition to follow an emotionally traumatizing conflict occurring before, during or shortly following a physically traumatic injury of the TMJ musculature or other associated structures. It has been postulated that the supraspinal structures learn on a subliminal level that it is easier to deal with physical pain than with emotional pain. Evidently, because of the mechanisms involved in attempts to compensate for or correct the damage done by the injury, the supraspinal structures learn to self-induce pain by producing selective muscular hypertonus.

Clinical research has verified that the gamma motor system (which is responsible for maintaining muscle tone) is so attuned to supraspinal structure influence that general muscle tone is directly affected by mood changes. Clinical observation seems to confirm that the degree of neuromuscular response is "learned" in association with particular environmental circumstances or stimuli, much as any other fine motor skill is "programmed." When emotional stress is somatized, it is simply the natural neuromuscular response to emotional tension, exaggerated and specified to a given muscle or muscle group.

Electromyometric feedback has proven to be effective in the treatment of imbalanced TMJ musculature (the single most common cause of the TMJ pain syndrome), which most often appears to result from the somatization of emotional stress, good or bad.

Procedures of Application (M18F2)

EMM instrumentation (M18F2A)

Instrumentation required for neuromuscular reeducation with electromyometric feedback of TMJ muscle imbalance includes one or two electromyometers with the most sophisticated feedback modes available, which can be preset to register different sensitivity levels. Common sensitivity settings include the 0 to 10, 0 to 50, 0 to 100, and 0 to 500 microvolt ranges. The basic electromyometer has both audio and visual (meter or light bar) feedback modes. These modes can be enhanced by built-in feedback mode variables (different sounds and multiple visual displays) or by external computer enhancement of both modes. Some types of EMM instrumentation also offer tactile feedback modes and/or robotic-mechanical displays.

Electrode placement (M18F2B) The EMM should be attached to the patient via electrode cables which terminate as they connect with surface electrodes which are affixed with taping materials, strapping, etc., to the skin above the various muscle groups.

If one-muscle or muscle group monitoring is utilized, both electrode cable pickup wires should be attached to surface electrodes affixed over the the muscle or muscle group, with the ground electrode in close association to them. If two-muscle monitry is used, the contralateral or homolateral muscle or muscle group pairs are simultaneously monitored with an electromyometer by having one of the electrode cable active pickup leads attached to a surface electrode affixed over one of the muscles or muscle groups while the other is attached over the second of the pair. The ground electrode is conventionally attached approximately one half the distance between the two.

Pterygoid placement (M18F2B01) The electrode placement for monitry of the pterygoid muscle group should be over the anterior distal portion of the temporalis muscle, just superior to the mandibular coronoid process. This placement will provide simultaneous pickup from the anterior temporalis and the pterygoids; however, clinical experience has demonstrated that the myoelectric activity picked up in this manner is adequately representative of pterygoid tonus or contraction to justify it.

Digastric and suprahyoid placement (M18F2B02) The electrode placement for monitry of the digastric and suprahyoid muscle group should be below and posterior to the lower border of the mandible, between midline and the angle of the mandible. While this placement will provide simultaneous pickup from the proximal portion of the platysma and the digastric and suprahyoid group, clinical experience has demonstrated that the myoelectric activity picked up in this manner is adequately representative of digastric and suprahyoid tonus or contraction to justify it.

Temporalis placement (M18F2B03) The electrode placement for monitry of the temporalis muscle should be superior to the zygomatic pro-

cess, posterior to the orbital process, and above the sphenoid and temporal fossa. One active pickup should be affixed over the sphenoid fossa and the other centrally located over the temporal fossa.

Masseter placement (M18F2B04) The electrode placement for monitry of the masseter muscle should be between the inferior aspect of the zygomatic process and the lower border of the mandible, one third of the distance from the angle of the mandible and the point of the mental eminence.

Evaluation (M18F2C)

The patient should be sitting in an upright position which will be comfortable for the duration of the evaluation session.

During the evaluation phase, the patient should remain unexposed to the feedback modes, unable to see, hear or feel changes in myoelectric production as reflected by EMM feedback.

Electrodes should be placed over each muscle or muscle group of interest. The electrodes should subsequently be attached to the EMM instrumentation. The electromyometer should then be turned on and each muscle monitored separately as the patient performs a few prescribed neuromuscular functions. These functions should include (1) relaxation of the TMJ musculature with the teeth of the upper and lower jaws meeting, (2) relaxation of the TMJ musculature with the jaw slack and the teeth apart, (3) biting down on a platform which is even on both sides of the jaw (two tongue depressors may serve), (4) and speaking (counting out loud from one to ten). The maximum myoelectric response to each function from each muscle or muscle group should be noted and recorded.

To establish the presence of hypertonicity in any particular muscle or muscle group, the myoelectric activity levels produced during each of the prescribed functions by each of the muscles or muscle groups should be compared to its contralateral counterpart and the other muscles on the homolateral side. Contralateral imbalances will become obvious as counterpart responses are compared, while homolateral imbalances will become obvious as the activity from homolateral synergists are compared. Normally, synergists produce relatively similar myoelectric activity levels when acting together.

Myoelectric activity produced by a particular TMJ muscle or muscle group during relaxation is between 3 to 6 microvolts; contracting against resistance, such muscles will produce myoelectric activity which may be in excess of 300 microvolts. During talking, myoelectric activity levels ranging between 10 and 50 microvolts may be produced.

Neuromuscular reeducation utilizing EMM feedback (M18F2D)

During the treatment phase, the patient should be sitting in an upright position which will remain comfortable for the duration of the session.

The process of neuromuscular reeducation utilizing EMM feedback is the same as that used for neuromuscular reeducation of pathological neuromuscular dysfunction (post CVA, post peripheral nerve injury and/or post spinal cord injury), and follows the basic learning phase pattern of facilitation, inhibition, balance and function.

Facilitation phase (M18F2D01) When the existence of relative hypertonus of a muscle or muscle group (balanced or imbalanced) has been established through EMM evaluation, the muscles to be monitored should be connected to the electromyometer(s) via electrode cables which terminate as they connect to surface electrodes affixed over the involved muscles.

If contralateral hypertonic imbalance has been established, the muscle or muscle group which is antagonistic to the hypertonic muscle should be monitored with the electromyometer. If ipsilateral hypertonic imbalance has been established, a muscle or muscle group which performs a clearly different function from that performed by the hypertonic muscle should be monitored. Utilizing EMM feedback to reflect myoelectric activity, the patient should be asked to voluntarily increase (facilitate) myoelectric activity from this relatively hypotonic muscle, and to maintain the resultant contraction at a maximum level for at least five or six seconds. If the contraction increases approximation of the jaw or teeth a thin soft bite plate may be supplied to the patient to provide a biting surface which protects the teeth.

During the facilitation stage the electromyometer should be set at a sensitivity level which best demonstrates positive or increasing myoelectric change. The patient must be able to see, hear and/or feel changes in myoelectric activity, during the entire procedure, to subliminally learn from them. Initially, the electromyometer may have to be set at the level of greatest sensitivity (0 to 6 or 0 to 10 microvolt ranges) so that any increase in myoelectric activity will be substantially demonstrated by all the feedback modes. If the muscle enjoys substantially more tone, the electromyometer should be set at a less sensitive level (0 to 30, 0 to 50, 0 to 100 microvolts or more) to provide a gauge of success as well as providing the patient with an opportunity to strive to produce more activity at observable levels.

Inhibition phase (M18F2D02) In the inhibition phase, the muscle or muscle group which has been established to be relatively hypertonic should be monitored by the electromyometer. Utilizing the EMM feedback to reflect myoelectric activity levels back to the patient, the patient should be asked to take several seconds to slowly increase the myoelectric activity produced by the muscle or muscle group to a small degree (less than 10 microvolts) and then to voluntarily inhibit or decrease the myoelectric activity below the previously established resting level. The myoelectric activity increase should last no longer than six seconds, but the suppression stage should last from six seconds to several minutes until adequate inhibition is attained or until failure must be admitted and another attempt made. The electromyometer should be set at a sensitivity level which will demonstrate the patient's successful inhibition of myoelectric activity (optimally at the 0 to 10 microvolt sensitivity level, but less sensitive levels may have to be initially used if resting levels exceed the 0 to 10 microvolt range).

Balance phase (M18F2D03) In the balance phase, two EMM setups are required to allow the patient the full opportunity of attempting to increase myoelectric activity produce by the muscle or muscle group to be facilitated while simultaneously decreasing or lowering myoelectric activity produced by the muscle or muscle group to be inhibited.

During the balance phase the electromyometer monitoring the muscle or muscle group to be facilitated should be set at the least sensitive level that best magnifies and reflects myoelectric increases (generally at the 0 to 50 microvolt level), while the unit monitoring the muscle or muscle group to be to be inhibited should be set at the sensitivity level that best magnifies and reflects decreases in myoelectric activity (shows the largest range of feedback, ideally at the 0 to 10 microvolt level).

The muscles most often requiring balancing are the contralateral masseter vs. masseter muscles, contralateral temporalis vs. temporalis muscles, contralateral pterygoid group vs. pterygoid group muscles, homolateral masseter vs. pterygoid muscles, homolateral masseter vs. temporalis muscles, pterygoid and contralateral masseter vs. masseter and contralateral pterygoid muscles, and bilateral masseter vs. the bilateral digastric-suprahyoid group muscles.

Functional phase (M18F2D04) In the functional phase, the hypertonic muscle or muscle group should be monitored. The patient should be asked to talk (an emotion-packed function) while keeping the myoelectric activity, as reflected by electromyometric feedback, from the muscle or muscle group below a given level arbitrarily determined by the practitioner. This level should be lowered as the patient's control increases.

During the talking (or functional) phase the patient should perform in speech categories which have increasingly greater emotional content while keeping myoelectric activity produced by the formerly hypertonic muscle below a set limit. Listed in order of increasing emotional content, speaking tasks from the different speech categories should include: (1) counting slowly out loud (whispering is initially permissible) from one to ten, (2) saying out loud the names of people the patient knows personally, and (3) talking conversationally to the practitioner regarding personal concerns or experiences. If the patient is able to successfully demonstrate control of the myoelectric activity produced by the formerly hypertonic muscle or muscle group by keeping the myoelectric activity produced below a preset level, the preset level should be lowered (made more difficult) during the following session. This process should progressively continue until the patient has reached the optimum level of control over myoelectric activity produced by the formerly hypertonic muscle or muscle group. Generally, the patient should be able to keep the myoelectric activity produced by the muscle or muscle group below 10 microvolts while engaging in conversation with the practitioner regarding topics of concern.

Computer enhancement of electromyometry (M18F2E)

Computer enhancement of electromyometric feedback has been shown to greatly add to patient appreciation and participation in the neuromuscular reeducation program by making feedback more graphic and easier to understand (for the patient as well as the practitioner) in terms of visual and auditory feedback modes. Some of the graphic designs currently in use for single electromyometer activity enhancement have shown themselves to be remarkably suited for increasing patient motivation and enthusiastic participation. One of the more effective graphic designs currently available requires the patient to keep the EMM activity line showing on the computer's video screen between parallel lines which cross the screen in various patterns or shapes. After the activity line has transversed the screen, a numerical readout in one corner of the screen tells the patient what percentage of the time the activity line was above, within, or below the parallel line pattern. The time it takes to cross the screen can be preset by the operator.

During the balance phase, computer enhancement is especially effective if the computer instrumentation allows representative activity from both electromyometers to appear on the video monitor simultaneously as distinctly different lines. Effectiveness increases if the patient can clearly relate the two to a graphic screen design which helps demonstrate their relative activity levels. Two fixed parallel lines across the screen placed relative to the action lines may be adequate.

During the functional phase, if computer enhancement is used, a single activity line from the monitored hypotonic muscle is compared to a preset line which represents the arbitrary level of control. If the activity level from the electromyometer exceeds or goes above the line during one of the talking exercises, the patient is said to be out of control and should be instructed to discontinue talking until the activity level is once again below the control line.

Session termination (M18F2F)

At the end of the treatment session the electromyometer and associated feedback mode instrumentation should be turned off.

The surface electrodes should be carefully removed and the electrode paste carefully cleansed from the patient's skin before the patient is allowed to leave the treatment area. A surface electrode count should be conducted after removal from the skin to prevent them from leaving when the patient does.

Partial List of Conditions Treated (M18F3)

Treatable Causes (M18F3A)

Dental pain (C22)
Extrafusal muscle spasm (C41)
Habitual joint positioning (C48)
Hypermobile/instable joint (C10)
Joint approximation (C47)
Ligamentous strain (C25)
Muscular weakness (C23)
Neuromuscular tonic imbalance (C30)
Psychoneurogenic neuromuscular
  hypertonus (C29)
Restricted joint range of motion (C26)
Trigger point formation (C01)

Syndromes (M18F3B)

CERVICAL (NECK) PAIN (S73)
EARACHE (S40)
HEADACHE PAIN (S02)
MIGRAINE (VASCULAR) HEADACHE (S18)
OSTEOARTHRITIS (S21)
POST IMMOBILIZATION (S36)
POST WHIPLASH (S55)
REFERRED PAIN (S01)
SINUS PAIN (S33)
TEMPOROMANDIBULAR JOINT (TMJ)
  PAIN (S06)
TIC DOULOUREUX (S53)
TINNITUS (S70)
TOOTHACHE/JAW PAIN (S41)

Precautions and Contraindications (M18F4)

No precautions or contraindications have been listed or detailed for electromyometric feedback, except in connection with the stimulation provided by tactile and proprioceptive feedback modes. For precautions and contraindications associated with these feedback modes refer to the Electrical Stimulation *Precautions and Contraindications* [M0204], Transcutaneous Nerve Stimulation *Precautions and Contraindications* [M0404], and Vibration *Precautions and Contraindications* [M0804] sections.

Notes Aside (M18F5)

Bruxism (M18F5A)

Neuromuscular reeducation of hypertonic TMJ musculature utilizing EMM feedback has shown itself to be coincidentally useful for the treatment of bruxism. It has been demonstrated through clinical experience that as the patient is successful in controlling a hypertonic neuromuscular imbalance associated with the TMJ syndrome, during the the functional phase of training, bruxism (nocturnal grinding of the teeth) seems to decrease in intensity and frequency, and sometimes becomes extinguished as an unconscious behavior.

Jaw range of motion (M18F5B)

The inhibitory phase described here has proven useful in the treatment of bilateral balanced acute hypertonicity of the jaw musculature with its consequence of restricted joint range of motion.

The treatment of choice is based upon the contract-relax Rood technique. The masseter muscles are generally monitored simultaneously, via split lead electrode attachment, with a single electromyometer. The patient should be provided with a bite plate (often made by stacking and taping together tongue depressors) which is just slightly too large to fit comfortably in the space between the front teeth. The patient should be asked to voluntarily increase the myoelectric activity produced by the hypertonic muscles by biting down, in a balanced manner, on the bite plate, for a six-second period. The patient should then attempt to relax the muscle and voluntarily inhibit the myoelectric activity produced by the muscle to a level below that previously demonstrated during rest. The inhibition attempt should continue until the desired level of myoelectric activity is attained or until failure must be admitted and the process repeated.

As the myoelectric hypertonic activity produced by the muscle slowly diminishes, with repeated attempts, and the mouth is able to open wide, the bite plate should be expanded or made thicker to promote a still wider aperture. The session should end when several repeated attempts fail to cause any further widening of the distance between the teeth of the upper and lower jaws.

Bibliography (M18F6)

Swaan, pp. 9–11.
Taylor, pp. 161–164.

ELECTROMYOMETRIC FEEDBACK NEUROMUSCULAR REEDUCATION OF LOW BACK MUSCULATURE
(M18G)

Definition (M18G1)

Neuromuscular sources of low back pain (M18G1A)

It has been clinically demonstrated that if the musculature supporting the spinal column becomes imbalanced (one side being tighter than the other), with one side continually producing relatively more myoelectric activity then the other, pain will result. The causes of the pain produced may include unequal pressure upon the involved vertebral disks, intrafusal muscle spasm (trigger point formations) accompanying the hypertonicity, extrafusal muscle spasm existing as an exaggeration of the hypertonicity, strain placed upon ligaments on the hypotonic side, excessive muscular pressure upon involved sensory nerve roots, or any combination of these causative agents. Low back pain may also be produced if any of these causes of pain result from *balanced* hypertonicity of the low back musculature (both sides of the back being too tight).

Acute neuromuscular imbalance of the back musculature generally occurs in response to muscle strain and/or other trauma to the muscles or associated structures (ligaments, facets, etc.). Such trauma may trigger an automatic protective mechanism which expresses itself as an unrelieved phasic stretch reflex, or as a balanced or imbalanced splinting of the musculature (general muscle tetanus of the involved muscles to prevent further injury). Such acute responses may be temporarily or permanently relieved by modes of treatment which usually do not include electromyometry (refer to Auricular Electroacustimulation, Pain Relief [M06A], Cryotherapy, Hypertonic Neuromuscular Management [M10B], Electrical Facilitation of Endorphin Production (Acusleep) [M03A], Electrical Stimulation, Muscle Lengthening [M02A], Vertebral Traction, Lumbar traction [M09B], and/or Vibration, Neuromuscular management [M08A]).

Chronic imbalance of the back musculature occurs when supraspinal structures fail to eventually correct neuromuscular imbalance(s) created during the acute phase or are provoked themselves to create the imbalance(s). Neuromuscular tonicity of the back musculature is generally controlled through supraspinal structure influence over the muscle spindle, or tonic stretch reflex (refer to Neuromuscular System [R002] and/or Muscle Spindle [R001]). If a chronic neuromuscular imbalance occurs in the back musculature, it is generally as an unconscious, subliminal, supraspinal response to a psychological need for a physical defense from an emotional stress. If this occurs, supraspinal structures unconsciously induce the muscle spindle system selectively maintain balanced or unbalanced hypertonicity of back musculature to create pain via the causes listed above or feeling of extreme tension.

Many who suffer from psychoneurogenic low back neuromuscular imbalance (hypotonus/ hypertonus) as an expression of somatized stress, report the onset of the condition following an emotionally traumatizing conflict occurring before, during or shortly following a physically traumatic injury of the low back musculature or related structures. It has been postulated that the supraspinal structures learn on a subliminal level that it is easier to deal with physical pain than with emotional pain. Evidently, because of the mechanisms involved in the system attempts to compensate for or correct the damage done by the injury, the supraspinal structures learn how to self-induce pain by tonically imbalancing opposing musculature. Clinical research has verified that the gamma motor system, which is responsible for maintaining muscle tone, is so attuned to supraspinal structure influence that general muscle tone is directly affected by mood changes. And clinical observation seems to confirm that the degree of neuromuscular response is "learned" in association with particular environmental circumstances or stimuli, much as any other fine motor skill is "programmed." When emotional stress is somatized, it is simply the natural neuromuscular response to emotional tension, exaggerated and specific to a given muscle or muscle group.

Procedures of Application (M18G2)

EMM instrumentation (M18G2A)

Instrumentation required for neuromuscular reeducation with electromyometric feedback of musculature associated with neuromuscular imbalance of back musculature includes one or two electromyometers with the most sophisticated feedback modes available, which can be preset to register different sensitivity levels. Common sensitivity settings include the 0 to 10, 0 to 50, 0 to 100, and 0 to 500 microvolt ranges. The basic electromyometer has both audio and visual (meter or light bar) feedback modes. These modes can be enhanced by built-in feedback mode variables (different sounds and multiple visual displays) or by external computer enhancement of both modes. Some types of EMM instrumentation also offer tactile feedback modes and/or robotic-mechanical displays.

Electrode placement (M18G2B)

The electromyometers should be attached to the patient via electrode cables which terminate by attaching to surface electrodes affixed to the skin over the muscles monitored.

The muscles most commonly involved in low back muscular imbalance and/or hypertonus are the iliocostalis, longissimus, and multifidus muscle groups.

Evaluation of the low back syndrome (M18G2C)

The patient should be sitting or lying in a comfortable position.

During the evaluation phase, the patient should remain unexposed to the feedback modes, unable to see, hear or feel changes in myoelectric production as reflected by EMM feedback.

Surface electrodes should be affixed over the back's dominant muscle groups. In general, electrode placement over back musculature should be in symmetrical fashion, the electrode placement on one side of the back mirroring the placement on the other side. On one side, one electrode should be placed over the prominent paraspinous muscles at the sixth thoracic vertebrae (T6) level, one at the tenth thoracic vertebrae (T10) level, and one at the fifth lumbar vertebrae (L5) level. For a general survey of myoelectric activity from the back musculature, the active pickup electrodes should be attached to the T6 and L5 sites, respectively, and the ground attached to T10. The convention is to call the placement arrangement on the right side Placement A, and that on the left side Placement A'. For a survey of the myoelectric activity from just the upper thoracic area, the active pickup electrodes should be attached over the T6 and T10 sites with the ground over the L5 site. Conventionally, this arrangement is called Placement B on the right side and Placement B' on the left. For a survey of the myoelectric activity from just the lower thoracic and lumbar area, the active pickup electrodes should be placed over the T10 and L5 and the ground over the T6 site. This is conventionally called Placement C on the right side and Placement C' on the left side.

In practice, with feedback modes withheld from the patient, the myoelectric activity from Placement A and A' should be monitored simultaneously. If an imbalance between A and A' or balanced hypertonicity is shown to exist, the Placements B and B' and Placements C and C' should be surveyed to establish the local area hypertonicity.

Subliminal learning pattern (M18G2D)

Initially the patient should be in an upright sitting position that is as comfortable as possible. The local muscular area which has been shown to be hypertonic (B and B' or C and C') should be monitored by electromyometry. A neuromuscular reeducation program should be instituted to correct imbalance or reduce general hypertonicity which follows the subliminal learning pattern of facilitation, inhibition, balance and function.

Facilitation phase (M18G2D01) Typically, in the facilitation phase the hypotonic muscle group (the anterior abdominal muscles if both sides of the back are hypertonic and balanced) is monitored, and the patient is asked to voluntarily increase its myoelectric activity as reflected by the available feedback modes. The patient is asked to maintain a contraction of the muscle, producing the maximum of feedback possible for a five- or six-second period, followed by at least a six-second rest period before another attempt is made.

During the facilitation stage the EMM should be set at a sensitivity level which best demonstrates positive or increasing myoelectric change. The patient must be able to see, hear and/or feel changes in myoelectric activity, during the entire procedure, to subliminally learn from any change which takes place in myoelectric activity from the muscle. If the muscle is severely under active, the electromyometer will most commonly have to be set initially at the level of greatest sensitivity (0 to 6 or 0 to 10 microvolt ranges) so that any increase in myoelectric activity will be substantially demonstrated by all the feedback modes. If the muscle is more active, the electromyometer should be set at a less sensitive level (0 to 30 or 0 to 50 microvolt ranges, for example) to provide a gauge of success as well as providing the patient with an opportunity to strive to produce more activity at observable levels.

Inhibition phase (M18G2D02) In the inhibition phase, the hypertonic muscle group(s) is monitored and the patient should be asked to increase activity from the muscle group(s) by a small amount (less than 10 microvolts) and then to actively suppress the activity level below the the previous resting level. The attempt to increase myoelectric activity from the muscle group(s) should last no longer than six seconds, but the suppression stage should last from six seconds to several minutes until adequate inhibition is attained or until failure must be admitted and another attempt made. The patient should be in a sitting position. The electromyometer should be set at a sensitivity level which will adequately demonstrate patient inhibitory success (optimally at the 0 to 10 microvolt sensitivity level, but less sensitive levels may be used initially if resting levels exceed the 0 to 10 microvolt range).

Balance phase (M18G2D03) In the balance phase, two EMM setups are required to allow the patient the opportunity of attempting to increase myoelectric activity from the facilitated muscle group(s) while simultaneously decreasing or lowering myoelectric activity from the inhibited muscle group(s) pair. This phase of training is generally conducted in the sitting position. The electromyometer monitoring the muscle group(s) to be facilitated should be set at the least sensitive level that best magnifies and reflects increases in myoelectric activity, without exceeding the maximum setting level (over 50 microvolts if the sensitivity level is set at a 0 to 50 microvolt range). The EMM monitoring the muscle group(s) to be inhibited should be set at the most sensitive level that best magnifies and reflects decreases in myoelectric activity (shows the largest range of feedback, ideally at the 0 to 10 microvolt sensitivity level).

The choice of feedback modes relative to the muscle group(s) and neuromuscular problem is of paramount importance. When in the balance phase of neuromuscular reeducation, the feedback modes that are the most attention-getting should be used in conjunction with the electromyometer monitoring the hardest muscle to control, which are usually the hypertonic muscle group(s).

Functional phase (M18G2D04) Initially, in the functional stage, the patient should be sitting while attempting to inhibit the myoelectric activity. The patient should be encouraged to find a position which causes an involuntary cessation or marked reduction of EMM activity from the muscle monitored (usually a slumped posture). The patient should then be instructed to slowly attempt to return to a more normal position while maintaining the low myoelectric activity. The position change should be gradual, with movement continuing only as long as the production of myoelectric activity remains low. If the myoelectric level rises, the patient should stop and hold the current position until the myoelectric activity level is reduced voluntarily to the previously low level.

When the patient is able to maintain the low myoelectric activity level in a regular sitting position, this process should then be repeated in the standing position. Usually, the standing position which involuntarily decreases myoelectric activity is with the legs straight and the patient's body weight shifted to the heels. The knees should not be allowed to flex. While doing so does reduce myoelectric activity from the back muscles, clinical experience has demonstrated that patients can not learn to keep low back myoelectric activity down when shifting to a straight knee position. The patient should attempt to gradually shift from the heel weight bearing position to a normal standing position while keeping the myoelectric activity from the back down.

A functional talking phase may also be used. The hypertonic muscle group(s) should be monitored while the patient attempts to keep the myoelectric activity below an arbitrarily set level while: (1) slowly counting out loud (whispering is initially permissible) from one to ten, (2) saying out loud the names of people involved in the patient's life, and (3) talking conversationally to someone in the same room. These are successive steps and the patient should be directed to proceed from one step to the next only as control has been demonstrated.

Computer enhancement of electromyometry (M18G2E)

Computer enhancement of electromyometric feedback has been shown to greatly add to patient appreciation and participation in the neuromuscular reeducation program by making feedback more graphic and easier to understand (for the patient as well as the practitioner) in terms of visual and auditory feedback modes. Some of the graphic designs currently in use for single electromyometer activity enhancement have shown themselves to be remarkably suited for increasing patient motivation and enthusiastic participation. One of the more effective graphic designs currently available requires the patient to keep the EMM activity line showing on the computer's video screen between parallel lines which cross the screen in various patterns or shapes. After the activity line has transversed the screen, a numerical readout in one corner of the screen tells the patient what percentage of the time the activity line was above, within, or below the parallel line pattern. The time it takes to cross the screen can be preset by the operator.

During the balance phase, computer enhancement is especially effective if the computer instrumentation allows representative activity from both electromyometers to appear on the video monitor simultaneously as distinctly different lines. Effectiveness increases if the patient can clearly relate the two to a graphic screen design which helps demonstrate their relative activity levels. Two fixed parallel lines across the screen placed relative to the action lines may be adequate.

During the functional phase, if computer enhancement is used, a single activity line from the monitored hypertonic muscle is compared to a preset line which represents the arbitrary level of control. If the activity level from the electromyometer exceeds or goes above the line during one of the talking exercises, the patient is said to be out of control and should be instructed to discontinue talking until the activity level is once again below the control line.

Termination of the treatment session (M18G2F)

At the end of the treatment session the electromyometer and associated feedback mode instrumentation should be turned off.

The surface electrodes should be carefully removed and the electrode paste carefully cleansed from the patient's skin before the patient is allowed to leave the treatment area. A surface electrode count should be conducted after removal from the skin to prevent them from leaving when the patient does.

Partial List of Conditions Treated (M18G3)

Treatable Causes (M18G3A)

Extrafusal muscle spasm (C41)
Habitual joint positioning (C48)
Hypermobile/instable joint (C10)
Interspinous ligamentous strain (C03)
Joint approximation (C47)
Ligamentous strain (C25)
Muscular weakness (C23)
Nerve root compression (C04)
Neuromuscular tonic imbalance (C30)
Pathological neuromuscular hypertonus (C28)
Restricted joint range of motion (C26)
Sympathetic hyperresponse (C38)
Trigger point formation (C01)

Syndromes (M18G3B)

FACET (S19)
HYSTERIA/ANXIETY REACTION (S59)
LOWER ABDOMINAL PAIN (S62)
LOW BACK PAIN (S03)
MENSTRUAL CRAMPING (S60)
OSTEOARTHRITIS (S21)
POST IMMOBILIZATION (S36)
REFERRED PAIN (S01)
SCIATICA (S05)

Precautions and Contraindications (M18G4)

No precautions or contraindications have been listed or detailed for electromyometric feedback, except in connection with the stimulation provided by tactile and proprioceptive feedback modes. For precautions and contraindications associated with these feedback modes refer to the Electrical Stimulation, *Precautions and Contraindications* [M0204], Transcutaneous Nerve Stimulation, *Precautions and Contraindications* [M0404], and/or Vibration, *Precautions and Contraindications* [M0804] sections.

Notes Aside (M18G5)

Reeducation of hypertonic back musculature utilizing EMM feedback for the relief of pain has proven to be effective, even when disc disease has been shown to be present (in some cases). In successful cases, it was apparent that muscle imbalance was responsible, at least in part, for excessive pressure which may have created disc distortion or greatly contributed to the ongoing pressure upon involved nerve roots.

Bibliography (M18G6)

Miller, pp. 1347–1354.
Swaan, pp. 9–11.
Taylor, pp. 164–169.

ELECTROMYOMETRIC FEEDBACK NEUROMUSCULAR REEDUCATION OF TORTICOLLIS (M18H)

Definition (M18H1)

The torticollis syndrome (M18H1A)

Electromyometric feedback has proven to be effective in the treatment of torticollis or wry neck. Torticollis is characterized by chronically imbalanced hypertonic neck musculature which effectively pulls the neck and head into flexion and lateral rotation either in a repetitive, rhythmic pattern of neck and head motion (spastic torticollis) or into a statically fixed position (wry neck).

Torticollis appears, in most cases, to have a psychoneurogenic origin (excluding those cases which have resulted from brain damage). It seems to stem from a somatic response to subliminal interpretations of externally or internally induced emotional stress and/or anxiety. Neuromuscular tonicity of the neck musculature is generally controlled through supraspinal structure influence over the muscle spindle, or tonic stretch reflex (refer to Neuromuscular System [R002] and/or Muscle Spindle [R001]).

Although the exact etiology of torticollis is unclear, it appears that if chronic neuromuscular hypertonicity (whether spastic or static) occurs in the neck musculature severely enough to establish the torticollis syndrome, it is generally as an unconscious, subliminal, supraspinal response to a psychological need for a physical defense against emotional stress. If this occurs, the supraspinal structures apparently subliminally induce the muscle spindle system to selectively maintain hypertonicity in one of the sternocleidomastoideus and/or the upper trapezius muscles on the same side.

Many who suffer from torticollis report the onset of the condition to follow an emotionally traumatizing conflict occurring before, during or shortly following a physically traumatic injury (sometimes no more serious then a stiff neck occurring after a restless night) of the sternocleidomastoideus and/or upper trapezius muscles. It has been postulated that the supraspinal structures involved seem somehow to conceptualize (on a subconscious level) that it is easier to deal with the discomfort and disfigurement of the twisted neck than to confront or deal directly with anxiety-provoking externally or internally induced stress. Apparently these supraspinal structures learn from the motor experiences consequent to the physical injury to self-induce neuromuscular imbalance or exaggerated phasic stretch responses between opposing neck muscles, some of which have been previously injured. The mind learns to distract itself with a physical reality of dysfunction from anxiety-provoking external or internal stimuli.

Clinical research has verified that the gamma motor system, which is responsible for maintaining muscle tone, is so attuned to supraspinal structure influence that general muscle tone is directly affected by mood changes. Clinical observation seems to confirm that the degree of neuromuscular response is "learned" in association with particular environmental circumstances or stimuli, much as any other fine motor skill is "programmed." When emotional stress is somatized, it is simply the natural neuromuscular response to emotional tension, exaggerated and specified to a given muscle or muscle group.

Procedures of Application (M18H2)

EMM instrumentation (M18H2A)

Instrumentation required for neuromuscular reeducation with electromyometric feedback of neck and/or shoulder (upper trapezius) muscular imbalance (torticollis) includes one or two electromyometers with the most sophisticated feedback modes available, which can be preset to register different sensitivity levels. Common sensitivity settings include the 0 to 10, 0 to 50, 0 to 100, and 0 to 500 microvolt ranges. The basic electromyometer has both audio and visual (meter or light bar) feedback modes. These modes can be enhanced by built-in feedback mode variables (different sounds and multiple visual displays) or by external computer enhancement of both modes. Some types of EMM instrumentation also offer tactile feedback modes and/or robotic- mechanical displays.

Electrode placement (M18H2B)

The EMM should be attached to the patient via electrode cables which terminate as they connect with surface electrodes which are affixed with taping materials, strapping, etc., to the skin above the various muscle groups.

If single muscle or muscle group monitoring is utilized, both electrode cable pickup wires should be attached to surface electrodes affixed over the the muscle or muscle group, with the ground electrode in close association to them. If two-muscle monitry is used, the contralateral or homolateral muscle or muscle group pairs are simultaneously monitored with an electromyometer by having one of the electrode cable active pickup leads attached to a surface electrode affixed over one of the muscles or muscle groups while the other is attached over the second of the pair. The ground electrode is conventionally attached approximately one half the distance between the two.

Sternocleidomastoideus (M18H2B01) The surface electrodes should be placed and affixed to the skin above and along the body of the sternocleidomastoideus, while the head is in the position in which it will be while testing takes place. The surface electrodes attached to the active pickup cable leads should be placed along the line of the sternocleidomastoideus, distal to the site of its mastoid insertion and proximal to its clavicular insertion. The ground electrode is conventionally placed in close proximity to the active electrodes, equidistant from each of them.

Upper trapezius (M18H2B02) Electrode placement on the skin over the upper trapezius muscle(s) should be on the skin above the site of the upper trapezius trigger point formation, which is approximately one half the distance along an imaginary line drawn from the acromioclavicular joint to the spinous process of the sixth cervical vertebrae (C7). The surface electrodes attached to the electrode cable active pickup leads should be at least an inch apart, along the imaginary line. The ground electrode placement should be in relatively close proximity to, but not touching, and equidistant from each of the active electrodes.

Evaluation of the torticollis (M18H2C)

The patient should be sitting in semi-reclined position with the head supported and turned toward the hypertonic side (the tight musculature being put on stretch).

During the evaluation phase, the patient should remain unexposed to the feedback modes, unable to see, hear or feel changes in myoelectric production as reflected by EMM feedback.

Each of the sternocleidomastoideus and upper trapezius muscles should be monitored separately. The myoelectric activity levels from each muscle should be noted and the maximum activity levels recorded. The patient should then be instructed to turn the head turned toward the hypotonic side while to attempt to keep it there while observations of myoelectric activity levels are repeated.

The hypertonic sternocleidomastoideus, when on stretch, may demonstrate a maximum activity level range between 150 and 300 or more microvolts. When it is contracting to turn the head in the opposite direction, myoelectric activity may approach or exceed 600 microvolts, which may be produced rhythmically, as in spastic torticollis, or maintained statically, as in the wry neck condition. If the upper trapezius is hypertonically involved, its myoelectric activity may be half that produced by the sternocleidomastoideus, whether at rest and being stretched or actively contracting.

The relatively hypotonic counterpart(s) to the hypertonic muscle(s) may show resting levels as low as 2 to 4 microvolts while resting and on stretch, and as little as 50 to 80 microvolts while attempting to contract to turn and hold the head toward the opposite side.

Neuromuscular reeducation utilizing EMM feedback (M18H2D)

Following evaluation, the patient should be sitting in a semi-reclined position with the head supported. The muscles to be monitored should be individually connected to electromyometers via electrode cables which are attached to surface electrodes affixed to the skin over the muscles.

The patient should be positioned to appreciate the available feedback modes, which may include audio, visual and (if available) tactile feedback. The electromyometers and accessory feedback modes should be turned and adjusted to acceptable levels.

A neuromuscular reeducation program utilizing EMM feedback, for the treatment or correction of torticollis should follow the learning phase pattern of facilitation, inhibition, balance and function.

Facilitation phase (M18H2D01) Typically, in the facilitation phase the hypotonic muscle group (the anterior abdominal muscles if both sides of the back are hypertonic and balanced) is monitored, and the patient is asked to voluntarily increase its myoelectric activity as reflected by the available feedback modes. The patient is asked to maintain a contraction of the muscle, producing the maximum of feedback possible for a five or six-second period, followed by at least a six-second rest period before another attempt is made.

During the facilitation stage the EMM should be set at a sensitivity level which best demonstrates positive or increasing myoelectric change. The patient must be able to see, hear and/or feel changes in myoelectric activity, during the entire procedure, to subliminally learn from any change which takes place in myoelectric activity from the muscle. If the muscle is severely under active, the electromyometer will most commonly have to be set initially at the level of greatest sensitivity (0 to 6 or 0 to 10 microvolt ranges) so that any increase in myoelectric activity will be substantially demonstrated by all the feedback modes. If the muscle is more active, the electromyometer should be set at a less sensitive level (0 to 30 or 0 to 50 microvolt ranges, for example) to provide a gauge of success as well as providing the patient with an opportunity to strive to produce more activity at observable levels.

Inhibition phase (M18H2D02) During the inhibition phase, the hypertonic muscle, with the head still turned toward the hypertonic side, should be monitored and the patient asked to increase activity from the muscle group(s) by a small amount (less than 10 microvolts) and then to actively suppress the activity level below the the previous resting level. The attempt to increase myoelectric activity from the muscle group(s) should last no longer than six seconds, but the suppression stage should last from six seconds to several minutes until adequate inhibition is attained or until failure must be admitted and another attempt made. The patient should be in a sitting position. The electromyometer should be set at a sensitivity level which will adequately demonstrate patient inhibitory success (optimally at the 0 to 10 microvolt sensitivity level, but less sensitive levels may be used initially if resting levels exceed the 0 to 10 microvolt range).

Balance phase (M18H2D03) During the balance phase, two EMM setups are required to allow the patient the opportunity of attempting to increase myoelectric activity from the facilitated muscle while simultaneously decreasing or lowering myoelectric activity from the inhibited muscle. The electromyometer monitoring the facilitated muscle should be set at the least sensitive level that best magnifies and reflects increases in myoelectric activity, without exceeding the maximum setting level (over 50 microvolts if the sensitivity level is set at 0 to 50 microvolts). The unit monitoring the inhibited muscle should be set at the most sensitive level which best magnifies and reflects decreases in myoelectric activity (shows the largest range of feedback, ideally at the 0 to 10 microvolt sensitivity level).

As the patient is able to successfully control the head in the fully rotated position toward the hypertonic side, the supported head position should be adjusted fifteen degrees toward the uninvolved side and the facilitation, inhibition and balance phases repeated. This process should be repeated until the patient can demonstrate control with the the head in midline. After head control has been demonstrated in the semi-reclined position, the entire reeducation process should be repeated with the patient sitting up and the head unsupported. Finally, the entire process should be repeated with the patient standing.

The choice of feedback modes relative to the muscle monitored and neuromuscular problem is of paramount importance. When in the balance phase of neuromuscular reeducation, the feedback modes that are the most attention-getting should be used in conjunction with the electromyometer monitoring the hardest muscle to control, which is usually the hypertonic muscle group(s).

Functional phase (M18H2D04) In the functional stage only the most hypertonic of the muscles is monitored. The patient is asked to talk (an emotion-packed exercise) while keeping the muscle's myoelectric activity, as reflected by electromyometric feedback, below an arbitrarily determined level. This level should be lowered as the patient's control increases. During the talking phase the patient should (1) slowly count out loud (whispering is initially permissible) from one to ten (the patient should be cautioned not to say more then a few numbers on a single breath because as the air runs out the shoulder and neck muscular tension increase), (2) say out loud the names of people involved in the patient's life, and (3) talk conversationally to someone in the same room. Each step should be attempted while striving to keep the feedback levels below the designated level. These are successive steps and the patient should be directed to proceed from one step to the next only as control has been demonstrated.

When control of the previously hypertonic musculature has been demonstrated during talking, the patient should be asked to demonstrate the same control while performing increasingly difficult or intricate hand functions, first while semireclining, then sitting upright, and finally while standing. Finally, it will be necessary to ask the patient to demonstrate control over the previously hypertonic musculature while performing such hand functions, without EMM feedback being present, to fully complete the neuromuscular reeducation process.

Computer enhancement of electromyometry (M18H2E)

Computer enhancement of electromyometric feedback has been shown to greatly add to patient appreciation and participation in the neuromuscular reeducation program by making feedback more graphic and easier to understand (for the patient as well as the practitioner) in terms of visual and auditory feedback modes. Some of the graphic designs currently in use for single electromyometer activity enhancement have shown themselves to be remarkably suited for increasing patient motivation and enthusiastic participation. One of the more effective graphic designs currently available requires the patient to keep the EMM activity line showing on the computer's video screen between parallel lines which cross the screen in various patterns or shapes. After the activity line has transversed the screen, a numerical readout in one corner of the screen tells the patient what percentage of the time the activity line was above, within, or below the parallel line pattern. The time it takes to cross the screen can be preset by the operator.

During the balance phase, computer enhancement is especially effective if the computer instrumentation allows representative activity from both electromyometers to appear on the video monitor simultaneously as distinctly different lines. Effectiveness increases if the patient can clearly relate the two to a graphic screen design which helps demonstrate their relative activity levels. Two fixed parallel lines across the screen placed relative to the action lines may be adequate.

During the functional phase, if computer enhancement is used, a single activity line from the monitored hypotonic muscle is compared to a preset line which represents the arbitrary level of control. If the activity level from the electromyometer exceeds or goes above the line during one of the talking exercises, the patient is said to be out of control and should be instructed to discontinue talking until the activity level is once again below the control line.

Session termination (M18H2F)

At the end of the treatment session the electromyometer and associated feedback mode instrumentation should be turned off.

The surface electrodes should be carefully removed and the electrode paste carefully cleansed from the patient's skin before the patient is allowed to leave the treatment area. A surface electrode count should be conducted after removal from the skin to prevent them from leaving when the patient does.

Partial List of Conditions Treated (M18H3)

Treatable Causes (M18H3A)

Extrafusal muscle spasm (C41)

Habitual joint positioning (C48)
Muscular weakness (C23)
Neuromuscular tonic imbalance (C30)
Psychoneurogenic neuromuscular hypertonus (C29)
Restricted joint range of motion (C26)
Sympathetic hyperresponse (C38)

Syndromes (M18H3B)

CERVICAL (NECK) PAIN (S73)
EARACHE (S40)
HEADACHE PAIN (S02)
HYSTERIA/ANXIETY REACTION (S59)
OSTEOARTHRITIS (S21)
TORTICOLLIS (WRY NECK) (S65)

Precautions and Contraindications (M18H4)

No precautions or contraindications have been listed or detailed for electromyometric feedback, except in connection with the stimulation provided by tactile and proprioceptive feedback modes. For precautions and contraindications associated with these feedback modes refer to the Electrical Stimulation *Precautions and Contraindications* [M0204], Transcutaneous Nerve Stimulation *Precautions and Contraindications* [M0404], and Vibration *Precautions and Contraindications* [M0804] sections.

Notes Aside (M18H5)

The success rate logged for the treatment with torticollis utilizing EMM feedback is fairly low. Treatment successes have occurred, but they are relatively rare and successful therapy has take many months (sometimes up to a year) to complete, and really depends upon the patient's unconscious being willing to not indulge in this symptom of a defense mechanism.

Bibliography (M18H6)

Braud, pp. 67–89.
Brudny, pp. 925–932.
Swaan, pp. 9–11.
Taylor, pp. 169–173.

ELECTROMYOMETRIC FEEDBACK POST IMMOBILIZATION NEUROMUSCULAR REEDUCATION
(M18I)

Definition (M18I1)

Electromyometric feedback has proven to be effective in the treatment of post immobilization syndrome. The post immobilization syndrome is characterized by tonically set muscular lengths, resulting from long-term splinting (casting) and/or joint and muscle disuse, which has the general effect of drastically restricting joint ranges of motion.

When a post immobilization syndrome occurs, the antagonistic musculature around the joint immobilized are universally affected. As one set of muscles loses the ability to shorten, the antagonistic musculature loses the ability to lengthen.

Procedures of Application (M18I2)

EMM instrumentation (M18I2A)

Instrumentation required for neuromuscular reeducation with electromyometric feedback of musculature which is affected by the post immobilization syndrome includes one or two electromyometers with the most sophisticated feedback modes available, which can be preset to register different sensitivity levels. Common sensitivity settings include the 0 to 10, 0 to 50, 0 to 100, and 0 to 500 microvolt ranges. The basic electromyometer has both audio and visual (meter or light bar) feedback modes. These modes can be enhanced by built-in feedback mode variables (different sounds and multiple visual displays) or by external computer enhancement of both modes. Some types of EMM instrumentation also offer tactile feedback modes and/or robotic-mechanical displays.

Electrode placement (M18I2B)

The muscles or muscle groups to be monitored should be connected to the electromyometers via electrode cable(s) terminating in surface electrodes affixed to the skin over the muscles or muscle group(s). Electrode placement should be with the active pickup electrodes placed over the body of the involved muscles, fairly close to the motor end plate, if it has been determined.

If single muscle or muscle group (in the same general area) monitoring is utilized, the muscle or muscle group is exclusively monitored, with both electrode cable active leads affixed over it. If dual muscle monitry of synergistic partners is used, each of the electrode cable active pickup leads should be attached to a surface electrode placed above each of the muscles monitored.

Evaluation of the post immobilization syndrome (M18I2C)

The patient should be sitting in an upright position which will be comfortable for the duration of the evaluation session.

During the evaluation phase, the patient should remain unexposed to the feedback modes, unable to see, hear or feel changes in myoelectric production as reflected by EMM feedback.

Each muscle or muscle group which articulates the involved joint should be monitored to determine how much it has been effected by prolonged disuse. In general, prolonged immobility promotes atrophy of the muscle, at least to the point of loss of tone and strength. The involved musculature usually develops hypotonicity and will be unable to voluntarily produce myoelectric activity at normal levels. To establish the norm for the involved muscle, its counterpart on the contralateral side should be monitored at rest and during isometric contraction. The activity levels should be noted, recorded, and contrasted with the myoelectric activity produced by the involved muscle or muscle group.

As a general rule, a normal muscle will produce myoelectric activity ranging between 150 and 300 or more microvolts during a sustained contraction against resistance. The activity level depends upon the size and strength of the muscle and the intensity of contraction. A muscle suffering from a post immobilization syndrome may, by contrast, produce a sustained level of myoelectric activity of only 50 to 120 microvolts (or less) when contracting against resistance.

Neuromuscular reeducation utilizing EMM feedback (M18I2D)

The neuromuscular reeducation of musculature affected by a post immobilization syndrome using-electromyometric feedback follows a learning pattern of facilitation, inhibition, balance and function in neuromuscular reeducation.

Facilitation phase (M18I2D01) Typically, in the facilitation phase, the muscle or muscle group which has been set long (relative to its antagonist) by the prolonged immobilization should be monitored. The patient should be asked to voluntarily increase the muscle or muscle group myoelectric activity, as reflected by the available feedback modes, by contracting the muscle. The patient should be asked to maintain the muscle contraction for at least a six-second period and then relaxing and resting for six seconds before attempting another contraction.

The electromyometers should be set at a sensitivity level which best demonstrates positive or increasing myoelectric change. The patient must be able to see, hear and/or feel changes in myoelectric activity, during the entire procedure, to subliminally learn from them. If the muscle or muscle group is severely underactive, the electromyometer will most commonly have to be set initially at the level of greatest sensitivity (0 to 10 microvolt range) so that any increase in myoelectric activity will be substantially demonstrated by all the feedback modes. If the muscle or muscle group is more active, the electromyometer should be set at a less sensitive level (0 to 30 or 0 to 50 microvolt ranges, for example) to provide a gauge of success as well as providing the patient with an opportunity to strive to produce more myoelectric activity at observable levels.

Inhibition phase (M18I2D02) In the inhibition phase, typically, the muscle or muscle group which has been set short (relative to its antagonist) should be monitored, and the patient should be asked to increase myoelectric activity from the muscle or muscle group by a small amount (less than 10 microvolts) and then to actively suppress the activity level below the the previous resting level. The attempt to increase myoelectric activity from the muscle group(s) should last no longer than six seconds, but the suppression stage should last from six seconds to several minutes until adequate inhibition is attained or until failure must be admitted and another attempt made. The patient should be in a sitting position. The electromyometer should be set at a sensitivity level which will adequately demonstrate patient inhibitory success (optimally at the 0 to 10 microvolt sensitivity level, but less sensitive levels may be used initially if resting levels exceed the 0 to 10 microvolt range).

Balance phase (M18I2D03) In the balance phase, two EMM setups are required to allow the patient the opportunity of attempting to increase myoelectric activity from the facilitated muscle or muscle group while simultaneously decreasing or lowering myoelectric activity from the inhibited muscle or muscle group. This phase of training is generally conducted in the sitting position. The electromyometer monitoring the muscle group to be facilitated should be set at the least sensitive level that best magnifies and reflects increases in myoelectric activity, without exceeding the maximum setting level (over 50 microvolts if the sensitivity level is set at a 0 to 50 microvolt range). The EMM monitoring the muscle to be inhibited should be set at the most sensitive level that best magnifies and reflects decreases in myoelectric activity (shows the largest range of feedback, ideally at the 0 to 10 microvolt sensitivity level).

The choice of feedback modes relative to the muscle and neuromuscular problem is of paramount importance. When in the balance phase of neuromuscular reeducation, the feedback modes that are the most attention-getting should be used in conjunction with the electromyometer monitoring the hardest muscle to control.

As the patient is successful in the balance phase, joint ranges should improve. Attempts should always be conducted at the maximum end of range.

Functional phase (M18I2D04) In the functional stage the patient must immediately use regained joint ranges, muscle tone and coordination in functional activities without benefit of EMM feedback, preferably during isotonic exercise (progression through range of motion against resistance) which not only maintains gains but also increases strength in the muscles exercised.

Computer enhancement (M18I2E)

Computer enhancement of electromyometric feedback has been shown to greatly add to patient appreciation and participation in the neuromuscular reeducation program by making feedback more graphic and easier to understand (for the patient as well as the practitioner) in terms of visual and auditory feedback modes. Some of the graphic designs currently in use for single electromyometer activity enhancement have shown themselves to be remarkably suited for increasing patient motivation and enthusiastic participation. One of the more effective graphic designs currently available requires the patient to keep the EMM activity line showing on the computer's video screen between parallel lines which cross the screen in various patterns or shapes. After the activity line has transversed the screen, a numerical readout in one corner of the screen tells the patient what percentage of the time the activity line was above, within, or below the parallel line pattern. The time it takes to cross the screen can be preset by the operator.

During the balance phase, computer enhancement is especially effective if the computer instrumentation allows representative activity from both electromyometers to appear on the video monitor simultaneously as distinctly different lines. Effectiveness increases if the patient can clearly relate the two to a graphic screen design which helps demonstrate their relative activity levels. Two fixed parallel lines across the screen placed relative to the action lines may be adequate.

Session termination (M1812F)

At the end of the treatment session the EMM and associated feedback mode instrumentation should be turned off.

The surface electrodes should be carefully removed and the electrode paste carefully cleansed from the patient's skin before the patient is allowed to leave the treatment area. A surface electrode count should be conducted after removal from the skin to prevent them from leaving when the patient does.

Partial List of Conditions Treated (M1813)

Treatable Causes (M1813A)

Habitual joint positioning (C48)
Joint ankylosis (C42)
Joint approximation (C47)
Muscular weakness (C23)
Neuromuscular tonic imbalance (C30)
Restricted joint range of motion (C26)

Syndromes (M1813B)

BURSITIS (S26)
CHONDROMALACIA (S15)
FROZEN SHOULDER (S64)
JOINT SPRAIN (S30)
MYOSITIS OSSIFICANS (S67)
OSTEOARTHRITIS (S21)
POST IMMOBILIZATION (S36)
POST WHIPLASH (S55)
SHOULDER/HAND (S57)

Precautions and Contraindications (M1814)

No precautions or contraindications have been listed or detailed for electromyometric feedback, except in connection with the stimulation provided by tactile and proprioceptive feedback modes. For precautions and contraindications associated with these feedback modes refer to the Electrical Stimulation *Precautions and Contraindications* [M0204], Transcutaneous Nerve Stimulation *Precautions and Contraindications* [M0404], and Vibration *Precautions and Contraindications* [M0804] sections.

Notes Aside (M1815)

Middle range immobilization (M1815A)

If the involved joint was immobilized in middle range, the muscle relationship may need to be reversed back and forth relative to the muscles facilitated and inhibited to regain range of motion in both directions.

Eclectic treatment of post immobilization syndrome (M1815B)

EMM feedback neuromuscular reeducation is generally effective, by itself, for regaining a portion of the range of motion lost through prolonged immobilization. It should generally be combined eclectically with other modalities which help increase muscle tone and strength as well as increase range of motion and relieve any accompanying pain (refer to Auricular Electroacustimulation, Pain Relief [M06A]; Cryotherapy, Analgesia/Anesthesia [M10C]; Electrical Facilitation of Endorphin Production (Acusleep) [M03]; Electrical Stimulation (ES), Muscle Lengthening [M02A] or Muscle Reeducation/Muscle Toning [M02B]; Exercise, Muscle Strengthening [M12A] or Muscle Relengthening [M12C]; Hydrotherapy, Muscle Tension Amelioration [M15C] or Range of Motion Enhancement and Physical Reconditioning [M15D]; Joint mobilization [M13]; Massage, Deep Soft Tissue Mobilization [M23F]; Ultrahigh Frequency Sound, Phonophoresis [M01A] or Scar Tissue Management [M01D]; and/or Vibration, Neuromuscular management [M08A]).

Bibliography (M1816)

Krebs, pp. 1017–1021.
Swaan, pp. 9–11.
Taylor, pp. 174–177.

ELECTROMYOMETRIC FEEDBACK BREATHING PATTERN CORRECTION (M18J)

Definition (M18J1)

Electromyometric feedback has proven to be effective in correcting abnormal breathing patterns which have naturally evolved and become habitual. The abnormal breathing pattern discussed here is commonly referred to as "the accessory breathing pattern".

In the accessory breathing pattern, the scaleni muscles (usurping the natural role of the diaphragm) are used to lift the rib cage to create negative pressure within the chest cavity, causing the lungs to fill. The drawback to the accessory breathing pattern is that it is only twenty-five percent as efficient as diaphragm breathing pattern, requiring a higher energy expenditure while providing a poorer tidal volume. Such inefficiency may greatly contribute to the effect of the various pathological conditions which restrict the passage of air into the lungs or interfere with gaseous exchange (asthma, bronchitis, emphysema, etc.) which leads to the symptomatology associated with oxygen deprivation (anxiety, fear, syncope, etc.).

The muscles most commonly involved in breathing pattern correction are the upper trapezius, scaleni and the diaphragm muscles. The treatment of choice for breathing pattern correction is neuromuscular reeducation utilizing electromyometric feedback.

Procedures of Application (M18J2)

EMM instrumentation (M18J2A)

Instrumentation required for neuromuscular reeducation of the accessory breathing pattern includes two electromyometers equipped with the most sophisticated feedback modes available, and the accessories required for their communication with the patient. The most suitable electromyometer will have the built-in option of being able to be preset to register different sensitivity levels (common EMM sensitivity settings include the 0 to 10, 0 to 50, 0 to 100, and 0 to 500 microvolt ranges).

The basic electromyometer has both audio and visual (meter or light bar) feedback modes. These modes can be enhanced by built in feedback mode variables (different sounds, multiple visual displays, and computer enhancement). Some electromyometry instrumentation may even provide the option of a tactile feedback mode and/or robotic-mechanical display modes.

Electrode placements (M18J2B)

The EMM's should be attached to the patient via electrode cables which terminate as they connect with surface electrodes affixed to the skin overlying the muscles monitored. These muscles usually include the diaphragm and scaleni muscles.

The electrode placement site for the scaleni muscle is at the approximate midpoint of an imaginary line drawn along the base of the neck from the origin of the sternocleidomastoideus muscle on the medial head of th sternum to the anterior edge of the upper trapezius muscle. If the scaleni muscle groups on either side of the neck are to be simultaneously monitored by the same electromyometer, the two active electrode cable leads should be split and each one attached over an opposing scaleni site, while the ground electrode is attached on the patient's chest and back, equidistant from each active electrode site.

For monitry of the diaphragm, the two active cable leads should be attached to surface electrodes attached to the anterior abdominal wall approximately three inches apart, in relatively symmetrical opposing positions across body midline, distal to the xiphoid process and just distal to the edge of the costal cartilage.

Evaluation of the breathing pattern (M18J2C)

Following electrode placement, the electromyometer should be turned on and the pattern reflected by the two electromyometers monitoring the scaleni and diaphragm muscles observed and noted.

During the evaluation phase, the patient should remain unexposed to the feedback modes, unable to see, hear or feel changes in myoelectric production as reflected by EMM feedback.

An accessory breathing pattern will be apparent if myoelectric activity from the scaleni muscle(s) increases appreciably (more then 5 to 10 microvolts) during inspiration while myoelectric activity from the diaphragm fails to increase above a few microvolts. If the more desirable diaphragm breathing pattern is being employed, the myoelectric activity from the diaphragm will increase more than 5 to 10 microvolts during inspiration, while that from the scaleni will increase only 2 or 3 microvolts.

Subliminal learning pattern (M18J2D)

If the accessory breathing pattern is positively established, neuromuscular reeducation is indicated which has the best chance of affecting a change in subliminal neuromuscular breathing patterns. The most efficient approach is to utilize electromyometric feedback training to establish a better coordination of the muscles involved. Clinical experience has taught that subliminal neuromuscular reeducation occurs best when the subliminal learning pattern employed for neuromuscular reeducation of central nervous system dysfunction is utilized. This process follows the procedural training phases of facilitation, inhibition, balance and function (refer to Electromyometric Feedback, Post Cerebral Accident Neuromuscular Reeducation [M18A] for further details).

Facilitation phase (M18J2D01) To affect breathing pattern correction and the establishment of the diaphragm breathing pattern the diaphragm is monitored during the facilitation phase. During this phase, the patient is asked to voluntarily increase myoelectric activity from the diaphragm muscle which is reflected by the available feedback modes. The electromyometers should be set at a sensitivity level which best demonstrates the myoelectric change that occurs. For subliminal learning to occur, the patient must be able to see, hear and/or feel changes in myoelectric activity during the entire procedure. If the diaphragm is severely under active, the electromyometer will have to be set initially at the level of greatest sensitivity (0 to 6 or 0 to 10 microvolt ranges) so that any increase in myoelectric activity will be demonstrated substantially for patient appreciation. If the diaphragm is active to an appreciable degree, the monitoring electromyometer should be set at a less sensitive level (0 to 30 or 0 to 50 microvolt ranges, for example) to provide a gauge of success as well as providing the patient with an opportunity to strive to produce more myoelectric activity at appreciable levels.

Inhibition phase (M18J2D02) During the inhibition phase, the scaleni muscle(s) should be monitored. The patient initially should be asked to voluntarily increase the myoelectric activity slightly (less than 10 microvolts), and then to voluntarily inhibit the myoelectric activity below the previous resting level. The attempt to increase myoelectric activity from the muscles should last no longer than six seconds, and the inhibition phase should last until adequate inhibition is attained (from six seconds to several minutes) or until failure must be admitted and another attempt made. The electromyometer should be set at a sensitivity level which will adequately demonstrate patient inhibitory success (optimally at the 0 to 10 microvolt level, but less sensitive levels may used if initial resting levels exceed the 0 to 10 microvolt range).

Balance phase (M18J2D03) In the balance phase, two electromyometers are required to allow the patient the opportunity of facilitating diaphragm myoelectric activity while simultaneously inhibiting scaleni myoelectric activity. The choice of feedback modes, relative to the muscle monitored, is of paramount importance. Success may be claimed when the patient demonstrates the ability to limit scaleni myoelectric activity to between 6 and 8 microvolts while simultaneously producing myoelectric activity from the diaphragm during inspiration in excess of 10 microvolts.

Functional phase (M18J2D04) During the functional phase, the patient should be instructed to engage in diaphragm breathing at home. As an aid to self awareness, the patient's hand may be placed over diaphragm to affirm rise and fall during the inspiration/expiration cycle. The patient should engage in this exercise at least ten minutes a day until the diaphragm breathing pattern has become habitual.

Computer enhancement of electromyometry (M18J2E)

Computer enhancement of electromyometric feedback has been shown to greatly add to patient appreciation of, and participation in, the neuromuscular reeducation program associated with the correction of the breathing pattern by making feedback more graphic and easier to understand for the patient as well as the practitioner. Some of the graphic designs currently in use for single electromyometer activity enhancement have shown themselves to be remarkably suited for increasing patient motivation and enthusiastic participation. One of the more effective graphic designs currently available requires the patient to keep the EMM activity line showing on the computer's video screen between parallel lines which cross the screen in various patterns or shapes. A multiple tooth spike shape is used for diaphragm facilitation to avoid hyperventilation. After the activity line has transversed the screen, a numerical readout in one corner of the screen tells the patient what percentage of the time the activity line was above, within, or below the parallel line pattern. The time it takes to cross the screen can be preset by the operator.

During the balance phase, computer enhancement is especially effective if the computer instrumentation allows representative activity from both electromyometers to appear on the video monitor simultaneously as distinctly different lines. Effectiveness increases if the patient can clearly relate the two to a graphic screen design which helps demonstrate their relative activity levels. Two fixed parallel lines across the screen placed relative to the action lines may be adequate.

Session termination (M18J2F)

At the end of the treatment session the EMM and associated feedback mode instrumentation should be turned off.

The surface electrodes should be carefully removed and the electrode paste carefully cleansed from the patient's skin before the patient is allowed to leave the treatment area. A surface electrode count should be conducted after removal from the skin to prevent them from leaving when the patient does.

Partial List of Conditions Treated (M18J3)

Conditions treated in the field (M18J3A)

Asthma

Treatable Causes (M18J3B)

Lung fluid accumulation (C13)
Psychoneurogenic neuromuscular
   hypertonus (C29)
Reduced lung vital capacity (C17)
Sympathetic hyperresponse (C38)
Visceral organ dysfunction (C02)

Syndromes (M18J3C)

CHRONIC BRONCHITIS/EMPHYSEMA (S13)
HYSTERIA/ANXIETY REACTION (S59)
PNEUMONIA (S58)

Precautions and Contraindications (M18J4)

No precautions or contraindications have been listed or detailed for electromyometric feedback, except in connection with the stimulation provided by tactile and proprioceptive feedback modes. For precautions and contraindications associated with these feedback modes refer to the Electrical Stimulation, *Precautions and Contraindications* [M0204], Transcutaneous Nerve Stimulation, *Precautions and Contraindications* [M0404], and/or Vibration, *Precautions and Contraindications* [M0804] sections.

Notes Aside (M18J5)

Patients who suffer from asthma may find that after having been successful in creating the diaphragm breathing pattern via EMM training, even before the pattern has become habitual, they may be able to avoid the symptoms of oxygen deprivation by sitting down and engaging in the functional exercise described above. This exercise increases the efficiency of the breathing pattern, thereby increasing the availability of oxygen and may aid in the avoidance of the anxiety and panic associated with an asthma attack.

Bibliography (M18J6)

Johnson, Lee, pp. 826–831.
Taylor, pp. 156–161.

ELECTROMYOMETRIC FEEDBACK MUSCLE TONING (M18K)

Definition (M18K1)

Electromyometry reflects myoelectric activity on a qualitative level, providing a measure of how much of the muscle is contracting at any given moment. By using EMM feedback to reflect that qualitative measure back to the patient, the patient may be aided to increase the general tone of (contract) the muscle monitored, even if the peripheral sensory nerve supply to the muscle has been impaired or if central nervous system damage has resulted in the muscle flaccidity.

Procedures of Application (M18K2)

EMM Instrumentation (M18K2A)

Instrumentation required for muscle toning with electromyometric feedback, includes one electromyometer (EMM), which communicates with the target muscle via an electrode cable. The electrode cable terminates in connection with surface electrodes coupled to the skin with electrode paste and secured in place by some form of strapping, tape, or other adhesive device. The electromyometer should have the option of being adjustable to register different sensitivity levels. Common EMM sensitivity settings include the 0 to 10, 0 to 50, 0 to 100, and 0 to 500 microvolt ranges).

The basic EMM has both audio and visual (meter or light bar) feedback modes. These modes can be enhanced by built-in feedback mode variables (different sounds and multiple visual displays) or by external computer enhancement of both these modes. Some electromyometry instrumentation also offers tactile feedback modes and/or robotic-mechanical displays, and may also be equipped with a graphic recorder to serve as a hard copy record of electromyometric activity.

Electrode placement (M18K2B)

The electromyometers are attached to the patient via electrode cables which terminate by attaching to surface electrodes affixed to the skin over the muscles to be monitored.

Evaluation (M18K2C)

The process of myoelectric assessment begins with the selection of a muscle to be monitored or tested. The patient should be placed in a position which allows easy access to the target muscle but prevents the patient from seeing the visual feedback. The audio and the tactile feedback modes (if available) should be turned off. The electromyometer should then be turned on.

The patient should be asked to relax and avoid voluntary contraction of the muscle while the resting myoelectric activity level is registered and noted. The involved joint should be fixed in place and no motion of it allowed during contraction. The patient should then be asked to isometrically contract the target muscle and attempt to hold the maximum contraction for six seconds, followed by relaxation. The level of myoelectric activity at the beginning and at the end of the sustained contraction, and the maximum level attained should be registered and recorded, as well as the level of myoelectric activity level attained at the beginning of the relaxation period. The time it takes for the myoelectric activity to reach the precontraction levels (or below) should be be noted (in minutes and/or seconds), as should the minimum level of activity attained.

Normal myoelectric activity from a muscle maintaining varies from 80 to 600 or more microvolts, depending upon the size and strength of the muscle. If possible, comparison of a muscle suspected of being weak or hypotonic should have its myoelectric activity level compared with that of its contralateral counterpart, or that of the same muscle belonging to a normal individual.

EMM muscle toning process (M18K2D)

Simply stated, the process of muscle toning with electromyometric feedback involves having the patient increase myoelectric activity reflected by the feedback modes from the electromyometer. A single muscle, muscle group, or synergistic muscle pair may be monitored.

The electromyometer should be set at a sensitivity level which best demonstrates positive or increasing myoelectric change. The patient must be able to see, hear and/or feel changes in myoelectric activity during the entire procedure to learn from them on a subliminal level. If a muscle is extremely weak, the electromyometer will most commonly have to be set at the level of greatest sensitivity (0 to 6 or 0 to 10 microvolt ranges), so that any increase in myoelectric activity will be substantially demonstrated by all the feedback modes. If the muscle is more active the unit should be set at a less sensitive level (0 to 30 or 0 to 50 microvolt ranges, for example) to provide a gauge of success, as well as providing the patient with an opportunity to strive to produce more activity at observable levels.

The patient should be asked to increase the myoelectric activity generated by the muscle or synergistic muscle pair, as reflected by the available feedback, to a level of myoelectric activity suggested by the practitioner as the goal to be reached. This can be set mechanically if a computer is being used, or as suggested arbitrary levels of activity measured by threshold responses, meter measurement marks, or frequency rates. The myoelectric goal should be set at a level which will adequately demonstrate failure or success. The myoelectric goal should be potentially attainable, but enough out of reach that the patient must exert a concentrated effort to attain it. The patient should be asked to hold the maximum level of activity for at least five or six seconds or more before relaxing and taking at least a six-second rest period before the next attempt.

Computer enhancement (M18K2E)

Computer enhancement of electromyometric feedback has been shown to greatly add to patient appreciation and participation in the neuromuscular reeducation program by making feedback more graphic and easier to understand (for the patient as well as the practitioner) in terms of visual and auditory feedback modes. Some of the graphic designs currently in use for single electromyometer activity enhancement have shown themselves to be remarkably suited for increasing patient motivation and enthusiastic participation. One of the more effective graphic designs currently available requires the patient to keep the EMM activity line showing on the computer's video screen between parallel lines which cross the screen in various patterns or shapes. After the activity line has transversed the screen, a numerical readout in one corner of the screen tells the patient what percentage of the time the activity line was above, within, or below the parallel line pattern. The time it takes to cross the screen can be preset by the operator.

During the balance phase, computer enhancement is especially effective if the computer instrumentation allows representative activity from both electromyometers to appear on the video monitor simultaneously as distinctly different lines. Effectiveness increases if the patient can clearly relate the two to a graphic screen design which helps demonstrate their relative activity levels. Two fixed parallel lines across the screen placed relative to the action lines may be adequate.

Session termination (M18K2F)

At the end of the treatment session the EMM and associated feedback mode instrumentation should be turned off.

The surface electrodes should be carefully removed and the electrode paste carefully cleansed from the patient's skin before the patient is allowed to leave the treatment area. A surface electrode count should be conducted after removal from the skin to prevent them from leaving when the patient does.

Partial List of Conditions Treated (M18K3)

Treatable Causes (M18K3A)

Habitual joint positioning (C48)
Hypermobile/instable joint (C10)
Muscular weakness (C23)
Neuromuscular tonic imbalance (C30)
Pathological neuromuscular hypotonus (C33)
Pathological neuromuscular dyscoordination (C35)
Peripheral nerve injury (C07)
Restricted joint range of motion (C26)
Spastic sphincter (C40)
Trigger point formation (C01)

Syndromes (M18K3B)

BELL'S PALSY (S66)
CEREBRAL PALSY (S25)
CERVICAL DORSAL OUTLET (SCALENICUS ATTICUS/CERVICAL RIB) (S09)
HYSTERIA/ANXIETY REACTION (S59)
MYOSITIS OSSIFICANS (S67)
NEUROMUSCULAR PARALYSIS (S22)
OSTEOARTHRITIS (S21)
PIRIFORMIS (S04)
POST CEREBRAL VASCULAR ACCIDENT (CVA) (S07)
POST IMMOBILIZATION (S36)
POST PERIPHERAL NERVE INJURY (S23)
POST SPINAL CORD INJURY (S24)
RADICULITIS (S29)
SCIATICA (S05)
SHOULDER/HAND (S57)

Precautions and Contraindications (M18K4)

No precautions or contraindications have been listed or detailed for electromyometric feedback, except in connection with the stimulation provided by tactile and proprioceptive feedback modes. For precautions and contraindications associated with these feedback modes refer to the Electrical Stimulation, *Precautions and Contraindications* [M0204], Transcutaneous Nerve Stimulation, *Precautions and Contraindications* [M0404], and/or Vibration, *Precautions and Contraindications* [M0804] sections.

Notes Aside (M18K5)

Toning of muscles paralyzed by peripheral nerve injury (M18K5A)

The use of electromyometry and its feedback appears to be indispensable for the treatment of paralysis which has occurred as the result of sensory nerve loss secondary to peripheral nerve injury. It has been demonstrated that patients may thus regain the ability to voluntarily contract muscles that are permanently lacking in sensory innervation (proprioception). This rehabilitation feat was thought to be absolutely impossible a few years ago, before Marinacci's studies of EMG training of paralyzed muscles resulting from polio induced peripheral nerve injury.

Muscle toning during joint immobilization (M18K5B)

Muscle toning of through the use of the electromyometer and its feedback modes has proven to be valuable in keeping muscles toned when an associated joint has been immobilized by prolonged splinting or casting. This may be accomplished by cutting a window in the casting or splinting material large enough to accommodate the placement of surface electrodes on the target muscle(s), and having the patient follow the protocol suggested above [M18K2].

Bibliography (M18K6)

Brudny, pp. 925–932.
Gosling, pp. 883–884.
LeVeau, pp. 1410–1415.
Lucca, pp. 200–203.
Marinacci, pp. 57–71.
Stratford, pp. 279–283.

ELECTROMYOMETRIC FEEDBACK ELECTROMYOMETRIC EVALUATION (M18L)

The evaluation of muscle states through the use of electromyometric feed back is based upon the assessment of myoelectric activity produced by the muscle monitored.

Procedures of Application (M18L01)

The process of myoelectric assessment begins with the selection of a muscle to be monitored or tested.

The patient should be placed in a position which allows easy access to the target muscle but prohibits the patient from seeing the EMM feedback of any myoelectric activity, and the audio and/or tactile feedback modes should be turned off.

The electrodes should be attached to the target muscle and the electromyometer turned on. The electromyometer should be set at a sensitivity level which best demonstrates the resting level of myoelectric activity, and will allow a clear assessment of any myoelectric increase which may accompany contraction. If a muscle is fairly inactive the unit should be set at the level of maximum sensitivity (0 to 10 microvolts for most machines) to best reflect myoelectric activity for adequate registration and recording. If a muscle is extremely active" at rest" the electromyometer may have to be set at a fairly insensitive level (0 to 50, 0 to 100 or even 0 to 500 microvolts) to both reflect and register resting activity variations, if any, as well as register and record myoelectric activity increases accompanying maximum contraction of the the muscle.

The maximum myoelectric activity level should be within the instrument range of measurement.

Instrumentation (M18L01A)

Myoelectric assessment is made through the use of an electromyometer (EMM), which communicates with the target muscle via an electrode cable. The electrode cable should terminate in connection with surface electrodes coupled to the skin with electrode paste, secured in place by some form of strapping, tape, or other adhesive device. The electromyometer should have the option of being adjustable to register different sensitivity levels. Common EMM sensitivity settings include the 0 to 10, 0 to 50, 0 to 100, and 0 to 500 microvolt ranges. It may be equipped with a graphic recorder for a hard copy record of electromyometric activity, or its feedback enhanced by computer to magnify and/or record such activity.

Session termination (M18L01B)

At the end of the treatment session the EMM and associated feedback mode instrumentation should be turned off.

The surface electrodes should be carefully removed and the electrode paste carefully cleansed from the patient's skin before the patient is allowed to leave the treatment area. A surface electrode count should be conducted after removal from the skin to prevent them from leaving when the patient does.

DETERMINATION OF NEUROMUSCULAR HYPERTONICITY (M18L1)

Neuromuscular Hypertonicity (M18L1A)

Neuromuscular hypertonicity is a condition in which a muscle enjoys relatively high tone when at rest and/or during phasic activity (voluntary contraction). Hypertonicity of the musculature is the result of the influence of supraspinal structures (cerebellum, cerebral motor cortex, neural pathways, etc.) over intrafusal and extrafusal muscle, via the tonic stretch reflex, to maintain hypertonus. It occurs because of impaired efferent or afferent communication between the supraspinal structures and the musculature, when the mechanisms responsible for maintaining the servo system or the tonic stretch reflex are caused to maintain a higher then normal muscle tone either because of a failure of a communication loop or group of loops to convey accurate neuromuscular information between the various structures comprising the neuromuscular system, or a failure of the supraspinal or infraspinal end organs themselves to operate correctly and to send the right afferent or efferent messages. The result is that the transported information is either inadequate and/or distorted. If communication loop damage is responsible, the impairment may be to the communication loop(s) which transport information from the muscles to the supraspinal structures, from one supraspinal structure to another, or from higher supraspinal structures to the muscles. Regardless of the cause, hypertonus occurs when the intrafusal structures (muscle spindles) are made to contract and maintain tension, thereby causing the extrafusal muscle to also contract and maintain tension (refer to the Neuromuscular System [R002] and/or the Muscle Spindle [R001] for further details).

EMM Determination of Neuromuscular Hypertonicity (M18L1B)

For the determination of myoelectric neuromuscular hypertonicity, The patient should be asked to relax and avoid voluntary contraction of the muscle while the resting myoelectric activity level is registered and noted. The involved joint should be fixed in place and no motion of it allowed during contraction. The patient should then be asked to isometrically contract the target muscle and attempt to hold the maximum contraction for six seconds, followed by relaxation. The level of myoelectric activity at the beginning and at the end of the sustained contraction, and the maximum level attained should be registered and recorded, as well as the level of myoelectric activity level attained at the beginning of the relaxation period. The time it takes for the myoelectric activity to reach the precontraction levels or below should be be noted in minutes and/or seconds.

Neuromuscular hypertonicity is defined in terms of a muscle's myoelectric activity at rest and the speed at which a return to resting levels is made after a sustained contraction. "Normal" resting myoelectric activity is generally in the range of 2 to 4 microvolts, depending on the muscle monitored, and a return to resting levels after a sustained contraction should take no more a second or two. The severity of the hypertonicity should be judged in comparison with these standards of measure. Neuromuscular spasticity is demonstrated by an involuntarily sustained level of myoelectric activity above precontraction resting levels, usually of several microvolts, which appears as a rhythmic fluctuation of feedback between resting maximum and minimum microvolt levels. This rhythmic activity may last for several seconds or minutes after suspension of the sustained isometric contraction.

Partial List of Conditions Evaluated (M18L1C)

Treatable Causes (M18L1C01)

Dental pain (C22)
Extrafusal muscle spasm (C41)
Habitual joint positioning (C48)
Interspinous ligamentous strain (C03)
Joint ankylosis (C42)
Joint approximation (C47)
Ligamentous strain (C25)
Menorrhalgia (C20)
Migraine (vascular headache) (C34)
Muscular weakness (C23)
Nerve root compression (C04)
Pathological neuromuscular hypertonus (C28)
Peripheral nerve injury (C07)
Psychoneurogenic neuromuscular hypertonus (C29)
Restricted joint range of motion (C26)

Tinnitus (C32)
Trigger point formation (C01)

Syndromes (M18L1C02)

ACHILLES TENDONITIS (S52)
BELL'S PALSY (S66)
BURSITIS (S26)
CALF PAIN (S50)
CAPSULITIS (S27)
CARPAL TUNNEL (S35)
CEREBRAL PALSY (S25)
CERVICAL DORSAL OUTLET (SCALENICUS ATTICUS/CERVICAL RIB) (S09)
CERVICAL (NECK) PAIN (S73)
CHEST PAIN (S14)
COLITIS (S42)
EARACHE (S40)
ELBOW PAIN (S47)
FACET (S19)
FASCIITIS (S20)
FOOT PAIN (S51)
FOREARM PAIN (S54)
FROZEN SHOULDER (S64)
HAND/FINGER PAIN (S17)
HEADACHE PAIN (S02)
HIGH THORACIC BACK PAIN (S48)
HYSTERIA/ANXIETY REACTION (S59)
KNEE PAIN (S46)
LOWER ABDOMINAL PAIN (S62)
LOW BACK PAIN (S03)
MENSTRUAL CRAMPING (S60)
MIGRAINE (VASCULAR) HEADACHE (S18)
NEUROMUSCULAR PARALYSIS (S22)
OSTEOARTHRITIS (S21)
PHOBIC REACTION (S44)
PIRIFORMIS (S04)
POST CEREBRAL VASCULAR ACCIDENT (CVA) (S07)
POST IMMOBILIZATION (S36)
POST PERIPHERAL NERVE INJURY (S23)
POST SPINAL CORD INJURY (S24)
POST WHIPLASH (S55)
RADICULITIS (S29)
REFERRED PAIN (S01)
SCIATICA (S05)
SHOULDER/HAND (S57)
SHOULDER PAIN (S08)
SINUS PAIN (S33)
TEMPOROMANDIBULAR JOINT (TMJ) PAIN (S06)
THIGH PAIN (S49)
TINNITUS (S70)
TOOTHACHE/JAW PAIN (S41)
TORTICOLLIS (WRY NECK) (S65)
WRIST PAIN (S16)

Notes Aside (M18L1D)

Preventing the Piriformis Syndrome (M18L1D01)

Electromyometric evaluation techniques have proven valuable for the evaluation of conditions which have resulted from muscular pressure upon peripheral nerve roots or nerves. Excessive myoelectric activity from a muscle associated with a particular joint motion, possibly responsible for nerve compression, and correlated to increased neuralgia, may provide valuable information to the patient as an indicator of what might be desirable or undesirable physical maneuvers. This process has shown itself particularly valuable in showing the patient which physical exercises to avoid in order to preclude the piriformis syndrome; for such determination, the piriformis should be monitored while the patient engages in various physical exercises.

DETERMINATION OF NEUROMUSCULAR HYPOTONICITY
(M18L2)

Neuromuscular Hypotonicity (M18L2A)

Neuromuscular hypotonicity is a condition in which the supraspinal central nervous system structures fail to maintain a normal resting muscle tone. This implies a general weakness or lack of contractile strength of the muscle. Hypotonicity of the muscle may stem from a damaged or impaired supraspinal structures, spinal cord injury (refer to Central Nervous System Dysfunction [R006]), peripheral nerve injury, or from simple disuse, as in the post immobilization syndrome (refer to Adverse Effects of Bed Rest [R008]).

EMM Determination of Neuromuscular Hypotonicity (M18L2B)

For the determination of myoelectric neuromuscular hypotonicity, the patient should be asked to relax and avoid voluntary contraction of the muscle while the resting myoelectric activity level(s) are registered and noted. The involved joint should be fixed in place and no motion of it allowed during contraction. The patient should then be asked to isometrically contract the target muscle and to attempt hold the maximum contraction for six seconds, followed by relaxation. The level of myoelectric activity at the beginning and at the end of the sustained contraction, and the maximum level attained should be registered and recorded, as should the level of myoelectric activity attained at the beginning of the relaxation period. The time it takes for the myoelectric activity to reach the precontraction levels or below should be be noted in minutes and/or seconds, as should the minimum level of activity attained.

Neuromuscular hypotonicity is defined in terms of a muscle's myoelectric activity at rest, the maximum level of myoelectric activity during contraction, and the ability to maintain near maximum myoelectric activity levels during a sustained contraction. "Normal" resting myoelectric activity is generally in the range of 2 to 4 microvolts, depending on the muscle, and the normal range of myoelectric activity during maximal contraction is between 80 and in excess of 500 microvolts; it is generally dependent upon the size of the muscle monitored. A "normal muscle" should be able sustain a myoelectric activity level which approaches the normal maximum level for that muscle for six seconds. The standard for myoelectric activity from a given muscle should be established relative to that muscle type by measuring the myoelectric activity from a normal muscle of that type, ideally on the patient's contralateral side, or on a more normal subject, under the test conditions suggested above, and comparing its activity levels with the activity levels from the target muscle.

Partial List of Conditions Evaluated (M18L2C)

Treatable Causes (M18L2C01)

Calcific deposit (C08)
Habitual joint positioning (C48)
Hypermobile/instable joint (C10)
Muscular weakness (C23)
Pathological neuromuscular hypotonus (C33)
Peripheral nerve injury (C07)
Restricted joint range of motion (C26)

Syndromes (M18L2C02)

ACHILLES TENDONITIS (S52)
ADHESION (S56)
BELL'S PALSY (S66)
BURNS (SECOND AND/OR THIRD DEGREE) (S72)
BURSITIS (S26)
CAPSULITIS (S27)
CARPAL TUNNEL (S35)
CEREBRAL PALSY (S25)
CERVICAL DORSAL OUTLET (SCALENICUS ATTICUS/CERVICAL RIB) (S09)
DUPYTREN'S CONTRACTURE (S34)
FROZEN SHOULDER (S64)
MYOSITIS OSSIFICANS (S67)
NEUROMUSCULAR PARALYSIS (S22)
OSTEOARTHRITIS (S21)
PIRIFORMIS (S04)
POST CEREBRAL VASCULAR ACCIDENT (CVA) (S07)
POST IMMOBILIZATION (S36)
POST PERIPHERAL NERVE INJURY (S23)
POST SPINAL CORD INJURY (S24)
RADICULITIS (S29)
SCIATICA (S05)
SHINGLES (HERPES ZOSTER) (S37)

SHOULDER/HAND (S57)
TORTICOLLIS (WRY NECK) (S65)

Notes Aside (M18L2D)

Spinal Cord vs. Peripheral Nerve Injury (M18L2D01)

The evaluation of hypotonus is one of the tests which may be utilized to help determine site and extent of neuromuscular paralysis associated with spinal cord injury. If hypotonicity is present six weeks after injury of subdural tissues within the spinal column, a peripheral nerve injury or cauda equina injury should be suspected. A transection of the spinal cord most often results in hypertonicity or spasticity of the muscles innervated by nerves which emerge from the spinal cord below the level of transection. If the injury is a cauda equina injury, the peripheral nerve degenerates back to originating ganglia before regeneration occurs. This causes neuromuscular flaccidity of the muscles innervated by the damaged nerve, which may eventually be reinnervated if the nerve regenerates to its end organs. EMM testing for hypertonus may be used to document such denervation and subsequent reinnervation over time by reflecting returned muscle control and the degree to which muscle contractibility exists.

DETERMINATION OF NEUROMUSCULAR IMBALANCE
(M18L3)

Neuromuscular Imbalance (M18L3A)

Neuromuscular imbalance is a relative concept, suggestive of a condition in which antagonistic muscles, which usually operate as reciprocal partners across a joint or joints, have established an abnormal hypertonic/hypotonic relationship. Muscles in opposition to one another, which normally work in balance (one normally increases myoelectric activity as its antagonist decreases myoelectric activity) to perform free, unhampered, painless motion of a commonly shared joint, develop an abnormal relationship in which one of these antagonists is perpetually too active (too tense) while the other is too under active to maintain a functional active balance, making reciprocal neuromuscular interactions between these muscles difficult or impossible. The result of such imbalances, if the nervous system is undamaged, is (often) pain or feelings of "tension" and/or curtailed active ranges of motion.

Neuromuscular imbalance may be neurogenic (resulting from nervous system injury), psychoneurogenic (resulting from a neurological response to emotional stress or other psychological influence, as in torticollis), or as an indirect result of trauma (excessive stress, strain or external blow) to the muscle or to related structures for which the system fails to tonically correct. In the latter case, muscle spasm or increased muscle tension may result from an unconscious protective reaction response designed to protect the musculature and related structures from further damage ("splinting"), but continue on as a neurological "habit" or symptom of a psychoneurogenic defense mechanism.

Determination of Neuromuscular Imbalance (M18L3B)

For the determination of myoelectric neuromuscular imbalance, antagonistic muscles across a joint or as synergistic opponents should be selected for monitry (refer to the Table of Antagonistic Muscles of Common Interest [T006]).

The patient should initially be asked to relax and avoid voluntary contraction of the muscles monitored while the resting myoelectric activity levels are registered and noted. The involved joint(s) should be fixed in place and no motion allowed during the contraction. The patient should then be asked to isometrically contract one of the two monitored muscles, and to attempt to hold the maximum contraction for six seconds, followed by relaxation. The level of myoelectric activity from each antagonist at the beginning and at the end of the sustained contraction, and the maximum myoelectric activity levels attained should be registered and recorded as well as the level of myoelectric activity levels attained at the beginning of the relaxation period. The time it takes for the myoelectric activity to reach precontraction levels or below should be be noted in minutes and/or seconds. This procedure should then be repeated with the contraction of the antagonist of the first muscle contracted.

Neuromuscular imbalance is defined in terms of relative and comparative myoelectric activity between antagonistic muscles. Agonistic contraction should reflect "normal" myoelectric activity levels as detailed above, while if the muscle balance is to be considered normal, its antagonist's myoelectric activity should remain low or slightly elevated above normal resting levels. An abnormal neuromuscular imbalance will be reflected by excessive myoelectric activity from the antagonist during agonist contraction and/or inappropriately low production of agonist myoelectric activity. Additionally, the hyperactive muscle of the antagonistic pair may be slow in the reduction of myoelectric activity to previous resting levels, following the beginning of the relaxation phase (whether or not it is the contracting agonist or not).

Refer to the description of hypertonicity and hypotonicity determination sections above for details on normal muscle myoelectric activity levels.

Partial List of Conditions Evaluated (M18L3C)

Treatable Causes (M18L3C01)

Extrafusal muscle spasm (C41)
Habitual joint positioning (C48)
Hypermobile/instable joint (C10)
Interspinous ligamentous strain (C03)
Joint ankylosis (C42)
Joint approximation (C47)
Ligamentous strain (C25)
Muscular weakness (C23)

Nerve root compression (C04)
Neuromuscular tonic imbalance (C30)
Pathological neuromuscular
  dyscoordination (C35)
Peripheral nerve injury (C07)
Psychoneurogenic neuromuscular
  hypertonus (C29)
Restricted joint range of motion (C26)
Sympathetic hyperresponse (C38)
Tinnitus (C32)
Trigger point formation (C01)

Syndromes (M18L3C02)

CALF PAIN (S50)
CEREBRAL PALSY (S25)
CERVICAL DORSAL OUTLET (SCALENICUS
  ATTICUS/CERVICAL RIB) (S09)
CERVICAL (NECK) PAIN (S73)
CHEST PAIN (S14)
EARACHE (S40)
ELBOW PAIN (S47)
FACET (S19)
FASCIITIS (S20)
FOOT PAIN (S51)
FOREARM PAIN (S54)
FROZEN SHOULDER (S64)
HAND/FINGER PAIN (S17)
HEADACHE PAIN (S02)
HIGH THORACIC BACK PAIN (S48)
HYSTERIA/ANXIETY REACTION (S59)
KNEE PAIN (S46)
LOWER ABDOMINAL PAIN (S62)
LOW BACK PAIN (S03)
MENSTRUAL CRAMPING (S60)
MIGRAINE (VASCULAR) HEADACHE (S18)
NEUROMUSCULAR PARALYSIS (S22)
OSTEOARTHRITIS (S21)
POST CEREBRAL VASCULAR ACCIDENT
  (CVA) (S07)
POST IMMOBILIZATION (S36)
POST PERIPHERAL NERVE INJURY (S23)
POST SPINAL CORD INJURY (S24)
POST WHIPLASH (S55)
RADICULITIS (S29)
REFERRED PAIN (S01)
SCIATICA (S05)
SHOULDER/HAND (S57)
SHOULDER PAIN (S08)
SINUS PAIN (S33)
TEMPOROMANDIBULAR JOINT (TMJ)
  PAIN (S06)
THIGH PAIN (S49)
TINNITUS (S70)
TOOTHACHE/JAW PAIN (S41)
TORTICOLLIS (WRY NECK) (S65)
WRIST PAIN (S16)

Notes Aside (M18L3D)

Muscle Imbalance and Low Back Pain (M18L3D01)

One of the classic "normal" forms of neuromuscular imbalance is that seen in some types of low back pain syndromes in which the pain stems from the muscles on one side of the spinal column being hypertonic when compared to its antagonist on the other side. This imbalance may occur in the middle thoracic, lower thoracic, lumbar, or combined areas of the back, involving single antagonistic muscles or larger antagonistic synergistic groups.

Neuromuscular Imbalance Associated With Post CVA (M18L3D02)

The most common pathologically induced neuromuscular imbalance is that associated with post cerebral vascular accident spastic neuromuscular patterns.

DETERMINATION OF NEUROMUSCULAR DYSCOORDINATION (M18L4)

Neuromuscular Dyscoordination (M18L4A)

Neuromuscular dyscoordination is a condition in which the supraspinal central nervous system structures fail to coordinate antagonistic and/or synergistic neuromuscular contractions during voluntary joint motion sufficiently to execute meaningful, planned, complicated movements. The extreme forms of neuromuscular dyscoordination stem from injury to supraspinal structures and/or to the communication pathways between the structures which make up the neuromuscular system (refer to the Table of Developmental Reflexes [T007] and/or the Table of Neuromuscular Synergistic Patterns [T008] for a description of some of the abnormal neuromuscular patterns which may be expected).

Determination of Neuromuscular Dyscoordination (M18L4B)

For the determination of neuromuscular dyscoordination, antagonistic muscles (across a joint or as synergistic opponents) should be selected for monitry.

The patient should initially be asked to relax and avoid voluntary contraction of the muscles monitored while the resting myoelectric activity levels are registered and noted. The patient should then be asked to isotonically contract one of the two monitored muscles and perform an appropriate simple joint motion, such as elbow flexion to bring hand from lap to chin while monitoring agonistic biceps and antagonistic triceps. The levels of myoelectric activity from each muscle should be registered and recorded during the attempted movement, and the range of motion measured by goniometry.

A neuromuscular dyscoordination is determined by the establishment of a failure of the agonist to produce an appropriate increase of myoelectric activity, with an appropriate change in the joint range of motion. This is simultaneously accompanied by an inappropriate increase in myoelectric activity (as representative of qualitative muscle contraction) from the antagonistic muscle, which may be substantially great enough to inhibit the desired motion.

A neuromuscular dyscoordination may occur when excessive myoelectric activity is produced from the antagonistic muscle while inappropriately low myoelectric activity is produced from the agonist. Additionally, the hyperactive muscle of an antagonistic pair may be slow in the reduction of myoelectric activity to resting levels following contraction, and spasticity may be evident in the myoelectric activity from the hyperactive muscle, which appears as a regular rhythmical vacillation of EMM feedback modes.

A coordinated antagonistic relationship between the monitored muscles will be reflected by an increase of myoelectric activity production from the agonist and little or no relative increase in myoelectric activity production from the antagonist and accompanied by an appropriate joint range of motion.

Partial List of Conditions Evaluated (M18L4C)

Treatable Causes (M18L4C01)

Dental pain (C22)
Extrafusal muscle spasm (C41)
Habitual joint positioning (C48)
Interspinous ligamentous strain (C03)
Joint approximation (C47)
Ligamentous strain (C25)
Muscular weakness (C23)
Nerve root compression (C04)
Neuromuscular tonic imbalance (C30)
Pathological neuromuscular dyscoordination (C35)
Peripheral nerve injury (C07)
Psychoneurogenic neuromuscular hypertonus (C29)
Restricted joint range of motion (C26)

Syndromes (M18L4C02)

BELL'S PALSY (S66)
CEREBRAL PALSY (S25)
CERVICAL DORSAL OUTLET (SCALENICUS ATTICUS/CERVICAL RIB) (S09)
FACET (S19)
HYSTERIA/ANXIETY REACTION (S59)
NEUROMUSCULAR PARALYSIS (S22)
PIRIFORMIS (S04)
POST CEREBRAL VASCULAR ACCIDENT (CVA) (S07)

POST IMMOBILIZATION (S36)
POST PERIPHERAL NERVE INJURY (S23)
POST SPINAL CORD INJURY (S24)
SCIATICA (S05)
TEMPOROMANDIBULAR JOINT (TMJ) PAIN (S06)
TOOTHACHE/JAW PAIN (S41)
TORTICOLLIS (WRY NECK) (S65)

Notes Aside (M18L4D)

Dyscoordination and Cerebral Palsy (M18L4D01)

The most classic examples of extreme neuromuscular dyscoordination are seen as part of the symptomatology accompanying various forms of cerebral palsy, especially if athetosis is present.

Dyscoordination as a Factor in the TMJ Syndrome (M18L4D02)

One of the most common examples of "normal" dyscoordination of neuromuscular activity may be seen in some cases of temporomandibular joint syndrome, where a masseter muscle may be in dyscoordination in relationship to the opposing masseter and/or the homolateral or contralateral temporalis muscles producing subluxation of the involved joint(s), joint excessive joint approximation, joint disarticulation during the functions of chewing or talking, and pain.

DETERMINATION OF MUSCULAR INNERVATION (M18L5)

Muscular Innervation (M18L5A)

In normal circumstance, a muscle must have both intact motor and sensory innervation. The motor nerve supplies efferent or positive feedback from supraspinal structures (brain and spinal cord) to the musculature, while the sensory nerve supplies proprioceptive afferent or negative feedback from the musculature to the spinal cord and the supraspinal structures. If the muscle is not provided with innervation neuromuscular dysfunction will occur. If motor innervation is not provided to the musculature, no tone or contraction can be developed, but if sensory innervation is deficient or missing muscular tone may develop (in time) and voluntary muscular contraction may eventually be possible.

Denervation of a muscle most often occurs as the result of traumatic peripheral nerve injury.

EMM Determination of Muscular Innervation (M18L5B)

For the assessment of muscular innervation or lack thereof, the patient should be asked to relax and avoid voluntary contraction of the muscle while the resting myoelectric activity level(s) are registered and noted. The involved joint should be fixed in place and no motion of it allowed during contraction. The patient should then be asked to isometrically contract the target muscle and to attempt hold the maximum contraction for six seconds, followed by relaxation. The level of myoelectric activity at the beginning and at the end of the sustained contraction, and the maximum level attained should be registered and recorded, as should the level of myoelectric activity attained at the beginning of the relaxation period. The time it takes for the myoelectric activity to reach the pre-contraction levels or below should be be noted in minutes and/or seconds, as should the minimum level of activity attained.

If a muscle is completely denervated, lacking both motor and proprioceptive nerve supply or just motor nerve supply, the patient will be unable to elicit voluntary increases or decreases in the production of myoelectric activity. If the muscle is only missing sensory nerve supply, the muscle may demonstrate increases and/or decreases in myoelectric activity production with patient effort.

EMM testing for innervation has been found useful for the evaluation of post peripheral nerve injury. "Normal" resting myoelectric activity is generally in the range of 2 to 4 microvolts, depending on the muscle, and the normal range of myoelectric activity during maximal contraction is between 80 and in excess of 500 microvolts; it is generally dependent upon the size of the muscle monitored. A "normal muscle" should be able sustain a myoelectric activity level which approaches the normal maximum level for that muscle for six seconds. The standard for myoelectric activity from a given muscle should be established relative to that muscle type by measuring the myoelectric activity from a normal muscle of that type, ideally on the patient's contralateral side, or on a more normal subject, under the test conditions suggested above, and comparing its activity levels with the activity levels from the target muscle.

Partial List of Conditions Evaluated (M18L5C)

Treatable Causes (M18L5C01)

Bacterial infection (C09)
Calcific deposit (C08)
Extrafusal muscle spasm (C41)
Habitual joint positioning (C48)
Muscular weakness (C23)
Nerve root compression (C04)
Neuroma formation (C14)
Neuromuscular tonic imbalance (C30)
Pathological neuromuscular hypertonus (C28)
Pathological neuromuscular hypotonus (C33)
Pathological neuromuscular dyscoordination (C35)
Peripheral nerve injury (C07)
Restricted joint range of motion (C26)

Syndromes (M18L5C02)

ADHESION (S56)
BACTERIAL INFECTION (S63)
BELL'S PALSY (S66)
CARPAL TUNNEL (S35)
CEREBRAL PALSY (S25)
CERVICAL DORSAL OUTLET (SCALENICUS ATTICUS/CERVICAL RIB) (S09)
HAND/FINGER PAIN (S17)
NEUROMUSCULAR PARALYSIS (S22)
PIRIFORMIS (S04)
POST IMMOBILIZATION (S36)
POST PERIPHERAL NERVE INJURY (S23)
POST SPINAL CORD INJURY (S24)
SCIATICA (S05)
SHOULDER/HAND (S57)

ELECTROMYOMETRIC FEEDBACK ELECTRODE PLACEMENTS (M18M)

List of Diagrams

Abdominal muscles	M18M49
Anterior deltoid muscle	M18M32
Anterior tibialis muscle	M18M16
Back muscles	M18M48
Biceps brachii muscle	M18M34
Biceps femoris muscle	M18M12
Brachioradialis muscle	M18M36
Corrugator supercilii muscle	M18M57
Depressor anguli oris & platysma muscles	M18M64
Depressor labii inferior & mentalis muscles	M18M63
Diaphragm muscle	M18M47
Finger extensor muscle group (extensor digitorum communis, extensor indicis proprius, & extensor digiti quinti proprius)	M18M39
Finger flexor muscle group (flexor digitorum sublimis & flexor digitorum profundus)	M18M40
Flexor digitorum brevis muscle	M18M20
Flexor pollicis brevis muscle	M18M44
Flexor pollicis longus muscle	M18M43
Frontalis muscle	M18M56
Gastrocnemius muscle	M18M14
Gluteus maximus muscle	M18M07
Gluteus medius muscle	M18M05
Gluteus minimus muscle	M18M06
Hamstring muscle group (Semitendinosus & semimembranosus)	M18M13
Hip adductor muscle group (adductor magnus, adductor brevis, adductor longus, pectineus, & gracilis)	M18M08
Hip external rotator muscle group (obturator externus, obturator internus, quadratus femoris, gemellus inferior, gemellus superior, & piriformis)	M18M02
Hip flexor muscle group (psoas major & iliacus)	M18M01
Infraspinatus muscle	M18M28
Levator scapulae muscle	M18M24
Lower trapezius muscle	M18M25
Lumbrical muscles	M18M46
Masseter muscle	M18M53
Masseter & pterygoid muscles	M18M55
Middle deltoid muscle	M18M31
Opponens muscle groups	M18M45
Orbicularis oculi muscle	M18M59
Orbicularis oris muscle	M18M60
Pectoralis major muscle	M18M29
Peroneal muscle	M18M19
Posterior deltoid muscle	M18M30
Posterior tibialis muscle	M18M17
Procerus muscle	M18M58
Pronator quadratus muscle	M18M41
Pterygoid muscle group	M18M54
Quadratus lumborum muscle	M18M04
Quadriceps muscle group (vastus intermedius & rectus femoris)	M18M11
Risorius & buccinator muscles	M18M62
Sartorius muscle	M18M03
Scaleni muscles	M18M51
Scapular adduction muscle group (rhomboids & middle trapezius)	M18M22
Serratus anterior muscle	M18M21
Shoulder extension muscle group (latissimus dorsi & teres major)	M18M26
Soleus muscle	M18M15
Sternocleidomastoideus muscle	M18M50
Supinator muscle	M18M35
Supraspinatus muscle	M18M27
Thumb extensor muscle group (extensor pollicis longus & extensor pollicis brevis)	M18M42
Toe extensor muscle group (extensor digitorum longus & extensor hallucis longus)	M18M18
Triceps muscles	M18M33
Upper trapezius muscle	M18M23
Upper trapezius & scaleni muscles	M18M52
Vastus lateralis muscle	M18M10
Vastus medialis muscle	M18M09
Wrist extensor muscle group (extensor carpi radialis longus, extensor carpi radialis brevis, & extensor carpi ulnaris)	M18M37
Wrist flexor muscle group (flexor carpi radialis & flexor carpi ulnaris)	M18M38
Zygomaticus major muscle	M18M65
Zygomaticus minor & levator anguli oris muscles	M18M61

electromyometric feedback electrode placements

Hip Flexor Group (Psoas Major and Iliacus) (M18M01)

ELECTRODE PLACEMENT

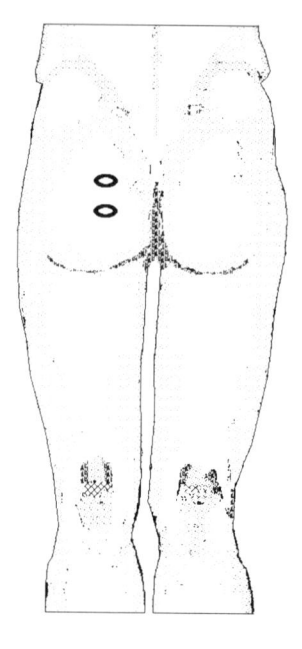

○ SURFACE ELECTRODE

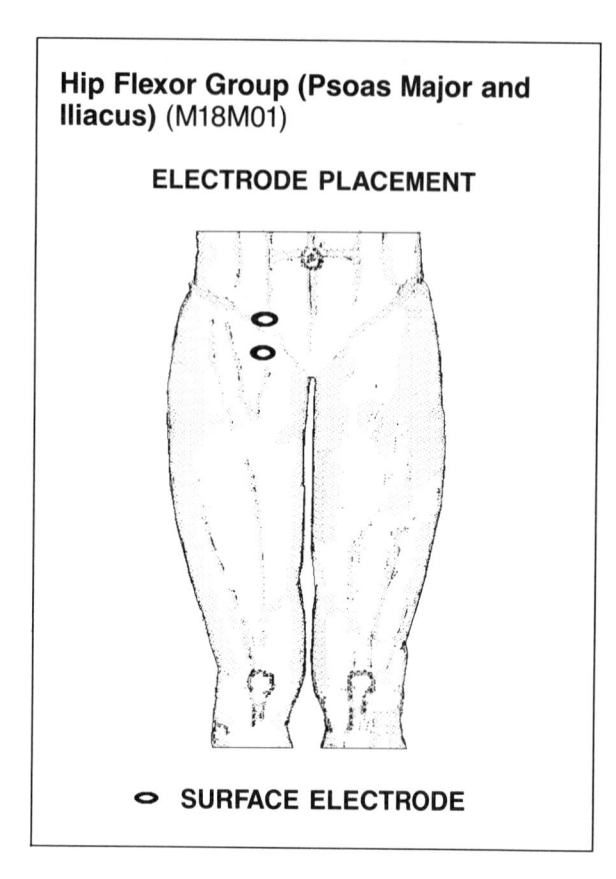

Hip External Rotator Group (Obturator, Externus, Obturator Internus, Quadratus Femoris, Gemellus Inferior, Gemellus Superior, and Piriformis) (M18M02)

ELECTRODE PLACEMENT

○ SURFACE ELECTRODE

electromyometric feedback electrode placements

Sartorius (M18M03)

ELECTRODE PLACEMENT

◯ SURFACE ELECTRODE

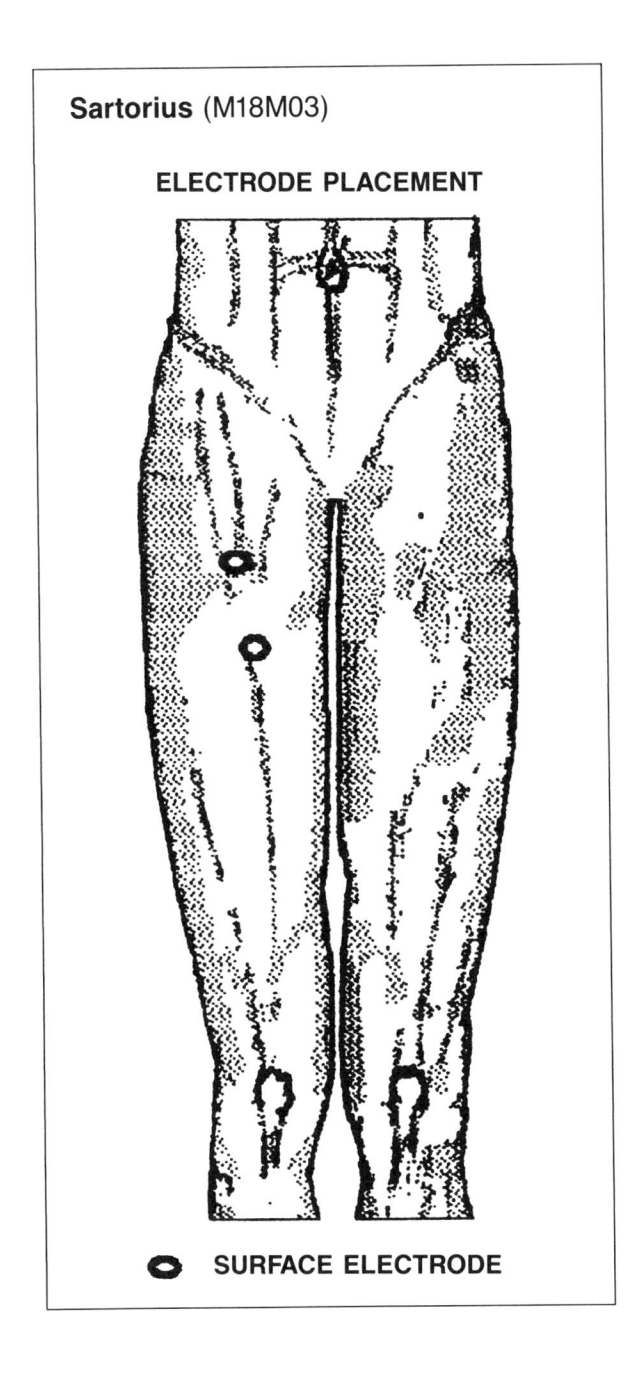

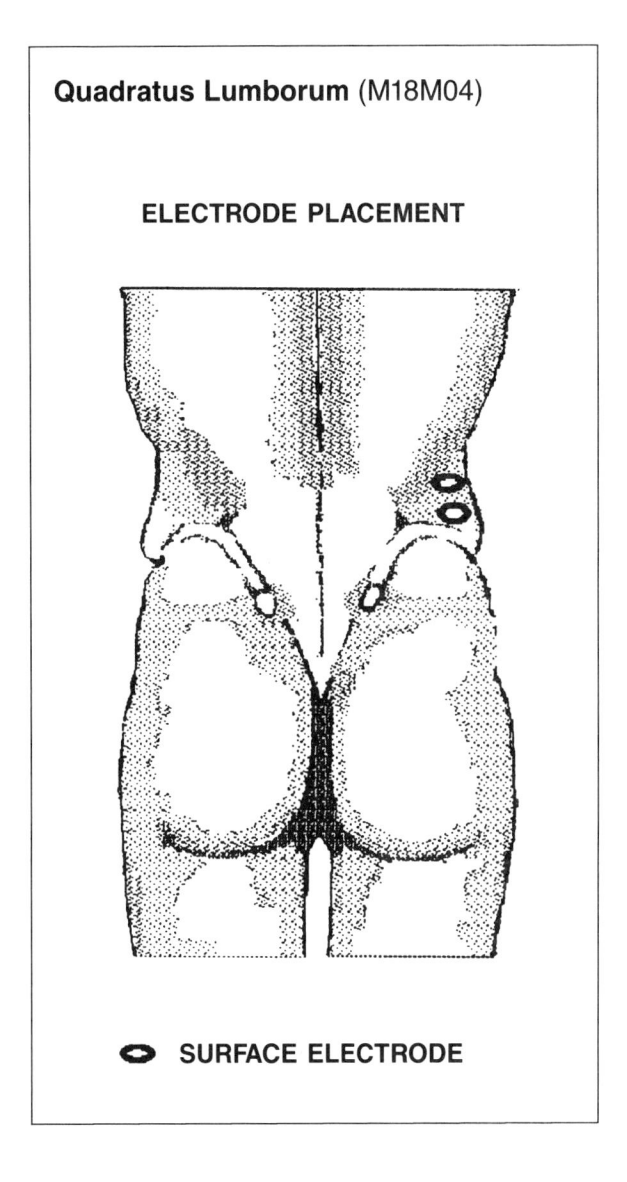

Quadratus Lumborum (M18M04)

ELECTRODE PLACEMENT

◯ SURFACE ELECTRODE

electromyometric feedback electrode placements

Gluteus Medius (M18M05)

ELECTRODE PLACEMENT

○ SURFACE ELECTRODE

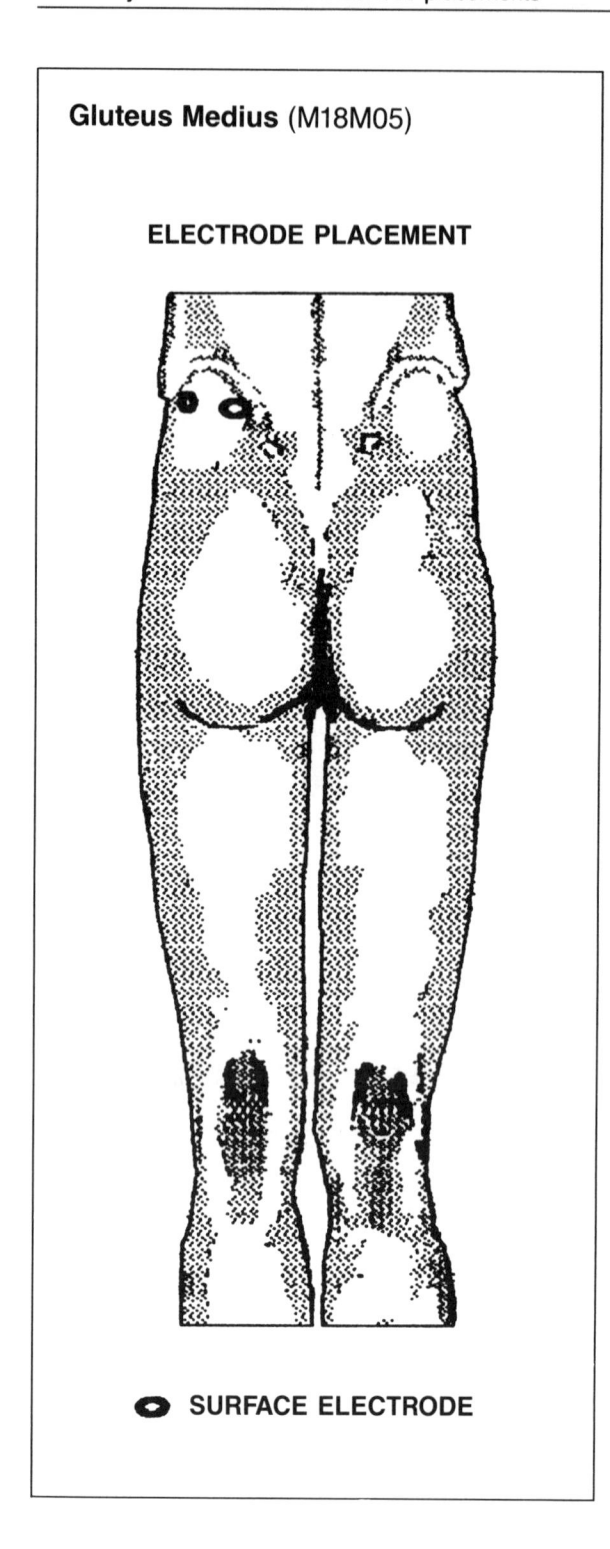

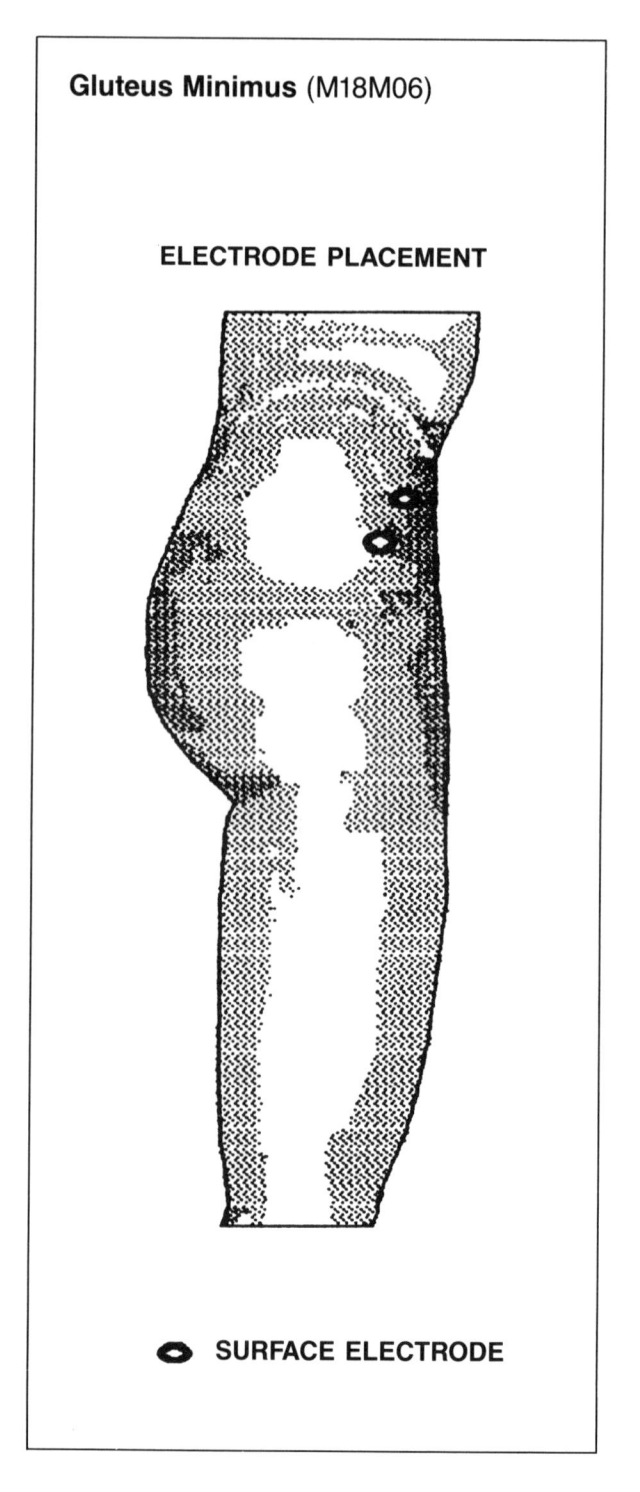

Gluteus Minimus (M18M06)

ELECTRODE PLACEMENT

○ SURFACE ELECTRODE

electromyometric feedback electrode placements

Gluteus Maximus (M18M07)

ELECTRODE PLACEMENT

⊙ SURFACE ELECTRODE

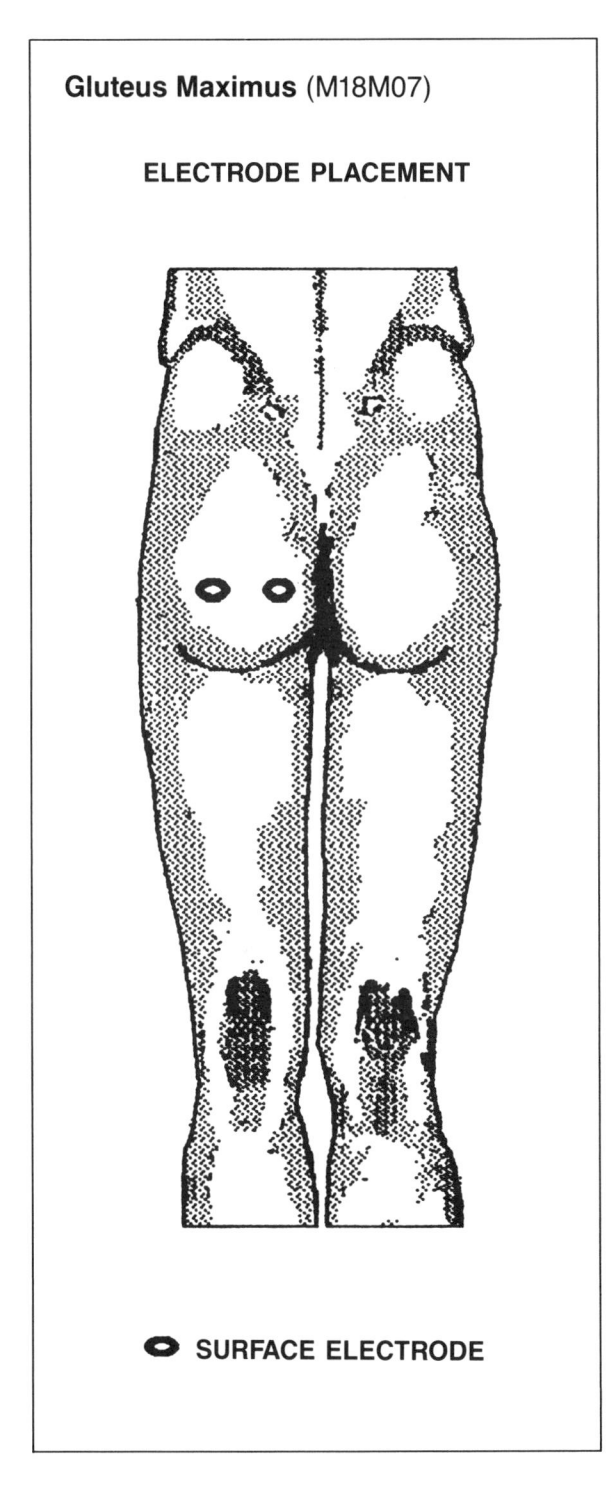

Hip Adductor Group (Adductor Magnus, Adductor Brevis, Adductor Longus, Pectineus, and Gracilis) (M18M08)

ELECTRODE PLACEMENT

⊙ SURFACE ELECTRODE

electromyometric feedback electrode placements

Vastus Medialis (M18M09)

ELECTRODE PLACEMENT

● SURFACE ELECTRODE

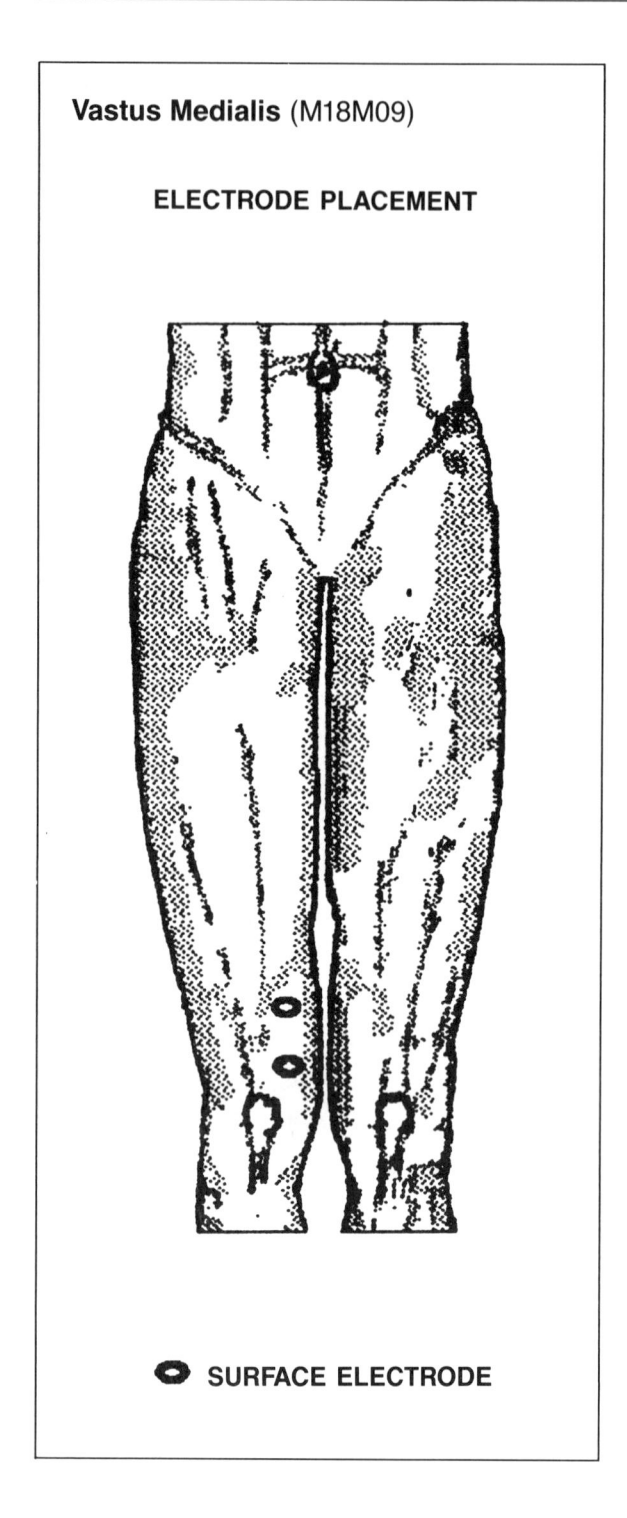

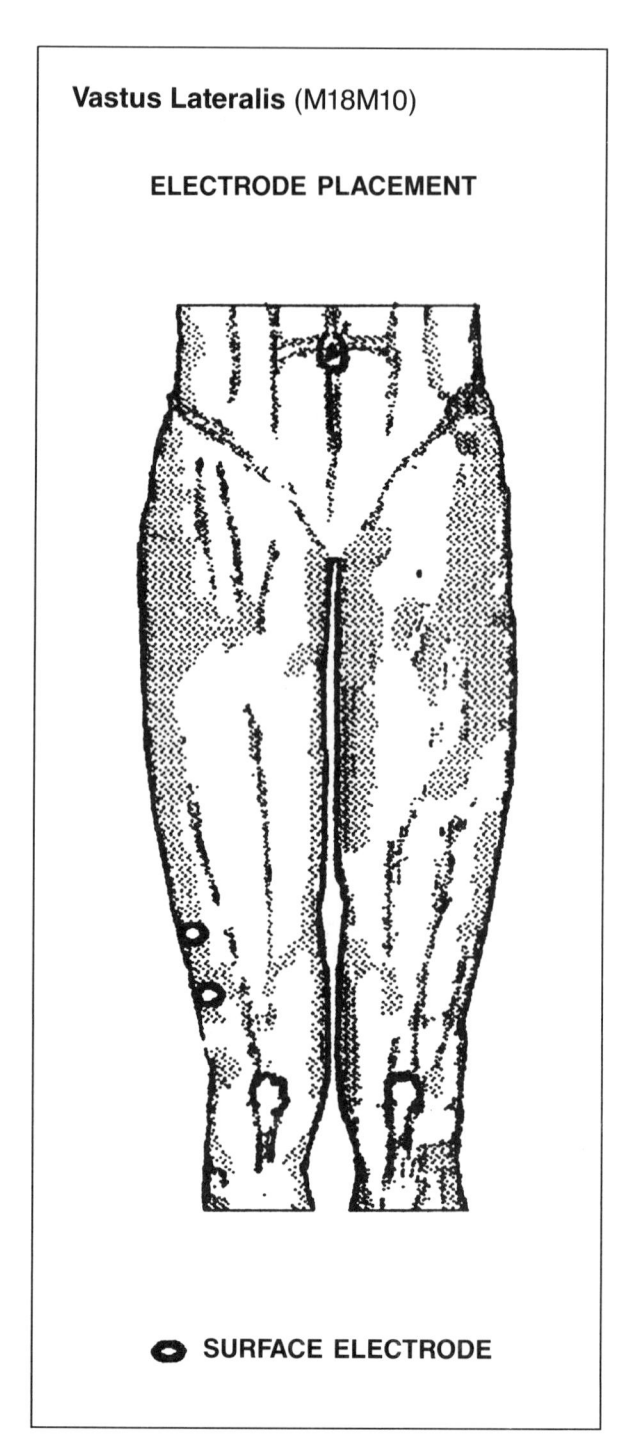

Vastus Lateralis (M18M10)

ELECTRODE PLACEMENT

● SURFACE ELECTRODE

electromyometric feedback electrode placements

Quadriceps Group (Rectus Femoris and Vastus Intermedius) (M18M11)

ELECTRODE PLACEMENT

○ SURFACE ELECTRODE

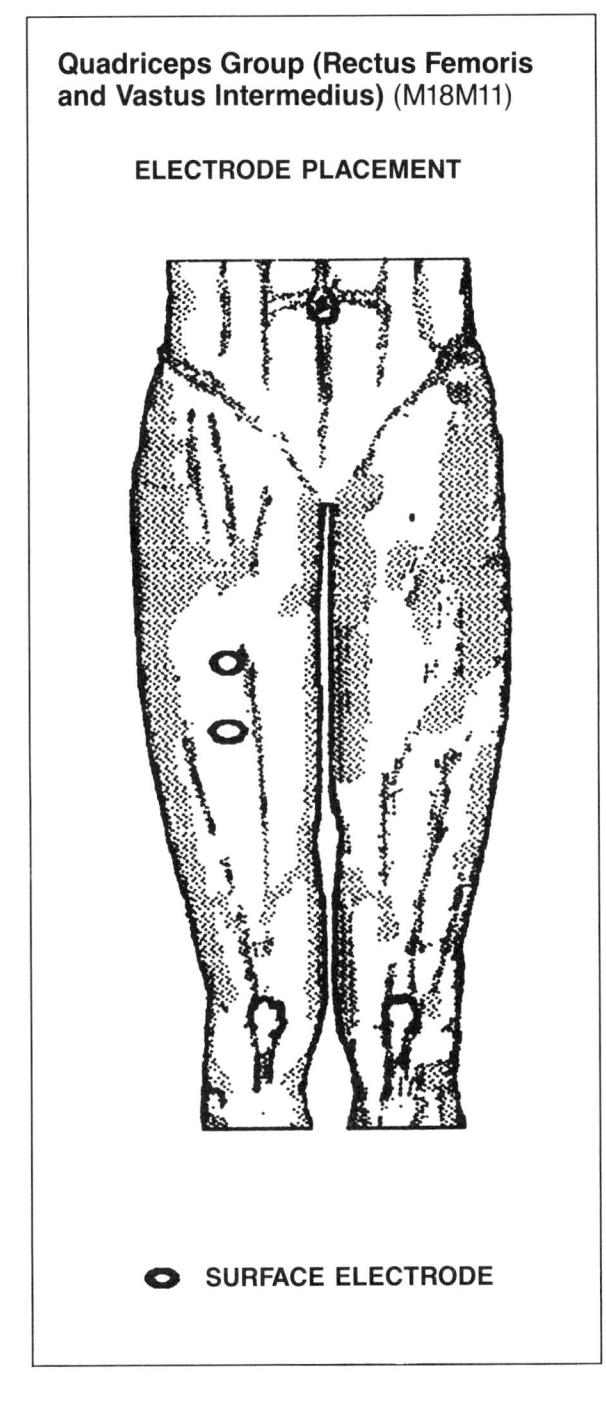

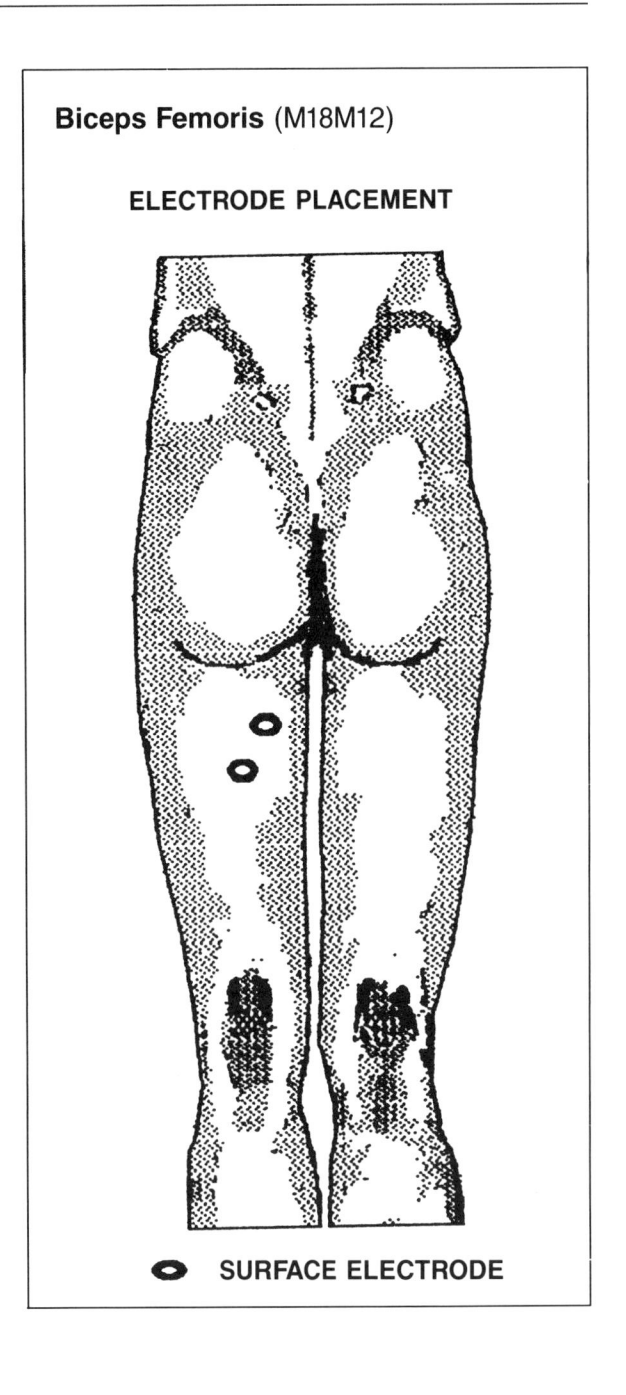

Biceps Femoris (M18M12)

ELECTRODE PLACEMENT

○ SURFACE ELECTRODE

electromyometric feedback electrode placements

Hamstring Group (Semitendinosus and Semimembranosus) (M18M13)

ELECTRODE PLACEMENT

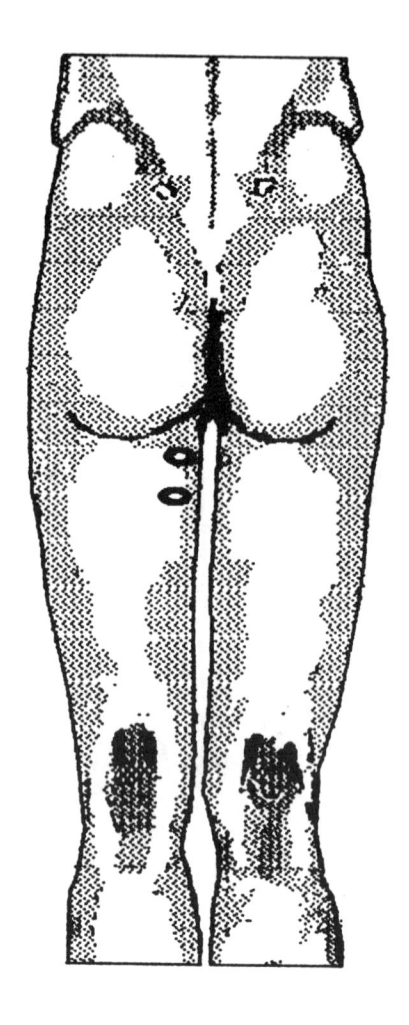

⊙ SURFACE ELECTRODE

Gastrocnemius (M18M14)

ELECTRODE PLACEMENT

⊙ SURFACE ELECTRODE

Anterior Tibialis (M18M16)

ELECTRODE PLACEMENT

o SURFACE ELECTRODE

Soleus Electrode Placement Surface Electrode (M18M15)

o SURFACE ELECTRODE

electromyometric feedback electrode placements

Posterior Tibialis (M18M17)

ELECTRODE PLACEMENT

○ SURFACE ELECTRODE

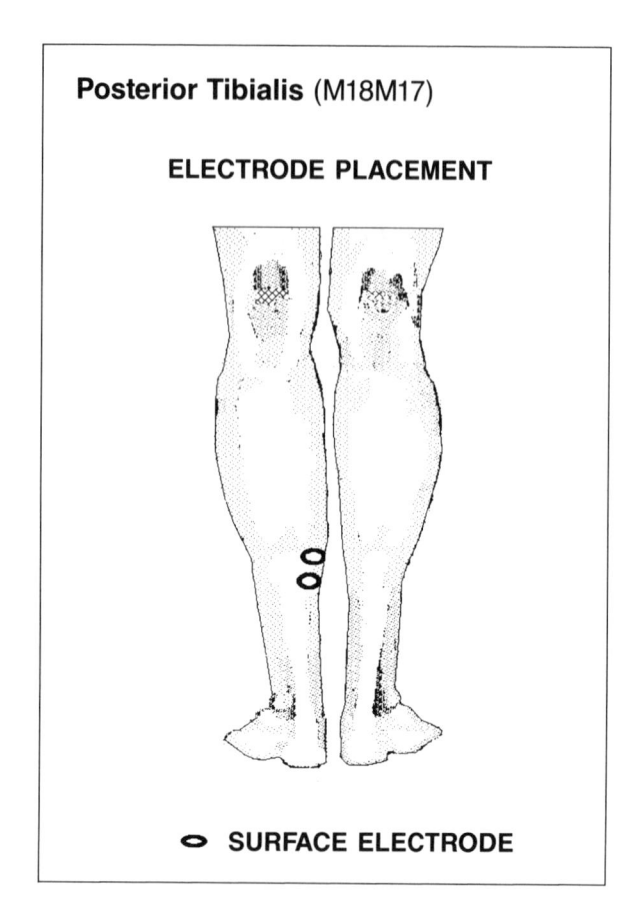

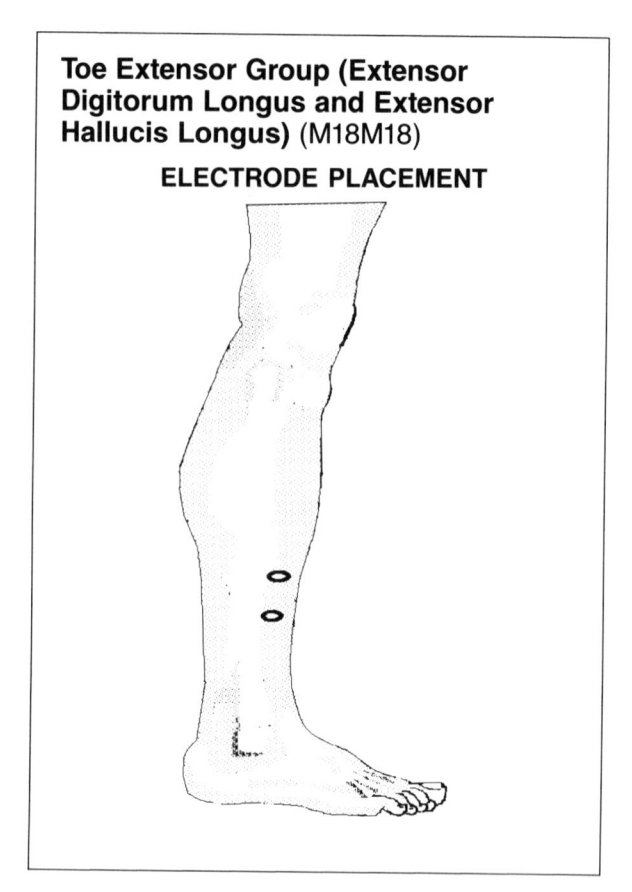

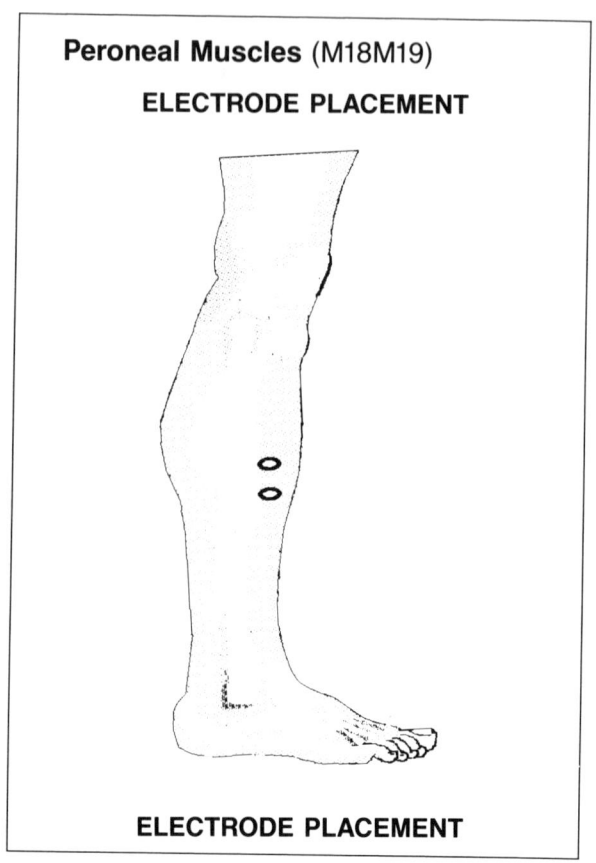

Toe Extensor Group (Extensor Digitorum Longus and Extensor Hallucis Longus) (M18M18)

ELECTRODE PLACEMENT

Peroneal Muscles (M18M19)

ELECTRODE PLACEMENT

ELECTRODE PLACEMENT

electromyometric feedback electrode placements

Flexor Digitorum Brevis (M18M20)

ELECTRODE PLACEMENT

○ SURFACE ELECTRODE

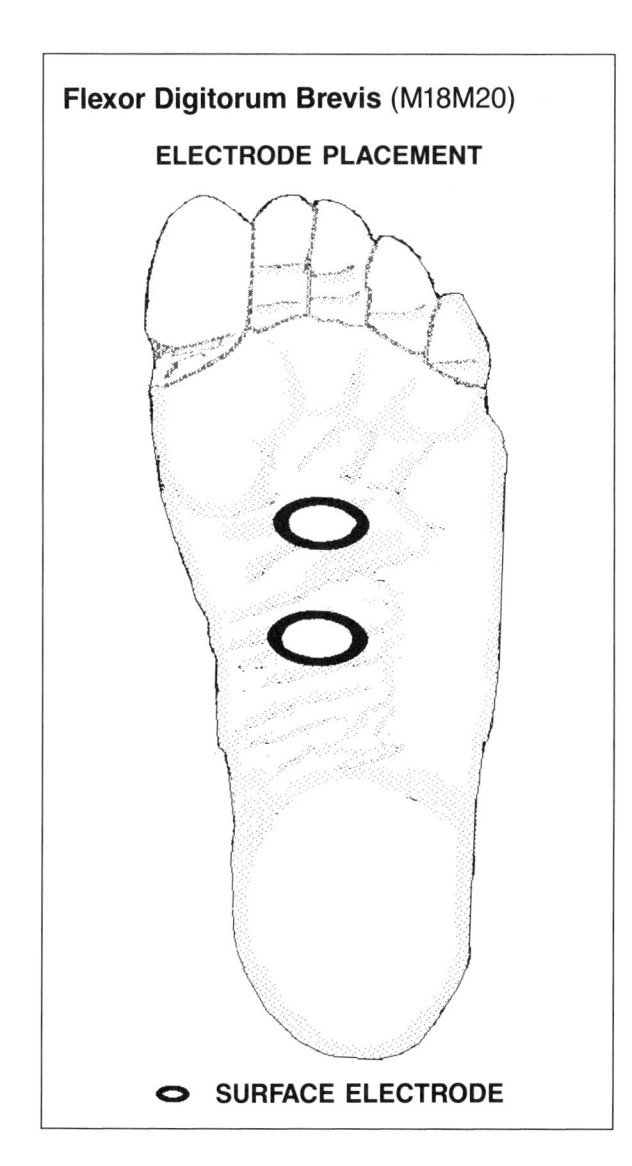

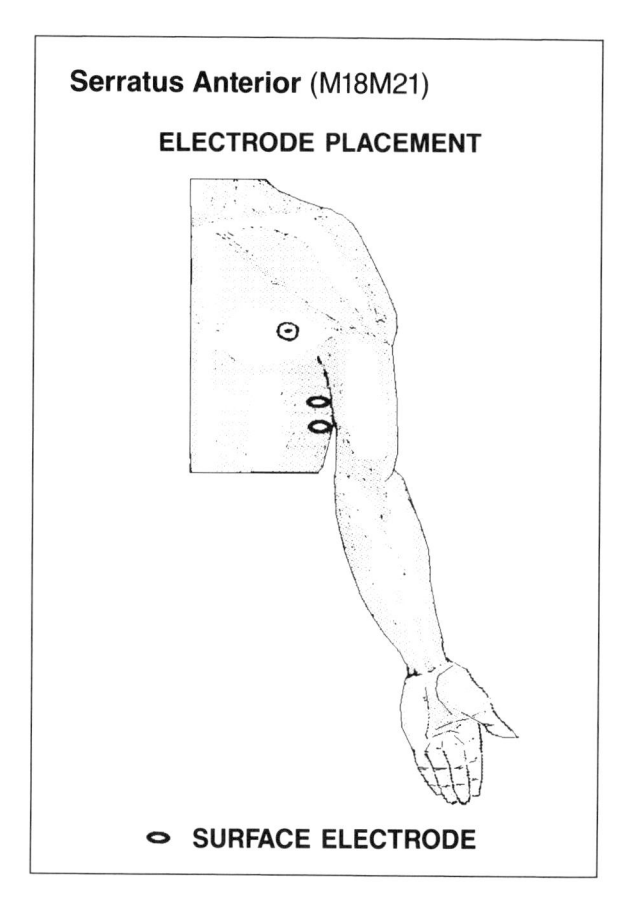

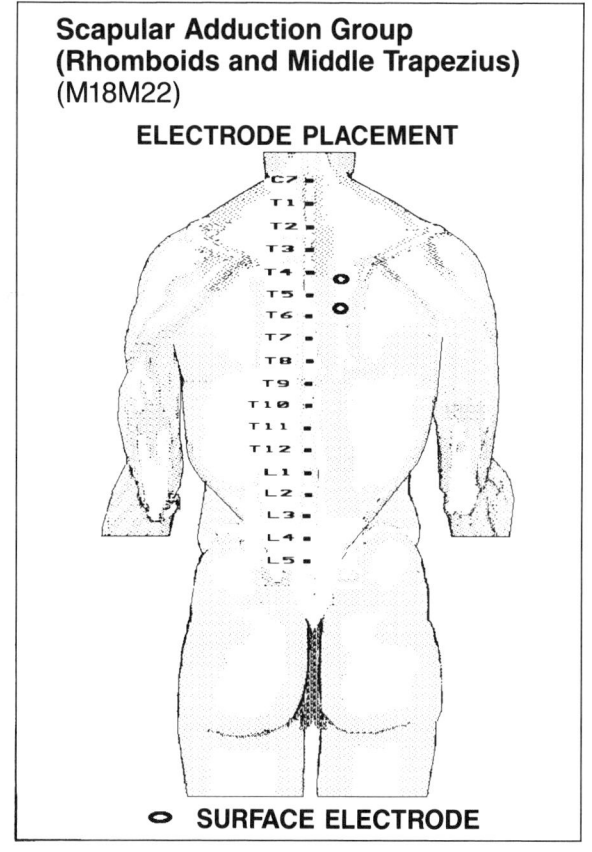

Serratus Anterior (M18M21)

ELECTRODE PLACEMENT

○ SURFACE ELECTRODE

Scapular Adduction Group (Rhomboids and Middle Trapezius) (M18M22)

ELECTRODE PLACEMENT

○ SURFACE ELECTRODE

electromyometric feedback electrode placements

Upper Trapezius (M18M23)
ELECTRODE PLACEMENT

○ SURFACE ELECTRODE

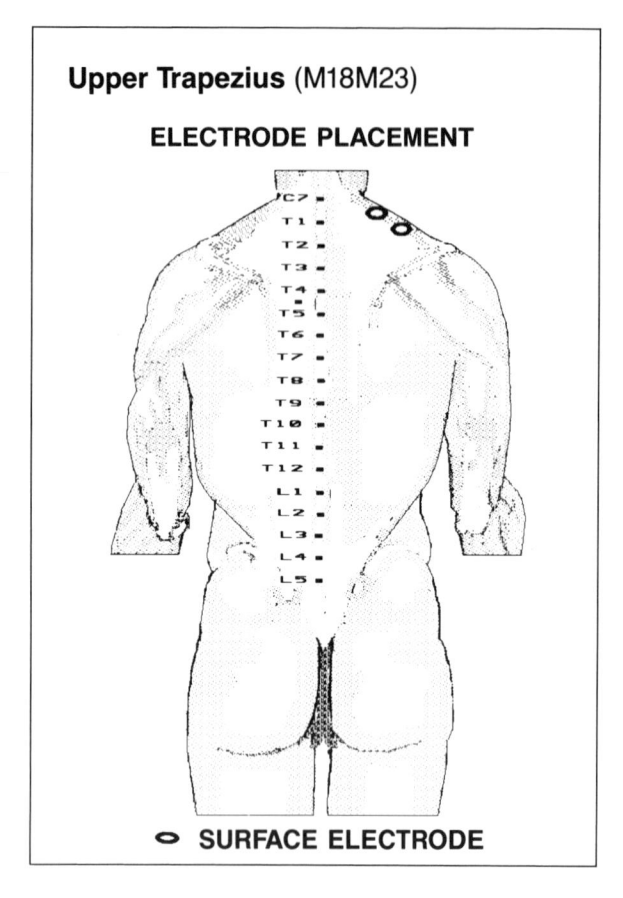

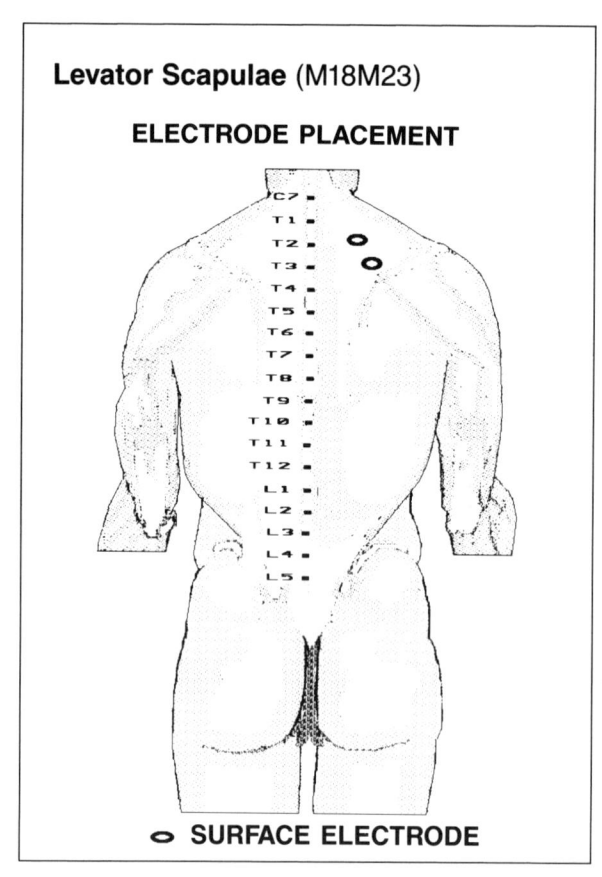

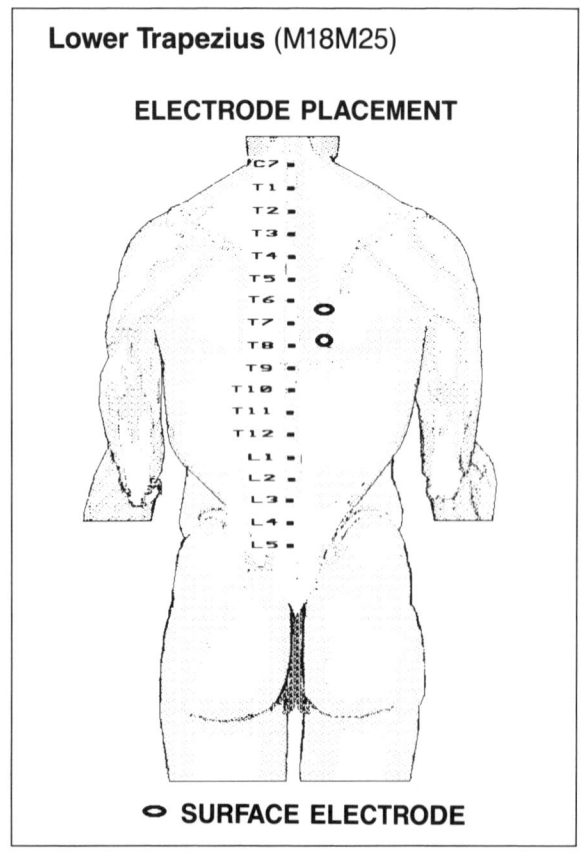

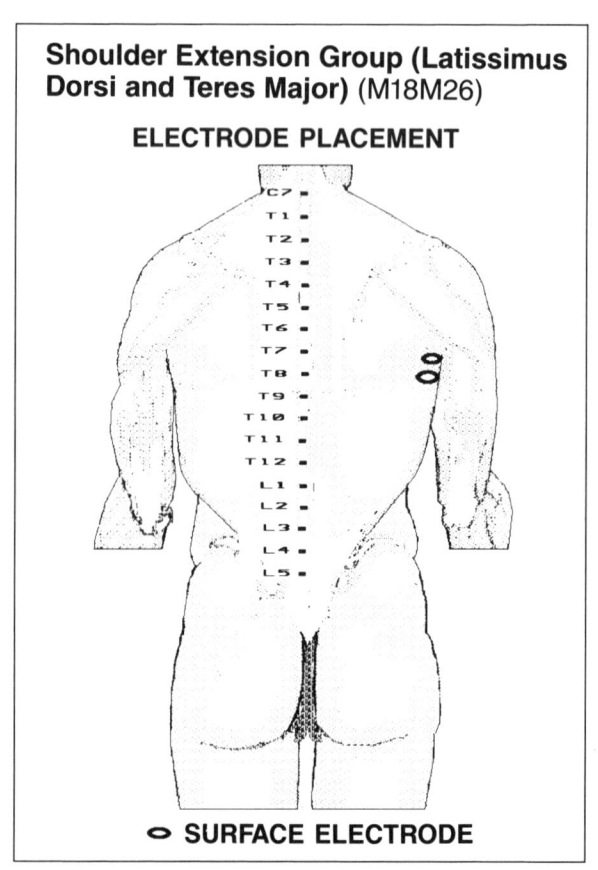

Levator Scapulae (M18M23)
ELECTRODE PLACEMENT

○ SURFACE ELECTRODE

Lower Trapezius (M18M25)
ELECTRODE PLACEMENT

○ SURFACE ELECTRODE

Shoulder Extension Group (Latissimus Dorsi and Teres Major) (M18M26)
ELECTRODE PLACEMENT

○ SURFACE ELECTRODE

electromyometric feedback electrode placements

Supraspinatus (M18M27)
ELECTRODE PLACEMENT

○ SURFACE ELECTRODE

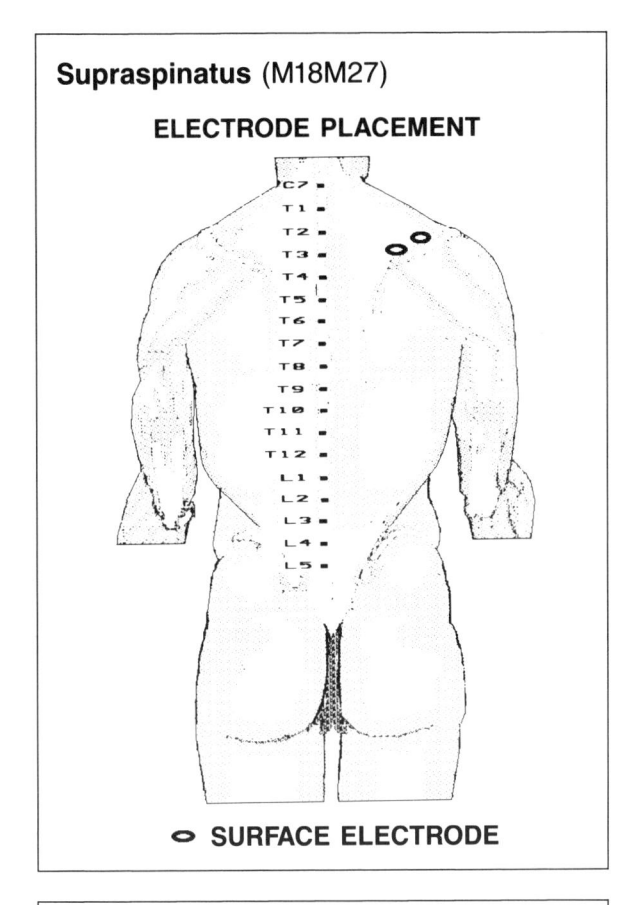

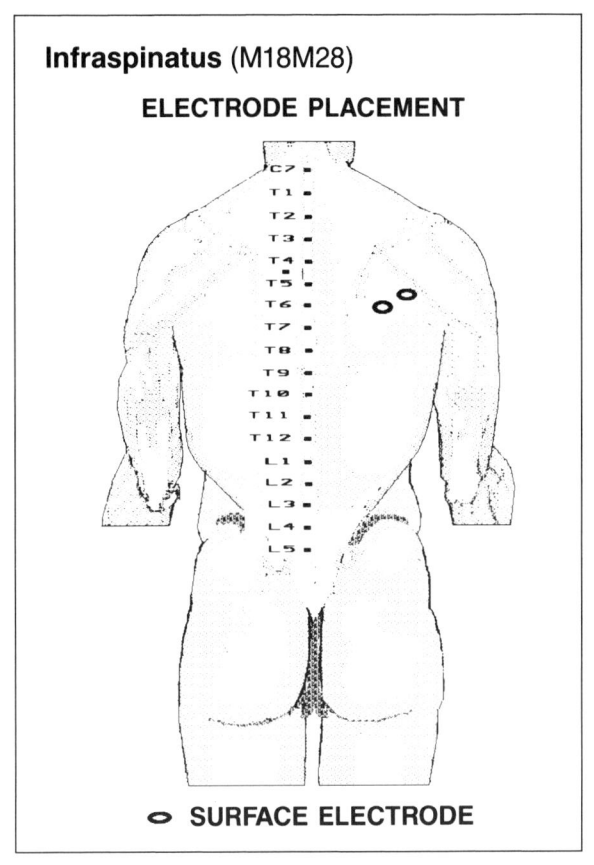

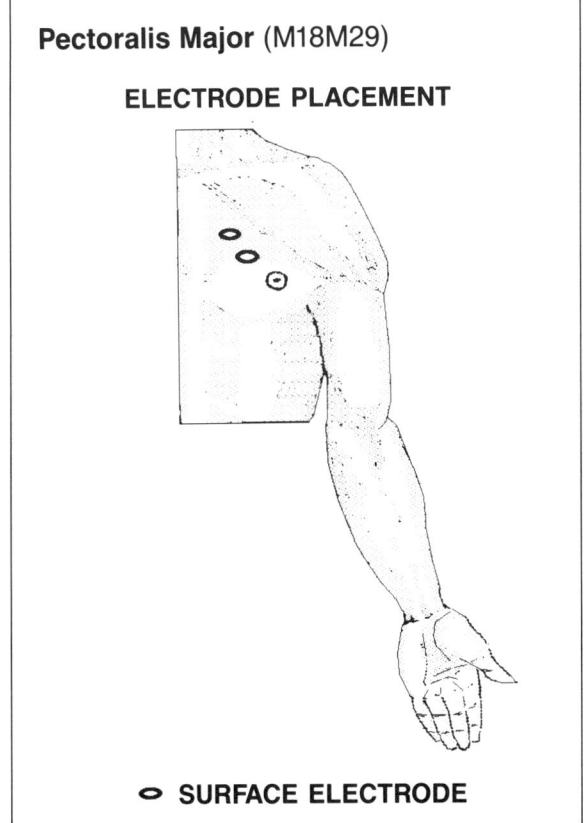

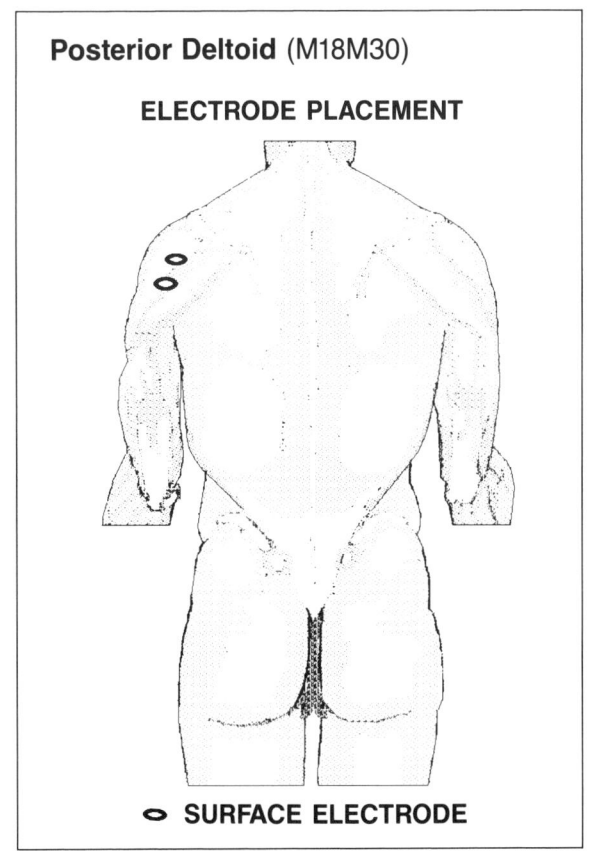

Infraspinatus (M18M28)
ELECTRODE PLACEMENT

○ SURFACE ELECTRODE

Pectoralis Major (M18M29)
ELECTRODE PLACEMENT

○ SURFACE ELECTRODE

Posterior Deltoid (M18M30)
ELECTRODE PLACEMENT

○ SURFACE ELECTRODE

electromyometric feedback electrode placements

Middle Deltoid (M18M31)
ELECTRODE PLACEMENT

○ SURFACE ELECTRODE

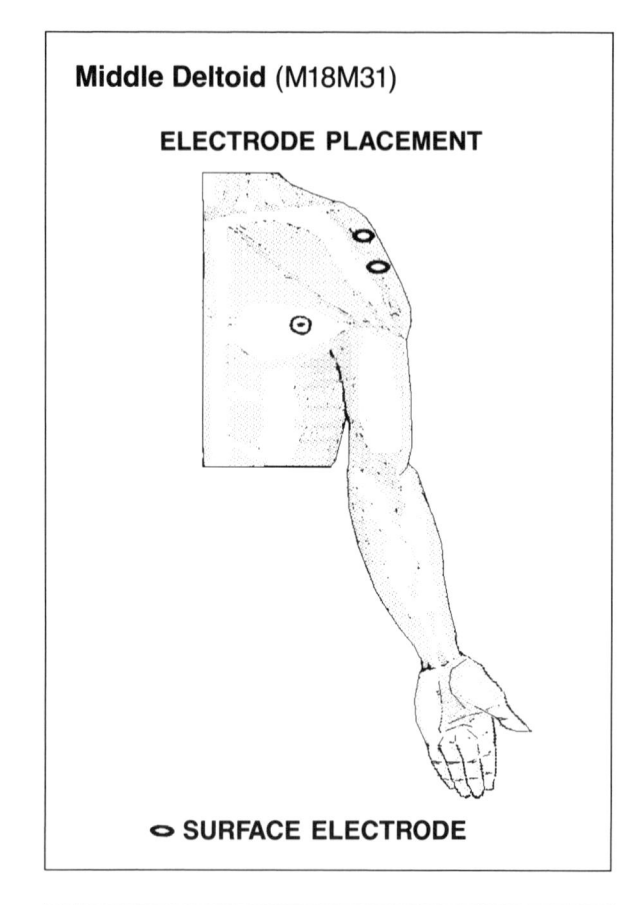

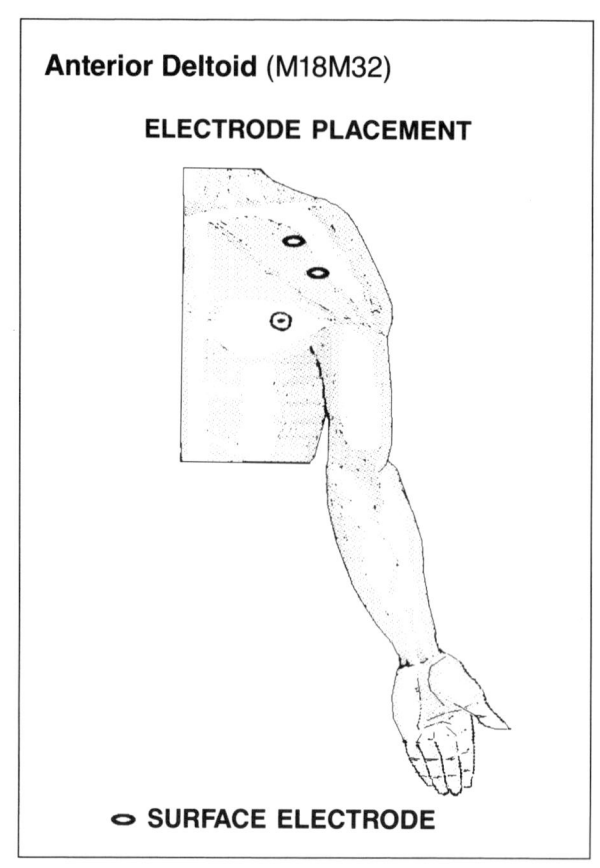

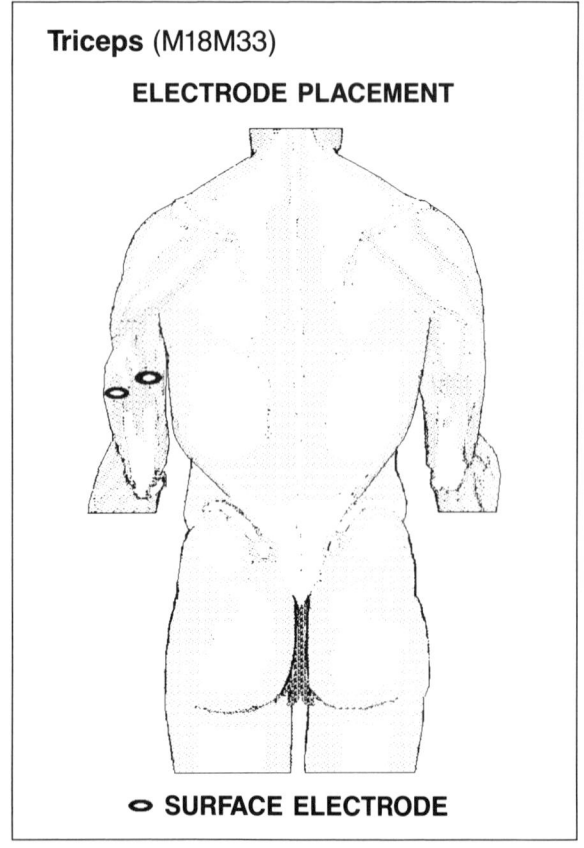

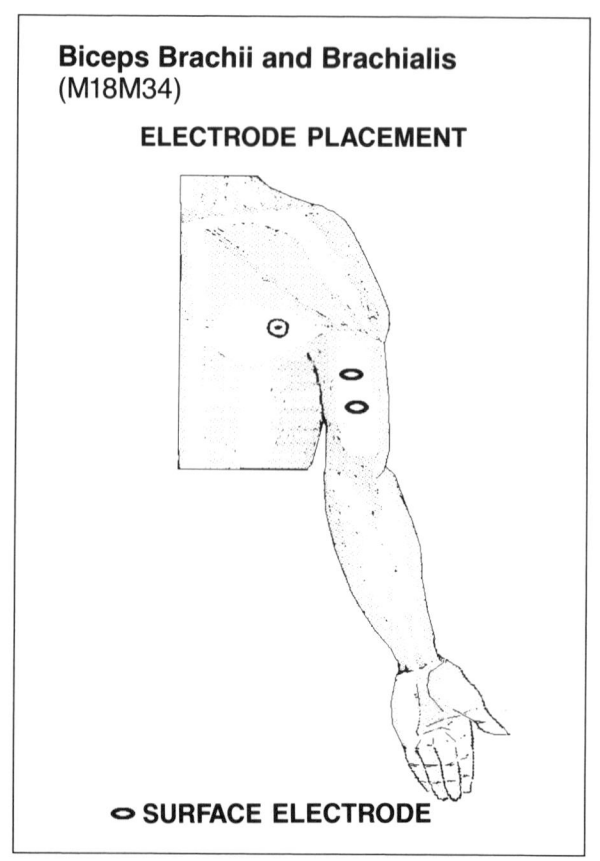

Anterior Deltoid (M18M32)
ELECTRODE PLACEMENT

○ SURFACE ELECTRODE

Triceps (M18M33)
ELECTRODE PLACEMENT

○ SURFACE ELECTRODE

Biceps Brachii and Brachialis (M18M34)
ELECTRODE PLACEMENT

○ SURFACE ELECTRODE

Supinator (M18M35)

ELECTRODE PLACEMENT

○ SURFACE ELECTRODE

Brachioradialis (M18M36)

ELECTRODE PLACEMENT

○ SURFACE ELECTRODE

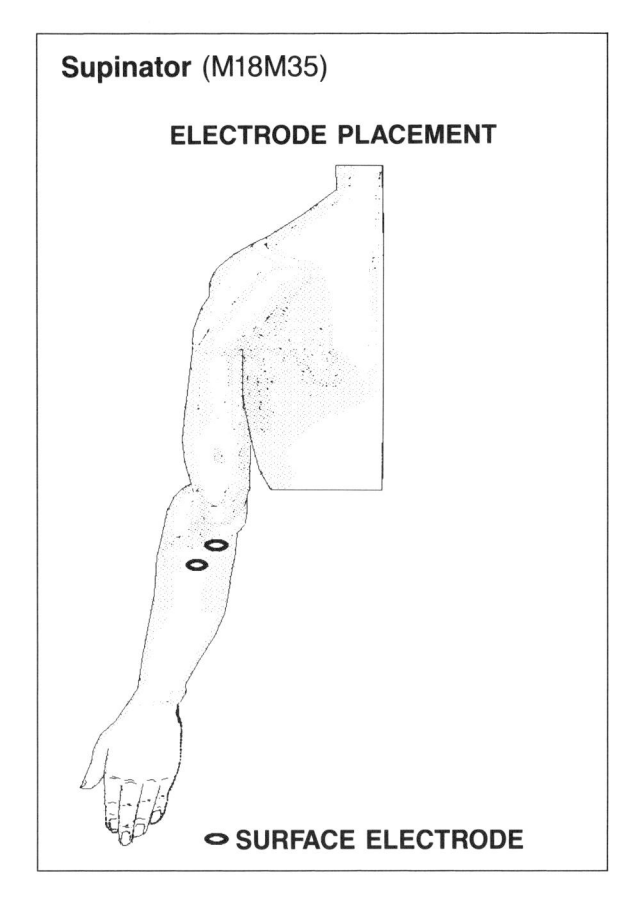

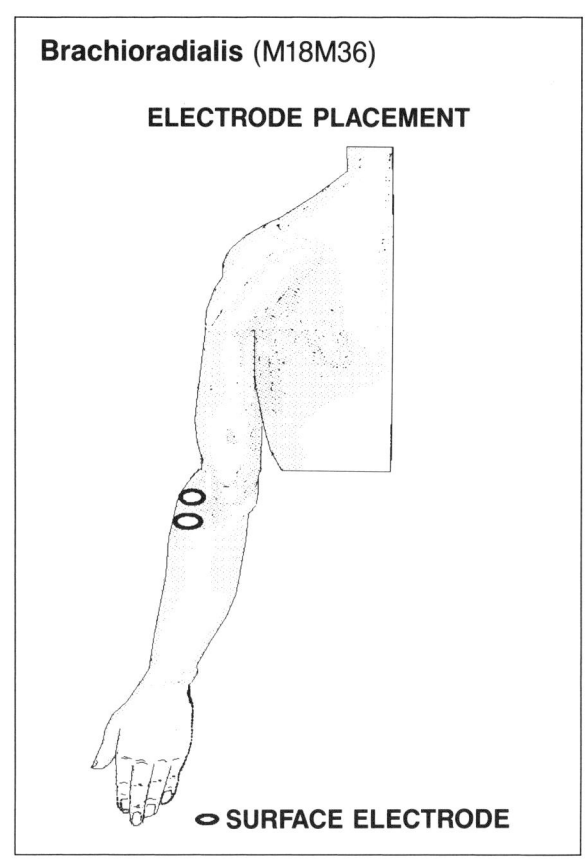

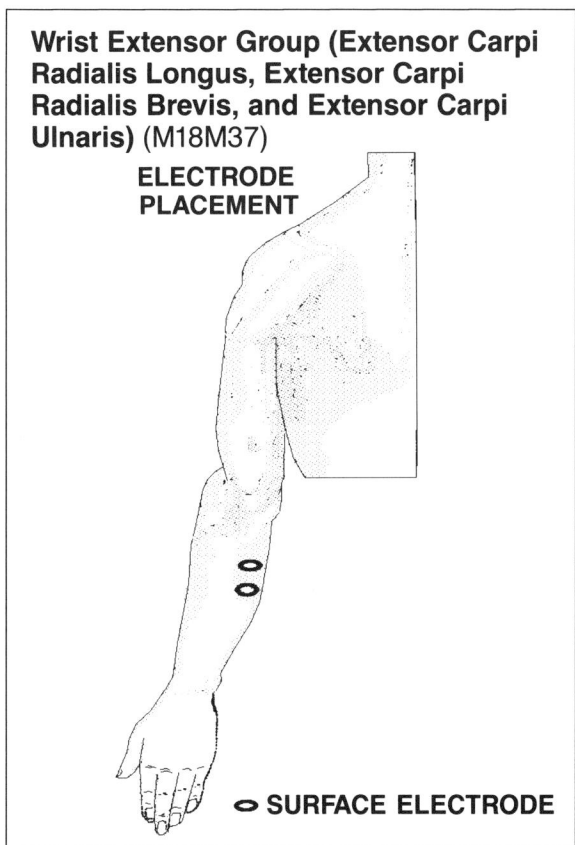

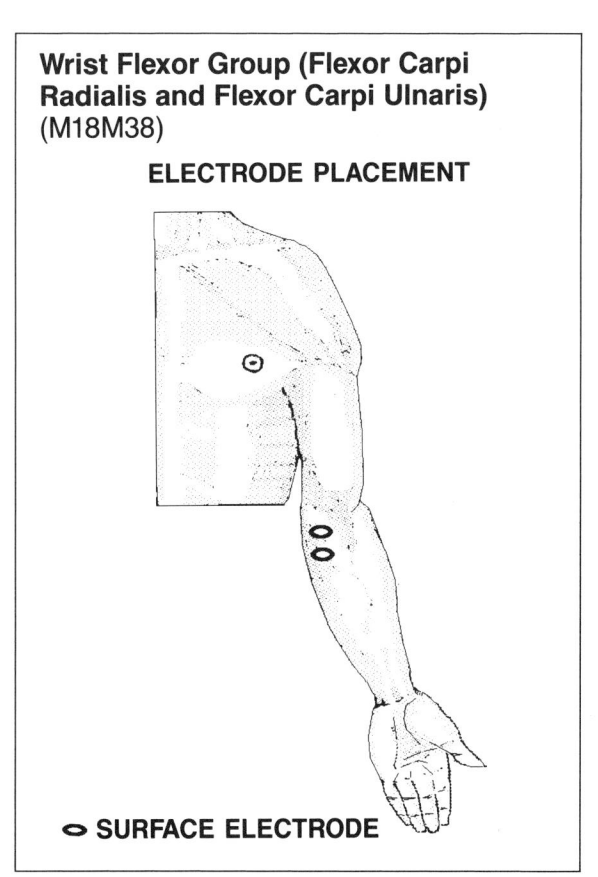

Wrist Extensor Group (Extensor Carpi Radialis Longus, Extensor Carpi Radialis Brevis, and Extensor Carpi Ulnaris) (M18M37)

ELECTRODE PLACEMENT

○ SURFACE ELECTRODE

Wrist Flexor Group (Flexor Carpi Radialis and Flexor Carpi Ulnaris) (M18M38)

ELECTRODE PLACEMENT

○ SURFACE ELECTRODE

electromyometric feedback electrode placements

Finger Extensor Group (Extensor Digitorum Communis, Extensor Indicis Proprius, and Extensor Digiti Quinti Proprius) (M18M39)

ELECTRODE PLACEMENT

○ SURFACE ELECTRODE

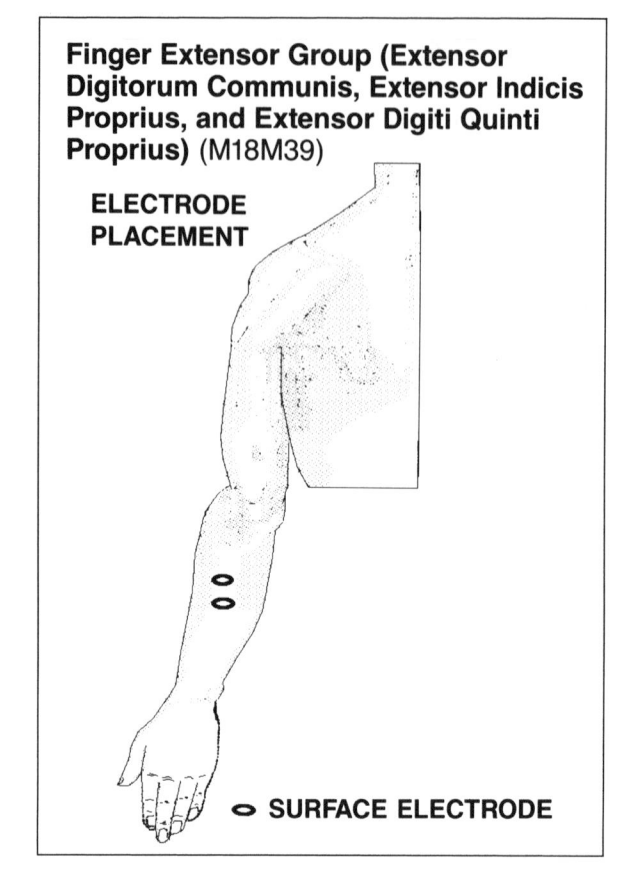

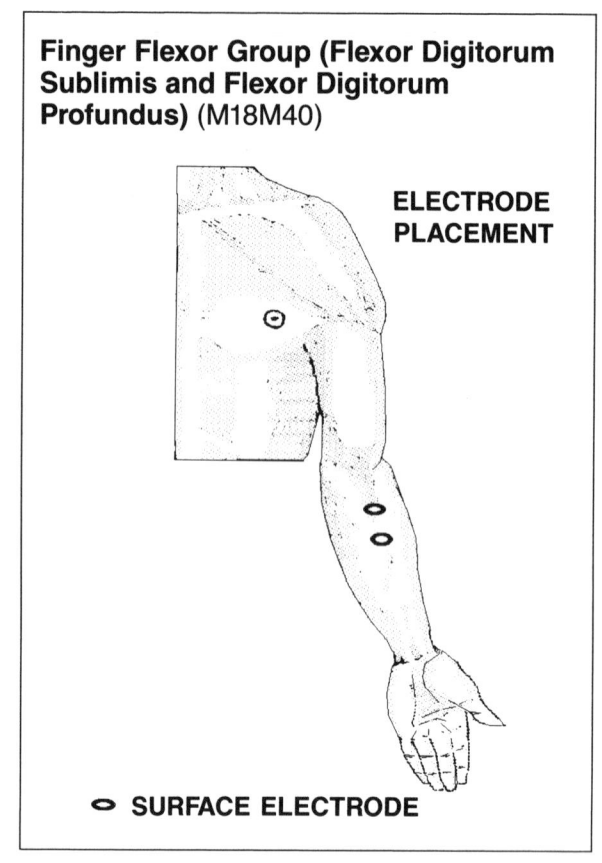

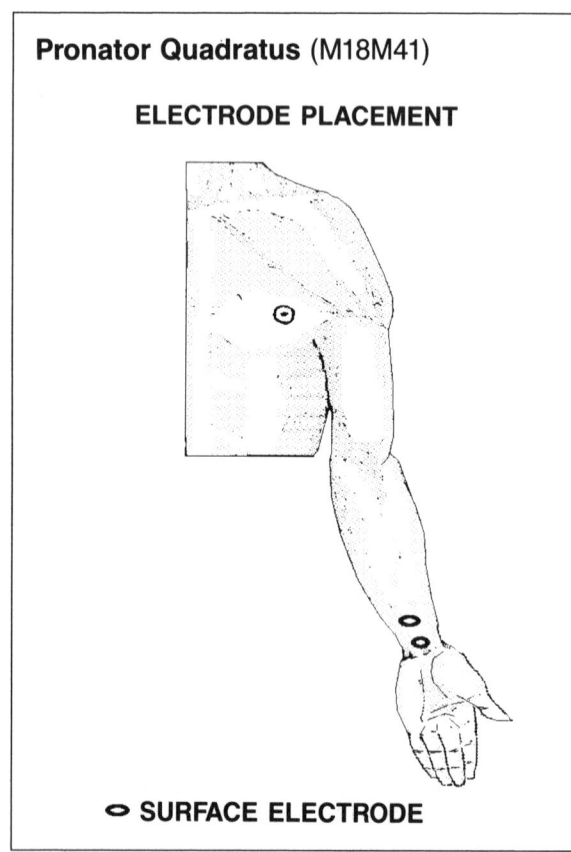

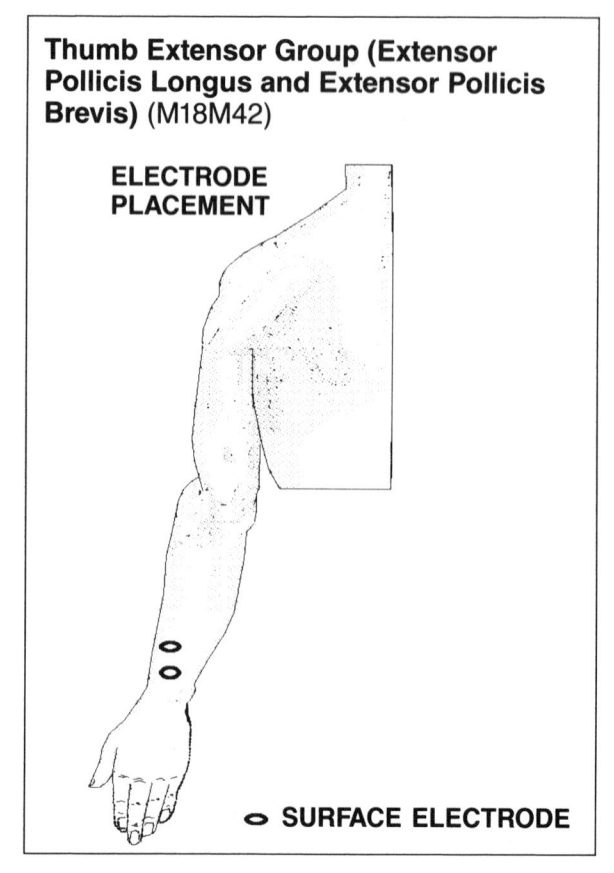

Finger Flexor Group (Flexor Digitorum Sublimis and Flexor Digitorum Profundus) (M18M40)

ELECTRODE PLACEMENT

○ SURFACE ELECTRODE

Pronator Quadratus (M18M41)

ELECTRODE PLACEMENT

○ SURFACE ELECTRODE

Thumb Extensor Group (Extensor Pollicis Longus and Extensor Pollicis Brevis) (M18M42)

ELECTRODE PLACEMENT

○ SURFACE ELECTRODE

electromyometric feedback electrode placements

Flexor Pollicis Longus (M18M43)

ELECTRODE PLACEMENT

○ SURFACE ELECTRODE

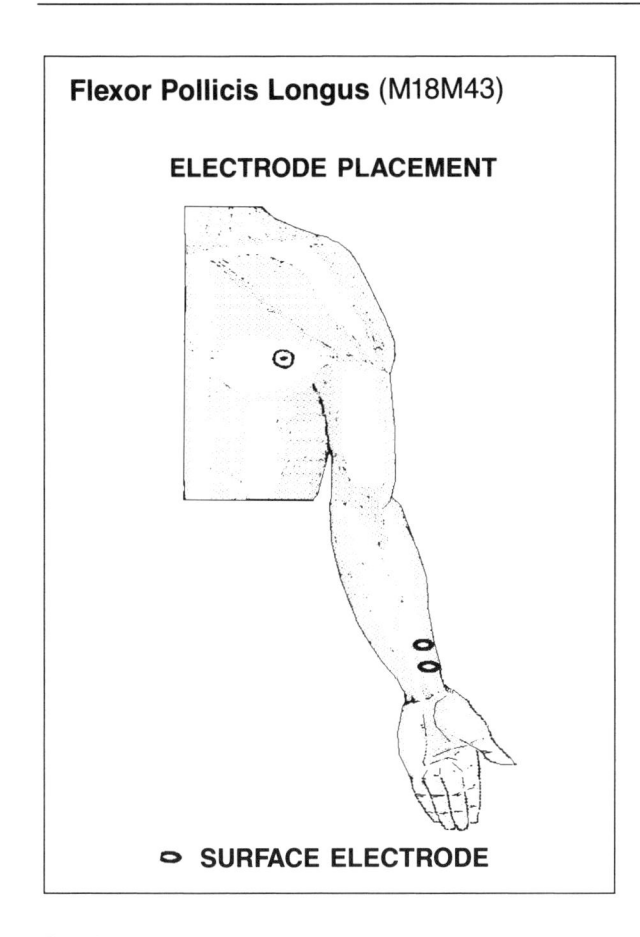

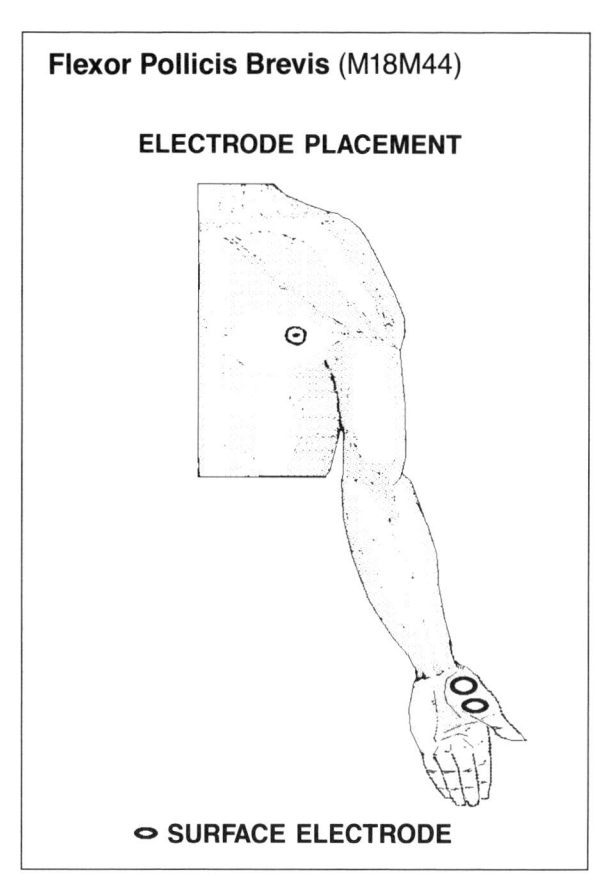

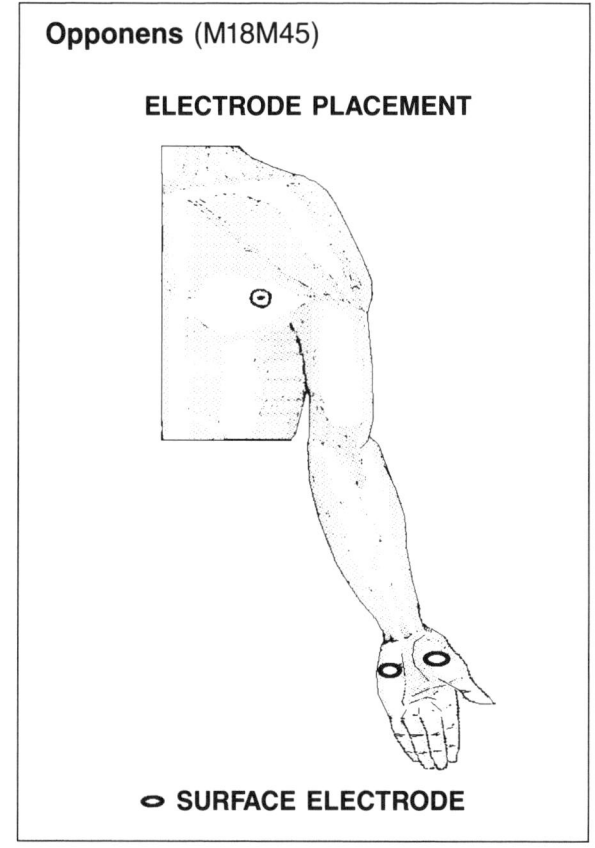

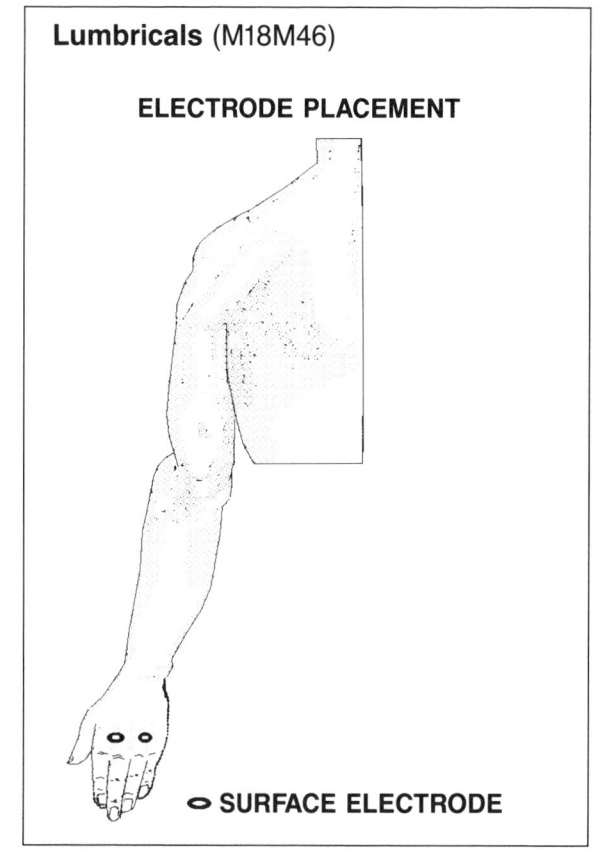

Flexor Pollicis Brevis (M18M44)

ELECTRODE PLACEMENT

○ SURFACE ELECTRODE

Opponens (M18M45)

ELECTRODE PLACEMENT

○ SURFACE ELECTRODE

Lumbricals (M18M46)

ELECTRODE PLACEMENT

○ SURFACE ELECTRODE

electromyometric feedback electrode placements

Diaphragm (M18M47)

ELECTRODE PLACEMENT

○ SURFACE ELECTRODE

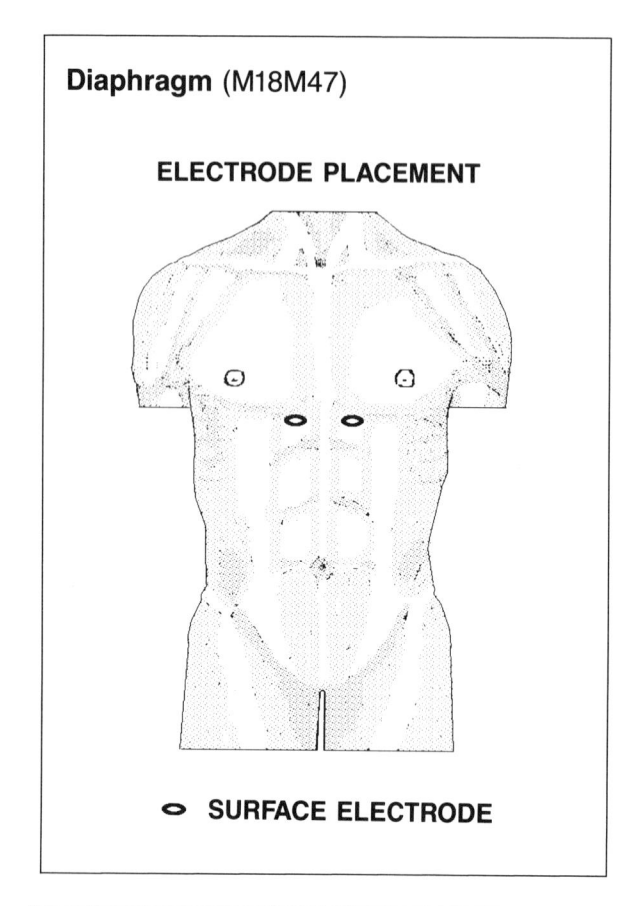

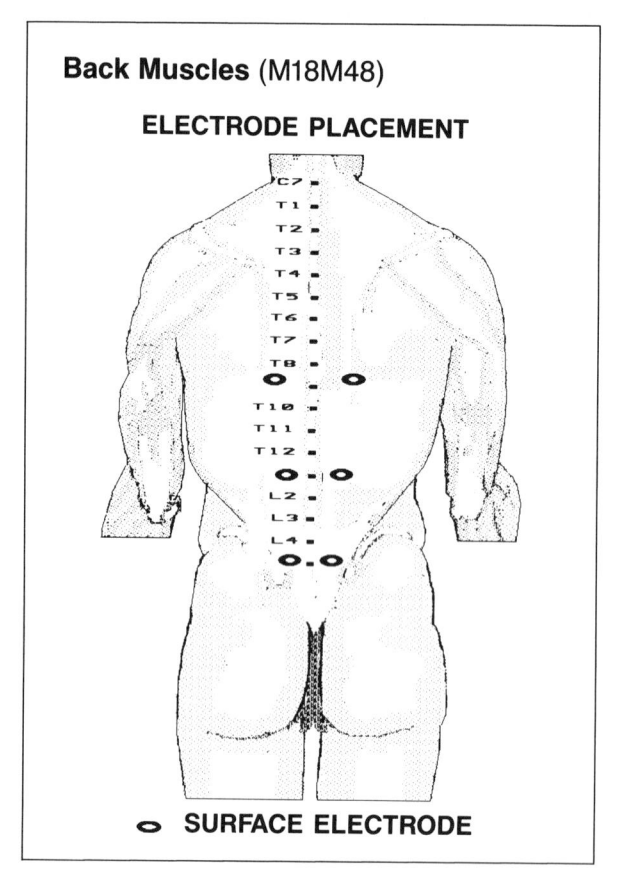

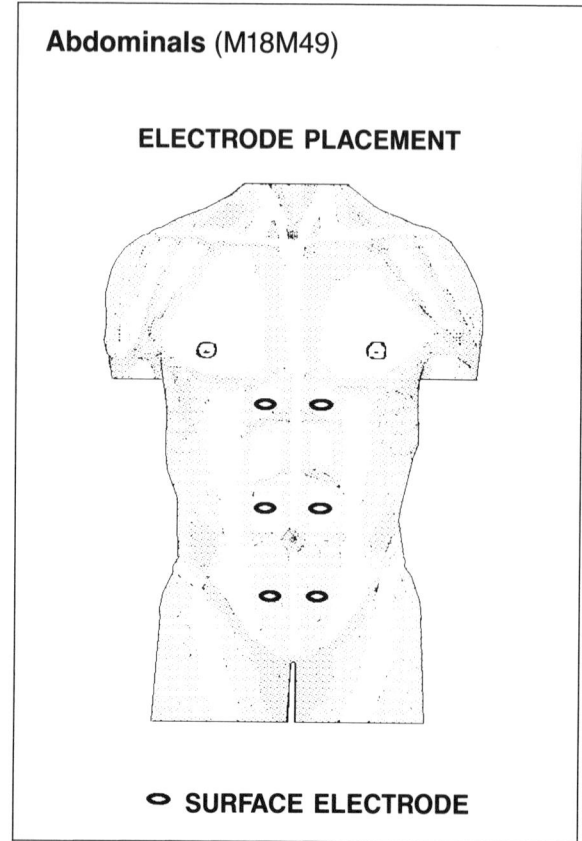

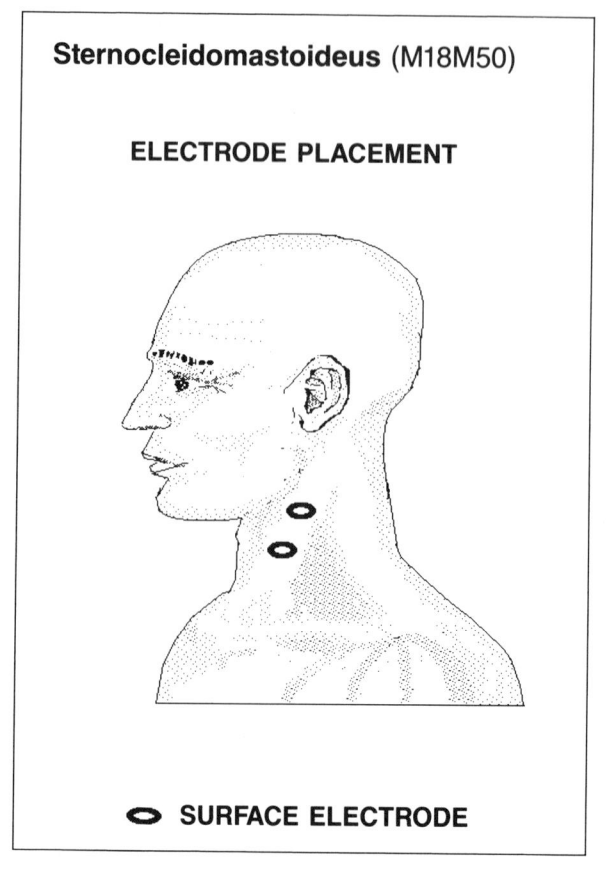

Back Muscles (M18M48)

ELECTRODE PLACEMENT

○ SURFACE ELECTRODE

Abdominals (M18M49)

ELECTRODE PLACEMENT

○ SURFACE ELECTRODE

Sternocleidomastoideus (M18M50)

ELECTRODE PLACEMENT

○ SURFACE ELECTRODE

Scaleni (M18M51)

ELECTRODE PLACEMENT

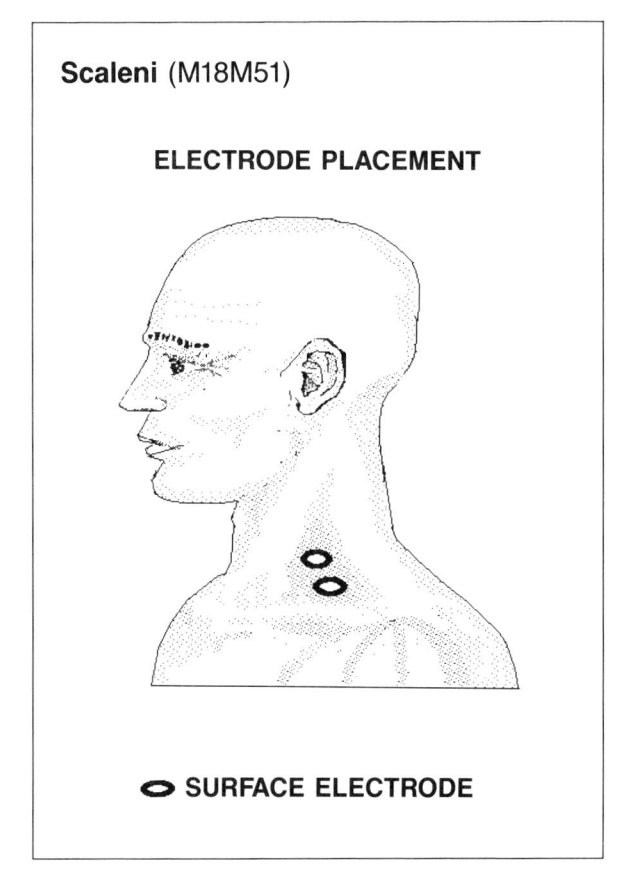

⬭ SURFACE ELECTRODE

Upper Trapezius and Scaleni (M18M52)

ELECTRODE PLACEMENT

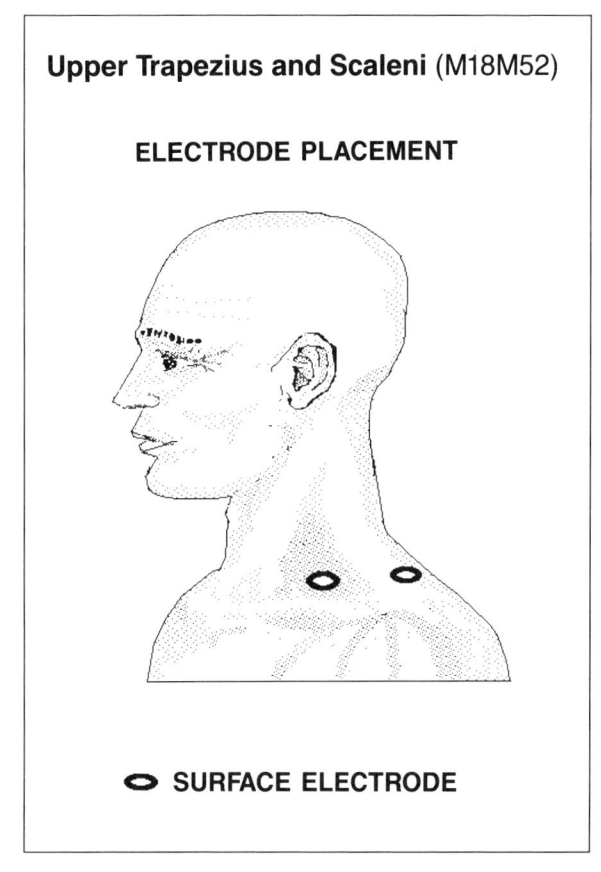

⬭ SURFACE ELECTRODE

Masseter Electrode Placement (M18M53)

ELECTRODE PLACEMENT

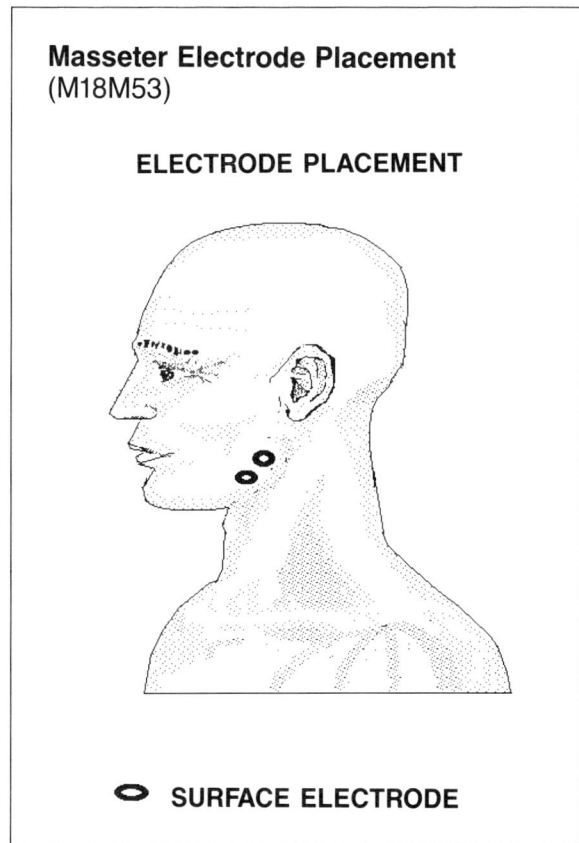

⬭ SURFACE ELECTRODE

Pterygoid Group (M18M54)

ELECTRODE PLACEMENT

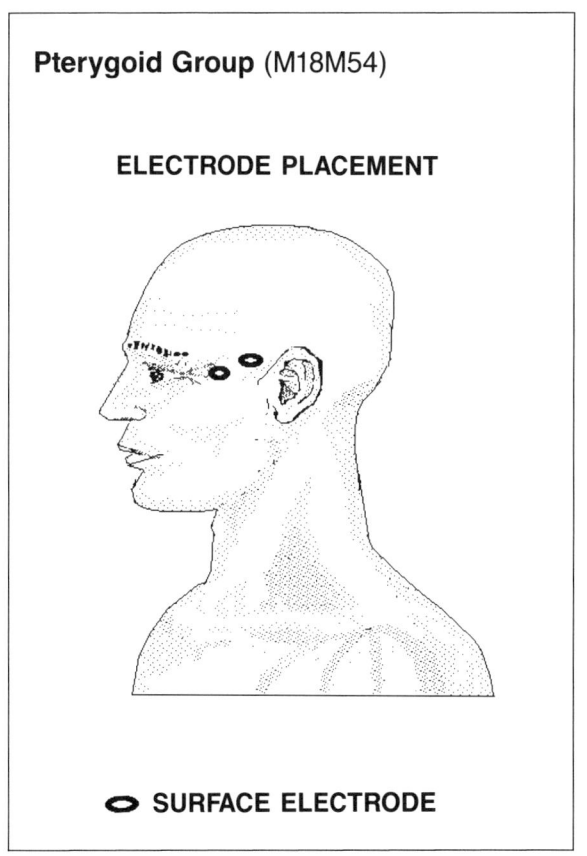

⬭ SURFACE ELECTRODE

electromyometric feedback electrode placements

Masseter and Pterygoid (M18M55)

ELECTRODE PLACEMENT

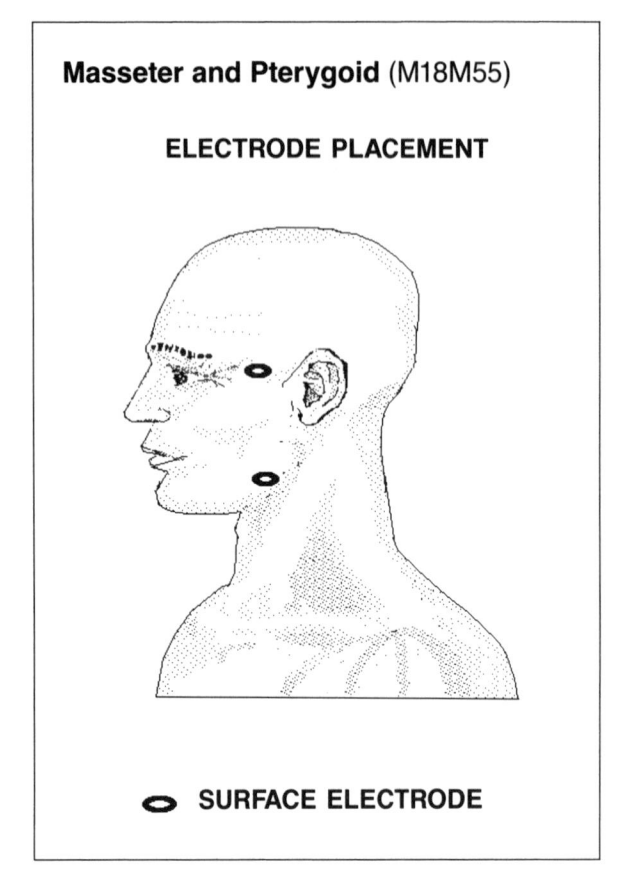

○ SURFACE ELECTRODE

Frontalis (M18M56)

ELECTRODE PLACEMENT

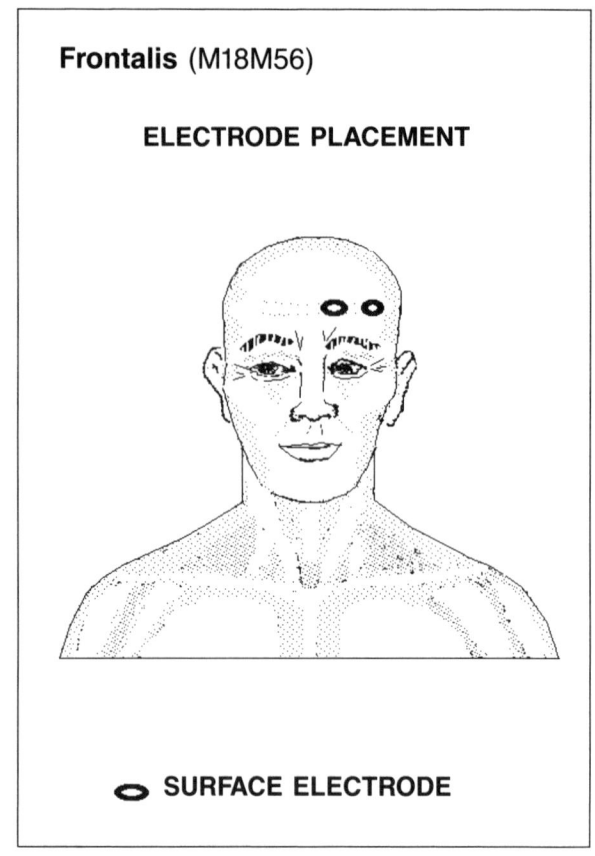

○ SURFACE ELECTRODE

Corrugator Supercilii (M18M57)

ELECTRODE PLACEMENT

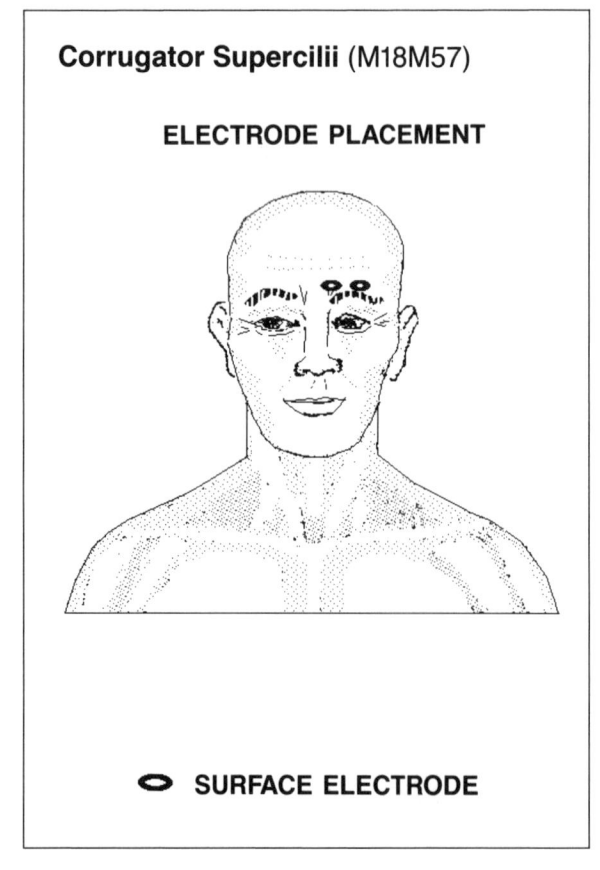

○ SURFACE ELECTRODE

Procerus (M18M58)

ELECTRODE PLACEMENT

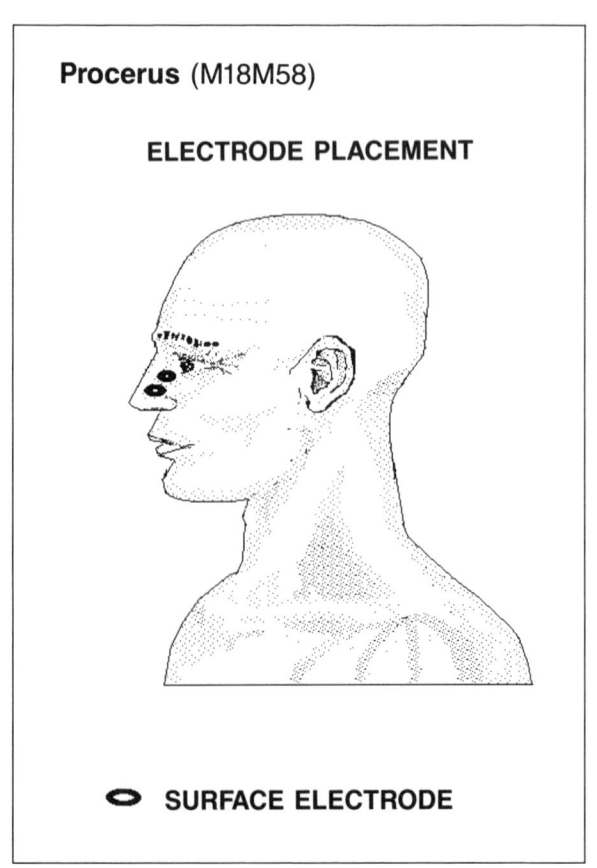

○ SURFACE ELECTRODE

Orbicularis Oculi (M18M59)

ELECTRODE PLACEMENT

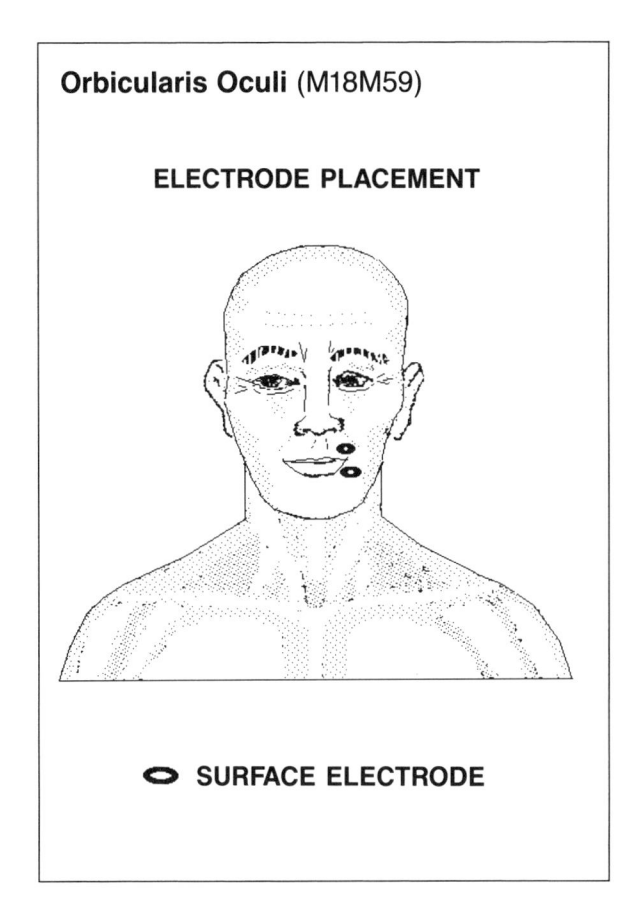

○ SURFACE ELECTRODE

Orbicularis Oris (M18M60)

ELECTRODE PLACEMENT

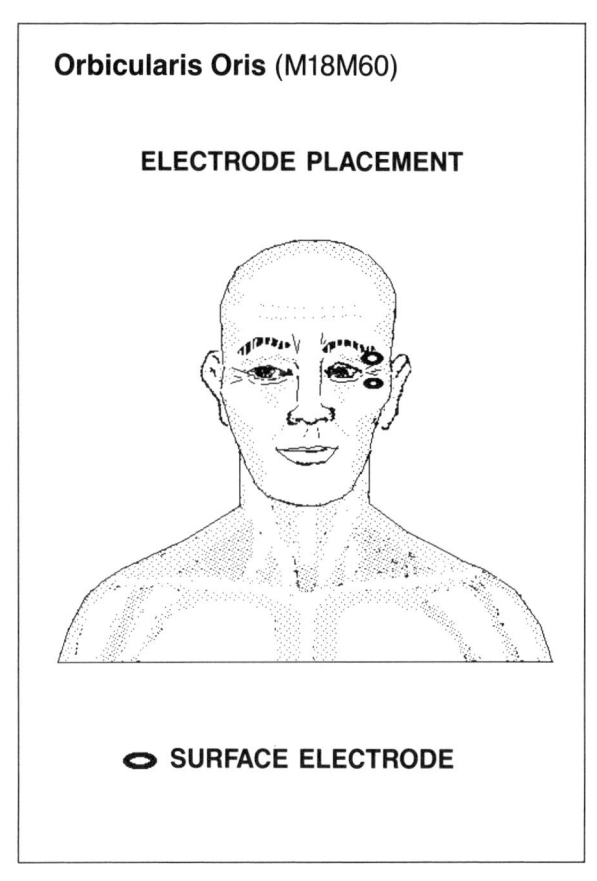

○ SURFACE ELECTRODE

Zygomaticus Minor and Levator Anguli Oris (M18M61)

ELECTRODE PLACEMENT

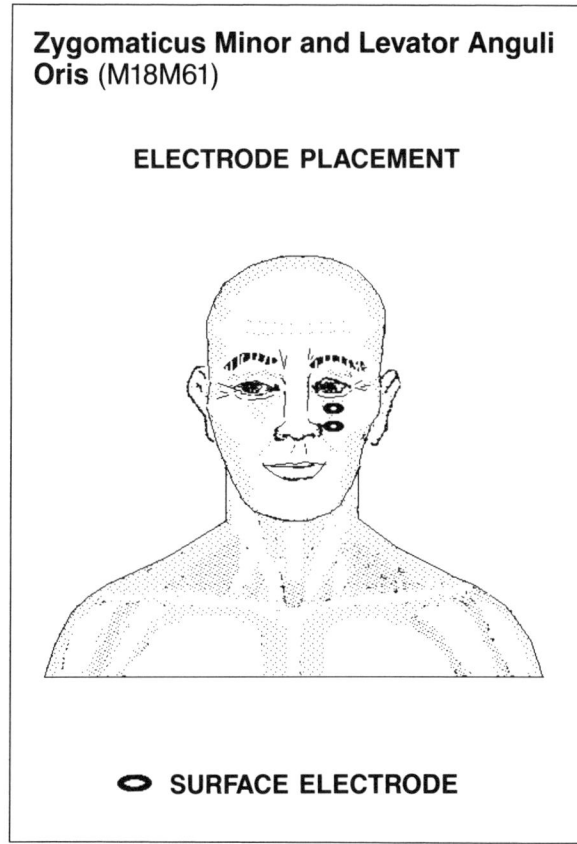

○ SURFACE ELECTRODE

Risorius and Buccinator (M18M62)

ELECTRODE PLACEMENT

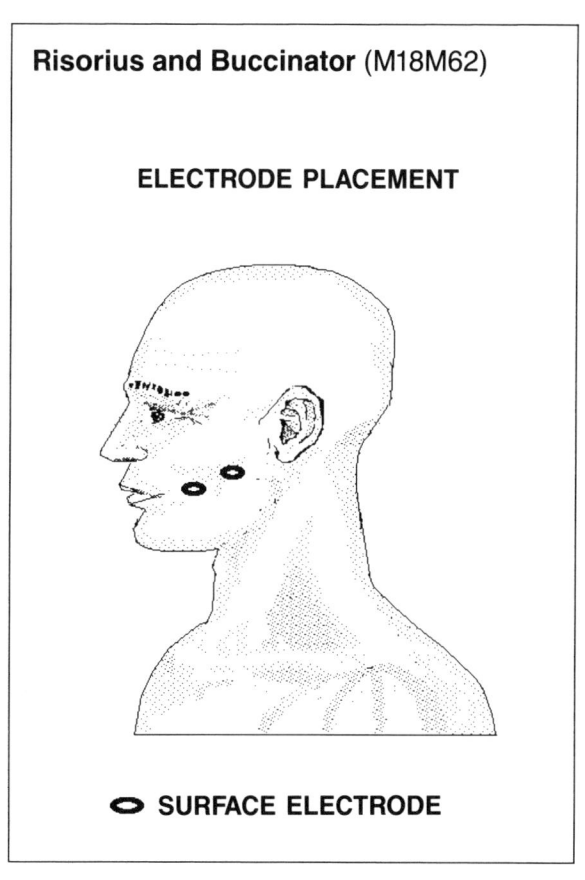

○ SURFACE ELECTRODE

electromyometric feedback electrode placements

Depressor Anguli Oris and Platysma
(M18M64)

ELECTRODE PLACEMENT

◯ SURFACE ELECTRODE

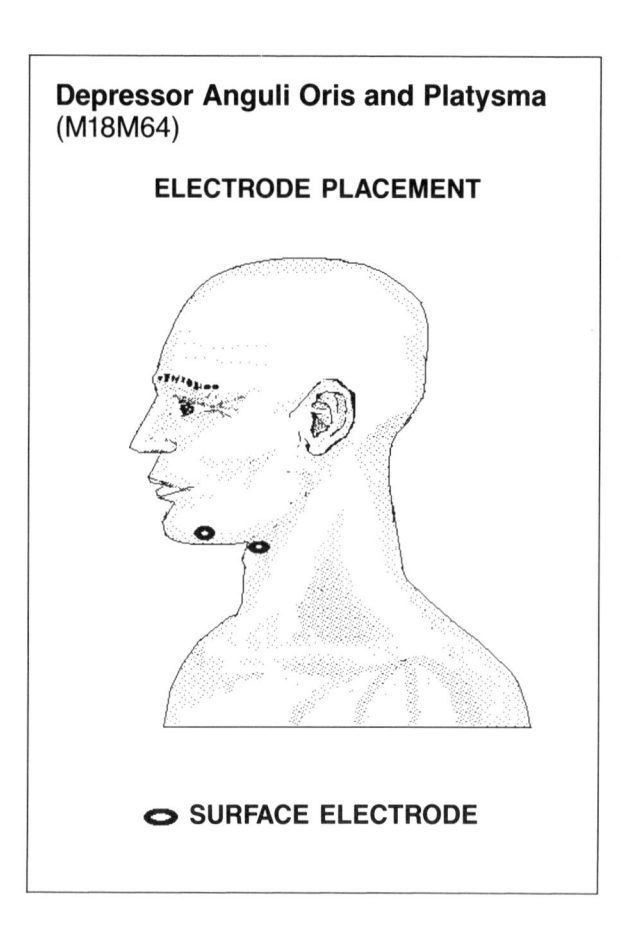

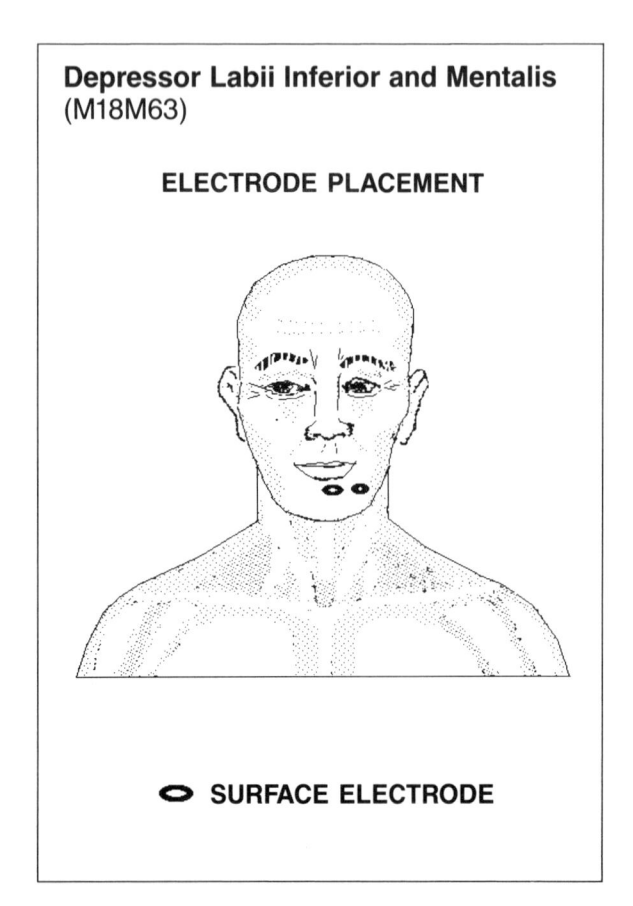

Depressor Labii Inferior and Mentalis
(M18M63)

ELECTRODE PLACEMENT

◯ SURFACE ELECTRODE

Zygomaticus Major (M18M65)

ELECTRODE PLACEMENT

⬭ SURFACE ELECTRODE

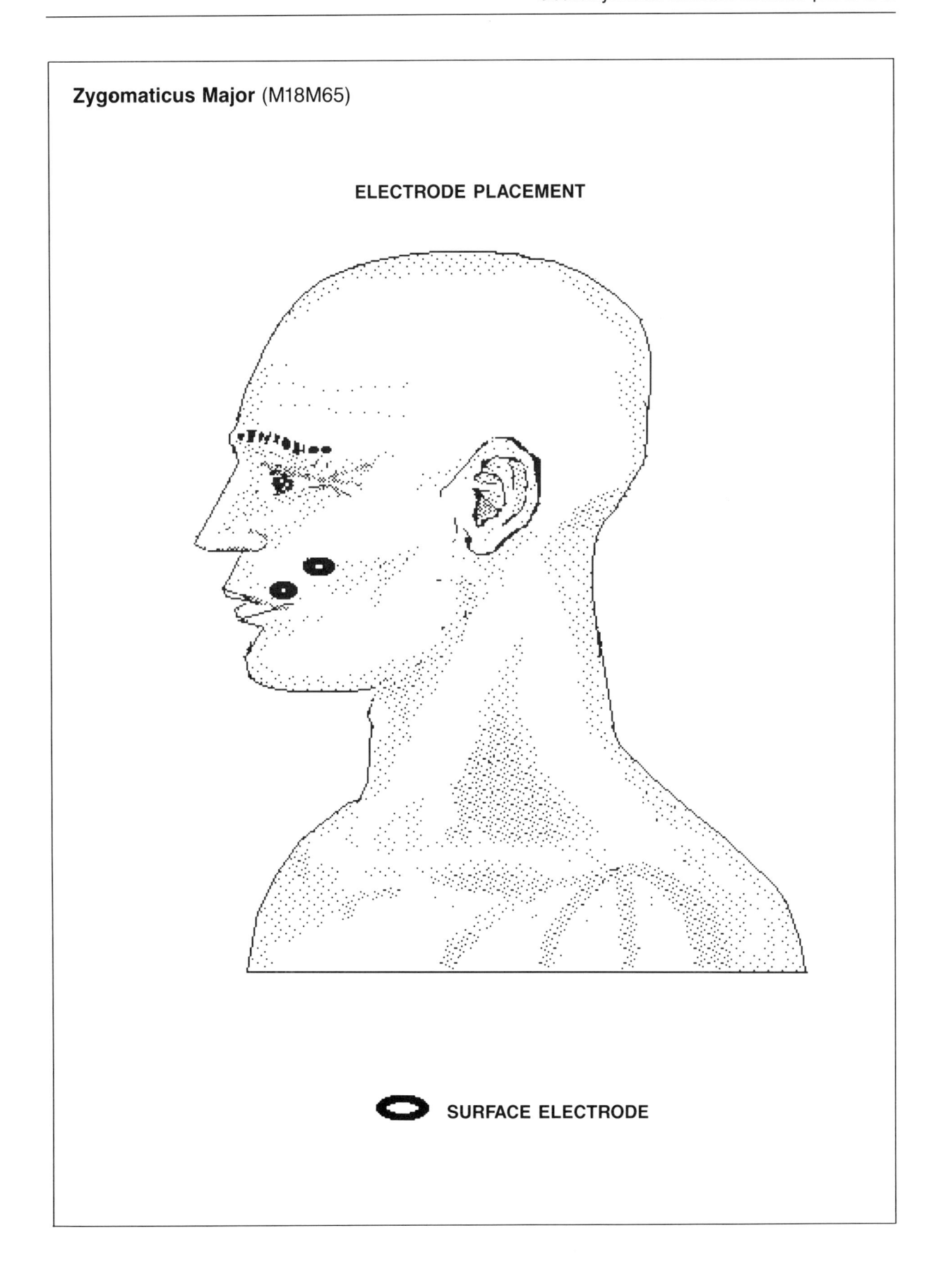

EXERCISE (M12)

Definition (M1201)

Exercise is defined here as the systematic use or operation of muscle and/or joint(s) to develop and/or maintain all or part of the musculoskeletal system in terms of dexterity, flexibility, strength, tone, endurance, coordination, and/or joint range of motion.

Most central to our interest here is the concept of therapeutic exercise, which is defined as exercise anatomically and physiologically based upon coordinated movement, designed to approximate or restore normal body function. It may involve the application of principles of proprioceptual-motor functioning to effect changes in the patient's neuromuscular response through the selective use of body posture and movement, or the selective use of sensorimotor stimulation (active or passive resistance, etc.) which is effective in restoring, improving or maintaining motor and/or other systemic function (joint range of motion included).

Metabolic rate (M1201A)

Strenuous exercise produces various changes in the body's physiological processes. Of those processes, the most dramatically effected is the metabolic rate. Even short bursts of maximal muscle contraction in any *single muscle* will cause an increase in the muscle metabolism sufficient to liberate one hundred times more heat than is normally produced during rest. Maximal muscle exercise can effect increases in heat production from the *entire body* in amounts twenty times above normal if the exercise is sustained for several minutes, or in amounts fifty times greater than normal if the exercise is performed in short bursts (performed for several seconds). Such increases may represent an increase in metabolic rate which may be two thousand percent above the normal rate measured during rest.

As the cellular metabolism increases occur with exercise the demand for metabolic fuel increases, along with a simultaneous increase in oxygen consumption. To fulfill these demands, glucose is caused to be transported from the blood into the muscle cells at an accelerated rate, even in the absence of insulin*, and under normal circumstances the respiratory rate is automatically increased to provide an adequate supply of oxygen. The pulse rate is consequently increased to both increase transport of fuel and oxygen to the area and to provide for the rapid elimination of metabolic by-products like carbon dioxide. During exercise, increased alveolar ventilation provides for the exchange of oxygen and carbon dioxide at a rate which generally keeps them in balance at "normal" levels.

During strenuous exercise, the muscle cells utilize oxygen at a rate which is as much as three times greater than is normal, and the full body requirement may rise to a level which is twenty times more than normal. The actual oxygen consumption rate may approach 2400 ml. per minute. The increased demand for oxygen may cause the interstitial fluid oxygen content to fall as low as 15 mm. Hg., which means that at this blood pressure only 4.4 ml. of oxygen may remain bound with the hemoglobin in each 100 ml. of blood, in each cycle through the tissues. Additionally, the increased tissue utilization of oxygen generally triggers an automatic increase in the rate of oxygen released from the hemoglobin to the extent that 75 to 100% of the hemoglobin may be caused to give up its oxygen, whereas only 25% is given up when the body is at rest.

Cardiac output (M1201B)

Normally, as an increase in the rate of oxygen consumption occurs, a simultaneous proportional increase in cardiac output is provoked. In youthful athletic individuals, the consequent rise in cardiac output may be five to six times the normal resting level, rising as high as 30 to 35 liters per minute. The compensatory heart stroke volume will vary from approximately 90 to just under 110 ml. per beat and the pulse rate may accelerate to as high as 195 beats per minute in the adult (the maximum usually decreasing with age). The pulse rate of children may reach as high as 200 beats per minute during exercise, including active play.

The direct causes for increases in cardiac output during exercise include: (1) intense sympathetic stimulation, which strengthens the force of heart contraction by as much as 70% with a consequent rise in the mean systemic pressure; (2) contraction of the muscles around associated blood vessels, which further increases the mean systemic pressure; and (3) the intense dilation of the resistance vessels in the muscles, facilitating venous return to the heart. Vasodilation in the active muscles will be compensated for by vasoconstriction in the viscera, including the kidneys and other organs, to maintain or allow a moderate rise in the arterial blood pressure levels. In a normal situation the diastolic pressure will tend to remain close to its normal level of around 80 mm. Hg., while the systolic pressure may gradually rise to approach 200 mm. Hg. The mean systemic pressure may rise from around 90 mm. Hg. to approach or exceed 120 mm. Hg.

* The insulin-like effect of exercise upon glucose transport into the muscle may provide the physically active diabetic patient (even those severely afflicted) with a better opportunity to control their diabetes then over protected sufferers by decreasing the need for the administration of supplementary insulin.

Procedures of Application (M1202)

Exercise is basically divided into five categories: (1) passive, (2) assistive, (3) active-assistive, (4) active, and (5) resistive.

Passive exercise (M1202A)

Passive exercise is applied to the patient, by the practitioner without the patient taking any active part. Passive exercise includes: (1) manual stretching, and (2) flail joint movement.

Manual stretching (M1202A01) Manual stretching entails the forcing of involved joint(s) through part or all of the normal range of motion. Manual stretching is generally performed on patients with nervous system impairment which has directly or indirectly affected the joints through neuromuscular hypertonicity, muscular atrophy and/or contracture (this may also occur as a result of simple muscular disuse). The joint is variously described as spastic, stiff, tight and/or rigid, and attempts to move the joint through the normal range of motion are actively (as in conditions stemming from neuromuscular hypertonicity) or passively resisted by atrophied and contracted musculature (refer to Central Nervous System Dysfunction [R006]).

Flail joint movement (M1202A02) The exercise of flail joint movement requires the practitioner to move the involved joint(s) through the normal range(s) of motion without patient participation, whether conscious or involuntary (i.e., neuromuscular hypertonic resistance). This type of exercise is usually performed on patients who suffer zero muscle grades (flaccidity) resulting from nervous system injury, and before muscular atrophy and/or contracture has occurred (refer to Central Nervous System Dysfunction [R006]).

Assistive exercise (M1202B)

Assistive exercise calls upon the the patient to actively attempt to move the involved joint(s) while practitioner assists this attempt by manually conducting the movement through the available or *full* joint range of motion. This type of exercise is generally conducted with patients who suffer trace to poor muscle grades, following peripheral nerve injury or as the result of disuse atrophy from enforced bed rest (refer to Adverse Effects of Bed Rest [R008]).

Active-assistive exercise (M1202C)

Active-assistive exercise takes place when the patient is able to actively move the involved joint(s) through some of the range(s) of motion and practitioner is only required to help the patient complete the movement. This type of exercise is usually performed with patients who suffer poor-plus to fair-minus muscle grade strengths.

During active-assistive exercise, a demand should be made of the patient that maximum effort be exerted during the exercise attempt. The practitioner should only help the patient complete the range. Assistance should be gradually decreased over time so that the patient eventually completes the range independently.

Active exercise (M1202D)

Active exercise takes place when the patient actively moves the involved joint(s) through the full range(s) of motion, unassisted by the therapist, but without external resistance. This type of exercise is usually performed by patients who suffer fair grade muscle strength, and relies upon the weight of the involved body part to tax the involved muscles. As strength improves, the patient should be progressed through various postures that sequentially increase the resistance of gravity upon the action of the muscle(s): (1) muscle contraction should move the joint across gravitational lines of force by movement parallel to the horizontal plane, or the floor (the extremity may have to be supported by a supporting surface (table) or by the practitioner's hands, supplied without increasing or decreasing resistance); (2) the muscle contraction should move the joint diagonally across gravitational lines at increasingly obtuse angles to the horizontal plane (incline boards may be helpful in defining and supporting extremity movement in these planes); (3) finally, the muscle contraction should move the joint parallel to the vertical plane against the resistance of gravity.

Resistive exercise (M1202E)

Resistive exercise occurs when the patient attempts to complete range(s) of motion against resistance, unassisted by the practitioner. This includes exercise in which joint motion against resistance is attempted but joint movement is disallowed, as in isometric exercise. Weight lifting and other forms of isotonic exercise are included in this classification. Resistance may be also be manually supplied by the practitioner, either isometrically or isotonically.

Partial List of Conditions Treated (M1203)

Conditions treated in the field (M1203A)

Amputations
Anterior poliomyelitis
Bunion (hallux valgus, prevention of)
Contractures (cicatricial, muscular, and/or tendinous)
Fibrosis ankylosis Fractures
Healed surgical repair Multiple sclerosis (selected cases)
Parkinsonism
Peripheral vascular diseases (Buerger's disease included)
Reduced epiphyseal separations
Rheumatoid conditions
Scoliosis
Sequelae of joint dislocations
Tabes dorsalis

Treatable Causes (M1203B)

Calcific deposit (C08)
Extrafusal muscle spasm (C41)
Habitual joint positioning (C48)
Hypermobile/instable joint (C10)
Hypertension (C39)
Hypotension (C45)
Insomnia (C46)
Interspinous ligamentous strain (C03)
Joint ankylosis (C42)
Joint approximation (C47)
Ligamentous strain (C25)
Lung fluid accumulation (C13)
Muscular weakness (C23)
Nerve root compression (C04)
Neuromuscular tonic imbalance (C30)
Pathological neuromuscular hypertonus (C28)
Pathological neuromuscular hypotonus (C33)
Pathological neuromuscular dyscoordination (C35)
Peripheral nerve injury (C07)
Psychoneurogenic neuromuscular hypertonus (C29)
Reduced lung vital capacity (C17)
Restricted joint range of motion (C26)
Soft tissue inflammation (C05)
Soft tissue swelling (C06)
Trigger point formation (C01)
Vascular insufficiency (C11)
Visceral organ dysfunction (C02)

Syndromes (M1203C)

ADHESION (S56)
BELL'S PALSY (S66)
BUERGER'S DISEASE (S11)
BURSITIS (S26)
CALF PAIN (S50)
CAPSULITIS (S27)
CARPAL TUNNEL (S35)
CEREBRAL PALSY (S25)
CERVICAL DORSAL OUTLET (SCALENICUS ATTICUS/CERVICAL RIB) (S09)
CERVICAL (NECK) PAIN (S73)
CHEST PAIN (S14)
CHONDROMALACIA (S15)
CHRONIC BRONCHITIS/EMPHYSEMA (OBSTRUCTIVE PULMONARY DISEASE) (S13)
DIABETES (S10)
EARACHE (S40)
ELBOW PAIN (S47)
FACET (S19)
FOOT PAIN (S51)
FOREARM PAIN (S54)
FROZEN SHOULDER (S64)
HAND/FINGER PAIN (S17)
HEADACHE PAIN (S02)
HIGH THORACIC BACK PAIN (S48)
HYPERTENSION (S43)
INSOMNIA (S71)
JOINT SPRAIN (S30)
KNEE PAIN (S46)
LOWER ABDOMINAL PAIN (S62)
LOW BACK PAIN (S03)
MENSTRUAL CRAMPING (S60)
MIGRAINE (VASCULAR) HEADACHE (S18)
MYOSITIS OSSIFICANS (S67)
NEUROMUSCULAR PARALYSIS (S22)
OSTEOARTHRITIS (S21)
PITTING (LYMPH) EDEMA (S31)
PNEUMONIA (S58)
POST CEREBRAL VASCULAR ACCIDENT (CVA) (S07)
POST IMMOBILIZATION (S36)
POST PERIPHERAL NERVE INJURY (S23)
POST SPINAL CORD INJURY (S24)
POST WHIPLASH (S55)
RADICULITIS (S29)
REFERRED PAIN (S01)
SCIATICA (S05)
SHOULDER/HAND (S57)
SHOULDER PAIN (S08)
SINUS PAIN (S33)
TEMPOROMANDIBULAR JOINT (TMJ) PAIN (S06)
TENDONITIS (S28)
TENNIS ELBOW (LATERAL

EPICONDYLITIS) (S32)
THIGH PAIN (S49)
TINNITUS (S70)
TOOTHACHE/JAW PAIN (S41)
TORTICOLLIS (WRY NECK) (S65)
WRIST PAIN (S16)

Precautions and Contraindications (M1204)

General rules for safe exercise (M1204A)

1. The spine should be maintained in an erect position (if the exercise is not conducted in prone or supine positions. Some deviation may be necessary during the exercise, but forward inclination of the trunk should be kept to a minimum.
2. Muscle(s) used in handling weight should be properly "set" before lifting is attempted. The muscle(s) must be prepared to lift the amount of weight to be lifted to avoid the common straining injury resulting from lifting a weight heavier than expected.
3. Motions should be smooth and steady when force is exerted, never jerky or sudden.
4. Muscles not required for an exercise should remain relatively relaxed to avoid unnecessary fatigue.
5. Exerting force when the trunk is twisted or the back hyperextended (abnormally arched with the lordotic curve increased beyond normal) should be avoided, especially when lifting.

Lifting compression force (M1204B)

Many exercises require the use of the arms in a relative position out in front of the body, as in lifting and reaching, with accompanying forward inclination of the trunk. While performing exercise in a standing position, forward trunk inclination should be kept at a minimum, because the farther the trunk is inclined forward, the greater the strain on the supporting back structures (muscles, discs, ligaments, etc.). As the trunk flexes, the rotational moment due to gravitational pull is increased, putting a greater load on the back extensor muscles and the spinal column supporting ligaments and intervertebral discs, while the angle of pull of the back extensor muscles becomes less favorable. This results in a decrease in the effectiveness of the muscles' angle-of-pull and necessitates an increase in the force of muscle contraction required to maintain posture or to produce back extension motion, thus causing the compression force acting downward along the spinal column to markedly increase. To illustrate, the effective angle-of-pull for the erector spinalis muscle group is approximately twelve degrees when the trunk is flexed forward sixty degrees from vertical. The tensile force in the erector spinalis group necessary to support the trunk of a man weighing 180 pounds in such a posture (with hands hanging freely) is approximately 450 pounds. If such a man is holding a fifty-pound weight, the muscle force must be approximately 750 pounds and the compression force on the L5-S1 joint is nearly 850 pounds, a potential threat to weakened or damaged discs and a possible source of strained back musculature.

Arteriosclerosis or atherosclerosis (M1204C)

Strenuous exercise may be contraindicated for those suffering from arteriosclerosis or atherosclerosis. A surge in blood pressure accompanying exercise or the strain of lifting a heavy weight may cause a "cracking" of a hardened cerebral artery and subsequent spurting of blood, under pressure, into vulnerable cerebral brain matter, causing destruction of any brain matter touched by the stream (a cerebral vascular accident).

Other peripheral vascular disease (M1204D)

Patients suffering from peripheral vascular disease should avoid any exercise which involves excessive use of an involved extremity, including forced walking, if there is lower extremity involvement, when symptoms of intermittent claudication are present. This is especially true for sufferers of Buerger's disease, Raynaud's disease, or diabetes.

Diabetes (M1204E)

An insulin reaction may occur in a diabetic patient who has taken an insulin supplement and has eaten too little food, failed to digest what has been eaten before beginning exercise, or taken too large a dose of insulin before performing strenuous exercise. Initial symptoms include extreme hunger, nervousness, heavy perspiration, palpitations of the heart, and occasionally double vision. These symptoms may be relieved by consumption of eight ounces of orange juice to which two teaspoonfuls of regular sugar have been added. If this is unavailable, any sugar or carbohydrate-containing food, like bread or crackers, may serve if it is given in sufficient quantities.

If initial symptoms are unrelieved, more serious symptoms may appear; these include delirium, convulsions, and/or unconsciousness (coma). Intravenous injection of glucose may be necessary to relieve these very dangerous symptoms.

Hyperventilation (alkalosis) (M1204F)

The respiratory rate increases with strenuous exercise, to increase the availability of oxygen introduced into the lungs, while at the same time providing a ready means for expulsion of the carbon dioxide from the blood, via the lungs, increasingly produced by stepped up metabolism. If the high respiration rate is not rapidly reduced following the termination of exercise, the continued high level uptake of oxygen and the continued high level of carbon dioxide elimination from the blood, without the balancing continued high level production of carbon dioxide from muscle metabolism, will cause a chemical imbalance to occur between oxygen and carbon dioxide in the blood. This may cause a transient alkalosis (depleted carbon dioxide) to develop which is commonly termed hyperventilation. The accompanying symptomatology includes: (1) the sensation of being "air hungry" (being unable to draw a satisfying breath), (2) dizziness or faintness, (3) nausea, (4) sensations of numbness, (5) sensations of tingling, (6) pounding of the heart, and (7) spasmodic muscle cramps (including of the muscles of the chest and lower abdomen).

The symptomatology of hyperventilation may be reduced by having the patient breathe into a paper bag for several minutes. As the carbon dioxide becomes concentrated in the bag, lung uptake of the carbon dioxide raises its level in the blood and the symptoms are relieved as the chemical imbalance between the oxygen and carbon dioxide is corrected. The symptoms of hyperventilation may additionally lead the patient to fear the onset of heart disease or other serious afflictions; verbal reassurance coupled by the demonstration of symptom relief may be necessary to still such anxieties.

Acidosis (M1204G)

If carbon dioxide elimination through respiration does not keep pace with carbon dioxide production during strenuous exercise, a transient systemic acidosis will be produced as a carbon dioxide excess builds up in the blood. The characteristic symptoms produced by acidosis is drowsiness, which may develop in extreme conditions into stupor and coma. If unrelieved spontaneously, treatment utilizing a mechanical respirator may be indicated.

Pulse rate regulation (M1204H)

To avoid exhaustion, the pulse rate of a patient engaged in strenuous exercise should be monitored and the exercise curtailed when the rate exceeds the heart's ability to maintain an adequate volume. As a general rule, exercise should be stopped when the patient's pulse exceeds the rate maximum, which is established by subtracting the patient's age from 220 (plus or minus 10) and multiplying the result by 0.85.

General contraindications (M1204I)

Therapeutic exercise is contraindicated for patients suffering from any depleting disease or neoplasm, certain pathological heart conditions, active systemic acute infections (bacterial or viral), acute inflammation, new fractures (post five weeks or less) or surgical orthopaedic correctional procedures, acute joint sprains, acute joint strains, and tuberculosis.

Exercise and menstruation (M1204J)

Regular intensive exercise has been noted to affect fertility by affecting the menstrual cycle. A common observation of women athletes who engage in long distance running and other physically demanding strenuous aerobic activities is that the menstrual cycle often becomes irregular, sometimes becoming nonexistent, while participating in the exercise program (exercise equivalent to running two miles a day, five days a week). Such exercise is reported to interfere with the normal production of estrogen, which has the direct effect of decreasing the fertility of the human female, while so engaged. Interestingly enough, some evidence has suggested that the effects of exercise on the estrogen levels may account for a general decrease in the occurrence of reported cases of cancer of the reproductive system in the population of current or former women athletes.

Notes Aside (M1205)

Calorie consumption of activities of daily living (M1205A)

To remain physiologically stable, the average human male weighing 70 kilograms lying in bed requires approximately 1600 Calories per day, with an additional 200-plus Calories required for the eating process. If the same human male sits in a chair all day, his energy requirements will be between 2000 and 2250 Calories. Walking up stairs requires approximately seventeen times more energy than lying in bed asleep, while a laborer may utilize as much as 6000 to 7000 Calories in a 24-hour period and may have a basal metabolism rate which is 3.5 times higher than is normal. The metabolic rate of a newborn child is almost 200%

higher then that of an elderly person because of the high energy cost high speed cellular reactions and rapid synthesis of cellular material required of the growth process (refer to the Table of Exercises - Energy Consumption Rates [T013]).

Bibliography (M1206)

Cooley, pp. 354–355, 238.
Ganong, pp. 504–506.
Guyton, pp. 158–159, 311–312, 314–315, 320, 482, 485–486, 502–503, 827, 925–926.
Holvey, pp. 133–135, 237.
Scott, pp. 1–13.
Tests Indicate Fertility Impeded by Regular and Intense Exercise, p. 9.
Williams, pp. 87–94.

EXERCISE MUSCLE STRENGTHENING
(M12A)

Definition (M12A1)

Muscle strength is here defined as the maximum tension a muscle may develop during a single contraction against resistance. Strenuous exercise has been shown to be useful for increasing the potential force a muscle is able to generate during contraction.

Muscle strengthening physiology (M12A1A)

When muscles become stronger, as the result of exercise, they become larger. Muscles are enlarged through a process called overloading, or exercising more than is normally required.

The physiological result of regularly overloading a muscle over time is an increase in the size and number of myofibrils (the protein cords inside the muscle fiber), an increase in size and number of capillaries supplying the muscle(s) (unless the exercise is isometric), an increase of myoglobin content of the muscle fibers, an increase in the abundance of enzyme rich mitochondria (for a more efficient transformation of fatty acids and glucose into energy) within the muscle fibers, and increases in ATP, creatine phosphate, glycogen and fat stored within the muscle (strong muscles have been shown to stockpile more of these materials than weak muscles). These changes, usually occurring after several weeks of regular strenuous exercise, ultimately result in larger, more efficient muscles which are capable of generating greater contractile force then similar muscles which have not been regularly exercised.

Strenuous exercise is usually performed with the goal of increasing muscle strength and/or muscle endurance. Although a general discourse on muscle anatomy and the physiology of muscle contraction and exercise is thought inappropriate here, it should be noted that the muscle is, for the most part, a collection of muscle fiber bundles which are composed of two basic muscle fiber types, fast-twitch and slow-twitch muscle fibers. There are actually three types, but we will only be concerning ourselves here with the two that are most common.

Fast-twitch muscle fibers (M12A1B)

Fast-twitch (phasic) muscle fibers perform anaerobically, having an oxygen-poor energy supply, and are associated with strong, powerful contractions produced by short bursts of energy and comparatively low stamina. They are sparsely supplied with capillaries, which makes them vulnerable to the build-up of metabolic by-products, including lactic acid, which have the general effect of driving the fibers to exhaustion.

Fast-twitch fibers are relatively light in color and are commonly referred to as "white" fibers. Muscles predominantly made up of fast-twitch fibers are often called "white" muscles or fast-twitch muscles. Fast-twitch muscles are most often anatomically superficial muscles, laterally located, and have, for the most part, tendinous origins and insertions, usually crossing two or more joints, with the distal segment generally being free; they are most often flexors.

Fast-twitch muscles in general have a relatively high threshold to tetany and a shorter duration of response to electrical stimuli then other muscle types. They enjoy relatively more muscle spindles and Golgi tendon organs, and are more sensitive to the influence of exteroceptor stimulation. Fast-twitch muscles are relatively more dependant upon cortical direction than other muscle types, and are generally used in coordinated movement to perform refined motor activities. Fast-twitch fibers often work in motor units containing many fibers innervated by a single motor nerve. Their collective contraction essentially magnifies the contractile force of the single fiber contraction. Fast-twitch muscle contraction is most often performed as part of a reciprocal inhibition relationship with an antagonist.

Fast-twitch fibers within a muscle are best tested and strengthened by high resistance exercise of short duration, involving a small number of repetitions of a stereotypic muscle contraction or sequence of muscle contractions. Such exercise includes isometric exercise, weightlifting, wrestling and some types of "calisthenic" exercise (like push-ups).

Slow-twitch muscle fibers (M12A1C)

Slow-twitch (tonic) muscle fibers perform aerobically, having an oxygen-rich energy supply. They have a rich supply of blood which, combined with plentiful myoglobin and thick intracellular fluid, lend the fibers color to the extent that they are commonly called "red" fibers. Slow-twitch fibers are associated with steady contractions which may last until the muscle's energy supply is exhausted, which means that they have a relative "high stamina."

Muscles composed primarily of slow-twitch fibers (slow-twitch muscles) are generally located anatomically deep. Their origins and insertions are usually fibrous, cross a single joint, and their distal segments are generally fixed; they are most often extensors. Slow-twitch muscles are primarily associated with cocontraction of agonists and antagonists, and their primary duty is generally to maintain postures, and to provide joint stabilization during refined motor functions.

Slow-twitch muscles are relatively more responsive to proprioception stimulation influence (stretch, resistance, etc.), and are more reflexly controlled. Motor units dominated by slow-twitch fibers are made up of relatively few muscle fibers.

Slow-twitch fibers are best strengthened and conditioned for endurance by exercise which involves low resistance, high repetition stereotypic muscle contractions exerted for a relatively prolonged period of time, sometimes for hours. Such exercises are usually called aerobics and include jogging, jumping rope, swimming, dancing, bicycle riding, Tai-Chi Chuan, yoga, and various types of "calisthenic" exercises (like jumping jacks).

Mixed fiber muscles (M12A1D)

In humans, all skeletal muscles possess both fast-twitch and slow-twitch fibers, but they occupy each muscle in varying proportions. The degree of dominance by a fiber type varies from muscle to muscle and individual to individual. When like muscles from different individuals are compared, the proportional content of each fiber may be found to be different. Additionally, in humans, the division between the fast-twitch and slow-twitch fibers is sometimes rather vague, with one type of fiber occasionally taking on some of the characteristics of the other. In general, however, one type of muscle fiber will predominate within a given muscle.

Resistive exercise (M12A1E)

Exercise which employs resistance is roughly classified as isometric and isotonic exercise.

Isometric exercise (M12A1E01) Isometric exercise employs the muscle in a contraction against resistance, during which the muscle does not shorten after the contraction begins, and no joint motion is supposed to take place. During an isometric exercise, both agonist and antagonist muscles contract together in cocontraction to maintain joint stability.

Isotonic exercise (M12A1E02) Isotonic exercise employs the muscle in a shortening contraction against resistance which moves an involved joint through part or all of its range of motion, while its tension is maintained. Typically, as the agonist contracts, the antagonistic muscle is called upon to perform an eccentric contraction (lengthening while maintaining tension), thereby allowing the in volved joint to move in a controlled manner in re sponse to agonist contraction. Isotonic exercise which employs slow controlled joint motion against resistance exercises both agonist and antagonist, allowing both to benefit from the exercise.

Blood pressure (M12A1F)

It has been shown that both isometric and isotonic exercise is normally capable of raising the patient's blood pressures (both systolic and diastolic) and heart rate between twenty and twenty-five percent. Isometric exercise has been shown to be significantly more effective at doing this than a comparable amount of isotonic exercise, in terms of the number of repetitions employed, the time spent, and similarity of grade of contraction.

Procedures of Application (M12A2)

Aerobic exercise (M12A2A)

Aerobic exercise regimens are not discussed here except to caution any who would engage in such exercise to periodically monitor the pulse rate and to use it as a gauge of endurance limits.

Isometric exercise (M12A2B)

Isometric exercise may be used to increase muscle tone, resetting the tonic fibers of the muscle spindle, and to strengthen muscle tissue without undue risk of tendon or muscle strain. Isometric exercise is generally easy to perform and is most convenient for patient home use since little or no equipment is required. It may be used as a preliminary toning exercise, preparatory for isotonic (progressive) resistive exercise.

exercise muscle strengthening

For appropriate isometric exercise regimes refer to Table of Exercises for the Lower Extremities — Isometrics [T014], Table of Exercises for the Neck — Isometrics [T015], and Table of Exercises for the Shoulder and Upper Extremities — Isometrics [T011].

Performing isometric exercise (M12A2B01)
(1) The patient should be instructed not to allow any motion of the involved joint(s) while contracting the muscle(s).
(2) The tension in the contracting muscle(s) should be increased slowly (taking as much as a full second) until a maximal contraction of the muscle is attained.
(3) The contraction should be held with unremitting intensity for six seconds without holding the breath and then the muscle(s) should be slowly and carefully relaxed, taking as much as a full second.
(4) The muscle(s) should be kept in a relaxed state for a full six seconds of rest (to allow an influx of blood into the exercised muscle fibers) before another isometric contraction is performed.

Frequency (M12A2B02) The number of repetitions of an isometric exercise of a given muscle or synergistic group usually ranges between five and ten.

Isometric exercise regimens are usually performed once or twice daily.

Isotonic exercise (M12A2C)

Since joint motion against resistance increases the possibility of muscle, tendon and joint ligament strain, isotonic exercise should be engaged in with care. It is often advisable for the patient to isometrically tone and strengthen the involved muscles, over several weeks (two may be enough), before isotonic exercise is begun. Most often, isotonic exercise in the clinical setting takes the form of weight lifting, or progressive resistive exercise.

For detailed descriptions of some of the isotonic exercises commonly recommended for patient treatment refer to Table of Exercise - Isotonic (Weight Lifting Protocols and Others) [T012].

Establishing appropriate lifting weight (M12A2C01) Before beginning a program of weight lifting (pulley weight exercises are included in this concept), the weight which may be safely lifted by the patient must be determined. The most appropriate starting weight for the patient is generally established through a process of trial and error. This is simply done by having the patient lift a particular weight in the correct exercise form with the appropriate equipment. If the patient can lift the selected weight with ease, but finds performing the lift increasingly difficult as consecutive lifts are performed, so that five repetitions are performed but the patient finds it increasingly difficult to reach ten, that is probably the weight that patient should be lifting.

Breathing during isotonic exercise (M12A2C02) While lifting, the patient should perform inspiration during the process of agonist muscular contraction (lifting the weight), completing the breath when the end of joint range has been reached. This approach helps prevent the patient from holding the breath. Expiration should be performed as the muscle is performing the eccentric contraction side of the exercise (lowering the weight).

Repetitions (M12A2C03) Weight lifting should be performed in a series of repetitions, called a set. Generally, a set is considered to be a series of ten consecutive lifts. Three sets of a given exercise per session are thought to be ideal for muscle strengthening. A rest of several minutes should take place between each set.

For best results, it is usually recommended that a particular weight lifting exercise (i.e., three sets) should be performed no more than three times a week, with at least a day between each session.

When the patient can easily perform ten repetitions, the weight should be increased by an amount following the procedure initially used to establish appropriate weight (increases usually vary from five to twenty pounds).

Partial List of Conditions Treated (M12A3)

Treatable Causes (M12A3A)

Blood sugar irregularities (C31)
Extrafusal muscle spasm (C41)
Habitual joint positioning (C48)
Hypermobile/instable joint (C10)
Hypotension (C45)
Insomnia (C46)
Interspinous ligamentous strain (C03)
Joint ankylosis (C42)
Ligamentous strain (C25)
Muscular weakness (C23)
Neuromuscular tonic imbalance (C30)
Pathological neuromuscular hypertonus (C28)
Pathological neuromuscular hypotonus (C33)
Pathological neuromuscular dyscoordination (C35)

Peripheral nerve injury (C07)
Psychoneurogenic neuromuscular hypertonus (C29)
Restricted joint range of motion (C26)
Trigger point formation (C01)

Syndromes (M12A3B)

BELL'S PALSY (S66)
BURSITIS (S26)
CALF PAIN (S50)
CAPSULITIS (S27)
CEREBRAL PALSY (S25)
CERVICAL (NECK) PAIN (S73)
CHEST PAIN (S14)
CHONDROMALACIA (S15)
DIABETES (S10)
EARACHE (S40)
ELBOW PAIN (S47)
FOOT PAIN (S51)
FOREARM PAIN (S54)
FROZEN SHOULDER (S64)
HAND/FINGER PAIN (S17)
HEADACHE PAIN (S02)
HIGH THORACIC BACK PAIN (S48)
INSOMNIA (S71)
JOINT SPRAIN (S30)
KNEE PAIN (S46)
LOWER ABDOMINAL PAIN (S62)
LOW BACK PAIN (S03)
MENSTRUAL CRAMPING (S60)
MIGRAINE (VASCULAR) HEADACHE (S18)
NEUROMUSCULAR PARALYSIS (S22)
OSTEOARTHRITIS (S21)
PIRIFORMIS (S04)
POST CEREBRAL VASCULAR ACCIDENT (CVA) (S07)
POST IMMOBILIZATION (S36)
POST PERIPHERAL NERVE INJURY (S23)
POST SPINAL CORD INJURY (S24)
POST WHIPLASH (S55)
RADICULITIS (S29)
REFERRED PAIN (S01)
SCIATICA (S05)
SHOULDER/HAND (S57)
SHOULDER PAIN (S08)
SINUS PAIN (S33)
TEMPOROMANDIBULAR JOINT (TMJ) PAIN (S06)
TENDONITIS (S28)
TENNIS ELBOW (LATERAL EPICONDYLITIS) (S32)
THIGH PAIN (S49)
TOOTHACHE/JAW PAIN (S41)
TORTICOLLIS (WRY NECK) (S65)
WRIST PAIN (S16)

Precautions and Contraindications (M12A4)

General rules for safe exercise (M12A4A)

1. The spine should be maintained in an erect position (if the exercise is not conducted in prone or supine positions. Some deviation may be necessary during the exercise, but forward inclination of the trunk should be kept to a minimum.
2. Muscle(s) used in handling weight should be properly "set" before lifting is attempted. The muscle(s) must be prepared to lift the amount of weight to be lifted to avoid the common straining injury resulting from lifting a weight heavier than expected.
3. Motions should be smooth and steady when force is exerted, never jerky or sudden.
4. Muscles not required for an exercise should remain relatively relaxed to avoid unnecessary fatigue.
5. Exerting force when the trunk is twisted or the back hyperextended (abnormally arched with the lordotic curve increased beyond normal) should be avoided, especially when lifting.

Lifting compression force (M12A4B)

Many exercises require the use of the arms in a relative position out in front of the body, as in lifting and reaching, with accompanying forward inclination of the trunk. While performing exercise in a standing position, forward trunk inclination should be kept at a minimum, because the farther the trunk is inclined forward, the greater the strain on the supporting back structures (muscles, discs, ligaments, etc.). As the trunk flexes, the rotational moment due to gravitational pull is increased, putting a greater load on the back extensor muscles and the spinal column supporting ligaments and intervertebral discs, while the angle of pull of the back extensor muscles becomes less favorable. This results in a decrease in the effectiveness of the muscles' angle-of-pull and necessitates an increase in the force of muscle contraction required to maintain posture or to produce back extension motion, thus causing the compression force acting downward along the spinal column to markedly increase. To illustrate, the effective angle-of-pull for the erector spinalis muscle group is approximately twelve degrees when the trunk is flexed forward sixty degrees from vertical. The tensile force in the erector spinalis group necessary to support the trunk of a man weighing 180 pounds in such a posture (with hands hanging freely) is approximately 450 pounds. If such a man is holding a fifty-pound weight, the muscle force must be approximately 750 pounds and the compression force on the L5-S1 joint is nearly 850 pounds, a potential threat to weakened or damaged discs and a possible source of strained back musculature.

Arteriosclerosis or atherosclerosis (M12A4C)

Strenuous exercise may be contraindicated for those suffering from arteriosclerosis or atherosclerosis. A surge in blood pressure accompanying exercise or the strain of lifting a heavy weight may cause a "cracking" of a hardened cerebral artery and subsequent spurting of blood, under pressure, into vulnerable cerebral brain matter, causing destruction of any brain matter touched by the stream (a cerebral vascular accident).

Other peripheral vascular disease (M12A4D)

Patients suffering from peripheral vascular disease should avoid any exercise which involves excessive use of an involved extremity, including forced walking, if there is lower extremity involvement, when symptoms of intermittent claudication are present. This is especially true for sufferers of Buerger's disease, Raynaud's disease, or diabetes.

Diabetes (M12A4E)

An insulin reaction may occur in a diabetic patient who has taken an insulin supplement and has eaten too little food, failed to digest what has been eaten before beginning exercise, or taken too large a dose of insulin before performing strenuous exercise. Initial symptoms include extreme hunger, nervousness, heavy perspiration, palpitations of the heart, and occasionally double vision. These symptoms may be relieved by consumption of eight ounces of orange juice to which two teaspoonfuls of regular sugar have been added. If this is unavailable, any sugar or carbohydrate-containing food, like bread or crackers, may serve if it is given in sufficient quantities.

If initial symptoms are unrelieved, more serious symptoms may appear; these include delirium, convulsions, and/or unconsciousness (coma). Intravenous injection of glucose may be necessary to relieve these very dangerous symptoms.

Hyperventilation (alkalosis) (M12A4F)

The respiratory rate increases with strenuous exercise, to increase the availability of oxygen introduced into the lungs, while at the same time providing a ready means for expulsion of the carbon dioxide from the blood, via the lungs, increasingly produced by stepped up metabolism. If the high respiration rate is not rapidly reduced following the termination of exercise, the continued high level uptake of oxygen and the continued high level of carbon dioxide elimination from the blood, without the balancing continued high level production of carbon dioxide from muscle metabolism, will cause a chemical imbalance to occur between oxygen and carbon dioxide in the blood. This may cause a transient alkalosis (depleted carbon dioxide) to develop which is commonly termed hyperventilation. The accompanying symptomatology includes: (1) the sensation of being "air hungry" (being unable to draw a satisfying breath), (2) dizziness or faintness, (3) nausea, (4) sensations of numbness, (5) sensations of tingling, (6) pounding of the heart, and (7) spasmodic muscle cramps (including of the muscles of the chest and lower abdomen).

The symptomatology of hyperventilation may be reduced by having the patient breathe into a paper bag for several minutes. As the carbon dioxide becomes concentrated in the bag, lung uptake of the carbon dioxide raises its level in the blood and the symptoms are relieved as the chemical imbalance between the oxygen and carbon dioxide is corrected. The symptoms of hyperventilation may additionally lead the patient to fear the onset of heart disease or other serious afflictions; verbal reassurance coupled by the demonstration of symptom relief may be necessary to still such anxieties.

Acidosis (M12A4G)

If carbon dioxide elimination through respiration does not keep pace with carbon dioxide production during strenuous exercise, a transient systemic acidosis will be produced as a carbon dioxide excess builds up in the blood. The characteristic symptoms produced by acidosis is drowsiness, which may develop in extreme conditions into stupor and coma. If unrelieved spontaneously, treatment utilizing a mechanical respirator may be indicated.

Pulse rate regulation (M12A4H)

To avoid exhaustion, the pulse rate of a patient engaged in strenuous exercise should be monitored and the exercise curtailed when the rate exceeds the heart's ability to maintain an adequate volume. As a general rule, exercise should be stopped when the patient's pulse exceeds the rate maximum, which is established by subtracting the patient's age from 220 (plus or minus 10) and multiplying the result by 0.85.

General contraindications (M12A4I)

Therapeutic exercise is contraindicated for patients suffering from any depleting disease or neoplasm, certain pathological heart conditions, active systemic acute infections (bacterial or viral), acute inflammation, new fractures (post 5 weeks or less) or surgical orthopaedic correctional procedures, acute joint sprains, acute joint strains and tuberculosis.

Valsalva maneuver (M12A4J)

The patient should be cautioned not to hold the breath while performing strenuous exercise, because of the danger of performing a Valsalva maneuver. A Valsalva maneuver is defined as a forced expiration against a closed glottis. It may occur following holding the breath while performing a strenuous isometric or isotonic exercise. The blood pressure rises at the outset of straining because the intrathoracic pressure is added to the pressure of the blood in the aorta. It then falls because the high intrathoracic pressure compresses the veins, decreasing venous return and cardiac output. The decrease in arterial pressure and pulse pressure may inhibit the baroreceptors to cause tachycardia and a rise in peripheral resistance. When the glottis is again open, the intrathoracic pressure will return to normal, restoring cardiac output without relieving peripheral vessel constriction; the blood pressure may then rise above normal. Eventually, the high blood pressure rise will stimulate baroreceptors to cause bradycardia and a drop in pressure to normal levels. So, the Valsalva maneuver may be a threat to the patient's well-being by posing the danger of cardiac malfunction as a result of internal attempts at system correction, as well as contributing to the possibility of cerebral vascular accident with the shifts in arterial pressure.

Notes Aside (M12A5)

Tonic resetting of muscle length with isometric exercise (M12A5A)

A tight muscle may not be a strong muscle; psychoneurogenic influences may tense a muscle utilizing the servo mechanism inherent in the central nervous system's relationship to the muscle spindle mechanisms. This mechanism is held by some to be one of the causes of the trigger point formations and, as a consequence, isometric exercise may be used to reset the tonic fibers of the muscle spindle (a common treatment of trigger point formations).

Muscle soreness following exercise (M12A5B)

Predictably, the patient may complain of a little additional soreness after the first one or two sessions of strenuous exercise, usually occurring within the the first two days after a workout. It results from the build-up of metabolites in muscles which are initially ill-supplied with circulation and from strains of unprepared or weak muscles. Preventative treatment is thought to be the best. A warm-up session of mild calisthenics before exercise and cooling down of mild calisthenics (or brief jog) afterward do much to prevent pain and stiffness by raising the temperature inside the muscles and drawing more blood into the muscle fibers to help enzyme action, which is responsible for the decay of inflammatory biochemicals resulting from stressed and strained soft tissues. As the exercise program has its effect and muscle strength increases, the efficiency of metabolic by-product elimination improves as capillary bed density increases, to provide better circulation and improve metabolite wash-out, and the patient will no longer experience soreness from strenuous exercise.

Rapid strength gain (M12A5C)

Clinical experience has demonstrated that strength gain, defined in terms or how much weight can be lifted, may be accelerated by decreasing the weight load to two thirds of the individual's current maximum and then having the individual increase the number of repetitions performed, during the set, to the maximum number which can be fully performed. Three sets should be performed.

Oxford technique (M12A5D)

For patients who are only able to lift a few pounds, the Oxford technique may be useful.

The weight to be initially lifted should be that which can be lifted a maximum of ten repetitions. In each successive set the load should be progressively lowered by about a pound. On each successive day the weight load of the first set should be raised a pound.

Bibliography (M12A6)

Delitto, Rose, Apts, pp. 1329–1348.
Ganong, pp. 481–482.
Greer, pp. 179–183.
Guyton, pp. 87–91.
Knapik, pp. 938–947.
Pelletier, pp. 1359–1364.
Selim, pp. 50–51, 128–131.
Taylor, pp. 274–284.

EXERCISE IMPROVING BREATHING FUNCTION (M12B)

Definition (M12B1)

Normal Breathing Pattern (M12B1A)

The muscles which provide the motor action required for normal inspiration ideally include mainly the diaphragm and intercostal muscles. This type of breathing is commonly called diaphragm breathing.

Diaphragm Muscle (M12B1A01) The diaphragm is anatomically attached inside the thorax, around the internal thoracic wall in such a way that it forms a domed-shaped structure which separates the upper abdominal area which houses the lungs from the abdominal viscera (liver, stomach, intestines, etc.) below. The diaphragm is designed to be the primary muscle responsible for the physical act of inspiration. When the diaphragm contracts, the "dome" shape flattens, creating negative pressure (or vacuum) within the upper thorax, which causes air to rush in and fill the lungs. Exhalation occurs when the diaphragm relaxes and returns to the dome shape, providing the positive pressure within the upper thorax which pushes the air out of the lungs.

Intercostal muscles (M12B1A02) The external intercostal muscles are thought by some authorities to aid in inspiration by tilting or lifting the ribs, while the internal intercostal muscles aid expiration by tilting the ribs back down. Other authorities believe that the intercostal muscles simply keep the intercostal spaces from being "sucked in" with inspiration and "bulged outward" with forced expiration.

Forced Inspiration Breathing Pattern (M12B1B)

Abnormal patterns may be observed which have resulted from a neurological deficit or as a "breathing habit" of unknown etiology; one of these abnormal breathing patterns is the forced inspiration pattern.

The muscles utilized for forced inspiration include the scaleni, sternocleidomastoid, pectoralis minor, pectoralis major, serratus anterior, sternohyoid, sternothyroid, serratus posterior superior, and levatores costarum. This type of breathing is commonly called accessory muscle breathing. To be used to produce inspiration the muscles involved are caused to reverse their primary motor functions, relative to the the shoulder girdle, neck and head, and/or the upper extremity. This is accomplished by having the body part upon which the muscle *inserts* (neck, head, clavicle, scapulae, etc.) neuromuscularly stabilized while the structure which normally serves as the muscle's *origin* is acted upon by the muscle's contraction to lift the rib cage (ribs, clavicle, scapulae, etc.).

Diaphragm breathing has been estimated to be seventy-five percent more efficient then breathing with the muscles of forced inspiration.

Forced Expiration (M12B1C)

The muscles utilized for forced expiration include the abdominal muscles (rectus abdominus, internal oblique and external oblique) and the latissimus dorsi. When these muscles contract together they create an intrathoracic positive pressure which pushes the viscera upward, causing the air in the lungs to be forced out. This same action may be used to breathe with, after a fashion, since the relaxation of those muscles after the united contraction causes a negative intrathoracic pressure to occur as the viscera drops back down, causing air to rush back into the lungs. This type of breathing is commonly called reversed breathing, and is the least efficient of the breathing techniques. It requires a relatively large amount of energy to force expiration, so the negative pressure caused by the following muscle relaxation is insufficient to cause complete filling of the lungs.

Thoracic Spine, Relative to Breathing Function (M12B1D)

Motion of the thoracic spine is interdependent with motion of the rib cage. When the thoracic spine is flexed, the ribs are depressed and their range of motion is mechanically limited; when the thoracic spine is extended, the ribs become elevated and the chest wall can move laterally, anteriorly and upward. Posture characterized by an habitually rounded back, hollow chest and and abducted scapulae (basketball player slump) may be expected to have an inflexible thorax. A patient suffering from this type of posture typically does not use the diaphragm to good advantage, and it will tend to be weak from under use Exercise which expands the rib cage through coordinated use of the musculature of the upper trunk, spine, scapulae, and humerus may be used to expand the rib cage and help to restore normal contour and mobility, while exercise which strengthens the diaphragm will improve the patients potential ability to breathe more efficiently.

Patients with asthma, chronic bronchitis, emphysema, or other chronic pulmonary obstructive disease may be plagued by chronic over-expansion of the chest, making it more difficult to enjoy normal expiration. If this condition is sufficiently severe and has existed for some time, the chest may become permanently inflated and take on the look of the "barrel chest". Patients suffering from this condition must develop control of the expiratory phase of breathing and to develop the muscles which are anatomically situated to provide compression of the thorax.

Procedures of Application (M12B2)

Exercises Which Promote Expansion of the Thorax (M12B2A)

Although a beginning supine position is described here (excepting the fourth exercise), the following exercises may also be performed with the patient in sitting, standing or semi-reclining positions.

Unresisted inspiration (M12B2A01)

(1) The patient should be lying in a supine position with the knees flexed, feet flat on the floor, and the the patient's arms against the patient's sides.
(2) One shoulder should be flexed and the extended arm lifted by the patient to the vertical position, as a deep breath is drawn on the ipsilateral side of the chest.
(3) The breath should then be slowly expelled as the arm is lowered to its resting position at the patient's side.
(4) This procedure should be repeated on the contralateral side.
(5) The patient should flex the shoulders and raise both extended arms simultaneously, while taking a deep breath.
(6) The breath should be slowly expelled as the arms are simultaneously lowered to their resting positions at the patient's sides.

(7) Finally, the patient should flex the shoulders and raise both extended arms over the head while taking a deep breath.
(8) The breath should be slowly expelled as the arms are simultaneously returned to their resting positions at the patient's sides.

Resisted inspiration (M12B2A02)

(1) The patient should be lying in a supine position with the knees flexed and the feet flat on the floor.
(2) A bath towel should be wrapped around the chest, the ends of the towel crossing over the anterior chest. The end of the towel proceeding around the chest from the right should be held in the left hand, and the end proceeding around from the left held in the right hand.
(3) The ends of the towel should be pulled in opposite directions to cause the chest to be "squeezed" as the patient forcefully inhales the resistance of the towel. Inspiration should take six or seven seconds.
(4) As the breath is slowly let out, the towel pressure should remain constant. Expiration should be controlled and last for ten seconds.

Chest elevation and forced expiration (M12B2A03)

(1) The patient should be lying in a supine position with the knees flexed, the feet flat on the floor, with the shoulders abducted and externally rotated to ninety degrees and the elbows flexed to ninety degrees, the whole arm lying on the resting surface.
(2) The patient should take a moderate breath, allowing the chest to rise to a moderate level.
(3) The moderately raised chest position should be maintained while the breath is forced out of the lungs by tightening and drawing in of the abdominal muscles.
(4) On the next inspiration, the ribs should be raised a little higher and that position held while forced expiration occurs as in the previous step. This process should be repeated until the chest is fully elevated and expanded, and no more air can be forced into the lungs.
(5) Finally, the patient should relax and the whole process begun again.

Panting with expanded chest (M12B2A04)

(1) The patient should be sitting upright with elbows out from the sides, with the hands on the hips.
(2) The patient should take a deep breath while pushing the chest up and forward.
(3) The expanded chest position should be held while exhalation is performed by drawing in the abdomen.

(4) A series of short inspirations should then be performed while the expanded chest position is being held.

Exercises Which Promote Contraction of the Thorax During Expiration (M12B2B)

In all exercises stressing expiration, the breathing cycle should take as long as possible. Inspiration should last from six to seven seconds and expiration should last for ten seconds.

Unresisted respiration with abdominal expiration (M12B2B01)

(1) The patient should be in a supine position with the knees flexed and the feet flat on the floor. One of the patient's hands should be placed on the upper abdomen, just below the lower ribs.

(2) The patient should expire slowly and completely by contracting the abdominal muscles; the hand on the abdomen serving to provide feedback to the patient regarding abdomen muscle action.

(3) The patient should then inspire gently as the preliminary step to repeating the process.

Forced mechanical expiration (M12B2B02)

(1) The patient should be sitting upright.

(2) A bath towel should be wrapped around the lower chest, the ends of the towel should cross over the anterior chest. The end of the towel proceeding around the chest from the right should be held in the left hand, and the end proceeding around from the left held in the right hand.

(3) The ends of the towel should be pulled in opposite directions to squeeze the chest as the patient expires by strongly contracting the abdominal muscles.

(4) At the end of chest motion, the patient should use the towel to squeeze the chest in an attempt to force as much air out of the lungs as possible and to exaggerate the movement of expiration.

Forced expiration with lateral flexion (M12B2B03)

(1) The patient should be sitting upright, with the knees abducted to increase stability.

(2) One of the patient's hands should be placed over the lower ribs on the ipsilateral side, with the fingers pointing forward.

(3) The patient should bend slightly forward and laterally flex the trunk toward the hand on the ribs while performing a strong expiration. Strong manual pressure from the hand should be exerted to increase the depression of the ribs on the side.

(4) The process should be repeated on the contralateral side.

Forced expiration with forward abdominal flexion (M12B2B04)

(1) The patient should be sitting upright with the arms relaxed at the sides.

(2) The upper trunk should be allowed to fall forward and the thoracic spine to flex while performing a strong expiration as the abdominal muscles are strongly contracted.

(3) The assumed flexed posture should be maintained during inspiration, which should be moderate and brief.

Shoulder Girdle and Upper Spine Mobilization Exercises (M12B2C)

Neck mobilization (M12B2C01)

(1) The patient should be sitting cross-legged with the arms hanging at the sides.

(2) The head should be allowed to fall forward on to the chest in a relaxed fashion.

(3) The neck should be extended to bring the head back up to the beginning position.

(4) While still maintaining good trunk position, the head should be allowed to fall forward onto the chest and the neck laterally flexed as the head is allowed to roll over.

(5) The head should be returned to its position on the chest, then the neck laterally flexed to the opposite side.

Shoulder mobilization (M12B2C02)

(1) The patient should be sitting cross-legged with the arms hanging at the sides.

(2) The patient should flex one shoulder forward, up, back and down; the thorax should not be allowed to turn with the movement.

(3) The patient should alternately circle the right and left shoulders in figure of eight patterns while emphasizing the backward and downward phase of the circles.

(4) This process should be repeated in a reversed direction, stressing the scapular adduction and suppression.

Partial List of Conditions Treated (M12B3)

Conditions Treated in the Field (M12B3A)

Asthma

Treatable Causes (M12B3B)

Habitual joint positioning (C48)
Lung fluid accumulation (C13)
Reduced lung vital capacity (C17)
Restricted joint range of motion (C26)

Syndromes (M12B3C)

CERVICAL DORSAL OUTLET (SCALENICUS ATTICUS/CERVICAL RIB) (S09)
CHEST PAIN (S14)
CHRONIC BRONCHITIS/EMPHYSEMA (OBSTRUCTIVE PULMONARY DISEASE) (S13)
HIGH THORACIC BACK PAIN (S48)
HYSTERIA/ANXIETY REACTION (S59)
POST IMMOBILIZATION (S36)

Precautions and Contraindications (M12B4)

Hyperventilation (M12B4A)

The respiratory rate increases with strenuous exercise, to increase the availability of oxygen introduced into the lungs, while at the same time providing a ready means for expulsion of the carbon dioxide from the blood, via the lungs, increasingly produced by stepped-up metabolism. If the high respiration rate is not rapidly reduced following the termination of exercise, the continued high level uptake of oxygen and the continued high level of carbon dioxide elimination from the blood, without the balancing continued high level production of carbon dioxide from muscle metabolism, will cause a chemical imbalance to occur between oxygen and carbon dioxide in the blood. This may cause a transient alkalosis (depleted carbon dioxide) to develop which is commonly termed hyperventilation. The accompanying symptomatology includes: (1) the sensation of being "air hungry" (being unable to draw a satisfying breath), (2) dizziness or faintness, (3) nausea, (4) sensations of numbness, (5) sensations of tingling, (6) pounding of the heart, and (7) spasmodic muscle cramps (including of the muscles of the chest and lower abdomen).

The symptomatology of hyperventilation may be reduced by having the patient breathe into a paper bag for several minutes. As the carbon dioxide becomes concentrated in the bag, lung uptake of the carbon dioxide raises its level in the blood and the symptoms are relieved as the chemical imbalance between the oxygen and carbon dioxide is corrected. The symptoms of hyperventilation may additionally lead the patient to fear the onset of heart disease or other serious afflictions; verbal reassurance coupled by the demonstration of symptom relief may be necessary to still such anxieties.

Notes Aside (M12B5)

If the patient suffers from pulmonary obstructive disease, exercises designed to improve lung function may be most effectively used when combined with a program of postural drainage and back pressure procedures (refer to Postural Drainage [M14]). Back pressure may include blowing water out of a disconnected 25-foot garden hose, blowing up balloons, blowing into a "blow bottle", or blowing out candles. If blowing out a candle is used, it should be lit and placed at mouth level. Through pursed lips, air from the lungs should be expelled to extinguish the flame. With success, the candle should be progressively placed at a greater and greater distances from the mouth; this exercise should follow the diaphragm breathing protocol. A positive pressure lung machine may also be used to provide mechanical back pressure.

Bibliography (M12B6)

Dull, pp. 655–659.
Sobush, pp. 542–544.
Williams, pp. 65–70.

EXERCISE MUSCLE RELENGTHENING
(M12C)

Definition (M12C1)

Tonically Shortened Musculature (M12C1A)

Neurogenic or psychoneurogenic influences of the supraspinal structure upon the muscle spindle may cause a shortening of the musculature (refer to Muscle Spindle [R001]). The muscle spindle has the ability to increase and/or maintain muscle tension (tonicity) by virtue of the tonic stretch reflex (refer to Neuromuscular System [R002]). Tonic shortening of the musculature may result from: (1) supraspinal central nervous system failure to inhibit muscle spindle tonic biasing, of which the spasticity that accompanies the post CVA syndrome is an illustration; (2) habitual maintenance of a muscle in its shortened range, as seen in the CDO syndrome, which develops out of habitually supporting the shoulder girdle so that the scaleni muscles tonically shorten over time as a function of the servo system; and (3) as the result of the somatization of psychological stress, self-induced unconscious neuromuscular tension produced when the patient is under emotional pressure.

Trigger Point Formation (M12C1B)

Shortened musculature, although tense, may not be strong. A weak, shortened muscle may be prone to injury if the muscle is inadvertently overstretched in attempts to reach its normal end of range or if called upon to suddenly perform an eccentric contraction to the limits of its normal range of motion. In either case the muscle exceeds its tonically set range. Additionally, tense, weak and shortened muscles have proven to be especially susceptible to trigger point formation development of the muscle spindle variety because of their general susceptibility to strain and their heightened sensitivity to psychoneurogenic influence which seems to result from the combination of on-going tonic biasing and a learned psychoneurogenic response, with selective muscle spindle spasm as the result.

Muscle Relengthening Through Reciprocal Inhibition (M12C1C)

Exercise may be utilized to increase muscle length by taking advantage of the reciprocal inhibitory relationship between antagonistic musculature. In general, when a muscle contracts, its antagonist is reciprocally inhibited to the degree that it remains relatively relaxed or engages in eccentric contraction, lengthening as the agonist shortens. This intermuscular relationship may be used to increase muscle length, if performed correctly. An exercise performed to lengthen a muscle must cause the agonist to contract sufficiently to trigger the reciprocal inhibition of the antagonist while compelling a change in the joint range of motion during or after agonist contraction which allows the tight muscle to lengthen.

Procedures of Application (M12C2)

The exercise techniques which seem to be most effectively utilized for increasing muscle length are the hold-relax, contract-relax, and the guided patterning employed in proprioceptive neuromuscular facilitation (PNF) approach.

The hold-relax and contract-relax techniques take advantage of the muscle spindle servomechanism function. When a contracting muscle is hyper-loaded (resisted), its muscle spindle and the associated extrafusal muscle tension, through the tonic stretch reflex, is reflexly increased and set shorter, while antagonistic muscle spindles and the housing extrafusal muscle is consequently reset to longer lengths by virtue of reciprocal inhibition relationship existing between antagonists.

Hold-relax Technique (M12C2A)

(1) The practitioner or the patient should take the involved joint to the end of the range that puts the tight (or short) muscle on stretch.

(2) The patient should be told to isometrically resist the practitioner's developing pressure without allowing any further joint motion.

(3) After gradually building it up to the maximum, the practitioner should exert constant pressure against the line pull of the tight muscle for five seconds.

(4) The patient should be told to gradually relax the contracting muscle as the practitioner's pressure is gradually decreased.

(5) When the pressure is fully relieved and the patient has relaxed the tight muscle as completely as possible for several seconds, the patient should

attempt to voluntarily increase the joint range, if possible, or the practitioner should take the joint into what new range may have been gained.

(6) After this process has been repeated several times, and no further range has been gained, the process should be repeated several times with the patient resisting pressure applied by the practitioner against the direction of pull of the tight muscle's antagonist. Additional range may be gained and the antagonistic muscle(s) may be tonically set in the new longer range.

Contract-relax Technique (M12C2B)

(1) The practitioner or the patient should take the involved joint to the end of the range that puts the tight (or short) muscle on stretch.

(2) The patient should be told to contract the tight muscle against the practitioner's pressure, which is sufficient to prevent any further joint motion (making the contraction an isometric one).

(3) After gradually building it up to a maximal contraction, the patient should perform a contraction of the muscle for five seconds.

(4) The patient should be told to gradually relax the contracting muscle as the practitioner's pressure is gradually decreased.

(5) When the practitioner's pressure is fully relieved, and the patient has relaxed the tight muscle as fully as possible for several seconds, the patient should voluntarily attempt to further increase the joint range as the tight muscle lengthens, if possible, or the practitioner should take the involved joint into what new range has been gained through muscle lengthening.

(6) After this process has been repeated several times, and no further range has been gained, the process should be repeated several times with the patient resisting pressure applied by the practitioner against the direction of pull of the tight muscle's antagonist. With this last step, additional range may be gained and the antagonistic muscles may be better tonically set in their new ranges.

PNF Muscle Lengthening (M12C2C)

(1) The practitioner starts the action by stretching the antagonist or antagonistic synergistic group of the tight muscle at the short end of the tight muscle's range of motion.

(2) The patient should be asked to contract the stretched muscle immediately following the practitioner's applied stretch, and to move the involved joint in the opposite direction of the tight muscle's line of pull as the practitioner resists the motion but allows it to occur.

(3) After a few degrees, the contracting muscle should be restretched and again resisted as the range increases.

(4) This process should be continued until the end of range is reached.

Partial List of Conditions Treated (M12C3)

Treatable Causes (M12C3A)

Extrafusal muscle spasm (C41)
Habitual joint positioning (C48)
Joint ankylosis (C42)
Joint approximation (C47)
Ligamentous strain (C25)
Nerve root compression (C04)
Neuromuscular tonic imbalance (C30)
Pathological neuromuscular hypertonus (C28)
Pathological neuromuscular dyscoordination (C35)
Psychoneurogenic neuromuscular hypertonus (C29)
Restricted joint range of motion (C26)
Trigger point formation (C01)

Syndromes (M12C3B)

ACHILLES TENDONITIS (S52)
CALF PAIN (S50)
CEREBRAL PALSY (S25)
CERVICAL DORSAL OUTLET (SCALENICUS ATTICUS/CERVICAL RIB) (S09)
CERVICAL (NECK) PAIN (S73)
CHEST PAIN (S14)
EARACHE (S40)
ELBOW PAIN (S47)
FACET (S19)
FOOT PAIN (S51)
FOREARM PAIN (S54)
FROZEN SHOULDER (S64)
HAND/FINGER PAIN (S17)
HEADACHE PAIN (S02)
HIGH THORACIC BACK PAIN (S48)
KNEE PAIN (S46)
LOWER ABDOMINAL PAIN (S62)
LOW BACK PAIN (S03)
MENSTRUAL CRAMPING (S60)
MIGRAINE (VASCULAR) HEADACHE (S18)
NEUROMUSCULAR PARALYSIS (S22)
OSTEOARTHRITIS (S21)
PIRIFORMIS (S04)
POST CEREBRAL VASCULAR ACCIDENT (CVA) (S07)
POST IMMOBILIZATION (S36)
POST WHIPLASH (S55)
REFERRED PAIN (S01)
SHOULDER/HAND (S57)

SHOULDER PAIN (S08)
SINUS PAIN (S33)
TEMPOROMANDIBULAR JOINT (TMJ)
  PAIN (S06)
TENDONITIS (S28)
TENNIS ELBOW (LATERAL
  EPICONDYLITIS) (S32)
THIGH PAIN (S49)
TOOTHACHE/JAW PAIN (S41)
TORTICOLLIS (WRY NECK) (S65)
WRIST PAIN (S16)

Precautions and Contraindications (M12C4)

Arteriosclerosis or atherosclerosis (M12C4A)

Strenuous exercise may be contraindicated for those suffering from arteriosclerosis or atherosclerosis. A surge in blood pressure accompanying exercise or the strain of lifting a heavy weight may cause a "cracking" of a hardened cerebral artery and subsequent spurting of blood, under pressure, into vulnerable cerebral brain matter, causing destruction of any brain matter touched by the stream (a cerebral vascular accident).

Other peripheral vascular disease (M12C4B)

Patients suffering from peripheral vascular disease should avoid any exercise which involves excessive use of an involved extremity, including forced walking, if there is lower extremity involvement, when symptoms of intermittent claudication are present. This is especially true for sufferers of Buerger's disease, Raynaud's disease, or diabetes.

Diabetes (M12C4C)

An insulin reaction may occur in a diabetic patient who has taken an insulin supplement and has eaten too little food, failed to digest what has been eaten before beginning exercise, or taken too large a dose of insulin before performing strenuous exercise. Initial symptoms include extreme hunger, nervousness, heavy perspiration, palpitations of the heart, and occasionally double vision. These symptoms may be relieved by consumption of eight ounces of orange juice to which two teaspoonfuls of regular sugar have been added. If this is unavailable, any sugar or carbohydrate-containing food, like bread or crackers, may serve if it is given in sufficient quantities.

If initial symptoms are unrelieved, more serious symptoms may appear; these include delirium, convulsions, and/or unconsciousness (coma). Intravenous injection of glucose may be necessary to relieve these very dangerous symptoms.

Hyperventilation (alkalosis) (M12C4D)

The respiratory rate increases with strenuous exercise, to increase the availability of oxygen introduced into the lungs, while at the same time providing a ready means for expulsion of the carbon dioxide from the blood, via the lungs, increasingly produced by stepped-up metabolism. If the high respiration rate is not rapidly reduced following the termination of exercise, the continued high level uptake of oxygen and the continued high level of carbon dioxide elimination from the blood, without the balancing continued high level production of carbon dioxide from muscle metabolism, will cause a chemical imbalance to occur between oxygen and carbon dioxide in the blood. This may cause a transient alkalosis (depleted carbon dioxide) to develop which is commonly termed hyperventilation. The accompanying symptomatology includes: (1) the sensation of being "air hungry" (being unable to draw a satisfying breath), (2) dizziness or faintness, (3) nausea, (4) sensations of numbness, (5) sensations of tingling, (6) pounding of the heart, and (7) spasmodic muscle cramps (including of the muscles of the chest and lower abdomen).

The symptomatology of hyperventilation may be reduced by having the patient breathe into a paper bag for several minutes. As the carbon dioxide becomes concentrated in the bag, lung uptake of the carbon dioxide raises its level in the blood and the symptoms are relieved as the chemical imbalance between the oxygen and carbon dioxide is corrected. The symptoms of hyperventilation may additionally lead the patient to fear the onset of heart disease or other serious afflictions; verbal reassurance coupled by the demonstration of symptom relief may be necessary to still such anxieties.

Valsalva maneuver (M12C4E)

The patient should be cautioned not to hold the breath while performing strenuous exercise, because of the danger of performing a Valsalva maneuver. A Valsalva maneuver is defined as a forced expiration against a closed glottis. It may occur following holding the breath while performing a strenuous isometric or isotonic exercise. The blood pressure rises at the outset of straining because the intrathoracic pressure is added to the pressure of the blood in the aorta. It then falls because the high intrathoracic pressure compresses the veins, decreasing venous return and cardiac output. The decrease in arterial pressure and pulse pressure may inhibit the baroreceptors to cause tachycardia and a rise in peripheral resistance. When the glottis is again open, the intrathoracic pressure will return to normal, restoring cardiac output without relieving peripheral vessel constriction; the blood pressure may then rise above normal. Eventually, the high blood pressure rise will stimulate baroreceptors to cause bradycardia and a drop in pressure to normal levels. So, the Valsalva maneuver may be a threat to the patient's well-being by posing the danger of cardiac malfunction as a result of internal attempts at system correction, as well as contributing to the possibility of cerebral vascular accident with the shifts in arterial pressure.

Acidosis (M12C4F)

If carbon dioxide elimination through respiration does not keep pace with carbon dioxide production during strenuous exercise, a transient systemic acidosis will be produced as a carbon dioxide excess builds up in the blood. The characteristic symptoms produced by acidosis is drowsiness, which may develop in extreme conditions into stupor and coma. If unrelieved spontaneously, treatment utilizing a mechanical respirator may be indicated.

Pulse rate regulation (M12C4G)

To avoid exhaustion, the pulse rate of a patient engaged in strenuous exercise should be monitored and the exercise curtailed when the rate exceeds the heart's ability to maintain an adequate volume. As a general rule, exercise should be stopped when the patient's pulse exceeds the rate maximum, which is established by subtracting the patient's age from 220 (plus or minus 10) and multiplying the result by 0.85.

General contraindications (M12C4H)

Therapeutic exercise is contraindicated for patients suffering from any depleting disease or neoplasm, certain pathological heart conditions, active systemic acute infections (bacterial or viral), acute inflammation, new fractures (post five weeks or less) or surgical orthopaedic correctional procedures, acute joint sprains, acute joint strains, and tuberculosis.

Stretch and spasticity (M12C4I)

Any attempt to stretch muscles across individual joints, whether manually or by bracing or splinting, does not change the condition of spasticity, but only shifts it to other joints, limiting the ability to make postural adjustments and movements elsewhere. Any accompanying pain or discomfort tends to increase spasticity throughout the involved extremity.

Notes Aside (M12C5)

The muscles most commonly associated with conditions resulting from shortened musculature are the hamstring, scaleni, upper trapezius, and piriformis muscles. Muscle lengthening through vibration utilizes the same relationship (refer to Vibration, Neuromuscular Management [M08A]).

The contract-relax technique is usually only applicable to patients who have not experienced nervous system injury, and usually only with patients who have (F+) grade muscle grades.

Some authorities advocate the use of the contract-relax technique for increasing range of motion by having the patient alternate muscle contractions, from one antagonist to the other, at the end of the tight muscle's range of motion. This process is called slow reversed- contract-relax technique.

Bibliography (M12C6)

Selim, p. 133.
An Exploratory and Analytical Survey of Therapeutic Exercise, pp. 868, 894, 996.

EXERCISE CARDIOVASCULAR CONDITIONING
(M12D)

Definition (M12D1)

Minimal Exercise and Its Effects (M12D1A)

Current research has demonstrated that basic on-going physical conditioning is necessary to maintain the cardiovascular system in good working order. The population of sedentary men has been shown to have suffered from a significantly greater rate of heart and circulatory disease then those men who employ a regular program of moderate exercise. When the two groups were compared, the sedentary group suffered twice the rate of cardiovascular disease and were statistically prone to die four or five years earlier than the exercising group.

Cardiovascular Conditioning Exercises (M12D1B)

Cardiovascular conditioning exercises include walking, climbing stairs, jogging, jump roping, and various sports. To be effective, the patient must burn at least 2,000 kilocalories of energy per week to be exercising at an adequate moderate level; an brisk walk lasting an hour will reportedly use up 450 kilocalories. The group of individuals that burn less then 500 kilocalories per week in leisure-time activities have been shown to enjoy an eighty-five percent higher risk of death from cardiovascular disease then those burning 2,000 (+) kilocalories per week, and their overall death rate was sixty-three percent higher.

Heavier forms of exercise have also been shown to be beneficial. Runners of ten miles a week showed a ten percent increase in high-density lipoproteins, which seem to scavenge cholesterol. The more miles run, within reason, the greater the reward. Chronic heavy exercise over a period of many weeks or months has been shown to produce hypertrophy of the cardiac muscle and to cause the enlargement of the ventricular chambers. As the overall strength of the heart becomes greatly enhanced and its effectiveness as a pump increases, potential cardiac output and efficiency may be increased from seventy to one hundred percent.

Effect of Exercise Upon Metabolism (M12D1C)

Strenuous exercise produces various changes in the body's physiological processes. Of the bodily processes, the most dramatically effected is the metabolic rate. Short bursts of maximal muscle contraction in any *single muscle*, and the consequent increases in the muscle metabolism, liberates as much as a hundred times the normal amount produced during rest. Maximal muscular exercise for extended periods of several minutes may effect increases in heat production from the *entire body* equal to an amount twenty times greater then that produced during normal rest, and fifty times that amount of heat may be produced during short bursts of several seconds of muscular exercise. Such heat production may represent increases in metabolic rate of 2000% above normal.

Procedures of Application (M12D2)

Metabolic Caloric Requirements (M12D2A)

A detailed description of cardiovascular reconditioning exercise regimens is thought to be inappropriate here. The general principles include the concept that the process of regularly using up metabolic calories at a rate of at least 2000 kilocalories per week is necessary for the patient to be able to enjoy the benefits of moderate exercise program. This may be accomplished, for example, by having the patient walk briskly for one hour, four times a week, or performing an alternative exercise program which burns up a comparable number of calories.

Bicycling consumes approximately 300 to 420 kilocalories per hour; golf, 210 to 300; swimming, 350 to 700; gardening, 220; and scull rowing at a racing pace, 840 calories per hour. An hour of racket sports three times a week should consume the equivalent of 2000 kilocalories. For a more complete listing of physical activities and their approximate of caloric consumption rates, refer to the Table of Exercises - Energy Consumption Rates [T013].

Cardiovascular Conditioning Through Weight Training (M12D2B)

Anaerobic exercise has also proven to be effective in improving cardiovascular function. A three-day per week weight training program carried on for at least four months has been noted to improve certain cardiovascular signs, including blood pressure and pulse rate. A review of available professional literature suggests that the effects of a regular exercise program of weight lifting upon the cardiovascular system has not been studied in depth, but apparently would be in order, especially in light of evidence that suggests that strenuous forms of cardiovascular conditioning exercises such as jogging, racket sports, bicycle riding and others may produce long term physical damage involving supportive joints and even the internal organs, which might be avoided if a well-regimented program of weight lifting could be substituted.

Cardiovascular Reconditioning Exercise (M12D2C)

Following myocardial infarction, it is commonly suggested that a program of bed rest and very restricted activity, excluding any therapeutic exercise, be adhered to for up to six weeks post incident. When an exercise regime is begun, sometimes as early as four weeks after onset, it should be developed slowly and progressively. Initially, the patient should be allowed to sit up, with support, for short periods of time, sometimes only minutes in the early stages, and to engage in self-feeding. The next step should entail passive leg raising. The patient should then be allowed to sit up without support. Finally, the patient should graduate to walking increasingly longer distances, and engaging in more strenuous exercise as the heart tolerates it. The patient's pulse rate should be monitored closely, with close attention paid to the speed and regularity of beat (refer to Precautions and Contraindications, Pulse rate regulation to avoid exhaustion [M12D4B]).

Partial List of Conditions Treated (M12D3)

Conditions Treated in the Field (M12D3A)

Post myocardial infarction reconditioning
General weakness following long periods of enforced bed rest

Treatable Conditions (M12D3B)

Blood sugar irregularities (C31)
Hypertension (C39)
Hypotension (C45)
Lung fluid accumulation (C13)
Reduced lung vital capacity (C17)
Soft tissue swelling (C06)
Vascular insufficiency (C11)
Visceral organ dysfunction (C02)

Syndromes (M12D3C)

BUERGER'S DISEASE (S11)
CHEST PAIN (S14)
CHRONIC BRONCHITIS/EMPHYSEMA
  (OBSTRUCTIVE PULMONARY DISEASE) (S13)
DIABETES (S10)
HYSTERIA/ANXIETY REACTION (S59)
INSOMNIA (S71)
PITTING (LYMPH) EDEMA (S31)
PNEUMONIA (S58)

Precautions and Contraindications (M12D4)

General rules for safe exercise (M12D4A)

1. The spine should be maintained in an erect position (if the exercise is not conducted in prone or supine positions. Some deviation may be necessary during the exercise, but forward inclination of the trunk should be kept to a minimum.
2. Muscle(s) used in handling weight should be properly "set" before lifting is attempted. The muscle(s) must be prepared to lift the amount of weight to be lifted to avoid the common straining injury resulting from lifting a weight heavier than expected.
3. Motions should be smooth and steady when force is exerted, never jerky or sudden.
4. Muscles not required for an exercise should remain relatively relaxed to avoid unnecessary fatigue.
5. Exerting force when the trunk is twisted or the back hyperextended (abnormally arched with the lordotic curve increased beyond normal) should be avoided, especially when lifting.

Lifting compression force (M12D4B)

Many exercises require the use of the arms in a relative position out in front of the body, as in lifting and reaching, with accompanying forward inclination of the trunk. While performing exercise in a standing position, forward trunk inclination should be kept at a minimum, because the farther the trunk is inclined forward, the greater the strain on the supporting back structures (muscles, discs, ligaments, etc.). As the trunk flexes, the rotational moment due to gravitational pull is increased, putting a greater load on the back extensor muscles and the spinal column supporting ligaments and intervertebral discs, while the angle of pull of the back extensor muscles becomes less favorable. This results in a decrease in the effectiveness of the muscles' angle-of-pull and necessitates an increase in the force of muscle contraction required to maintain posture or to produce back extension motion, thus causing the compression force acting downward along the spinal column to markedly increase. To illustrate, the effective angle-of-pull for the erector spinalis muscle group is approximately twelve degrees when the trunk is flexed forward sixty degrees from vertical. The tensile force in the erector spinalis group necessary to support the trunk of a man weighing 180 pounds in such a posture (with hands hanging freely) is approximately 450 pounds. If such a man is holding a fifty-pound weight, the muscle force must be approximately 750 pounds and the compression force on the L5-S1 joint is nearly 850 pounds, a potential threat to weakened or damaged discs and a possible source of strained back musculature.

Arteriosclerosis or atherosclerosis (M12D4C)

Strenuous exercise may be contraindicated for those suffering from arteriosclerosis or atherosclerosis. A surge in blood pressure accompanying exercise or the strain of lifting a heavy weight may cause a "cracking" of a hardened cerebral artery and subsequent spurting of blood, under pressure, into vulnerable cerebral brain matter, causing destruction of any brain matter touched by the stream (a cerebral vascular accident).

Other peripheral vascular disease (M12D4D)

Patients suffering from peripheral vascular disease should avoid any exercise which involves excessive use of an involved extremity, including forced walking, if there is lower extremity involvement, when symptoms of intermittent claudication are present. This is especially true for sufferers of Buerger's disease, Raynaud's disease, or diabetes.

Diabetes (M12D4E)

An insulin reaction may occur in a diabetic patient who has taken an insulin supplement and has eaten too little food, failed to digest what has been eaten before beginning exercise, or taken too large a dose of insulin before performing strenuous exercise. Initial symptoms include extreme hunger, nervousness, heavy perspiration, palpitations of the heart, and occasionally double vision. These symptoms may be relieved by consumption of eight ounces of orange juice to which two teaspoonfuls of regular sugar have been added. If this is unavailable, any sugar or carbohydrate-containing food, like bread or crackers, may serve if it is given in sufficient quantities.

If initial symptoms are unrelieved, more serious symptoms may appear; these include delirium, convulsions, and/or unconsciousness (coma). Intravenous injection of glucose may be necessary to relieve these very dangerous symptoms.

Hyperventilation (alkalosis) (M12D4F)

The respiratory rate increases with strenuous exercise, to increase the availability of oxygen introduced into the lungs, while at the same time providing a ready means for expulsion of the carbon dioxide from the blood, via the lungs, increasingly produced by stepped-up metabolism. If the high respiration rate is not rapidly reduced following the termination of exercise, the continued high level uptake of oxygen and the continued high level of carbon dioxide elimination from the blood, without the balancing continued high level production of carbon dioxide from muscle metabolism, will cause a chemical imbalance to occur between oxygen and carbon dioxide in the blood. This may cause a transient alkalosis (depleted carbon dioxide) to develop which is commonly termed hyperventilation. The accompanying symptomatology includes: (1) the sensation of being "air hungry" (being unable to draw a satisfying breath), (2) dizziness or faintness, (3) nausea, (4) sensations of numbness, (5) sensations of tingling, (6) pounding of the heart, and (7) spasmodic muscle cramps (including of the muscles of the chest and lower abdomen).

The symptomatology of hyperventilation may be reduced by having the patient breathe into a paper bag for several minutes. As the carbon dioxide becomes concentrated in the bag, lung uptake of the carbon dioxide raises its level in the blood and the symptoms are relieved as the chemical imbalance between the oxygen and carbon dioxide is corrected. The symptoms of hyperventilation may additionally lead the patient to fear the onset of heart disease or other serious afflictions; verbal reassurance coupled by the demonstration of symptom relief may be necessary to still such anxieties.

Acidosis (M12D4G)

If carbon dioxide elimination through respiration does not keep pace with carbon dioxide production during strenuous exercise, a transient systemic acidosis will be produced as a carbon dioxide excess builds up in the blood. The characteristic symptoms produced by acidosis is drowsiness, which may develop in extreme conditions into stupor and coma. If unrelieved spontaneously, treatment utilizing a mechanical respirator may be indicated.

Pulse rate regulation (M12D4H)

To avoid exhaustion, the pulse rate of a patient engaged in strenuous exercise should be monitored and the exercise curtailed when the rate exceeds the heart's ability to maintain an adequate volume. As a general rule, exercise should be stopped when the patient's pulse exceeds the rate maximum, which is established by subtracting the patient's age from 220 (plus or minus 10) and multiplying the result by 0.85.

General contraindications (M12D4I)

Therapeutic exercise is contraindicated for patients suffering from any depleting disease or neoplasm, certain pathological heart conditions, active systemic acute infections (bacterial or viral), acute inflammation, new fractures (post five weeks or less) or surgical orthopaedic correctional procedures, acute joint sprains, acute joint strains, and tuberculosis.

Running and injury (M12D4J)

Long distance running has long been equated with injuries to the weight bearing joints and other associated soft tissues. Evidence has been put forward that the rate of injury is markedly increased when runners exceed fifteen miles per week.

Notes Aside (M12D5)

Effect of cardiovascular exercise upon the diabetic (M12D5A)

Incidentally, as cellular metabolism increases during strenuous exercise like that performed for cardiovascular conditioning, the demand for metabolic fuel increases along with a simultaneous need for an increase in the available oxygen so that metabolism can take place. Consequently, a physiological process occurs which allows glucose to be transported into the muscle cells at an accelerated rate, even in the absence of insulin. This insulin-like effect of exercise upon glucose transport into the muscle may provide the physically active diabetic patient, even those severely afflicted, with a better opportunity to control their diabetes than over-protected sufferers by decreasing the need for the administration of supplementary insulin.

Bibliography (M12D6)

Gold, p. 16.
Guyton, pp. 158–159.
Holvey, pp. 133–135.
Wade, pp. 202–204.

EXERCISE GAIT TRAINING (M12E)

Definition (M12E1)

Functional Bipedal Ambulation (Gait) (M12E1A)

Functional bipedal ambulation (gait) depends upon the ability to (1) independently support the body in an upright position, (2) balance the body in that upright position, and (3) to execute the stepping motion required of the legs.

Normal ambulation depends upon a cycle of "walking" maneuvers performed by the two lower extremities, providing continuous locomotion in a linear direction. These maneuvers include (1) swing phase, (2) heel strike (initial contact of the foot with the floor), (3) double support (when both feet are simultaneously in contact with the floor), and (4) stance phase, sometimes called the propulsion phase (when one foot is in full contact with the floor and the opposite leg is in swing phase taking the next step).

Periods of stance and swing occur alternately and continuously, with one lower extremity required to provide support and balance while allowing the opposite lower extremity to move freely. Within each walking cycle there are two periods of single lower extremity support and two periods of double lower extremity support, when one lower extremity is in the beginning of the stance phase and the other is ending a stance phase. During normal ambulation in the forward direction, at the beginning of stance phase, coordinated synergistic muscular activity directs the lower extremity obliquely forward to provide restraint from excessive forward motion while the opposite lower extremity is directed obliquely backward to provide propulsion in a forward direction. During this process the trunk is continuously translated forward over alternating bases of support, usually at a constant linear horizontal velocity. The speed of translation from one base of support to the other depends upon the rapidity of steps and the stride length, which is the distance between successive contact points of the opposite feet.

Characteristics of Normal Ambulation (M12E1B)

Normal ambulation depends upon the coordination of antagonistic muscles within synergistic patterns which together produce the combination of proximal stabilization with distal joint motion required by the neuromuscular complexity of ambulation. For a more detailed discussion of muscular interplay during ambulation refer to the Kinesiology of Ambulation [R004]. Normal ambulation is characterized by (1) a wide range of rapid and comfortable walking speeds, (2) smooth forward translation of the trunk, and (3) a rhythmicity of step length and duration of the temporal components of the walking cycle.

Disruptive Influences Upon Normal Ambulation (M12E1C)

Various physical conditions may result in a disruption or impairment of the execution of the various operational components required for normal ambulation. The impediments of normal ambulation may include: (1) pain (especially of the joints), (2) structural malformation, (3) weakness, (4) appliances (casting or splinting of the lower extremity, (5) central neuromuscular injury or dysfunction, and (6) peripheral nerve injury. If normal ambulation has been impaired or precluded by any of these impediments, gait training (ambulation reeducation) may be indicated.

Gait Training Objectives (M12E1D)

The object of gait training is to develop, refine and/or perfect the patient's ambulatory skills to the degree that the patient may realize what potential for functional gait still remains following impairment. This may mean (1) the training of balance skills, (2) modifying motor responses of the involved musculature to obtain neuromuscular synergies or muscular associations resembling those required for normal ambulation, and/or (3) the training of alternate responses of involved muscular antagonists to permit a rapid release of tension in muscles or muscle groups following their activation.

Procedures of Application (M12E2)

As indicated above, there is a basic sequence of training steps inherent in a gait training (ambulation reeducation) program. This sequence follows the subliminal learning pattern of (1) facilitation, (2) inhibition, (3) balance, and (4) function found to be effective in programs of neuromuscular reeducation directed at correcting dyscoordinated or nonfunctional musculature produced by nervous system injury and/or malfunction.

Facilitation (M12E2A)

During the facilitation phase, the muscles necessary for balance and body support should be strengthened (if necessary) to the degree of being able to perform their required functions of maintaining erect standing and walking postures. This may entail the toning of muscles of the legs, hips and trunk. If nervous system function has been impaired, electromyometric feedback may prove beneficial (indeed, indispensable) by providing a means through which the patient may be shown on a subliminal level how to control affected musculature without adequate endogenous proprioceptive perception. If the nervous system is intact, the muscles may be selectively toned through the use of isometric and/or isotonic exercise (refer to the Table of Exercise for the Lower Extremities—Isometrics [T014] and Table of Exercises—Isotonic [T012]).

When sufficient muscle toning has occurred, exercise which fosters balance should be instituted to facilitate coordinated action of the muscles required to maintain erect postures. Exercise designed to foster balance is generally based upon directly challenging the vestibular phasic reflex by testing the patient's ability to compensate for sudden changes in spatial orientation; this involves pushing the patient off balance and waiting for the patient to recover erect posture. A series of sudden shoves, insufficiently violent to throw the subject down or to snap the subject's neck, should be administered from various directions and with varying degrees of force, or a series of steady pressures and sudden releases may be applied to the patient's upper trunk from different directions to provide the same neuromuscular demand as shoving. Resistance should be administered as a shove or steady pressure applied first to one side of the body and then the other, with varying frequency and duration; progression should be made from sitting and standing postures.

Sitting (M12E2A01) The patient should be sitting with head and back unsupported, the arms and hands at the sides. The patient should then be encouraged to attempt to maintain the erect sitting posture without reaching out with the hand(s) to gain steadying support when the pressure or shoving sequences are performed by the practitioner.

Standing (M12E2A02) The patient should be standing with the feet no more than six inches apart, arms at the sides and the head erect, and without support of any kind. The patient should then be encouraged to attempt to maintain the erect standing posture without changing foot position when the pressure or shoving sequences are performed by the practitioner.

Inhibition (M12E2B)

If hypertonicity of lower extremity musculature is present to the extent that it interferes with standing or walking, efforts should be made to help the patient learn to reduce it. If nervous system damage has resulted in hypertonicity, electromyometric feedback may prove to be useful as a means of teaching the patient how to voluntarily reduce hypertonicity in spite of reduced supraspinal structure appreciation and/or reception of proprioception (refer to Electromyometric Feedback, Post Cerebral Accident Neuromuscular Reeducation [M18A]). Various sensory stimulation techniques may also prove useful for the involuntary reduction of spasticity or hypertonicity as well (refer to Sensory Stimulation, Neuromuscular Spasticity Control [M17A]).

If the patient's nervous system is intact and functioning normally, exercise utilizing the principles of reciprocal inhibition to reduce hypertonicity and/or to lengthen involved musculature may be useful (refer to Exercise, Muscle Relengthening [M12C]). Other techniques which have proven useful for the reduction of hypertonicity, when the nervous system is functioning normally, include electrical stimulation (refer to Electrical Stimulation, Muscle Lengthening [M02A]) and/or vibration (refer to Vibration, Neuromuscular Management [M08A]).

Balance (M12E2C)

During the balance phase, the selective and timed facilitation and inhibition of antagonistic synergistic muscle and muscle groups, if not already present, must be developed. It is an essential element of lower extremity swing phase motion and the smooth process of agonist muscular contraction with simultaneous antagonist eccentric contraction seen during stance phase. Electromyometric feedback may prove useful for teaching the patient how to coordinate lower extremity musculature, whether the nervous system has been affected by previous injury or not (refer to Electromyometric Feedback, Post Cerebral Accident Neuromuscular Reeducation [M18A]). Such training should begin with the patient in lying or sitting positions, with the lower extremities bearing no weight, and progress first to standing and finally to walking. A similar process may prove useful utilizing guided pattern techniques, if electromyometry is unavailable (refer to Exercise, Guided Patterning [M12H]).

Function (M12E2D)

Assistive devices (M12E2D01) During the functional phase, the patient should be directed and tutored in the act of ambulation. This training may involve first teaching the patient to ambulate with supportive devices, progressing through a series of assistive devices which successively provide less and less support. These assistive devices may include parallel bars, pickup and/or rolling walkers, forearm crutches (Canadian canes), hemi-walkers and/or quadcanes (both of which can stand alone), regular canes or a single cane. In some cases, leg bracing with short or long leg braces and/or splinting may be necessary until sufficient strength has been developed in weak muscles, hypertonicity sufficiently inhibited, or dyscoordinated musculature has been sufficiently reeducated. Optimally, the patient may be progressed to the point where no assistive device support is necessary (refer to Ambulation with Assistive Devices [R005]).

Ambulation stage training sequence (M12E2D02) During the functional walking stage, it may be necessary to help the patient through a training process which includes having the patient proceed from (1) sitting to standing, (2) standing to walking, (3) walking to standing, and (4) standing to sitting. During this process, the practitioner may have to aid the patient by (1) supporting, assisting and directing the patient, (2) assisting and directing the unsupported patient, (3) directing the unassisted, unsupported patient, and finally (4) observing the independent ambulation of the patient.

Training should progressively proceed from flat surfaces to stairways, rampways, and finally to broken uneven surfaces. Distances should be progressively increased, and the line of ambulation should be made progressively more complex. The patient should first ambulate in a straight line, then proceed to going up and back, and then be directed through a series of turns, and around and over obstacles of various heights and sizes.

In the supportive and assistive stages of training, a belt around the patient's waist should be used to help the practitioner to hold and direct the patient. The practitioner should hold the belt from behind the patient, being careful only to provide support and physical direction which is absolutely necessary. Support and physical direction should be progressively decreased to the point that the patient is absolutely free of practitioner physical contact.

Partial List of Conditions Treated (M12E3)

Treatable Causes (M12E3A)

Habitual joint positioning   (C48)
Hypermobile/instable joint   (C10)
Joint ankylosis   (C42)
Muscular weakness   (C23)
Neuromuscular tonic imbalance   (C30)
Pathological neuromuscular hypertonus   (C28)
Pathological neuromuscular hypotonus   (C33)
Pathological neuromuscular dyscoordination   (C35)
Peripheral nerve injury   (C07)
Restricted joint range of motion   (C26)

Syndromes (M12E3B)

BUERGER'S DISEASE   (S11)
CEREBRAL PALSY   (S25)
KNEE PAIN   (S46)
NEUROMUSCULAR PARALYSIS   (S22)
OSTEOARTHRITIS   (S21)
POST CEREBRAL VASCULAR ACCIDENT (CVA)   (S07)
POST IMMOBILIZATION   (S36)
POST PERIPHERAL NERVE INJURY   (S23)
POST SPINAL CORD INJURY   (S24)

Precautions and Contraindications (M12E4)

Providing appropriate support (M12E4A)

Of paramount importance is that the practitioner prevent the gait training patient from falling at any stage in the process. Acute judgement must be used to gauge when the patient's ability is sufficient to allow safe progression from one stage to another. Especially important is the judgement and decision as to when it is appropriate to give up being physically in contact with by supporting and/or assisting the patient.

Pulse regulation to avoid exhaustion (M12E4B)

To avoid exhaustion, the pulse rate of a patient engaged in gait training should be monitored and training curtailed when the rate exceeds the heart's ability to maintain an adequate volume. As a general rule, exercise should be stopped when the patient's pulse exceeds the rate maximum, which is established by subtracting the patient's age from 220 (plus or minus 10) and multiplying the result by 0.85.

Peripheral vascular disease (M12E4C)

Patients suffering from peripheral vascular disease should avoid any exercise which involves excessive use of an involved extremity, including forced walking, if there is lower extremity involvement, when symptoms of intermittent claudication are present. This is especially true for sufferers of Buerger's disease, Raynaud's disease, or diabetes.

Hyperventilation (alkalosis) (M12E4D)

The respiratory rate increases with strenuous exercise, to increase the availability of oxygen introduced into the lungs, while at the same time providing a ready means for expulsion of the carbon dioxide from the blood, via the lungs, increasingly produced by stepped-up metabolism. If the high respiration rate is not rapidly reduced following the termination of exercise, the continued high level uptake of oxygen and the continued high level of carbon dioxide elimination from the blood, without the balancing continued high level production of carbon dioxide from muscle metabolism, will cause a chemical imbalance to occur between oxygen and carbon dioxide in the blood. This may cause a transient alkalosis (depleted carbon dioxide) to develop which is commonly termed hyperventilation. The accompanying symptomatology includes: (1) the sensation of being "air hungry" (being unable to draw a satisfying breath), (2) dizziness or faintness, (3) nausea, (4) sensations of numbness, (5) sensations of tingling, (6) pounding of the heart, and (7) spasmodic muscle cramps (including of the muscles of the chest and lower abdomen).

The symptomatology of hyperventilation may be reduced by having the patient breathe into a paper bag for several minutes. As the carbon dioxide becomes concentrated in the bag, lung uptake of the carbon dioxide raises its level in the blood and the symptoms are relieved as the chemical imbalance between the oxygen and carbon dioxide is corrected. The symptoms of hyperventilation may additionally lead the patient to fear the onset of heart disease or other serious afflictions; verbal reassurance coupled by the demonstration of symptom relief may be necessary to still such anxieties.

Acidosis (M12E4E)

If carbon dioxide elimination through respiration does not keep pace with carbon dioxide production during strenuous exercise, a transient systemic acidosis will be produced as a carbon dioxide excess builds up in the blood. The characteristic symptoms produced by acidosis is drowsiness, which may develop in extreme conditions into stupor and coma. If unrelieved spontaneously, treatment utilizing a mechanical respirator may be indicated.

Diabetes (M12E4F)

An insulin reaction may occur in a diabetic patient who has taken an insulin supplement and has eaten too little food, failed to digest what has been eaten before beginning exercise, or taken too large a dose of insulin before performing strenuous exercise. Initial symptoms include extreme hunger, nervousness, heavy perspiration, palpitations of the heart, and occasionally double vision. These symptoms may be relieved by consumption of eight ounces of orange juice to which two teaspoonfuls of regular sugar have been added. If this is unavailable, any sugar or carbohydrate-containing food, like bread or crackers, may serve if it is given in sufficient quantities.

If initial symptoms are unrelieved, more serious symptoms may appear; these include delirium, convulsions, and/or unconsciousness (coma). Intravenous injection of glucose may be necessary to relieve these very dangerous symptoms.

Notes Aside (M12E5)

One of the most important concepts to be considered during gait training or any sort of functional training, is the idea that the patient should be asked to perform tasks which are progressively more difficult. If this is not done, the patient will tend to plateau at the current level of competence. However, the human system tends to respond to demand as long as the demand is reasonable and the system has the potential to comply with the demand and is capable of performing at the level of competence consistent with the demand. The system also has the facility of adjusting its perceptions of difficulty, so that when a task is attempted which has a higher level of difficulty than previously accomplished tasks, the system will perceive those tasks as relatively easy so that confidence at that level is increased. Consequently, a patient who perceives walking on flat surfaces with an assistive device such as a cane as difficult will consider flat surface walking relatively easy and non-threatening, even without the assistive device, when stair climbing without an assistive device has been attempted and some success safely achieved.

Bibliography (M12E6)

Arnheim, pp. 28, 203–204, 220.
Brunnstrom, pp. 111–128.
An Exploratory and Analytical Survey of Therapeutic Exercise, pp. 290–333.

EXERCISE MUSCLE STRETCHING (M12F)

Definition (M12F1)

It has been demonstrated that a limitation of joint range resulting from of tight (abnormally shortened) musculature may be corrected or ameliorated through exercise which stretches the muscle(s) sufficiently long enough to cause the muscle(s) involved to assume a greater length (tonically biased to a longer length).

In general, when a muscle is mechanically stretched, it tends to pull back by virtue of the phasic and tonic stretch reflexes and the servo mechanism built into the muscle spindle system (refer to Neuromuscular System [R002] and/or Muscle Spindle [R001]). If, however, the muscle is stretched for a sufficient length of time, the muscle spindles housed in the involved muscle(s) will gradually reset at an extended length, which causes the extrafusal muscle to reflexly assume a longer length, by virtue of the tonic stretch reflex mechanism.

In the clinical setting, muscle stretching exercise has proven to be of most value for the relief of myogenic pressure upon nerve roots and/or peripheral nerves. Such cases include nerve root pressure in the lumbosacral region where shortened muscles may contribute to pressure supplied by narrowed or ruptured disks (sciatica), the cervical dorsal outlet (CDO) or scalenicus atticus syndrome, and the piriformis syndrome (small sciatica). The latter two cases are examples of peripheral nerves being affected by shortened muscles. Muscle stretching exercise has also been found to be valuable for increasing joint range of motion to enhance flexibility often required for the proficient performance of physical sports and occupations which require greater then normal hamstring and hip adductor muscles may be required to have greater then normal length (as in dance).

Procedures of Application (M12F2)

General Rules for Muscle Stretching (M12F2A)

(1) The body part providing the site of insertion or origin for the muscle or muscle group at one end must be in position to provide the force of pull exerted upon the muscle the opportunity for direct communication with the stabilized site of attachment at the muscle's other end.

(2) Stretching should occur in the opposite direction of the natural line of muscle pull.

(3) Ideally, the body parts involved should be placed in such a position that gravity will be helpful and the patient's body weight may be used to exert or add to the force of stretch.

General Principles (M12F2B)

(1) Muscle stretching is most effective when rhythmical spring stretching (small bouncing motions forcing more and more stretch with each motion) is additionally applied to the main stretch at the full end of range.

(2) After each prolonged stretch or series of stretches the muscles involved should be relaxed for several seconds and then "shaken out" to prevent the stretched muscles from cramping (an involuntary sustained spasm of the muscle).

(3) To have the desired effect of achieving relatively permanent greater muscle length, prolonged stretch should last for between five to thirty minutes.

(4) Passive and passive/assistive stretching applied by the practitioner may be entirely necessary at regular intervals, if nervous system injury has occurred, to prevent contractures.

For detailed descriptions of commonly utilized stretching exercise regimens, refer to the Table of Stretching Exercises for Muscle Lengthening [T016].

Partial List of Conditions Treated (M12F3)

Treatable Causes (M12F3A)

Extrafusal muscle spasm (C41)
Habitual joint positioning (C48)
Joint ankylosis (C42)
Joint approximation (C47)
Nerve root compression (C04)
Peripheral nerve injury (C07)
Psychoneurogenic neuromuscular hypertonus (C29)
Restricted joint range of motion (C26)
Scar tissue formation (C15)
Trigger point formation (C01)

Syndromes (M12F3B)

ACHILLES TENDONITIS (S52)
ADHESION (S56)

CEREBRAL PALSY (S25)
CERVICAL DORSAL OUTLET (SCALENICUS ATTICUS/CERVICAL RIB) (S09)
FROZEN SHOULDER (S64)
LOW BACK PAIN (S03)
NEUROMUSCULAR PARALYSIS (S22)
PIRIFORMIS (S04)
POST CEREBRAL VASCULAR ACCIDENT (CVA) (S07)
POST IMMOBILIZATION (S36)
POST PERIPHERAL NERVE INJURY (S23)
POST SPINAL CORD INJURY (S24)
TEMPOROMANDIBULAR JOINT (TMJ) PAIN (S06)
TENDONITIS (S28)
TENNIS ELBOW (LATERAL EPICONDYLITIS) (S32)
TORTICOLLIS (WRY NECK) (S65)

Precautions and Contraindications (M12F4)

No stretching procedure should be carried out for the benefit of one muscle, muscle group, or joint which will cause the over-stretching or deformity of another muscle, muscle group, or joint.

The stretching of a given muscle or muscle group should be avoided if it can be demonstrated that the abnormally short (tight) muscle or muscle group contributes to the stability of a given joint and that stability would be put in jeopardy if that muscle or muscle group were to be lengthened (stretched).

Gastrocnemius muscle and Achilles tendon stretching should be limited to maintaining ankle neutral position, especially when performed on a comatose or cerebral palsy patient. Nothing can be gained by producing an unstable joint through over-stretching a muscle and/or tendon.

When a patient is engaged in the simultaneous stretching of the hip adductor and hamstring muscles in a sitting position, the patient should be informed of the possibility and dangers of hip dislocation which may occur if hip flexion is attempted at the end of range and the hip is externally rotated.

Notes Aside (M12F5)

Muscle stretching techniques are often initially uncomfortable for the patient to engage in, especially if an entrapped nerve is involved. When a muscle is pulled it pulls back, when a hypertonic muscle is pulled it pulls back even harder.

It has often been reported that if a patient engages in the regimen utilized to stretch and ultimately provide for lengthening of scaleni muscles (Hale's regime) the discomfort associated with the CDO syndrome being treated is increased and remains so for the first week of daily exercise, gradually decreasing toward the end of the second week. Because of this, Hale's regime (refer to Table of Stretching Exercises for Muscle Lengthening [T016]) is now only recommended for patient's who are unable to avail themselves of treatment designed to relengthen the tight muscles through the use of electrical stimulation and vibration (refer to Electrical Stimulation, Muscle Lengthening [M02A] and Vibration, Neuromuscular Management [M08A]).

Bibliography (M12F6)

Williams, p.36.
Starring, pp. 314–320.

EXERCISE DIRECTED FUNCTIONAL ACTIVITY (M12G)

Definition (M12G1)

Directed functional activity exercises are designed to make the patient use muscles which need to be exercised in a coordinated functional manner. Functional exercises usually involve the patient in performing an activity of daily living which the patient would normally be inclined to perform on the job or in the home, and thereby may not seem esoteric, like turning a door knob to promote supinator/pronator muscle group control of the wrist. (This exercise was reported by Margaret Rood to serve a supraspinal integrative function which may lead to further neuromuscular learning.)

Functional exercise has been shown to complete the process of neuromuscular reeducation (an especially important part of neuromuscular reeducation utilizing electromyometric feedback) by placing a demand upon the central nervous system to use gained muscle control in an integrated, coordinated manner. This apparently has the effect of "putting it all together" on the subliminal neuromuscular level, and the patient is much more likely to keep what has been learned or gained in terms of neuromuscular control.

Functional exercise includes both gross motor and refined motor functional activities. Examples of a gross motor functional activity include walking and carrying an object in an involved hand (usually fairly heavy, relative to the functional ability of the patient), or flipping a light switch on and/or off. Examples of a refined motor functional activity include picking up, lifting, placing and releasing utensils or other small objects of common use, eating skills (like cutting or spooning up), or tying a shoe.

Procedures of Application (M12G2)

Functional exercise most usually arises out of the imagination of the practitioner; the exercise design should depend upon the muscles to be used, the relative functional level of the patient, and the equipment at hand. When designing an functional exercise, consider the following:

General Principles of Functional Exercise (M12G2A)

1) The exercise should be performable in the patient's home or living environment.

2) The exercise should employ the muscles needing exercise.

3) The exercise should be performable by the patient, but it should be hard enough to challenge to the patient's ability.

Neuromuscular Principles of Functional Exercise (M12G2B)

1) An unresisted isotonic contraction, not followed by resistance, results in inhibition of the muscle contracting (this response is related to golgi tendon function).

2) A quick stretch not followed by resistance will ultimately result in inhibition of the muscle stretched and facilitation of its antagonist.

3) Contraction and/or quick stretch followed by resistance will result in facilitation of the muscle resisted.

4) Stimulation of the dermatome over a muscle group generally facilitates the agonist and inhibits the antagonist.

5) Bilateral exercise should occur before unilateral exercise. In unilateral exercises, work with the good extremity first and then with the affected extremity.

6) Keep the developmental sequence in mind when dealing with post CVA, alcoholic, or dementia patients.

7) Following nervous system injury, flexion develops before extension.

8) Reciprocal activity will generally develop in the extremities between antagonists before cocontraction of antagonists develops.

9) Mild compression or approximation (body weight or less) applied into a joint tends to inhibit tight muscles and relieve arthrogenic pain while promoting joint motion.

10) Traction of a joint tends to inhibit tight muscle, relieve arthrogenic pain and facilitate joint motion.

11) An isometric contraction serves to inhibit an agonist and facilitate the antagonist following contraction.

Partial List of Conditions Treated   (M12G3)

Treatable Causes   (M12G3A)

Habitual joint positioning   (C48)
Joint ankylosis   (C42)
Muscular weakness   (C23)
Neuromuscular tonic imbalance   (C30)
Pathological neuromuscular hypertonus   (C28)
Pathological neuromuscular hypotonus   (C33)
Pathological neuromuscular dyscoordination   (C35)
Peripheral nerve injury   (C07)
Psychoneurogenic neuromuscular hypertonus   (C29)
Restricted joint range of motion   (C26)

Syndromes   (M12G3B)

BELL'S PALSY   (S66)
CEREBRAL PALSY   (S25)
NEUROMUSCULAR PARALYSIS   (S22)
POST CEREBRAL VASCULAR ACCIDENT (CVA)   (S07)
POST IMMOBILIZATION   (S36)
POST PERIPHERAL NERVE INJURY   (S23)
POST SPINAL CORD INJURY   (S24)
SHOULDER/HAND   (S57)
TEMPOROMANDIBULAR JOINT (TMJ) PAIN   (S06)
TORTICOLLIS (WRY NECK)   (S65)

Precautions and Contraindications   (M12G4)

Functional activity exercise design must be such that the ability of the patient will eventually allow successful performance; unattainable goals will extinguish motivation and the will of the patient to perform.

Notes Aside   (M12G5)

In one of her lectures at the University of Southern California, Physical Therapy Professor Margaret Rood made the statement that the most effective exercises for use in a program of neuromuscular rehabilitation were functional exercises. Muscles and the neuromuscular system seemed to benefit most from exercise that a patient could use in common activities of daily living. Additionally, as a rule of thumb, muscles seemed to build tone faster if they were exercised in positions relative to the pull of gravity that best imitated the position they were most commonly used in and if the exercise employed a coordination of synergistic muscles.

Bibliography   (M12G6)

Taylor, pp. 56–57.

EXERCISE GUIDED PATTERNING (M12H)

Definition (M12H1)

Guided patterning is a process in which a practitioner systematically utilizes muscle stretch, manual resistance, joint approximation, joint positioning, and/or body posture, sometimes combined with other sensory and proprioceptive mechanisms, to influence neuromuscular control or activity. Such techniques are generally utilized to change the uncoordinated motor behavior of patients suffering from central nervous system dysfunction. For a detailed discussion of central nervous system function and/or dysfunction refer to the Neuromuscular System [R002], Muscle Spindle [R001], and/or Central Nervous System Dysfunction [R006].

The problem produced by supraspinal neuromuscular dysfunction is usually manifests itself as the incoordination of motor function, resulting in abnormal muscle tone (spasticity, flaccidity, rigidity or tonic spasm, and hypotonicity), which results in abnormal muscular interrelationships. The abnormal neuromuscular relationships which are produced express themselves as pathological neuromuscular synergies which create abnormal extremity movement patterns and postures. These patterns often resemble neuromuscular responses produced by the development reflexes and the contralateral and homolateral reactions associated with infant development (refer to the Table of Developmental Reflexes [T007] and the Table of Neuromuscular Synergistic Patterns [T008]).

The formation of pathological neuromuscular postural patterns often precludes the possibility of selective movement or a variability of joint motion outside of the patterns produced by uninhibited tonic and spinal reflexes. Sensory loss or distortion caused by deprivation of proprioceptive input (lack of motor experience) may cause additional incoordination or further loss of the ability to produce directed movement, thereby complicating or distorting body image. If damage has occurred to an immature central nervous system, motor development may be retarded.

The techniques of guided patterning are generally based upon the theory that by externally causing a change in the patterns of muscular stabilization and/or motion as an automatic and/or voluntary response the central nervous system will learn how to produce more normal postural patterns or synergies. It is hoped that this may lead to the development of selective and variable voluntary motion and the possible facilitation of "more normal" neuromuscular development. Indeed, only by changing existing abnormal postural patterns can normal postural tone be obtained, and only on the basis of more normal postural tone may movement be obtained that is adequate, relative to variety and quality, to produce normal function.

The most commonly known and utilized techniques of guided patterning include the Temple Fay [M12H2A], Brunnstrom [M12H2B], Bobath [M12H2C], Proprioceptive Neuromuscular Facilitation (PNF) [M12H2D], and Rood [M12H2E] approaches. This listing does not include all of the guided patterning techniques, but those listed are generally the most used and best known to Physical and Occupational Therapists.

It is noted that competent application of any of these guided patterning approaches described here is dependant upon the practiced skill of a practitioner who has not only studied theoretical material regarding a particular guided pattern technique, but who has generally apprenticed under the watchful eye of a talented and highly trained technical specialist in the application of the particular treatment technique. It is only intended here to supply the reader with some of the bare bones of the most popular or guided pattern approaches. A full detailed description of any of these approaches is beyond the scope of this work.

All discussion upon any of these approaches is made relative to the belief system incorporated within the approach being discussed and does not necessarily reflect the views of the author, nor is it meant to be put forward as a statement of absolute truth. Much of what is discussed here is based upon theory and the empirical observation of the developers of the particular approach discussed.

Procedures of Application (M12H2)

The Temple Fay Approach (M12H2A)

Fay referred to reflexes, spinal automatisms and tonic responses as "built in exercises," suggesting that they are fragments of ancient motor behaviors which still persist or exhibit themselves to a degree that higher cortical control is impaired or lost. The treatment of neuromuscular dysfunction must begin with the simple elements of movement and must be built upon the patterns of reflexes which prevail.

Initially, through evaluation and observation the functional level of the patient must be established and the existing pathological reflexes and automatic responses identified (refer to Exercise, Evaluation of Developmental Reflexes [M12I6] and Exercise, Pathological Neuromuscular Synergistic Pattern Evaluation [M12I7]).

Principles (M12H2A01)

1) Treatment must begin with the simple elements of motion and build upon the reflex patterns which predominate.

2) It is essential that the mobility skills at the patient's lowest level of competency be learned before learning the skills required of higher level. For instance, if the patient is dominated by some of the reflexes associated with the first stage of infant development, the patient must develop all of the reflexes at that level, and then learn to break those reflexes and to produce movement not dictated by those reflexes as a step in toward developing the reflexes associated with the next higher developmental stage; i.e., the patient must crawl and creep before standing and walking.

3) By utilizing existing intrinsic reflex mechanisms (triggered by appropriate stimulation, including postures), reflexes and tonic responses may become integrated and subsequently suppressed to promote a more mature neuromotor system. In other words, reflex responses should be used to develop muscles, inhibit hypertonic antagonists, and to promote balanced neuromuscular relationships between antagonistic muscles and synergists, relative to tone, to improve or partially coordinate movement.

4) Passive exercise, in the form of total patterns (not isolated joint by joint), should be properly timed to effect the sensory feedback mechanisms in a manner designed to provide the central mechanisms with the perception of more normal motion and joint-muscle relationships.

5) Continued active or passive practice of the reflex patterns may lead spontaneously to the next higher developmental level of movement; i.e., the use of the asymmetrical tonic neck reflex as an active forced pattern of activity should lead to the development of the tonic lumbar reflex.

Developmental patterns of posture and movement (M12H2A02) The developmental reflexes and tonic patterns may be used to play a progressive role in a program of neuromuscular re-education with the end being the patient's assumption of erect posture and free, self-directed selective mobility of the extremities. The sequence of development should begin with the patient in the prone position and progressing to the upright position of full vertebral extension.

1) The most elemental pattern of coordinated movement is called the homolateral pattern. In this pattern, the patient is first placed in a prone posture. The head, thorax and pelvis are then turned to varying degrees toward the advancing or actively flexing ipsilateral upper and lower extremities (referring to shoulder and hip), while contralateral extremities are extending. In a serialized motion, beginning with the eyes and the head rotating toward the the opposite side, the thorax, upper extremities, the pelvis, and lower extremities reverse position. This pattern (rudimentary creeping) is an elementary form of forward propulsion (seen in amphibians and reptilian life forms) and is seen in the normally developing human infant. In a stationary position, this process may be used as basic exercise to develop the coordination necessary for effective motion of the extremities.

2) The homologous pattern is a bilateral symmetrical movement pattern. The head is maintained in a midline position (some neck flexion or extension may present) while the upper extremities perform the flexion sequence seen in the homolateral pattern while the lower extremities perform the extension phase, these movements are then reversed rhythmically. This pattern is seen in normal developing infants, persisting for only a short period without providing effective propulsion in the prone position.

3) The crossed-diagonal pattern, a step higher in the developmental sequence then the homologous pattern, should be performed by turning the eyes and head to the left as the thorax is rotated to the left while the left upper extremity moves into flexion and the right upper extremity moves into extension. This occurs as the pelvis is rotated to the right as the right lower extremity moves into flexion and the left lower extremity is extended. This pattern of counter-rotation of the upper and lower trunk segments is essential for coordinated motion in the erect posture.

4) At the next postural level an antigravity all-fours position should be used in which the neck is extended and the trunk is nearly horizontal. The body's weight is borne on the hands and knees. Movement of the extremities should occur in the crossed-diagonal prone progression for the most efficient and effective means of balance and propulsion. This occurs almost spontaneously if the automatic responses at the previous levels have become integrated.

5) The plantigrade posture should be assumed next and all-fours walking performed utilizing crossed-diagonal prone progression, with the body weight being borne on the hands and feet. This posture is not frequently observed in children, but it is a stage in the progression toward erect posture and walking.

6) In the full erect posture, initially either the homolateral or crossed-diagonal pattern of progression may be used for walking, but the crossed-

diagonal pattern is always the eventual goal with an accompanying integration of the coordinated of counter-rotation of the trunk segments and the inclusion of the natural contralateral arm swing.

Procedures (M12H2A03)

1) Repetitive use of the flexor-withdrawal response of the lower extremities may be used to aid in achieving a release of the extensor thrust in preparation for reciprocal action of the lower extremities.

2) Unlocking the flexed hand may be achieved by placing the patient in a prone position, and putting the upper extremity in the extended, adducted and internally rotated position, with the dorsum of the hand resting on the buttocks. The thumb should be passively moved to the opposable position to cause the flexor tone of the other digits to diminish. Repetitive use of the procedure from thirty to fifty times may aid the patient in gaining awareness of the more normal hand position as well as a reduction of hypertonicity and may serve to prepare the patient for the neuromuscular activity required for the total patterns of movement.

3) Total patterns of motion depend upon the level of function which the patient is able to demonstrate. For example, when efforts at forward propulsion in the prone posture appear to be random and ineffectual, passive imposition of the homolateral pattern should serve as the starting point. The patient should then be assisted in rhythmical repetition of the movements of the head and extremities as previously described to aid the patient in the development of the coordinated movement necessary to effectively progress to the next level.

4) As the patient gains the ability to self-elicit the homolateral pattern effectively, without assistance, progression to the crossed-diagonal pattern should be made. Once again, it may be necessary to assist the patient with the passive patterning in the prone position to improve the coordination of the movements of the crossed-diagonal pattern, as a preliminary step to self-generation of the pattern.

5) The patient should be encouraged to perform an active role in the process of pattern utilization and control, and be made aware that without motivated volitional participation progress will be minimal (you can lead a horse to water but you can't make it float on its back). The skill of the practitioner to psychologically manipulate or foster the patient's motivational drive will be very important to the patient.

6) Graduation of demand is inherent in the progression of the patient to the antigravity positions. Propulsion in the prone position, for example, may be graded by variations in surface textures and the supporting surface angle; such variations are dependent upon the ingenuity and imagination of the practitioner.

7) Timing control (the sequence of joint motion) may be affected by the rhythm of the performed pattern. For best results, practitioner-assisted patterning should occur at a rate and rhythm that closely imitates reality.

8) Directional control of the extremities may develop during patterning if the movements are properly guided, in imitation of natural, normal muscle/joint behaviors, since the sequence follows a natural progression. As a consequence, it may be assumed that body awareness and perception of position and space may be developed as progression takes place from total body support to antigravity postures and movement.

9) The effectiveness of any of the above treatment procedures will depend upon the initiation of guided pattern therapy at the developmental stage appropriate attained by the patient and not upon the patient's chronological age.

The Brunnstrom Approach (M12H2B)

The Brunnstrom approach to guided patterning was primarily developed as a means of assisting in the recovery of upper extremity neuromuscular control by the post cerebral vascular accident (CVA) victims, but it has proven to be valuable for the treatment of central neurogenic muscular dysfunctions which have arisen from other sources of injury. The Brunnstrom approach is based upon the influence of neuromuscular reflexes upon motor behavior, and is commonly called reflex training.

Stages of post CVA recovery (M12H2B01) The stages of post CVA recovery normally follow a definite pattern of reflex and/or neuromuscular control development, which each patient naturally reaches at a particular and relatively different stage of development.

1) In the first stage of development, the post CVA victim will experience flaccidity, with no voluntary motion. If the patient does not suffer from functional flaccidity (refer to Central Nervous System Dysfunction [R006]), spasticity will eventually begin to develop in the form of flexion and extension pathological stereotypic synergies, appearing initially as antagonistic cocontractions. As development progresses, the pathological stereotypic synergistic patterns become increasingly distinct and may be elicited as synergistic motion patterns in response to triggered neuromuscular reflexes. In the upper extremity, pathological flexion synergies usually predominate over extension patterns, and in the lower extremity pathological extension synergies tend to predominate over flexion patterns.

2) In the second stage, spasticity is usually quite severe, but the patient should generally be able to voluntarily initiate stereotypic synergistic motion. All attempts at voluntary motion of the involved extremity will result in the dominating pathological stereotypic synergistic patterns.

3) In the third stage, spasticity begins to decrease, and the patient should be able to perform a few movements which deviate from the stereotypic pathological synergistic patterns. Such movements will initially be relatively uncoordinated, and will be necessarily slow and deliberate. At this stage, reciprocal, more normal, synergistic movement may begin to appear.

4) In the fourth stage, a relative neuromuscular independence from the dominance of the stereotypic pathological synergistic patterns should have developed, with selective joint motion becoming increasingly possible. At this stage, spasticity should gradually decline and the patient may be able to perform simple functional activities without the interference of previously dominating stereotypic patterned motion. Such independent motion will be necessarily slow and deliberate.

5) In the fifth stage, spasticity will have almost completely disappeared, with individual joint motion occurring more freely and with a better control of speed and direction. Some incoordination may still be existent, especially when rapid reciprocal movements are employed.

6) In the sixth stage, normal motor function should have been restored.

General considerations (M12H2B02)

1) Evaluation is the first step in developing a program of reflex training, primarily to establish the patient's present stage of recovery and to determine the starting point and direction of therapy.

2) Wrist and finger control development does not follow a well-defined pattern of recovery because the movements of these segments involve a high degree of cortical control. In the early stages of development, only stereotypic pathological synergistic reactions triggered by reflex activity may be apparent. Such reflex activity should be utilized to promote gross motor function control as an essential step toward more refined cortical neuromuscular control. Stereotypic pathological synergistic finger and wrist flexion patterns generally precede the development of extension patterns, and independent thumb motion will generally precede finger motion.

3) The lower extremity advances through the stages of recovery faster then the upper extremity because the required cortical control is less for walking and leg control than is necessary for arm-wrist-finger coordination. Basic extension synergies will usually predominate in the lower extremity in early stages of development.

4) Once the patient's stage of recovery has been determined, the goal of treatment is to help the patient progress from the level of function present and, through reflex training, to overcome subcortical dominance of motor activity by increasing cortical control of muscle function.

5) Essentially, the general pattern of therapy is (a) the facilitation, (b) the utilization, and (c) the breaking of a reflex, preliminary to beginning the facilitation of a reflex one step higher in the developmental sequence.

Posture and positioning (M12H2B03)

Integral to the Brunnstrom approach is the use of posture or positioning to facilitate or inhibit reflex activity and response:

Supine positions

1) The supine position requires little patient stabilization and may be used in the earliest stages of development.

2) The supine position facilitates extension synergies and inhibits flexion synergies.

3) In the supine position, a triggered withdrawal reflex should result in facilitation of flexion synergies in both the upper and lower extremities, but they may be inhibited from full response by the tonic labyrinthine reflex.

4) Teaching the patient to roll from supine to prone promotes the development of neck and trunk neuromuscular control.

Prone positions

The prone position facilitates flexion synergies and inhibits extensor synergies as a response to the labyrinthine prone reflex.

Side-lying positions

Side-lying positions facilitate flexion synergies in the upper extremities and extension synergies in the lower extremities, by virtue of the thalamic posture response, thereby promoting righting reflexes.

Sitting positions

1) Sitting positions, which may be considered as neutral, should be used when facilitation of either flexion or extension synergies is desired, without the inhibitory presence of either the tonic labyrinthine prone or supine reflexes.

2) A hard-surfaced, armless, straight-backed chair is considered ideal for therapeutic sitting position postures, providing the sensory cues of touch, pressure, and vision which the patient may be able to use as a substitute for lost neuromuscular proprioception. The sitting position in the chair described is additionally important because it provides for good trunk position while providing for functional upper extremity positions.

3) The tonic neck and tonic lumbar reflexes are able to demonstrate themselves in the sitting position, allowing the patient the opportunity of developing the head and trunk control necessary for balancing the unsupported trunk.

4) Practicing trunk balance in the sitting position will help develop the kinesthetic sense necessary for walking.

Standing postures

1) Standing postures are often possible in early stages of development if the positive supporting reaction is present, but walking attempts may not be safe until a logical progression of exercise through the supine, sitting, and standing positions has occurred.

2) The voluntary responses, relative to motion synergies, learned and developed in the supine positions, must be subsequently learned and developed in the sitting and finally the standing posture.

Coordinated movement (M12H2B04) Coordinated movement is the primary goal of almost all therapeutic exercise, and coordinated movement is dependent upon supraspinal structure control. It has been demonstrated that supraspinal control of the components of coordinated movement may be fostered via appropriate reflex mechanisms, sensory cues, patient volitional effort and the gradation of task demands applied throughout the various stages of patient recovery. This belief system is the essence of the Brunnstrom approach to guided patterning.

Reflex mechanisms are used quite extensively in the quest for coordinated movement in the early stages of development. For example, the tonic neck, tonic labyrinthine and tonic lumbar reflexes are used extensively to achieve both mobility and stability, as well as providing the patient with the means of controlling timing and direction of motion. In practice, after motion has been reflexly and involuntarily elicited, the patient should attempt to voluntarily repeat the same motion. Repetition of motions reflexively evoked will help to increase patient appreciation of self-generated motion.

Sensory cues (M12H2B05) Sensory cues are sometimes used to promote coordinated motion.

Deep stroking

Deep stroking of the skin over the antagonist of a spastic muscle, proceeding from the proximal portion of the muscle to the distal extent, will provoke the neuromuscular phasic stretch reflex from the muscle so stroked, which increase its tone and by virtue of the process of reciprocal inhibition, inhibits tone within the spastic agonist. Ideally, this may serve to consequently increase joint mobility and to permit more voluntary motion and stability as stretch reflex sensitivity decreases in the spastic agonist. The same response may be elicited if the antagonist of the spastic agonist is appropriately vibrated (refer to Vibration, Neuromuscular Management [M08A]).

Brisk tapping

As with deep stroking, brisk tapping or slapping (a crude form of vibration) of a muscle or muscle group will facilitate involuntary tone within the stimulated agonist and inhibit tone in the antagonist if used frequently.

Resistance

Resistance to muscular contraction may be used to increase joint stability by fostering muscular recruitment, and by incidentally increasing proprioceptive input (both muscle spindle functions) as an aid in teaching the patient both timing and directional control (refer to Muscle Spindle [R001] and/or Neuromuscular System [R002]).

Resistance applied against the contraction of counterpart muscles on the uninvolved, contralateral side will facilitate an associated reaction from the muscles on the involved, ipsilateral side. This reaction may be utilized in early stages to initiate response and to reinforce a voluntary patient effort. Resistance to contraction of the stronger components of a synergistic muscle group may be used to reinforce the contraction of weaker components. For example, resistance of hip and knee flexion will facilitate dorsiflexion of the ankle.

Voluntary effort (M12H2B06) The use of voluntary action or volitional effort on the part of the patient is necessary if acquired muscle control is to become functional. Voluntary effort may be reinforced through reflex activation and sensory cues. Task grading should be utilized, with the number of reinforcing movements decreased as control of voluntary motion improves. Resistance provided by the practitioner should progressively decrease as voluntary control is achieved, and the difficulty of functional activity increased as the patient replaces stereotypic pathological synergistic motions with selective voluntary motion.

Progress in treatment and functional carry-over may occur over a periods measured in weeks or months, and visible improvements in the patient's functional condition may be very slow to appear.

The Bobath Approach (M12H2C)

Preliminary assessment (M12H2C01) Preparatory to treatment, a preliminary assessment of the patient's postural patterns must be made to determine the sources of disability and which reflex-inhibiting patterns might be effectively applied.

A survey to determine the distribution of hypertonus should be made in a systematic manner, with emphasis put upon the discovery of pathological postural patterns and not upon the determination of specific muscle spasticity.

To test neuromuscular hypertonic response, the patient should be placed in various positions and the patient's extremities maneuvered into various positions as changes in neuromuscular tone are observed. The test positions include the (1) supine, (2) the prone, (3) sitting, (4) kneeling, and (5) standing positions. The test maneuvers include (1) external rotation and flexion of the arms into the over-head position, (2) hyperextension and turning from one side to the other of the head, (3) external rotation and abduction of the legs, and finally (4) dorsiflexion and then extension of the feet. Any immediate alteration in muscle tone or unusual joint movement should be noted. Special attention should be paid to muscle hypotonicity which appears relative to joint is hyperextension, and any unusual resistance to testing maneuvers in terms of degree, pattern of resistance and the presence of spasticity, rigidity or intermittent resistance.

Techniques of handling (M12H2C02) Central to the theory of the Bobath approach is the idea that the patient must be provided with the experience of movement in a normal functional pattern before it possible for abnormal postural patterns or synergies to be inhibited and/or extinguished. Coupled with this is the concept that normal motion will only develop out of a normal posture and that normal motion cannot develop out of abnormal postures. Postures are generally not fixed, without joint ankylosis, and patterns must be changed so that postural tone may be adjusted to the requirements of the activity at hand. The experience of normal motion should be provided for the patient by changing the relationship between the parts of the body to each other by *handling* the patient.

Handling simply means that the practitioner is required to physically manipulate or change the positional relationship between the patient's extremities while maintaining the patient in a particular "normal" posture. Handling should be performed in such a way that abnormal patterns are blocked while more normal postures or movements are facilitated or promoted. This can be accomplished by controlling patient postures and motion from *key points* from which it has been found possible to control the state of tone in other body parts (trunk and/or extremities), while simultaneously stimulating or facilitating desired activity performed by a particular body part.

Key points are sites from which the practitioner may control patient movement. They are generally sites over proximal joints which may be used to control the movement of the head, shoulder girdle, trunk and/or pelvis. When stimulated, the sensory receptors associated with these particular joints are thought to effect vestibular and midbrain nuclei and to effect the developmental reflexes (refer to Table of Developmental Reflexes [T007]). Skillful control of the *key points* (through handling) may affect dominating developmental reflexes and the various stereotypic synergistic patterns to such an extent that they may be made to effect each other to produce more normal motions. For instance, lifting and turning the head may be used to elicit a series of righting sequences or reactions, which may be used to progress the patient to roll over, sitting up, go to hands and knees, rise to kneeling, and finally come to full standing.

Rotational manipulations of key points is especially effective at interchanging or melding developmental and other synergistic patterns since they tend to inhibit both flexor and extensor hypertonus. Manually turning the head facilitates the extremity musculature responsible for the pattern of extension/abduction/internal rotation on the side the face is turned to and the flexion/adduction/external rotation on the contralateral side, while simultaneously inhibiting the antagonistic musculature in opposition to those patterns.

The shoulder girdle is a very effective key point; control of it allows stabilization of the head, thorax, upper extremities and pelvis (by virtue of linking muscular connections) while leaving the head free to move or adjust itself as various handling manipulations take place. Other key points include the arm and forearm. Upward pressure under the chest may be used to assist trunk extension, and downward pressure on the sternum may be used to inhibit extensor spasm. Control of the pelvis and knees may be used to assist rising from sitting to standing.

Reflex inhibition patterns (M12H2C03) Reflex inhibiting patterns are used to interrupt abnormal motor responses and to redirect impulses into more desirable motion patterns which may be subsequently learned. Reflex-inhibition alone, through key point handling and pattern control, should be used only in exceptional circumstances, as when hypertonus is so severe that it permits little or no movement, or when severe tonic spasms distort movement (dyskinesia). In such cases, inhibition of hypertonus should precede attempts to facilitate motion. The patient should not be held statically, but should be constantly moved in "normal" patterns to impart the idea of movement and to prevent the patient from becoming stiff in newly assumed positions. Any reflex inhibition procedure which inhibits abnormal activity makes other and more normal reactions easier or possible, and in this sense may be said to facilitate or at least make possible other reactions.

Reflex inhibiting patterns include:

Neck flexion/hyperextension

Manually raising the head, by hyperextending the neck, usually facilitates extension in the rest of the body, while flexion of the neck generally inhibits extensor tonus. If a strong symmetrical tonic neck reflex is present, the response pattern in the lower extremities will be opposite from that in the upper extremities.

Lumbar lordosis/kyphosis

Manually forced lordosis of the lower spine may be used to facilitate hip flexion and inhibit extension, while manually forced kyphosis may be used to facilitate hip extension and inhibit flexion.

Shoulder rotation

Manual internal rotation of the upper extremities tends to inhibit extension patterns, especially for those suffering from athetosis, while external rotation tends to inhibit flexion.

Shoulder horizontal abduction/extension

Manual horizontal abduction and extension of the shoulder directly diagonally backward, tends to inhibit flexion in the neck, arms and hands, and adduction and internal rotation of the lower extremities.

Shoulder flexion

Manual flexion of the upper extremities into the overhead position tends to facilitate extension of the hips and trunk while inhibiting hip and trunk extension.

Shoulder elevation & lateral abdominal stretch

Manual lifting of the shoulder girdle while laterally stretching the ipsilateral lateral trunk flexors tends to inhibit spasticity in the upper extremity.

Weight bearing

Weight bearing on the heel of the hand tends to facilitate extension patterns and promote postural support.

Lower extremity extension/external rotation

Manual external rotation and extension of the lower extremities tends to facilitate hip abduction and dorsiflexion of feet while inhibiting the synergies responsible for the positive supporting reaction.

Dorsiflexion of the toes

Manual dorsiflexion of the toes tends to inhibit extensor spasticity and facilitates the flexor patterns in the lower extremity.

Hip flexion or abduction

Manually forced hip flexion and hip abduction are associated movements and manual performance of one facilitates the other and inhibits their respective antagonists.

Trunk rotation

Manually forced trunk rotation between the pelvis and shoulder girdle tends to inhibit both flexor and extensor tonus and to facilitate the functional patterns required to turn over and go to standing.

Thoracic spine extension

Manual extension of the thoracic spine tends to inhibit hip flexion and flexor synergies of the upper extremities.

Tactile stimulation (M12H2C04) Proprioceptive and tactile stimuli may be used in conjunction with, and to enhance handling. The tactile stimuli traditionally recommended for use with this approach is generally *tapping* (manual, low frequency vibration) applied to muscle belly and/or tendon.

Tapping may be used to increase involuntary muscle tone if the patient has become neuromuscularly too relaxed from the application of reflex inhibiting patterns, the musculature is hypotonic or flaccid, or fluctuating muscular tone requires modulation. The most effective method of tapping is believed to be through the stimulation of receptor tendons as they cross the joint space. When tapping is performed it is important to control the patient's body via a reflex-inhibiting pattern to prevent the development or heightening of spasticity or spasm which may be present in the involved musculature. Tapping should be halted if spasticity does develop.

Tactile stimulation techniques

1) To reduce tone in a spastic muscle or synergistic muscle group, the spastic muscle should be put on stretch and the tendon(s) of its antagonistic muscle or muscle group should be repeatedly tapped, at a rate fast enough to avoid a rebound phasic stretch reflex from the indirectly stimulated agonist(s). Tapping should continue until the spastic muscle(s) reduces tone and permits passive lengthening. Thirty hertz linear vibration applied to the muscle's tendon for one minute followed by a brief rest period may serve (refer to Vibration, Neuromuscular Management [M08A]).

2) The involuntary postural tone generated by tapping may be sufficient to provide support for functional movement and body postures held against gravity. To facilitate muscle tone, fairly heavy percussion (or vibration) should be applied to the muscle to be facilitated. Applied to the paravertebral and gluteal muscles, such percussion may be utilized to facilitate trunk extension.

3) To increase muscle tone generated by tapping, joint approximation may be applied during or shortly following the tapping (or vibration) procedure. To apply approximation, the involved extremity should be driven directly into the joint with care taken to avoid tangential force lines. Muscle tone will be involuntarily increased in all the muscles surrounding and supporting the approximated joint. Pressure applied to the soles of the feet or palms of the hands on a supporting surface provokes the approximation response in the involved extremity.

4) Tapping of antagonists, alternately or simultaneously (from a midpoint position), may be used to promote postural cocontractions This technique may be used to promote sitting, standing, or kneeling balance. Tapping should be applied in a slow and firm manner for gross postural enhancement (a vibrator may be used), or applied lightly and tentatively with the fingers to cause slight displacements of the trunk and to promote fine adjustments in balance.

5) Brief muscle stretch should be used to further increase muscle tone. For example, head righting may be facilitated by gently tossing the head up (a push being applied to the underside of the chin) and caught again as it begins the drop back.

6) Light touch or brushing may be applied to facilitate the muscles used for extremity withdrawal. Fingers and toes may be made to extend with wrist and ankle dorsiflexion by light touch or brushing (tickling) be applied to the palmar surfaces of the tips of the fingers or toes.

7) Joint shaking may be used to inhibit muscular hypertonus. A spastically closed hand will tend to open when the arm is abducted and the forearm shaken.

Perceptual-motor learning (M12H2C05) Position sense derived from afferent sensory input combined with the observation and appreciation of temporal sequencing of physical events is at the base of perceptual-motor learning. Position sense, relative to the physical environment, helps produce our body image and provides us with a conceptual relationship between external objects, and the appreciation of temporal sequences allows us the opportunity to motor plan. Normally, a child utilizes these components of perceptual-motor learning in conjunction with a program of trial and error movement to learn fine motor control necessary for most functional activities. Patients who have sustained central nervous system damage which interferes with afferent sensory input or processing are often deprived of the opportunity of engaging in this trial and error experimentation since free motion is generally precluded for such individuals by the fixed postures or movement patterns dictated by dominant pathological developmental reflex patterns and/or other stereotypic synergistic patterns. For further neuromuscular voluntary control to be allowed to developed, these individuals need to be provided with external help to experience the sensory input provided by motion. Such movement may be initiated through the stimulation of exteroceptors (proprioception) provoked by the complex pattern of sensory input which results from the handling and reflex-inhibiting patterns utilized in the Bobath approach to guided patterning. During such treatment, error should not be permitted to occur, until the patient has experienced the sensations of a normal movement pattern. Unless the sensations of normal motion are appreciated by the patient, movement errors will go undetected and abnormal sensory input will be assumed to be normal by the previously damaged nervous system.

To make perceptual-motor learning possible, the patient should not be asked to perform a task which is patently impossible, or one which can not be performed without undue effort or by a grossly abnormal way performance A reasonable effort should be met with reasonable success, and undue effort is undesirable because it will increase general tone in musculature already prone to hypertonus or trigger associative reactions, and consequently distort proprioceptive feedback. Immature developmental reflex patterns may be used initially to prepare for more mature developmental reflex patterns and selective motion. These patterns should be made up of a variety of postural and motion patterns required for functional activities. Patterns which may be useful in this regard are those that (1) utilize automatic movements like righting, protective and/or equilibrium (balance) maintaining reactions, (2) utilize movements which would normally be performed by infants for self-exploration and relating to others and/or objects, (3) utilize basic postural patterns required for sitting, standing and/or the reciprocal movements associated with progression, and (4) utilize voluntary functional movements, as well as combinations of automatic and voluntary movements, controlled by key point handling. Ideally, the patient should experience easy movement, performing activities which provide motivation by being fun and/or interesting to do.

exercise guided patterning

The Proprioceptive Neuromuscular Facilitation (PNF) Approach (M12H2D)

Of the approaches discussed here, the PNF approach to guided patterning is the most demanding upon the practitioner in that it requires the most hands-on skill to apply. It depends upon the practitioner's learned ability to appreciate what physical resistance means relative to movement patterns, an exact knowledge of the sequence of patterned muscular contractions which produce synergistic motion, and an integration of the knowledge of the synergistic patterns and timed application of resistance to provoke desired neuromuscular response.

Basic technique (M12H2D01) The PNF approach to guided patterning requires that during treatment the practitioner must be in direct, almost continuous contact with the patient. Essentially, the PNF approach is an exercise program which employs the patient's proprioceptive response to quick stretch to facilitate synergistic muscle activity in the muscles stretched. These contractions are subsequently guided via directed manual resistance to complete an appropriate range of motion.

Provoking synergistic patterns

Whole synergistic pattern facilitation generally requires the practitioner to take the following steps:

1) The patient's extremity should be grasped by taking hold of the distal segment(s) of the extremity with one hand while supporting and holding the extremity, usually near the central joint interposed between the extremity's proximal and distal extremes.

2) All of the muscles which comprise a synergistic muscle group (or pattern) should be put on stretch simultaneously, by taking all joints to the end of their respective ranges. For example, to facilitate the leg extension synergy the hip should be flexed, abducted and externally rotated, while the knee is fully flexed and the ankle dorsiflexed to the end of range.

3) A quick-stretch should then be simultaneously applied to the entire muscle group comprising the synergy by suddenly pushing against the resistance of the already taut muscles, the practitioner's pressure following the path of the antagonistic synergy.

4) When the synergistic muscle group responds to the quick stretch with an involuntary phasic contraction, the patient, having been previously warned of the eventuality, should be cued to attempt to voluntarily contract the synergistic muscle group and move against the practitioner's resistance.

5) The practitioner should manually resist the patient's contraction and movement through the range of motion, until the patient's contraction begins to lose force. At that point, quick stretch should once again be applied in the direction of the antagonistic synergistic pattern, to facilitate a renewal of the force of muscle contraction, and the resulting contraction allowed to move the extremity further along the pattern of motion.

6) This pattern of stimulation and response should continue until the entire range of motion of all the involved joints is completed, the practitioner having truly guided the patient's motion to the zero degree end of range. For the leg extension synergy, this means that the hip has been fully extended, adducted and internally rotated to zero ranges, the knee extended to zero, and the ankle and toes plantar flexed to the end of range.

7) If any particular muscle or muscle group in the synergy fails to maintain the force of contraction, the practitioner should require the patient stop motion at the point where force is being lost and to maintain tone in the stronger proximal muscles as the related proximal joints are stabilized, while the practitioner repeatedly applies quick stretch to the weak muscle(s) to elicit or pump up contractile force. When contractile force in the weak muscle is sufficient, the synergy should be allowed to continue and the range of motion eventually completed.

Principles of treatment (M12H2D02)

1) The patient's abilities must be harnessed and channeled to reduce inabilities. Therapy should be centered upon the patient's existing ability and not upon the patient's disability.

2) Normal neuromuscular development proceeds caudally, from proximal to distal. Consequently, therapy should initially be directed at the development of neck and trunk skills (especially in the area of stabilization) before attention is paid to improving extremity function.

3) Normally, early motor behavior is dominated by developmental reflex activity while more mature motor behavior is reinforced by postural reflexes. Developmental reflex activity may be harnessed to reinforce voluntary effort, and the use of total patterns of motion and posture may be used to promote and then to balance reflex activity.

4) Early neuromuscular development is characterized by spontaneous motion which oscillates between extremes of flexion and extension in a rhythmically alternating (vibration-like) pattern of motion. Alternating movement may be used therapeutically to establish or reestablish interaction between antagonists. The ability to rhythmically reverse the direction of motion should be an objective of treatment, if the patient lacks that ability.

5) Neuromuscular development is characterized by an orderly sequence of total patterns of motion and posture. Each total pattern of motion must integrate the component motion patterns of the head, neck and trunk with those of the extremities to be complete. Normal motion is made up of component patterns combined to produce bilateral symmetrical, asymmetrical, ipsilateral, and alternating reciprocal and diagonally reciprocal patterns. It is believed that patients who have failed to experience these patterns must be made to experience them if further neuromuscular control is to occur.

6) Total patterns of motion may be initiated with either flexor or extensor dominance. Rocking, the alteration of flexion and extension movements alternately between diagonal flexion and extension, as well as in other directions, may be used to develop balance and to establish reciprocal relationships between antagonistic motions.

7) The overlapping of total pattern performance is used to enhance or speed the development of diagonal patterns.

8) Locomotion depends upon the reciprocal contraction of flexors and extensors, while the maintenance of an erect posture during ambulation requires continual neuromuscular adjustments to compensate for the constantly created neuromuscular imbalances. These functions are all dependant upon synergistic action of component muscle groups, and a balanced interaction occurring between antagonistic muscles and synergistic muscle groups. One therapeutic goal is to prevent and/or correct imbalances between antagonistic muscles and/or synergistic muscle groups.

9) Along with the physical techniques of neuromuscular facilitation, appropriate sensory cues (including vision and hearing), which clearly express demand, should be selected to enhance the patient's motor learning. The technique used for the enhancement of motor learning should be superimposed upon present motion and posture.

10) The techniques designed to facilitate agonistic neuromuscular activity, and conversely to inhibit antagonistic neuromuscular activity, include: (a) position, for optimal support of postural reflexes during voluntary effort and the stimulation of the equilibrium saving reactions; (b) manual contact with the patient, which permits the administration of direct pressure, stretch, and resistance to oppose and guide the patient's efforts; and (c) maximal resistance, applied for increased proprioceptive response through the processes of *irradiation* (energy channeled from stronger to weaker muscle groups or synergistic patterns), *successive induction* (alternate application to agonistic and antagonistic patterns to increase agonistic response), and *reciprocal inhibition* (application which inhibits the antagonist).

11) The frequency of sensory stimulation and repetitive activity should be used to promote and retain motor learning, and to promote the development of additional muscle strength and endurance.

12) Supplementary activities should include supervised mat exercise, the use of weights and pulleys to provide resistance of spiral and diagonal patterns, and pattern combinations which the patient can successfully perform in a coordinated manner.

13) Goal-directed activities coupled with techniques of neuromuscular facilitation should be used to hasten learning of the total patterns of walking, climbing up and down stairs, and other activities of daily living, including self-care.

14) Maximal resistance and the selective techniques of slow reversal and rhythmic stabilization should be superimposed upon one another for optimal results.

15) Recapitulation should be used to the couple postural reflexes and voluntary motion through the balancing of reflex activity with of antagonistic synergistic muscle action. Recapitulation is defined as the process of developing a readiness for learning an activity at a higher skill level by utilizing exercises which employ reflexes or movement at a lower level of skill. This process is used when primitive movement associated with one stage of reflex development is used to facilitate the development of a reflex pattern at a higher developmental reflex level.

Postural reflexes generally promote development of voluntary motion. The asymmetrical tonic neck reflex with its promotion of extremity movement and synergistic tone, for example, may be used to reinforce the total pattern of rolling, promote eye-hand coordination, and/or strengthen a synergistic muscle group which contributes to the whole pattern. When correctly utilized, the process of recapitulation may be used to create a progression of motor acts which may be used to assess a patient's level of development, revealing the patient's level of motor competence and helping establish the level from which the rebuilding of total patterns of movement and posture may begin. In therapy, the patient is led to recapitulate the developmental sequence insofar as is possible.

The Rood Approach (M12H2E)

The Rood approach is based upon a philosophy of treatment which is concerned with the interaction of somatic, autonomic, and psychological factors and their combined role in the regulation of motor behavior. It is based upon specific concepts about the development of human motion and its relationship to dysfunction. Motor function is thought to be inseparable from sensory mechanisms and therefore a strong emphasis is placed upon it in the design of treatment procedures. The goal of the Rood approach is to activate patient movement and postural response in the same automatic manner in which they occur in normal subjects, without the need of conscious patient awareness of what the appropriate neuromuscular response should be.

Philosophy (M12H2E01) The Rood approach is built upon the concept that all functions and structures of the neuromuscular system can be related to (1) the biological survival of the individual through protection and mobility and (2) the growth of the individual through pursuit and adaptation which is aimed at maintaining contact with the environment, as a mirror or proof of self control. These factors play a major role in the development and functional design of neuromuscular stability.

There are two general reaction systems at play in any human nervous system, the autonomic and the somatic. The autonomic nervous system is made up of the sympathetic and parasympathetic divisions; sympathetic activity is generally brief and selective in response, being devoted to responding to changes in the internal and external environment. Parasympathetic functions, on the other hand, are generally continuous and ongoing in nature, usually providing a steady background of somatic stabilizing functions. The somatic reaction system has a functional duality, contributing either to mobility or to the stability of a response. The more complex a motor activity, the more complex will be the variations of interaction, in respect to degree of response and timing of the two functions. Muscles differ structurally and kinesiologically depending on which function they most often fulfill (refer to Exercise, Muscle Strengthening [M12A] for details regarding muscle fiber variables).

There are two somatic sensory functions, one related to specific sensory reception and the other to nonspecific or general sensory reception. The somatic sensory specific mechanisms are initially protective and indiscriminate in their functional response, but with continued stimulation become discriminatory. They include free pain nerve endings and the sensory receptors designed to respond selectively to cold, heat, light touch, deep touch, skin stretch, muscle stretch, and others. The somatic sensory nonspecific mechanisms are static and maintain the quality of function as a part of the regulation of response. The most important of these are muscle spindle trail nerve endings, and golgi tendon organs.

Principles (M12H2E02)

1) The achievement of a desirable response is considered to be related to the influence of appropriate sensory factors upon the neuromuscular system. Normal movement is associated with stimuli present prior to, and sensory feedback provided during muscular response. Most proprioception before and during motor activity is provided by the muscle spindle (refer to The Muscle Spindle [R001] for details of muscle spindle function); these stimuli may activate, facilitate or inhibit the involved motor components. Natural or therapeutically-devised stimuli which imitate nature should be an integral and carefully considered part of the treatment of neuromuscular dysfunction. Such stimuli should be used according to their ability to facilitate and/or inhibit neuromuscular responses.

2) Motor responses should be sought in the order in which they appear in the developmental reflex sequence. Coordinated motion is normally acquired during the different and overlapping stages in the sequence of sensorimotor reflex development This sequence forms the comparative basis for identifying the level at which the patient is functioning, the components found wanting and necessary for the development of coordinated motion, and the order in which further pattern development should be encouraged.

3) Coordinated movement develops out of the interaction of neuromuscular mobilizing and stabilizing actions. As the relationship between these two actions develop, greater refinement of response develops accompanied by the acquisition of effective control of the direction and speed of neuromuscular action.

4) There is an intimate ongoing interaction between autonomic, somatic, and psychological functions. Consideration should be given to the influence which therapeutic stimuli may have upon autonomic and/or psychological functions as well as upon somatic responses, and a careful selection of stimuli made based upon their direct effect upon one system or their indirect effect (via effects upon the stimulated system) upon another.

5) Individual muscle functions should be identified relative to the type of developmental reflex or synergistic pattern in which the muscle participates. The individual muscle should be described, studied, and reeducated according to whether it works in (a) a non-weight-bearing reciprocal pattern moving a small lever, (b) a weight-bearing co-contraction pattern, (c) a weight-bearing pattern moving a large lever, or (d) a coordinated pattern.

6) Developmental reflex patterns should be built upon in the order of their ascending complexity, as occurs in normal developmental reflex development. The most complex patterns cannot be fully

developed if not preceded by the development of the more elementary pattern.

7) In all conditions of neuromuscular dysfunction, concern must given to the present state of the nervous system and the means available to influence it, since it is only through the nervous system that the musculoskeletal system is able to function dynamically.

8) An exercise is not considered therapeutic unless the provoked neuromuscular pattern of response is correct and results in proprioceptive feedback from the involved muscles which will enhance learning from that response. Therapy is not defined by the neuromuscular action alone, but in terms of the application of stimuli to elicit a response, the afferent sensory (proprioceptive) response to a correct neuromuscular reaction, and the additional stimuli applied to facilitate or inhibit components within the provoked neuromuscular pattern. Sensory factors are essential for the achievement and maintenance of normal motor function since the development of coordinated motion are normally dependent upon them. Therefore, sensory stimuli may have to be supplied if a reflex pattern is to be produced.

Developmental patterns of movement and posture (M12H2E03) There are two major sequences of motor development, the vital functions sequence and the skeletal, ontogenetic motor patterns functions sequence. Functions which are considered vital include food and water consumption, waste product elimination (not considered here), respiration, and speech. Skeletal functions include neck, trunk and extremity activities.

Vital functions development sequence

1) *Inspiration* is the intake of air into the lungs.
2) *Expiration* is the output of air from the lungs. Other functions in this category include crying, sneezing and coughing.
3) *Sucking*, which is a grasping or pursuit response. The sucking center sets the pattern for respiration by coordinating sucking, swallowing and breathing.
4) *Swallowing* is activated by the stimulus of food, or any sharp stimulus to areas supplied by the trigeminal nerve.
5) *Phonation* (or controlled expiration) normally requires the prerequisite development of the respiratory control developed during sucking.
6) *Chewing* occurs as masticators work against resistance in rhythmic contractions. Chewing is coordinated with tongue action and the gag reflex must have been inhibited for it to occur successfully.
7) *Speech* or articulation is a motor act dependent upon the contributions of coordinated motion developed during the previous steps, in the designated order.

Skeletal functions developmental sequence

1) *Withdrawal*, which is performed in the supine position and requires heavy work effort of the musculature of the trunk, neck, and proximal joints of the extremities, with motion occurring in a reciprocal innervation pattern.
2) *Roll-over* is performed by flexion of the upper and lower extremities on the superior ipsilateral side and extension of the extremities on the contralateral inferior side.
3) *Pivot prone* is lying prone with full extension of the neck, trunk, and lower extremities. Once this pattern has been established, the patient may be advanced to weight-bearing patterns.
4) *Cocontraction of the neck musculature* occurs when the neck extensors and flexors cocontract with thoracic extension.
5) *On-elbows* is produced with cocontraction of the scapular and glenohumeral joint musculature and involves a pushing off or pushing back tendency of the upper extremities.
6) *All-fours* occurs when body weight is shifted backward and forward and side to side on the knees and palms of the hands, ultimately progressing to the alternate arm and leg motions of creeping.
7) *Standing* is a static posture; the body's weight is generally shifted from foot to foot.
8) *Walking*, involving stance, push off, pick up, and heel strike phases. Each step in the previous sequence is built on or develops out of the previous step.

Procedures for achieving coordinated motion (M12H2E04) Therapeutic procedures for achieving coordinated motion involve the application of stimuli to facilitate, inhibit, and/or activate neuromuscular responses. These procedures should be applied to activate automatic (reflex) responses without conscious effort on the part of the patient until the patient is able to perform the pattern or function in a relatively normal manner.

Evaluation and treatment planning begin with the identification of the pattern to be activated and determination of the stimuli to be used to elicit the pattern. This may mean placing the patient in a position which will spontaneously activate the desired pattern or the application of a stimulus such as resistance, vibration or manipulation which may serve as an activator.

Procedures

1) *Quick stretch* (or muscle stretch part way into range) may be used to facilitate the mobilizing component of a muscle and to inhibit its antagonist.

exercise guided patterning

2) *Prolonged stretch* or *heavy resistance* of a stabilizer will facilitate flexor activity while dampening facilitation of extensor activity, so that cocontraction may be initiated.

3) *Fast brushing* of the skin over all stabilizers involving the affected area to bias the spindle to a new shorter length, making it more sensitive to stretch, so that normal stretch or resistance will facilitate cocontraction. Fast brushing should occur thirty minutes before the rest of the treatment. This procedure has been found most helpful in the facilitation of the pivot prone response. The tonic biasing of the agonist to shorten by fast brushing influences the antagonist muscle spindles to lengthen, as an expression of the reciprocal inhibitory relationship of antagonists; used for this effect, fast brushing has proven to be effective for the reduction and inhibition of spasticity, especially for acute post CVA syndrome victims.

4) *Quick-ice* (12 to 17° C. for three to five seconds) should be applied to the skin area overlying stabilizing muscles of the extremities only, to facilitate contraction. Quick-ice applied over the trunk has visceral effects and should usually be avoided on both anterior and posterior aspects.

5) *Intermittent quick-stretch* (or pressure) applied to a muscle or muscle group may be used to facilitate contraction in that muscle, but the contraction elicited must be immediately resisted as it is produced, or the response will be phasic and too short lived to produce tonic biasing of the muscle spindle to a shortened length. Resistance of moderate force may be used to facilitate a stabilizer if a reciprocal holding contraction is sought (as in the pivot prone pattern).

6) *Joint compression*, adding of compression to a weight bearing joint or a shift of weight, results in cocontraction of the stabilizers which have gone through the stage of holding against resistance in the shortened range.

7) *Heavy resistance* applied against the pull of a stabilizer of sufficient force to increase the lever arm length will result in a cocontraction rather than a reciprocal contraction.

8) *Slow repetitive stroking* lasting for three minutes or *warm to neutral applications* applied to the skin overlying the area of the back innervated by the posterior rami will reduce sympathetic outflow and will have an inhibitory effect on somatic motor functions.

9) The *quadruped position*, which activates the carotid sinus reflex, may be used to reduce general tension without inhibiting postural functions.

10) *Quick tactile heavy stroking or rubbing* or *cold stimuli* are appreciated by the sensory system as nociceptive, especially when applied to the soles of the feet, palms of the hand, or the lips, and may be used to mobilize responses. Nociceptive stimuli are generally avoided unless a response can not be produced by other stimuli.

11) *Repeated unresisted contractions* (produced without stretch) will have an inhibitory effect upon muscle contraction and a facilitatory effect on its antagonist.

12) The mobility of the tongue is facilitated by *taste stimuli*; bitter substances facilitate active protrusion, while sweet or pleasant substances facilitate retraction, especially where protrusion is excessive, and may be used in preparation for sucking.

13) *Quick touch* to the mucosa or alveolar ridge may be used to stimulate tongue movement.

14) *Quick ice* to the lips may be used to facilitate mouth opening.

15) *Fast brushing* or *quick icing* for three to five seconds over the skin area of the stabilizer may be used in preparation for muscular facilitation.

16) In the stability developing patterns of sucking, swallowing and chewing, *resistance* to those actions and/or *pressure* over the muscle bellies increases facilitation.

17) *Low threshold stimuli* are related to reciprocal mobilizing functions which provide range and speed in early developmental reflex patterns and provide distal skilled movement of more advanced patterns. Such stimuli include *quick tactile* and *quick cold* stimuli, *acceleration* and *deceleration* during movement of the head, *tendon tapping*, *quick stretch*, and *moderate resistance* during motion.

18) *High threshold stimuli* are related to cocontraction stabilizing responses which are maintained and static in nature. Such stimuli include *fast brushing* or *maintained touch*, three to five second *quick icing*, *heavy resistance* to maintained contraction in the shortened range, *maintained stretch* of a stabilizer at one joint, *constant pressure* on the muscle belly, *static position* of the head, and *weight bearing* on one extremity which will require constant holding.

Partial List of Conditions Treated   (M12H3)

Treatable Causes   (M12H3A)

Extrafusal muscle spasm   (C41)
Habitual joint positioning   (C48)
Hypermobile/instable joint   (C10)
Joint approximation   (C47)
Muscular weakness   (C23)
Nerve root compression   (C04)
Neuromuscular tonic imbalance   (C30)
Pathological neuromuscular hypertonus   (C28)
Pathological neuromuscular hypotonus   (C33)

Pathological neuromuscular dyscoordination (C35)
Peripheral nerve injury (C07)
Psychoneurogenic neuromuscular hypertonus (C29)
Restricted joint range of motion (C26)
Trigger point formation (C01)

Syndromes (M12H3B)

CEREBRAL PALSY (S25)
CERVICAL (NECK) PAIN (S73)
FROZEN SHOULDER (S64)
HAND/FINGER PAIN (S17)
HEADACHE PAIN (S02)
NEUROMUSCULAR PARALYSIS (S22)
PIRIFORMIS (S04)
POST CEREBRAL VASCULAR ACCIDENT (CVA) (S07)
POST IMMOBILIZATION (S36)
POST PERIPHERAL NERVE INJURY (S23)
POST SPINAL CORD INJURY (S24)
POST WHIPLASH (S55)
REFERRED PAIN (S01)
SCIATICA (S05)
SHOULDER/HAND (S57)
SHOULDER PAIN (S08)
TORTICOLLIS (WRY NECK) (S65)

Precautions and Contraindications (M12H4)

Arteriosclerosis or atherosclerosis (M12H4A)

Strenuous exercise may be contraindicated for those suffering from arteriosclerosis or atherosclerosis. A surge in blood pressure accompanying exercise or the strain of lifting a heavy weight may cause a "cracking" of a hardened cerebral artery and subsequent spurting of blood, under pressure, into vulnerable cerebral brain matter, causing destruction of any brain matter touched by the stream (a cerebral vascular accident).

Other peripheral vascular disease (M12H4B)

Patients suffering from peripheral vascular disease should avoid any exercise which involves excessive use of an involved extremity, including forced walking, if there is lower extremity involvement, when symptoms of intermittent claudication are present. This is especially true for sufferers of Buerger's disease, Raynaud's disease, or diabetes.

Diabetes (M12H4C)

An insulin reaction may occur in a diabetic patient who has taken an insulin supplement and has eaten too little food, failed to digest what has been eaten before beginning exercise, or taken too large a dose of insulin before performing strenuous exercise. Initial symptoms include extreme hunger, nervousness, heavy perspiration, palpitations of the heart, and occasionally double vision. These symptoms may be relieved by consumption of eight ounces of orange juice to which two teaspoonfuls of regular sugar have been added. If this is unavailable, any sugar or carbohydrate-containing food, like bread or crackers, may serve if it is given in sufficient quantities.

If initial symptoms are unrelieved, more serious symptoms may appear; these include delirium, convulsions, and/or unconsciousness (coma). Intravenous injection of glucose may be necessary to relieve these very dangerous symptoms.

Hyperventilation (alkalosis) (M12H4D)

The respiratory rate increases with strenuous exercise, to increase the availability of oxygen introduced into the lungs, while at the same time providing a ready means for expulsion of the carbon dioxide from the blood, via the lungs, increasingly produced by stepped-up metabolism. If the high respiration rate is not rapidly reduced following the termination of exercise, the continued high level uptake of oxygen and the continued high level of carbon dioxide elimination from the blood, without the balancing continued high level production of carbon dioxide from muscle metabolism, will cause a chemical imbalance to occur between oxygen and carbon dioxide in the blood. This may cause a transient alkalosis (depleted carbon dioxide) to develop which is commonly termed hyperventilation. The accompanying symptomatology includes: (1) the sensation of being "air hungry" (being unable to draw a satisfying breath), (2) dizziness or faintness, (3) nausea, (4) sensations of numbness, (5) sensations of tingling, (6) pounding of the heart, and (7) spasmodic muscle cramps (including of the muscles of the chest and lower abdomen).

The symptomatology of hyperventilation may be reduced by having the patient breathe into a paper bag for several minutes. As the carbon dioxide becomes concentrated in the bag, lung uptake of the carbon dioxide raises its level in the blood and the symptoms are relieved as the chemical imbalance between the oxygen and carbon dioxide is corrected. The symptoms of hyperventilation may additionally lead the patient to fear the onset of heart disease or other serious afflictions; verbal reassur-

Valsalva maneuver (M12H4E)

The patient should be cautioned not to hold the breath while performing strenuous exercise, because of the danger of performing a Valsalva maneuver. A Valsalva maneuver is defined as a forced expiration against a closed glottis. It may occur following holding the breath while performing a strenuous isometric or isotonic exercise. The blood pressure rises at the outset of straining because the intrathoracic pressure is added to the pressure of the blood in the aorta. It then falls because the high intrathoracic pressure compresses the veins, decreasing venous return and cardiac output. The decrease in arterial pressure and pulse pressure may inhibit the baroreceptors to cause tachycardia and a rise in peripheral resistance. When the glottis is again open, the intrathoracic pressure will return to normal, restoring cardiac output without relieving peripheral vessel constriction; the blood pressure may then rise above normal. Eventually, the high blood pressure rise will stimulate baroreceptors to cause bradycardia and a drop in pressure to normal levels. So, the Valsalva maneuver may be a threat to the patient's well-being by posing the danger of cardiac malfunction as a result of internal attempts at system correction, as well as contributing to the possibility of cerebral vascular accident with the shifts in arterial pressure.

Acidosis (M12H4F)

If carbon dioxide elimination through respiration does not keep pace with carbon dioxide production during strenuous exercise, a transient systemic acidosis will be produced as a carbon dioxide excess builds up in the blood. The characteristic symptoms produced by acidosis is drowsiness, which may develop in extreme conditions into stupor and coma. If unrelieved spontaneously, treatment utilizing a mechanical respirator may be indicated.

Pulse rate regulation (M12H4G)

To avoid exhaustion, the pulse rate of a patient engaged in strenuous exercise should be monitored and the exercise curtailed when the rate exceeds the heart's ability to maintain an adequate volume. As a general rule, exercise should be stopped when the patient's pulse exceeds the rate maximum, which is established by subtracting the patient's age from 220 (plus or minus 10) and multiplying the result by 0.85.

General contraindications (M12H4H)

Therapeutic exercise is contraindicated for patients suffering from any depleting disease or neoplasm, certain pathological heart conditions, active systemic acute infections (bacterial or viral), acute inflammation, new fractures (post five weeks or less) or surgical orthopaedic correctional procedures, acute joint sprains, acute joint strains, and tuberculosis.

Stretch and spasticity (M12H4I)

Any attempt to stretch muscles across individual joints, whether manually or by bracing or splinting, does not change the condition of spasticity, but only shifts it to other joints, limiting the ability to make postural adjustments and movements elsewhere. Any accompanying pain or discomfort tends to increase spasticity throughout the involved extremity.

Sensory stimulation (M12H4J)

There is a danger of over-stimulation when any of the sensory or proprioceptive facilitatory or inhibitory procedures noted above are utilized. Soft tissues may be irritated or injured, and a risk exists of sensory elements accommodating to stimulation, thereby defeating the procedure's beneficial effect.

Accommodation dangers (M12H4K)

Patients with mild to moderate hypertonus who have considerable voluntary motion may have a tendency to use abnormal patterns (stereotypic pathological synergistic or developmental reflexes) to perform functional activities, thus reinforcing them. Such patterns are unsightly and lead to deformities such as equinus with internal rotation or hip flexion with lumbar lordosis. Once abnormal patterns are used successfully to perform functional activities they become well established, increasing the likelihood that they may not be brought under control.

Although some abduction is desirable in the lower extremities as a joint position for lateral stability and as a base for the muscle action required for lateral weight shifting and maintaining balance, if the early linkage of flexion with abduction is too strong, the ensuing flexor pattern will interfere with standing if the legs are abducted too far apart.

In the treatment of the post CVA syndrome, in the acute stage, attempts to train a sound upper extremity to substitute for the involved upper extremity for self-help activities should be discouraged so that as much relearning as possible may take place, in respect to the use of the involved upper extremity, before progress is halted by adaptive compensatory actions of the uninvolved side.

Gait training   (M12H4L)

Gait training should not be attempted until balance reactions have been developed in standing postures and in the forward step positions. Early steps may be facilitated as automatic movements to assist in the recovery of lost balance.

Notes Aside   M12H5)

Wrist and elbow flexion spastic pattern   (M12H5A)

To defeat the wrist and elbow flexion spastic pattern, the head and thoracic spine should be extended, then the pectoralis muscles should be inhibited by shoulder external rotation which will decrease the whole pattern of spasticity so that the wrist and elbow may be extended.

Athetoid cerebral palsy patients   (M12H5B)

Athetoid cerebral palsy victims are less likely to have perceptual motor problems than spastic ones. Owing to the intermittent state of abnormal tone in athetosis, there are moments of normal tone when the patient may experience the sensory feedback from normal movement patterns.

Hyperkinetic children   (M12H5C)

Hyperkinetic children may be placed in total reflex inhibiting postures and held for a considerable time while engaged in mental or communicative activity, which may help them to gain control of their physical movements which seem to be associated with short attention spans. Drills which are comprised of precise commands and games may help them to organize and control sequential movement. Immature patterns in equilibrium reactions and persistence of righting reactions often characterize their motor behavior. As gross motor control improves, mental function has reportedly improved.

Abnormal postural patterns   (M12H5D)

Certain postural patterns are always abnormal. In lower extremities, adduction with internal rotation is always abnormal, whether combined with flexion or extension; equinus with internal rotation of the hip is another example of pattern interference with propulsion at the end of stance and with heel strike.

Bibliography   (M12H6)

Blakely, pp. 1224–1227.
Dickstein, pp. 1233–1238.
Markos, pp. 1366–1373.
Page, pp. 816–837.
Perry, pp. 789–815.
Pink, pp. 1158–1162.
Prevost, pp. 228–232.
Semans, pp. 732–788.
Stockmeyer, pp. 900–961.
Sullivan, pp. 283–288, 1980.
Surburg, pp. 1413–1517.
Voss, pp. 838–899.

EXERCISE FUNCTIONAL TESTING (M12I)
EVALUATION OF JOINT RANGE OF MOTION (M12I1)

Definition (M12I1A)

An anatomical joint is the site of juncture, or union, between two or more bones. Of interest here are the joints which allow motion of one or more of its component parts. The joint is generally described as being made up of two lever arms, a stable arm and a mobile arm (each a bone structure), related to each other via a central axis. The measurement of how much motion the joint allows is made in terms of degrees of deviation existing between the mobile arm and the stable arm relative to their common axis; this is called the joint range of motion.

Joint classification (M12I1A01)

Joints may be classified according to the number of planes of motion they afford:

Uniplane (single plane) joints, like the knee and elbow, are true hinge joints. They have a natural range of motion for the mobile arm in only one direction, beginning from an ideal zero starting point and flexing to the closest possible proximity to the stable arm, while conversely providing for extension from maximum flexion back to the zero point (flexion and extension). Some joints of this type have what is sometimes referred to as an unnatural range of motion which takes the mobile arm from zero and extends it in the opposite direction from its natural flexion range into what is called hyperextension.

Biplane (two plane) joints, like the wrist, have two natural ranges of motion, in two different planes. The biplane joint provides the same flexion-extension range of motion of the uniplane joint while additionally providing movement of the mobile arm which begins from the ideal zero starting point and proceeds to move into another plane of deviation (as into abduction or adduction and back).

Compound (three dimensional or ball and socket type) joints like the shoulder and hip, have very complex ranges of motion from the zero point. These ranges may include flexion/extension, abduction/adduction, internal/external rotation, and various combinations and permutations of these possible ranges which may provide deviations into other planes of motion.

Procedures of Application (M12I1B)

Principles of joint range of motion evaluation (M12I1B01)

1) The method of measuring joint range of motion is based upon the Neutral Zero Method.

2) All joint ranges of motion are measured from the defined zero starting point, and the degrees of motion performed by the joint, usually provided by the mobile arm, are counted in the direction that the joint moves from the zero point to where mobile arm motion is halted.

3) The ideal extended anatomical position of an extremity is accepted as the zero point and not the 180 degree point.

4) When defining the "normal" range of motion for a given joint, the measured range should be compared (whenever possible) with its counterpart on the subject's contralateral side. The differences between the ranges of the two joints may be expressed in comparative degrees of motion or in the percentage of lost range of motion of one joint compared to that of the counterpart joint on the contralateral side.

5) If a contralateral extremity joint is unavailable for joint range comparison, the joint range of motion should be compared with the average joint range of motion established from measurement of the same joint belong to "normal" individuals of the same age and similar build.

6) Motion should be described as *active* (voluntarily performed by the subject using muscles associated with the joint) or passive (the range of motion performed by the tester, without assistance from the subject).

7) Extension is the term designated for the motion performed in the opposite direction from flexion. Hyperextension is the range of motion in the extension direction beyond the zero point. Hyperextension is often called an unnatural range of mo-

tion, but it is a normal range of motion for the wrist, elbow, shoulder, and knee joints.

8) Any limitation of joint range of motion should be simply described in terms of missing degrees of motion and any deformity that may be present.

9) If joint motion is painful, the description of range of motion should be put in terms of ranges of motion limited by pain, or ranges of motion which are pain free.

10) The term *ankylosis* may be used to denote complete loss of joint range of motion.

11) The use of a goniometer is considered optional by many, but accuracy would seem to demand its use since it is usually the only "objective" measurement of range of motion available.

12) Joint range of motion measurement should be accurately and clearly recorded.

13) A function of range is defined as the direction in which the mobile element of a joint is moved. As such, functions are usually considered to be a basic pair of directional movements: flexion/extension and abduction/adduction. Measurement of such function of range pairs generally assumes that the function which is described to occur from zero to the end of range is the opposite function of its converse function; for example, if the optimal range for flexion is 0 to 180 degrees, the optimal extension range will be 180 to 0 degrees.

Use of the goniometer (M12I1B02)

The goniometer is made up of two arms which are attached (riveted) to each other at a pivotal axis which is designed and premarked to register the angles formed by the two arms, from zero to 180 or 360°.

When performing goniometry measurement, the axis should be placed above the approximate site of the joint, with one arm extending in a line parallel with the stable arm (S.Arm) of the joint and the other extending in a line parallel with the mobile arm (M.Arm), in consistent alignment with the line of motion being measured.

Measurements should be made from the beginning of range, as close to the zero point as possible, to the maximum end range, and both measurements should be recorded.

Goniometry measurement always depends upon the skill and accuracy of the tester. Measurements taken by two testers of the same joint at relatively the same time are basically incomparable, since the two measurements will generally vary from one another. Range of motion of a particular joint should, over time, be measured by the same given tester if statements of change in the joint range of motion measurements are to be deemed relatively accurate.

Normal Joint Ranges of Motion (M12I1B03)

Joint	Type	Function	Goniometric Measurement
Shoulder	compound	flexion (+ internal rot.)	Axis : Acromium
			S.Arm : Head of the trochanter
			M.Arm : Lateral Epicondyle
			Range : 0 to 180°
		hyperextension	Axis : Acromium
			S.Arm : Head of the trochanter
			M.Arm : Lateral Epicondyle
			Range : 0 to 60°
		abduction (+ external rot.)	Axis : Head of Humerus
			S.Arm : Parallel to sagittal plane
			M.Arm : Olecranon Process
			Range : 0 to 180°
		abduction (+ int.rot.90°)	Axis : Head of Humerus
			S.Arm : Parallel to sagittal plane
			M.Arm : Olecranon Process
			Range : 0 to 90°
		hyperadduction (across body)	Axis : Acromium
			S.Arm : Parallel to sagittal plane
			M.Arm : Olecranon Process
			Range : 0 to 75°
		internal rot. (with elbow flexed to 90°, shoulder at 0°)	Axis : Olecranon Process
			S.Arm : Parallel to support surface
			M.Arm : Parallel to ulna
			Range : 0 to 90° (inward)

Joint	Type	Function	Goniometric Measurement
		external rot. (with elbow flexed to 90°, shoulder at 0°)	Axis : Olecranon Process S.Arm : Parallel to support surface M.Arm : Parallel to ulna Range : 0 to 90° (outward)
		internal rot. (shoulder abd. to 90°, elbow flexed to 90°)	Axis : Olecranon Process S.Arm : Parallel to support surface M.Arm : Parallel to ulna Range : 0 to 90° (inward)
		external rot. (shoulder abd. to 90°, elow flexed to 90°)	Axis : Olecranon Process S.Arm : Parallel to support surface M.Arm : Parallel to ulna Range : 0 to 90° (outward)
		horizontal add. (abd. to 90°, internal rot. to 90°)	Axis : Acromium S.Arm : Parallel to support surface M.Arm : Parallel to humerus Range : 0 to 135°
		horiz. hyperabd. (abd. to 90°, internal rot. to 90°)	Axis : Acromium S.Arm : Parallel to support surface M.Arm : Parallel to humerus Range : 0 to 50°
Elbow	uniplane	flexion	Axis : Lateral Humeral Epicondyle S.Arm : Parallel to humerus M.Arm : Parallel to radius Range : 0 to 150°
		hyperextension (highly variable)	Axis : Lateral Humeral Epicondyle S.Arm : Parallel to humerus M.Arm : Parallel to radius Range : 0 to 10°
Forearm	uniplane	supination (thumb up, perform external rotation)	Axis : Shaft of the forearm S.Arm : Parallel to the humerus M.Arm : Parallel to volar wrist Range : 0 to 90°
		pronation (thumb up, perform internal rotation)	Axis : Shaft of the forearm S.Arm : Parallel to the humerus M.Arm : Parallel to volar wrist Range : 0 to 90°
Wrist	Biplane	dorsiflexion (from neutral, fingers slightly flexed)	Axis : Carpal region of the wrist S.Arm : Parallel to the ulna M.Arm : Parallel to 5th metacarpal Range : 0 to 80°
		palmar flexion (from neutral, fingers slightly flexed)	Axis : Carpal region of the wrist S.Arm : Parallel to the ulna M.Arm : Parallel to 5th metacarpal Range : 0 to 80°
		radial deviation (wrist neutral)	Axis : Carpal bones S.Arm : Line between radius & ulna M.Arm : Parallel 3rd metacarpal Range : 0 to 20°
		ulna deviation (wrist neutral)	Axis : Carpal bones S.Arm : Line between radius & ulna M.Arm : Parallel 3rd metacarpal Range : 0 to 40°

Joint	Type	Function	Goniometric Measurement
Finger metacarpo-phalangeal (M.P.)	biplane	flexion	Axis : Just dorsal to M.P. joint S.Arm : Parallel metacarpal dorsum M.Arm : Parallel proximal phalange Range : 0 to 90°
		ulnar deviation	Axis : M.P. joint S.Arm : Parallel to metacarpal M.Arm : Parallel proximal phalange Range : 0 to 20° (5th 0 to 30°)
index & 3rd M.P.		radial deviation	Axis : M.P. joint S.Arm : Parallel metacarpal M.Arm : Parallel proximal phalange Range : 0 to 20°
proximal interpha-langeal (PIP)	uniplane	flexion	Axis : Just dorsal to PIP joint S.Arm : Parallel proximal phalange M.Arm : Parallel to middle phalange Range : 0 to 110°
distal interpha-langeal (DIP)	uniplane	flexion	Axis : Just dorsal to DIP joint S.Arm : Parallel middle phalange M.Arm : Parallel to distal phalange Range : 0 to 90°
Thumb carpal-metacarpal (saddle)	compound	palmar abduction (circumduction)	Axis : Base of the saddle joint S.Arm : Parallel to 2nd metacarpal M.Arm : Parallel to 1st metacarpal Range : 0 to 50°
		abduction	Axis : Base of the saddle joint S.Arm : Parallel to 2nd metacarpal M.Arm : Parallel to 1st metacarpal Range : 0 to 60°
metacarpal interpha-langeal (M.P.)	uniplane	flexion	Axis : M.P. joint S.Arm : Parallel to 1st metacarpal M.Arm : Parallel proximal phalange Range : 0 to 50 to 60°
interpha-langeal (I.P.)	uniplane	flexion	Axis : I.P. joint S.Arm : Parallel proximal phalange M.Arm : Parallel to distal phalange Range : 0 to 80°
Hip	compound	flexion (with knee flexed)	Axis : Greater Trochanter S.Arm : Parallel to support surface M.Arm : Parallel to femur Range : 0 to 125°
		flexion (with knee straight)	Axis : Greater Trochanter S.Arm : Parallel to support surface M.Arm : Parallel to femur Range : 0 to 90°
		hyperextension	Axis : Greater Trochanter S.Arm : Parallel to support surface M.Arm : Parallel to femur Range : 0 to 15°
		abduction	Axis : Ant. Superior Iliac Crest S.Arm : Line between ant. crests M.Arm : Parallel to femur Range : 0 to 45°

Joint	Type	Function	Goniometric Measurement
		hyperadduction	Axis : Ant. Superior Iliac Crest
			S.Arm : Line between ant. crests
			M.Arm : Parallel to femur
			Range : 0 to 35°
		internal rot. (from neutral, foot drawn outward)	Axis : Base of patella
			S.Arm : Parallel to support surface
			M.Arm : Parallel to femur
			Range : 0 to 45°
		external rot. (from neutral, foot drawn inward)	Axis : Base of patella
			S.Arm : Parallel to support surface
			M.Arm : Parallel to femur
			Range : 0 to 45°
Knee	uniplane	flexion	Axis : Lateral Epicondyle of femur
			S.Arm : Parallel to femur
			M.Arm : Parallel to fibula
			Range : 0 to 140°
		hyperextension	Axis : Lateral Epicondyle of femur
			S.Arm : Parallel to femur
			M.Arm : Parallel to fibula
			Range : 0 to 10°
Ankle	uniplane	dorsiflexion	Axis : Just distal-lat. malleolus
			S.Arm : Parallel to fibula
			M.Arm : Parallel to sole of foot
			Range : 0 to 25°
		plantar flexion	Axis : Just distal-lat. malleolus
			S.Arm : Parallel to fibula
			M.Arm : Parallel to sole of foot
			Range : 0 to 45°
Fore foot	uniplane	inversion	Axis : Line between malleoli
			S.Arm : Parallel to tibia
			M.Arm : Parallel to 3rd metatarsal
			Range : 0 to 40°
		eversion	Axis : Line between malleoli
			S.Arm : Parallel to tibia
			M.Arm : Parallel to 3rd metatarsal
			Range : 0 to 20°
Hind foot (subtalar)	uniplane	inversion	Axis : Line between malleoli
			S.Arm : Parallel to tibia
			M.Arm : Parallel - Achilles tendon
			Range : 0 to 5°
		eversion	Axis : Line between malleoli
			S.Arm : Parallel to tibia
			M.Arm : Parallel - Achilles tendon
			Range : 0 to 5°
Great Toe metatarsal-phalangeal (M.P.)	uniplane	plantar flexion	Axis : M.P. joint
			S.Arm : Parallel to Metatarsal
			M.Arm : Parallel to prox. phalange
			Range : 0 to 30 to 45°

Joint	Type	Function	Goniometric Measurement
		dorsiflexion	Axis : M.P. joint S.Arm : Parallel to Metatarsal M.Arm : Parallel to prox. phalange Range : 0 to 50 to 80°
interpha- langeal (I.P.)	uniplane	plantar flexion	Axis : I.P. joint S.Arm : Parallel to Metatarsal M.Arm : Parallel to prox. phalange Range : 0 to 30 to 90°
Toes (2-5) metatarsal- phalangeal (M.P.)	uniplane	dorsiflexion	Axis : M.P. joint S.Arm : Parallel to metatarsal M.Arm : Parallel- proximal phalange Range : 0 to 40°
		plantar flexion	Axis : M.P. joint S.Arm : Parallel to metatarsal M.Arm : Parallel- proximal phalange Range : 0 to 30 to 40°
interpha- langeal (I.P.)	uniplane	plantar flexion	Axis : I.P. joint S.Arm : Parallel- proximal phalange M.Arm : Parallel to medial phalange Range : 0 to 40°
distal interpha- langeal (DIP)	uniplane	plantar flexion	Axis : DIP joint S.Arm : Parallel to medial phalange M.Arm : Parallel to distal phalange Range : 0 to 60°
Cervical Spine	compound	flexion	Axis : Sagittal & frontal cross S.Arm : Parallel to vertical M.Arm : Parallel C7 to caudal apex Range : 0 to 45°
		hyperextension	Axis : Sagittal & frontal cross S.Arm : Parallel to vertical apex M.Arm : Parallel C7 to caudal apex Range : 0 to 45°
		lateral flexion (midline toward one shoulder)	Axis : C7 posterior process S.Arm : Parallel to vertical plane M.Arm : C7 post.process-caudal apex Range : 0 to 45°
		rotation (midline to one side)	Axis : Sagittal & frontal cross S.Arm : Parallel to frontal plane M.Arm : Parallel line ear to ear Range : 0 to 60°
Thoracic & Lumbar Spine	compound	flexion	Axis : Greater Trochanter S.Arm : Parallel to femur M.Arm : Parallel line to C7 Range : 0 to 90°
		hyperextension	Axis : Greater Trochanter S.Arm : Parallel to femur M.Arm : Parallel line to C7 Range : 0 to 30°
		lateral flexion	Axis : Cross midline L-S joint S.Arm : Parallel to frontal plane M.Arm : L-S joint to C7 Range : 0 to 35°

exercise functional testing

Joint	Type	Function	Goniometric Measurement
		rotation (midline to one side)	Axis : Sagittal & frontal cross S.Arm : Parallel of frontal plane M.Arm : Parallel line ear to ear Range : 0 to 45°

Key:
S.Arm—stable arm
d.—degree
+—plus
rot.—rotation

horiz.—horizontal
L-S—lumbosacral
M.Arm—mobile arm
int.—internal

ext.—external
abd.—abduction
post.—posterior
C7—seventh cervical vertebrae posterior process

Partial List of Conditions Evaluated (M12l1C)

Treatable Causes (M12l1C01)

Calcific deposit (C08)
Extrafusal muscle spasm (C41)
Habitual joint positioning (C48)
Hypermobile/instable joint (C10)
Interspinous ligamentous strain (C03)
Joint ankylosis (C42)
Joint approximation (C47)
Ligamentous strain (C25)
Muscular weakness (C23)
Nerve root compression (C04)
Neuromuscular tonic imbalance (C30)
Peripheral nerve injury (C07)
Psychoneurogenic neuromuscular hypertonus (C29)
Restricted joint range of motion (C26)
Soft tissue inflammation (C05)
Soft tissue swelling (C06)

Syndromes (M12l1C02)

ADHESION (S56)
BURSITIS (S26)
CALF PAIN (S50)
CAPSULITIS (S27)
CARPAL TUNNEL (S35)
CERVICAL DORSAL OUTLET (SCALENICUS ATTICUS/CERVICAL RIB) (S09)
CERVICAL (NECK) PAIN (S73)
DUPYTREN'S CONTRACTURE (S34)
ELBOW PAIN (S47)
FACET (S19)
FOOT PAIN (S51)
FOREARM PAIN (S54)
FROZEN SHOULDER (S64)
HAND/FINGER PAIN (S17)
HIGH THORACIC BACK PAIN (S48)
JOINT SPRAIN (S30)
KNEE PAIN (S46)
LOW BACK PAIN (S03)
MYOSITIS OSSIFICANS (S67)
NEUROMUSCULAR PARALYSIS (S22)
OSTEOARTHRITIS (S21)
PIRIFORMIS (S04)
PITTING (LYMPH) EDEMA (S31)
POST CEREBRAL VASCULAR ACCIDENT (CVA) (S07)
POST IMMOBILIZATION (S36)
POST PERIPHERAL NERVE INJURY (S23)
POST SPINAL CORD INJURY (S24)
POST WHIPLASH (S55)
RADICULITIS (S29)
SCIATICA (S05)
SHOULDER/HAND (S57)
SHOULDER PAIN (S08)
TEMPOROMANDIBULAR JOINT (TMJ) PAIN (S06)
THIGH PAIN (S49)
TOOTHACHE/JAW PAIN (S41)
TORTICOLLIS (WRY NECK) (S65)
WRIST PAIN (S16)

Bibliography (M12l1D)

Pandya, pp. 1339–1342.

EVALUATION OF MUSCLE STRENGTH (M1212)

Definition (M1212A)

Muscle strength evaluation, commonly referred to as muscle testing, is the determination and grading of a muscle's ability to contract. Muscle testing must be performed in a manner which isolates the muscle's action from that of other muscles which have a similar function. Patient position (body part position relative to gravity), muscle line of pull, line of practitioner resistance, and joint ranges of motion affected by the contracting muscle are all important components of the process.

Procedures of Application (M1212B)

Muscle strength is graded on how much range of motion a muscle can move its respective joint(s) through, relative to how much resistance it is able to overcome in doing so.

Muscle strength grading (M1212B01)

Muscle strength grading is based upon the muscle's ability to complete or demonstrate a range of motion, relative to how much resistance is overcome in the performance:

Muscle grade	Motor task performed
Zero	No contraction of the muscle can be felt or observed when contraction is attempted by the patient (palpation of the muscle's belly or tendon is indicated)
Trace	A contraction of the muscle is felt during palpation but no joint motion is produced
Poor (−)	A contraction is produced which is sufficient to produce movement in the related joint(s) of a few degrees, with gravity eliminated and friction and other resistance factors reduced to near zero levels
Poor	A contraction is produced sufficient to perform full range of motion with gravity eliminated and friction reduced to minimum levels
Poor (+)	A contraction is produced sufficient to perform full range of motion with gravity eliminated but against slight resistance applied by the practitioner
Fair (−)	A contraction is produced sufficient to cause a movement of a few degrees of the range of motion against gravity
Fair	A contraction is produced sufficient to perform full range of motion against gravity; the contraction will be unsustainable against even minor resistance applied at the end of range of motion
Fair (+)	A contraction is produced sufficient to perform full range of motion against gravity and may be sustained against slight resistance applied by the practitioner
Good	A contraction is produced sufficient to perform full range of motion against gravity and may be sustained against moderate resistance applied by the practitioner
Normal	A contraction is produced sufficient to perform full range of motion against gravity and may be sustained against considerable resistance; to be graded normal, the strength of the contraction must at least match that of its counterpart on the contralateral side (if one is available), assuming the contralateral muscle to be "normal"

evaluation of muscle strength

Muscles and Testing Parameters   (M12I2B02)

Muscle(s)		Test Parameters
Sternocleido-mastoideus	Innervation:	C2, C3; spinal accessory
	Action:	1) *Cervical flexion*;
		2) *cervical rotation* toward the opposing side
	Non-gravity position:	Lying on involved side, head supported by tester
	Gravity position:	Back lying
	Resistance:	Applied to the forehead against flexion and rotation toward the uninvolved side or, with the neck flexed and rotated toward the opposite side, above the superior ear (the one on top); the lower thorax should be stabilized.
Upper trapezius	Innervation:	C3, C4; spinal accessory
Semispinalis capitis		Posterior rami of spinal nerves
Splenius capitis		Posterior rami of the middle
Splenius cervicis		and lower cervical nerves
	Action:	*Neck extension*
	Non-gravity position:	Side lying with head supported
	Gravity position:	Face lying with neck flexed over supporting table edge
	Resistance:	At the posterior base of the skull against extension; the upper thoracic and scapular areas should be stabilized
Rectus abdominis	Innervation:	Lower intercostal nerves
	Action:	*Anterior flexion of the thorax*
	Non-gravity position:	Back lying, knees and hips flexed to reduce lumbar lordotic curve; arms at sides
	Gravity position:	Back lying, knees and hips flexed to reduce lumbar lordotic curve; arms at sides
	Resistance:	Applied over sternum; knees stabilized
External oblique	Innervation:	Lower intercostals
Internal oblique		Lower intercostals; iliohypogastic and occasionally the ilioinguinal nerves
	Action:	Left external oblique and right internal oblique combine to produce *trunk rotation* to the left; right external oblique and left internal oblique produce trunk rotation to the right.
	Non-gravity position:	Sitting, pelvis stabilized
	Gravity position:	Back lying with pelvis and straight legs stabilized
	Resistance:	Obliquely to the shoulder being lifted
Iliocostalis dorsi	Innervation:	Adjacent spinal nerves
Longissimus dorsi		T12, L1, L2
Spinalis dorsi	Action:	*Trunk extension*
Iliocostalis lumborum	Non-gravity position:	Face lying, straight legs and pelvis stabilized
Quadratus lumborum	Gravity position:	Face lying, straight legs and pelvis stabilized
	Resistance:	Applied to area between the scapulae, pelvis stabilized
Quadratus lumborum	Innervation:	T12, L1, L2
Iliocostalis lumborum		Adjacent spinal nerves
	Action:	*Pelvis elevation*
	Non-gravity position:	Back lying, leg and pelvis supported by tester
	Gravity position:	Standing, thorax stabilized by holding lower rib area
	Resistance:	Ankle held and pulled in the direction of pelvis depression; the patient should be back lying

Muscle(s)		Test Parameters
Psoas major	Innervation:	L2, L3
Iliacus		L2, L3; femoral nerve
	Action:	*Hip flexion*
	Non-gravity position:	Side lying on side to be tested, uninvolved leg supported from the rear at the knee; the pelvis is supported by reaching across the patient's pelvis and slightly lifting the pelvis from under gluteus medius area; flexion of the lower leg is then attempted, allowing the knee to flex to eliminate hamstring resistance
	Gravity position:	Sitting with the knees over the edge
	Resistance:	Applied in a downward direction just proximal to the knee joint
Gluteus maximus	Innervation:	L4-S2; inferior gluteal nerve
Semitendinosus		L4-S3; sciatic (tibial) nerve
Semimembranosus	Action:	*Hip extension*
Biceps femoris (long head)	Non-gravity position:	Side lying on side to be tested, the lower leg flexed, with the uninvolved leg extended and supported from the rear at the knee, and the pelvis stabilized; extension of the lower leg is attempted
	Gravity position:	Face lying with legs extended, pelvis stabilized above the hip; knee may be flexed or extended
	Resistance:	Applied downward just proximal to flexed knee; pelvis stabilized over iliac crest
Gluteus medius	Innervation:	L4-S1; superior gluteal nerve
	Action:	*Hip abduction*
	Non-gravity position:	Back lying with legs extended, tested leg supported and involved side of the pelvis stabilized just proximal to hip joint
	Gravity position:	Side lying on the uninvolved side, pelvis stabilized just proximal to the hip joint
	Resistance:	Applied downward on the lateral aspect of the knee joint
Adductor magnus	Innervation:	L3,L4; obturator nerve
Adductor brevis		L2-L4; femoral nerve
Adductor longus	Action:	*Hip adduction*
Pectineus	Non-gravity position:	Back lying, involved leg supported
Gracilis	Gravity position:	Side lying with the tested leg on the bottom; uninvolved leg supported at the medial aspect of the knee and along the calf
	Resistance:	Applied downward, just above the knee on the medial aspect
Obturator externus	Innervation:	L3,L4; obturator nerve
Obturator internus		S1-S3
Quadratus femoris		L5,S1
Piriformis		S1,S2
Gemellus superior		S1-S3
Gemellus inferior		L5,S1
Gluteus maximus		L5-S2; inferior gluteal nerve
	Action:	*Hip external rotation*
	Non-gravity position:	Back lying with the leg internally rotated, pelvis stabilized on tested side just proximal to hip joint

evaluation of muscle strength

Muscle(s)		Test Parameters
	Gravity position:	Back lying with the tested knee dangling over the edge of the table and the uninvolved foot resting on the table surface; the thigh should be stabilized (but not resisted) just proximal to the knee
	Resistance:	Applied just proximal to the ankle in the lateral direction; thigh stabilized just proximal to the knee
Gluteus minimus	Innervation:	L4-S1; superior gluteal nerve
Tensor fasciae latae	Action:	*Hip internal rotation*
	Non-gravity position:	Back lying with the tested leg externally rotated; pelvis stabilized just proximal to the involved hip joint
	Gravity position:	Back lying with the tested knee dangling over the edge of the table and the uninvolved foot resting on the table surface; the thigh should be stabilized (but not resisted) just proximal to the knee
	Resistance:	Applied just proximal to the ankle in a medial direction; the thigh is stabilized just proximal to the knee
Sartorius	Innervation:	L2-L4; femoral nerve
	Action:	*Hip flexion, abduction and external rotation with knee flexion*
	Non-gravity position:	Back lying, the tested leg supported at lateral aspect of knee and under the calf, the pelvis supported just proximal to the involved hip joint
	Gravity position:	Sitting with the knees over the table edge, pelvis stabilized just proximal to involved hip joint
	Resistance:	Applied medially at the knee and laterally at the ankle
Tensor fasciae latae	Innervation:	L4-S1; superior gluteal nerve
	Action:	*Hip abduction with hip flexed*
	Non-gravity position:	Sitting on a table with the knees extended and with the trunk supported at a 45° angle with the hip, the pelvis stabilized just proximal to the involved hip joint
	Gravity position:	Side lying on the uninvolved side with the knees flexed, the tested hip at a 45° angle; the pelvis stabilized just proximal to the hip joint
	Resistance:	Applied in a downward direction just proximal to the involved knee joint; hip stabilized just proximal to the involved hip joint
Biceps femoris (long head)	Innervation:	L4-S3; sciatic (tibial) nerve L4-S2; sciatic(common peroneal)
Biceps femoris (short head)	Action:	*Knee flexion*
Semitendinosus Semimembranosus	Non-gravity position:	Side lying on the tested side, the knees extended with the uninvolved side supported at the medial knee and calf
	Gravity position:	Face lying, thigh stabilized just proximal to posterior aspect of the knee
	Resistance:	Applied just proximal to the heel in a horizontal distal direction
Rectus femoris	Innervation:	L2-L4; femoral nerve
Vastus intermedius	Action:	*Knee extension*

Muscle(s)		Test Parameters
Vastus medialis Vastus lateralis	Non-gravity position:	Back lying, the tested hip flexed to 45 degrees, the foot resting on the support surface, with the knee supported posteriorly; the uninvolved knee extended
	Gravity position:	Back lying with the tested knee over the table's edge and the knee stabilized just proximal to the involved knee joint
	Resistance:	Applied just proximal to the ankle joint in a downward direction, knee stabilized just proximal to the knee joint
Gastrocnemius Soleus	Innervation:	S1,S2; tibial nerve
	Action:	*Ankle flexion*
	Non-gravity position:	Side lying on the involved side, the tested ankle dorsiflexed
	Gravity position:	1) Face lying with the tested foot over the table edge, the calf supported over the belly of the gastrocnemius; 2) Standing
	Resistance:	1) Applied in a distal direction to the calcaneus, the calf stabilized over the belly of the gastrocnemius; 2) the body weight lifted on the ball of the foot
Anterior tibialis	Innervation:	L4-S1; deep peroneal nerve
	Action:	*Dorsiflexion and inversion of the foot*
	Non-gravity position:	Side lying with the tested foot resting on its lateral border and passively inverted, the ankle supported and the knee stabilized just proximal to the patella
	Gravity position:	Sitting with the legs dangling over the table's edge, the calf stabilized just proximal to the malleoli
	Resistance:	Applied to the dorsum of the foot in the distal direction, with the calf stabilized just proximal to the malleoli
Posterior tibialis	Innervation:	L5,S1; tibial nerve
	Action:	*Foot inversion from plantar flexion*
	Non-gravity Position:	Back lying with the foot over the table edge in plantar flexion, calf stabilized just proximal to malleoli
	Gravity Position:	Side lying on the involved side, with the foot plantar flexed and resting its lateral border on the support surface
	Resistance:	Applied to the dorsum of the foot over the first metatarsal tendon in a lateral direction
Peroneus longus Peroneus brevis	Innervation:	L4-S1; superficial peroneal N.
	Action:	*Foot eversion from plantar flexion*
	Non-gravity Position:	Back lying with the foot over the table edge in plantar flexion, calf stabilized just proximal to malleoli
	Gravity Position:	Side lying on the uninvolved side, with the foot plantar flexed and resting its medial border on the support surface
	Resistance:	Applied downward to lateral aspect of the fifth metacarpal
Lumbricals (1st)	Innervation:	S1,S2; medial plantar nerve
Lumbricals (2nd-4th)		S1,S2; Lateral plantar nerve
Flexor hallucis brevis		S1,S2; medial plantar nerve
	Action:	*Flexion of metatarsophalangeal joints of toes*

Muscle(s)		Test Parameters
	Non-gravity Position:	Back lying with the metatarsals stabilized just proximal to the plantar surfaces of their heads
	Gravity Position:	Back lying with the metatarsals stabilized just proximal to the plantar surfaces of their heads
	Resistance:	Applied horizontal in a proximal direction to the proximal phalange(s)
Flexor digitorum longus	Innervation:	L5,S1; tibial nerve S1,S2; medial plantar
Flexor digitorum brevis		L5-S2; tibial nerve
	Action:	*Flexion of the interphalangeal joints of the toes*
Flexor hallucis longus	Non-gravity Position:	Back lying with the proximal phalanges stabilized by pressure on their plantar surfaces
	Gravity Position:	Back lying with the proximal phalanges stabilized by pressure on their plantar surfaces
	Resistance:	Applied to the second row of phalanges with the proximal phalanges stabilized by pressure on their plantar surfaces
Extensor digitorum brevis	Innervation:	L4-S1; deep peroneal nerve
	Action:	*Extension of the metatarsophalangeal joints of the toes*
Extensor digitorum longus	Non-gravity Position:	Back lying, the metatarsals stabilized just proximal of the metatarsal heads
Extensor hallucis longus	Gravity Position:	Back lying, the metatarsals stabilized just proximal of the metatarsal heads
	Resistance:	Applied in a distal direction over the proximal phalanges
Dorsal interossei	Innervation:	S1,S2; lateral plantar nerve
Abductor hallucis		S1,S2; medial plantar nerve
Abductor digiti quinti		S1,S2; lateral plantar nerve
	Action:	*Toe abduction*
	Non-gravity Position:	Back lying
	Gravity Position:	Back lying
	Resistance:	Applied in a lateral direction to the medial aspect of one toe and simultaneously in a medial direction to the lateral aspect of the toe laterally adjacent
Plantar interossei	Innervation:	S1,S2; lateral plantar nerve
Adductor hallucis	Action:	*Toe adduction*
	Non-gravity Position:	Back lying
	Gravity Position:	Back lying
	Resistance:	Applied in a medial direction to the lateral aspect of one toe and simultaneously in a lateral direction to the medial aspect of the toe laterally adjacent
Serratus anterior	Innervation:	C5-C7; long thoracic nerve
	Action:	*Scapular abduction and upward rotation*
	Non-gravity Position:	Sitting at a shoulder high table with the shoulder flexed to 90 degrees and resting on the table; the thorax should be stabilized across the upper trapezius area
	Gravity Position:	Back lying with the shoulder flexed to 90 degrees and the scapulae resting on the table surface
	Resistance:	Applied in a downward direction along the line of the humerus
Upper trapezius	Innervation:	C3,C4; spinal accessory nerves
Levator scapulae		C3-C5; dorsal scapular nerve
	Action:	*Scapular elevation*

evaluation of muscle strength

Muscle(s)		Test Parameters
	Non-gravity Position:	Face lying with shoulders supported from underneath by the tester and forehead resting on the table
	Gravity Position:	Sitting with the arms down at the patient's sides
	Resistance:	Applied downward into the patient's shoulders
Middle trapezius	Innervation:	C3,C4; spinal accessory nerves
Rhomboid major		C4,C5; dorsal scapular
	Action:	*Scapular adduction*
	Non-gravity position:	Sitting with the shoulder flexed to 90 degrees, horizontally abducted to 45 degrees, and resting on a shoulder high table; the thorax stabilized across upper trapezius area of the opposite side
	Gravity position:	Face lying with the shoulder flexed to 90 degrees, abducted to 90 degrees, and hyperexternally rotated to 90 degrees; thorax stabilized over opposing rhomboid area
	Resistance:	Applied downward to the lateral angle and posterior aspect of the scapula
Lower trapezius	Innervation:	C3,C4; spinal accessory nerves
	Action:	*Scapular depression and adduction*
	Non-gravity position:	Face lying with the patient's forehead resting on the table and the tested shoulder flexed to 180 degrees; an attempt is made to lift arm straight up off table
	Gravity position:	Face lying with the patient's forehead resting on the table and the tested shoulder flexed to 180 degrees; the patient to lift arm straight up off the table
	Resistance:	Applied to the lateral angle of the scapula in a proximal (toward head) and outward direction
Rhomboid major	Innervation:	C4,C5; dorsal scapular
Rhomboid minor	Action:	*Scapular adduction and downward rotation*
	Non-gravity position:	Sitting with arms at sides and shoulders relaxed
	Gravity position:	Face lying with the arm internally rotated and resting on the back; opposite shoulder stabilized over opposite scapula
	Resistance:	Applied over the medial border of the scapulae in a lateral and downward direction
Anterior deltoid	Innervation:	C5,C6; axillary nerve
Coracobrachialis		C5-C7; musculocutaneous
	Action:	*Shoulder flexion to 90 degrees*
	Non-gravity Position:	Side lying on untested side with the tested shoulder slightly flexed and arm supported by an anteriorly positioned horizontal platform at shoulder level; scapula stabilized over the upper trapezius area
	Gravity Position:	Sitting with tested arm at the patient's side with the palm down; scapula stabilized over the upper trapezius area
	Resistance:	Applied downward just proximal to elbow; scapula stabilized over the upper trapezius area
Latissimus dorsi	Innervation:	C5-C8; subscapular nerve
Teres major		C5,C6; subscapular nerve
Posterior deltoid		C5,C6; axillary nerve
	Action:	*Shoulder hyperextension*

evaluation of muscle strength

Muscle(s)		Test Parameters
	Non-gravity Position:	Side lying on untested side with the tested shoulder slightly extended and arm supported by a horizontal platform positioned posteriorly at shoulder level; scapula stabilized over the upper trapezius area
	Gravity Position:	Face lying with the tested arm at the patient's side with the palm up; scapula stabilized over the upper trapezius area
	Resistance:	Applied downward just proximal to the elbow; scapula stabilized over the upper trapezius area
Middle deltoid	Innervation:	C5,C6; axillary nerve
Supraspinatus		C4-C6; suprascapular
	Action:	*Shoulder abduction to 90 degrees*
	Non-gravity Position:	Back lying with the arm at the patient's side; scapula stabilized over the upper trapezius area
	Gravity Position:	Sitting with the arm at the patient's side; scapula stabilized over the upper trapezius area
	Resistance:	Applied downward just proximal to the elbow; scapula stabilized over the upper trapezius area
Posterior deltoid	Innervation:	C5,C6; axillary nerve
	Action:	*Shoulder horizontal abduction*
	Non-gravity Position:	Sitting with the patient's arm supported at 90 degrees of flexion by a horizontal platform positioned at shoulder height which could provide support through the horizontal abduction range of motion; scapula stabilized over the upper trapezius area
	Gravity Position:	Face lying with the patient's shoulder abducted to 90 degrees and the forearm dangling passively over the table edge; scapula stabilized over the upper trapezius area
	Resistance:	Applied downward just proximal to the elbow; scapula stabilized over the upper trapezius area
Pectoralis major	Innervation:	C5-T1; thoracic anterior nerve
	Action:	*Shoulder horizontal adduction*
	Non-gravity Position:	Sitting with the patient's arm resting upon a shoulder height horizontal platform, positioned laterally, with the shoulder abducted to 90 degrees; scapula stabilized over the upper trapezius area
	Gravity Position:	Back lying with the patient's shoulder abducted to 90 degrees; scapula stabilized over the upper trapezius area
	Resistance:	Applied to the inside and just proximal to the elbow in a lateralward direction; scapula stabilized over the upper trapezius area
Infraspinatus	Innervation:	C4-C6; suprascapular nerve
Teres minor		C4,C5; axillary nerve
	Action:	*Shoulder external rotation*
	Non-gravity Position:	Face lying with the shoulder dangling vertically over table's edge; scapula stabilized over the upper trapezius area
	Gravity Position:	Face lying with the shoulder abducted to 90 degrees, arm supported by the table, forearm dangling vertically over table's edge; scapula stabilized over the upper trapezius area

evaluation of muscle strength

Muscle(s)	Test Parameters	
	Resistance:	Applied just proximal to the dorsal surface of the wrist, in a downward direction; the arm stabilized just proximal to the elbow
Subscapularis	Innervation:	C5-C8; subscapular nerve
Pectoralis major		C5-T1; thoracic anterior nerve
Latissimus dorsi		C5-C8; subscapular nerve
Teres major	Action:	*Shoulder internal rotation*
	Non-gravity Position:	Face lying with the shoulder dangling vertically over table's edge; scapula stabilized over the upper trapezius area
	Gravity Position:	Face lying with the shoulder abducted to 90 degrees, arm supported by the table, forearm dangling vertically over table's edge; scapula stabilized over the upper trapezius area
	Resistance:	Applied just proximal to the palmar surface of the wrist, in a downward direction; the arm stabilized just proximal to the elbow
Biceps brachii	Innervation:	C5,C6; musculocutaneous nerve
Brachialis	Action:	*Elbow flexion*
	Non-gravity Position:	Back lying with the shoulder externally rotated and abducted to 90 degrees; shoulder should be stabilized just distal to shoulder joint
	Gravity Position:	Sitting with forearm fully supinated and arm at side; shoulder stabilized over shoulder joint and over triceps areas
	Resistance:	Applied just proximal to the supinated wrist in an anteriorly-downward direction; the arm stabilized over the biceps
Triceps	Innervation:	C6-C8; radial nerve
	Action:	*Elbow extension*
	Non-gravity Position:	Back lying with the shoulder externally rotated and abducted to 90 degrees, elbow flexed; shoulder should be stabilized just distal to shoulder joint
	Gravity Position:	Face lying, shoulder abducted to 90 degrees, the arm supported on the table, with the forearm dangling over the table's edge; the arm should be stabilized just proximal to the elbow under the biceps area
	Resistance:	Applied just proximal to the dorsum of the wrist in a downward direction; the arm should be stabilized just proximal to the elbow under the biceps area
Biceps brachii	Innervation:	C5,C6; musculocutaneous nerve
Supinator		C5-C7; radial nerve
	Action:	*Forearm supination*
	Non-gravity Position:	Sitting with arm down at the patient's side, the elbow flexed to 90 degrees, forearm pronated and supported by the tester
	Gravity Position:	Sitting with arm down at the patient's side, the elbow flexed to 90 degrees; arm stabilized just proximal to the elbow joint and over distal triceps area
	Resistance:	Applied to the gripped supinated wrist in a pronating rotary direction; arm stabilized just proximal to the elbow joint and over distal triceps area

evaluation of muscle strength

Muscle(s)		Test Parameters
Pronator teres	Innervation:	C6,C7; median nerve
Pronator quadratus		C8,T1; median nerve
	Action:	*Forearm pronation*
	Non-gravity Position:	Sitting with arm down at the patient's side, the elbow flexed to 90 degrees, forearm supinated and supported by the tester
	Gravity Position:	Sitting with arm down at the patient's side, the elbow flexed to 90 degrees; arm stabilized just proximal to the elbow joint and over distal triceps area
	Resistance:	Applied to the gripped supinated wrist in a supinating rotary direction; arm stabilized just proximal to the elbow joint and over distal triceps area
Flexor carpi radialis	Innervation:	C6,C7; median nerve
Flexor carpi ulnaris		C8,T1; ulnar nerve
	Action:	*Wrist flexion*
	Non-gravity Position:	Sitting with the hand resting on its lateral border, tester supporting the forearm under and just proximal to the wrist
	Gravity Position:	Sitting with the supinated forearm resting upon the table; forearm stabilized distal to elbow
	Resistance:	Applied to palm of the hand in a downward direction; forearm stabilized just proximal to wrist
Extensor carpi radialis longus	Innervation:	C6-C8; radial nerve
	Action:	*Wrist extension*
Extensor carpi radialis brevis	Non-gravity Position:	Sitting with the hand resting on its lateral border, tester supporting the forearm under and just proximal to the wrist
Extensor carpi ulnaris	Gravity Position:	Sitting with the pronated forearm resting upon the table; forearm stabilized proximal to wrist
	Resistance:	Applied to dorsum of hand in a distal-downward direction; forearm stabilized proximal to wrist
Lumbricales (medial 2)	Innervation:	C8,T1; ulnar
Lumbricales (lateral 2)		C8,T1; median
Dorsal interossei		C8; ulnar
Palmar interossei	Action:	*Flexion of the Metacarpophalangeal joints of the fingers*
	Non-gravity Position:	Sitting with the hand at the zero point of forearm rotation and supported by the tester's fingers and thumb clamping the lateral and medial edges of the hand across the dorsum
	Gravity Position:	Sitting with the dorsum of the hand resting supinated on the table surface; stabilized by the tester's fingers and thumb clamping the dorsal and palmar surfaces of the metacarpal area between them
	Resistance:	Applied to the palmar surfaces of the proximal row of phalanges in a distal direction; the hand stabilized by the tester's fingers and thumb clamping the dorsal and palmar surfaces of the metacarpal area between them
Flexor digitorum sublimis	Innervation:	C7-T1; median nerve
		C7-T1; ulnar nerve

evaluation of muscle strength

Muscle(s)		*Test Parameters*
Flexor digitorum profundus	Action:	*Flexion of the Proximal and distal interphalangeal joints of the fingers*
	Non-gravity Position:	Sitting with palm up and the dorsum of the hand resting on the table surface; each finger is tested separately, with first the proximal phalange stabilized by the tester along its medial and lateral surfaces and then the middle phalange is stabilized
	Gravity Position:	Sitting with palm up and the dorsum of the hand resting on the table surface; each finger is tested separately, with first the proximal phalange stabilized by the tester along its medial and lateral surfaces and then the middle phalange is stabilized
	Resistance:	Applied first to the palmer surface of the middle phalange, and then to the distal phalange in a distal direction; each finger is tested separately, with first the proximal phalange stabilized by the tester along its medial and lateral surfaces and then the middle phalange is stabilized
Extensor digitorum communis Extensor indicis proprius Extensor digiti quinti proprius	Innervation:	C6-C8; radial nerve
	Action:	*Extension of the metacarpophalangeal joints of the fingers*
	Non-gravity position:	Sitting with the hand supported at the forearm rotary zero point and the wrist in neutral, the tester clasping across the palmar aspect to support and stabilize the metacarpals with fingers and thumb, the patient's fingers flexed at the middle interphalangeal joints
	Gravity position:	Sitting with the forearm internally rotated to 90 degrees and wrist neutral, the hand supported and stabilized by the tester clasping across the palmar aspect to the metacarpals with fingers and thumb, the patient's fingers flexed at the middle interphalangeal joints
	Resistance:	Applied to the dorsal surface of the proximal row of finger phalanges; the hand supported and stabilized by the tester clasping across the palmar aspect to the metacarpals with fingers and thumb, the patient's fingers flexed at the middle interphalangeal joints
Dorsal interossei Abductor digiti quinti	Innervation:	C8,T1; ulnar
	Action:	*Finger abduction*
	Non-gravity Position:	Sitting with the palm resting on the table with the fingers fully adducted
	Gravity Position:	Sitting with the palm resting on the table with the fingers fully adducted; the hand should be stabilized with pressure applied above the carpal bone area
	Resistance:	Applied to lateral and medial aspects of adjacent fingers in respective medial and lateral directions; the hand should be stabilized with pressure applied above the carpal bone area
Palmar interossei	Innervation:	C8,T1; ulnar
	Action:	*Finger adduction*
	Non-gravity Position:	Sitting with the palm resting on the table surface, the fingers fully abducted

Muscle(s)		Test Parameters
	Gravity Position:	Sitting with the palm resting on the table with the fingers fully adducted; the hand should be stabilized with pressure applied above the carpal bone area
	Resistance:	Applied when the attempt is made to prize adjoining fingers apart
Flexor pollicis brevis	Innervation:	C7-T1; median N. (lateral portion) C8-T1; ulnar N. (medial portion)
	Action:	*Flexion of the metacarpophalangeal joint of the thumb*
	Non-gravity Position:	Sitting with the hand resting palm up on the table
	Gravity Position:	Sitting with the hand resting palm up on the table; the first metacarpal stabilized
	Resistance:	Applied to the palmar surface of the proximal phalange in a lateral direction; the first metacarpal stabilized
Flexor pollicis longus	Innervation:	C7-T1; median nerve
	Action:	*Flexion of the Interphalangeal joint of the thumb*
	Non-gravity Position:	Sitting with the hand resting palm up on the table
	Gravity Position:	Sitting with the hand resting palm up on the table; the first phalange stabilized
	Resistance:	Applied to the palmar surface of the distal phalange of the thumb in a lateral direction; the first phalange stabilized
Extensor pollicis brevis	Innervation:	C7,C8; radial nerve
	Action:	*Extension of the metacarpophalangeal joint of the thumb*
	Non-gravity Position:	Sitting with the hand resting palm up on the table
	Gravity Position:	Sitting with the hand resting palm up on the table; the first metacarpal stabilized
	Resistance:	Applied to the dorsal surface of the proximal phalange in a medial direction; the first metacarpal stabilized
Extensor pollicis longus	Innervation:	C6-C8; radial nerve
	Action:	*Extension of the interphalangeal joint of the thumb*
	Non-gravity Position:	Sitting with the hand resting palm up on the table
	Gravity Position:	Sitting with the hand resting palm up on the table; the first proximal phalanges stabilized
	Resistance:	Applied to the dorsal surface of the distal phalange in a medial direction; the proximal phalange stabilized
Abductor pollicis longus	Innervation:	C7,C8; radial nerve C7-T1; median nerve
Abductor pollicis brevis	Action:	*Abduction of the thumb*
	Non-gravity Position:	Sitting with the hand resting palm up on the table
	Gravity Position:	Sitting with the hand resting palm up on the table; the metacarpals and wrist stabilized
	Resistance:	Applied to the lateral border of the first phalange of the thumb; the metacarpals and wrist stabilized
Adductor pollicis obliquus	Innervation:	C8,T1; ulnar nerve
	Action:	*Adduction of the thumb*
Adductor pollicis transversus	Non-gravity Position:	Sitting with the hand resting palm up on the table
	Gravity Position:	Sitting with the hand resting on its medial edge, thumb up and metacarpals stabilized

Muscle(s)

Opponens pollicis
Opponens digiti
  quinti

Test Parameters

Resistance: Applied to the medial border of the first phalange in a lateral direction
Innervation: C8,T1; median nerve
C7-T1; ulnar nerve
Action: *Opposition of the thumb and fifth finger*
Non-gravity Position: Sitting with the hand resting palm up on the table
Gravity Position: Sitting with the hand resting palm up on the table
Resistance: Applied by attempting to pry the first and fifth metacarpals further apart

Partial List of Conditions Evaluated (M12I2C)

Treatable Causes (M12I2C01)

Calcific deposit (C08)
Extrafusal muscle spasm (C41)
Habitual joint positioning (C48)
Hypermobile/instable joint (C10)
Interspinous ligamentous strain (C03)
Joint ankylosis (C42)
Joint approximation (C47)
Ligamentous strain (C25)
Muscular weakness (C23)
Nerve root compression (C04)
Neuromuscular tonic imbalance (C30)
Peripheral nerve injury (C07)
Psychoneurogenic neuromuscular hypertonus (C29)
Restricted joint range of motion (C26)

Syndromes (M12I2C02)

ADHESION (S56)
BURSITIS (S26)
CALF PAIN (S50)
CAPSULITIS (S27)
CARPAL TUNNEL (S35)
CERVICAL DORSAL OUTLET (SCALENICUS ATTICUS/CERVICAL RIB) (S09)
CERVICAL (NECK) PAIN (S73)
DUPYTREN'S CONTRACTURE (S34)
ELBOW PAIN (S47)
FACET (S19)
FOOT PAIN (S51)
FOREARM PAIN (S54)
FROZEN SHOULDER (S64)
HAND/FINGER PAIN (S17)
HEADACHE PAIN (S02)
HIGH THORACIC BACK PAIN (S48)
JOINT SPRAIN (S30)
KELOID FORMATION (S38)
KNEE PAIN (S46)
LOWER ABDOMINAL PAIN (S62)
LOW BACK PAIN (S03)
MENSTRUAL CRAMPING (S60)
MIGRAINE (VASCULAR) HEADACHE (S18)
MYOSITIS OSSIFICANS (S67)
NEUROMUSCULAR PARALYSIS (S22)
OSTEOARTHRITIS (S21)
PIRIFORMIS (S04)
POST IMMOBILIZATION (S36)
POST PERIPHERAL NERVE INJURY (S23)
POST SPINAL CORD INJURY (S24)
POST WHIPLASH (S55)
RADICULITIS (S29)
SCIATICA (S05)
SHOULDER/HAND (S57)
SHOULDER PAIN (S08)
THIGH PAIN (S49)
TORTICOLLIS (WRY NECK) (S65)
WRIST PAIN (S16)

Bibliography (M12I2E)

Chusid, pp. 171–184.
Daniels, pp. 2–161.

EVALUATION OF FUNCTIONAL MOTOR SKILLS (M12I3)

Definition (M12I3A)

Physical growth may occur without a corresponding development of motor skills. The normal development of motor skills is achieved through the individual's neuromuscular experience and experimentation with movement and patterns of movement. Evaluation of functional motor skills may be made by comparing demonstrations of motor skills with skills thought to be appropriate for the patient's chronological age group. Failure to demonstrate functional motor skills thought appropriate for the age group may indicate a failure to develop normal neuromuscular control.

Sensory motor development sequence (M12I3A01)

Generally, normal sensorimotor development occurs in the following sequence:

1) Early movements involve activity of the whole organism.
2) Specific movement develops out of such generalized patterns. Neuromuscular integration must be learned and controlled.
3) Development of controlled motion of the head and trunk precedes such development in the extremities. Movement control of large muscle groups proximal to the spine precedes motor control of muscles occurring more distally.
4) Eye-hand coordination develops with the internal awareness of (a) the two sides of the body and their differences (which is necessary for balance, complex motor pattern, and direction sense development), (b) the spacial relationships between objects, and (c) the body image relative to spatial relations and position in space (orientation).

Procedures of Application (M12I3B)

To evaluate sensorimotor development the following functional assessment may be useful.

Assessment of functional skills (M12I3B01)

To evaluate basic skill development, an assessment of the following functional skills should be administered:

1) Balancing: resistance applied first to one side of the body and then the other, with varying frequency, duration of resistance and the force applied. This should first be administered with the subject sitting and then while standing.
2) Bending: the attempt to flex, extend, abduct, adduct, circumduct, evert, and invert all jointed body parts available.
3) Bouncing: the performance a series of ten short bipedal jumps.
4) Carrying: the lifting of various objects of varying weights in different positions relative to the subject's base of support.
5) Catching: the hands and eyes used in a coordinated fashion to grasp an object flying through space in the subject's vicinity.
6) Climbing: the arms and legs used to raise the body against the pull of gravity, as in ascending and/or descending a ladder or incline.
7) Crawling: self-propelling by using opposing hands and legs in a prone position
8) Creeping: self-propelling by using the opposing hands and legs in a hands and knees position.
9) Crouching: concerted flexion of the subject's ankles, knees, hips and trunk.
10) Dodging: the ability to stop and change direction upon visual and/or auditory cue.
11) Dribbling: the ability to repetitively bounce a ball on the floor and catch it while moving the entire body.
12) Falling: the ability to drop to the ground in a controlled manner.
13) Galloping: running with long steps.
14) Hanging: the ability to resist gravitational pull when suspended by the arms or legs.
15) Hitting: the ability to hit an external object with the hand.
16) Hopping: propulsion of the body up into the air from one foot and landing on the same foot.
17) Jumping: propulsion of the body into the air from one foot or both, landing on both feet simultaneously.
18) Kicking: the ability to hit an external object with the foot in a forward direction.
19) Leaping: propulsion of the body forward from one foot to land on the other.
20) Lifting: overcoming the gravitational pull of an object by exertions of muscular force.
21) Passing: throwing an object to another individual.
22) Pivoting: turning the body with one foot while standing on the other.
23) Pulling: moving an external object toward the self.
24) Punching: hitting an external object with a closed hand.

25) Pushing: repelling an external object.
26) Reaching: attempting to touch an object away from the body.
27) Rising: raising the center of gravity from lying to sitting positions.
28) Rocking: rhythmical shifting to the center of gravity in opposing directions.
29) Rolling: causing an object to spin across a horizontal surface.
30) Running: ambulation in which a phase occurs when both feet are simultaneously off the supporting surface.
31) Skipping: continuous forward motion in which a hop is followed by a leap.
32) Starting: overcoming inertia to set the body in motion.
33) Stopping: overcoming momentum to bring the body to rest.
34) Stretching: moving a joint to the end of its full range of motion.
35) Striking: using an implement to hit an external object.
36) Swinging: moving the suspended body back and forth.
37) Tagging: running, reaching, and hitting simultaneously.
38) Throwing: performing a pushing motion of the arm to impart force to a hand-held object, which flies away from the body when released at the end of the pushing range by the hand opening.
39) Tossing: a short throw (usually underhanded) of graded impetus.
40) Touching: voluntary contact of the finger(s) with an external surface to discern texture, size and shape.
41) Trotting: a relatively slow run.
42) Tumbling: a continuous transfer of body weight from one body part to another in a smooth fashion, as in rolling.
43) Twisting: rotation of a body part.
44) Vaulting: the transfer of body weight from the feet to hands to feet, moving in a continuous direction.
45) Walking: continuous locomotion of the body in an upright posture, in a linear direction, provided by the leg and foot sequential phases of swing, heel strike, stance and propulsion.
46) Stooping: Hip, knee and ankle flexion to lower the center of gravity over the base of support.
47) Holding: grasping an object and preventing its vertical downward motion.

Motor Development Scale (M12I3B02)

Ambulation

Age	Function
13 months	Walking
16.5 months	Walking sideways
18–24 months	Uniform rate of walking
31.3 months	Walking a straight line
37 months	Walking a one-inch path ten feet long
50 months	Graceful walking with a smooth transfer of weight
5–6 years	Walking in an adult manner

Stair Climbing

18–20 months	Ascends staircase with help
	Descends backwards or sit-slides downstairs
	Marking time (one stair at a time), precedes alternating feet
24 months	Ascends three steps without support, marking time (one step at a time)
28 months	Descends three steps without support, marking time
29 months	Ascends eleven steps without support, marking time
34 months	Descends eleven steps without support, marking time
29 months	Ascends three steps, alternate feet with support
48 months	Descends three steps, alternate feet with support
31 months	Ascends eleven steps, alternated feet with support
48 months	Descends eleven steps, alternated feet with support
31 months	Ascends three steps, alternated feet without support
49 months	Descends three steps, alternated feet without support
41 months	Ascends eleven steps, alternated feet without support
55 months	Descends eleven steps, alternated feet without support

Ladder Climbing

24 months	Marking time, cautiously descends a large ladder
27 months	Marking time, ascends a large ladder with difficulty
33 months	Marking time, ascends a large ladder with facility
34 months	Cautiously ascends a small ladder, alternating feet
38 months	Marking time, descends a large ladder with facility
38 months	Easily ascends a small ladder, alternating feet
45 months	Ascends a large ladder cautiously, alternating feet
47 months	Ascends a large ladder with facility, alternating feet
51 months	Cautiously descends a small ladder, alternating feet
53 months	Descends a small ladder, alternating feet with facility
56 months	Cautiously descends a large ladder, alternating feet
62 months	Descends a large ladder, alternating feet with facility

Jumping

	Jumping starts by dropping downward from a height
	The two-legged jump precedes the hop
24 months	Jumps down with help from an eighteen-inch high box
24 months	Jumps alone from a twelve-inch high box
28 months	Jumps off floor using both feet
31 months	Jumps alone from an eighteen-inch high box, one foot forward
32.1 months	Jumps from a chair seat
33 months	Jumps alone from an eight-inch box with feet together
34 months	Jumps alone from a twelve-inch box with feet together
36 months	Jumps with help from a height of 28 inches
37 months	Jumps alone from an eighteen-inch box with feet together
39.7 months	Jumps a distance of 24 inches.
41.5 months	Jumps over a rope less than 8 inches high.
43 months	Jumps alone from a 28-inch box with one foot forward
46 months	Jumps alone from a 28-inch box with feet together
48.4 months	Jumps a distance of 24 to 32 inches

Hopping

38 months	Hops 1 to 3 steps on both feet
40 months	Hops 4 to 6 steps on both feet
41 months	Hops 7 to 9 steps on both feet
42 months	Hops 10 or more steps on both feet
43 months	Hops 1 to 3 steps on one foot
46 months	Hops 4 to 6 steps on one foot
55 months	Hops 7 to 9 steps on one foot
60 months	Hops 10 or more steps on one foot

Ball Throwing

	Arm in the anterioposterior plane with no trunk rotation or weight transfer
	Arm and body movements occur in a horizontal plane, with hip and trunk rotation
	Arm is held upward and obliquely with whole body rotation
	Opposition of movement with a mature throwing pattern
30 months	A small or large ball 4–5 feet
33 months	A small ball 6–7 feet
43 months	A small ball 8–9 feet
52 months	A small ball 10–11 feet
53 months	A large ball 8–9 feet
57 months	A small ball 12–13 feet
63 months	A large ball 10–11 feet
65 months	A small ball 14–15 feet
72 months	A large ball 12–13 feet, a small ball 16–17 feet

Catching

Arms are initially held stiff with no adjustment to the moving ball

Arms and hands work as a unit to corral or basket the ball against the chest

Synchronization of the arms and hands with the ball

The body is moved to advantageous catching position

Partial List of Conditions Evaluated (M12I3C)

Treatable Causes (M12I3C01)

Extrafusal muscle spasm (C41)
Habitual joint positioning (C48)
Muscular weakness (C23)
Neuromuscular tonic imbalance (C30)
Pathological neuromuscular hypertonus (C28)

Pathological neuromuscular hypotonus (C33)
Pathological neuromuscular dyscoordination (C35)
Peripheral nerve injury (C07)
Restricted joint range of motion (C26)

Syndromes (M12I3C02)

CEREBRAL PALSY (S25)
CERVICAL DORSAL OUTLET (SCALENICUS ATTICUS/CERVICAL RIB) (S09)
HAND/FINGER PAIN (S17)
NEUROMUSCULAR PARALYSIS (S22)
POST CEREBRAL VASCULAR ACCIDENT (CVA) (S07)
POST IMMOBILIZATION (S36)
POST PERIPHERAL NERVE INJURY (S23)
POST SPINAL CORD INJURY (S24)
SHOULDER/HAND (S57)

Bibliography (M12I3D)

Arnheim, pp. 26–30.
Chusid, pp. 427–428.
Connolly, pp. 1505–1512.

EVALUATION OF AMBULATION SKILLS (M1214)

Definition (M1214A)

Ambulation is made up of coordinated lower extremity movements and functions which generally follow the sequence described as swing phase, heel strike (contact), double support phase, and push-off (propulsion). These coordinated movements and functions are designed to allow the human subject to maintain an erect posture and to provide for simultaneous propulsion of the body in a linear direction. Ambulation is a bit more complicated than an exercise of the lower extremities and, in fact, involves patterns and postures of all the moveable joints. For a more complete discussion of ambulation refer to Exercise, Gait Training [M12E]. To evaluate a patient's ambulation skills, the gait pattern must be compared against the gait pattern considered to be normal, and any abnormalities observed and noted.

Procedures of application (M1214B)

The evaluation of a patient's ambulation should take the following considerations into account.

Ambulation analysis check list (M1214B01)

1) Is drop foot present? A drop foot occurs as the result of weak dorsiflexors which fail, through weakness or flaccidly, to dorsiflex the foot during swing phase. A clear differentiation should be made between the true drop foot which occurs as the result of muscle hypotonicity (as in peripheral nerve injury and flaccid post CVA syndromes), and dynamic plantar flexion which occurs as the dorsiflexors are being overwhelmed by hypertonic (often spastic) plantar flexors (as in the spastic post CVA syndrome).

Drop foot often results in a steppage gait, which characteristically presents the anterior portion of the foot striking the floor first. This gait pattern is often associated with equinus deformity and accompanying genu recurvatum (especially when the flexors of the knee are paralyzed).

2) Pelvic levels should be assessed and any abnormality investigated. If one iliac crest is noticeably higher than the other, leg length from the point of the anterior superior iliac crest to the prominence of the medial malleolus should be comparatively measured. It is normal for legs to vary in length from one to three quarters of an inch; as the discrepancy widens, the chances of gait being affected, creating a limp, become greater.

3) Visual and palpation appraisal of muscle tone should be made with the aim of establishing any muscle atrophy that could be affecting the gait pattern.

4) The pelvis should be watched for evidence of weak hip abductors or hip extensors. Weak hip abductors characteristically allow the pelvis to drop on the opposite side during stance phase (the Trendelenburg gait). Weak hip extensors characteristically cause the patient to lean over in the direction of the involved hip during stance phase (the gluteus maximus limp).

5) The relative position of the feet to one another should be noted. The toed-in posture is abnormal for Caucasians without American Indian progenitors and the toeing-out posture is abnormal for Caucasians if it exceeds 15 to 20 degrees of external rotation of the hip during stance phase.

6) Scaring of the hip, leg, ankle or foot should be noted to establish any previous surgical procedure or traumatic accident which may have resulted in nervous or other soft tissue damage which may hamper normal function. Soft tissues should be inspected under and around scaring for adhesion to lower structures.

7) Comparison of arm swing should be made to establish discrepancies between the length of the arcs and any other regularity of motion. Such discrepancies may indicate irregularities in hip structure or function.

8) Compare leg stride lengths and timing; normally they are the same.

9) Shoulder levels should be checked to establish the presence of scoliosis or differences in leg lengths.

10) The width of the base of support should be inspected to help establish the presence of knee and hip joint structural and functional regularity. Is it too wide or too narrow?

11) Ascertain whether the pelvis tilts forward or backward; lordosis may be indicative of weak abdominal muscles, short hip flexor muscles, spinal column structural abnormality or hip joint malformation.

12) Knee structures should be observed for deformities such as genu recurvatum, hyperextension during stance phase and bow-leggedness.

13) The degree of kyphosis and lordosis of the spinal column should be noted along with the di-

rection and degree of head tilt (side to side, forward to back).

14) Pathological synergistic gait patterns should be noted and the influence of developmental reflexes and/or pathological synergies of the lower extremities determined (refer to Exercise, Evaluation of Developmental Reflexes [M12I6], Pathological Neuromuscular Synergistic Pattern Evaluation [M12I7], and/or Kinesiology of Ambulation [R004]).

Partial List of Conditions Evaluated (M12I4C)

Treatable Causes (M12I4C01)

Calcific deposit (C08)
Extrafusal muscle spasm (C41)
Habitual joint positioning (C48)
Hypermobile/instable joint (C10)
Joint ankylosis (C42)
Joint approximation (C47)
Ligamentous strain (C25)
Muscular weakness (C23)
Nerve root compression (C04)
Neuromuscular tonic imbalance (C30)
Pathological neuromuscular hypertonus (C28)
Pathological neuromuscular hypotonus (C33)
Pathological neuromuscular dyscoordination (C35)
Peripheral nerve injury (C07)
Restricted joint range of motion (C26)
Scar tissue formation (C15)

Syndromes (M12I4C02)

ACHILLES TENDONITIS (S52)
ADHESION (S56)
CALF PAIN (S50)
CEREBRAL PALSY (S25)
FACET (S19)
FOOT PAIN (S51)
JOINT SPRAIN (S30)
KELOID FORMATION (S38)
KNEE PAIN (S46)
LOWER ABDOMINAL PAIN (S62)
LOW BACK PAIN (S03)
NEUROMUSCULAR PARALYSIS (S22)
OSTEOARTHRITIS (S21)
PIRIFORMIS (S04)
POST CEREBRAL VASCULAR ACCIDENT (CVA) (S07)
POST IMMOBILIZATION (S36)
POST PERIPHERAL NERVE INJURY (S23)
POST SPINAL CORD INJURY (S24)
REFERRED PAIN (S01)
SCIATICA (S05)
THIGH PAIN (S49)

Bibliography (M12I4D)

Arnheim, pp. 28, 203–204, 220.
Brunnstrom, pp. 101–110.
Rasch, pp. 290–291, 386–387.

EVALUATION OF BALANCE SKILLS (M1215)

Definition (M1215A)

Balance (equilibrium) is the ability to physically maintain the head in space without external support, i.e., to hold the head up through the body's own mechanisms without having to have it propped up for us, whether in a sitting or standing position. The ability to balance depends upon a highly sophisticated integration of sensory (afferent) input from the various sense organs (golgi tendon organ, muscle spindle, etc.), the vestibular apparati (semicircular canals, utricle, and saccule), and the central nervous system efferent control of antagonistic supporting musculature.

Even the eyes may play a part in maintaining equilibrium (balance). Visual clues may provide a convenient (though not absolutely essential) means of spatial orientation. Indeed, nature has constructed the visual system to play a role in spatial orientation. The eye is constructed so that the rod cells are arranged in vertical and horizontal patterns. When stimulated, impulses from the rods are interpreted by the visual cortex in terms of vertical and horizontal planes to help the individual appreciate spatial relationships relative to the position of the head.

Obviously, the process of maintaining equilibrium is highly complex, and a full description of the process is beyond the scope of our discussion here. However, simply stated, when a subject is maintaining an unsupported head position in space, continuous impulses (the resulting coordinated response to integrated data originating from the sensory organs) are transmitted from the reticular formation via the spinal cord (primarily along reticulospinal and vestibulospinal tracts) to the supporting musculature with the purpose of maintaining appropriate tone. Changes in this tone depends upon changes in sensory input which varies as the physical environment and the body's relationship to it changes. Thus, when the "normal" subject begins to fall to one side, the supporting musculature on one side of the involved joints shortens while the opposing musculature lengthens (in an eccentric contraction) so that the head may maintain its position relative to the horizontal and vertical planes. This system is so sophisticated that vectors of motion may be imposed upon it (permutations of up and down, side to side and forward and back body movements, as in ambulating) and a continuum of adjustments will continue to occur to maintain head position.

Injury or impairment of the elements within the central nervous system, or of the sensory mechanisms responsible for the transmission of essential data to maintain equilibrium, may result in a loss of the ability to balance. This loss may be qualitative or quantitative, and the subject's ability to maintain a "heads up" position while semireclined, sitting (supported or unsupported), standing and/or walking, may be adversely affected to varying degrees, depending on the extent of injury and which mechanisms have been damaged. Balance testing is designed to evaluate an individual's ability to balance and the extent to which balance may have been affected by system damage.

Procedures of Application (M1215B)

Evaluation of vestibular phasic reflexes (M1215B01)

Vestibular phasic reflex testing is designed to test the subject's ability to compensate for sudden changes in spatial orientation. Simply stated, vestibular phasic reflexes may be tested by pushing the subject off balance and observing the subject's ability to recover.

This test of vestibular reactivity is performed as a series of sudden shoves from various directions and with varying degrees of force. The shoves should not be violent enough to throw the subject down or to snap the subject's neck. Similar reactions may be provoked by applying a steady pressure to the subjects upper trunk and suddenly removing it. The response should be tested with the pressure applied from different directions around the trunk.

Sitting If the subject is sitting, the head and back should be unsupported, with the arms and hands at the sides. A normal response to an imbalancing shove will be for the sitting subject to respond by reaching out the arm and hand to for steadying support.

Standing The standing subject should have the feet no more than six inches apart, arms at the sides and the head erect, and no part of the body supported. A normal response to a shove is for the subject to respond by stepping out in the direction of the shove and to use the force exerted by foot and leg to oppose momentum and to halt motion.

Abnormal responses If the subject is be unable to recover as efficaciously as a normal subject, taking longer to recover or not at all, the subject has demonstrated an abnormal response and, if further such testing is intended, measures should be taken to protect the subject from a fall to the supporting surface.

Testing the integrity of equilibrium mechanism (M1215B02)

The integrity of the equilibrium mechanism may be tested by having the subject stand perfectly still, with the eyes closed. Normally the subject will have no trouble maintaining position; but if the equilibrium mechanism (utricle function) has been impaired, the subject will be unable to maintain the body in this standing static posture, and will be seen to waver from side to side, and may even begin to fall.

Testing of bilateral semicircular canal function (M1215B03)

Semicircular canal function may be tested through the *Barany test*.

The subject should be placed in a Barany (swivel) chair and rotated rapidly. The subject should be asked to maneuver the head so that each pair of semicircular canals are placed successively in the horizontal plane of rotation. This may be accomplished by having the subject (1) flex the neck and hold the head forward and (2) angulation of the head on one side and then the other. After these head postures have been performed successively, the chair should be suddenly stopped and the resultant nystagmus of the eyes observed.

The normal response is for nystagmus to result, with the eyes moving as a slow component of nystagmus toward the direction the chair was rotated in and a fast correcting component in the opposite direction. Nystagmus should last for from ten to twenty seconds. Normally, the patient should report having the sensation of rotating in the direction opposite of the chair rotation. This test serves as a check on the semicircular canals on both sides of the head simultaneously.

Testing of unilateral semicircular canal function (M1215B04)

A small amount of ice water should be placed in one of the subject's ears. The cooling caused by the ice water should serve to increase the density of the fluid (endolymph) in the cooled semicircular canal, causing the endolymph to sink downward, resulting in slight movement of the fluid in the semicircular canal. This causes the normal subject to experience the sensation of rotating and also initiates nystagmus similar to that seen in Barany test described above.

Partial List of Conditions Evaluated (M1215C)

Treatable Causes (M1215C01)

Bacterial infection (C09)
Habitual joint positioning (C48)
Hypertension (C39)
Hypotension (C45)
Localized viral infection (C27)
Neuromuscular tonic imbalance (C30)
Pathological neuromuscular hypertonus (C28)
Pathological neuromuscular hypotonus (C33)
Pathological neuromuscular dyscoordination (C35)
Peripheral nerve injury (C07)
Tinnitus (C32)

Syndromes (M1215C02)

BACTERIAL INFECTION (S63)
BELL'S PALSY (S66)
CEREBRAL PALSY (S25)
DIABETES (S10)
EARACHE (S40)
HYPERTENSION (S43)
HYSTERIA/ANXIETY REACTION (S59)
NEUROMUSCULAR PARALYSIS (S22)
POST CEREBRAL VASCULAR ACCIDENT (CVA) (S07)
POST IMMOBILIZATION (S36)
POST PERIPHERAL NERVE INJURY (S23)
TINNITUS (S70)

Notes Aside (M1215D)

Epileptic subject reaction to the Barany test (M1215D01)

Historically some fears have been expressed that vestibulogenic seizures might result from the spinning integral to the Barany test. A clinical study of the EEG activity of seizure prone children five to fifteen years old subjected to such testing did not support such fears. In fact, evidence suggested that EEG activity associated with seizure activity was showed a significant decrease in the paroxysmal abnormalities of more then half the subjects tested and the others showed no change. It is still safe to say, however, that exceptions do occur and there may be a remote possibility that a vestibulogenic seizure might occur to the exceptional subject. It is, of course, advisable to restrict a seizure-prone subject's vision during and shortly after spinning to preclude the development of photogenic seizures; also, seizure prone subjects should be monitored for signs of hyperventilation during and shortly after spinning.

Bibliography (M12I5E)

Guyton, pp. 662–668.
Kantner, pp. 16–21, 1982.
Rivlin, pp. 100–103.

EVALUATION OF DEVELOPMENTAL REFLEXES (M12I6)

Definition (M12I6A)

When supraspinal neuromuscular dysfunction occurs, abnormal muscle tone may result which is organized in postural patterns made up of abnormal neuromuscular synergies, including developmental reflexes and associated and homolateral reactions (refer to Table of Developmental Reflexes [T007]).

Evaluation performed to establish the presence of dominating developmental reflexes is generally conducted upon subjects who are thought to be influenced by neuromuscular reflex activity which is influenced by developmental reflexes. These reflexes may be appropriately present in the infant if their dominance is demonstrated at the developmental stage appropriate for the age of the subject. If developmental reflexes are demonstrated to dominate or inordinately influence subject neuromuscular behavior at an inappropriate age (as an adult, for example), an abnormal central nervous condition may be postulated to exist.

Developmental reflexes play an integral role in the preliminary stages of many approaches to neuromuscular reeducation (rehabilitation) of victims of supraspinal structure injury and/or dysfunction, especially those approaches utilizing guided patterning (refer to Exercise, Guided Patterning [M12H]).

Procedures of Application (M12I6B)

Flexor Withdrawal (M12I6B01)

1) The patient should be placed in a supine position with the head in midline and the lower extremities fully extended.
2) The sole of the foot should be stimulated with a thumb stroke drawn from the heel base, across the plantar surface of the arch of the foot, to just proximal to the distal metatarsal heads (the ball of the foot).

Positive Response : An uncontrolled flexion pattern of the stimulated lower extremity and/or an apparent increase of the myoelectric (myogenic electrical) activity from the hamstring muscle group, hip flexor muscle group, tibialis anterior and long toe extensor muscles.

Extensor Thrust (M12I6B02)

1) The patient should be placed in a supine position with the head in midline.
2) The uninvolved lower extremity should be extended and the involved lower extremity hip and knee should be flexed to ninety degrees, respectively.
3) The sole of the foot on the involved side should be stimulated with a thumb stroke delivered across the arch, from the heel, across the arch to the base of the distal metatarsal heads.

Positive Response : An uncontrolled extension pattern of the flexed leg with marked plantar flexion and inversion of the ankle, and/or an increase of the myoelectric activity from the quadriceps muscle group, gluteus maximus, gastrocnemius, and posterior tibialis muscles.

Crossed Extension 1 (M12I6B03)

1) The patient should be placed in a supine position with the head in midline.
2) One lower extremity should be extended and the involved lower extremity hip and knee flexed to ninety degrees, respectively.
3) The opposite lower extremity should be manually flexed by the tester.
4) This procedure should be repeated with the lower extremities reversing roles.

Positive Response : An involuntary extension of the flexed lower extremity, and/or an increase of myoelectric activity from the quadriceps muscle group, gluteus maximus, gastrocnemius, and anterior tibialis muscles.

Crossed Extension 2 (M12I6B04)

1) The patient should be placed in a supine position with the head in midline.
2) The lower extremities should be extended, externally rotated, and abducted to fifteen degrees.
3) The medial surface of one of the thighs should be tapped by the tester.
4) This procedure should be repeated and the other thigh tapped.

Positive Response : The opposing hip (opposite from that tapped) will spontaneously adduct and internally rotate while the knee extends and the ankle plantar flexes and inverts (scissor response), and/or an increase of myoelectric activity from the quadriceps muscle group, hip adductor group, gluteus minimus, gastrocnemius and posterior tibialis muscles.

Tonic Labyrinthine Supine (M12I6B05)

1) The patient should be placed in a supine position with the head in midline.

2) The upper and lower extremities should be extended and an attempt should be made by the tester to manually and passively flex the major joints (elbow, hip and knee).

Positive Response : A spontaneous increase of extensor muscle tone in upper and lower extremities and/or an increase of myoelectric activity from the quadriceps muscle group, gluteus maximus, triceps, and wrist extensor muscles.

Tonic Labyrinthine Prone (M12I6B06)

1) The patient should be placed in a prone position with the head in midline.

2) The lower extremities should be extended while the upper extremities are brought up over the head with the elbows extended.

Positive Response : An involuntary spontaneous increase of muscle tone and/or an increase of myoelectric activity from the biceps muscle group, hamstring muscle group, wrist flexor, finger flexor, thumb flexor, and hip flexor muscles.

Positive Supporting Reaction (M12I6B07)

1) If the patient is a child, it should be held up in a standing position. The adult patient should be supported in a standing position.

2) The child should be bounced several times up and down on the soles of the feet. If the patient is an adult, the body weight should be involuntarily shifted from one foot to the other.

Positive Response : An involuntary extension (and often genu recurvatum) of the knee, plantar flexion and inversion of the ankle, and clawing (sustained flexion) of the toes, and/or an increase of myoelectric activity from the quadriceps muscle group, gastrocnemius, posterior tibialis, anterior tibialis, and the long and short toe flexor muscles.

Negative Supporting Reaction (M12I6B08)

1) If the patient is a child, it should be held up in a standing position. The adult patient should be supported in the standing position.

2) The child should be bounced up and down, several times on the soles of the feet. The adult should attempt to put the involved foot on the floor and to bear weight on it.

Positive Response : Involuntary flexion with internal or external rotation of the hip, flexion of the knee, and ankle plantar flexion and inversion, and/or an increase of myoelectric activity from the hip external rotator muscle group or gluteus minimus, hip flexor, gastrocnemius, posterior tibialis, and anterior tibialis muscles.

Asymmetric Tonic Neck (M12I6B09)

1) The patient should be placed in a supine position with the head in midline.

2) The upper and lower extremities should be extended and slightly abducted.

3) The patient's face should be voluntarily or involuntarily turned to one side.

Positive Response : Involuntary extension (or a spontaneous increase of extensor tone) of the upper and lower extremities with lower extremity internal rotation on the side the face is turned toward, with external rotation and abduction of the shoulder, and flexion of the elbow, hip and knee, inversion and plantar flexion of the ankle. On the extending side, an increase of myoelectric activity from quadriceps muscle group, triceps, gluteus minimus, and gluteus maximus muscles may occur, while on the flexing side myoelectric activity may increase from the hamstring muscle group, biceps muscle group, infraspinatus, teres minor, hip flexor, posterior tibialis, and anterior tibialis muscles.

Tonic Lumbar (M12I6B10)

1) The patient should be placed in a supine position with the head in midline.

2) The lumbar spine should be voluntarily or involuntarily rotated toward one side (the involved side if unilaterally affected).

Positive Response : Involuntary abduction and flexion of the upper extremity and lower extremity extension on the side the lumbar spine is rotated toward, and/or an increase of myoelectric activity from the affected biceps muscle group, quadriceps muscle group, middle deltoid, supraspinatus, gluteus maximus, gastrocnemius, posterior tibialis and anterior tibialis muscles.

Symmetrical Tonic Neck 1 (M12I6B11)

1) The patient should be sitting with the hips and knees flexed to ninety degrees, respectively, with the feet off the floor and the head in midline.

2) The patient should be flexed forward at the waist to full flexion of the hips while the neck is anteriorly flexed.

Positive Response : Involuntary flexion (or flexor muscle tone) of the elbow, wrist, fingers and thumb, with extension of the knee, plantar flexion and inversion of the ankle, and/or an increase of myoelectric activity from the biceps muscle group, quadriceps muscle group, wrist flexor, finger flexor, thumb flexor, gastrocnemius, posterior tibialis, and anterior tibialis muscles.

Symmetrical Tonic Neck 2 (M12I6B12)

1) The patient should be placed on a table or plinth in a prone position with the head in midline, or sitting and bending forward at the waist with the feet off the floor.

2) The upper extremities should be distended over the edge of the supporting platform if the patient is lying, or out over the knees if sitting.

3) The neck should be posteriorly flexed.

Positive Response : Involuntary flexion of the shoulder, extension of the elbow, wrist, finger and thumb, and flexion of the knee, and/or an increase of myoelectric activity from the hamstring muscle group, anterior deltoid, triceps, wrist extensor, finger extensor, thumb extensor, and hip flexor muscles.

Contralateral Associative Reaction (M12I6B13)

1) The patient should be placed in the position appropriate for the testing of a particular muscle's strength (refer to Exercise, Evaluation of Muscle Strength [M12I2]).

2) The patient should contract the same muscle on the opposing side or contralateral side, first without resistance and then against isometric resistance.

Positive Response : Involuntary contraction (or increased tone) of the muscle being tested, and/or an increase of myoelectric activity from the muscle tested.

Homolateral Associative Reaction (M12I6B14)

1) The patient should be placed in a position appropriate for the testing of a particular muscle's strength (refer to the Exercise, Evaluation of Muscle Strength [M12I2]).

2) The patient should attempt to contract a muscle in the same synergistic group on the same (homolateral) side, first without and then against isometric resistance.

Positive Response : An involuntary contraction of the muscle tested, and/or an increase of its myoelectric activity.

Partial List of Conditions Evaluated (M12I6C)

Treatable Causes (M12I6C01)

Habitual joint positioning (C48)
Muscular weakness (C23)
Neuromuscular tonic imbalance (C30)
Pathological neuromuscular hypertonus (C28)
Pathological neuromuscular hypotonus (C33)
Pathological neuromuscular
 dyscoordination (C35)
Peripheral nerve injury (C07)
Restricted joint range of motion (C26)

Syndromes (M12I6C02)

BELL'S PALSY (S66)
CEREBRAL PALSY (S25)
CERVICAL DORSAL OUTLET (SCALENICUS
 ATTICUS/CERVICAL RIB) (S09)
NEUROMUSCULAR PARALYSIS (S22)
POST CEREBRAL VASCULAR ACCIDENT
 (CVA) (S07)
POST IMMOBILIZATION (S36)
POST PERIPHERAL NERVE INJURY (S23)
POST SPINAL CORD INJURY (S24)
SHOULDER/HAND (S57)
TORTICOLLIS (WRY NECK) (S65)

Bibliography (M12I6D)

Brunnstrom, pp. 158–178.
Chusid, pp. 427–428.
Taylor, pp. 119–137.

EVALUATION OF PATHOLOGICAL NEUROMUSCULAR SYNERGISTIC PATTERNS (M12I7)

Definition (M12I7A)

Pathological synergistic neuromuscular patterns often occur as the result of supraspinal central nervous system insult (or damage) and are commonly seen to accompany spastic forms of the post cerebral vascular and cerebral palsy syndromes. Neuromuscular synergies are, in effect, groups of muscles combining to produce a stereotypic sequence of joint movements or a to maintain a pattern of muscle contraction. They are viewed by many to result from uninhibited primitive spinal cord reflex patterns and play a role in the motor behaviors exhibited when many of the developmental reflexes are activated. The evaluation of pathological synergistic patterns is dependent upon direct observation of joint movement and/or position, and/or the measurement of myoelectric activity produced by the muscles suspected of being part of such a synergy and their antagonists by electromyography or electromyometry; muscles which are prime movers in a pathological synergy will typically demonstrate higher then normal myoelectric (demonstrating hypertonicity) while their antagonists may demonstrate relatively low myoelectric activity (refer to Electromyometric Feedback, Electromyometric Evaluation [M18L]).

Procedures of Application (M12I7B)

Evaluation or assessing the influence of synergistic patterns is generally performed through visual observation of the patient's patterns of motion or extremity postures, or through the use of electromyometric evaluation of myoelectric activity from the muscles affected. Following are descriptions of the most common synergies (exceptions are seen) in terms of the joint functions involved and the prime moving muscles most responsible:

Upper extremity flexion synergy (M12I7B01)

Scapular adduction: middle trapezius, rhomboid major, rhomboid minor
Scapular elevation: upper trapezius, levator scapulae
Shoulder external rotation: infraspinatus, teres minor
Shoulder abduction: middle deltoid, supraspinatus
Elbow flexion: biceps brachii, brachialis
Forearm supination:biceps brachii, supinator
Wrist flexion: flexor carpi radialis, flexor carpi ulnaris
Finger flexion: lumbricales, dorsal interossei, flexor digitorum sublimis, flexor digitorum profundus
Finger adduction: palmar interossei
Thumb flexion: flexor pollicis brevis, flexor pollicis longus
Thumb adductor: adductor pollicis

Upper extremity extension synergy (M12I7B02)

Scapular abduction: serratus anterior
Shoulder internal rotation: subscapularis, latissimus dorsi, teres major
Shoulder adduction: pectoralis major, latissimus dorsi
Elbow extension: triceps
Forearm pronation: pronator teres, pronator quadratus
Wrist extension: extensor carpi radialis longus, extensor carpi radialis brevis, extensor carpi ulnaris

Lower extremity flexion synergy (M12I7B03)

Hip abduction: gluteus medius, sartorius, tensor fasciae latae
Hip flexion: psoas major and iliacus, sartorius
Hip external rotation: obturator externus, obturator internus, quadratus femoris, gemellus inferior, gemellus superior, piriformis, gluteus maximus, sartorius
Knee flexion: biceps femoris, semitendinosis, semimembranosus, sartorius
Ankle dorsiflexion: anterior tibialis, extensor digitorum longus, extensor hallucis longus
Toe dorsiflexion: extensor digitorum longus, extensor digitorum brevis, extensor hallucis longus

Lower extremity extension synergy (M12I7B04)

Hip adduction: adductor magnus, adductor longus, adductor brevis, pectineus, gracilis

Hip extension: gluteus maximus, semimembranosus, biceps femoris, semitendinosus
Hip internal rotation: gluteus minimus, tensor fasciae latae
Knee extension: rectus femoris, vastus intermedius, vastus medialis, vastus lateralis
Ankle plantar flexion: gastrocnemius, soleus
Ankle inversion: tibialis posterior
Toe plantar flexion: lumbricales, flexor hallucis brevis, flexor hallucis longus, flexor digitorum longus, flexor digitorum brevis

Partial List of Conditions Evaluated (M1217C)

Treatable Causes (M1217C01)

Habitual joint positioning (C48)
Joint ankylosis (C42)
Joint approximation (C47)
Muscular weakness (C23)
Neuromuscular tonic imbalance (C30)
Pathological neuromuscular hypertonus (C28)
Pathological neuromuscular hypotonus (C33)
Pathological neuromuscular dyscoordination (C35)
Restricted joint range of motion (C26)

Syndromes (M1217C02)

CEREBRAL PALSY (S25)
NEUROMUSCULAR PARALYSIS (S22)
POST CEREBRAL VASCULAR ACCIDENT (CVA) (S07)
POST IMMOBILIZATION (S36)
POST SPINAL CORD INJURY (S24)

Bibliography (M1217D)

Brunnstrom, pp. 7–22.
Taylor, pp. 146–147.

TABLE OF EXERCISE — ENERGY CONSUMPTION RATES (T013)

Activity	Kilocalories Consumed per Hour* (based upon the consumption rate of a person weighing 150 lbs.)
Badminton	350
Bicycling (5.5 mph)	210
Bicycling (13 mph)	660
Bowling	270
Canoeing (2.5 mph)	230
Climbing (hills at 100 ft/h)	490
Dancing	470–700
Dancing (square dancing)	350
Digging (hand shovel)	400
Driving an Automobile	120
Fencing	300
Gardening	220–350
Golf	210–300
Handball (or squash)	600
Horse-back riding	180–480
Horse-back riding (trotting)	350
Housework	180–240
Lawn mowing (hand mower)	270
Lawn mowing (power mower)	250
Lying down or sleeping	80
Ping pong (table tennis)	360
Rowing (2.5 mph)	300
Rowing (racing pace)	840
Running (10 mph)	800–1000
Sitting	100
Skating (ice skating 10 mph)	360–400
Skating (roller skating)	300–700
Snow skiing (10 mph)	600
Standing	140
Swimming (0.25 mph)	300–350
Swimming (0.50 mph)	600–700
Tennis	400–500
Walking (2.5 mph)	210
Walking (3.75 mph)	210–330
Water skiing	480
Wood chopping	400
Volleyball	210–350

*Consumption rates vary with the weight of the patient and the vigorousness of participation

Bibliography

Wade, pp. 202–204.

TABLE OF EXERCISES FOR ABDOMINAL MUSCLE TONING (T019)

Anterior Abdominal Muscle Strengthening (T019A)

(1) Lie in a supine position on the floor with the calves of the leg supported by a chair so that the hips and knees are in 90 degrees of flexion. The arms should be folded across the chest, and the chin should be tucked in.

(2) Raise the head and shoulders off the ground until the distal tips of the shoulder blades are one inch off the floor; do not try to sit all the way up.

(3) Repeat this exercise as rapidly as possible for the number of repetitions specified (usually fifty).

(4) After two weeks of performing this exercise on a daily basis, alternately lead with each shoulder toward the contralateral knee.

Repetitions _____

Back Extension Exercise (T019B)

WARNING : During isometric exercise the breath should *not* be held, under any circumstances. Holding the breath during exertion may provoke a Valsalva maneuver (a forced expiration against a closed glottis), which may cause tachycardia and a subsequent rise in blood pressure, followed by bradycardia and a sudden drop in blood pressure pressure. Such shifts in heart rate and blood pressure may precipitate cardiac malfunction or a cerebral vascular accident. The Valsalva maneuver may be avoided by the simple procedure of counting the seconds out loud during exertion.

(1) Lie in a prone position over two pillows (positioned under the pelvic area to reduce the lordotic curve) with the head turned to one side and the arms down at your sides.

(2) Lift the turned head and the anterior surfaces of the shoulders one inch above the surface of the floor, but do not arch the back; hold that position for six seconds.

(3) Allow the head and shoulders to relax down to the beginning position.

(4) Turn the head to the contralateral side and lift the turned head and the anterior surfaces of the shoulders one inch above the surface of the floor, but do not arch the back; hold that position for six seconds.

Repetitions_____

TABLE OF FACIAL MUSCLE EXERCISES (T010)

Each exercise listed below should be performed so that the muscular contraction is held for six seconds, and then followed by a six-second rest period to allow the musculature to relax. Each exercise should be performed ten times, several times a day (three times is often recommended).

It may be helpful for the very involved patient to use the fingers to push the appropriate facial skin in the direction of the muscle pull, while attempting to contract the appropriate musculature. After several attempts, the fingers should be taken away, just following the attempted contraction, and an attempt made to hold the contraction for a full six seconds.

A mirror may be helpful as a of visual feedback source in the home environment. The first evidence of success may occur with the observation of voluntary musculature relaxation. A voluntary contraction may develop so slowly that the contraction cannot be visually observed and the only evidence of it may come when it suddenly relaxes.

Facial Exercises

1) Raise the eyebrows and wrinkle the forehead.
2) Draw the eyebrows downward and toward one another in a frown.
3) Open the eyes as widely as is possible.
4) Wrinkle the nose.
5) Flare the nostrils, spreading them as wide as possible.
6) Lift and protrude the upper lip in a sneer.
7) Pucker the lips as if kissing and attempt to whistle.
8) Compress the lips together as firmly as possible.
9) Blow air into the cheeks while attempting to keep the lips compressed together.
10) Smile while keeping the lips compressed together.
11) Smile with the lips apart.
12) Wrinkle the point of the chin.

TABLE OF EXERCISES FOR THE LOWER EXTREMITIES — ISOMETRICS (T014)

General Considerations

Isometric exercise is generally performed twice daily, one set in the morning and one in the afternoon. The number of repetitions in each set generally varies between five and ten.

During an isometric contraction no joint motion should be allowed if muscle, tendon or joint ligament strains are to be avoided.

When performing an isometric contraction, the tension in the contracting muscle(s) should be increased slowly until a maximal contraction of the muscle is attained. The maximal contraction should be held with unremitting intensity for six seconds without holding the breath and then the muscle(s) should be slowly and carefully relaxed. The muscle(s) should be kept in a relaxed state for a full six seconds of rest, which allows an influx of blood into the exercised muscle fibers, before performing another isometric contraction with the same muscles.

WARNING: During isometric exercise the breath should *not* be held, under any circumstances. Holding the breath during exertion may provoke a Valsalva maneuver (a forced expiration against a closed glottis), which may cause tachycardia and a subsequent rise in blood pressure, followed by bradycardia and a sudden drop in blood pressure pressure. Such shifts in heart rate and blood pressure may precipitate cardiac malfunction or a cerebral vascular accident. The Valsalva maneuver may be avoided by the simple procedure of counting the seconds out loud during exertion.

Hip Abduction & Internal Rotation (T014A)

(1) Lying supine with the feet spread shoulder width apart (or closer together if comfort dictates), and a pillow or two placed under the knees to decrease the lumbar lordotic curve, a nonstretchable belt or strap be placed to encircle the ankles.

(2) The legs should be spread against the resistance of the restraining belt for six seconds. Care should be taken not to lift the feet or legs or to drive them down into the resting surface while performing the exercise.

(3) The hips and legs should be completely relaxed for six seconds before repeating the exercise.

Repetitions _____

Hip Abduction and External Rotation (T014B)

(1) Lying supine with the feet spread shoulder width apart (or closer together if comfort dictates), and a pillow or two placed under the knees to decrease the lumbar lordotic curve, a nonstretchable belt or strap should be placed to encircle the knees.

(2) The legs should be spread against the restraining resistance of the belt for six seconds. Care should be taken not to lift the feet or legs or to drive them down into the resting surface while performing the exercise.

(3) The hips and legs should be completely relaxed for six seconds before repeating the exercise.

Repetitions _____

Hip Adduction (T014C)

(1) In a sitting or semi-reclining position, place a pillow or two between the knees.

(2) Squeeze the knees together against the resistance of the pillow(s) for six seconds.

(3) Completely relax the hips for six seconds before repeating the exercise.

Repetitions _____

Quadriceps and Hamstring Muscle Groups (T014D)

(1) With the shoulders and buttocks resting up against a smooth wall (or door) surface, the feet should be parallel to one another and spread approximately six inches apart, six inches in front of the wall.

(2) While keeping the buttocks and shoulders against the wall, bend the knees and slide down the wall six to eight inches.

(3) Hold the squatting position for six seconds before sliding up and standing erect.

(4) Relax for six seconds before repeating the exercise.

Repetitions _____

Quadriceps Muscle Group (T014E)

(1) Sitting on a chair or other high platform, one leg should dangle over the edge while the other is

table of exercises for the lower extremities–isometrics

supported by a stool or other short platform, so that the knee is flexed from five to ten degrees. The ankle or foot supported by the stool may be weighted as is judged necessary. Five to ten pounds is most common, but it may be as much as fifty pounds.

(2) The knee should be straightened so that the foot is lifted from its supporting platform and held up for six seconds.

(3) The foot should be placed back on the low support and the thigh muscles relaxed for six seconds before repeating the exercise.

       Repetitions _____

Hamstring Muscle Group (T014F)

(1) Sitting in a chair or couch with a solid base board, place a pillow behind the heel.

(2) Pull the heel back against the resistance the pillow for six seconds.

(3) The hamstrings should be relaxed for six seconds before repeating the exercise.

       Repetitions _____

Gastrocnemius and Soleus Muscles (T014G)

(1) Stand erect with the knees straight, feet parallel and six inches apart. Use the wall or a stable table to maintain balance.

(2) Rise up on the balls of the feet and hold that position for six seconds.

(3) Lower the heels to the floor and relax the calf muscles for six seconds before repeating the exercise.

       Repetitions _____

Dorsiflexors of the Ankle (T014H)

(1) Standing with the knees straight, feet parallel and six inches apart, use the wall or a stable table to maintain balance.

(2) Lift the balls of the feet off the floor and hold that position for six seconds.

(3) Lower the balls of the feet to the floor and relax the dorsiflexors of the foot for six seconds before repeating the exercise.

       Repetitions _____

TABLE OF EXERCISES FOR THE NECK — ISOMETRICS (T015)

General Considerations

Isometric exercise is generally performed twice daily, one set in the morning and one in the afternoon. The number of repetitions in each set generally varies between five and ten.

During an isometric contraction no joint motion should be allowed if muscle, tendon or joint ligament strains are to be avoided.

When performing an isometric contraction, the tension in the contracting muscle(s) should be increased slowly until a maximal contraction of the muscle is attained. The maximal contraction should be held with unremitting intensity for six seconds without holding the breath and then the muscle(s) should be slowly and carefully relaxed. The muscle(s) should be kept in a relaxed state for a full six seconds of rest, which allows an influx of blood into the exercised muscle fibers, before performing another isometric contraction with the same muscles.

WARNING: During isometric exercise the breath should not be held, under any circumstances. Holding the breath during exertion may provoke a Valsalva maneuver (a forced expiration against a closed glottis), which may cause tachycardia and a subsequent rise in blood pressure, followed by bradycardia and a sudden drop in blood pressure pressure. Such shifts in heart rate and blood pressure may precipitate cardiac malfunction or a cerebral vascular accident. The Valsalva maneuver may be avoided by the simple procedure of counting the seconds out loud during exertion.

Lateral Rotation (T015A)

(1) Sitting upright, looking straight ahead, without projecting the chin forward (the head in neutral), place the palm of one hand against the ipsilateral facial cheek.

(2) Attempt to rotate the head against the resistance of the palm for six seconds, without allowing any rotational motion of the head.

(3) The neck muscles should be relaxed for six seconds before repeating the exercise.

Repetitions _____

Forward Flexion (T015B)

(1) Sitting upright, looking straight ahead, without projecting the chin forward (the head in neutral), one or both hands should be placed against the forehead.

(2) Attempt to tuck the chin, pushing the forehead against the resistance of the hand(s) for six seconds, without allowing any forward flexion of the neck.

(3) The neck muscles should be relaxed for six seconds before repeating the exercise.

Repetitions _____

Lateral Flexion (T015C)

(1) Sitting upright, looking straight ahead, without projecting the chin forward (i.e., the head held in neutral), place a hand just superior to (above) one ear.

(2) Attempt to laterally tilt the head toward the ipsilateral side against the resistance of the hand for six seconds, without allowing any lateral flexion of the neck.

(3) The neck muscles should be relaxed for six seconds before repeating the exercise.

Repetitions _____

Posterior Flexion (T015D)

(1) Sitting upright, looking straight ahead, without projecting the chin forward (i.e., the head in neutral), place the hands behind the head over the occiput.

(2) Push the head back against the resistance of the hands, without raising the chin (push straight back), for six seconds.

(3) The neck muscles should be relaxed for six seconds before repeating the exercise.

Repetitions _____

TABLE OF EXERCISES FOR THE SHOULDERS AND UPPER EXTREMITIES — ISOMETRICS
(T011)

General Considerations

Isometric exercise is generally performed twice daily, one set in the morning and one in the afternoon. The number of repetitions in each set generally varies between five and ten.

During an isometric contraction no joint motion should be allowed if muscle, tendon or joint ligament strains are to be avoided.

When performing an isometric contraction, the tension in the contracting muscle(s) should be increased slowly until a maximal contraction of the muscle is attained. The maximal contraction should be held with unremitting intensity for six seconds without holding the breath and then the muscle(s) should be slowly and carefully relaxed. The muscle(s) should be kept in a relaxed state for a full six seconds of rest, which allows an influx of blood into the exercised muscle fibers, before performing another isometric contraction with the same muscles.

WARNING: During isometric exercise the breath should *not* be held, under any circumstances. Holding the breath during exertion may provoke a Valsalva maneuver (a forced expiration against a closed glottis), which may cause tachycardia and a subsequent rise in blood pressure, followed by bradycardia and a sudden drop in blood pressure pressure. Such shifts in heart rate and blood pressure may precipitate cardiac malfunction or a cerebral vascular accident. The Valsalva maneuver may be avoided by the simple procedure of counting the seconds out loud during exertion.

Shoulder Adduction   (T011A)

(1) The hands should be clasped together, with overlapping thumbs, and held in close to the chest. The elbows should be pointed out.

(2) The palms of the hands should be pushed together for six seconds.

(3) The hands should be lowered into the lap; the arms and shoulders should be relaxed for six seconds before repeating the exercise.

Repetitions _____

Shoulder Abduction   (T011B)

(1) The arms should be folded and held down against the chest, while the elbows should be grasped and held by the hands.

(2) The elbows should be pushed out against the resistance of the hands for six seconds.

(3) The hands should be lowered into the lap. The arms and shoulders should be relaxed for six seconds before repeating the exercise.

Repetitions _____

Shoulder Elevation   (T011C)

(1) While sitting erect in an armless chair, the bottom of the chair seat should be grasped by the hands. The elbows should be pointed straight out from the body.

(2) An attempt should be made to lift the bottom of the chair, which coincidentally drives the body down into the chair seat. The attempt should last for six seconds.

(3) The hands and arms should allowed to relax and dangle at the sides for six seconds before repeating the exercise.

Repetitions _____

Shoulder Horizontal Adduction   (T011D)

(1) The palmar surfaces of the hands (including the fingers) should be placed in direct opposition, and held out from the body at shoulder level. The elbows should be slightly flexed.

(2) The hands should be pressed together for six seconds.

(3) The hands should be lowered into the lap, and the arms and shoulders allowed to relax for six seconds before repeating the exercise.

Repetitions _____

Finger Flexion   (T011E)

(1) The elbow should be flexed to approximately ninety degrees and the forearm supported.

(2) A hard baseball should be grasped between the fingers and thumb and squeezed for six seconds. The force applied to the ball by the fingers and thumb should be evenly distributed.

(3) The fingers and thumb should be relaxed for six seconds before repeating the exercise.

Repetitions _____

TABLE OF EXERCISES — ISOTONIC (WEIGHT LIFTING PROTOCOLS) (T012)

General Principles

Before beginning a program of weight lifting, the weight which may be safely lifted should be determined. The most appropriate weight with which to begin is generally established through a process of trial and error. A particular weight should be lifted using the correct form with the appropriate equipment. If the weight can be lifted with ease, but each successive lift is increasingly difficult and the tenth repetition is very difficult to perform, that is probably the correct starting weight.

Weight lifting should be performed as a series of repetitions, called a set. Generally, a set is considered to be a series of ten consecutive lifts. Three sets of a given exercise per session are thought to be ideal for muscle strengthening. A rest of several minutes should be take place between each set.

For best results, it is usually recommended that a particular weight lifting exercise (i.e., three sets) should be performed no more than three times a week, with at least a day between each session.

When ten repetitions can be easily performed, the weight should be increased by an amount determined by following the procedure initially used to establish appropriate weight (increases usually vary from five to twenty pounds).

While lifting, the patient should perform inspiration during the process of agonist muscular contraction (lifting the weight), completing the breath when the end of joint range has been reached. This approach helps prevent the patient from holding the breath. Expiration should be performed as the muscle is performing the eccentric contraction side of the exercise (lowering the weight).

WARNING : While doing a lift, don't hold your breath, under any circumstances. Holding the breath during exertion may provoke a Valsalva maneuver (a forced expiration against a closed glottis), which may cause tachycardia and a subsequent rise in blood pressure, followed by bradycardia and a sudden drop in blood pressure pressure. Such shifts in heart rate and blood pressure may precipitate cardiac malfunction or a cerebral vascular accident. The Valsalva maneuver may be avoided by making the inspiration of breath last from the beginning of the lift to when the top of the lift is reached.

Bench Press (T012A)

(1) The equipment required includes a weight lifting bench (approximately forty-eight inches long and twelve inches wide), a foot support (part of the bench or not), weight stand attached to the bench, and a barbell loaded with the requisite weights. A weight lifting machine and bench which is set up for bench pressing is ideal.

(2) The barbell having previously been placed on the weight stand with its load of weights, assume a supine position on the bench with the knees flexed to approximately forty-five degrees and the feet flat on top of the foot support.

(3) The barbell bar should be grasped with both pronated hands, equidistant from the weighted ends, and lifted off the weight stand and held above the chest so that the shoulders are abducted to ninety degrees and with no horizontal abduction, elbows flexed to ninety degrees.

(4) As a breath is taken in (inspired), the lift should be executed by extending the elbows to zero degrees and horizontally adducting the shoulders (relative to body position) to ninety degrees. Essentially, the weight should be lifted vertically, and the movement being completed when the inspired breath is complete.

(5) The weight should be vertically lowered to the beginning position, with the elbows being allowed to go into ninety degrees of flexion and the shoulders horizontally abducted back to zero degrees. The downward movement of the weight should be accompanied by the slow expiration of the breath, and should come to a halt over the chest. The weight should not be returned to the weight stand until the last lift in the set is complete (the weight should be settled onto the stand on the way down from the last lift).

Repetitions _____

Weight Rowing (T012B)

(1) The equipment required for weight rowing includes a platform for sitting and a pulley setup, weighted with the requisite weight, and provided with a double-handed handle.

(2) The sitting platform should be situated relative to the pulley handle so that the weight lifter

should have to flex at the waist to forty-five degrees to grasp the handle with both hands, with the elbows fully extended and the shoulders flexed from 105 to 120 degrees.

(3) The pulley's line of pull, for a sitting weight lifter flexed at the waist to forty-five degrees, should be in the horizontal plane or along an inclined plane as much as thirty degrees below the horizontal plane.

(4) Sitting on the platform with the feet flat on the floor, flex forward at the waist and grasp the pulley handle at either end with pronated hands.

(5) To lift the weight, lean back against the pull of the handle, extending the back to from 35 to 45 degrees as the elbows, which are pointed out and shoulders flexing to compensate, are simultaneously flexed to at least 120 degrees. Inspiration of a breath should accompany the action, terminating at the end of the motion.

(6) The weight should be lowered back to the beginning position, through the extension of the elbows and flexion of the waist, as the breath is slowly let out.

The entire action is one of rowing and should be performed rhythmically in a slow steady pattern of motion.

Repetitions _____

Pull Down (T012C)

(1) The equipment required for the pull down exercise includes a platform for sitting and a pulley setup, weighted with the requisite weight, and provided with a double-handed handle.

(2) The sitting platform should be situated relative to the pulley handle so that the weight lifter should have to reach out and up, so that the shoulders are flexed to 160 degrees and the elbows are fully extended, to grasp the handle with both hands.

(3) The pulley's line of pull (relative to a weight lifter sitting erect) should be nearly vertical, descending along an oblique angled plane of approximately 100 to 110 degrees.

(4) Sitting on the platform with the feet flat on the floor, reach up and grasp the pulley handle at either end with pronated hands.

(5) Keeping the elbows down, pull against the weight of the handle by extending and hyperextending the shoulders while allowing the elbows to flex to around forty-five degrees. This action should be accompanied by an intake of breath which lasts until the movement ends.

(6) Slowly lower the weight by allowing the shoulders to go into flexion and the elbows to extend back to the beginning position, slowly letting out the breath as the action takes place.

Repetitions _____

Curls (T012D)

(1) The equipment required to perform curls includes a barbell loaded with the requisite weights, and a curler's yoke. The curler's yoke is a steel band formed to fit across the lifter's chest and behind the back of the upper arms, above the elbows, which hangs from the lifter's neck by a strap which is affixed to the steel band one quarter of the length from each end. The curler's yoke prevents hyperextension of the shoulders during flexion of the elbows.

(2) Standing and slightly leaning with the back and buttocks up against a wall, the curler's yoke hung across the neck and fitted across the chest and behind the elbows, the barbell should be grasped in both supinated hands, equidistant from the weighted ends and shoulder width apart.

(3) Being sure to keep the buttocks against its support, the lift should be performed by slowly flexing the elbows to the maximum, bringing the bar up in front of the face. A breath should be taken as flexion occurs, ending when full flexion has been achieved.

(4) The barbell should be slowly lowered as the elbows are allowed to fully extend back to the beginning position. The breath should be let out slowly as the weight is lowered.

Repetitions _____

EXTRAFUSAL MUSCLE SPASM (C41)

Definition (C41A)

An extrafusal muscle spasm is an involuntary contraction of extrafusal muscle fibers (most certainly including and affecting intrafusal or muscle spindle structures) which is usually sudden, severe, often extremely painful, and interferes with the normal motor function of the involved muscle(s). Not included in our discussion here are involuntary spasm-like muscle contractions resulting from supraspinal lesion or malfunction.

An extrafusal muscle spasm may involve the whole muscle (a "cramp") or just a few muscle fiber bundles to form a palpable *knot* in the muscle (sometimes forming around an intrafusal structure spasm, creating an exaggerated myositic nodule). If the contraction is steady and prolonged it may be called a tonic spasm. If the contractions are steadily intermittent, alternating contraction with brief periods of relaxation, the spasms are said to be clonic and phasic (an uninhibited series of phasic stretch reflexes).

Evaluation (C41B)

The effect of extrafusal muscle spasm may be evaluated through range of motion measurement (tonic muscle spasm will limit range(s) of motion), palpation to assess gross muscle tension and to discern or find *knots*, and/or with electromyometry to establish the presence of muscular hypertonicity.

Etiology (C41C)

Extrafusal muscle spasms are often associated with changes in mineral content of the muscles (calcium and/or potassium imbalance), idiopathic supraspinal influence of tonic muscle control to maintain in a painful state of contraction, trauma to the muscle tissue sufficient to induce a muscle to "splint" (as seen in post long bone fracture or from a direct blow to the muscle), excessive muscle strain, severe cold, lack of blood flow to the muscle and/or muscle fatigue (excessive exercise).

Of special interest is the mechanism behind the tonic muscle spasm. Any irritation of muscle tissue or metabolic abnormality produced by severe cold, lack of blood flow, or fatigue can elicit pain or other types of sensory response that, transmitted from the muscle to the spinal cord, cause a reflex hypertonic contraction of the muscle. The elicited involuntary contraction in turn stimulates the same sensory receptors to a further degree to promote a spinal cord response which produces an escalation of muscle contraction, which finally culminates in a full blown muscle cramp or tonic muscle spasm.

Treatment Notes (C41D)

Treatment of the extrafusal muscle spasm (including the tonic variety) should be directed at taking advantage of the reciprocal inhibitory relationship between antagonistic muscles. A muscle spasm may be completely relieved by inducing a voluntary or involuntary contraction of a muscle or muscles antagonistic to the one in spasm. Linear vibration, various forms of electrical stimulation and electromyometric feedback may by used to this end (especially if employed to also promote lengthening of the involved muscle).

Extrafusal muscle spasm may also be treated by promoting lengthening of the involved muscle through the use of ice packing, or ice massage applied to the involved musculature while the muscle is on stretch.

Bibliography (C41E)

Cooley, pp. 363, 988, 1031.
Guyton, pp. 659–660.
Lehmann, pp. 410–413, 563–517, 580–582.

TREATMENT OF EXTRAFUSAL MUSCLE SPASM (C41F)

Technique Options (C41F1)

Evaluation techniques (C41F1A)

Auxiliary evaluation techniques	*M26*
Auscultation	M26C
Joint and/or tendon sounds	M26C2C
Cryotherapy	*M10*
Evaluation of capillary response to cold	M10H
Determination of hyperreactivity to cold	M10H2A
Electromyometric feedback	*M18*
Electromyometric evaluation	M18L
Determination of neuromuscular hypertonicity	M18L1
Exercise	*M12*

extrafusal muscle spasm

Functional testing	M12I
Evaluation of joint range of motion	M12I1
Galvanic skin response monitry	*M20*
GSR assessment	M20C
Hydrotherapy	*M15*
Hydrotherapeutic Tests	M15F
Test for hypersensitivity to cold	M15F3
Massage	*M23*
Palpation evaluation	M23G
Determination of extrafusal muscle spasm or hypertonicity	M23G2
Peripheral electroacustimulation	*M05*
Peripheral acupoint survey	M05D

Treatment techniques (C41F1B)

Auricular electroacustimulation	*M06*
Pain relief	M06A
Cryotherapy	*M10*
Suppression of sympathetic autonomic hyperreactivity	M10A
Hypertonic neuromuscular management	M10B
Extrafusal muscle spasm treatment	M10B2B
Coolant sprays	M10B2C
Analgesia/anesthesia	M10C
Electrical facilitation of endorphin production (acusleep)	*M03*
Increasing pain tolerance	M0305A
Electrical stimulation (ES)	*M02*
Muscle lengthening	M02A
Muscle lengthening	M02A5A
Muscle reeducation/muscle toning	M02B
Increasing blood flow	M02C
Increasing capillary density	M02C2B
Electromyometric feedback	*M18*
Neuromuscular reeducation of neck and shoulder musculature	M18E (1)
Neuromuscular reeducation of temporomandibular musculature	M18F (2)
Bruxism	M18F5A
Jaw range of motion	M18F5B
Neuromuscular reeducation of low back musculature	M18G (3)
Exercise	*M12*
Muscle strengthening	M12A
Muscle relengthening	M12C
Tonically shortened musculature	M12C1A
Muscle relengthening through reciprocal inhibition	M12C1C
Galvanic skin response monitry	*M20*
General relaxation	M20A
Desensitization	M20B
Hydrotherapy	*M15*
Muscle tension amelioration	M15C
Range of motion enhancement and physical reconditioning	M15D
Suppression of sympathetic hyperresponsiveness	M15E
Iontophoresis	*M07*
Pain relief	M07A
Muscle spasm relief	M07F
Hypertonicity relief	M07G
Joint mobilization	*M13*
Laser stimulation	*M27*
Massage	*M23*
Muscle fatigue (and/or tension) relief	M23A
Acupressure	M23C
Deep soft tissue mobilization	M23F
Muscles	M23F2A
Ligaments	M23F2B
Tendon sheaths	M23F2C
Tendons	M23F2D
Peripheral electroacustimulation	*M05*
Analgesia	M05B
Sensory stimulation	*M17*
Neuromuscular spasticity control	M17A
Provoking reflex physiological responses	M17C
Prolonged heating for reflex physiological responses	M17C2A
Prolonged cooling for reflex physiological responses	M17C2B
Brief cooling for reflex physiological responses	M17C2C
Superficial thermal (heat) therapy	*M11*
Lower abdominal pain relief	M11B
Muscle relaxation (wet and dry sauna)	M11C
Taping	*M16*
Transcutaneous nerve stimulation	*M04*
Prolonged stimulation of acupoints	M04B
Ultrahigh frequency sound (ultrasound)	*M01*
Increasing circulation	M01B
Acupoint stimulation	M01K
Autonomic sympathetic hyperactivity suppression	M01N
Vertebral Traction	*M09*
Cervical traction	M09A (1)
Lumbar traction	M09B (3)
Vibration	*M08*
Neuromuscular management	M08A
Muscle lengthening	M08A2B
Cervical traction enhancement	M08A2C
Lumbar traction enhancement	M08A2D

(1) If cervical or shoulder musculature is involved
(2) If jaw musculature is involved
(3) If midthoracic to lumbar musculature is involved

Techniques Suggested for the Acute Condition (C41F2)

Evaluation techniques (C41F2A)

Cryotherapy	*M10*
Evaluation of capillary response to cold	M10H
Determination of hyperreactivity to cold	M10H2A

extrafusal muscle spasm

Exercise	*M12*
Functional testing	M12I
Evaluation of joint range of motion	M12I1
Massage	*M23*
Palpation evaluation	M23G
Determination of extrafusal muscle spasm or hypertonicity	M23G2

Treatment techniques (C41F2B)

Auricular electroacustimulation	*M06*
Pain relief	M06A
Cryotherapy	*M10*
Suppression of sympathetic autonomic hyperreactivity	M10A
Hypertonic neuromuscular management	M10B
Extrafusal muscle spasm treatment	M10B2B
Electrical facilitation of endorphin production (acusleep)	*M03*
Increasing pain tolerance	M0305A
Electrical stimulation (ES)	*M02*
Muscle lengthening	M02A
Muscle lengthening	M02A5A
Exercise	*M12*
Muscle relengthening	M12C
Muscle relengthening through reciprocalinhibition	M12C1C
Taping	*M16*
Ultrahigh frequency sound (ultrasound)	*M01*
Autonomic sympathetic hyperactivity suppression	M01N
Vertebral Traction	*M09*
Cervical traction	M09A (1)
Lumbar traction	M09B (3)
Vibration	M08
Neuromuscular management	M08A
Muscle lengthening	M08A2B
Cervical traction enhancement	M08A2C
Lumbar traction enhancement	M08A2D

(1) If cervical or shoulder musculature is involved
(2) If jaw musculature is involved
(3) If midthoracic to lumbar musculature is involved

Techniques Suggested for the Chronic Condition (C41F3)

Evaluation techniques (C41F3A)

Auxiliary evaluation techniques	*M26*
Auscultation	M26C
Joint and/or tendon sounds	M26C2C
Electromyometric feedback	*M18*
Electromyometric evaluation	M18L
Determination of neuromuscular hypertonicity	M18L1
Exercise	*M12*
Functional testing	M12I
Evaluation of joint range of motion	M12I1
Galvanic skin response monitry	*M20*
GSR assessment	M20C
Massage	*M23*
Palpation evaluation	M23G
Determination of extrafusal muscle spasm or hypertonicity	M23G2
Peripheral electroacustimulation	*M05*
Peripheral acupoint survey	M05D

Treatment techniques (C41F3B)

Electrical facilitation of endorphin production (acusleep)	*M03*
Increasing pain tolerance	M0305A
Electrical stimulation (ES)	*M02*
Muscle lengthening	M02A
Muscle lengthening	M02A5A
Muscle reeducation/muscle toning	M02B
Increasing blood flow	M02C
Increasing capillary density	M02C2B
Electromyometric feedback	*M18*
Neuromuscular reeducation of neck and shoul-der musculature	M18E (1)
Neuromuscular reeducation of temporomandibular musculature	M18F (2)
Bruxism	M18F5A
Jaw range of motion	M18F5B
Neuromuscular reeducation of low back musculature	M18G (3)
Exercise	*M12*
Muscle strengthening	M12A
Muscle relengthening	M12C
Tonically shortened musculature	M12C1A
Galvanic skin response monitry	*M20*
Desensitization	M20B
Massage	*M23*
Deep soft tissue mobilization	M23F
Muscles	M23F2A
Ligaments	M23F2B
Tendon sheaths	M23F2C
Tendons	M23F2D
Peripheral electroacustimulation	*M05*
Analgesia	M05B
Transcutaneous nerve stimulation	*M04*
Prolonged stimulation of acupoints	M04B
Vibration	*M08*
Neuromuscular management	M08A
Muscle lengthening	M08A2B

(1) If cervical or shoulder musculature is involved
(2) If jaw musculature is involved
(3) If midthoracic to lumbar musculature is involved

FACET SYNDROME (S19)

Definition (S19A)

The posterior portion of the vertebra is composed of two vertebral arches, two transverse processes, a central posterior spinous process and inferior and superior articular surfaces called facets. The facets pilot the direction of movement between two adjacent vertebrae; their directional planes prevent or restrict movement in a direction contrary to the planes of articulation. Vertebral planes of articulation change depending on where in the spinal column the vertebrae exists. In the lumbar area the facets lie in a vertical sagittal plane and allow flexion and extension, but severely limit lateral flexion or rotation. In the thoracic area the facets are convex-concave in a horizontal plane and allow lateral flexion and rotation.

The facets are lined with synovial tissue and are separated by synovial fluid which is contained within an articular capsule. Should the facets impinge and compress upon the adjacent synovial tissues, acute synovitis may result. If this happens, an immediate protective spasm of adjacent musculature will occur which will increase compressive force upon the interposed discs and will simultaneously produce further impaction of the posterior facets. This whole process is called the *facet syndrome*.

Related Syndromes (S19B)

CERVICAL (NECK) PAIN (S73)
CHEST PAIN (S14)
FASCIITIS (S20)
HIGH THORACIC BACK PAIN (S48)
LOWER ABDOMINAL PAIN (S62)
LOW BACK PAIN (S03)
OSTEOARTHRITIS (S21)
POST WHIPLASH (S55)
RADICULITIS (S29)
REFERRED PAIN (S01)
SCIATICA (S05)
SHOULDER/HAND (S57)
SHOULDER PAIN (S08)

Treatable Causes Which May Contribute To The Syndrome (S19C)

Calcific deposit (C08)
Extrafusal muscle spasm (C41)
Habitual joint positioning (C48)
Hypermobile/instable joint (C10)
Interspinous ligamentous strain (C03)
Joint ankylosis (C42)
Joint approximation (C47)
Ligamentous strain (C25)
Localized viral infection (C27)
Nerve root compression (C04)
Neuromuscular tonic imbalance (C30)
Peripheral nerve injury (C07)
Psychoneurogenic neuromuscular hypertonus (C29)
Restricted joint range of motion (C26)
Soft tissue inflammation (C05)
Soft tissue swelling (C06)
Trigger point formation (C01)

Treatment Notes (S19D)

Treatment of the facet syndrome should be directed at reducing extrafusal muscle spasm in the area adjacent to the facet impingement in order to reduce pressure on the interposing disk, and increasing the distance between the facets. Electrical stimulation, ice packing, vibration and positional (gravity) traction may be used most effectively to these ends, especially when in appropriate combination with each other.

The following is a list of trigger point formations which may, singly or in combination, refer pain into the cervical thoracic or lumbar paraspinous areas. It should be noted that it is possible for the given patient to experience pain throughout the entire stereotypical referred pain pattern or only in part or parts of it. Opposite to each listing is a description of the parts of the stereotypical referred pain pattern most commonly experienced by previous patients, relative to the given trigger point formation. Any descriptive reference which is underlined is a part of the pattern which has *not* been commonly experienced by previous patients. For complete description of each stereotypical referred pain pattern, refer to the Table of Referred Pain Patterns of Trigger Point Formation Origin [T005]. Please note the key to abbreviated words at the bottom of the list.

LOCATION (MUSCLE)		REFERRED PAIN ZONES (GENERAL)
Upper trapezius [A]	(2)	Posterior neck, occiput, temple, *eye*
Posterior cervical group (semispinalis capitus, semispinalis cervicis, C4 or C5 multifidus)	(3)	Paraspinous area from nuchal line to mid thoracic area

Sternocleido-mastoideus (superficial)	(3)	Sterno., occiput, across forehead & temple, *face, chin, superior lateral throat, top of head, behind & in eye*	
Sternocleido-mastoid (deep)	(3)	Sterno., ear, occiput, bil. forehead	
Levator scapulae	(2)	*Med. & l. bord. of scapula, joint,* para. C2-T2, upper trapezius	
Lower splenius cervicus		*Para. C1-C7,* upper trapezius	
Upper trapezius [B]		Subocciput, *upper trapezius, l.p.neck*	
Lower trapezius [A]		Subocciput, *para. C1-C6,* upper trapezius, A/C joint	
Cervical multifidus (C4-C5)		Subocciput, para. *C1-C7, upper trapezius*	
Supraspinatus (muscle)	(2)	*Upper trapezius,* deltoids, *l.arm*	
Serratus anterior		Lower a.l.chest, *med. a.arm & hand,* d.med. bord. of scapula, a.f.s 4 & 5	
Rhomboids	(3)	Med. bord. of scapula, suprasp inatus	
Serratus posterior inferior		Para. T9-L2 & l.over p.lower ribs	
Multifidus (T4-T5)		Bil. para. T2-T7, l.over med. bord. of scapula	
Multifidus (S4)		Bil. para. S2-coccyx, & bil. med. prox. gluteus maximus	
Longissimus thoracis (L1)		*Para. from L1* to & over p.iliac crest	
Longissimus thoracis (T10-T11)		*Para. from T10,* gluteus maximus to gluteal fold	
Multifidi (L2)		Bil. para. T12-L4, *to p.iliac crest a.upper quadrant of the abdomen*	
Multifidi (S1)		Bil. para. L4-S5, med. gluteus maximus, coccyx, *prox. p.hamstrings, a.lower abdominal quadrant*	
Iliocostalis lumborum (L1)		L. from L1 iliocostalis, *down over* central gluteus maximus & *l.to head* of trochanter	
Iliocostalis thoracis (T6)		*Para. T4-T10,* med. scapula, *a.lower med. chest*	
Iliocostalis thoracis (T11)		*Para. T11-S1,* l.bottom ribs, *l.bord. of the scapula,* a.lower abdominal quadrant	
Upper rectus abdominus		Bil. across the p. lower thoracic area (T8-T11) i. to the scapulae	
Caudal rectus abdominus		Bil. across the low back (L4-S4) from l. extreme to l. extreme	
Gluteus minimus		Gluteus maximus (excepting central area) & gluteus medius, p.l.leg, l.calf to p.of l. malleolus	
Gluteus medius		Para. L4-Coccyx, *gluteus medius, gluteus maximus to head of trochanter*	

Key:
TMJ—temporomandibular joint
A/C—acromioclavicular
med.—medial
brach.—brachioradialis
para.—paraspinous
MP—metacarpal phalangeal
sterno.—sternocleidomastoideus
pect.—pectoralis
a.—anterior
s.—superior
&—and
maj.—major
bord.—border
f.—finger
l.—lateral
m.—middle
d.—distal
p.—posterior
prox.—proximal
bil.—bilateral
f.s—fingers
i.—inferior

Bibliography (S19E)

Cailliet, pp. 6–7, 68–71.
Goss, pp. 115–125.
Scott, pp. 218–219.
Shands, p. 330.

TABLE OF FACIAL MUSCLE FUNCTION (T009)

Muscle	Muscle Function	Expression Produced
Occipito frontalis	Raises the eyebrows to form horizontal forehead lines	Surprise
Corrugator supercilii	Draws the eyebrows medially and down, to form vertical lines between the eyerows	The eyebrow component of frowning
Procerus	Lifts the lateral borders of the nostrils to form diagonal lines across the bridge of the nose	Wrinkles the nose as if sniffing
Nasalis	Dilates (flap fibers) and compresses (transverse fibers) nostrils	
Orbicularis oculi	Draws the eyelids together	Squeezes the eyes closed
Levator palpebrae superioris	Lifts the eye lids	Opens the eyes
Orbicularis oris	Approximates and compresses the lips	Purses the lips together
Zygomaticus minor	Protrudes the upper lip	One of the components of kissing or lip puckering
Caninus	Lifts the upper border of the lip without raising the lateral angle of the mouth.	Brings the lip up in a snarl or sneer
Zygomaticus major	Raises the lateral angle of the mouth upward and latterally	Smile
Risorius	Approximates the lips and draws the corners of the mouth laterally	Grimace
Buccinator	Approximates the lips and compresses the cheeks	Blowing
Depressor labii inferior	Protrudes the lower lip	Pout
Depressor anguli oris and platysma	Draws the corners of the mouth downward	Frown
Mentalis	Wrinkles the chin and draws the tip of the chin upward	Associated with forced protrusion of the lower lip

Bibliography

Goss, pp. 382–390.
Daniels, pp. 162–171.

FASCIITIS SYNDROME (S20)

Definition (S20A)

Fascia is a sheet of fibrous tissue which envelops much of the body below the skin. Fascia encloses the muscles and groups of muscles, and separates their several layers or groups.

Fasciitis is an inflammation of a fascial layer. The fascial layers pertinent here are those inclosing muscular tissues. The patient suffering from the *fasciitis syndrome* complains of a burning, aching pain directly over the inflamed zones. Patients sometimes complain of referred pain, but such pain is usually found, upon examination, to originate from trigger point formations housed in musculature adjacent to or associated with the inflamed soft tissues.

Related Syndromes (S20B)

ACHILLES TENDONITIS (S52)
BURSITIS (S26)
CALF PAIN (S50)
CAPSULITIS (S27)
CARPAL TUNNEL (S35)
CERVICAL DORSAL OUTLET (SCALENICUS ATTICUS/CERVICAL RIB) (S09)
CERVICAL (NECK) PAIN (S73)
CHEST PAIN (S14)
ELBOW PAIN (S47)
FACET (S19)
FOOT PAIN (S51)
FOREARM PAIN (S54)
HAND/FINGER PAIN (S17)
HIGH THORACIC BACK PAIN (S48)
JOINT SPRAIN (S30)
KNEE PAIN (S46)
LOWER ABDOMINAL PAIN (S62)
LOW BACK PAIN (S03)
OSTEOARTHRITIS (S21)
PIRIFORMIS (S04)
POST IMMOBILIZATION (S36)
POST WHIPLASH (S55)
RADICULITIS (S29)
REFERRED PAIN (S01)
SCIATICA (S05)
SHOULDER/HAND (S57)
SHOULDER PAIN (S08)
TEMPOROMANDIBULAR JOINT (TMJ) PAIN (S06)
TENDONITIS (S28)
TENNIS ELBOW (LATERAL EPICONDYLITIS) (S32)
THIGH PAIN (S49)
WRIST PAIN (S16)

Treatable Causes Which May Contribute To The Syndrome (S20C)

Calcific deposit (C08)
Extrafusal muscle spasm (C41)
Habitual joint positioning (C48)
Interspinous ligamentous strain (C03)
Joint ankylosis (C42)
Joint approximation (C47)
Ligamentous strain (C25)
Localized viral infection (C27)
Muscular weakness (C23)
Nerve root compression (C04)
Restricted joint range of motion (C26)
Soft tissue inflammation (C05)
Soft tissue swelling (C06)
Tactile hypersensitivity (C24)
Trigger point formation (C01)
Vascular insufficiency (C11)

Treatment Notes (S20D)

Fasciitis is generally established clinically through differential skin resistance (DSR) survey (refer to Differential Skin Resistance Survey [M22]). It responds well to treatment by the phonophoresis of nonsteroid antiinflammatories (refer to Ultrahigh Frequency Sound, Phonophoresis [M01A]) combined with peripheral electroacustimulation (refer to Peripheral Electroacustimulation, Acupoint Electrical Stimulation Within an Inflamed Zone [M05A]), multiple frequency pulsed galvanic electrical stimulation (refer to Electrical Stimulation, Inflammation Control [M02G]) and/or ice packing (refer to Cryotherapy, Soft Tissue Inflammation Control [M10E]), with almost immediate decreases in pain sensation and increases in functional joint motion.

FOOT PAIN SYNDROME (S51)

Definition (S51A)

The foot is designed to serve as a support for the weight of the body when standing and to act as a lever for the raising and propelling the body forward in walking and running. The muscles in the leg supply the power and the heads of the metatarsals provide the fulcrum on which the body weight is lifted. The foot contains two main arches formed by the bones and supported by ligaments, and indirect support provided by the associated tendons and muscles. Motion and elasticity (to a degree) are required from the arches if the foot is to function properly.

Foot pain may result from flat feet (fallen transverse arch), fractures of the metatarsal bones, hallux valgus (bunion), corns or callus formation, plantar warts, diseases that cause inflammation and/or swelling of the joints, aseptic soft tissue inflammation, soft tissue swelling, strain of associated ligament(s) and/or tendon(s), extrafusal muscle spasm, muscle strain, abnormal bony or joint formation (including clawfoot), osteochondritis of the navicular bone or metatarsal head, anterior metatarsalgia, plantar neuroma, hallux varus (medial angulation of the big toe), hallus rigidus, hammer toe, overlapping or dorsal displacement of the toes, subungual exostosis, accessory bone formation, displacement of peroneal tendons, various inflammatory disorders of the Achilles tendon and its attachments, intermittent claudication, and/or referred pain from trigger point formations.

Related Syndromes (S51B)

ACHILLES TENDONITIS (S52)
ACNE (S45)
ADHESION (S56)
BACTERIAL INFECTION (S63)
BURSITIS (S26)
CAPSULITIS (S27)
DIABETES (S10)
FASCIITIS (S20)
HYSTERIA/ANXIETY REACTION (S59)
JOINT SPRAIN (S30)
OSTEOARTHRITIS (S21)
PHLEBITIS (S68)
PHOBIC REACTION (S44)
PITTING (LYMPH) EDEMA (S31)
PLANTAR WART (S69)
POST CEREBRAL VASCULAR ACCIDENT (CVA) (S07)
POST IMMOBILIZATION (S36)
POST PERIPHERAL NERVE INJURY (S23)
POST SPINAL CORD INJURY (S24)
REFERRED PAIN (S01)
SCIATICA (S05)
SHINGLES (HERPES ZOSTER) (S37)
TENDONITIS (S28)
UNHEALED DERMAL LESION (S39)

Treatable Causes Which May Contribute To The Syndrome (S51C)

Bacterial infection (C09)
Blood sugar irregularities (C31)
Calcific deposit (C08)
Extrafusal muscle spasm (C41)
Gel-foam implant (C21)
Habitual joint positioning (C48)
Hypermobile/instable joint (C10)
Interspinous ligamentous strain (C03)
Joint ankylosis (C42)
Joint approximation (C47)
Ligamentous strain (C25)
Localized viral infection (C27)
Muscular weakness (C23)
Nerve root compression (C04)
Neuroma formation (C14)
Neuromuscular tonic imbalance (C30)
Pathological neuromuscular hypertonus (C28)
Pathological neuromuscular hypotonus (C33)
Pathological neuromuscular dyscoordination (C35)
Peripheral nerve injury (C07)
Psychoneurogenic neuromuscular hypertonus (C29)
Restricted joint range of motion (C26)
Soft tissue inflammation (C05)
Soft tissue swelling (C06)
Sympathetic hyperresponse (C38)
Tactile hypersensitivity (C24)
Trigger point formation (C01)
Unhealed dermal lesion (C12)
Vascular insufficiency (C11)

Treatment Notes (S51D)

The following is a list of trigger point formations which may, singly or in combination, refer pain into the area of the foot. It should be noted that it is possible for the given patient to experience pain throughout the entire stereotypical referred pain pattern or only in part or parts of it. Opposite to each listing is a description of the parts of the stereotypical referred pain pattern most commonly experienced by previous patients, relative to the given trigger point formation. Any descriptive reference which is underlined is a part of the pattern which has *not* been commonly experienced by previous patients. For complete description of each stereotypical referred pain pattern, refer to the Table of Referred Pain Patterns of Trigger Point Formation Origin [T005]. Please note the key to abbreviated words at the bottom of the list.

LOCATION (MUSCLE)	REFERRED PAIN ZONES (GENERAL)
Gastrocnemius	P. med. calf, central thigh, behind med. malleolus, arch of the foot
Anterior tibialis	A. calf just l.to tibia, *over metatarsal 1, big toe*
Long toe extensors	A. l.distal 2/3 of calf, over metatarsals & toes 2-5
Soleus	Achilles tendon & heel
Short toe extensors	L. instep, metatarsals & toes 2-5
Abductor hallucis	Metatarsals & toes 1-2

Key:
med.—medial
a.—anterior
&—and
l.—lateral
p.—posterior

Bibliography (S51E)

Goss, pp. 262–276, 507–526.
Scott, pp. 358–361.
Shands, pp. 390–411.

FOREARM PAIN SYNDROME (S54)

Definition (S54A)

Forearm pain is produced by radiohumeral bursitis (tennis elbow, lateral epicondylitis), tendonitis (tenosynovitis, inflammation of the synovial sheaths covering the tendons) of the various tendons which run along the forearm, referred pain from other structures, peripheral nerve impingement, fractures of either the radius or ulnar shafts, muscle strain and/or extrafusal muscle spasm, tendon ruptures, calcific tenosynovitis, myositis ossificans, and/or ganglion formation.

Related Syndromes (S54B)

ADHESION (S56)
BACTERIAL INFECTION (S63)
BUERGER'S DISEASE (S11)
BURSITIS (S26)
CAPSULITIS (S27)
CARPAL TUNNEL (S35)
CERVICAL DORSAL OUTLET (SCALENICUS ATTICUS/CERVICAL RIB) (S09)
CERVICAL (NECK) PAIN (S73)
DIABETES (S10)
ELBOW PAIN (S47)
FASCIITIS (S20)
FROZEN SHOULDER (S64)
HAND/FINGER PAIN (S17)
HYSTERIA/ANXIETY REACTION (S59)
JOINT SPRAIN (S30)
MYOSITIS OSSIFICANS (S67)
OSTEOARTHRITIS (S21)
PHOBIC REACTION (S44)
POST CEREBRAL VASCULAR ACCIDENT (CVA) (S07)
POST IMMOBILIZATION (S36)
POST PERIPHERAL NERVE INJURY (S23)
POST SPINAL CORD INJURY (S24)
RADICULITIS (S29)
RAYNAUD'S DISEASE (S12)
REFERRED PAIN (S01)
SHINGLES (HERPES ZOSTER) (S37)
SHOULDER/HAND (S57)
TENDONITIS (S28)
TENNIS ELBOW (LATERAL EPICONDYLITIS) (S32)
UNHEALED DERMAL LESION (S39)
WRIST PAIN (S16)

Treatable Causes Which May Contribute To The Syndrome (S54C)

Bacterial infection (C09)
Calcific deposit (C08)
Extrafusal muscle spasm (C41)
Habitual joint positioning (C48)
Hypermobile/instable joint (C10)
Interspinous ligamentous strain (C03)
Joint ankylosis (C42)
Joint approximation (C47)
Ligamentous strain (C25)
Localized viral infection (C27)
Muscular weakness (C23)
Nerve root compression (C04)
Neuroma formation (C14)
Neuromuscular tonic imbalance (C30)
Pathological neuromuscular hypertonus (C28)
Pathological neuromuscular hypotonus (C33)
Peripheral nerve injury (C07)
Psychoneurogenic neuromuscular hypertonus (C29)
Restricted joint range of motion (C26)
Soft tissue inflammation (C05)
Soft tissue swelling (C06)
Sympathetic hyperresponse (C38)
Tactile hypersensitivity (C24)
Trigger point formation (C01)
Unhealed dermal lesion (C12)
Vascular insufficiency (C11)
Visceral organ dysfunction (C02)

Treatment Notes (S54D)

The following is a list of trigger point formations which may, singly or in combination, refer pain into the forearm. It should be noted that it is possible for the given patient to experience pain throughout the entire stereotypical referred pain pattern or only in part or parts of it. Opposite to each listing is a description of the parts of the stereotypical referred pain pattern most commonly experienced by previous patients, relative to the given trigger point formation. Any descriptive reference which is underlined is a part of the pattern which has *not* been commonly experienced by previous patients. For complete description of each stereotypical referred pain pattern, refer to the Table of Referred Pain Patterns of Trigger Point Formation Origin [T005]. Please note the key to abbreviated words at the bottom of the list.

LOCATION (MUSCLE)	REFERRED PAIN ZONES (GENERAL)
Scalenus	(4) *Shoulder*, med. & l.pect. maj., med. bord. of scapula, deltoids, *l.arm*, p.thumb, *p.index & m.finger*

Scalenus (minimus)	L. triceps, *p.forearm & hand & f.s*	Supinator	L. elbow, *p.brach.*, p.metacarpals 1-2
Infraspinatus	(2) *Subocciput*, deltoids, l.arm, palm, p.hand A/C joint, p.& m.deltoids, *p.forearm*	Extensor carpi radialis brevis	P. d.2/3 forearm, p.wrist & hand
Lateral teres major		Extensor carpi ulnaris	P. med. d.forearm & wrist & prox. hand
Coracobrachialis	A. deltoid, shoulder joint, l. triceps, *p.arm & hand, p.m.finger* Goose flesh of l.triceps & brach.	Middle finger extensor	P. central forearm & middle finger
		Ring finger extensor	P. origin of brach., p.med. forearm, p.finger 4
Middle trapezius [C]			
Supraspinatus (muscle)	*Upper trapezius,* (2) deltoids, *l.arm*	Palmaris longus	A. med. forearm, palm
Upper latissimus dorsi	Latissimus dorsi, *med. arm, l.hand*, fingers 4 & 5	Flexor carpi radialis	L. a.forearm, a.wrist, *palm*
		Flexor carpi ulnaris	Med. a.d.forearm, a.med. wrist-hand
Serratus posterior superior (under the scapula)	Scapula, p.deltoid, med. arm, wrist, med. hand, finger 5	Brachioradialis	P. l.elbow, *l.forearm,* p.metacarpals 1 & 2
		Pronator teres	A. l.forearm & wrist, a.metacarpal #1
Serratus anterior	Lower a.l.chest, *med. a.arm & hand*, d.med. bord. of scapula, *a.f.s 4 & 5*	Extensor indicus	P. wrist, *p.metacarpals 2-4 & second MP joint*
		Extensor carpi radialis longus	P. l.elbow, p.l.forearm & wrist, p.carpals, p.metacarpals 1-4
Subclavius	Under clavicle, *a.deltoid*, biceps, brach., *l.hand*, a.& *p.f.s 1-3*		
Subscapularis	*Scapula, axilla, p.deltoid, med. triceps, a.p.wrist*	Key: TMJ—temporomandibular joint A/C—acromioclavicular med.—medial brach.—brachioradialis para.—paraspinous MP—metacarpal phalangeal sterno.—sternocleidomastoideus pect.—pectoralis a.—anterior m.—middle s.—superior &—and maj.—major bord.—border f.—finger l.—lateral d.—distal p.—posterior prox.—proximal bil.—bilateral f.s—fingers i.—inferior	
Pectoralis major (sternal)	Pectoralis major, (3) a.deltoid, *med. a. upper arm*, prox. a.forearm, *m.palm, a.f.s 3-5*		
Pectoralis minor	*Pectoralis major*, a.deltoid, *med. a. arm, med. a.hand, a.f.s 3-5*		
Medial triceps (deep fibers)	Med. a.elbow & *forearm, a.f.s 3-5*		
Medial triceps (l. fibers)	P. l.elbow, posterior brach.		
Lateral triceps	L. triceps, *p.forearm, p.f.s 4-5*		
Longhead of the triceps	*Upper trapezius,* (2) p.deltoid, *p.upper arm, l.p.forearm, p.wrist*		
Distal medial triceps	P. elbow		
Brachialis (superior & inferior)	A. deltoid, a.half m.deltoid, a.elbow, a.& (4) p.first metacarpal, l.hand, p.second metacarpal		

Bibliography (S54E)

Goss, pp. 220–227, 464–475.
Scott, pp. 140–155.
Shands, pp. 438–451.

FROZEN SHOULDER SYNDROME (S64)

Definition (S64A)

The *frozen shoulder syndrome* (periarthritis, adhesive capsulitis) is characterized by severe limitation of shoulder motion which has resulted from degenerative changes involving the musculo-tendinous cuff, synovial membrane, articular cartilage, bicipetal tendon and tendon sheath, glenohumeral joint, and/or acromioclavicular joint. Round cell infiltration, indicative of low grade inflammatory process, may usually be demonstrated histologically, with attendant edema and developing fibrosis. Elasticity of the periarticular tissues is progressively lost as they become progressively shortened and fibrotic. These changes firmly fix the humeral head within the glenoid cavity, greatly reducing the range of motion and eventually resulting in marked atrophy of affected musculature. Contracture of the coracohumeral ligament and the subscapularis tendon additionally limits range as external rotation of the head of the humerus is progressively lost. Range of motion loss may be subsequently so severe that there may be a complete loss of scapulohumeral motion.

Onset of the *frozen shoulder syndrome* may be insidious and follow on the heels of a direct or indirect local trauma to the shoulder structures and/or tissues, or as a sequel to injuries sustained by structures and/or tissues of the elbow, wrist or hand. It may also arise indirectly from cerebrovascular accident (CVA) paralysis, from reduced shoulder motion provoked by referred pain from the structures associated with the chest or pressure on a cervical nerve root, or from trigger point formation referred pain. The frozen shoulder syndrome may also arise as a direct consequence of subacromial bursitis, calcific tendonitis, and/or tenosynovitis of the long head of the biceps muscle.

Pain associated with the frozen shoulder syndrome may be produced and increased by attempts at scapulohumeral abduction, and external rotation and/or extension of the shoulder. Pain has also been reported to radiate from the soft tissues surrounding the shoulder joint into the anterolateral aspect of the shoulder, the biceps muscle belly, to the flexor surface of the forearm, and to the inferior angle of the scapula. Palpation tenderness of the intertubercular sulcus and the biceps tendon may be evident. The pain may contribute to the frozen shoulder syndrome as the voluntary range of motion is limited by the intensity of the resultant increases in pain caused by attempts to move. Characteristically, the pain is reported to be worse at night, even without attempts to move the shoulder.

Of special note is that individuals beyond the age of forty are generally subject to degenerative changes in the rotator cuff and long head of the biceps. Any painful afflictions of the upper extremity that cause an adducted, internally rotated position of the shoulder may trigger pathologic changes leading to frozen shoulder.

Related Syndromes (S64B)

ADHESION (S56)
BACTERIAL INFECTION (S63)
BURSITIS (S26)
CAPSULITIS (S27)
CERVICAL DORSAL OUTLET (SCALENICUS ATTICUS/CERVICAL RIB) (S09)
FASCIITIS (S20)
HIGH THORACIC BACK PAIN (S48)
HYSTERIA/ANXIETY REACTION (S59)
JOINT SPRAIN (S30)
MYOSITIS OSSIFICANS (S67)
NEUROMUSCULAR PARALYSIS (S22)
OSTEOARTHRITIS (S21)
PHOBIC REACTION (S44)
PITTING (LYMPH) EDEMA (S31)
POST CEREBRAL VASCULAR ACCIDENT (CVA) (S07)
POST IMMOBILIZATION (S36)
POST PERIPHERAL NERVE INJURY (S23)
POST SPINAL CORD INJURY (S24)
RADICULITIS (S29)
REFERRED PAIN (S01)
SHOULDER/HAND (S57)
SHOULDER PAIN (S08)
TENDONITIS (S28)

Treatable Causes Which May Contribute To The Syndrome (S64C)

Bacterial infection (C09)
Calcific deposit (C08)
Extrafusal muscle spasm (C41)
Habitual joint positioning (C48)
Interspinous ligamentous strain (C03)
Joint ankylosis (C42)
Joint approximation (C47)
Ligamentous strain (C25)
Muscular weakness (C23)

Nerve root compression (C04)
Neuromuscular tonic imbalance (C30)
Pathological neuromuscular hypertonus (C28)
Pathological neuromuscular dyscoordination (C35)
Peripheral nerve injury (C07)
Psychoneurogenic neuromuscular hypertonus (C29)
Restricted joint range of motion (C26)
Scar tissue formation (C15)
Soft tissue inflammation (C05)
Soft tissue swelling (C06)
Trigger point formation (C01)
Visceral organ dysfunction (C02)

Treatment Notes (S64D)

Treatment of the frozen shoulder should first be directed at relieving the soft tissue inflammatory process, if it is present. Soft tissue inflammation associated with the frozen shoulder syndrome may be determined through differential skin resistance survey (refer to Differential Skin Resistance [M22]). Soft tissue inflammation may be treated with the phonophoresis of nonsteroidal antiinflammatories (refer to Ultrahigh Frequency Sound, Phonophoresis [M01A]), ten to fifteen minutes of continuous electrical stimulation of acupoints within the inflamed zone (refer to Peripheral Electroacustimulation, Acupoint Electrical Stimulation Within an Inflamed Zone [M05A]), and/or multifrequency pulsed galvanic electrical stimulation of the inflamed zone (refer to Electrical Stimulation, Inflammation Control [M02G]).

The patient should be instructed in non-gravity range of motion exercises for the shoulder (refer to Table of Stretching Exercises for Muscle Lengthening, Shoulder Muscle Lengthening [T016C]), and shoulder muscle isometrics (refer to Table of Exercises for the Shoulders and Upper Extremities - Isometrics [T011]).

A treatment program of joint manipulation may be necessary to regain lost range of motion in the various involved joints.

The following is a list of trigger point formations which may, singly or in combination, refer pain into the shoulder area (and may be sometimes responsible for the development of the frozen shoulder). It should be noted that it is possible for the given patient to experience pain throughout the entire stereotypical referred pain pattern or only in part or parts of it. Opposite to each listing is a description of the parts of the stereotypical referred pain pattern most commonly experienced by previous patients, relative to the given trigger point formation. Any descriptive reference which is underlined is a part of the pattern which has *not* been commonly experienced by previous patients. For complete description of each stereotypical referred pain pattern, refer to the Table of Referred Pain Patterns of Trigger Point Formation Origin [T005]. Please note the key to abbreviated words at the bottom of the list.

LOCATION (MUSCLE)		REFERRED PAIN ZONES (GENERAL)
Posterior cervical group (Semispinalis capitus, semispinalis cervicis, C4 or C5 multifidus)	(3)	Paraspinous area from nuchal line to mid thoracic area
Levator scapulae	(2)	*Med. & l. bord. of scapula, joint, para. C2-T2, upper trapezius*
Scalenus	(4)	*Shoulder, med. & l.pect. maj., med. bord. of scapula, deltoids, l.arm, p.thumb, p.index & m.finger*
Scalenus (minimus)		*L. triceps, p.forearm & hand & f.s*
Infraspinatus	(2)	*Subocciput, deltoids, l.arm, palm, p.hand*
Infraspinatus (abnormal)		Med. bord. & i.angle of scapula
Medial teres major		*A/C joint, p.& m.deltoids, p.forearm*
Lateral teres major		*A/C joint, p.& m.deltoids, p.forearm*
Teres minor		P. & m.deltoid, *med. & l.triceps*
Coracobrachialis		A. deltoid, shoulder joint, *l. triceps, p.arm & hand, p.m.finger*
Lower splenius cervicus		Para. *C1-C7, upper trapezius*
Upper trapezius [B]		Subocciput, *upper trapezius, l.p. neck*
Middle trapezius [A]		Para. C7-T4, *upper trapezius*
Middle trapezius [B]		S. scap.border (supraspinatus)
Middle trapezius [C]		Goose flesh of l.triceps & brach.
Lower trapezius [A]		Subocciput, *para. C1-C6, upper trapezius,* A/C joint
Lower trapezius [B]		M. scapula bord.
Cervical multifidus (C4-C5)		Subocciput, para. C1-C7, *upper trapezius*
Supraspinatus (muscle)	(2)	*Upper trapezius, deltoids, l.arm*

Supraspinatus (tendon)	Deltoids	Brachialis (superior & inferior)	A. deltoid, a.half m.deltoid, a.elbow, a.&
Latissimus dorsi (abnormal)	A. delt., l.lower abdomen		(4) p.first metacarpal, l.hand, p.second metacarpal
Serratus posterior superior (under the scapula)	Scapula, p.deltoid, med. arm, wrist, med. hand, finger 5	Multifidus (T4-T5) Iliocostalis thoracis (T6)	Bil. para. T2-T7, l.over med. bord. of scapula Para. T4-T10, med. scapula, a.lower med. chest
Subclavius	Under clavicle, a.deltoid, biceps, brach., l.hand, a.& p.f.s 1-3		
Subscapularis	Scapula, axilla, p.deltoid, med. triceps, a.p.wrist		
Posterior deltoid	P. deltoid, a.& m.deltoid, l.triceps		
Anterior deltoid	A. deltoid, m.deltoid, l.upper arm		
Pectoralis major (costal)	Breast, chest, axilla (2)		
Pectoralis major (clavicular)	Subclavical, anterior (2) deltoid		
Pectoralis major (sternal)	Pectoralis major, (3) a.deltoid, med. a. upper arm, prox. a.forearm, m.palm, a.f.s 3-5		
Pectoralis minor	Pectoralis major, a.deltoid, med. a. arm, med. a.hand, a.f.s 3-5		
Sternalis;	(3) Bil. parasternum, med. subclavicle, a.deltoid, med. a.upper arm & elbow		
Rhomboids	(3) Med. bord. of scapula, supraspinatus		
Longhead of the triceps	Upper trapezius, (2) p.deltoid, p.upper arm, l.p.forearm, p.wrist		
Biceps brachii	Upper trapezius, a.& (2) m.deltoid, biceps, a.elbow		

Key:
TMJ—temporomandibular joint
A/C—acromioclavicular
med.—medial
brach.—brachioradialis
para.—paraspinous
MP—metacarpal phalangeal
sterno.—sternocleidomastoideus
pect.—pectoralis
a.—anterior
s.—superior
&—and
maj.—major
bord.—border
f.—finger
l.—lateral
m.—middle
d.—distal
p.—posterior
prox.—proximal
bil.—bilateral
f.s—fingers
i.—inferior

Bibliography (S64E)

Cailliet, pp. 98–101.
Basmajian, pp. 135–137.
Salter, pp. 244–245.
Scott, pp. 110–115.
Shands, pp. 426–427.

GALVANIC SKIN RESPONSE MONITRY (M20)

Definition (M2001)

Galvanic skin response monitry is the measurement of skin resistance via an ohm meter especially designed to monitor the resistance of the skin to the passage of an electrical current.

There are three basic types of these skin resistance monitors used in the clinical setting: (1) the acupoint finder, (2) the differential skin resistance (DSR) measurement device and (3) the galvanic skin response (GSR) feedback monitor. These devices are all mentioned together because they are really the same instrument adjusted and accessorized to perform different tasks; each of these instruments may be changed to perform the functions of the other two.

The acupoint finder is primarily used to discern sites of low skin resistance and high conductance, commonly termed acupoints. The use of this variation of the galvanic skin response monitor is discussed in detail in Peripheral Electroacustimulation [M05] and Auricular Electroacustimulation [M06].

The DSR measurement device is utilized to find zones of high skin resistance. These zones have been correlated to soft tissue inflammation, both on the surface of the skin and at the much deeper levels occupied by tendons, fascial layers, muscles, ligaments and joints. The use of the DSR measurement device is discussed in detail in Differential Skin Resistance Survey [M28]).

To be discussed here is the GSR feedback monitor, which is clinically utilized to: (1) provide a base line indicator of autonomic activity or status, (2) provide a gauge of autonomic response to emotional stress and/or (3) to serve as a feedback device for patient training in GSR control.

Normally, the human autonomic nervous system expresses its response to emotional stress, or lack of it, by inducing changes (increases and/or decreases) in the rate at which sweat is produced by the sweat glands in the palms of the hands, the palms of the feet, and temporal areas. When a person is emotionally stressed, the sweat production in these areas is normally increased, thereby decreasing the skin resistance, but if emotional stress is reduced, the sweat production is decreased and the skin resistance is consequently increased. Why these correlates exist and which neuroautogenic mechanisms and/or relationships are responsible for this phenomena remains to be fully explained (and lies beyond the scope of this discussion). It is sufficient to note that the phenomena exists and may be utilized as an indicator of autonomic reactivity to emotional stress.

It has been demonstrated that when provided with adequate negative feedback relative to GSR activity, a patient may be able to learn to reduce sympathetic autonomic reactivity to emotional stress voluntarily. A reduction in sympathetic autonomic reactivity to emotional stress coincidentally causes decreases in the production of adrenaline and insulin and other related endocrine substances which are hormones directly related to the autonomic "fight or flight" mechanism. The patient may be able to learn to do this by learning to voluntarily keep GSR instrument feedback low while being emotionally stressed.

The GSR instrumentation utilized in the clinical setting may employ audio, visual and/or tactile modes of feedback. Instruments utilized for monitoring patient autonomic activity exclusively, without providing feedback to the patient, need only employ visual modes (meters, light bars, video graphs, strip recorders, etc.) and are usually kept out of the patient's line of sight. GSR instruments utilized for feedback training procedures usually employ audio and visual feedback modes for presentation to the patient, but only audio feedback is considered to be absolutely essential. The most efficient instrumentation provides audio, visual and tactile feedback to the patient, in ranges discernible to the patient.

Physically, the GSR instrument is generally attached to the patient's skin via two insulated wire electrode cables, attached to the patient's skin via surface electrodes, coupled by electrode cream, affixed in place with tape, strapping or other vehicle.

When the GSR instrument has been turned on, a small electrical current (called a trickle current) is passed from one of the electrodes to the other. The instrument registers how much of the current is passed between the two electrodes, thereby providing a measure of skin resistance.

GSR instruments generally measure resistance levels ranging from 0 to 50 or more micro ohms (mohms). To be most effective the GSR instruments should be set to monitor different ranges of sensitivity. One such instrument, for example, may be set to monitor a full range of 0 to 50 mohms or a limited range of 10 mohm increments within the general 50 mohm range (0 to 10, 10 to 20, 20 to 30, 30 to 40, and 40 to 50 mohms), with the additional facility of being adjustable to monitor a 2 mohm range, found in the mid range of any of the 10 mohm incremental ranges.

Procedures of Application (M2002)

1) The patient should be sitting (or semi-reclining) in a position which will remain comfortable for the duration of the session.

2) The GSR instrument should be attached to the patient via the electrode cables and surface electrodes affixed over the site to be monitored (generally to the palmer surfaces of two fingers or on either side of the palm of the hand).

3) If a tactile feedback mode is available, the stimulation electrodes or the vibrating element should be affixed over an appropriate site. Stimulation elements are generally affixed to skin over the stomach area or to the anterior surface of the forearm, so that the stimulation is perceived as noxious, as a stimulus to be avoided, promoting decreases in GSR feedback activity levels, thereby increasing skin resistance.

4) The GSR instrument and feedback modes should be turned on.

5) If GSR training is to be attempted, the patient should be prevented from perceiving the feedback modes (prevented from hearing, seeing or feeling changes in GSR activity) until after the feedback modes have stopped changing and a base line GSR activity may be noted and recorded, if visual feedback is provided. If monitry without training is to be engaged in, the visual feedback should be out of the patient's line of vision and no other feedback provided to the patient.

6) If feedback training is to take place, after the base line of GSR activity has been established, the patient should be exposed to the feedback modes, which have been adjusted to provide feedback at pitch levels or meter needle positions which will allow the patient to perceive changes in GSR feedback activity.

7) The patient should be instructed in the procedure to be followed, and treatment goals clearly explained.

8) At the end of the treatment session the GSR level of activity should be noted and recorded, if visual feedback is available.

9) The instrumentation should be turned off and the electrodes removed from the patient's skin.

10) All electrode cream or paste should be cleansed from the patient's skin.

Partial List of Conditions Treated (M2003)

Treatable Causes (M2003A)

Extrafusal muscle spasm (C41)
Hyperhidrosis (C37)
Hypertension (C39)
Insomnia (C46)
Migraine (vascular headache) (C34)
Nausea/vomiting (C36)
Neurogenic dermal disorders (psoriasis, hives) (C49)
Psychoneurogenic neuromuscular hypertonus (C29)
Sympathetic hyperresponse (C38)
Trigger point formation (C01)
Vascular insufficiency (C11)
Visceral organ dysfunction (C02)

Syndromes (M2003B)

BUERGER'S DISEASE (S11)
CERVICAL (NECK) PAIN (S73)
CHEST PAIN (S14)
COLITIS (S42)
EARACHE (S40)
ELBOW PAIN (S47)
FOREARM PAIN (S54)
HAND/FINGER PAIN (S17)
HEADACHE PAIN (S02)
HIGH THORACIC BACK PAIN (S48)
HYPERTENSION (S43)
HYSTERIA/ANXIETY REACTION (S59)
INSOMNIA (S71)
LOWER ABDOMINAL PAIN (S62)
LOW BACK PAIN (S03)
MIGRAINE (VASCULAR) HEADACHE (S18)
PHOBIC REACTION (S44)
POST WHIPLASH (S55)
RADICULITIS (S29)
RAYNAUD'S DISEASE (S12)
REFERRED PAIN (S01)
SCIATICA (S05)
SHOULDER/HAND (S57)
SHOULDER PAIN (S08)
TEMPOROMANDIBULAR JOINT (TMJ) PAIN (S06)
TINNITUS (S70)
TOOTHACHE/JAW PAIN (S41)
TORTICOLLIS (WRY NECK) (S65)
WRIST PAIN (S16)

Precautions and Contraindications (M2004)

There are no precautions or contraindications directly associated with GSR monitry or GSR training in the literature reviewed. However, caution should be observed when GSR general relaxation or desensitization training is performed with patients suffering from diabetes. The depression of sympathetic activity may decrease the patient's metabolism and reduce demand for insulin. This increases the possibility of an insulin reaction occurring, especially if the diabetic patient has (1) taken an insulin supplement just prior to GSR training, (2) failed to previously consume adequate amounts of food or failed to digest what has been eaten or (3) taken too large a dose of insulin before beginning training.

Initial symptoms of an insulin reaction include extreme hunger, nervousness (general agitation), heavy perspiration, palpitations of the heart, and (occasionally) double vision. These symptoms may be relieved by consumption of eight ounces of orange juice to which to two teaspoonfuls of regular sugar have been added, or an orange "instant" breakfast beverage. If unavailable, any sugar or carbohydrate-containing food (bread, crackers, bananas, etc.), will serve.

If initial symptoms are unrelieved, more serious symptoms may progressively appear, which may include delirium, convulsions and/or unconsciousness (coma). To relieve these symptoms, intravenous injection of glucose may be necessary.

Notes Aside (M2005)

Limited patient use of the GSR (M2005A)

GSR monitry may not be possible if the patient is physiologically depressed and autonomically induced sweat production is suppressed. Such depression and/or sweat suppression may be seen in patients who suffer from diabetes, hypoglycemia, psychopathy, primary (endogenous) depression, long-term chronic pain (the long term pain causing an antidiuretic hormone to be produced, for unknown reasons), or are taking antidiuretic medication; such patients are said to have flat affect, and their GSR levels may be extremely low and/or will not vary when the patient is emotionally stressed.

Hand temperature and the GSR (M2005B)

It is estimated that for seventy-five percent of the population, there exists a reciprocal correlation between the GSR and the temperature of the hand; if the GSR increases (sweat production goes up) the temperature of the hand decreases, and if the GSR decreases the temperature of the hand will rise. For patients with this reciprocal correlations, the GSR relaxation technique may be used to treat peripheral circulatory disorders (increase capillary dilation) and vascular headaches.

Bibliography (M2006)

Green, pp. 1–26.
Klinge, pp. 305–317.
Miller, pp. 443–445.

GALVANIC SKIN RESPONSE MONITRY GENERAL RELAXATION (M20A)

Definition (M20A1)

General relaxation implies the depression of sympathetic autonomic activity. Utilizing galvanic skin resistance (GSR) monitry, this means learning voluntarily increasing the skin resistance of the palm of the hand, palm of the foot, or the temporal area by decreasing the GSR monitor feedback levels.

Procedures of Application (M20A2)

1) The patient should be sitting (or semi-reclining) in a position which will remain comfortable for the duration of the session.

2) The GSR instrument should be attached to the patient via the electrode cables and surface electrodes affixed over the site to be monitored (generally to the palmer surfaces of two fingers or on either side of the palm of the hand).

3) If a tactile feedback mode is available, the stimulation electrodes or the vibrating element should be affixed over an appropriate site. Stimulation elements are generally affixed to skin over the stomach area or to the anterior surface of the forearm, so that the stimulation is perceived as noxious, as a stimulus to be avoided, promoting decreases in GSR feedback activity levels, thereby increasing skin resistance.

4) The GSR instrument and feedback modes should be turned on.

5) The patient should be prevented from perceiving the feedback modes (from hearing, seeing or feeling changes in GSR activity) until after the feedback modes have stopped changing and the base line GSR activity may be noted and recorded, if visual feedback is provided.

6) The sensitivity of the instrument should be set at an appropriate sensitivity level: the patient should be instructed to take a deep breath or the patient's anterior forearm (or thigh) briefly stroked. If the visual feedback mode demonstrates a large deflection (20 to 30 mohms) the instrument should remain set in a broad range (0 to 50 mohms), but if the patient response is less than 10 mohms the instrument should be set in the appropriate 10 mohm range (0 to 10, 10 to 20, 20 to 30, 30 to 40 or 40 to 50 mohm range), and if the deflection is just under 2 mohms the instrument should be set in the appropriate 10 mohm increment range and sensitized to monitor the 2 mohm range. If a meter is being used as one of the feedback modes, the meter needle should be set near the end of the selected meter range.

7) After the base line of GSR activity has been established and the feedback instrument set at the appropriate sensitivity level, the patient should be exposed to the feedback modes, which have been adjusted to provide feedback at pitch levels or meter needle positions that, when changes in GSR activity occur, will allow the patient to perceive the change.

8) The patient should be instructed to attempt to *let* the feedback be depressed to increasingly lower levels, thereby progressively increasing skin resistance and reducing sympathetic autonomic activity.

9) At the end of the treatment session the GSR level of activity should be noted and recorded, if visual feedback is available.

10) The instrumentation should be turned off and the electrodes removed from the patient's skin.

11) All electrode cream or paste should be cleansed from the patient's skin.

This technique is often utilized as a precursor to the desensitization technique.

Partial List of Conditions Treated (M20A3)

Treatable Causes (M20A3A)

Blood sugar irregularities (C31)
Extrafusal muscle spasm (C41)
Hyperhidrosis (C37)
Hypertension (C39)
Insomnia (C46)
Migraine (vascular headache) (C34)
Nausea/vomiting (C36)
Psychoneurogenic neuromuscular hypertonus (C29)
Sympathetic hyperresponse (C38)
Tinnitus (C32)
Trigger point formation (C01)
Vascular insufficiency (C11)
Visceral organ dysfunction (C02)

Syndromes (M20A3B)

BUERGER'S DISEASE (S11)
CERVICAL (NECK) PAIN (S73)

CHEST PAIN (S14)
COLITIS (S42)
DIABETES (S10)
EARACHE (S40)
ELBOW PAIN (S47)
FOREARM PAIN (S54)
HAND/FINGER PAIN (S17)
HEADACHE PAIN (S02)
HIGH THORACIC BACK PAIN (S48)
HYPERTENSION (S43)
HYSTERIA/ANXIETY REACTION (S59)
INSOMNIA (S71)
LOWER ABDOMINAL PAIN (S62)
LOW BACK PAIN (S03)
MIGRAINE (VASCULAR) HEADACHE (S18)
PHOBIC REACTION (S44)
POST WHIPLASH (S55)
RADICULITIS (S29)
RAYNAUD'S DISEASE (S12)
REFERRED PAIN (S01)
SCIATICA (S05)
SHOULDER/HAND (S57)
SHOULDER PAIN (S08)
TEMPOROMANDIBULAR JOINT (TMJ) PAIN (S06)
TINNITUS (S70)
TOOTHACHE/JAW PAIN (S41)
WRIST PAIN (S16)

Precautions and Contraindications (M20A4)

There are no precautions or contraindications directly associated with GSR monitry or GSR training in the literature reviewed. However, caution should be observed when GSR general relaxation or desensitization training is performed with patients suffering from diabetes. The depression of sympathetic activity may decrease the patient's metabolism and reduce demand for insulin. This increases the possibility of an insulin reaction occurring, especially if the diabetic patient has (1) taken an insulin supplement just prior to GSR training, (2) failed to previously consume adequate amounts of food or failed to digest what has been eaten or (3) taken too large a dose of insulin before beginning training.

Initial symptoms of an insulin reaction include extreme hunger, nervousness (general agitation), heavy perspiration, palpitations of the heart, and (occasionally) double vision. These symptoms may be relieved by consumption of eight ounces of orange juice to which to two teaspoonfuls of regular sugar have been added, or an orange "instant" breakfast beverage. If unavailable, any sugar or carbohydrate-containing food (bread, crackers, bananas, etc.), will serve.

If initial symptoms are unrelieved, more serious symptoms may progressively appear, which may include delirium, convulsions and/or unconsciousness (coma). To relieve these symptoms, intravenous injection of glucose may be necessary.

Notes Aside (M20A5)

Limited patient use of the GSR (M20A5A)

GSR monitry may not be possible if the patient is physiologically depressed and autonomically induced sweat production is suppressed. Such depression and/or sweat suppression may be seen in patients who suffer from diabetes, hypoglycemia, psychopathy, primary (endogenous) depression, long-term chronic pain (the long term pain causing an antidiuretic hormone to be produced, for unknown reasons), or are taking antidiuretic medication; such patients are said to have flat affect, and their GSR levels may be extremely low and/or will not vary when the patient is emotionally stressed.

Hand temperature and the GSR (M20A5B)

It is estimated that for seventy-five percent of the population, there exists a reciprocal correlation between the GSR and the temperature of the hand; if the GSR increases (sweat production goes up) the temperature of the hand decreases, and if the GSR decreases the temperature of the hand will rise. For patients with this reciprocal correlations, the GSR relaxation technique may be used to treat peripheral circulatory disorders (increase capillary dilation) and vascular headaches.

Bibliography (M20A6)

Green, pp. 1–26.
Klinge, pp. 305–317.

GALVANIC SKIN RESPONSE MONITRY DESENSITIZATION (M20B)

Definition (M20B1)

Clinical research and experience has demonstrated that some patients may be taught to prevent sympathetic hyperresponse to emotional stimuli by learning to control the galvanic skin response (GSR) associated with it. The process which has been found to be the most successful is called GSR desensitization. During this process the patient is taught to control autonomic hyperresponses to emotional stress by learning to control GSR monitor feedback responses at progressively greater sensitivity levels.

GSR training for desensitization is best performed with a GSR feedback unit which employs audio, visual and tactile feedback modes and may be preset at increasingly higher sensitive levels.

Procedures of Application (M20B2)

1) The patient should be sitting (or semi-reclining) in a position which will remain comfortable for the duration of the session.

2) The GSR instrument should be attached to the patient via the electrode cables and surface electrodes affixed over the site to be monitored (generally to the palmer surfaces of two fingers or on either side of the palm of the hand).

3) If a tactile feedback mode is available, the stimulation electrodes or the vibrating element should be affixed over an appropriate site. Stimulation elements are generally affixed to skin over the stomach area or to the anterior surface of the forearm, so that the stimulation is perceived as noxious, as a stimulus to be avoided, promoting decreases in GSR feedback activity levels, thereby increasing skin resistance.

4) The GSR instrument and feedback modes should be turned on.

5) The patient should be prevented from perceiving the feedback modes (from hearing, seeing or feeling changes in GSR activity) until after the feedback modes have stopped changing and the base line GSR activity may be noted and recorded, if visual feedback is provided.

6) The sensitivity of the instrument should be set at an appropriate sensitivity level: the patient should be instructed to take a deep breath. If the visual feedback mode demonstrates a large deflection (20 to 30 mohms) the instrument should remain set in a broad range (0 to 50 mohms), but if the patient response is less than 10 mohms the instrument should be set in the appropriate 10 mohm range (0 to 10, 10 to 20, 20 to 30, 30 to 40 or 40 to 50 mohm range), and if the deflection is just under 2 mohms the instrument should be set in the appropriate 10 mohm increment range and sensitized to monitor the 2 mohm range. This step is usually taken after the patient has been successful in controlling GSR responses at less sensitive levels. If a meter is being used as one of the feedback modes, the meter needle should be set near the end of the selected meter range.

7) After the base line of GSR activity has been established and the feedback instrument set at the appropriate sensitivity level, the patient should be exposed to the feedback modes, which have been adjusted to provide feedback at pitch levels or meter needle positions that, when changes in GSR activity occur, will allow the patient to perceive the change.

8) The patient should be encouraged to talk to the practitioner, initially about topics that are non-threatening or non-stimulating to the patient. The patient should be allowed to continue to talk as long as the GSR activity is below a prescribed level (usually the upper end of the preset sensitivity range, or just below it).

9) If the GSR activity exceeds the maximum allowed ceiling level, the patient should be instructed to stop talking and not to begin again until the activity level has been made to drop below the ceiling level. If the feedback level exceeds the ability of the patient to perceive change in it, it should be adjusted down to the point where perception of change is possible, or the instrument reset to a more appropriate range.

10) When the patient is able to keep the GSR activity below the acceptable ceiling level, while talking about non-threatening subjects, the practitioner should suggest that the one-sided conversation be directed into areas that may be more stimulating and/or threatening to the patient (the patient can usually suggest the best topics to explore, if asked).

11) As the conversation continues, the practitioner should make note of the areas which provoke the greatest autonomic response. As the patient progresses, these topics should be brought up occasionally to test the patient's sympathetic autonomic reactivity.

12) When the patient is able to maintain control (keep GSR feedback activity below the designated

level), within a given sensitivity range, the instrument should be reset at a more sensitive level. This process should eventually progress to the most sensitive level.

13) At the end of each session the final GSR level should be noted and recorded, as a spot gauge of success or failure.

14) The instrumentation should be turned off and the surface electrodes removed from the patient's skin.

15) All electrode cream or paste should be thoroughly cleansed from the patient's skin.

When the patient is able to maintain control at the most sensitive GSR instrument level while talking about formerly sympathetic autonomic arousing topics, the patient may said to be *desensitized* and may no longer suffer from the symptomatology that was the basic reason for beginning such therapy.

Partial List of Conditions Treated   (M20B3)

Treatable Causes   (M20B3A)

Hyperhidrosis (C37)
Hypertension (C39)
Insomnia (C46)
Migraine (vascular headache) (C34)
Nausea/vomiting (C36)
Neurogenic dermal disorders (psoriasis, hives) (C49)
Neuromuscular tonic imbalance (C30)
Psychoneurogenic neuromuscular hypertonus (C29)
Sinusitis (C16)
Sympathetic hyperresponse (C38)
Tactile hypersensitivity (C24)
Tinnitus (C32)
Trigger point formation (C01)
Visceral organ dysfunction (C02)

Syndromes   (M20B3B)

CERVICAL (NECK) PAIN (S73)
CHEST PAIN (S14)
COLITIS (S42)
EARACHE (S40)
ELBOW PAIN (S47)
FOREARM PAIN (S54)
HAND/FINGER PAIN (S17)
HEADACHE PAIN (S02)
HIGH THORACIC BACK PAIN (S48)
HYPERTENSION (S43)
HYSTERIA/ANXIETY REACTION (S59)
INSOMNIA (S71)
LOWER ABDOMINAL PAIN (S62)
LOW BACK PAIN (S03)
MIGRAINE (VASCULAR) HEADACHE (S18)
PHOBIC REACTION (S44)
POST WHIPLASH (S55)
RADICULITIS (S29)
REFERRED PAIN (S01)
SCIATICA (S05)
SHOULDER/HAND (S57)
SHOULDER PAIN (S08)
SINUS PAIN (S33)
TEMPOROMANDIBULAR JOINT (TMJ) PAIN (S06)
TINNITUS (S70)
TOOTHACHE/JAW PAIN (S41)
TORTICOLLIS (WRY NECK) (S65)
WRIST PAIN (S16)

Precautions and Contraindications   (M20B4)

There are no precautions or contraindications directly associated with GSR monitry or GSR training in the literature reviewed. However, caution should be observed when GSR general relaxation or desensitization training is performed with patients suffering from diabetes. The depression of sympathetic activity may decrease the patient's metabolism and reduce demand for insulin. This increases the possibility of an insulin reaction occurring, especially if the diabetic patient has (1) taken an insulin supplement just prior to GSR training, (2) failed to previously consume adequate amounts of food or failed to digest what has been eaten or (3) taken too large a dose of insulin before beginning training.

Initial symptoms of an insulin reaction include extreme hunger, nervousness (general agitation), heavy perspiration, palpitations of the heart, and (occasionally) double vision. These symptoms may be relieved by consumption of eight ounces of orange juice to which to two teaspoonfuls of regular sugar have been added, or an orange "instant" breakfast beverage. If unavailable, any sugar or carbohydrate-containing food (bread, crackers, bananas, etc.), will serve.

If initial symptoms are unrelieved, more serious symptoms may progressively appear, which may include delirium, convulsions and/or unconsciousness (coma). To relieve these symptoms, intravenous injection of glucose may be necessary.

Notes Aside   (M20B5)

Practitioner discipline   (M20B5A)

Clinical experience suggests that the administering practitioner is harder to discipline than the patient in regards to having the patient stop talking when the ceiling level is exceeded. Practitioners seem to have a need to let the patient cathart (probably out of curiosity) and often seem to forget that the object of the exercise is autonomic desensitization of the patient and not an opportunity for the patient to bare the soul. The disciplined GSR process predictably produces autonomic desensitization for the patient and is far more important than a practitioner's apparent belief that an emotional "breakthrough" has an over-riding importance. Allowing the patient to cathart without maintaining control of GSR activity may destroy the process the patient has contracted to obtain.

Limited patient use of the GSR (M20B5B)

GSR monitry may not be possible if the patient is physiologically depressed and autonomically induced sweat production is suppressed. Such depression and/or sweat suppression may be seen in patients who suffer from diabetes, hypoglycemia, psychopathy, primary (endogenous) depression, long-term chronic pain (the long term pain causing an antidiuretic hormone to be produced, for unknown reasons), or are taking antidiuretic medication; such patients are said to have flat affect, and their GSR levels may be extremely low and/or will not vary when the patient is emotionally stressed.

Behavioral change (M20B5C)

It has been noted that those patients who have gone through the GSR desensitization process have been observed to change some of their behavior patterns as they gain autonomic control. Patients who apparently have spent their lives being victims to everyone in their lives (husbands, wives, children, parents, friends, associates, etc.) often begin to demonstrate assertive behavior toward former persecutors, making their own decisions and often becoming so self-directing that jobs and even marital relationships have been changed. New styles of dress and social interaction patterns seem to spontaneously appear. It seems that such patients seem to come to realize that if they can control themselves to such an extreme degree they should be able to control some of the influences in their environment, or at least their reactions to them, and go out to prove it.

Bibliography (M20B6)

Green, pp. 1–26.
Klinge, pp. 305–317.
Miller, pp. 443–445.

GALVANIC SKIN RESPONSE MONITRY GSR ASSESSMENT (M20C)

Definition (M20C1)

Evaluation of a patient's baseline galvanic skin response (GSR) activity may be useful in helping define patient states of low or diminished affect as associated with primary (endogenous) depression, hypoglycemia, diabetes, or psychopathic states (if the latter is suspected psychometric testing is advised). GSR monitry may also help evaluate patients who may be suffering from hyperhidrosis, and/or states associated with sympathetic autonomic hyperreactivity to emotional stimuli, which may be reflected by exceedingly high, maintained or widely varying GSR responses. GSR assessment may also be useful in evaluating patient response during therapy, as an augmentation of more conventional psychotherapeutic or psychiatric protocols.

Procedures of Application (M20C2)

1) The patient should be sitting (or semi-reclining) in a position which will remain comfortable for the duration of the session.

2) The GSR instrument should be attached to the patient via the electrode cables and surface electrodes affixed over the site to be monitored (generally to the palmer surfaces of two fingers or on either side of the palm of the hand).

3) If a tactile feedback mode is available, the stimulation electrodes or the vibrating element should be affixed over an appropriate site. Stimulation elements are generally affixed to skin over the stomach area or to the anterior surface of the forearm, so that the stimulation is perceived as noxious, as a stimulus to be avoided, promoting decreases in GSR feedback activity levels, thereby increasing skin resistance.

4) The GSR instrument and feedback modes should be turned on.

5) The patient should be prevented from perceiving (seeing, hearing or feeling) feedback responses, with visual feedback remaining out of the patient's line of site.

6) When the base line GSR activity has been established (the feedback from the GSR instrument stabilizing) it should be noted and recorded.

7) The sensitivity of the instrument should be set at an appropriate sensitivity level: the patient should be instructed to take a deep breath or the patient's anterior forearm (or thigh) briefly stroked. If the visual feedback mode demonstrates a large deflection (20 to 30 mohms) the instrument should remain set in a broad range (0 to 50 mohms), but if the patient response is less than 10 mohms the instrument should be set in the appropriate 10 mohm range (0 to 10, 10 to 20, 20 to 30, 30 to 40 or 40 to 50 mohm range), and if the deflection is just under 2 mohms the instrument should be set in the appropriate 10 mohm increment range and sensitized to monitor the 2 mohm range. A lack of a significant GSR response to the deep breath when the instrument is set at the most sensitive level is indicative of a lack of affect or sympathetic autonomic responsiveness.

8) If the patient's GSR activity is to be observed during a psychotherapeutic or psychiatric session, the session should begin and the patient's most notable GSR responses should be noted relative to the conversational content.

9) At the end of the treatment session the GSR level of activity should be noted and recorded, if visual feedback is available.

10) The instrumentation should be turned off and the electrodes removed from the patient's skin.

11) All electrode cream or paste should be cleansed from the patient's skin.

Partial List of Conditions Evaluated (M20C3)

Treatable Causes (M20C3A)

Blood sugar irregularities (C31)
Hyperhidrosis (C37)
Hypertension (C39)
Insomnia (C46)
Migraine (vascular headache) (C34)
Nausea/vomiting (C36)
Neurogenic dermal disorders (psoriasis, hives) (C49)
Psychoneurogenic neuromuscular hypertonus (C29)
Sinusitis (C16)
Sympathetic hyperresponse (C38)
Tinnitus (C32)
Trigger point formation (C01)
Visceral organ dysfunction (C02)

Syndromes (M20C3B)

CERVICAL (NECK) PAIN (S73)
CHEST PAIN (S14)
COLITIS (S42)
EARACHE (S40)
ELBOW PAIN (S47)
FOREARM PAIN (S54)
HAND/FINGER PAIN (S17)
HEADACHE PAIN (S02)
HIGH THORACIC BACK PAIN (S48)
HYPERTENSION (S43)
HYSTERIA/ANXIETY REACTION (S59)
INSOMNIA (S71)
LOWER ABDOMINAL PAIN (S62)
LOW BACK PAIN (S03)
MIGRAINE (VASCULAR) HEADACHE (S18)
NEUROMUSCULAR PARALYSIS (S22)
PHOBIC REACTION (S44)
POST WHIPLASH (S55)
RADICULITIS (S29)
REFERRED PAIN (S01)
SCIATICA (S05)
SHOULDER/HAND (S57)
SHOULDER PAIN (S08)
SINUS PAIN (S33)
TEMPOROMANDIBULAR JOINT (TMJ) PAIN (S06)
TINNITUS (S70)
TOOTHACHE/JAW PAIN (S41)
TORTICOLLIS (WRY NECK) (S65)
WRIST PAIN (S16)

Notes Aside (M20C5)

Limited patient use of the GSR (M20C5A)

It is estimated that for seventy-five percent of the population, there exists a reciprocal correlation between the GSR and the temperature of the hand; if the GSR increases (sweat production goes up) the temperature of the hand decreases, and if the GSR decreases the temperature of the hand will rise. For patients with this reciprocal correlations, the GSR relaxation technique may be used to treat peripheral circulatory disorders (increase capillary dilation) and vascular headaches.

GEL-FOAM IMPLANT (C21)

Definition (C21A)

Gel-foam is a man-made substance which is implanted in a surgically excised lesion and is meant to be left in the wound after incision closure. Gel-foam is designed to remain in the wound, absorb blood, and fill the tissue void created by the excision until it is reabsorbed and slowly replaced by healing body tissue.

Evaluation (C21B)

Evaluation of an over-large gel-foam implant is dependent upon visual observation and palpation examination of the implant site, and the patient's report of uncomfortable pressure and/or pain with or without compression of the body part supporting the implant.

Etiology (C21C)

Should too much gel-foam be implanted, as the natural swelling of tissue trauma diminishes a discernible swelling at the implant site may become apparent. The swelling may be accompanied by consequent pressure pain. Because of its size and the failure of the body to reabsorb it rapidly, replacement of the gel-foam implant may be an extended process and the patient may have difficulty using the involved extremity because of pain and discomfort.

Treatment Notes (C21D)

Implanted gel-foam has proven to be susceptible to the effects of ultrahigh frequency sound (ultrasound), and will apparently break down under the pressure of ultrasound waves.

To facilitate the dissolution or the break down of the gel-foam implant, ultrasound should be applied in a continuous wave form at .8 w/cm^2 from four to six minutes, once daily until the gel-foam dissolves. Usually only two or three applications are necessary to precipitate dissolution or break down.

Implant swelling will continue unchanged through the first or second ultrasound application until a critical amount of energy has been absorbed by the gel-foam. When the critical amount has been absorbed, the gel-foam will seem to "melt" under the sound head. The implant swelling will be almost entirely relieved, with the patient reporting an immediate relief to pain and a loss of the feeling of internal pressure.

TREATMENT OF GEL-FOAM SWELLING (C21F)

Technique Options (C21F1)

Evaluation techniques (C21F1A)

Massage	*M23*
Palpation evaluation	M23G
Determination of soft tissue tenderness	M23G3
Determination of soft tissue swelling	M23G6

Treatment techniques (C21F1B)

Ultrahigh frequency sound (ultrasound)	*M01*
Gel-foam dissolution	M01L

Techniques Suggested for the Chronic Condition (C21F2)

Evaluation techniques (C21F2A)

Massage	*M23*
Palpation evaluation	M23G
Determination of soft tissue tenderness	M23G3
Determination of soft tissue swelling	M23G6

Treatment techniques (C21F2B)

Ultrahigh frequency sound (ultrasound)	*M01*
Gel-foam dissolution	M01L

HABITUAL JOINT POSITIONING (C48)

Definition (C48A)

Habitual joint positioning is the voluntary or involuntary maintenance of a limited joint range of motion (allowing only a few degrees of motion) for an extended period of time.

Evaluation (C48B)

Evaluation of habitual joint positioning should be based upon range of motion measurement and muscle strength evaluation. Differential skin resistance (DSR) survey should be performed to establish the presence of any soft tissue inflammation.

Etiology (C48C)

Habitual joint positioning results when a joint is splinted or casted, for a period of four to eight weeks, as part of a therapeutic procedure or when a joint is kept within a narrow range of motion as a "habit" or in an attempt to "protect" or ignore an extremity, especially if the extremity or joint has been associated by the patient with previous injury or other pathology such as post CVA syndrome, peripheral nerve injury, or hysteria.

Adverse effects of prolonged immobilization (C48C)

(1) Osteoporosis of the bones associated with the involved joint(s) may occur if immobilization continues for more than two months if the muscles which have anchorage along the bone are prevented from stressing the bone. Lack of stress upon the structure of the bone, especially by attached muscles, promotes osteoclastic changes in the bone and inhibits osteoblastic activity. This phenomena can be summed up with the expression, "if you don't use it, you lose it".

(2) Disuse atrophy, weakness and contracture of the involved muscles may occur upon muscles both medial and distal to the immobilized joint(s).

(3) Congestion of venous and/or lymph circulation consequent to muscular disuse may foster lymph edema and create a condition conducive to thrombus formation.

(4) Long term immobilization may also promote joint capsule contractures and the formation of adhesions within involved joints and along associated tendons.

Treatment Notes (C48D)

Neuromuscular reeducation utilizing electromyometric feedback, electrical "toning", ice packing and linear vibration may be used to good effect in concert with joint manipulation and manual stretching in restoring range of motion lost from habitual joint position.

One of the most common features of habitual joint positioning is a heightened sensitivity of the skin around the joint and sometimes of the entire associated extremity. If this condition is present, a program of sensory desensitization may be necessary as a preliminary step to more extensive therapy directed at treating underlying causes, or simply to get the patient's cooperation (refer to Sensory Stimulation, Desensitization [M17D]).

Bibliography (C48E)

Basmajian, pp. 160–161.
Salter, pp. 80, 406–407.
Scott, pp. 178–181.
Shands, pp. 469–472.

TREATMENT OF HABITUAL JOINT POSITIONING (C48F)

Technique Options (C48F1)

Evaluation techniques (C48F1A)

Auxiliary evaluation techniques	**M26**
Auscultation	M26C
Joint and/or tendon sounds	M26C2C
Cryotherapy	**M10**
Evaluation of capillary response to cold	M10H
Determination of hyperreactivity to cold	M10H2A
Differential skin resistance survey	**M22**
Electrical stimulation (ES)	**M02**
Electrical evaluation	M02H
Galvanic testing	M02H2A
Faradic testing	M02H2B
Electromyometric feedback	**M18**
Electromyometric evaluation	M18L
Determination of neuromuscular hypertonicity	M18L1

habitual joint positioning

Determination of neuromuscular hypotonicity	M18L2
Determination of neuromuscular imbalance	M18L3
Determination of muscular innervation	M18L5
Exercise	*M12*
Functional testing	M12I
Evaluation of joint range of motion	M12I1
Evaluation of muscle strength	M12I2
Evaluation of functional motor skills	M12I3
Evaluation of ambulation skills	M12I4
Evaluation of balance skills	M12I5
Evaluation of developmental reflexes	M12I6
Pathological neuromuscular synergistic pattern evaluation	M12I7
Massage	*M23*
Palpation evaluation	M23G
Determination of trigger point formations	M23G1
Determination of extrafusal muscle spasm or hypertonicity	M23G2
Determination of soft tissue tenderness	M23G3
Determination of fibroid, scar, or calcific formation	M23G4
Determination of soft tissue swelling	M23G6
Sensory stimulation	*M17*
Neurological evaluation	M17E
Sensory evaluation	M17E2

Treatment techniques (C48F1B)

Cryotherapy	*M10*
Hypertonic neuromuscular management	M10B
Extrafusal muscle spasm treatment	M10B2B
Analgesia/anesthesia	M10C
Electrical facilitation of endorphin production (acusleep)	*M03*
Increasing pain tolerance	M0305A
Electrical stimulation (ES)	*M02*
Muscle lengthening	M02A
Muscle lengthening	M02A5A
Muscle reeducation/muscle toning	M02B
Increasing blood flow	M02C
Facilitation of venous blood flow	M02C2A
Increasing capillary density	M02C2B
Electromyometric feedback	*M18*
Post cerebral accident neuromuscular reeducation	M18A
Spastic hemiplegic EMM neuromuscular reeducation	M18A1C
Flaccid hemiplegic EMM neuromuscular reeducation	M18A1D
Post peripheral nerve injury neuromuscular reeducation	M18B
Post spinal cord injury neuromuscular reeducation	M18C
Neuromuscular reeducation of cerebral palsy conditions	18D
Post immobilization neuromuscular reeducation	M18I
Muscle toning	M18K
Exercise	*M12*
Muscle strengthening	M12A
Muscle relengthening	M12C
Tonically shortened musculature	M12C1A
Muscle relengthening through reciprocal inhibition	M12C1C
Guided patterning	M12H
Hydrotherapy	*M15*
Muscle tension amelioration	M15C
Ice bath treatment of rigidity & semicomatose states	M15C5A
Range of motion enhancement and physical reconditioning	M15D
Hydrotherapeutic tests	M15F
Gibbons-Landis procedure	M15F1
Test for hypersensitivity to cold	M15F3
Iontophoresis	*M07*
Pain relief	M07A
Edema control	M07C
Ischemia relief	M07D
Muscle spasm relief	M07F
Hypertonicity relief	M07G
Joint mobilization	*M13*
Massage	*M23*
Muscle fatigue (and/or tension) relief	M23A
Circulation enhancement and reduction of edema	M23B
Deep soft tissue mobilization	M23F
Muscles	M23F2A
Ligaments	M23F2B
Tendon sheaths	M23F2C
Tendons	M23F2D
Positive pressure	*M24*
Edema reduction	M24A
Sensory stimulation	*M17*
Provoking reflex physiological responses	M17C
Prolonged heating for reflex physiological responses	M17C2A
Prolonged cooling for reflex physiological responses	M17C2B
Brief cooling for reflex physiological responses	M17C2C
Desensitization	M17D
Ultrahigh frequency sound (ultrasound)	*M01*
Desensitization	M01C
Scar tissue management	M01D
Vibration	*M08*
Neuromuscular management	M08A
Muscle lengthening	M08A2B
Spasticity reduction	M08B
Inhibition of spasticity via reciprocal innervation	M08B2A
Desensitization of soft tissue	M08E

Techniques Suggested for the Acute Condition (C48F2)

Evaluation techniques (C48F2A)

Cryotherapy	**M10**
Evaluation of capillary response to cold	M10H
Determination of hyperreactivity to cold	M10H2A
Differential skin resistance survey	**M22**
Electrical stimulation (ES)	**M02**
Electrical evaluation	M02H
Galvanic testing	M02H2A
Faradic testing	M02H2B
Exercise	**M12**
Functional testing	M12I
Evaluation of joint range of motion	M12I1
Evaluation of muscle strength	M12I2
Evaluation of functional motor skills	M12I3
Evaluation of ambulation skills	M12I4
Evaluation of balance skills	M12I5
Evaluation of developmental reflexes	M12I6
Pathological neuromuscular synergistic pattern evaluation	M12I7
Massage	**M23**
Palpation evaluation	M23G
Determination of trigger point formations	M23G1
Determination of extrafusal muscle spasm or hypertonicity	M23G2
Determination of soft tissue tenderness	M23G3
Determination of fibroid, scar, or calcific formation	M23G4
Determination of soft tissue swelling	M23G6
Sensory stimulation	**M17**
Neurological evaluation	M17E
Sensory evaluation	M17E2

Treatment techniques (C48F2B)

Cryotherapy	**M10**
Hypertonic neuromuscular management	M10B
Extrafusal muscle spasm treatment	M10B2B
Electrical facilitation of endorphin production (acusleep)	**M03**
Increasing pain tolerance	M0305A
Electrical stimulation (ES)	**M02**
Muscle lengthening	M02A
Muscle lengthening	M02A5A
Muscle reeducation/muscle toning	M02B
Electromyometric feedback	**M18**
Post immobilization neuromuscular reeducation	M18I
Exercise	**M12**
Muscle strengthening	M12A
Muscle relengthening	M12C
Tonically shortened musculature	M12C1A
Muscle relengthening through reciprocal inhibition	M12C1C
Guided patterning	M12H
Positive pressure	**M24**
Edema reduction	M24A
Ultrahigh frequency sound (ultrasound)	**M01**
Desensitization	M01C
Vibration	**M08**
Neuromuscular management	M08A
Muscle lengthening	M08A2B
Desensitization of soft tissue	M08E

Techniques Suggested for the Chronic Condition (C48F3)

Evaluation techniques (C48F3A)

Auxiliary evaluation techniques	**M26**
Auscultation	M26C
Joint and/or tendon sounds	M26C2C
Cryotherapy	**M10**
Evaluation of capillary response to cold	M10H
Determination of hyperreactivity to cold	M10H2A
Differential skin resistance survey	**M22**
Electrical stimulation (ES)	**M02**
Electrical evaluation	M02H
Galvanic testing	M02H2A
Faradic testing	M02H2B
Electromyometric feedback	**M18**
Electromyometric evaluation	M18L
Determination of neuromuscular hypertonicity	M18L1
Determination of neuromuscular hypotonicity	M18L2
Determination of neuromuscular imbalance	M18L3
Determination of muscular innervation	M18L5
Exercise	**M12**
Functional testing	M12I
Evaluation of joint range of motion	M12I1
Evaluation of muscle strength	M12I2
Evaluation of functional motor skills	M12I3
Evaluation of ambulation skills	M12I4
Evaluation of Balance skills	M12I5
Evaluation of developmental reflexes	M12I6
Pathological neuromuscular synergistic pattern evaluation	M12I7
Massage	**M23**
Palpation evaluation	M23G
Determination of trigger point formations	M23G1
Determination of extrafusal muscle spasm or hypertonicity	M23G2

Determination of soft tissue tenderness	M23G3		Muscle strengthening	M12A
			Muscle relengthening	M12C
Determination of fibroid, scar, or calcific formation	M23G4		Tonically shortened musculature	M12C1A
Determination of soft tissue swelling	M23G6		Muscle relengthening through reciprocal inhibition	M12C1C

Sensory stimulation **M17**
   Neurological evaluation   M17E
   Sensory evaluation   M17E2

Treatment techniques (C48F3B)

Cryotherapy   **M10**
   Analgesia/anesthesia   M10C

Electrical facilitation of endorphin production (acusleep)   **M03**
   Increasing pain tolerance   M0305A

Electrical stimulation (ES)   **M02**
   Muscle lengthening   M02A
     Muscle lengthening   M02A5A
   Muscle reeducation/muscle toning   M02B
   Increasing blood flow   M02C
     Facilitation of venous blood flow   M02C2A
     Increasing capillary density   M02C2B

Electromyometric feedback   **M18**
   Post immobilization neuromuscular reeducation   M18I
   Muscle toning   M18K

Exercise   **M12**

Joint mobilization   **M13**

Massage   **M23**
   Deep soft tissue mobilization   M23F
     Muscles   M23F2A
     Ligaments   M23F2B
     Tendon sheaths   M23F2C
     Tendons   M23F2D

Sensory stimulation   **M17**
   Provoking reflex physiological responses   M17C
   Prolonged heating for reflex physiological responses   M17C2A
   Prolonged cooling for reflex physiological responses   M17C2B
   Brief cooling for reflex physiological responses   M17C2C
   Desensitization   M17D

Ultrahigh frequency sound (ultrasound)   **M01**
   Desensitization   M01C

Vibration   **M08**
   Neuromuscular management   M08A
     Muscle lengthening   M08A2B
   Desensitization of soft tissue   M08E

HAND/FINGER PAIN SYNDROME (S17)

Definition (S17A)

Pain in the hand and/or fingers may be a direct result of trauma to the soft tissues or joints, including burns and bacterial infection. Pain may be referred to the hand or fingers from trigger point formations occurring in other areas and/or from strained interspinous ligaments (C6, C7 and C8). Hand and/or finger pain may result from a disease process which interferes with blood supply or as a consequence of injury to the peripheral or central nervous systems from direct trauma such as nerve severance, cerebral concussion, and diseases which block nerve impulses. Pain in the hand and/or fingers may also have psychogenic origins (hysteria, psychosis, etc.), and may even be the symptom of a psychosomatic defense mechanism, manufactured or created for the purposes of hysterical conversion. In such cases the pain may be imagined but is usually real. Such pain may also be part of an anxiety reaction, in which case the pain may be induced voluntarily from trigger point formations or normal nonpainful sensory feedback may be subconsciously distorted and perceived as painful.

Related Syndromes (S17B)

ADHESION (S56)
BACTERIAL INFECTION (S63)
BUERGER'S DISEASE (S11)
BURNS (SECOND AND/OR THIRD DEGREE) (S72)
CAPSULITIS (S27)
CARPAL TUNNEL (S35)
CERVICAL DORSAL OUTLET (SCALENICUS ATTICUS/CERVICAL RIB) (S09)
CERVICAL (NECK) PAIN (S73)
DIABETES (S10)
DUPYTREN'S CONTRACTURE (S34)
FOREARM PAIN (S54)
HAND/FINGER PAIN (S17)
HYSTERIA/ANXIETY REACTION (S59)
JOINT SPRAIN (S30)
OSTEOARTHRITIS (S21)
PHOBIC REACTION (S44)
PITTING (LYMPH) EDEMA (S31)
POST IMMOBILIZATION (S36)
POST PERIPHERAL NERVE INJURY (S23)
RADICULITIS (S29)
RAYNAUD'S DISEASE (S12)
REFERRED PAIN (S01)
SHOULDER/HAND (S57)
TENDONITIS (S28)
UNHEALED DERMAL LESION (S39)

Treatable Causes Which May Contribute To The Syndrome (S17C)

Bacterial infection (C09)
Blood sugar irregularities (C31)
Calcific deposit (C08)
Extrafusal muscle spasm (C41)
Habitual joint positioning (C48)
Hypermobile/instable joint (C10)
Interspinous ligamentous strain (C03)
Joint ankylosis (C42)
Joint approximation (C47)
Ligamentous strain (C25)
Muscular weakness (C23)
Necrotic soft tissue (C18)
Nerve root compression (C04)
Neuroma formation (C14)
Neuromuscular tonic imbalance (C30)
Peripheral nerve injury (C07)
Psychoneurogenic neuromuscular hypertonus (C29)
Restricted joint range of motion (C26)
Soft tissue inflammation (C05)
Soft tissue swelling (C06)
Sympathetic hyperresponse (C38)
Tactile hypersensitivity (C24)
Trigger point formation (C01)
Unhealed dermal lesion (C12)
Vascular insufficiency (C11)
Visceral organ dysfunction (C02)

Treatment Notes (S17D)

Nerve root impingement of the C7 spinal nerve root by calcific bony formation, as is usually associated with osteoarthritic changes, or ruptured or bulging disc may be the source of pain in the hand and/or fingers. Treatment should be directed at reducing any associated inflammation, relieving any coincidental contributory trigger point formations and/or extrafusal muscle spasm and increasing vertebral joint space and/or reducing calcific deposit. Neuromuscular hypertonicity of associated musculature has often been found to be contributory to nerve root impingement occurring in the cervical area and electromyometric neuromuscular reeducation directed at teaching the patient how to reduce muscular tension associated with peripheral nerve root compression has proven useful as a treatment technique.

The following is a list of trigger point formations which may, singly or in combination, refer pain into the fingers and/or hand. It should be noted that it is possible for the given patient to experience pain throughout the entire stereotypical referred pain pattern or only in part or parts of it. Opposite to each listing is a description of the parts of the stereotypical referred pain pattern most commonly experienced by previous patients, relative to the given trigger point formation. Any descriptive reference which is underlined is a part of the pattern which has *not* been commonly experienced by previous patients. For complete description of each stereotypical referred pain pattern, refer to the Table of Referred Pain Patterns of Trigger point Formation Origin [T005]. Please note the key to abbreviated words at the bottom of the list.

LOCATION (MUSCLE)		REFERRED PAIN ZONES (GENERAL)
Scalenus	(4)	*Shoulder*, med. & l.pect. maj., med.bord. of scapula, deltoids, *l.arm*, p.thumb, *p.index & m.finger*
Scalenus (minimus)		L. Triceps, *p. forearm & hand & f.s*
Infraspinatus	(2)	*Subocciput*, deltoids, *l.arm*, palm, *p.hand*
Coracobrachialis		A. deltoid, shoulder joint, *l.triceps, p.arm & hand, p.m.finger*
Upper latissimus dorsi		Latissimus dorsi, *med. arm, l.hand*, fingers 4 & 5
Serratus posterior superior (under the scapula)		Scapula, p.deltoid, med. arm, wrist, med. hand, finger 5
Serratus anterior		Lower a.l.chest, *med. a.arm & hand*, d.med. bord. of scapula, *a.f.s 4 & 5*
Subclavius		Under clavicle, *a.deltoid*, biceps, brach., *l.hand, a.& p.f.s 1–3*
Pectoralis major (sternal)	(3)	Pectoralis major, a.deltoid, *med. a. upper arm*, prox. a.forearm, m.palm, *a.f.s 3–5*
Pectoralis minor		*Pectoralis major*, a.deltoid, *med. a. arm*, med. a.hand, *a.f.s 3–5*
Medial triceps (deep fibers)		Med. a.elbow & *forearm*, *a.f.s 3–5*
Lateral triceps		L. triceps, *p.forearm*, *p.f.s 4–5*
Brachialis (superior & inferior)	(4)	A. deltoid, *a.half m.deltoid, a.elbow*, a.& p.first metacarpal, l.hand, p.second metacarpal
Extensor carpi radialis brevis		P. d.2/3 *forearm*, p.wrist & hand
Extensor carpi ulnaris		P. med. d.forearm & wrist & prox. hand
Middle finger extensor		P. *central forearm* & middle finger
Ring finger extensor		P. origin of brach., p.med. forearm, p.finger 4
Palmaris longus		A. *med. forearm*, palm
Flexor carpi radialis		L. *a.forearm*, a.wrist, *palm*
Flexor carpi ulnaris		Med. a.d.forearm, a.med. wrist-hand
Brachioradialis		P. l.elbow, *l.forearm*, p.metacarpals 1 & 2
Pronator teres		A. *l.forearm* & wrist, *a.metacarpal #1*
Extensor indicus		P. wrist, *p.metacarpals 2-4 & second* MP joint
Radial head of the flexor digitorum sublimis		A. *l.wrist, central palm*, a.m.finger
Humeral head of the flexor digitorum sublimis		A. of *carpals & metacarpals* & f.s 4–5
Flexor pollicis longus		A. *thenar eminence* & thumb
Abductor digiti minimi		P. *l.metacarpal* & finger 5
Second dorsal interosseus	(2)	P. *l.finger 3*
Opponens pollicis		A. l.wrist, a.metacarpal & finger 1
Adductor pollicis		L. metacarpal 1, a.& *p.of metacarpals 1–2 & thumb*
First dorsal interosseus		*Palm*, f.2, p.metacarpals 2-5, l.p.f.5
Extensor carpi radialis longus		P. l.elbow, p.l.forearm & wrist, p.carpals, p.metacarpals 1–4

Key:
TMJ—temporomandibular joint
A/C—acromioclavicular
med.—medial
brach.—brachioradialis
para.—paraspinous

MP—metacarpal phalangeal
sterno.—sternocleidomastoideus
a.—anterior
s.—superior
&—and
maj.—major
bord.—border
f.—finger
l.—lateral
m.—middle
d.—distal
p.—posterior

prox.—proximal
bil.—bilateral
f.s—fingers
i.—inferior
pect.—pectoralis

Bibliography (S17E)

Rivlin, pp. 29–49.
Salter, pp. 244–248.
Scott, pp. 156–211.
Shands, pp. 444–457.

HEADACHE PAIN SYNDROME (S02)

Definition (S02A)

Headache (cephalgia) is a diffuse pain which may occur in various parts of the head and is not generally confined to the area of distribution of any given peripheral nerve.

Headaches vary between patients relative to the precise location, time of occurrence, duration, frequency and accompanying symptoms.

Headaches may be accompanied by malaise, disorientation, photophobia, irritability, sleeplessness, fever, nausea, vomiting, nervousness and/or vertigo. Some patients may additionally have pain in other areas of the body which they commonly associate with the headache pain (shoulder pain, stomach cramps, back pain, etc.).

Headache without fever may be an accompanying symptom of central nervous system tumors and abscesses, middle ear disease, sinusitis, the late stage of syphilis, cysts, simple or malignant encephalitis, multiple sclerosis, cephalic thrombosis, embolism, aneurysm and arteriosclerosis, mastoiditis, osteoarthritis, herpes zoster, spondylitis, iritis, glaucoma, uremia, gout, diabetes, high blood pressure, low blood pressure, hypoglycemia ("hunger headaches"), motion sickness, sunstroke, allergies and as a sequelae of episodic malaria.

Headache with fever may accompany kidney disease, cephalic or meningeal disease, influenza, measles, gastroenteritis, typhoid fever, paratyphoid fever, tropical illnesses of various types and the common cold.

Related Syndromes (S02B)

BACTERIAL INFECTION (S63)
CERVICAL (NECK) PAIN (S73)
EARACHE (S40)
HYPERTENSION (S43)
HYSTERIA/ANXIETY REACTION (S59)
MIGRAINE (VASCULAR) HEADACHE (S18)
OSTEOARTHRITIS (S21)
PHOBIC REACTION (S44)
POST IMMOBILIZATION (S36)
POST PERIPHERAL NERVE INJURY (S23)
POST WHIPLASH (S55)
REFERRED PAIN (S01)
SHINGLES (HERPES ZOSTER) (S37)
SINUS PAIN (S33)
TEMPOROMANDIBULAR JOINT (TMJ) PAIN (S06)
TIC DOULOUREUX (S53)
TOOTHACHE/JAW PAIN (S41)

Treatable Causes Which May Contribute to the Syndrome (S02C)

Bacterial infection (C09)
Blood sugar irregularities (C31)
Calcific deposit (C08)
Dental pain (C22)
Extrafusal muscle spasm (C41)
Habitual joint positioning (C48)
Hypermobile/instable joint (C10)
Hypertension (C39)
Insomnia (C46)
Interspinous ligamentous strain (C03)
Joint ankylosis (C42)
Joint approximation (C47)
Ligamentous strain (C25)
Localized viral infection (C27)
Migraine (vascular headache) (C34)
Muscular weakness (C23)
Nerve root compression (C04)
Neuromuscular tonic imbalance (C30)
Peripheral nerve injury (C07)
Psychoneurogenic neuromuscular hypertonus (C29)
Restricted joint range of motion (C26)
Sinusitis (C16)
Soft tissue inflammation (C05)
Soft tissue swelling (C06)
Sympathetic hyperresponse (C38)
Tactile hypersensitivity (C24)
Tinnitus (C32)
Trigger point formation (C01)
Visceral organ dysfunction (C02)

Treatment Notes (S02D)

The most important step in the treatment of the headache pain syndrome is the evaluation. The headache can have so many causes which produce similar symptomatology that without careful assessment it can be very easy to presume a cause or causes which may in fact not be responsible.

Special attention should be paid to the evidence supplied by electromyometry and trigger point surveys regarding the role(s) that various neck, shoulder and/or jaw muscles may be playing in the production of the headache syndrome.

The most common mistake made when establishing the source of the headache is the assumption that the first treatable cause established to be present is the only cause. It is true that most commonly a headache will have a single source, but a not inconsiderable number have multiple causes.

headache pain syndrome

For example, a patient may have a migraine (vascular) headache and a tension or referred pain (trigger point) headache at the same time, with one syndrome (or set of symptoms) overlapping the other. Typically, these patients will not respond well to vasoconstrictor medication (which is remarkably effective in the treatment of the vascular headaches) because the vasoconstrictor medication has little or no effect upon the trigger point formation or the perception of its referred pain. In such cases, both sources of the headache must be treated if the patient is to be entirely relieved of the pain.

The following is a list of trigger point formations which may, singly or in combination, refer pain into the area of the head concern. It should be noted that it is possible for the given patient to experience pain throughout the entire stereotypical referred pain pattern or only in part or parts of it. Opposite to each listing is a description of the parts of the stereotypical referred pain pattern most commonly experienced by previous patients, relative to the given trigger point formation. Any descriptive reference which is underlined is a part of the pattern which has *not* been commonly experienced by previous patients. For complete description of each stereotypical referred pain pattern, refer to the Table of Referred Pain Patterns of Trigger point Formation Origin [T005]. Please note the key to abbreviated words at the bottom of the list.

LOCATION (MUSCLE)		REFERRED PAIN ZONES (GENERAL)
Masseter (deep)		TMJ, lateral jaw, ear, *suboccipital*
Masseter (superficial B)	(2)	Mandible, TMJ, eyebrow, *cheek bone*
Temporalis (anterior)		Upper incisors, a.temple, eyebrow
Temporalis (middle A)		Upper bicuspids, m.temple
Temporalis (middle B)		Upper bicuspids & molars, p.temple
Temporalis (posterior)		Posterior & superior to the ear
Medial pterygoid		*Ear*, TMJ, *mandible, lateral throat*
Lateral pterygoid	(2)	Anterior maxilla, TMJ
Posterior digastric		Mastoid process, *jaw line, occiput*
Frontalis		Ipsilateral forehead
Suboccipital	(2)	Lower hemicranium, *eye*
Occipitalis		Central cranium, superior eye orbit
Semispinalis capitus		*Central cranium,* a.temple, *forehead*
Semispinalis cervicis		*Hemiocciput*
Upper trapezius [A]	(2)	Posterior neck, occiput, temple, *eye*
Posterior cervical group (Semispinalis capitus, semispinalis cervicis, C4 or C5 multifidus)	(3)	Paraspinous area from nuchal line to mid thoracic area
Splenius capitus [A]		*Lower hemicranium, ear,* behind eye
Sternocleido-mastoideus (superficial)	(3)	Sterno., occiput, across forehead & temple, *face, chin, superior lateral throat, top of head, behind & in eye*
Sternocleido-mastoid (deep)	(3)	Sterno., ear, occiput, bil. forehead
Obicularis oris (orbital)		Eyebrow, nose, *a.face, upper lip*
Zygomaticus (major)		Medial forehead, l.of nose & mouth
Splenius capitus [B]		Top of the head

Key:
TMJ—temporomandibular joint
A/C—acromioclavicular
med.—medial
brach.—brachioradialis
para.—paraspinous
MP—metacarpal phalangeal
sterno.—sternocleidomastoideus
pect.—pectoralis
a.—anterior
s.—superior
&—and
maj.—major
bord.—border
f.—finger
l.—lateral
m.—middle
d.—distal
p.—posterior
prox.—proximal
bil.—bilateral
f.s—fingers
i.—inferior

Bibliography (S02E)

Basmajian, pp. 99–105.
Chusid, pp. 391–397, 394–397.
Gardner, pp. 108–110.
Scott, pp. 56–58.
Travell, Simons, pp. 207–209, 236–238.

HICCUP (HICCOUGH) OR DIAPHRAGM SPASM (C44)

Definition (C44A)

A hiccup (hiccough) is an involuntary spasmodic contraction of the diaphragm which results in the spasmodic intake of small amount of air, followed by the sudden closure of the glottis which checks the inflow of air, producing a characteristically peculiar gasping sound.

Evaluation (C44B)

The severity of hiccups depends upon the frequency (occurrences per minute) and the duration of the episode. Frequency may be determined by observing the rate of diaphragm spasm.

Etiology (C44C)

The hiccup may result from irritation of the afferent or efferent nerve pathways to the diaphragm, or nervous centers that control the muscles of respiration, especially the diaphragm. Hiccups may be caused when the afferent nerve pathways are stimulated by swallowing hot foods or other irritating substances, including roughly textured foods like popcorn or caustic substances like raw alcohol, and by disorders of the stomach and/or esophagus such as gastric dilatation, gastritis, or esophagospasm.

Pathological sources of hiccup include intestinal obstruction, pancreatitis, strangulated hernia, peritonitis, intraperitoneal tumor, amebic, typhoid or other types of enteritis, hepatitis, pregnancy and bladder irritation. Another source may be a serious, life-threatening, post operative condition, interfering with eating, drinking and even breathing. Hiccups may be caused by vascular or inflammatory lesions or tumors of the medulla oblongata (a control center of the respiratory muscles), inflammation or tumors of the mediastinum, enlargement of the heart and adherent pericarditis; they may accompany diaphragmatic pleurisy, pneumonia, uremia or alcoholism, and may result from psychogenic sources.

Treatment Notes (C44D)

Continuous peripheral electroacustimulation (fifteen to twenty-minutes), and auricular electroacustimulation of appropriate acupoints has proven to be notably effective (in some cases) at relieving prolonged episodes of hiccups which have been resistant to other remedies, as long as seven days post onset.

Bibliography (C44E)

Cooley, pp. 238–239, 953.
Holvey, pp. 1020–1022.

TREATMENT OF HICCUP (HICCOUGH) OR DIAPHRAGM SPASM (C44F)

Technique Options (C44F1)

Evaluation techniques (C44F1A)

Auricular electroacustimulation	*M06*
Auricular acupoint evaluative survey	M06C
Determination of pathological symptom source	M06C5B
Peripheral electroacustimulation	*M05*
Peripheral acupoint survey	M05D

Treatment techniques (C44F1B)

Auricular electroacustimulation	*M06*
Pathology control	M06B
Electromyometric feedback	*M18*
Breathing pattern correction	M18J
Laser stimulation	*M27*
Massage	*M23*
Acupressure	M23C
Peripheral electroacustimulation	*M05*
Pathology control	M05C
Physiological dysfunction	M05C5A
Transcutaneous nerve stimulation	*M04*
Prolonged stimulation of acupoints	M04B
Ultrahigh frequency sound (ultrasound)	*M01*
Acupoint stimulation	M01K

Techniques Suggested for the Acute Condition (C44F2)

Evaluation techniques (C44F2A)

Auricular electroacustimulation	*M06*
Auricular acupoint evaluative survey	M06C
Determination of pathological symptom source	M06C5B

Peripheral electroacustimulation **M05**
  Peripheral acupoint survey   M05D

Treatment techniques (C44F2B)

Auricular electroacustimulation   **M06**
  Pathology control   M06B
Peripheral electroacustimulation   **M05**
  Pathology control   M05C
    Physiological dysfunction   M05C5A
Transcutaneous nerve stimulation   **M04**
  Prolonged stimulation of acupoints   M04B

Techniques Suggested for the Chronic Condition (C44F3)

Evaluation techniques (C44F3A)

Auricular electroacustimulation   **M06**
  Auricular acupoint evaluative survey   M06C
  Determination of pathological symptom source   M06C5B
Peripheral electroacustimulation   **M05**
  Peripheral acupoint survey   M05D

Treatment techniques (C44F3B)

Auricular electroacustimulation   **M06**
  Pathology control   M06B
Electromyometric feedback   **M18**
  Breathing pattern correction   M18J
Peripheral electroacustimulation   **M05**
  Pathology control   M05C
    Physiological dysfunction   M05C5A
Transcutaneous nerve stimulation   **M04**
  Prolonged stimulation of acupoints   M04B

HIGH THORACIC BACK PAIN SYNDROME (S48)

Definition (S48A)

Pain in the high thoracic back area may result from soft tissue inflammation and/or swelling, muscle strain or spasm, pain referred from other parts and areas of the body or from peripheral nerve impingement. High thoracic back pain may also arise from pathological relationships between the vertebral bodies, intervertebral discs and/or costal articulations and joint distortion from various diseases of the joint.

Malformations of the spine which may lead to joint or muscular problems include scoliosis, kyphosis and lordosis. These conditions may eventually lead to pain because of abnormal pressure exerted upon intervertebral discs, calcific deposits encroaching upon soft tissues and/or abnormal muscle balance between stabilizing antagonists resulting from structural anomaly or lack of structural symmetry.

Related Syndromes (S48B)

ACNE (S45)
BACTERIAL INFECTION (S63)
CERVICAL DORSAL OUTLET (SCALENICUS ATTICUS/CERVICAL RIB) (S09)
CERVICAL (NECK) PAIN (S73)
CHEST PAIN (S14)
CHRONIC BRONCHITIS/EMPHYSEMA (OBSTRUCTIVE PULMONARY DISEASE (S13)
FACET (S19)
FASCIITIS (S20)
FROZEN SHOULDER (S64)
HAND/FINGER PAIN (S17)
HYSTERIA/ANXIETY REACTION (S59)
MYOSITIS OSSIFICANS (S67)
OSTEOARTHRITIS (S21)
PHOBIC REACTION (S44)
PNEUMONIA (S58)
REFERRED PAIN (S01)
SHINGLES (HERPES ZOSTER) (S37)
SHOULDER/HAND (S57)
SHOULDER PAIN (S08)
UNHEALED DERMAL LESION (S39)

Treatable Causes Which May Contribute To The Syndrome (S48C)

Bacterial infection (C09)
Calcific deposit (C08)
Extrafusal muscle spasm (C41)
Habitual joint positioning (C48)
Hypermobile/instable joint (C10)
Interspinous ligamentous strain (C03)
Joint ankylosis (C42)
Joint approximation (C47)
Ligamentous strain (C25)
Localized viral infection (C27)
Lung fluid accumulation (C13)
Muscular weakness (C23)
Necrotic soft tissue (C18)
Nerve root compression (C04)
Neurogenic dermal disorders (psoriasis, hives) (C49)
Neuroma formation (C14)
Neuromuscular tonic imbalance (C30)
Peripheral nerve injury (C07)
Psychoneurogenic neuromuscular hypertonus (C29)
Reduced lung vital capacity (C17)
Restricted joint range of motion (C26)
Scar tissue formation (C15)
Soft tissue inflammation (C05)
Soft tissue swelling (C06)
Sympathetic hyperresponse (C38)
Trigger point formation (C01)
Unhealed dermal lesion (C12)
Visceral organ dysfunction (C02)

Treatment Notes (S48D)

Muscular imbalance of the paraspinous and/or rhomboid (including lower trapezius) muscles has been successfully treated utilizing electromyometry (EMM) in a program of neuromuscular reeducation. Balancing antagonistic paraspinous thoracic musculature or balancing paraspinous thoracic musculature against homolateral pectoralis minor musculature is the basis for this procedure. The subliminal learning theme of facilitation, inhibition, balance and function (in this case, electromyometric inhibition of previous hyperactive musculature during the patient's speech is performed during the functional phase) is generally utilized.

The following is a list of trigger point formations which may, singly or in combination, refer pain into the upper thoracic area. It should be noted that it is possible for the given patient to experience pain throughout the entire stereotypical referred pain pattern or only in part or parts of it. Opposite to each listing is a description of the parts of the stereotypical referred pain pattern most commonly experienced by previous patients, relative to the given trigger point formation. Any descriptive reference which is underlined is a part of the pattern which has *not* been commonly experienced by previous patients. For complete description of each stereotypical referred pain pattern, refer to the Table of Referred Pain Patterns of Trigger point Formation Origin [T005]. Please note the key to abbreviated words at the bottom of the list.

LOCATION (MUSCLE)		REFERRED PAIN ZONES (GENERAL)
Posterior cervical group (Semispinalis capitus, semispinalis cervicis, C4 or C5 multifidus)	(3)	Paraspinous area from nuchal line to mid thoracic area
Levator scapulae	(2)	*Med. & l. bord. of scapula, joint, para. C2-T2, upper trapezius*
Scalenus	(4)	*Shoulder*, med. & l.pect. maj., med. bord. of scapula, deltoids, *l.arm, p.thumb, p.index & m.finger*
Infraspinatus (abnormal)		Med bord. & i.angle of scapula
Lower trapezius [A]		Subocciput, *para. C1-C6, upper trapezius*, A/C joint
Lower trapezius [B]		M. scapula bord.
Upper latissimus dorsi		Latissimus dorsi, *med. arm, l.hand, fingers 4 & 5*
Serratus posterior superior (under the scapula)		Scapula, p.deltoid, med. arm, wrist, med. hand, finger 5
Serratus anterior		Lower a.l.chest, *med. a.arm & hand*, d.med. bord. of scapula, *a.f.s 4 & 5*
Subscapularis		*Scapula, axilla, p.deltoid, med. triceps, a.p.wrist*
Rhomboids	(3)	Med. bord. of scapula, *supraspinatus*
Upper rectus abdominus		Bil. across the p. lower thoracic area (T8-T11) i. to the scapulae

Key:
TMJ—temporomandibular joint
A/C—acromioclavicular
med.—medial
brach.—brachioradialis
para.—paraspinous
MP—metacarpal phalangeal
sterno.—sternocleidomastoideus
pect.—pectoralis
a.—anterior
s.—superior
&—and
maj.—major
bord.—border
f.—finger
l.—lateral
m.—middle
d.—distal
p.—posterior
prox.—proximal
bil.—bilateral
f.s—fingers
i.—inferior

Bibliography (S48E)

Basmajian, pp. 91–99.
Salter, pp. 310–319.
Scott, pp. 233–239.
Shands, pp. 299–320.
Travell, Simons, pp. 614–656.

HYDROTHERAPY (M15)

Definition (M1501)

Hydrotherapy, or medical hydrology, has traditionally been defined as the external or internal application of water in its solid, liquid, or vapor forms for promotion of healing processes. Here we will limit our definition to the external application of water in its liquid or vapor forms for the production of physically beneficial physiological changes.

Hydrotherapy, in its various forms, is usually applied to help relieve pain, restore function or to accelerate the repair of diseased or injured tissues. The forms of hydrotherapy considered here include: (1) immersion baths such as Hubbard tanks, whirlpool or oxygen baths, contrast baths, etc., (2) moist hot air (steam) baths, and (3) hydrosprays (showers or douches). These basic forms of hydrotherapy assume different names and are used for different purposes as the temperature of the water utilized is changed, the manner of application is varied, and as other modalities are applied in conjunction with the hydrotherapy application.

Procedures of Application (M1502)

Immersion baths (M152A)

1) Immersion baths require a vessel large enough to contain the portion of the patient to be immersed and the volume of water necessary for immersion.

2) The surrounding room should be warm and free from draft.

3) Before use, the vessel to be used should be thoroughly cleansed with an antiseptic solution, well enough that the practitioner would be willing to be similarly immersed in it.

4) Before filling the vessel, determine where and how it should be situated to allow for comfortable body part immersion without the vessel rim impinging upon any other body part, such as the inside of the arm (axilla) or the patient's trunk.

5) Before filling the vessel, situate it so that the body part may be immersed and retrieved easily without undue strain upon the patient or the practitioner. If full body immersion is practiced, the vessel should be easy to get into and out of.

6) Before immersion, remove all clothing, except for bathing attire, as well as all dressings which may be removed safely. Sanitary precautions on the part of the practitioner should be observed, for self-protection and to avoid undue contamination of the bath.

7) Before immersion, the vessel should be filled to the desired level.

8) Before immersion, any electrical appliances, such as whirlpools, ultrasound or electrical stimulation units, which are to be used while immersion is taking place should be adequately grounded by three-way plug or separate ground wire attached to water pipe or radiator and should be safety tested by the practitioner. To do this, the electrical equipment should be turned on and applied to the practitioner, or at least to a part of the practitioner, as a test of equipment and set-up integrity. After testing, electrical equipment should be turned off until after the patient's body part has been immersed.

9) Immersion should not commence until the prescribed water temperature has been reached (refer to *Notes Aside* [1505B] for water temperature values).

10) Immerse the patient's body part for the time required for completion of the treatment procedure.

11) Immersion in warm water baths should be followed, as a general rule (if the patient's constitution allows it), by a cold shower for the vasomotor response.

12) Following immersion, thoroughly dry the body part and any other area which has been moistened. Apply dry dressings to wounds when appropriate. The patient should not be allowed to leave with damp clothes or skin.

13) If the patient has engaged in exercise during immersion, the patient should be made to rest for ten to fifteen minutes before being allowed to leave the treatment area.

14) Thoroughly cleanse and rinse vessel before using it again.

Moist hot air (steam) baths (M1502B)

1) Lying or sitting platforms should be cleansed thoroughly before use.

2) Before the patient enters the steam room or heat cabinet, it should be steamed or heated to desired levels. The temperature should be checked with thermometry by the practitioner.

3) Before the patient enters, the steam (if piped in) should be turned off.

4) Before entering the bath, the patient's pulse and respiratory count should be taken.

5) Upon entering the room or cabinet, the patient should assume a lying or sitting position.

6) During the bath, the patient's pulse and respiratory count should be taken every five minutes.

hydrotherapy

If a cabinet is utilized and the patient's head extrudes through the cabinet top, the carotid pulse may be taken. If pulse rate exceeds 100, dyspnea occurs. If the patient feels faint, the treatment should be terminated.

7) During the treatment session, damp cool towels to be wrapped around the head and tepid liquids for drinking may be offered to the patient to prevent heat exhaustion.

8) During the treatment session, the steam may be turned on or water poured over hot rocks or pipes several times to maintain the steam concentration. The patient should be kept away from steam vents or the vapor source to guard against being scalded.

9) After the prescribed period of treatment, upon leaving the bath, the patient should shower or sponge down with tepid or cool water.

10) The patient should towel down thoroughly and leave the treatment area with dry skin and in warm dry clothing.

11) All support apparatus (stools, benches, etc.) should be thoroughly cleansed and dried before being used again.

Hydrosprays (showers or douches) (M1502C)

1) Hydrospray application (shower or douche) requires a multiple needle spray head; the water is generally conveyed to the spray head by water pipe (showers) or by hose.

2) Sitting or lying platforms should be cleansed thoroughly before use. The hydrospray may be applied with the patient in lying, sitting or standing positions.

3) Before application, all clothing except for appropriate bathing costume should be removed from the patient. A bathing cap may be worn if the head is to remain dry. Waterproof thong sandals or clogs should be worn if the patient is to be standing during application.

4) Before application, the water temperature should be tested and adjusted to the required temperature (98° F. if used simply for cleansing).

5) During treatment, the spray should generally be applied to strike the entire body surface. If applied via hose, it should be applied horizontally from four directions.

6) Application of warm water sprays or douches are often followed by a tepid or cool shower or douche for the vasomotor response.

7) Following application, the spray should be turned off.

8) Following application, the patient should be aided, if necessary, to dry off completely by toweling, with special attention to drying off the feet.

9) The patient should leave the treatment area with dry skin and in warm, dry clothing.

10) All support apparatus (floor, stool, bench, table, etc.) should be thoroughly cleansed and dried before being used again.

Partial List of Conditions Treated (M1503)

Conditions treated in the field (M1503A)

Amputated stumps
Sprains
Contusions
Fibrositis
Burns
Lymphangitis
Rheumatoid arthritis
Fever convalescence
Hyperthyroidism
Nervous irritability
Poliomyelitis
Parkinson's disease
Multiple sclerosis
Postoperative orthopaedic conditions
Decubitus ulcers
Mild fever induction
Increase pulse rate
Increase metabolism
Gout
Obesity
Alcoholism and other substance abuse
Sweat induction

Treatable Causes (M1503B)

Bacterial infection (C09)
Blood sugar irregularities (C31)
Calcific deposit (C08)
Extrafusal muscle spasm (C41)
Habitual joint positioning (C48)
Hiccup (C44)
Hypermobile/instable joint (C10)
Hypertension (C39)
Hypotension (C45)
Insomnia (C46)
Interspinous ligamentous strain (C03)
Joint ankylosis (C42)
Joint approximation (C47)
Ligamentous strain (C25)
Localized viral infection (C27)
Lung fluid accumulation (C13)
Migraine (vascular headache) (C34)
Muscular weakness (C23)
Nausea/vomiting (C36)
Necrotic soft tissue (C18)
Nerve root compression (C04)
Neurogenic dermal disorders
   (psoriasis, hives) (C49)

Pathological neuromuscular hypertonus (C28)
Pathological neuromuscular hypotonus (C33)
Pathological neuromuscular dyscoordination (C35)
Psychoneurogenic neuromuscular hypertonus (C29)
Reduced lung vital capacity (C17)
Restricted joint range of motion (C26)
Sinusitis (C16)
Soft tissue inflammation (C05)
Soft tissue swelling (C06)
Sympathetic hyperresponse (C38)
Tactile hypersensitivity (C24)
Trigger point formation (C01)
Unhealed dermal lesion (C12)
Vascular insufficiency (C11)
Visceral organ dysfunction (C02)

Syndromes (M1503C)

ACHILLES TENDONITIS (S52)
ACNE (S45)
ADHESION (S56)
BACTERIAL INFECTION (S63)
BUERGER'S DISEASE (S11)
BURNS (SECOND AND/OR THIRD DEGREE) (S72)
CALF PAIN (S50)
CEREBRAL PALSY (S25)
CERVICAL (NECK) PAIN (S73)
CHEST PAIN (S14)
CHONDROMALACIA (S15)
CHRONIC BRONCHITIS/EMPHYSEMA (OBSTRUCTIVE PULMONARY DISEASE) (S13)
COLITIS (S42)
DIABETES (S10)
ELBOW PAIN (S47)
FACET (S19)
FASCIITIS (S20)
FOOT PAIN (S51)
FOREARM PAIN (S54)
FROZEN SHOULDER (S64)
HAND/FINGER PAIN (S17)
HEADACHE PAIN (S02)
HIGH THORACIC BACK PAIN (S48)
HYPERTENSION (S43)
HYSTERIA/ANXIETY REACTION (S59)
INSOMNIA (S71)
JOINT SPRAIN (S30)
KNEE PAIN (S46)
LOWER ABDOMINAL PAIN (S62)
LOW BACK PAIN (S03)
MIGRAINE (VASCULAR) HEADACHE (S18)
NEUROMUSCULAR PARALYSIS (S22)
OSTEOARTHRITIS (S21)
PIRIFORMIS (S04)
PITTING (LYMPH) EDEMA (S31)
PNEUMONIA (S58)
POST CEREBRAL VASCULAR ACCIDENT (CVA) (S07)
POST IMMOBILIZATION (S36)
POST WHIPLASH (S55)
RADICULITIS (S29)
RAYNAUD'S DISEASE (S12)
REFERRED PAIN (S01)
SCIATICA (S05)
SHOULDER/HAND (S57)
SHOULDER PAIN (S08)
SINUS PAIN (S33)
TENDONITIS (S28)
TENNIS ELBOW (LATERAL EPICONDYLITIS) (S32)
THIGH PAIN (S49)
UNHEALED DERMAL LESION (S39)
WRIST PAIN (S16)

Precautions and Contraindications (M1504)

Water temperature (M1504A)

For immersion baths or sprays, water temperatures should not exceed 110° F. and steam room temperatures should not exceed 110° F.

Chilling (M1504B)

Extreme care should be taken to avoid chilling the patient both before and after treatment. Signs of chilling include bluish lips, undue pallor, cutis anserina (goose flesh) and shivering.

Hypersensitivity to cold (M1504C)

Individuals who are hypersensitive to cold may develop severe vasospasms upon immersion in low temperature water. If the hand of such a patient is immersed in low temperature water for a few minutes it will produce redness, swelling, uticaria, and increased local temperature. After a latent period of three to six minutes, a systemic reaction will occur with the symptoms of flushing of the face, a sharp fall in blood pressure, a rise in pulse rate, and occasionally syncope. Recovery from such symptoms generally takes from five to ten minutes.

Frostbite (M1504D)

A sustained immersion in water below 43° F. may produce cell damage and frostbite. Frostbite may cause blueness, mottled appearance, or redness of the skin, with areas of anesthesia and stiffness of the involved part.

Peripheral vascular disease (M1504E)

Avoid overheating the patient who suffers from peripheral vascular disease or peripheral nerve injury.

Prolonged local heat application to areas with impaired blood supply may produce tissue damage. Heat application increases tissue temperature and causes a rise in local metabolic rate which results in waste product accumulation, causing increased cell membrane permeability and increasing the possibility of edema. Varicose veins, arteriosclerosis, thromboangiitis obliterans and other diseases of the circulatory system will limit the ability of the body to rapidly dissipate heat applied to it. An extremity suffering from vascular disease should not be used to elicit a reflex vasodilation because of the danger of local tissue damage and the desired effect may not be elicited since the vessels may not dilate or constrict normally.

Heat exhaustion (M1504F)

Heat exhaustion is due to circulatory collapse following excessive loss of fluids and chlorides due to sweating. It may occur if the bath temperature remains or exceeds 98° F. for a prolonged period. A hot, humid atmosphere such as is found in a moist air sauna may produce the same results, which are giddiness, faintness, nausea, shallow or gasping breath, weak, rapid or fluttering pulse, moist cool skin, profuse sweating, facial pallor, drooping of the corners of the mouth, anoxemia and tissue anoxia, cramps, tetany, low blood pressure, a drop in systemic temperature and/or dehydration. Heat exhaustion should be treated by keeping the patient warm by covering with blankets, placing the head in a low position, and administering cool water with .5 grams of salt per pint as tolerated.

Heat stroke (M1504G)

Heat stroke may result from a failure of the body to regulate heat and exhaustion of the sweat glands. Symptoms may include suppression of perspiration (hot dry skin), rapid pulse, red face, increased blood pressure, increased systemic temperature, increased urination, pulmonary edema, loss of consciousness and/or delirium. The treatment of heat stroke should include elevating the patient's head, cooling the patient's body with cold applications (ice water bath) or fanning. The patient should be kept quiet until the body temperature is stable.

Hypersensitivity to heat (M1504H)

Heat sensitivity is a rare condition but occasionally occurs and results from a derangement of the heat-regulating mechanism and may produce exhaustion. Symptoms include weakness, dizziness, fainting spells, palpitations of the heart, and temporary anorexia (loss of appetite and nausea). Victims of heat hypersensitivity may be easily burned.

Cerebral edema (M1504I)

Cerebral edema may occur among patients who suffer from organic diseases of the brain and experience an application of heat. Symptoms include pallor around the mouth, tremor of the lips and fingers, vomiting and convulsions. It may be prevented by applying cool wet towels to the head, renewing every two minutes.

Bath rash and abrasion (M1504J)

Bath rash and abrasion (corrugated or macerated skin) frequently occur with prolonged wet applications; the skin should be oiled if moist application is to be lengthy. Adjacent skin surfaces should be separated by cloth. If a patient is to be in a bath longer than four hours the skin should be lubricated with a lanolin cream preceding the bath.

Preclusions (M1504K)

Patients with communicable diseases, skin rash or the common cold should have immersion baths postponed until the disease is gone.

Immersion baths may be contraindicated for patients suffering acute kidney disease, advanced cardiac disease, anemia, hypotension, asthenia, dermatitis, subnormal body temperature, debilitation and tuberculosis.

Pool therapy is contraindicated in any febrile condition. The patient's temperature should be normal 72 hours before treatment. It is contraindicated for patients suffering from cardiac decompensation, very high or low blood pressure, acute inflammation or active disease of the joints, acute painful neuritis, nephritis or kidney infection, acute systemic infection, infections of the eye, ear, nose or throat, active pulmonary tuberculosis, acute poliomyelitis and/or infective skin conditions. It is not recommended during menstruation or for incontinent patients. Pool therapy should be discontinued if the patient experiences undue fatigue or loss of appetite following therapy.

Contrast baths are contraindicated in cases of malignancies or where there is a tendency to hemorrhage.

Percussion (Scotch) douche (a column of water directed against some portion of the patient's body) should be avoided in the treatment of patient's with cardiac and renal disease, arteriosclerosis, exophthalmic goiter, hypertension, nervous irritability, neuritis, varicose veins (avoid pressure on involved areas) and severe illness.

During percussion douches, the stream should be fanned over the popliteal area and over the front of the body, except for the thighs. Application should begin and end with the feet. The breasts should be protected by the patient's arm.

If the treated area suffers tactile hypersensitivity, any water agitation or spray force should be reduced to tolerable levels.

Steam baths are contraindicated for patients with hypertension, circulatory difficulties arising from diabetes, exophthalmic goiter, arteriosclerosis, respiratory diseases, epilepsy, mental illness, neurasthenia, poor tolerance to the bath, advanced pulmonary tuberculosis, acute infection, acute inflammation of the eyes or nose, acute kidney involvement, and cardiac impairment.

Thermal reaction (M1504L)

Brief exposure to a hot and cold stimulus may produce a *thermal reaction* which begins to occur when the stimulus is discontinued. The thermal reaction occurs as the body attempts to maintain thermal and circulatory equilibrium. A thermal reaction has three phasic components: (1) the thermic phase, during which there is an increase in tissue heat production to combat the cold stimulus; (2) the circulatory phase, during which a primary cutaneous vasoconstriction is produced followed very quickly by vasodilation accompanied by a warm tingling feeling and rubor of the skin, slowing heart rate and a slight rise in blood pressure; and (3) the nervous phase, during which there is an increase in muscle tone and a general feeling of increased vigor and well-being.

Incomplete thermal reaction (M1504M)

An *incomplete thermal reaction* is less desirable, producing cyanosis of the skin, chilliness, shivering, cold hands and feet, the sensation of having fullness of the head, lethargy and/or malaise. If it should develop, the patient should be encouraged to lie down and be covered lightly with blankets to prevent further cooling and to be gently warmed. Recovery from the incomplete thermic reaction may take over an hour. It is signalled by a halting of the shivering, an increase in the patient's hand temperature and a decrease in the sensations of malaise and lethargy.

General precautions (M1504N)

Before steam baths, the patient should have had a bowel movement in the preceding 24 hours; the patient should void before treatment.

The patient should *not* be left alone during any hydrotherapy procedure.

Avoid applying moist cold over sinus areas.

Notes Aside (M1505)

Physiological effects of heat or cold hydrotherapy (M1505A)

A prolonged intense application of either cold or heat may lead to tissue damage, while neutral temperature applications produce only minor physiological effects, including a slight lowering of blood pressure and a state of mild sedation.

The physiological effects of hydrotherapeutic procedures vary a good deal and depend directly upon the temperature of the water used:

1) Metabolism is first decreased by the application of cold and then increased as the cooling continues. General heating of the body increases metabolism.

2) Perspiration is decreased by the application of cold and increased by application of heat.

3) "Gooseflesh" results as a primary response to cold stimulation, but flushing of the skin follows as cooling continues. Heat stimulation causes flushing of the skin.

4) Pulse rate is first increased and then decreased as cooling continues. Heating increases the pulse rate.

5) Blood pressure is first increased and then lowered, though it will remain above normal, by continued cold application. Application of heat lowers blood pressure.

6) Superficial blood vessels are first constricted by the application of cold in a physiological attempt to maintain core temperature, and then dilated in the body's attempt to save cooled areas as cooling continues. Heat application first causes a brief vasoconstriction and then vasodilation as heating continues, the blood shunted to the skin from deeper tissues.

7) Sudden application of cold causes an initial respiratory gasping followed by quick, deep breaths which slow as cooling continues. Application of heat causes increased shallow respiration.

8) Muscles are excited by short applications of cold, while prolonged applications decrease muscle tone. Prolonged applications of heat diminish muscle excitability.

hydrotherapy

Temperature ranges for hydrotherapy (M1505B)

Cold : 65° F. (18.3° C.) and below; lower safe limit is 40° F.
Cool : 65 – 75° F. (18.3 – 23.9° C.)
Tepid : 75 – 92° F. (23.9 – 33.3° C.)
Neutral : 92 – 97° F. (33.3 – 36.1° C.)
Warm to hot : 98 – 104° F. (36.7 – 40° C.)
Hot to very hot : 104° F. (40° C.) and up; upper safe temperature limits are 110 – 115° F. (43 to 47° C.), but dry air saunas can be tolerated at much higher levels.

Bibliography (M1506)

Lehmann, pp. 172–197, 404–602.
Michlovitz, pp. 73–97, 99–117, 119–134.

HYDROTHERAPY WOUND (AND BURN) DEBRIDEMENT (M15A)

Definition (M15A1)

Soft tissue damage (M15A1A)

Open lesions (wounds resulting from cell death) may be produced by circulatory insufficiency, sudden and complete (or near-complete) ischemia (creating an infarct), mycotic infection, as the result of abcess forming collection of neutrophils causing indiscriminate digestion of tissue, mesenchymal autoimmune disease, ischemic necrosis with superimposed saprophytic or putrefying infection (gangrene) or from the effects of extreme heat or dehydration (burns).

Essentially, necrotic soft tissue results from tissue death caused by an underlying failure of the cell or tissue to compensate for intense or unusual damage, which is as a consequence subject to the process of necrosis.

Wound healing (M15A1B)

Wound healing normally follows a pattern of stages: (1) acute inflammation, with accompanying inflammatory exudate production to clear wound site of microorganisms and tissue debris, (2) the contraction of wound edges, (3) the ingrowth of capillary buds, (4) fibroblast proliferation, (5) synthesis and aggregation of tropocollagen, (6) reepithelialization and (7) the maturation and contraction of scar tissue.

Because of the extensive tissue damage resulting from a third degree burn, complete healing may ultimately depend upon skin grafting to provide the cells necessary for reepithelialization.

Treatment of severe wounds (M15A1C)

The treatment of severe wounds (including burns) may include and revolve around the careful debridement of necrotic tissue from the wound site. Successful debridement depends upon removing the necrotic tissue without causing additional hemorrhage or damaging granulating tissue. Debridement is performed to reduce the warm debris cover, which may serve to enhance bacteria incubation, and to facilitate the healing process by assisting the second stage of wound healing.

Evaluation of the stages of necrosis (M15A1D)

Evaluation of the condition of necrotic soft tissue depends upon the observation of the involved tissue and identifying the stages of necrosis. Of special importance is the ability to identify granulation tissue and to differentiate it from surrounding necrotic tissue, especially before debridement.

Granulation tissue develops as an intermediate stage between blood clotting and scar tissue formation. Distributed over its uncovered surface are tiny red granules (bumps) that are actually capillary loops and associated tissue pushing above the surface layer; they are fragile, highly vascular, easily injured and copiously hemorrhagic when damaged.

Necrotic soft tissue which is ready for debridement is usually pale and pasty looking and is easily lifted away from the granulating tissue.

Wound immersion (M15A1E)

Wound debridement may be enhanced by immersion of the burn site in neutral temperature (or tepid) water. If the wound is extensive, involving large areas of the body (as sometimes occurs in burn injury), full body immersion may be advisable. The water serves as a mild analgesic and sedative, improving patient comfort and decreasing the pain of the disturbance of sensitive, newly forming or exposed corium, which is unavoidable during debridement.

Procedures of Application (M15A2)

1) Immersion baths require a vessel large enough to contain the portion of the patient to be immersed and the volume of water necessary for immersion.

2) The surrounding room should be warm and free from draft.

3) Before use, the vessel to be used should be thoroughly cleansed with an antiseptic solution, well enough that the practitioner would be willing to be similarly immersed in it.

4) Before filling the vessel, determine where and how it should be situated to allow for comfortable body part immersion without the vessel rim impinging upon any other body part, such as the inside of the arm (axilla) or the patient's trunk.

5) Before filling the vessel, situate it so that the body part may be immersed and retrieved easily without undue strain upon the patient or the practitioner.

6) The vessel should be filled with aseptic water at a neutral temperature (92–97° F.).

7) Before immersion, the portion of the patient's body to be immersed should be stripped of clothing and any dressings which may be removed painlessly.

8) The patient should be assisted to fully immerse the portion of the body to be debrided.

9) If dressings are still present over the wound, time should be allowed for dressings to soak and loosen, and then they should be removed as gently as is possible.

10) When the patient has gotten used to the water and become as comfortable as possible, the wound area should be brought to the surface and debridement begun.

11) Only that necrotic tissue which readily comes up when touched with a blunt instrument should be removed. In some cases, a whirlpool may be used to facilitate debridement.

12) During the entire debridement procedure, water temperature must be carefully regulated. The temperature must remain tepid as long as debridement takes place.

13) Following debridement, the patient should be helped to remove the immersed portion of the body from the vessel.

14) The portion of the body which has been immersed should be carefully dried off, excluding the wound area.

15) If appropriate, the wound should be redressed.

16) The patient should leave the treatment area in warm, dry clothing.

17) The immersion vessel should be thoroughly cleansed.

Partial List of Conditions Treated (M15A3)

Treatable Causes (M15A3A)

Bacterial infection (C09)
Necrotic soft tissue (C18)
Unhealed dermal lesion (C12)

Syndromes (M15A3B)

BACTERIAL INFECTION (S63)
BURNS (SECOND AND/OR THIRD DEGREE) (S72)
UNHEALED DERMAL LESION (S39)

Precautions and Contraindications (M15A4)

General precautions (M15A4A)

Debridement of necrotic tissue should be performed with care. Attention should be directed at not disturbing the granulating tissue, especially the red granules; only loose necrotic tissue from over and between the red granules should be removed. If care is not taken, and granular tissue is damaged, healing may be delayed and scar tissue formation may be more severe.

Great care must be taken to avoid contamination of the wound. Aseptic conditions should be maintained within the immersion vessel, the water and instruments used and the dressings applied.

If a whirlpool is utilized, care should be taken to retrieve any dressing material that may float free before it enters the whirlpool intake valve.

The patient should not be left alone during debridement.

Water temperature (M15A4B)

Water temperature should not exceed 92° F. during debridement.

Chilling (M15A4C)

Extreme care should be taken to avoid chilling the patient both before and following treatment. Signs of chilling include bluish lips, undue pallor, cutis anserina (goose flesh), and shivering.

Bath rash and abrasion (M15A4D)

Bath rash and abrasion (corrugated or macerated skin) frequently occur with prolonged wet applications; the skin should be oiled if the moist application is to be lengthy. Adjacent skin surfaces should be separated by cloth. If a patient is to be in a bath longer than four hours the skin should be lubricated with a lanolin cream preceding the bath, taking care to avoid contaminating the wound with the cream.

Preclusions (M15A4E)

Immersion baths may be contraindicated for patients suffering from acute kidney disease, advanced cardiac disease, anemia, hypo-tension, asthenia (weakness), dermatitis, subnormal body temperature, debilitation and tuberculosis.

Notes Aside (M15A5)

The painstaking debridement with the involved extremity or body part immersed in tepid water discussed here is far less painful and safer than the widely accepted, efficient, but brutal and dangerous method of sponging or towelling to strip necrotic tissue from the debridement site.

Bibliography (M15A6)

Bickley, pp. 48–52.
Finnerty, pp. 33–39, 48–51, 55–57, 61–79.
Moor, pp. 39–41, 82–86, 90–92, 92–94.
Salter, pp. 391.
Shands, pp. 472–473.

HYDROTHERAPY CIRCULATION ENHANCEMENT
(M15B)

Definition (M15B1)

Hydrotherapeutic applications of hot and cold have been noted to have a facilitatory effect on the circulation of body parts stimulated, by promoting reflex dilation of arterioles and/or capillary beds.

Immersion baths which employ contrasting temperatures (heat and cold) have been shown to markedly increase the blood flow in the body parts so immersed. Blood flow in the lower extremities has been shown to increase as much as 95% when they were stimulated by a thirty-minute contrast bath. When the same procedure was performed with simultaneous immersion of both upper and lower extremities, an increased blood flow of 100% was demonstrated in the upper extremities and a 70% increase in the lower extremities.

Less dramatically, warm whirlpool baths are also noted to increase circulation in the body parts immersed.

Procedures of Application (M15B2)

Contrast Baths (M15B2A)

1) Immersion baths for circulation enhancement require two vessels large enough to contain the body part and purified water of sufficient amount to provide complete immersion.

2) The surrounding room should be warm and free from draft.

3) Before use, the vessels to be used should be thoroughly cleansed with an antiseptic solution, well enough that the practitioner would be willing to be similarly immersed in them.

4) Before filling the vessels, determine where and how they should be situated to allow for comfortable body part immersion without vessel rims impinging upon any other body part, such as the inside of the arm (axilla) or the patient's trunk.

5) Before filling the vessels, situate them so that the body part may be immersed and retrieved easily without undue strain upon the patient or the practitioner.

6) One vessel should be filled to the required level with hot pure (or disinfected) water (100 to 115° F.), while the other is filled with cold pure water (65 to 50° F.).

7) Before immersion, the body part to be immersed should be stripped of all clothing and dressings which are easily, safely and painlessly removed.

8) Before immersion, the patient's pulse should be taken.

9) The patient should assisted, if necessary, to immerse the body part in the vessel containing the hot water.

10) The body part should be immersed for one to four minutes before being transferred to the other vessel.

11) The body part should be immersed in the cold water and kept there for twenty to sixty seconds before transfer back to the vessel containing hot water.

12) Six to eight transfers should be made during the procedure, which should take between twenty and thirty minutes to complete. The last immersion should be in the cold water, unless the condition being treated is rheumatoid arthritis, in which case, the last immersion should be in the warm water.

13) During the procedure, the patient's pulse should be taken every five minutes. The procedure should be halted if the pulse exceeds 100 beats per minute.

14) During the procedure, the water contained in each vessel should be monitored and maintained at beginning temperatures.

15) When the procedure is complete, the body part should be removed from the vessels and thoroughly dried off. If necessary, dressings should be reapplied.

16) Observe the patient for any signs of development of an incomplete thermic reaction; if such should develop, the patient should be encouraged to lie down and blankets used to prevent further cooling and to gently warm the patient. Recovery may take up to an hour.

17) The patient should rest for several minutes before leaving in warm, dry clothing.

18) Each of the vessels used for immersion should be thoroughly cleansed before being used again.

Whirlpool Baths (M15B2B)

Whirlpool baths utilized to enhance circulation should be applied with the water between 105 and 110° F., following the protocol described in the Hydrotherapy, *Procedures of Application* [M1502A] section. The treatment should last from ten to twenty-five minutes.

Partial List of Conditions Treated (M15B3)

Treatable Causes (M15B3A)

Bacterial infection   (C09)
Hypermobile/instable joint   (C10)
Ligamentous strain   (C25)
Migraine (vascular headache)   (C34)
Restricted joint range of motion   (C26)
Soft tissue inflammation   (C05)
Soft tissue swelling   (C06)
Tactile hypersensitivity   (C24)
Unhealed dermal lesion   (C12)
Vascular insufficiency   (C11)

Syndromes (M15B3B)

ACHILLES TENDONITIS   (S52)
BACTERIAL INFECTION   (S63)
CALF PAIN   (S50)
ELBOW PAIN   (S47)
FOOT PAIN   (S51)
FOREARM PAIN   (S54)
HAND/FINGER PAIN   (S17)
HEADACHE PAIN   (S02)
JOINT SPRAIN   (S30)
MIGRAINE (VASCULAR) HEADACHE   (S18)
OSTEOARTHRITIS   (S21)
PITTING (LYMPH) EDEMA   (S31)
POST IMMOBILIZATION   (S36)
TENDONITIS   (S28)
TENNIS ELBOW (LATERAL EPICONDYLITIS)   (S32)
THIGH PAIN   (S49)
UNHEALED DERMAL LESION   (S39)
WRIST PAIN   (S16)

Precautions and Contraindications (M15B4)

General precautions (M15B4A)

Contrasted baths should not be used if the patient suffers from arterial insufficiency, advanced arteriosclerosis, advanced peripheral vascular diseases, diabetes or from conditions for which hot or cold applications are contraindications.

The patient should not be left alone during any hydrotherapy procedure.

Avoid moist cold applications over the sinus areas.

Water temperature (M15B4B)

For immersion baths or sprays, water temperatures should not exceed 110° F. and steam room temperatures should not exceed 110° F.

Chilling (M15B4C)

Extreme care should be taken to avoid chilling the patient both before and after treatment. Signs of chilling include bluish lips, undue pallor, cutis anserina (goose flesh) and shivering.

Hypersensitivity to cold (M15B4D)

Individuals who are hypersensitive to cold may develop severe vasospasms upon immersion in low temperature water. If the hand of such a patient is immersed in low temperature water for a few minutes it will produce redness, swelling, uticaria, and increased local temperature. After a latent period of three to six minutes, a systemic reaction will occur with the symptoms of flushing of the face, a sharp fall in blood pressure, a rise in pulse rate, and occasionally syncope. Recovery from such symptoms generally takes from five to ten minutes.

Frostbite (M15B4E)

A sustained immersion in water below 43° F. may produce cell damage and frostbite. Frostbite may cause blueness, mottled appearance, or redness of the skin, with areas of anesthesia and stiffness of the involved part.

Peripheral vascular disease (M15B4F)

Avoid overheating the patient who suffers from peripheral vascular disease or peripheral nerve injury.

Prolonged local heat application to areas with impaired blood supply may produce tissue damage. Heat application increases tissue temperature and causes a rise in local metabolic rate which results in waste product accumulation, causing increased cell membrane permeability and increasing the possibility of edema. Varicose veins, arteriosclerosis, thromboangiitis obliterans and other diseases of the circulatory system will limit the ability of the body to rapidly dissipate heat applied to it. An extremity suffering from vascular disease should not be used to elicit a reflex vasodilation because of the danger of local tissue damage and the desired effect may not be elicited since the vessels may not dilate or constrict normally.

Heat exhaustion (M15B4G)

Heat exhaustion is due to circulatory collapse following excessive loss of fluids and chlorides due to sweating. It may occur if the bath temperature remains or exceeds 98° F. for a prolonged period. A hot, humid atmosphere such as is found in a moist air sauna may produce the same results, which are giddiness, faintness, nausea, shallow or gasping breath, weak, rapid or fluttering pulse, moist cool skin, profuse sweating, facial pallor, drooping of the corners of the mouth, anoxemia and tissue anoxia, cramps, tetany, low blood pressure, a drop in systemic temperature and/or dehydration. Heat exhaustion should be treated by keeping the patient warm by covering with blankets, placing the head in a low position, and administering cool water with .5 grams of salt per pint as tolerated.

Heat stroke (M15B4H)

Heat stroke may result from a failure of the body to regulate heat and exhaustion of the sweat glands. Symptoms may include suppression of perspiration (hot dry skin), rapid pulse, red face, increased blood pressure, increased systemic temperature, increased urination, pulmonary edema, loss of consciousness and/or delirium. The treatment of heat stroke should include elevating the patient's head, cooling the patient's body with cold applications (ice water bath) or fanning. The patient should be kept quiet until the body temperature is stable.

Hypersensitivity to heat (M15B4I)

Heat sensitivity is a rare condition but occasionally occurs and results from a derangement of the heat-regulating mechanism and may produce exhaustion. Symptoms include weakness, dizziness, fainting spells, palpitations of the heart, and temporary anorexia (loss of appetite and nausea). Victims of heat hypersensitivity may be easily burned.

Cerebral edema (M15B4J)

Cerebral edema may occur among patients who suffer from organic diseases of the brain and experience an application of heat. Symptoms include pallor around the mouth, tremor of the lips and fingers, vomiting and convulsions. It may be prevented by applying cool wet towels to the head, renewing every two minutes.

Preclusions (M15B4K)

Contrast baths are contraindicated in cases of malignancies or where there is a tendency to hemorrhage.

Percussion (Scotch) douche (a column of water directed against some portion of the patient's body) should be avoided in the treatment of patient's with cardiac and renal disease, arteriosclerosis, exophthalmic goiter, hypertension, nervous irritability, neuritis, varicose veins (avoid pressure on involved areas) and severe illness.

During percussion douches, the stream should be fanned over the popliteal area and over the front of the body, except for the thighs. Application should begin and end with the feet. The breasts should be protected by the patient's arm.

If the treated area is sensitive, any water agitation or spray force should be reduced to tolerable levels.

Immersion baths may be contraindicated for patients suffering acute kidney disease, advanced cardiac disease, anemia, hypotension, asthenia, dermatitis, subnormal body temperature, debilitation and tuberculosis.

Thermal reaction (M15B4L)

Brief exposure to a hot and cold stimulus may produce a *thermal reaction* which begins to occur when the stimulus is discontinued. The thermal reaction occurs as the body attempts to maintain thermal and circulatory equilibrium. A thermal reaction has three phasic components: (1) the thermic phase, during which there is an increase in tissue heat production to combat the cold stimulus; (2) the circulatory phase, during which a primary cutaneous vasoconstriction is produced followed very quickly by vasodilation accompanied by a warm tingling feeling and rubor of the skin, slowing heart rate and a slight rise in blood pressure; and (3) the nervous phase, during which there is an increase in muscle tone and a general feeling of increased vigor and well-being.

Incomplete thermal reaction (M15B4M)

An *incomplete thermal reaction* is less desirable, producing cyanosis of the skin, chilliness, shivering, cold hands and feet, the sensation of having fullness of the head, lethargy and/or malaise. If it should develop, the patient should be encouraged to lie down and be covered lightly with blankets to prevent further cooling and to be gently warmed. Recovery from the incomplete thermic reaction may take over an hour. It is signalled by a halting of the shivering, an increase in the patient's hand temperature and a decrease in the sensations of malaise and lethargy.

Notes Aside (M15B5)

Contrast bath for evaluation of vascular headache (M15B5A)

Contrast baths was one of the first treatment techniques used for the management of the vascular (migraine) headache. Contrast baths applied to the hands give almost immediate temporary relief of pain caused by the true, pure vascular headache by causing reflex vasodilation. Because of this predictable response, contrast baths may be utilized to help evaluate the headache syndrome. Failure to provide relief of the headache for more then a few seconds is indicative of the presence of trigger point formation referred pain and/or sinusitis components, pointing to a mixed treatable cause syndrome (several causative components) or a clear case of misdiagnoses (i.e., the headache is caused by an undiagnosed cause).

Bibliography (M15B6)

Finnerty, pp. 57–58.
Lehmann, pp. 436, 590–594.
Moor, pp. 53–55.

HYDROTHERAPY
MUSCLE TENSION AMELIORATION
(M15C)

Definition (M15C1)

Local and general applications of cold have been demonstrated to temporarily reduce neuromuscular hypertonicity. Immersion of the patient to the waist in a cool or cold (ice) bath has been especially impressive for the reduction of spasticity accompanying multiple sclerosis (sited in connection are also improvements in vision, dysarthria, impaired sensation, depressed mood, ophthalmoplegia and ataxic spastic paraparesis), spasticity resulting from post cerebral vascular accident (CVA) syndromes or brain surgery and rigidity resulting from head trauma. Some of the cited beneficial effects have also been associated with cool or cold sprays.

Muscle hypertonicity (tension) found in otherwise normal subjects has been found to be reduced by warm steam baths or hot dry air sauna.

Procedures of Application (M15C2)

Ice bath (M15C2A)

1) The tub to be used should be thoroughly cleansed before use.
2) The room used should be warm and without drafts.
3) 50 to 100 pounds of ice should be deposited in the tub.
4) Pure or disinfected water should be poured into the tub, sufficient to float the ice and to reach the patient's belly button level when sitting.
5) The patient's clothes should be removed; appropriate swim wear may be worn.
6) The patient should be helped to assume a sitting position in the tub.
7) The patient should remain sitting in the tub for five minutes.
8) The patient should be assisted from the tub.
9) The patient's skin should be thoroughly dried.
10) The patient should leave the area in warm, dry clothing.
11) The tub should be drained and thoroughly cleansed before being used again.

Cold spray (M15C2B)

Cold spray applied for five minutes will have similar effects on spasticity as those produced by an ice bath.

For specifics of the spray application refer to Hydrotherapy, *Procedures of Application* [M1502C].

Steam bath (M15C2C)

A steam bath, used to reduce muscle tension, should be applied with the room heated to 120° F., for a duration of five to thirty minutes, repeated two or three times. It may be utilized daily.

For specifics of steam bath application refer to Hydrotherapy, *Procedures of Application* [M1502B] section.

Partial List of Conditions Treated (M15C3)

Treatable Causes (M15C3A)

Extrafusal muscle spasm (C41)
Habitual joint positioning (C48)
Joint ankylosis (C42)
Neuromuscular tonic imbalance (C30)
Pathological neuromuscular hypertonus (C28)
Pathological neuromuscular dyscoordination (C35)
Psychoneurogenic neuromuscular hypertonus (C29)
Restricted joint range of motion (C26)

Syndromes (M15C3B)

CEREBRAL PALSY (S25)
CERVICAL (NECK) PAIN (S73)
CHEST PAIN (S14)
FACET (S19)
FOOT PAIN (S51)
FOREARM PAIN (S54)
HIGH THORACIC BACK PAIN (S48)
KNEE PAIN (S46)
LOWER ABDOMINAL PAIN (S62)
LOW BACK PAIN (S03)
NEUROMUSCULAR PARALYSIS (S22)
OSTEOARTHRITIS (S21)
POST CEREBRAL VASCULAR ACCIDENT (CVA) (S07)
POST IMMOBILIZATION (S36)
POST PERIPHERAL NERVE INJURY (S23)
POST SPINAL CORD INJURY (S24)
POST WHIPLASH (S55)
REFERRED PAIN (S01)
SCIATICA (S05)

SHOULDER PAIN (S08)
THIGH PAIN (S49)
WRIST PAIN (S16)

Precautions and Contraindications (M15C4)

General precautions (M15C4A)

Before steam baths, the patient should have had a bowel movement in the preceding 24 hours; the patient should void prior to treatment.

If the treated area is sensitive, any water agitation or spray force should be reduced to tolerable levels.

Moist cold applied over the sinus areas should be avoided.

The patient should not be left alone during any hydrotherapy procedure.

Water temperature (M15C4B)

Ice bath and cold shower temperatures should be in the area of 50° F.

Steam room temperatures should not exceed 110° F. but dry sauna temperatures may be between 110 and 120° F.

Chilling (M15C4C)

Extreme care should be taken to avoid chilling the patient both before and after treatment. Signs of chilling include bluish lips, undue pallor, cutis anserina (goose flesh) and shivering.

Hypersensitivity to cold (M15C4D)

Individuals who are hypersensitive to cold may develop severe vasospasms upon immersion in low temperature water. If the hand of such a patient is immersed in low temperature water for a few minutes it will produce redness, swelling, uticaria, and increased local temperature. After a latent period of three to six minutes, a systemic reaction will occur with the symptoms of flushing of the face, a sharp fall in blood pressure, a rise in pulse rate, and occasionally syncope. Recovery from such symptoms generally takes from five to ten minutes.

Frostbite (M15C4E)

A sustained immersion in water below 43° F. may produce cell damage and frostbite. Frostbite may cause blueness, mottled appearance, or redness of the skin, with areas of anesthesia and stiffness of the involved part.

Peripheral vascular disease (M15C4F)

Avoid overheating the patient who suffers from peripheral vascular disease or peripheral nerve injury.

Prolonged local heat application to areas with impaired blood supply may produce tissue damage. Heat application increases tissue temperature and causes a rise in local metabolic rate which results in waste product accumulation, causing increased cell membrane permeability and increasing the possibility of edema. Varicose veins, arteriosclerosis, thromboangiitis obliterans and other diseases of the circulatory system will limit the ability of the body to rapidly dissipate heat applied to it. An extremity suffering from vascular disease should not be used to elicit a reflex vasodilation because of the danger of local tissue damage and the desired effect may not be elicited since the vessels may not dilate or constrict normally.

Heat exhaustion (M15C4G)

Heat exhaustion is due to circulatory collapse following excessive loss of fluids and chlorides due to sweating. It may occur if the bath temperature remains or exceeds 98° F. for a prolonged period. A hot, humid atmosphere such as is found in a moist air sauna may produce the same results, which are giddiness, faintness, nausea, shallow or gasping breath, weak, rapid or fluttering pulse, moist cool skin, profuse sweating, facial pallor, drooping of the corners of the mouth, anoxemia and tissue anoxia, cramps, tetany, low blood pressure, a drop in systemic temperature and/or dehydration. Heat exhaustion should be treated by keeping the patient warm by covering with blankets, placing the head in a low position, and administering cool water with .5 grams of salt per pint as tolerated.

Heat stroke (M15C4H)

Heat stroke may result from a failure of the body to regulate heat and exhaustion of the sweat glands. Symptoms may include suppression of perspiration (hot dry skin), rapid pulse, red face, increased blood pressure, increased systemic temperature, increased urination, pulmonary edema, loss of consciousness and/or delirium. The treatment of heat stroke should include elevating the patient's head, cooling the patient's body with cold applications (ice water bath) or fanning. The patient should be kept quiet until the body temperature is stable.

Hypersensitivity to heat (M15C4I)

Heat sensitivity is a rare condition but occasionally occurs and results from a derangement of the heat-regulating mechanism and may produce exhaustion. Symptoms include weakness, dizziness, fainting spells, palpitations of the heart, and temporary anorexia (loss of appetite and nausea). Victims of heat hypersensitivity may be easily burned.

Cerebral edema (M15C4J)

Cerebral edema may occur among patients who suffer from organic diseases of the brain and are exposed to prolonged general heating. Symptoms include pallor around the mouth, tremor of the lips and fingers, vomiting and convulsions. It may be prevented by using cool wet towels to the head, renewed every two minutes.

Preclusions (M15C4K)

Ice baths and sprays are contraindicated for women experiencing menstruation and for individuals suffering from cystitis and acute pelvic, pulmonary and/or abdominal inflammations.

Steam baths are contraindicated for individuals suffering from diabetes, serious cardiac conditions, respiratory diseases, exophthalmic goiter, arteriosclerosis and hypertension.

Immersion baths may be contraindicated for patients suffering from acute kidney disease, advanced cardiac disease, anemia, hypo-tension, asthenia (weakness), dermatitis, subnormal body temperature, debilitation and tuberculosis.

Percussion (Scotch) douche (a column of water directed against some portion of the patient's body) should be avoided in the treatment of patient's with cardiac and renal disease, arteriosclerosis, exophthalmic goiter, hypertension, nervous irritability, neuritis, varicose veins (avoid pressure on involved areas), and severe illness.

During percussion douches, the stream should be fanned over the popliteal area and over the front of the body, except for the thighs. Application should begin and end with the feet. The breasts should be protected by the patient's arm.

Steam baths are contraindicated for patients with hypertension, circulatory difficulties arising from diabetes, exophthalmic goiter, arteriosclerosis, respiratory diseases, epilepsy, mental illness, neurasthenia, poor tolerance to the bath, advanced pulmonary tuberculosis, acute infection, acute inflammation of the eyes or nose, acute kidney involvement and cardiac impairment.

Thermal reaction (M15C4L)

Brief exposure to a hot and cold stimulus may produce a *thermal reaction* which begins to occur when the stimulus is discontinued. The thermal reaction occurs as the body attempts to maintain thermal and circulatory equilibrium. A thermal reaction has three phasic components: (1) the thermic phase, during which there is an increase in tissue heat production to combat the cold stimulus; (2) the circulatory phase, during which a primary cutaneous vasoconstriction is produced followed very quickly by vasodilation accompanied by a warm tingling feeling and rubor of the skin, slowing heart rate and a slight rise in blood pressure; and (3) the nervous phase, during which there is an increase in muscle tone and a general feeling of increased vigor and well-being.

Incomplete thermal reaction (M15C4M)

An *incomplete thermal reaction* is less desirable, producing cyanosis of the skin, chilliness, shivering, cold hands and feet, the sensation of having fullness of the head, lethargy and/or malaise. If it should develop, the patient should be encouraged to lie down and be covered lightly with blankets to prevent further cooling and to be gently warmed. Recovery from the incomplete thermic reaction may take over an hour. It is signalled by a halting of the shivering, an increase in the patient's hand temperature and a decrease in the sensations of malaise and lethargy.

Notes Aside (M15C5)

Ice bath treatment of rigidity and semicomatose states (M15C5A)

Semicomatose patients suffering from rigidity have been noted to benefit from ice baths. The beneficial effects of the ice bath, in such a case, may not only include a decrease in rigidity but may also help to relieve the semicomatose condition. The sudden cold appears to have the effect of "jarring" the patient back into conscious awareness in some cases.

Bibliography (M15C6)

Finnerty, pp. 130–131, 137–141.
Moor, pp. 22–23, 94–95.

HYDROTHERAPY RANGE OF MOTION ENHANCEMENT AND PHYSICAL RECONDITIONING (M15D)

Definition (M15D1)

Range of motion may be enhanced and physical reconditioning begun in large vessels containing water which allow full immersion of the patient and additionally enough room for the patient to be able to move around in. This is possible in the Hubbard tank, which allows the patient to fully recline and potentially utilize full upper and lower extremity ranges of motion, and in the swimming or therapeutic pool.

The buoyancy or upward force exerted on an immersed body by the containing water (the larger the body of water the greater the buoyancy) has the effect of decreasing the effects of gravity upon the patient, allowing motion of the extremities which would otherwise be impossible. Exercise of weak musculature is made possible in ranges which would otherwise be made impossible by the force of gravity. The number and type of exercises available in the therapeutic tank or pool is limited only by the depth and size of the vessel and the skill and imagination of the practitioner.

Procedures of Application (M15D2)

Hubbard Tank (M15D2A)

1) Immersion baths require a vessel large enough to contain the patient and the volume of water necessary for immersion.

2) The surrounding room should be warm and free from draft.

3) Before use, the vessel to be used should be thoroughly cleansed with an antiseptic solution, well enough that the practitioner would be willing to be similarly immersed in it.

4) Before filling the vessel, situate it so that the body may be immersed and retrieved easily without undue strain upon the patient or the practitioner; the vessel should be easy to get into and out of.

5) Before immersion, remove all clothing, except for bathing attire, as well as all dressings which may be removed safely. Sanitary precautions on the part of the practitioner should be observed, for self-protection and to avoid undue contamination of the bath.

6) Before immersion, the vessel should be filled to the desired levels.

7) Before immersion, the water should have reached the prescribed temperature. The temperature of the water should be maintained from 90 to 104° F.; the lower end of the temperature range should be used for muscle reeducation or active exercise.

8) The patient should be helped or hoisted into the tank.

9) The patient should generally be reclined with the head and upper trunk supported by an apparatus secured to the side of the tank.

10) The patient should be engaged in a series of exercises. If assistance is required, the practitioner should manually assist over the side of the tank.

11) Generally, the treatment session should last from ten to thirty minutes.

12) Following the treatment session, the patient should be assisted leaving the tank.

13) The patient's skin should be dried thoroughly.

14) The patient should leave the area in warm, dry clothing.

Therapeutic pool (M15D2B)

Therapeutic pools generally vary in size from 10 × 12 to 15 × 30 feet with increasing depths from 2.5 feet at the shallow ends to 5 feet or more at the deep end. The main advantages of the pool, when compared with the standard Hubbard tank, is that the pool depth allows the patient to sit and/or stand while being immersed and exercising. The practitioner may also be able to better assist the patient perform the required exercises by being beside the patient within the pool itself. Additionally pool equipment may be used in the pool which space would not permit in the Hubbard tank. Such equipment includes plinths, walking aids (crutches, splints, pickup walkers, and parallel bars) and floating devices which may be used in the exercise program (inflated balls, inner tubes, paddle boards, etc.).

Pool temperatures range from 85 to 100° F. and depend upon the functions taking place in the pool. Temperatures ranging from 98 to 100° F. have been recommended for patients suffering from spastic paralysis, while 92 to 95° F. has been recommended for general exercise and 85 to 90° F. is recommended if the main activity in the pool is to be swimming.

Partial List of Conditions Treated (M15D3)

Treatable Causes (M15D3A)

Blood sugar irregularities (C31)
Calcific deposit (C08)
Extrafusal muscle spasm (C41)
Habitual joint positioning (C48)
Hypermobile/instable joint (C10)
Interspinous ligamentous strain (C03)
Ligamentous strain (C25)
Muscular weakness (C23)
Nerve root compression (C04)
Neuromuscular tonic imbalance (C30)
Pathological neuromuscular hypertonus (C28)
Pathological neuromuscular hypotonus (C33)
Pathological neuromuscular dyscoordination (C35)
Peripheral nerve injury (C07)
Psychoneurogenic neuromuscular hypertonus (C29)
Restricted joint range of motion (C26)
Unhealed dermal lesion (C12)

Syndromes (M15D3B)

BURNS (SECOND AND/OR THIRD DEGREE) (S72)
BURSITIS (S26)
CALF PAIN (S50)
CEREBRAL PALSY (S25)
CERVICAL DORSAL OUTLET (SCALENICUS ATTICUS/CERVICAL RIB) (S09)
ELBOW PAIN (S47)
FACET (S19)
FROZEN SHOULDER (S64)
HIGH THORACIC BACK PAIN (S48)
KNEE PAIN (S46)
LOW BACK PAIN (S03)
NEUROMUSCULAR PARALYSIS (S22)
OSTEOARTHRITIS (S21)
POST CEREBRAL VASCULAR ACCIDENT (CVA) (S07)
POST IMMOBILIZATION (S36)
POST PERIPHERAL NERVE INJURY (S23)
POST SPINAL CORD INJURY (S24)
RADICULITIS (S29)
REFERRED PAIN (S01)
SCIATICA (S05)
SHOULDER/HAND (S57)
SHOULDER PAIN (S08)
THIGH PAIN (S49)

Precautions and Contraindications (M15D4)

General considerations (M15D4A)

Hubbard tank and therapeutic pool programs are contraindicated if there is any increase in pain or if the patient becomes emotionally disturbed by the procedure. They are also contraindicated for patients suffering from acute infection and/or febrile diseases, acute inflamma-tion of the joints or other active joint diseases (tuberculosis of the joints), acute neuritis, kidney infection, cardiac lesions (decompen-sated heart conditions included), ear/eye/nose and throat conditions (including colds), excessive patient fatigue, acute poliomyelitis (especially in the presence of severe pain, nephritis or considerable respiratory weakness. Therapeutic pools are contraindicated in the presence of discharging wounds or sores, contagious skin infections (athlete's foot, etc.), active tuberculosis (pulmonary or otherwise) and other conditions from which there may be danger of cross-infection from one patient to another.

The patient should not be left alone during any hydrotherapy procedure.

Avoid moist cold applications over the sinus areas.

Water temperature (M15D4B)

For immersion baths water temperatures should not exceed 110° F.

Chilling (M15D4C)

Extreme care should be taken to avoid chilling the patient both before and after treatment. Signs of chilling include bluish lips, undue pallor, cutis anserina (goose flesh), and shivering.

Heat exhaustion (M15D4D)

Heat exhaustion is due to circulatory collapse following excessive loss of fluids and chlorides due to sweating. It may occur if the bath temperature remains or exceeds 98° F. for a prolonged period. A hot, humid atmosphere such as is found in a moist air sauna may produce the same results, which are giddiness, faintness, nausea, shallow or gasping breath, weak, rapid or fluttering pulse, moist cool skin, profuse sweating, facial pallor, drooping of the corners of the mouth, anoxemia and tissue anoxia, cramps, tetany, low blood pressure, a drop in systemic temperature and/or dehydration. Heat exhaustion should be treated by keeping the patient warm by covering with blankets, placing the head in a low position, and administering cool water with .5 grams of salt per pint as tolerated.

Heat stroke (M15D4E)

Heat stroke may result from a failure of the body to regulate heat and exhaustion of the sweat glands. Symptoms may include suppression of perspiration (hot dry skin), rapid pulse, red face, increased blood pressure, increased systemic temperature, increased urination, pulmonary edema, loss of consciousness and/or delirium. The treatment of heat stroke should include elevating the patient's head, cooling the patient's body with cold applications (ice water bath) or fanning. The patient should be kept quiet until the body temperature is stable.

Hypersensitivity to heat (M15D4F)

Heat sensitivity is a rare condition but occasionally occurs and results from a derangement of the heat-regulating mechanism and may produce exhaustion. Symptoms include weakness, dizziness, fainting spells, palpitations of the heart, and temporary anorexia (loss of appetite and nausea). Victims of heat hypersensitivity may be easily burned.

Cerebral edema (M15D4G)

Cerebral edema may occur among patients who suffer from organic diseases of the brain and experience an application of heat. Symptoms include pallor around the mouth, tremor of the lips and fingers, vomiting and convulsions. It may be prevented by applying cool wet towels to the head, renewing every two minutes.

Bath rash and abrasion (M15D4H)

Bath rash and abrasion (corrugated or macerated skin) frequently occur with prolonged wet applications; the skin should be oiled if moist application is to be lengthy. Adjacent skin surfaces should be separated by cloth. If a patient is to be in a bath longer than four hours the skin should be lubricated with a lanolin cream preceding the bath.

Preclusions (M15D4I)

Patients with communicable diseases, skin rash, or the common cold should have immersion baths postponed.

Immersion baths may be contraindicated for patients suffering from acute kidney disease, advanced cardiac disease, anemia, hypotension, asthenia (weakness), dermatitis, subnormal body temperature, debilitation and tuberculosis.

Pool therapy is contraindicated in any febrile condition. The patient's temperature should be normal 72 hours before treatment. It is contraindicated for patients suffering from cardiac decompensation, very high or low blood pressure, acute inflammation or active disease of the joints, acute painful neuritis, nephritis or kidney infection, acute systemic infection, infections of the eye, ear, nose or throat, active pulmonary tuberculosis, acute poliomyelitis and/or infective skin conditions. It is not recommended during menstruation or for incontinent patients. Pool therapy should be discontinued if the patient experiences undue fatigue or loss of appetite following therapy.

Contrast baths are contraindicated in cases of malignancies or where there is a tendency to hemorrhage.

Thermal reaction (M15D4J)

Brief exposure to a hot and cold stimulus may produce a *thermal reaction* which begins to occur when the stimulus is discontinued. The thermal reaction occurs as the body attempts to maintain thermal and circulatory equilibrium. A thermal reaction has three phasic components: (1) the thermic phase, during which there is an increase in tissue heat production to combat the cold stimulus; (2) the circulatory phase, during which a primary cutaneous vasoconstriction is produced followed very quickly by vasodilation accompanied by a warm tingling feeling and rubor of the skin, slowing heart rate and a slight rise in blood pressure; and (3) the nervous phase, during which there is an increase in muscle tone and a general feeling of increased vigor and well-being.

Incomplete thermal reaction (M15D4M)

An *incomplete thermal reaction* is less desirable, producing cyanosis of the skin, chilliness, shivering, cold hands and feet, the sensation of having fullness of the head, lethargy and/or malaise. If it should develop, the patient should be encouraged to lie down and be covered lightly with blankets to prevent further cooling and to be gently warmed. Recovery from the incomplete thermic reaction may take over an hour. It is signalled by a halting of the shivering, an increase in the patient's hand temperature and a decrease in the sensations of malaise and lethargy.

Notes Aside (M15D5)

In general, exercise programs utilized in Hubbard tanks and therapeutic pools fall into categories of exercise, listed below in the order of increasing difficulty or resistance:

Passive exercise (M15D5A)

Shortened muscles are passively stretched while immersed.

Active assistive exercise (M15D5B)

Movement in any and all of the planes of motion available under the water's surface; the practitioner may assist the patient in completing attempted ranges.

Active buoyed exercise (M15D5C)

Movement beginning from below and progressing toward the surface, to the buoyant level of the extremity.

Active lateral exercise (M15D5D)

Movement parallel to the water's surface (at the buoyant level of the extremity), with the patient sitting, standing, supine or prone.

Active decreasingly-buoyant exercise (M15D5E)

Movement from the extremity's buoyant level up toward the water's surface.

Active counter-buoyant exercise (M15D5F)

Movement away from the extremity's buoyant level downward away from the surface.

Active antigravity exercise (M15D5G)

Extremity movement taking place above the water's surface.

Active counter-buoyant resistive exercise (M15D5H)

Movement away from the extremity's buoyant level downward away from the surface against the resistance of a float device such as an inflated ball held in the hand, an inflated collar around the distal end of the extremity, etc.

The greater room provided by the therapeutic pool allows the patient the opportunity to engage in a greater range of exercise than available in the Hubbard tank. These include bicycle action with the legs, flexion of the hips and trunk simultaneously, performed in sitting, side-lying or standing starting positions, and standing and walking with or without assistive devices such as crutches, pickup walker, parallel bars, splints, etc.

Bibliography (M15D6)

Finnerty, pp. 71–79, 161–184.
Moor, pp. 83–94, 114–122.

HYDROTHERAPY SUPPRESSION OF SYMPATHETIC HYPERACTIVITY (M15E)

Definition (M15E1)

Some hydrotherapeutic applications have been noted to have a sedative effect upon the patient when they are applied at appropriate temperatures, including the immersion bath (whirlpool included) and the various hydrosprays (neutral spray, neutral douche, etc.). The sedative effect seems to come from an apparent suppression of sympathetic activity which produces a slowing of the pulse rate, relaxation and dilation of the cutaneous blood vessels, reduction of muscular rigidity, decreased metabolism (a reduction in body heat production), a lowering of blood pressure and a general state of emotional relaxation with accompanying drowsiness. These effects are noted to occur when the temperature of the water used in the hydrotherapeutic applications mentioned is within the tepid range (75 to 92° F.). Historically such applications have proven useful for the treatment of hysterical states, anxiety reactions, nervous insomnia, functional neuroses, spastic paralysis, renal insufficiency, pyrexia in infants and muscle cramps or irregular muscular twitching.

Procedures of Application (M15E2)

The water used in the treatment should be maintained at a constant 92° F., while neutral sprays or douches should last for 5 minutes.

Tepid immersion baths (M15E2A)

1) Immersion baths require a vessel large enough to contain the patient and the volume of water necessary for immersion.
2) The surrounding room should be warm and free from draft.
3) Before use, the vessel to be used should be thoroughly cleansed with an antiseptic solution, well enough that the practitioner would be willing to be similarly immersed in it.
4) Before filling the vessel, situate it so that the body may be immersed and retrieved easily without undue strain upon the patient or the practitioner. The vessel should be easy to get into and out of.
5) Before immersion, remove all clothing, except for bathing attire, as well as all dressings which may be removed safely. Sanitary precautions on the part of the practitioner should be observed, for self-protection and to avoid undue contamination of the bath.
6) Before immersion, the vessel should be filled to the desired levels.
7) Before immersion, the water should have reached the prescribed temperature of between 90 and 92° F.
8) In general, immersion tepid baths should last from thirty to sixty minutes, but may last longer if the condition of the patient warrants it.
9) For the duration of the treatment session, the temperature of the water should be maintained between 90 and 92° F.
10) Following immersion, thoroughly dry the body part and any other area which has been moistened. Dry dressings should be applied to wounds when appropriate. The patient should not be allowed to leave with damp clothes or skin.
11) Thoroughly cleanse and rinse vessel before using it again.

Tepid hydrosprays (showers or douches) (M15E2B)

1) Hydrospray applications (shower or douche) require a multiple needle spray head. The water is generally conveyed to the spray head by water pipe (showers) or by hose.
2) Sitting or lying platforms should be cleansed thoroughly before use. The hydrospray may be applied with the patient in lying, sitting or standing positions.
3) All clothing except for appropriate bathing costume should be removed from the patient. A bathing cap may be worn if the head is to remain dry. Waterproof thong sandals or clogs should be worn if the patient is to be standing during application.
4) Before application, the water temperature should be tested and adjusted to between 90 and 92° F.
5) The spray should generally be applied to strike the entire body surface; if applied via hose, it should be applied horizontally from four directions.
6) Following application, the spray should be turned off.
7) The patient should be aided, if necessary, to dry off completely, with special attention to drying off the feet.

8) The patient should leave the treatment area with dry skin and in warm, dry clothing.

9) All support apparatus (floor, stool, bench, table, etc.) should be thoroughly cleansed and dried before being used again.

Partial List of Conditions Treated (M15E3)

Treatable Causes (M15E3A)

Blood sugar irregularities (C31)
Extrafusal muscle spasm (C41)
Hypertension (C39)
Insomnia (C46)
Migraine (vascular headache) (C34)
Nausea/vomiting (C36)
Neurogenic dermal disorders
 (psoriasis, hives) (C49)
Neuromuscular tonic imbalance (C30)
Pathological neuromuscular hypertonus (C28)
Psychoneurogenic neuromuscular
 hypertonus (C29)
Spastic sphincter (C40)
Sympathetic hyperresponse (C38)
Tinnitus (C32)
Trigger point formation (C01)
Visceral organ dysfunction (C02)

Syndromes (M15E3B)

CALF PAIN (S50)
CHEST PAIN (S14)
COLITIS (S42)
DIABETES (S10)
ELBOW PAIN (S47)
FOOT PAIN (S51)
FOREARM PAIN (S54)
HAND/FINGER PAIN (S17)
HEADACHE PAIN (S02)
HIGH THORACIC BACK PAIN (S48)
HYPERTENSION (S43)
HYSTERIA/ANXIETY REACTION (S59)
INSOMNIA (S71)
KNEE PAIN (S46)
LOWER ABDOMINAL PAIN (S62)
LOW BACK PAIN (S03)
MIGRAINE (VASCULAR) HEADACHE (S18)
PHOBIC REACTION (S44)
POST CEREBRAL VASCULAR ACCIDENT
 (CVA) (S07)
POST IMMOBILIZATION (S36)
REFERRED PAIN (S01)
SCIATICA (S05)
SHOULDER PAIN (S08)
THIGH PAIN (S49)
TINNITUS (S70)
WRIST PAIN (S16)

Precautions and Contraindications (M15E4)

General considerations (M15E4A)

The patient should not be left alone during any hydrotherapy procedure.

Water temperature (M15E4B)

For tepid immersion baths or sprays, water temperatures should not exceed 92° F.

Chilling (M15E4C)

Extreme care should be taken to avoid chilling the patient both before and after treatment. Signs of chilling include bluish lips, undue pallor, cutis anserina (goose flesh) and shivering.

Heat exhaustion (M15E4D)

Heat exhaustion is due to circulatory collapse following excessive loss of fluids and chlorides due to sweating. It may occur if the bath temperature remains or exceeds 98° F. for a prolonged period. A hot, humid atmosphere such as is found in a moist air sauna may produce the same results, which are giddiness, faintness, nausea, shallow or gasping breath, weak, rapid or fluttering pulse, moist cool skin, profuse sweating, facial pallor, drooping of the corners of the mouth, anoxemia and tissue anoxia, cramps, tetany, low blood pressure, a drop in systemic temperature and/or dehydration. Heat exhaustion should be treated by keeping the patient warm by covering with blankets, placing the head in a low position, and administering cool water with .5 grams of salt per pint as tolerated.

Heat stroke (M15E4E)

Heat stroke may result from a failure of the body to regulate heat and exhaustion of the sweat glands. Symptoms may include suppression of perspiration (hot dry skin), rapid pulse, red face, increased blood pressure, increased systemic temperature, increased urination, pulmonary edema, loss of consciousness and/or delirium. The treatment of heat stroke should include elevating the patient's head, cooling the patient's body with cold applications (ice water bath) or fanning. The patient should be kept quiet until the body temperature is stable.

Bath rash and abrasion (M15E4F)

Bath rash and abrasion (corrugated or macerated skin) frequently occur with prolonged wet applications; the skin should be oiled if the moist application is to be lengthy. Adjacent skin surfaces should be separated by cloth. If a patient is to be in a bath longer than four hours the skin should be lubricated with a lanolin cream preceding the bath.

Preclusions (M15E4G)

Neutral baths are contraindicated for patients suffering from advanced cardiac diseases, the very young and the very old.

Patients with communicable diseases, skin rash, or the common cold should have immersion baths postponed.

Immersion baths may be contraindicated for patients suffering from acute kidney disease, advanced cardiac disease, anemia, hypotension, asthenia (weakness), dermatitis, subnormal body temperature, debilitation and tuberculosis.

Percussion (Scotch) douche (a column of water directed against some portion of the patient's body) should be avoided in the treatment of patient's with cardiac and renal disease, arteriosclerosis, exophthalmic goiter, hypertension, nervous irritability, neuritis, varicose veins (avoid pressure on involved areas) and severe illness.

During percussion douches, the stream should be fanned over the popliteal area and over the front of the body, except for the thighs. Application should begin and end with the feet. The breasts should be protected by the patient's arm.

If the treated area is sensitive, any water agitation or spray force should be reduced to tolerable levels.

Notes Aside (M15E5)

Tepid baths have been historically important in psychiatric settings having proven effective for reducing patient agitation. Prolonged sessions, sometimes lasting from four to six hours, have been known to be extremely useful as a treatment of psychiatric conditions which require sedation and may sometimes be used in place of tranquilizing medication.

Bibliography (M15E6)

Finnerty, pp. 39–42, 81–82.
Moor, pp. 39–40, 71–72.

HYDROTHERAPY HYDROTHERAPEUTIC TESTS
(M15F)

GIBBON-LANDIS PROCEDURE
(M15F1)

Definition (M15F1A)

The Gibbons-Landis procedure is essentially used to test the ability of hot water to induce reflex vasodilation in the extremities. The normal response to having an extremity immersed in hot water is for an increase in vasodilation to occur in the immersed extremity which is accompanied by increased perspiration production, a sharp rise in skin temperature and the patient report of a general sensation of warmth. If the extremity has been adversely affected by peripheral vascular disease, to any great extent, none of these responses will occur.

Warming a large area of skin on one extremity will normally produce vasodilation in the other three extremities. Warming of two extremities simultaneously normally results in the complete vasodilation in the digits of the other extremities. If circulation is normal, the temperature of the immersed extremities also rises.

Procedures of Application (M15F1B)

1) The Gibbons-Landis procedure requires a vessel large enough to contain the extremity to be immersed and the volume of water required for immersion.

2) The surrounding room should be warm and free from draft.

3) Before use, the vessel to be used should be thoroughly cleansed with an antiseptic solution, well enough that the practitioner would be willing to be similarly immersed in it.

4) Before filling the vessel, determine where and how it should be situated to allow for comfortable body part immersion without the vessel rim impinging upon any other body part, such as the inside of the arm (axilla) or the patient's trunk.

5) Before filling the vessel, situate it so that the body part may be immersed and retrieved easily without undue strain upon the patient or the practitioner. If full body immersion is practiced, the vessel should be easy to get into and out of.

6) Before immersion, remove all clothing, except for bathing attire, as well as all dressings which may be removed safely. Sanitary precautions on the part of the practitioner should be observed, for self-protection and to avoid undue contamination of the bath.

7) Before immersion, the vessel should be filled to the desired levels.

8) Before immersion, the temperature of the water should have reached 113° F.

9) The skin temperature of an extremity to remain unimmersed should be taken and recorded.

10) One extremity should be immersed in the water; if the feet are to be tested a hand should be immersed, and if the hands are to be tested a foot should be immersed. The forearm should be immersed to a point just above the elbow, and the leg immersed to midway between the ankle and the knee.

11) The extremity should be immersed from fifteen to forty-five minutes, to allow time for the reflex to occur.

12) For the duration of the procedure, the temperature of the monitored extremity should be taken every five minutes until the reaction occurs or until 45 minutes has elapsed without the reflex occurring.

13) Following immersion, thoroughly dry the body part and any other area which has been moistened. Dry dressings should be applied to wounds when appropriate. The patient should not be allowed to leave with damp clothes or skin.

14) Thoroughly cleanse and rinse the vessel before using it again.

Partial List of Conditions Evaluated (M15F1C)

Treatable Causes (M15F1C01)

Bacterial infection (C09)
Blood sugar irregularities (C31)
Nerve root compression (C04)
Peripheral nerve injury (C07)
Phlebitis (C43)
Soft tissue inflammation (C05)
Soft tissue swelling (C06)
Unhealed dermal lesion (C12)

Vascular insufficiency (C11)
Visceral organ dysfunction (C02)

Syndromes (M15F1C02)

BACTERIAL INFECTION (S63)
BUERGER'S DISEASE (S11)
CALF PAIN (S50)
CERVICAL DORSAL OUTLET (SCALENICUS ATTICUS/CERVICAL RIB) (S09)
DIABETES (S10)
ELBOW PAIN (S47)
FOOT PAIN (S51)
FOREARM PAIN (S54)
HAND/FINGER PAIN (S17)
KNEE PAIN (S46)
PHLEBITIS (S68)
POST PERIPHERAL NERVE INJURY (S23)
RAYNAUD'S DISEASE (S12)
SCIATICA (S05)
SHOULDER/HAND (S57)
THIGH PAIN (S49)
UNHEALED DERMAL LESION (S39)
WRIST PAIN (S16)

Precautions and Contraindications (M15F1D)

Vascular occlusion (M15F1D01)

An extremity which is afflicted with a circulatory disease should never have heat applied to it for therapeutic or testing purposes because of the danger of burns and the fact that the extremity will not respond favorably to the heating.

An occluded leg should never be tested with the Gibbons-Landis procedure.

Prolonged local heat application to areas with impaired blood supply may produce tissue damage. Heat application increases tissue temperature and a rise in local metabolic rate which results in waste product accumulation, causing increased cell membrane permeability and increasing the possibility of edema. Varicose veins, arteriosclerosis, thromboangiitis obliterans and other diseases of the circulatory system will limit the ability of the body to rapidly dissipate heat applied to it. An extremity suffering from vascular disease should not be used to elicit a reflex vasodilation because of the danger of local tissue damage and the desired effect may not be elicited since the vessels may not dilate or constrict normally.

Heat exhaustion (M15F1D02)

Heat exhaustion is due to circulatory collapse following excessive loss of fluids and chlorides due to sweating. It may occur if the bath temperature remains or exceeds 98° F. for a prolonged period. A hot, humid atmosphere such as is found in a moist air sauna may produce the same results, which are giddiness, faintness, nausea, shallow or gasping breath, weak, rapid or fluttering pulse, moist cool skin, profuse sweating, facial pallor, drooping of the corners of the mouth, anoxemia and tissue anoxia, cramps, tetany, low blood pressure, a drop in systemic temperature and/or dehydration. Heat exhaustion should be treated by keeping the patient warm by covering with blankets, placing the head in a low position, and administering cool water with .5 grams of salt per pint as tolerated.

Heat stroke (M15F1D03)

Heat stroke may result from a failure of the body to regulate heat and exhaustion of the sweat glands. Symptoms may include suppression of perspiration (hot dry skin), rapid pulse, red face, increased blood pressure, increased systemic temperature, increased urination, pulmonary edema, loss of consciousness and/or delirium. The treatment of heat stroke should include elevating the patient's head, cooling the patient's body with cold applications (ice water bath) or fanning. The patient should be kept quiet until the body temperature is stable.

Hypersensitivity to heat (M15F1D04)

Heat sensitivity is a rare condition but occasionally occurs and results from a derangement of the heat-regulating mechanism and may produce exhaustion. Symptoms include weakness, dizziness, fainting spells, palpitations of the heart, and temporary anorexia (loss of appetite and nausea). Victims of heat hypersensitivity may be easily burned.

Cerebral edema (M15F1D05)

Cerebral edema may occur among patients who suffer from organic diseases of the brain and experience an application of heat. Symptoms include pallor around the mouth, tremor of the lips and fingers, vomiting and convulsions. It may be prevented by applying cool wet towels to the head, renewing every two minutes.

Thermal reaction (M15F1D06)

Brief exposure to a hot and cold stimulus may produce a *thermal reaction* which begins to occur when the stimulus is discontinued. The thermal reaction occurs as the body attempts to maintain thermal and circulatory equilibrium. A thermal reaction has three phasic components: (1) the thermic phase, during which there is an increase in tissue heat production to combat the cold stimulus; (2) the circulatory phase, during which a primary cutaneous vasoconstriction is produced followed very quickly by vasodilation accompanied by a warm tingling feeling and rubor of the skin, slowing heart rate and a slight rise in blood pressure; and (3) the nervous phase, during which there is an increase in muscle tone and a general feeling of increased vigor and well-being.

Incomplete thermal reaction (M15F1D07)

An *incomplete thermal reaction* is less desirable, producing cyanosis of the skin, chilliness, shivering, cold hands and feet, the sensation of having fullness of the head, lethargy and/or malaise. If it should develop, the patient should be encouraged to lie down and be covered lightly with blankets to prevent further cooling and to be gently warmed. Recovery from the incomplete thermic reaction may take over an hour. It is signalled by a halting of the shivering, an increase in the patient's hand temperature and a decrease in the sensations of malaise and lethargy.

General precaution (M15F1D08)

The patient should *not* be left alone during any hydrotherapy procedure.

Notes Aside (M15F1E)

If an immersion bath is inconvenient, other sources of heat may be substituted (heat packs, pulsed hot air, etc.) following the protocol detailed above.

Bibliography (M15F1F)

Finnerty, pp. 204–205.
Moor, pp. 12–14.

COLD PRESSOR TEST (M15F2)

Definition (M15F2A)

The Cold Pressor Test is based upon the fact that arterial blood pressure will go up in response to a prolonged cold stimulus. It is used to detect instability of the vasomotor mechanism. Patient response to cold falls into three groups:

1) *Normal reactors*, who respond to the cold stimulus by producing a diastolic blood pressure increase of approximately 15 mm. Hg. and a systolic pressure increase of approximately 20 mm. Hg.

2) *Hypo-reactors*, who experience a rise in blood pressure which is less than that of normal reactors.

3) *Hyper-reactors*, who experience a blood pressure rise which exceeds that experienced by normal reactors. They have a tendency toward arterial spasm and are prime candidates for hypertensive diseases.

The Cold Pressor Test produces reflex vasoconstriction to produce the heightened blood pressures, which usually occur shortly after the cold stimulus is applied.

Procedures of Application (M15F2B)

1) The patient is required to rest quietly in a supine position for from fifteen to sixty minutes before the test is begun.

2) After the rest period, the patient's blood pressure is measured for one arm and recorded.

3) The hand on the opposing arm is immersed above the wrist in a vessel containing cold water (40–59° F.) for five minutes. The temperature of the water should be prevented from rising above the maximum by adding ice when necessary.

4) The blood pressure should be taken and recorded (by the same tester) every minute until the hand is removed from the water.

5) After the hand is removed from the water, the blood pressure should be taken and recorded every two minutes until the pressure returns to normal.

6) The quantitative measure of the response is equal to the difference between the blood pressures taken at the beginning of the procedure and the maximum pressure taken.

Partial List of Conditions Evaluated (M15F2C)

Treatable Causes (M15F2C01)

Blood sugar irregularities (C31)
Hypertension (C39)
Hypotension (C45)
Vascular insufficiency (C11)
Visceral organ dysfunction (C02)

Syndromes (M15F2C02)

BUERGER'S DISEASE (S11)
CARPAL TUNNEL (S35)
CERVICAL DORSAL OUTLET (SCALENICUS ATTICUS/CERVICAL RIB) (S09)
DIABETES (S10)
ELBOW PAIN (S47)
FOOT PAIN (S51)
FOREARM PAIN (S54)
HAND/FINGER PAIN (S17)
HYPERTENSION (S43)
KNEE PAIN (S46)
PHLEBITIS (S68)
POST PERIPHERAL NERVE INJURY (S23)
RAYNAUD'S DISEASE (S12)
SHOULDER/HAND (S57)
THIGH PAIN (S49)
UNHEALED DERMAL LESION (S39)
WRIST PAIN (S16)

Precautions and Contraindications (M15F2D)

Chilling (M15F2D01)

Extreme care should be taken to avoid chilling the patient both before and after treatment. Signs of chilling include bluish lips, undue pallor, cutis anserina (goose flesh) and shivering.

Hypersensitivity to cold (M15F2D02)

Individuals who are hypersensitive to cold may develop severe vasospasms upon immersion in low temperature water. If the hand of such a patient is immersed in low temperature water for a few minutes it will produce redness, swelling, uticaria, and increased local temperature. After a latent period of three to six minutes, a systemic reaction will occur with the symptoms of flushing of the face, a sharp fall in blood pressure, a rise in pulse rate, and occasionally syncope. Recovery from such symptoms generally takes from five to ten minutes.

Frostbite (M15F2D03)

A sustained immersion in water below 43° F. may produce cell damage and frostbite. Frostbite may cause blueness, mottled appearance, or redness of the skin, with areas of anesthesia and stiffness of the involved part.

Thermal reaction (M15F2D04)

Brief exposure to a hot and cold stimulus may produce a *thermal reaction* which begins to occur when the stimulus is discontinued. The thermal reaction occurs as the body attempts to maintain thermal and circulatory equilibrium. A thermal reaction has three phasic components: (1) the thermic phase, during which there is an increase in tissue heat production to combat the cold stimulus; (2) the circulatory phase, during which a primary cutaneous vasoconstriction is produced followed very quickly by vasodilation accompanied by a warm tingling feeling and rubor of the skin, slowing heart rate and a slight rise in blood pressure; and (3) the nervous phase, during which there is an increase in muscle tone and a general feeling of increased vigor and well-being.

Incomplete thermal reaction (M15F2D05)

An *incomplete thermal reaction* is less desirable, producing cyanosis of the skin, chilliness, shivering, cold hands and feet, the sensation of having fullness of the head, lethargy and/or malaise. If it should develop, the patient should be encouraged to lie down and be covered lightly with blankets to prevent further cooling and to be gently warmed. Recovery from the incomplete thermic reaction may take over an hour. It is signalled by a halting of the shivering, an increase in the patient's hand temperature and a decrease in the sensations of malaise and lethargy.

Bibliography (M15F2E)

Finnerty, pp. 205–206.

TEST FOR HYPERSENSITIVITY TO COLD (M15F3)

Definition (M15F3A)

Some individuals have such a violent response to the application of the cold stimulus that they are often said to be allergic to cold. They experience many symptoms common to histamine poisoning including urticaria, headaches, swelling of the lips and/or hands, and/or syncope.

Procedures of Application (M15F3B)

1) The patient's pulse rate and blood pressure should be checked before beginning the procedure.
2) One of the patient's hands should be immersed in cold water (47 to 50° F.) for five to six minutes.
3) The patient's pulse rate and blood pressure should be checked at one minute intervals during and up to twenty minutes following removal of the hand from the water.
4) The formerly immersed hand should be inspected for symptoms of hypersensitivity.

An adverse reaction to the cold stimulus will include swelling of the hand in addition to the normal physiological responses of initial pallor followed be rubor and increased calor upon removal of the stimulus. Three to six minutes later, the hypersensitive patient may suffer flushing of face, a fall in blood pressures, increased pulse rate, and possibly syncope.

Partial List of Conditions Evaluated (M15F3C)

Treatable Causes (M15F3C01)

Blood sugar irregularities (C31)
Peripheral nerve injury (C07)
Soft tissue swelling (C06)
Unhealed dermal lesion (C12)
Vascular insufficiency (C11)
Visceral organ Dysfunction (C02)

Syndromes (M15F3C02)

BUERGER'S DISEASE (S11)
CERVICAL DORSAL OUTLET
  (SCALENICUS ATTICUS/CERVICAL RIB) (S09)
DIABETES (S10)
FOOT PAIN (S51)
FOREARM PAIN (S54)
HAND/FINGER PAIN (S17)
KNEE PAIN (S46)
PITTING (LYMPH) EDEMA (S31)
POST PERIPHERAL NERVE INJURY (S23)
RAYNAUD'S DISEASE (S12)
SHOULDER/HAND (S57)
THIGH PAIN (S49)
UNHEALED DERMAL LESION (S39)
WRIST PAIN (S16)

Precautions and Contraindications (M15F3D)

Chilling (M15F3D01)

Extreme care should be taken to avoid chilling the patient both before and after treatment. Signs of chilling include bluish lips, undue pallor, cutis anserina (goose flesh) and shivering.

Hypersensitivity to cold (M15F3D02)

Individuals who are hypersensitive to cold may develop severe vasospasms upon immersion in low temperature water. If the hand of such a patient is immersed in low temperature water for a few minutes it will produce redness, swelling, uticaria, and increased local temperature. After a latent period of three to six minutes, a systemic reaction will occur with the symptoms of flushing of the face, a sharp fall in blood pressure, a rise in pulse rate, and occasionally syncope. Recovery from such symptoms generally takes from five to ten minutes.

Frostbite (M15F3D03)

A sustained immersion in water below 43° F. may produce cell damage and frostbite. Frostbite may cause blueness, mottled appearance, or redness of the skin, with areas of anesthesia and stiffness of the involved part.

Thermal reaction (M15F3D04)

Brief exposure to a hot and cold stimulus may produce a *thermal reaction* which begins to occur when the stimulus is discontinued. The thermal reaction occurs as the body attempts to maintain thermal and circulatory equilibrium. A thermal reaction has three phasic components: (1) the thermic phase, during which there is an increase in tissue heat production to combat the cold stimulus; (2) the circulatory phase, during which a primary cutaneous vasoconstriction is produced followed very quickly by vasodilation accompanied by a warm tingling feeling and rubor of the skin, slowing heart rate and a slight rise in blood pressure; and (3) the nervous phase, during which there is an increase in muscle tone and a general feeling of increased vigor and well-being.

Incomplete thermal reaction (M15F3D05)

An *incomplete thermal reaction* is less desirable, producing cyanosis of the skin, chilliness, shivering, cold hands and feet, the sensation of having fullness of the head, lethargy and/or malaise. If it should develop, the patient should be encouraged to lie down and should be covered lightly with blankets to prevent further cooling and to gently warm the patient. Recovery from the incomplete thermic reaction may take over an hour, and is signalled by a halting of the shivering, increase in hand temperature, and a decrease in the feelings of malaise and lethargy.

Notes Aside (M15F3E)

If the cold immersion bath is inconvenient to apply, a towel soaked in cold water and then wrung out may be wrapped around the patient's forearm. After fifteen seconds the towel should be removed; if the skin is not pinkish in color the patient has had an abnormal reaction to the cold stimulus.

Bibliography (M15F3F)

Finnerty, pp. 206–207.

HYPERHIDROSIS (C37)

Definition (C37A)

One of the mechanisms the body uses to reduce the effects of overheating is to produce sweat which is subsequently evaporated to provide rapid cooling of the skin. The sweat is produced by the *coiled portion* of eccrine sweat glands, distributed unevenly beneath the surface layers of the skin and which communicate with the surface via a *duct portion*, which penetrates through the dermis of the skin to terminate as a opening in the skin called the *pore*.

Eccrine glands in general are innervated and stimulated by cholinergic nerve fibers which make up part of the autonomic nervous system. Eccrine glands found in the palms of the hands, soles of the feet, and the temporal areas of the head are additionally innervated by adrenergic nerve fibers controlled by sympathetic activity, which may stimulate sweat production in these glands in response to emotional stress or physical exertion.

Sweat production in response to general body heating normally varies from 1.5 to 4 liters per hour. Sweat is basically water with various impurities added, including sodium chloride, urea, lactic acid, and potassium ions. The concentrations of urea and lactic acid sometimes found within excreted sweat has led many to classify the whole system as an important excretory system; urea concentration during high levels of sweat production may be as high as 100% greater than that found in blood plasma, and lactic acid may be 400% higher in sweat than in the plasma.

Abnormally high rates of sweat production, without appropriate precipitating stimulus (atmospheric temperature, physical exertion, obvious emotional stress, fever or other disease elements) is considered pathological and is called *hyperhidrosis*.

Hyperhidrosis may afflict the entire sweat production system or be confined to the palms of the hands, soles of the feet, axillas, inframammary regions, and/or groin. The sweat produced may be normally composed and simply be annoying relative to the problems of excessive water loss (wet clothes, dripping puddles, etc.), or the exudate may be extremely malodorous (bromhidrosis) and create major social problems for the patient. Additionally, hyperhidrosis may cause or contribute to the development of pyogenic infection, contact dermatitis, fungal infection, allergic dermatosis, or seborrhoica dermatitis.

Evaluation (C37B)

Hyperhidrosis occurs within a wide range of severity. Mild cases may only demonstrate the phenomena of clammy palms, dripping temples, sweaty feet when in stressful situations, but in severe cases the patient's skin may suffer the extremity of maceration, fissuring and/or scaling as the result of profuse and continuous sweat production. The skin affected by severe hyperhidrosis is often bluish white in color (erythematosis).

True hyperhidrosis will occur regardless of environmental temperature.

Galvanic skin response monitry may be useful in establishing the severity of hyperhidrosis.

Etiology (C37C)

Hyperhidrosis of the palms of the hands and/or soles of the feet may be largely psychogenic in origin.

Generalized hyperhidrosis is often a consequence of fever, certain systemic diseases (anemia or diabetes), obesity, endocrine gland dysfunction (hyperthyroidism), and occasionally as the result of nervous system abnormality or disease. It is sometimes seen to be an accompanying symptom of the post CVA syndrome (on the involved side) and as a sequelae of spinal cord injury.

Localized hyperhidrosis is generally idiopathic, occurring in otherwise normal individuals, but some cases are thought to be psycho-neurogenically triggered by subliminal tension resulting from anxiety.

The fetid odor associated with bromhidrosis results from yeast and bacterial cellular debris and the decomposition of elements within the sweat.

Treatment Notes (C37D)

Galvanic skin response (GSR) monitry may be used in the treatment of hyperhidrosis, especially in mild to moderate psychoneurogenic cases. The GSR desensitization process has proven to be the most successful technique utilized to date.

Zinc iontophoresis has been historically utilized successfully for the treatment of localized and generalized hyperhidrosis of idiopathic nature.

Bibliography (C37E)

Abell, pp. 87–91.
Guyton, pp. 832–836.
Hill, pp. 69–72.
Levit, pp. 505–507.
Holvey, pp. 1385–1386.
Morgan, p. 45.

TREATMENT OF HYPERHIDROSIS (C37F)

Technique Options (C37F1)

Evaluation techniques (C37F1A)

Cryotherapy	*M10*
Evaluation of capillary response to cold	M10H
Determination of hyperreactivity to cold	M10H2A
Galvanic skin response monitry	*M20*
GSR assessment	M20C

Treatment techniques (C37F1B)

Cryotherapy	*M10*
Suppression of sympathetic autonomic hyperreactivity	M10A
Electrical facilitation of endorphin production (acusleep)	*M03*
Galvanic skin response monitry	*M20*
Desensitization	M20B
Hydrotherapy	*M15*
Suppression of sympathetic hyperresponsiveness	M15E
Iontophoresis	*M07*
Hyperhidrosis control	M07K
Ultrahigh frequency sound (ultrasound)	*M01*
Autonomic sympathetic hyperactivity suppression	M01N

Techniques Suggested for the Acute Condition (C37F2)

Evaluation techniques (C37F2A)

Cryotherapy	*M10*
Evaluation of capillary response to cold	M10H
Determination of hyperreactivity to cold	M10H2A
Galvanic skin response (GSR) monitry	*M20*
GSR assessment	M20C

Treatment techniques (C37F2B)

Cryotherapy	*M10*
Suppression of sympathetic autonomic hyperreactivity	M10A
Electrical facilitation of endorphin production (acusleep)	*M03*
Ultrahigh frequency sound (ultrasound)	*M01*
Autonomic sympathetic hyperactivity suppression	M01N

Techniques Suggested for the Chronic Condition (C37F3)

Evaluation techniques (C37F3A)

Galvanic skin response (GSR) monitry	*M20*
GSR assessment	M20C

Treatment techniques (C37F3B)

Galvanic skin response monitry	*M20*
Desensitization	M20B

HYPERMOBILE/INSTABLE JOINT (C10)

Definition (C10A)

A hypermobile or instable joint is a joint which has lost some or all of its ligamentous and/or tendinous support.

Evaluation (C10B)

Evaluation of the hypermobile or instable joint is dependent upon range of motion evaluation and draw signs present upon lateral manipulation of the joint.

Etiology (C10C)

Joint hypermobility or instability is usually resultant from stretching or tearing of the supportive ligament. It can also occasionally result from cartilage attachment tearing or avulsion of bony attachment without damage to the ligament. Joint stability may also be lost if tendon support of the joint which had been compensatory for previously over-stretched ligaments is lost. Joint instability often stems from dislocation or subluxation of the involved joint.

Treatment Notes (C10D)

Treatment of the hypermobile or instable joint should be directed at decreasing attendant soft tissue components of inflammation, swelling and pain and providing support for the joint to facilitate healing and to prevent further injury by casting, strapping, taping, and/ or bracing.

Bibliography (C10E)

Salter, pp. 103–104, 418–487.

TREATMENT OF HYPERMOBILE/UNSTABLE JOINT (C10F)

Technique Options (C10F1)

Evaluation techniques (C10F1A)

Auxiliary evaluation techniques	M26
Auscultation	M26C
Joint and/or tendon sounds	M26C2C
Differential skin resistance survey	M22
Exercise	M12
Functional testing	M12I
Evaluation of joint range of motion	M12I1
Evaluation of muscle strength	M12I2
Massage	M23
Palpation evaluation	M23G
Determination of soft tissue tenderness	M23G3

Treatment techniques (C10F1B)

Auricular electroacustimulation	M06
Pain relief	M06A
Electrical facilitation of endorphin production (acusleep)	M03
Increasing pain tolerance	M0305A
Electrical stimulation (ES)	M02
Muscle reeducation/muscle toning	M02B
Exercise	M12
Muscle strengthening	M12A
Positive pressure	M24
Edema reduction	M24A
Taping	M16
Ultrahigh frequency sound (ultrasound)	M01
Desensitization	M01C

Techniques Suggested for the Acute Condition (C10F2)

Evaluation techniques (C10F2A)

Auxiliary evaluation techniques	M26
Auscultation	M26C
Joint and/or tendon sounds	M26C2C
Differential skin resistance survey	M22
Exercise	M12
Functional testing	M12I
Evaluation of joint range of motion	M12I1
Evaluation of muscle strength	M12I2
Massage	M23
Palpation evaluation	M23G
Determination of soft tissue tenderness	M23G3

Treatment techniques (C10F2B)

Positive pressure	M24
Edema reduction	M24A
Taping	M16

Techniques Suggested for the Chronic Condition (C10F3)

Evaluation Techniques (C10F3A)

Auxiliary evaluation techniques	*M26*
Auscultation	M26C
Joint and/or tendon sounds	M26C2C
Differential skin resistance survey	*M22*
Exercise	*M12*
Functional testing	M12I
Evaluation of joint range of motion	M12I1
Evaluation of muscle strength	M12I2
Massage	*M23*
Palpation evaluation	M23G
Determination of soft tissue tenderness	M23G3

Treatment techniques (C10F3B)

Electrical stimulation (ES)	*M02*
Muscle reeducation/muscle toning	M02B
Exercise	*M12*
Muscle strengthening	M12A
Taping	*M16* (1)

(1) If joint actively in jeopardy

HYPERTENSION (C39)

Definition (C39A)

Hypertension is defined as the sustained elevation of a person's systemic blood pressure at a level exceeding that which is normal for the given age and sex of the individual. In general, systemic blood pressure which exceeds 140/90 mm. Hg. is considered hypertension (for more detailed discussions refer to Hypertension Syndrome [S43] and Blood Pressure Control with Physiological Feedback Training [R007]).

Hypertension is generally divided etiologically into three primary groups: (1) essential hypertension, (2) renal hypertension, and (3) correctable hypertension.

Evaluation (C39B)

Hypertension may be evaluated through the mechanical assessment of the patient's blood pressure through the use of a sphygmomanometer, comparing patient readings with "normal" standards for age and sex. Readings of 140/90 mm. Hg. systolic/diastolic pressures and above are considered to represent a state of hypertension, regardless of the age or sex of the patient.

Etiology (C39C)

Essential hypertension occurs as an isolated symptom of idiopathic origin in an otherwise healthy individual, and it is not part of another disease process. Most common among women, it usually occurs in middle to late age groups, and may lead to cardiac hypertrophy, renal vascular deterioration, and/or an acceleration of atherosclerosis if not treated adequately or in time.

Renal hypertension occurs as the result of the renal angiotensin mechanism which is triggered when blood flow is restricted through the kidneys from a variety of causes. Other forms of hypertension may cause renal hypertension and, in many cases, it may be debatable to whether renal or primary hypertension was precedent. Renal hypertension may result directly from malignant nephrosclerosis, producing "malignant hypertension." It is usually fatal, as the blood pressure continues to rise until the patient dies of vascular failure, usually a cerebral hemorrhage. Fibrinoid necrosis in the walls of the arterioles, especially in the renal arterial system, can usually be found by autopsy in victims of "malignant" hypertension.

Correctable hypertension may result from pheochromocytoma (a functional tumor of the renal medulla), adrenocortical adenoma producing aldosterone, idiosyncratic response to the metabolism of certain foods (cheese, licorice, and tyrosine-containing foods like beer), and systemic responses to some medications. Monoamine oxidase inhibitors predispose a patient to tyramine-provoked hypertension since both agents occupy the body's monoamine oxidases and prevents them from detoxifying endogenous catecholamines. Correction may come from simply removing the triggering cause, whether tumor, chemical or food source.

Treatment Notes (C39D)

Temperature monitry and galvanic skin response monitry in a program of general relaxation and autonomic desensitization have proven effective in ameliorating or controlling chronic essential and psychogenic hypertension.

Acute episodes of hypertension, or dangerous surges of high blood pressure above already high levels (nonmalignant, excluding sudden blood pressure rise from bladder and/or rectal distention) have proven to be controllable and reducible with ice packing of the C7 vertebral paraspinous areas, electrical facilitation of endorphin production, tepid immersion baths and/or auricular electroacustimulation of appropriate acupoints.

Bibliography (C39E)

Bali, pp. 637–646.
Bickley, p. 211.
Blanchard, pp. 241–245.
Brener, pp. 1063–1064.
Elder, pp. 377–382.
Frankel, pp. 276–293.
Kristt, pp. 370–378.
Holvey, pp. 216–223.
Richter–Heinrich, pp. 251–258.

TREATMENT OF HYPERTENSION (C39F)

Technique Options (C39F1)

Evaluation techniques (C39F1A)

Auricular electroacustimulation	*M06*
Auricular acupoint evaluative survey	M06C
Determination of pathological symptom source	M06C5B
Auxiliary evaluation techniques	*M26*

hypertension

Auscultatory method of measuring blood pressure	M26A
Upper extremity	M26A2A
Lower extremity	M26A2B
Cryotherapy	*M10*
Evaluation of capillary response to cold	M10H
Determination of hyperreactivity to cold	M10H2A
Electroencephalometry	*M21*
Electroencephalometric evaluation	M21C
Electromyometric feedback	*M18*
Electromyometric evaluation	M18L
Determination of neuromuscular imbalance	M18L3
Galvanic skin response monitry	*M20*
GSR assessment	M20C
Hydrotherapy	*M15*
Hydrotherapeutic Tests	M15F
Cold pressor	M15F2
Peripheral electroacustimulation	*M05*
Peripheral acupoint survey	M05D

Treatment techniques (C39F1B)

Auricular electroacustimulation	*M06*
Pathology control	M06B
Cryotherapy	*M10*
Suppression of sympathetic autonomic hyperreactivity	M10A
Electrical facilitation of endorphin production (acusleep)	*M03*
Electroencephalometry	*M21*
Hypertension control	M21B
Electromyometric feedback	*M18*
Breathing pattern correction	M18J
Exercise	*M12*
Cardiovascular conditioning	M12D
Cardiovascular conditioning weight training	M12D2B
Galvanic skin response monitry	*M20*
General relaxation	M20A
Desensitization	M20B
Hydrotherapy	*M15*
Suppression of sympathetic hyperresponsiveness	M15E
Laser stimulation	*M27*
Massage	*M23*
Muscle fatigue (and/or tension) relief	M23A
Acupressure	M23C
Peripheral electroacustimulation	*M05*
Pathology control	M05C
Physiological dysfunction	M05C5A
Sensory stimulation	*M17*
Provoking reflex physiological responses	M17C
Prolonged heating for reflex physiological responses	M17C2A
Prolonged cooling for reflex physiological responses	M17C2B
Brief cooling for reflex physiological responses	M17C2C
Superficial thermal (heat) therapy	*M11*
Muscle relaxation (wet and dry sauna)	M11C
Excretory (sweat) facilitation	M11D
Temperature monitry	*M19*
Hypertension control	M19B
Transcutaneous nerve stimulation	*M04*
Prolonged stimulation of acupoints	M04B
Ultrahigh frequency sound (ultrasound)	*M01*
Acupoint stimulation	M01K
Autonomic sympathetic hyperactivity suppression	M01N

Techniques Suggested for the Acute Condition (C39F2)

Evaluation techniques (C39F2A)

Auricular electroacustimulation	*M06*
Auricular acupoint evaluative survey	M06C
Determination of pathological symptom source	M06C5B
Auxiliary evaluation techniques	*M26*
Auscultatory method of measuring blood pressure,	M26A
Upper extremity	M26A2A
Lower extremity	M26A2B
Cryotherapy	*M10*
Evaluation of capillary response to cold	M10H
Determination of hyperreactivity to cold	M10H2A

Treatment techniques (C39F2B)

Auricular electroacustimulation	*M06*
Pathology control	M06B
Cryotherapy	*M10*
Suppression of sympathetic autonomic hyperreactivity	M10A
Electrical facilitation of endorphin production (acusleep)	*M03*
Sensory stimulation	*M17*
Provoking reflex physiological responses	M17C
Prolonged heating for reflex physiological responses	M17C2A
Prolonged cooling for reflex physiological responses	M17C2B
Brief cooling for reflex physiological responses	M17C2C
Ultrahigh frequency sound (ultrasound)	*M01*
Autonomic sympathetic hyperactivity suppression	M01N

hypertension

Techniques Suggested for the Chronic Condition (C39F3)

Evaluation techniques (C39F3A)

Auricular electroacustimulation — **M06**
  Auricular acupoint evaluative survey — M06C
  Determination of pathological
    symptom source — M06C5B
Auxiliary evaluation techniques — **M26**
  Auscultatory method of measuring
    blood pressure — M26A
    Upper extremity — M26A2A
    Lower extremity — M26A2B
Electroencephalometry — **M21**
  Electroencephalometric evaluation — M21C

Galvanic skin response monitry — **M20**
  GSR assessment — M20C

Treatment techniques C39F3B)

Electroencephalometry — **M21**
  Hypertension control — M21B
Exercise — **M12**
  Cardiovascular conditioning — M12D
    Cardiovascular conditioning weight
      training — M12D2B
Galvanic skin response monitry — **M20**
  General relaxation — M20A
  Desensitization — M20B
Temperature monitry — **M19**
  Hypertension control — M19B

HYPERTENSION SYNDROME (S43)

Definition (S43A)

Primary hypertension (benign essential hypertension or hypertensive vascular disease) is characterized by an elevated diastolic blood pressure associated with chronically sustained arteriolar vaso-constriction and is commonly termed diastolic hypertension. Systolic blood pressure elevation usually accompanies diastolic hypertension. Approximately ninety percent of the high blood pressure population suffer from primary hypertension; the remaining ten percent suffer from systolic hypertension or diastolic hypertension resulting as a secondary symptom of kidney disease (glomerulonephritis, pylonephritis, polycystic disease and narrowing of arteries into the kidneys), tumors of the adrenal glands (as in Cushing's disease, pheochromocytoma, and aldosteronism), and disorders of the nervous system (including encephalitis and tumors of the brain).

Primary hypertension occurs idiopathically, but is often associated with an hereditary predisposition, emotional stress, obesity, excessive dietary salt, and excessive smoking. Primary hypertension develops insidiously, its symptoms appear as detrimental effects upon various organs, sometimes as long as twenty years after hypertension has long been present. The heart is the organ most commonly damaged by hypertension since it is required to expend greater then normal energy in the course of circulating the blood against the back pressure of constricted arteries. The heart muscle generally enlarges to meet this demand, but if blood pressure elevation is sustained for too long the heart eventually becomes overly fatigued and essentially wears out and congestive heart failure results. Additionally, the heart may not be able to meet additional requirements of supplying its enlarged self with blood, leading to myocardial infarction and its accompanying symptom of angina pectoris.

The brain is another organ which may be affected by long standing or severe high blood pressure. Prolonged hypertension may harden and weaken cerebral arteries, eventually leading to cerebral vascular hemorrhage possibly as a result of episodic increases in intra-arterial pressure. The kidneys may also be affected by prolonged high blood pressure, to the extent that the arteriole supply to the kidneys may be damaged and kidney function and renal capacity reduced. If unrelieved, the consequence may be uremia and/or kidney failure. Additionally, the result on the large arteries of the body may prolong hypertension by hardening (arteriosclerosis), loss of elasticity, and narrowed channels; the ultimate effect of which is that the general blood flow to the tissues may be compromised.

Obviously, severe hypertension, uncontrolled and unsuccessfully treated, may lead to fatal complications and the life span of the patient may be greatly shortened. The prognosis for the patient suffering from primary hypertension is greatly improved with early discovery and blood pressure reduction. Rapid developing, uncontrolled "malignant" hypertension may shorten the patient's life span to as little as two years after onset.

Related Syndromes (S43B)

BACTERIAL INFECTION (S63)
DIABETES (S10)
HEADACHE PAIN (S02)
HYSTERIA/ANXIETY REACTION (S59)
INSOMNIA (S71)
PHOBIC REACTION (S44)
TINNITUS (S70)

Treatable Causes Which May Contribute To The Syndrome (S43C)

Hypertension (C39)
Insomnia (C46)
Localized viral infection (C27)
Lung fluid accumulation (C13)
Nausea/vomiting (C36)
Reduced lung vital capacity (C17)
Sympathetic hyperresponse (C38)
Visceral organ dysfunction (C02)

Treatment Notes (S43D)

Hypertension may be successfully treated (if arteriosclerosis has not yet occurred to the extent of destroying arterial vessel elasticity) through the use of feedback training, desensitization to sphygmography, and a home program of self-regulation.

Temperature monitry (feedback) is initially utilized to teach the patient to voluntarily raise the skin temperature of the hands. As the hand temperature rises voluntarily, most people experience a decrease in autonomic sympathetic activity with a decrease in the blood pressure. This is a highly successful short-term approach, but the instrumentation (by its nature) tends to encourage the patient to compete with the self to see just how high the skin temperature can be driven; consequently, the patient's blood pressure may inadvertently begin to rise with the attempt. At this point, galvanic skin response (GSR) monitry or electro myometric (EMM) desensitization should be begun to teach the patient how not to over-respond, either autonomically or neuromuscularly to emotional stimuli. Or, brainwave (EEM) training (alpha or alpha/theta) should be started (if the patient shows no signs of epilepsy or primary endogenous depression) as a form of meditation, which is famous for reducing blood pressure levels.

The patient's blood pressure is taken at the beginning of each session and after each procedure. Doing so serves to verify patient success or failure and provides graphable datum for the record, and provides the means of helping to desensitize the patient to the process of blood pressure taking.

To help develop the patient develop a home program of self-regulation, after the patient has demonstrated the ability to make hand skin temperature rise on the temperature monitor, which usually in the first couple of sessions, the patient is encouraged to try to perform the same exercise at home for ten minutes in the mid-morning and mid-afternoon, utilizing a mood ring, temperature sensitive color changing strip or thermometer for visual feedback measurement of success or failure. This procedure is meant to serve as a mental "coffee-break".

Experience has shown that "normal" hypertensive patients, without manic/depression components or severe arteriosclerosis, respond to the above therapeutic approach with fair predictability, with the following episodes occurring three to four sessions apart, one after another: (1) a sharp decline of pre-session blood pressure levels occurs within the first few sessions; (2) a crisis and sharp rise in pre-session blood pressure levels occurs; (3) another drop in the pre-session blood pressure levels occurs, more extensive than its predecessor; (4) another crisis occurs, with its sharp rise in blood pressure, higher than its predecessor; and finally (5) a gradual decrease of pre-session pressure occurs, eventually leveling off at close to or lower than normal levels. The systolic and diastolic pressures simultaneously follow approximately the same response pattern. In all, sixteen to eighteen sessions, occurring on a once or twice a week basis, must take place before the blood pressure levels are reduced and can be maintained at normal levels of in and around 120 mm. Hg. systolic and 80 mm. Hg. diastolic, regardless of the patient's age.

Systolic hypertension occurs frequently in the absence of diastolic blood pressure elevation; it is most often caused by arteriosclerosis and it is a fairly common malady of advancing age.

For further discussion refer to Blood Pressure Control with Physiological Feedback Training [R007].

Bibliography (S43E)

Bali, pp. 637–646.
Barr, pp. 339–342.
Blanchard, pp. 241–245.
Brener, pp. 1063–1064.
Elder, pp. 377–382.
Frankel, pp. 276–293.
Kristt, pp. 370–378.
Holvey, pp. 216–223.
Richter–Heinrich, pp. 251–258
Shapiro, pp. 296–304.

HYPOTENSION (C45)

Definition (C45A)

Hypotension is defined as low blood pressure; it is sometimes significant as a health problem, but it may also promise extended longevity.

Evaluation (C45B)

Hypotension may be determined by taking the patient's blood pressure. Hypotension is usually considered to exist at pressure levels at or below 90/60 mm. Hg. (systolic/diastolic).

Etiology (C45C)

Causes of low blood pressure may include malnutrition, hormonal deficiencies (adrenal insufficiency), various cardiovascular disorders, spinal cord injury and other causes of orthostatic hypotension (lowered blood pressure brought on by going from lying to an upright position).

Treatment Notes (C45D)

Patients who suffer from hypotension who are underweight or physically inactive may benefit from an improved diet and graduated isometric and isotonic exercise. Static hypotensive patients may benefit from a process of desensitization to position changes by graduated increase in tilt table angulation from 0 to 90 degrees (from supine to upright position). Careful attention should be paid to arm and leg blood pressures during the process: the extent of physical tolerance is marked by a sudden drop below beginning pressure.

Bibliography (C45E)

Chusid, pp. 387.
Cooley, pp. 122, 331, 958, 995.

TREATMENT OF HYPOTENSION (C45F)

Technique Options (C45F1)

Evaluation techniques (C45F1A)

Auricular electroacustimulation	**M06**
Auricular acupoint evaluative survey	M06C
Determination of pathological symptom source	M06C5B
Auxiliary evaluation techniques	**M26**
Auscultatory method of measuring blood pressure	M26A
Upper extremity	M26A2A
Lower extremity	M26A2B
Hydrotherapy	**M15**
Hydrotherapeutic Tests	M15F
Cold pressor	M15F2
Peripheral electroacustimulation	**M05**
Peripheral acupoint survey	M05D

Treatment techniques (C45F1B)

Auricular electroacustimulation	**M06**
Pathology control	M06B
Electrical stimulation (ES)	**M02**
Muscle reeducation/muscle toning	M02B
Exercise	**M12**
Muscle strengthening	M12A
Cardiovascular conditioning	M12D
Cardiovascular conditioning weight training	M12D2B
Cardiovascular reconditioning exercise	M12D2C
Laser stimulation	**M27**
Massage	**M23**
Acupressure	M23C
Peripheral electroacustimulation	**M05**
Pathology control	M05C
Physiological dysfunction	M05C5A
Positive pressure	**M24**
Phlebitis and/or edema control	M24B
Sensory stimulation	**M17**
Provoking reflex physiological responses	M17C
Prolonged heating for reflex physiological responses	M17C2A
Prolonged cooling for reflex physiological responses	M17C2B
Brief cooling for reflex physiological responses	M17C2C
Transcutaneous nerve stimulation	**M04**
Prolonged stimulation of acupoints	M04B
Ultrahigh frequency sound (ultrasound)	**M01**
Acupoint stimulation	M01K

Techniques Suggested for the Acute Condition (C45F2)

Evaluation techniques (C45F2A)

Auricular electroacustimulation	**M06**
Auricular acupoint evaluative survey	M06C
Determination of pathological symptom source	M06C5B

Auxiliary evaluation techniques   *M26*
  Auscultatory method of measuring
    blood pressure   *M26A*
      Upper extremity   M26A2A
      Lower extremity   M26A2B
Peripheral electroacustimulation   *M05*
  Peripheral acupoint survey   M05D

Treatment techniques (C45F2B)

Auricular electroacustimulation   *M06*
  Pathology control   M06B
Electrical stimulation (ES)   *M02*
  Muscle reeducation/muscle
    toning   M02B
Peripheral electroacustimulation   *M05*
  Pathology control   M05C
    Physiological dysfunction   M05C5A
Positive pressure   *M24*
  Phlebitis and/or edema control   M24B
Sensory stimulation   *M17*
  Provoking reflex physiological
    responses   M17C
    Prolonged heating for reflex
      physiological responses   M17C2A
    Prolonged cooling for reflex
      physiological responses   M17C2B
    Brief cooling for reflex physiological
      responses   M17C2C
Transcutaneous nerve stimulation   *M04*
  Prolonged stimulation of acupoints   M04B

Techniques Suggested for the Chronic Condition (C45F3)

Evaluation techniques (C45F3A)

Auricular electroacustimulation   *M06*
  Auricular acupoint evaluative survey   M06C
    Determination of pathological
      symptom source   M06C5B
Auxiliary evaluation techniques   *M26*
  Auscultatory method of measuring
    blood pressure   M26A
      Upper extremity   M26A2A
      Lower extremity   M26A2B
Peripheral electroacustimulation   *M05*
  Peripheral acupoint survey   M05D

Treatment techniques (C45F3B)

Auricular electroacustimulation   *M06*
  Pathology control   M06B
Electrical stimulation (ES)   *M02*
  Muscle reeducation/muscle toning   M02B
Exercise   *M12*
  Muscle strengthening   M12A
  Cardiovascular conditioning   M12D
    Cardiovascular conditioning weight
      training   M12D2B
    Cardiovascular reconditioning
      exercise   M12D2C
Peripheral electroacustimulation   *M05*
  Pathology control   M05
    Physiological dysfunction   M05C5A
Positive pressure   *M24*
  Phlebitis and/or edema control   M24B
Transcutaneous nerve stimulation   *M04*
  Prolonged stimulation of acupoints   M04B

HYSTERIA/ANXIETY REACTION SYNDROME (S59)

Definition (S59A)

Hysteria, for discussion here, is defined as an *anxiety reaction*, *anxiety hysteria*, or *anxiety neurosis*, which produces an unconscious conversion of emotional stress into a physical symptom. Hysteria, as such, may characteristically involve the patient in a direct, subjective experience of emotional tension, apprehension, the often sudden sensation of general fatigue, and/or panic. These subjective feelings may be accompanied by excessive production of sweat, palpitations of the heart, vomiting, abdominal cramps, hyperventilation, diarrhea, urinary urgency, and/or widespread vasomotor disturbances. Hysteria may aggravate or initiate organic disorders including bronchial asthma, peptic ulcer, chronic colitis, enteritis, psychoneurogenic pain syndromes, and heart disease (especially coronary artery disease).

Conversion hysteria (reaction) may be characterized by *symptoms* of dysfunction of an organ or organs, or parts of the body, in an unconscious response to emotional conflict. Physical symptoms usually involve the voluntary muscles or the special sense organs (sight, smell, etc.). Typically, the patient will appear relatively unconcerned by the sensory loss or motor paralysis. The conversion symptomatology involving the neuromuscular system will predictably fail to follow the anatomic distribution of the sensorimotor nerves, and is determined instead by the patient's unconscious. This results in "stocking" or "glove" anesthesia, tunnel vision, blindness, loss of the sense of smell, and/or bilateral paralysis without pathological basis. The secondary gain and the form of the sensory or motor loss may give important clues as to the nature of the unconscious conflict responsible for the conversion symptomatology. Sufferers are most often from relatively unsophisticated social groups.

Conversion reactions are distinctly different from *psychophysiological reactions* described in the previous paragraph by virtue of the fact that the latter usually occur from organic processes dominated by the autonomic nervous system.

Related Syndromes (S59B)

CALF PAIN (S50)
CERVICAL DORSAL OUTLET (SCALENICUS ATTICUS/CERVICAL RIB) (S09)
CERVICAL (NECK) PAIN (S73)
CHEST PAIN (S14)
COLITIS (S42)
ELBOW PAIN (S47)
FOOT PAIN (S51)
FOREARM PAIN (S54)
FROZEN SHOULDER (S64)
HAND/FINGER PAIN (S17)
HEADACHE PAIN (S02)
HIGH THORACIC BACK PAIN (S48)
HYPERTENSION (S43)
INSOMNIA (S71)
KNEE PAIN (S46)
LOWER ABDOMINAL PAIN (S62)
LOW BACK PAIN (S03)
MENSTRUAL CRAMPING (S60)
MIGRAINE (VASCULAR) HEADACHE (S18)
NEUROMUSCULAR PARALYSIS (S22)
PHOBIC REACTION (S44)
RADICULITIS (S29)
REFERRED PAIN (S01)
SCIATICA (S05)
SHOULDER/HAND (S57)
SHOULDER PAIN (S08)
SINUS PAIN (S33)
TEMPOROMANDIBULAR JOINT (TMJ) PAIN (S06)
THIGH PAIN (S49)
TINNITUS (S70)
TOOTHACHE/JAW PAIN (S41)
TORTICOLLIS (WRY NECK) (S65)
WRIST PAIN (S16)

Treatable Causes Which May Contribute To The Syndrome (S59C)

Blood sugar irregularities (C31)
Dental pain (C22)
Extrafusal muscle spasm (C41)
Hyperhidrosis (C37)
Hypertension (C39)
Insomnia (C46)
Irregular menses (C19)
Menorrhalgia (C20)
Migraine (vascular headache) (C34)
Nausea/vomiting (C36)
Neurogenic dermal disorders (psoriasis, hives) (C49)
Neuromuscular tonic imbalance (C30)
Psychoneurogenic neuromuscular hypertonus (C29)
Reduced lung vital capacity (C17)
Sympathetic hyperresponse (C38)
Trigger point formation (C01)
Visceral organ dysfunction (C02)

Treatment Notes (S59D)

Immediate treatment of the psychophysiological hysteria/anxiety reaction syndrome is to treat the sympathetic autonomic hyperresponse. Ice packing of the dermatome over the sites of the C7/T1 nerve root ganglion (refer to Cryotherapy, Suppression of Sympathetic Autonomic Hyperreactivity [M10A]), electrical facilitation of prostaglandin production (refer to Electrical Facilitation of Endorphin Production [M03]), and short term (fifteen to sixty seconds/per acupoint) auricular electroacustimulation of appropriate acupoints (refer to Auricular Electroacustimulation, Pathology Control [M06B]) may be effective, either singly or in combination, in temporarily arresting hysteria reaction symptoms.

Trigger point formations may be psychoneurogenically produced and play a role in the hysteria/anxiety reaction syndrome by providing various symptoms for an unconscious defense mechanism. For a detailed description of possible contributory trigger point formations, refer to Table of Trigger Point Formation Locations & Descriptions of Their Referred Pain Patterns [T005].

Bibliography (S59E)

Bickley, pp. 320–321.
Chusid, pp. 388, 415–417.
Holvey, pp. 1101–1105.
Shands, pp. 203, 415–416.

INSOMNIA (C46)

Definition (C46A)

Insomnia is a state of wakefulness, or inability to sleep, in the absence of external impediments such as noise or bright light during the period when sleep should normally occur for the patient in question. It may vary from restlessness (light sleep) or the shortening of the sleep period to absolute wakefulness.

Evaluation (C46B)

The characteristic pattern of insomnia is difficulty in getting to sleep, light sleeping in the early part of the period, and difficulty staying asleep. Eventually exhaustion will take its toll and the victim will sleep.

Depression-provoked insomnia is marked by early waking. Mild depression may cause the patient to awaken slightly earlier then usual, while deep depression may cause the patient to awaken very much earlier than normal after which the patient may not be able to return to sleep.

Etiology (C46C)

The causes of insomnia include disturbances of regular routine, irregular work patterns, pain or discomfort, anxiety, emotional upset or disorders, and psychiatric disorders (including depression).

Treatment Notes (C46D)

Electrical facilitation of endorphin production (acusleep) has proven effective in restoring disturbed sleep patterns, and effectively relieving acute cases of insomnia.

Bibliography (C46E)

Gardner, p. 84.

TREATMENT OF INSOMNIA (C46F)

Technique Options (C46F1)

Evaluation techniques (C46F1A)

Auricular electroacustimulation	*M06*
Auricular acupoint evaluative survey	M06C
Determination of pathological symptom source	M06C5B
Auxiliary evaluation techniques	*M26*
Auscultatory method of measuring blood pressure	M26A
Upper extremity	M26A2A
Cryotherapy	*M10*
Evaluation of capillary response to cold	M10H
Determination of hyperreactivity to cold	M10H2A
Electroencephalometry	*M21*
Electroencephalometric evaluation	M21C
Galvanic skin response monitry	*M20*
GSR assessment	M20C
Hydrotherapy	*M15*
Hydrotherapeutic Tests	M15F
Test for hypersensitivity to cold	M15F3
Peripheral electroacustimulation	*M05*
Peripheral acupoint survey	M05D

Treatment techniques (C46F1B)

Auricular electroacustimulation	*M06*
Pathology control	M06B
Cryotherapy	*M10*
Suppression of sympathetic autonomic hyperreactivity	M10A
Electrical facilitation of endorphin production (acusleep)	*M03*
Relieving the insomnia syndrome	M0305C
Electroencephalometry	*M21*
Meditation facilitation	M21A
Exercise	*M12*
Cardiovascular conditioning	M12D
Cardiovascular conditioning weight training	M12D2B
Cardiovascular reconditioning exercise	M12D2C
Galvanic skin response monitry	*M20*
General relaxation	M20A
Desensitization	M20B
Hydrotherapy	*M15*
Suppression of sympathetic hyperresponsiveness	M15E
Laser stimulation	*M27*
Massage	*M23*
Muscle fatigue (and/or tension) relief	M23A
Acupressure	M23C
Peripheral electroacustimulation	*M05*
Pathology control	M05C
Physiological dysfunction	M05C5A

insomnia

Sensory stimulation	*M17*
Provoking reflex physiological responses	M17C
Prolonged heating for reflex physiological responses	M17C2A
Prolonged cooling for reflex physiological responses	M17C2B
Brief cooling for reflex physiological responses	M17C2C
Superficial thermal (heat) therapy	*M11*
Muscle relaxation (wet and dry sauna)	M11C
Excretory (sweat) facilitation	M11D
Temperature monitry	*M19*
General relaxation (sympathetic hyper-response suppression)	M19A
Transcutaneous nerve stimulation	*M04*
Prolonged stimulation of acupoints	M04B
Ultrahigh frequency sound (ultrasound)	*M01*
Acupoint stimulation	M01K
Autonomic sympathetic hyperactivity suppression	M01N

Techniques Suggested for the Acute Condition (C46F2)

Evaluation techniques (C46F2A)

Auxiliary evaluation techniques	*M26*
Auscultatory method of measuring blood pressure	M26A
Upper extremity	M26A2A

Treatment techniques (C46F2B)

Electrical facilitation of endorphin production (acusleep)	*M03*
Relieving the insomnia syndrome	M0305C

Techniques Suggested for the Chronic Condition (C46F3)

Evaluation techniques (C46F3A)

Auricular electroacustimulation	*M06*
Auricular acupoint evaluative survey	M06C
Determination of pathological symptom source	M06C5B
Auxiliary evaluation techniques	*M26*
Auscultatory method of measuring blood pressure	M26A
Upper extremity	M26A2A
Electroencephalometry	*M21*
Electroencephalometric evaluation	M21C
Galvanic skin response monitry	*M20*
GSR assessment	M20C
Peripheral electroacustimulation	*M05*
Peripheral acupoint survey	M05D

Treatment techniques (C46F3B)

Auricular electroacustimulation	*M06*
Pathology control	M06B
Electrical facilitation of endorphin production (acusleep)	*M03*
Relieving the insomnia syndrome	M0305C
Electroencephalometry	*M21*
Meditation facilitation	M21A
Exercise	*M12*
Cardiovascular conditioning	M12D
Cardiovascular conditioning weight training	M12D2B
Cardiovascular reconditioning exercise	M12D2C
Galvanic skin response monitry	*M20*
Desensitization	M20B
Peripheral electroacustimulation	*M05*
Pathology control	M05C
Physiological dysfunction	M05C5A
Temperature monitry	*M19*
General relaxation (sympathetic hyperresponse suppression)	M19A
Transcutaneous nerve stimulation	*M04*
Prolonged stimulation of acupoints	M04B

INSOMNIA SYNDROME (S71)

Definition (S71A)

The individual who suffers from insomnia has sleep patterns which are characteristically disturbed and is subject to wakefulness, or an inability to sleep, during the period when sleep should normally occur. This condition is generally thought to be psychoneurogenic, usually occurring in the absence of external impediments such as noise, or bright light. Causes usually sited include disturbances of regular routines, irregular work patterns, pain or discomfort, anxiety, emotional stress or disorder, or psychiatric disorders, including functional or endogenous depression. Insomnia may vary in degree from restlessness (light sleep) or shortened sleep periods to absolute wakefulness.

Related Syndromes (S71B)

COLITIS (S42)
HYPERTENSION (S43)
HYSTERIA/ANXIETY REACTION (S59)
MIGRAINE (VASCULAR) HEADACHE (S18)
PHOBIC REACTION (S44)
SINUS PAIN (S33)
TINNITUS (S70)

Treatable Causes Which May Contribute To The Syndrome (S71C)

Bacterial infection (C09)
Calcific deposit (C08)
Dental pain (C22)
Extrafusal muscle spasm (C41)
Hypertension (C39)
Insomnia (C46)
Interspinous ligamentous strain (C03)
Joint approximation (C47)
Ligamentous strain (C25)
Menorrhalgia (C20)
Migraine (vascular headache) (C34)
Nausea/vomiting (C36)
Nerve root compression (C04)
Neuroma formation (C14)
Peripheral nerve injury (C07)
Phlebitis (C43)
Psychoneurogenic neuromuscular hypertonus (C29)
Sinusitis (C16)
Sympathetic hyperresponse (C38)
Tactile hypersensitivity (C24)
Tinnitus (C32)
Trigger point formation (C01)
Visceral organ dysfunction (C02)

Treatment Notes (S71D)

Insomnia may be treatable with relaxation therapy utilizing temperature monitry (refer to Temperature Monitry, General Relaxation [M19A]) and galvanic skin resistance monitry (refer to Galvanic Skin Response Monitry, General Relaxation [M20A], and Desensitization [M20B]). Symptomatic relief may be attained through modality therapy which decreases sympathetic autonomic activity (refer to Hydrotherapy, Suppression of Sympathetic Hyperresponsiveness [M15E]) and/or increases in central nervous system endorphin production (refer to Electrical Facilitation of Endorphin Production [M03]). The pain which is produced by trigger point formations may be a source of "functional" insomnia. For a description of the trigger point formations which may be responsible refer to Table of Trigger Point Formation Locations & Descriptions of Their Referred Pain Patterns [T005].

Bibliography (S71E)

Gardner, p. 84.

INTERSPINOUS LIGAMENTOUS STRAIN (C03)

Definition (C03A)

Interspinous ligaments are thin membranous ligaments which connect adjoining vertebral spinous processes and extend from the root to the apex of each process. They join the ligamenta flava ventrally and the supraspinal ligament dorsally. Minimally developed in the cervical region, the interspinous ligament is narrow and elongated in the thoracic region, and broader, thicker and quadrilateral in the lumbar region.

Strain, excessive shearing force or traction (stretching), of an interspinous ligament is reported to cause localized and referred pain patterns, without signs or symptoms of trigger point formation referred pain, extrafusal muscle strain, visceral referred pain or nerve root irritation.

Evaluation (C03B)

The patient typically reports a central locus of pain along the vertebral column, accompanied by a stereotypic referred pain pattern in a more distal area. The centralized pain along the vertebral column is usually constant but the referred pain pattern is affected by changes in ligamentous stretch. There is significant increase of referred pain intensity if the ligament is mechanically stretched (often used diagnostically through the use of a round keyring applied over the ligament site), and a decrease in the intensity of pain when the stretch is decreased. There will be palpation tenderness over the involved interspinous ligament but not in the areas exhibiting the referred pain.

Etiology (C03C)

Commonly, interspinous ligamentous strain arises from acts of lifting, twisting, catching a falling heavy object, or falling in an awkward manner and landing in a twisted posture.

Some controversy revolves around whether the interspinous ligament is provided with sensory nerves capable of producing the sensation of pain upon stimulation (R. Cailliet, *Low Back Pain Syndrome*), but the work of Kellgren and Lewis (described in *Pain*), however, was able to demonstrate (and thereby graphically represent) referred pain patterns produced by interspinous ligament stretch. The mechanical stretch was provided by injection of saline solution under the various interspinous ligaments. The interspinous ligament is noted as a source of pain by other more recent authors.

Treatment Notes (C03D)

Ultrasound may prove useful in desensitizing sensitive interspinous ligamentous tissue as well as a providing the means of phonophoresing nonsteroidal antiinflammatories into inflamed tissues commonly associated with this condition. Supporting the trunk into the extension ranges, forcing associated vertebral joints out of flexion ranges, may provide pressure relief for the involved interspinous ligament.

Bibliography (C03E)

Basmajian, pp. 83–86.
Cailliet, pp. 17–19.
Chusid, p. 205.
Goss, pp. 310–311

TREATMENT OF INTERSPINOUS LIGAMENTOUS STRAIN (C03F)

Technique Options (C03F1)

Evaluation techniques (C03F1A)

Auricular electroacustimulation	**M06**
Auricular acupoint evaluative survey	M06C
Determination of referred pain source	M06C5A
Cryotherapy	**M10**
Evaluation of capillary response to cold	M10H
Determination of hyperreactivity to cold	M10H2A
Differential skin resistance survey	**M22**
Hydrotherapy	**M15F**
Hydrotherapeutic Tests	M15
Test for hypersensitivity to cold	M15F3
Massage	**M23**
Palpation evaluation	M23G
Determination of trigger point formations	M23G1
Determination of soft tissue tenderness	M23G3

interspinous ligamentous strain

Peripheral electroacustimulation	*M05*
Peripheral acupoint survey	M05D

Treatment techniques C03f1B

Auricular electroacustimulation	*M06*
Pain relief	M06A
Cryotherapy	*M10*
Hypertonic neuromuscular management	M10B
Extrafusal muscle spasm treatment	M10B2B
Analgesia/anesthesia	M10C
Soft tissue inflammation control	M10E
Electrical facilitation of endorphin production (acusleep)	*M03*
Increasing pain tolerance	M0305A
Electrical stimulation (ES)	*M02*
Muscle lengthening	M02A
Muscle lengthening	M02A5A
Muscle reeducation/muscle toning	M02B
Increasing blood flow	M02C
Increasing capillary density	M02C2B
Inflammation control	M02G
Decreasing soft tissue inflammation	M02G5A
Exercise	*M12*
Muscle strengthening	M12A
Muscle relengthening	M12C
Tonically shortened musculature	M12C1A
Muscle relengthening through reciprocal inhibition	M12C1C
Hydrotherapy	*M15*
Range of motion enhancement and physical reconditioning	M15D
Iontophoresis	*M07*
Pain relief	M07A
Inflammation amelioration	M07B
Muscle spasm relief	M07F
Joint mobilization	*M13*
Laser stimulation	*M27*
Massage	*M23*
Circulation enhancement and reduction of edema	M23B
Acupressure	M23C
Deep soft tissue mobilization	M23F
Muscles	M23F2A
Ligaments	M23F2B
Peripheral electroacustimulation	*M05*
Acupoint electrical stimulation within an inflamed zone	M05A
Analgesia	M05B
Sensory stimulation	*M17*
Provoking reflex physiological responses	M17C
Prolonged heating for reflex physiological responses	M17C2A
Prolonged cooling for reflex physiological responses	M17C2B
Brief cooling for reflex physiological responses	M17C2C
Desensitization	M17D
Superficial thermal (heat) therapy	*M11*
Muscle relaxation (wet and dry sauna)	M11C
Transcutaneous nerve stimulation	*M04*
Prolonged stimulation of acupoints	M04B
Ultrahigh frequency sound (ultrasound)	*M01*
Phonophoresis	M01A
The inflammatory process: control with phonophoresis	M01A5A
Desensitization	M01C
Acupoint stimulation	M01K
Vibration	*M08*
Neuromuscular management	M08A
Muscle lengthening	M08A2B
Desensitization of soft tissue	M08E

Technique Suggestions for the Acute Condition (C03F2)

Evaluation techniques (C03F2A)

Auricular electroacustimulation	*M06*
Auricular acupoint evaluative survey	M06C
Determination of referred pain source	M06C5A
Cryotherapy	*M10*
Evaluation of capillary response to cold	M10H
Determination of hyperreactivity to cold	M10H2A
Differential skin resistance survey	*M22*
Massage	*M23*
Palpation evaluation	M23G
Determination of trigger point formations	M23G1
Determination of soft tissue tenderness	M23G3
Peripheral electroacustimulation	*M05*
Peripheral acupoint survey	M05D

Treatment techniques (C03F2B)

Auricular electroascustimulation	*M06*
Pain Relife	M06A
Cryotherapy	*M10*
Hypertonic neuromuscular management	M10B
Extrafusal muscle spasm treatment	M10B2B
Electrical facilitation of endorphin production (acusleep)	*M03*
Increasing pain tolerance	M0305A
Electrical stimulation (ES)	M02
Muscle lengthening	M02A
Muscle lengthening	M02A5A
Inflammation control	M02G
Decreasing soft tissue inflammation	M02G5A
Exercise	*M12*
Muscle relengthening	M12C

Tonically shortened musculature	M12C1A
Muscle relengthening through reciprocal inhibition	M12C1C
Joint mobilization	*M13*
Massage	*M23*
Deep soft tissue mobilization	M23F
Muscles	M23F2A
Ligaments	M23F2B
Peripheral electroacustimulation	*M05*
Acupoint electrical stimulation within an inflamed zone	M05A
Analgesia	M05B
Transcutaneous nerve stimulation	*M04*
Prolonged stimulation of acupoints	M04B
Ultrahigh frequency sound (ultrasound)	*M01*
Phonophoresis	M01A
The inflammatory process: control with phonophoresis	M01A5A
Vibration	*M08*
Neuromuscular management	M08A
Muscle lengthening	M08A2B

Techniques Suggested for the Chronic Condition (C03F3)

Evaluation techniques (C03F3A)

Auricular electroacustimulation	*M06*
Auricular acupoint evaluative survey	M06C
Determination of referred pain source	M06C5A
Cryotherapy	*M10*
Evaluation of capillary response to cold	M10H
Determination of hyperreactivity to cold	M10H2A
Differential skin resistance survey	*M22*
Massage	*M23*
Palpation evaluation	M23G
Determination of soft tissue tenderness	M23G3
Peripheral electroacustimulation	*M05*
Peripheral acupoint survey	M05D

Treatment techniques (C03F3B)

Auricular electroacustimulation	*M06*
Pain relief	M06A
Cryotherapy	*M10*
Soft tissue inflammation control	M10E
Electrical facilitation of endorphin production (acusleep)	*M03*
Increasing pain tolerance	M0305A
Electrical stimulation (ES)	*M02*
Muscle reeducation/muscle toning	M02B
Increasing blood flow	M02C
Increasing capillary density	M02C2B
Inflammation control	M02G
Decreasing soft tissue inflammation	M02G5A
Exercise	*M12*
Muscle strengthening	M12A
Muscle relengthening	M12C
Tonically shortened musculature	M12C1A
Muscle relengthening through reciprocal inhibition	M12C1C
Massage	*M23*
Deep soft tissue mobilization	M23F
Muscles	M23F2A
Ligaments	M23F2B
Peripheral electroacustimulation	*M05*
Acupoint electrical stimulation within an inflamed zone	M05A
Transcutaneous nerve stimulation	*M04*
Prolonged stimulation of acupoints	M04B
Ultrahigh frequency sound (ultrasound)	*M01*
Phonophoresis	M01A
The inflammatory process: control with phonophoresis	M01A5A
Vibration	*M08*
Neuromuscular management	M08A
Muscle lengthening	M08A2B

IONTOPHORESIS (M07)

Definition (M0701)

Iontophoresis is the name of an electrical process through which ions of a chosen medication or chemical may be introduced into the tissues by means of the passage of an electrical current.

Historically, iontophoresis is very old treatment technique dating from the middle of the eighteenth century. It was discovered that certain chemicals, when exposed to a constant direct current, seemed to travel through the tissues from one electrode pole to the other, as if attracted to a particular pole by some mysterious force. The "mysterious force" was ion transfer. Ion transfer essentially means that an ion (an electrically charged particle or chemical radical, with either a positive or negative polarity) will be attracted by and travel through an essentially liquid medium (the soft tissues of the human body) toward an electrical pole (electrode) of the opposite charge.

As different chemicals were found to be subject to the influence of ion transfer, often simply through trial and error, therapeutic attempts to utilize the iontophoresis process in the treatment of various ailments became fairly common. Some of the ions utilized included aconite, adrenaline, arsenic, bee venom, calcium, carbaine, chloride, cocaine, copper, curare, dibenamine, dionin, iodide, iron, lithium, magnesium, mecholyl, mercury, morphine, novocaine, prostaglandin, pyribenzamine, salicylate, silver, sulphanamide, thallium, uranium and zinc. When histamines, antihistamines, sulfones and antibiotics joined this group of recognized substances susceptible to "electrical pressure", the number of conditions treated with iontophoresis mushroomed. Over fifty conditions were reported to have been successfully cured or improved through iontophoresis of various chemicals or other medications, differently designated. They included acne, adhesions, arthritis, asthma, Bell's palsy, boils, bleeding, bursitis, cancer, cervicitis, colitis, corneal ulcer, cystitis, dermatophytosis, Dupytren's contracture, ecchymosis, gonorrhea, gout, hay fever, hemiplegia, hyperhidrosis, hypertension, hyperthyroidism, hypertrophied prostate, keloid formation, keratitis, laryngeal tuberculosis, lupus, migraine, multiple sclerosis, neuralgia, ophthalmia, otitis, ozena, pelvic inflammation, phlebitis, poliomyelitis, pruritis ani, pneumonia, rectal ulcer, root canal infection, sciatica, sinusitis, psychosis, thrombophlebitis, scar tissue, scleroderma, ulcer, vaginitis and warts.

Medical use of iontophoresis reached its zenith between 1940 and 1950, but its use quickly fell off in the early 1950's, for uncertain reasons. It currently seems to be enjoying a modest resurgence in use in the application of a few select substances for a few specific conditions. The heavy metals ions of zinc and copper, applied under the positive pole, are currently being employed in the treatment of various skin infections and chronic infections of the sinuses. Chlorine ions, applied under the negative pole, are being applied to loosen superficial scars and reduce adhesions. Vasodilating drugs are being used to treat some rheumatic afflictions, while anesthetic drugs, applied under the positive pole, are being applied to produce anesthesia of the skin, and various other medications are being applied to produce general metabolic or specific hormonal effects.

It should be noted, as a general rule, that medicinal ions cannot be made to migrate far below the surface of the skin and mucous membranes, and usually precipitate or bind in the local tissues associated with the site of application; therefore, iontophoresis is often most effective in treating localized conditions which may have the appropriate electrode applied over or near its location. Scars, inflammations and sclerotic or calcific deposits are examples.

It has been demonstrated that the lighter the atomic weight of the ions introduced, the further the ions will penetrate into the tissues. If heavy enough, some ions will reach the bloodstream, and even travel into and be found in the hair follicles. The number of ions introduced into the tissues depends upon (1) the intensity of the direct current driving them, (2) the duration of the treatment session, and (3) the strength of the solution used. If the solution is too strong, however, the binding effect may be increased and the penetrating amount decreased.

For further discussion of therapeutic electrical stimulation refer to Electrical Stimulation [M02].

Procedures of Application (M0702)

General principles (M0702A)

Essential to iontophoresis is an electrical generator which can supply a continuous direct, or monophasic current. Electrode cables, attached to the electrical stimulator at the positive anode and negative cathode poles, should terminate in attachment to surface electrodes, which are affixed to the patient's skin.

If the ion rich substance to be applied is a solution, the active, ion driving electrode pad should be saturated by the solution at a strength of 1% or less and firmly secured over the treatment site via taping, strapping or weighting. If a bath is utilized, the active electrode should be placed in a vessel adequately filled with enough water or saline solution to cover the treatment site; the body part should then be placed in the vessel as well. If the ion rich substance is in a lotion or ointment form, it should be spread evenly over the treatment site and the water or saline-saturated active electrode pad placed directly over it. In either of these cases, the water or saline-saturated, nonactive, ion attracting electrode pad should be placed over an area of skin distal to the active electrode. A foot or forearm bath may be used as a dispersive if conveniently distant from the active electrode pad or bath.

Solutions containing ions with a positive charge (zinc, copper, histamine, and the vasodilating alkaloids) should be introduced into the skin and mucal membranes from under the positive pole, while those containing ions with a negative charge should be introduced from under the negative pole.

Technical procedure (M0702B)

1) The patient should be placed in a comfortable sitting, lying, or semi-reclining position.

2) The electrodes should be secured over the appropriate sites on the patient's skin.

3) The electrical stimulator should be preset at zero intensity and a timer set for the prescribed time for the treatment protocol.

4) The electrical stimulator should be turned on and the intensity slowly increased until the patient feels a slight burning sensation. The current intensity should be slowly reduced if patient discomfort increases. Current strengths should generally not exceed two milliamperes per square inch of active electrode.

5) After the treatment time has elapsed, the electrical stimulation, if not terminated by an automatic timer, should be reduced slowly back to zero and switched off.

6) The electrodes should be removed and the patient allowed to rest for a few minutes before leaving.

7) The electrodes, if they are to be used again, should be thoroughly cleansed and rinsed before reusing, to get rid of the secondary chemical products which may have accumulated near the metal component of the electrode.

Partial List of Conditions Treated (M0703)

Conditions Treated in the Field (M0703A)

Indolent ulcers
Chronic fungus infections
Vasodilation
Ischemia
Gouty tophi
Sclerotic deposit
Vascular congestion
Peyronie's disease

Treatable Causes (M0703B)

Bacterial infection (C09)
Calcific deposit (C08)
Extrafusal muscle spasm (C41)
Hyperhidrosis (C37)
Hypertension (C39)
Interspinous ligamentous strain (C03)
Joint ankylosis (C42)
Ligamentous strain (C25)
Migraine (vascular headache) (C34)
Necrotic soft tissue C18)
Nerve root compression (C0)
Neurogenic dermal disorders (psoriasis, hives) (C49)
Restricted joint range of motion (C26)
Scar tissue formation (C15)
Sinusitis (C16)
Soft tissue inflammation (C05)
Soft tissue swelling (C06)
Sympathetic hyperresponse (C38)
Tactile hypersensitivity (C24)
Unhealed dermal lesion (C12)
Vascular insufficiency (C11)

Syndromes (M0703C)

ACHILLES TENDONITIS (S52)
ACNE (S45)
ADHESION (S56)
BACTERIAL INFECTION (S63)
BUERGER'S DISEASE (S11)
BURSITIS (S26)
CALF PAIN (S50)
CAPSULITIS (S27)
CARPAL TUNNEL (S35)

DIABETES (S10)
DUPYTREN'S CONTRACTURE (S34)
ELBOW PAIN (S47)
FACET (S19)
FASCIITIS (S20)
FOOT PAIN (S51)
FOREARM PAIN (S54)
HAND/FINGER PAIN (S17)
HEADACHE PAIN (S02)
HIGH THORACIC BACK PAIN (S48)
HYPERTENSION (S43)
KELOID FORMATION (S38)
KNEE PAIN (S46)
LOW BACK PAIN (S03)
MIGRAINE (VASCULAR) HEADACHE (S18)
MYOSITIS OSSIFICANS (S67)
OSTEOARTHRITIS (S21)
PITTING (LYMPH) EDEMA (S31)
POST IMMOBILIZATION (S36)
RADICULITIS (S29)
RAYNAUD'S DISEASE (S12)
SHOULDER/HAND (S57)
SHOULDER PAIN (S08)
SINUS PAIN (S33)
TENDONITIS (S28)
TENNIS ELBOW (LATERAL EPICONDYLITIS) (S32)
THIGH PAIN (S49)
UNHEALED DERMAL LESION (S39)
WRIST PAIN (S16)

Precautions and Contraindications (M0704)

Galvanism (M0704A)

Direct current has effects not only on the free ions introduced into the tissues but has a physiological effect on the tissues themselves: an affect of "galvanism". Nondisassociated and charged colloid molecules from the adsorption of ions, such as droplets of fat, albumin, particles of starch, blood cells, bacteria and other single cells will be made to move toward the negative pole along with the ions (cataphoresis). Additionally, there are direct effects upon the chemicals which combine to make up and are contained within the various soft tissues, especially in the area of the electrodes. Essentially, the positive anode electrode will have the physiochemical effect of producing an acid reaction, repelling alkaloids and trace metals which have a positive charge, while producing the general physiological effects of "hardening" the soft tissues and decreasing nerve irritability. The negative cathode electrode will have the physiochemical effect of precipitating alkaline reactions while repelling acids and acidic radicals, "softening" the tissues and increasing nerve irritability. Both electrodes have the general effect of producing a mild heating of the tissues directly under them and stimulating vasomotor activity. These effects indicate the need for care when choosing electrode sites, and for consideration of the degree to which they may offset the beneficial effects of the iontophoresis process proposed.

Burn dangers (M0704B)

The chief danger inherent in iontophoresis is burns of the skin and excessive destruction of mucous membrane resulting from excessive current density. There is also danger of chemical burns resulting from acid and/or alkaloid reactions taking place below them, regardless of the chemical applied. Electrical burn danger may be reduced by taking care to avoid direct contact between the metal portion of the surface electrode and the patient's skin by covering all metal surfaces of the electrode with padding. Current strength must be regulated relative to the size of the electrodes, and current levels reduced according to electrode size.

Specific precautions (M0704C)

Excessive dosage or individual sensitivity may cause a systemic or autonomic reaction with mecholyl iontophoresis.

In general, to qualify for iontophoresis, the patient's skin sensation should be normal. If it is not, tissue damage which ordinarily would be avoided by patient report of developing discomfort must be precluded by keeping current levels low.

Current should not be applied over denuded areas or over a recently developed scar.

Electrode pads should be evenly soaked with saline, water, or the ion-rich solution and must remain wet during the treatment session; a drying electrode may cause skin irritation or a burn.

The patient must be encouraged to report any uncomfortable burning sensation. Current levels should be adjusted to suit patient tolerance.

The patient should be allowed to leave the treatment area only with dry skin and clothing.

Bibliography (M0705)

Bertolucci, pp. 103–108.
Delacerda, Iontophoresis for Treatment of Shinsplints: pp. 183–185.
Delacerda, A Comparative Study of Three Methods of Treatment for Shoulder Girdle Myofascial Syndrome: pp. 51–54.
Glass, pp. 519–525.
Gordon, pp. 869–870.
Harris, pp. 109–112.

Kahn, A Case Report: Lithium Iontophoresis for Gouty Arthritis: pp. 113–114. Kahn, Use of Iontophoresis in Peyronie's Disease: pp. 995–996.

Kahn, Iontophoresis and Ultrasound for Postsurgical Temporomandibular Trismus and Paresthesia: pp. 307–309.

Langley, pp. 1395–1396.
Shriber, pp. 124–135.
Stillwell, pp. 34–40.
Tannenbaum, pp. 792–793.
Watkins, pp. 114–127.

IONTOPHORESIS PAIN RELIEF (M07A)

Definition (M07A1)

Iontophoresis of either lidocaine or trolamine salicylate has been reported to be effective as a means of relieving pain from an anatomically superficial and localized source.

Lidocaine (M07A1A)

Lidocaine introduced through iontophoresis may provide a temporary relief of pain by producing a local analgesia or anesthesia of the skin area treated. The analgesia produced may be useful for the temporary relief of pain accompanying tendonitis, superficial fasciitis, or other superficial inflammation.

Trolamine salicylate (M07A1B)

Iontophoresis of trolamine salicylate has been noted to be effective for the relief of mild, localized joint and/or myogenic pain, providing a mild analgesia by filling receptor sites on the free pain nerve endings, while simultaneously working to inhibit at or near the location of treatment the production of prostaglandins and the other inflammatories which play a role in the inflammation process (refer to Soft Tissue Inflammation [C05] for further discussion).

Procedures of Application (M07A2)

Concentrations (M07A2A)

Both the lidocaine (at a 5% concentration) and trolamine salicylate (ranging in strength from 5 to 15%) come in the form of ointments. Lidocaine is also available in solution.

Patient preparation (M07A2B)

1) The patient should be placed in a comfortable position which allows easy access to the treatment site.
2) The treatment site should be stripped of covering material and cleansed with a mild alcohol rub or saline solution.

Procedure utilized in application of ointment (M07A2C)

1) The ointment of choice should be spread evenly over the selected treatment site.
2) The active surface electrode pad, generally made of a sponge-like material, should be (a) saturated by water or a saline solution, (b) connected via electrode cable to the appropriate pole of the electrical stimulation (positive for lidocaine, negative for trolamine salicylate) and (c) placed directly over the ointment and firmly secured in place over the skin via taping, strapping or weighting.
3) The inactive surface electrode pad, also soaked with water or saline solution, should be placed over an area of skin distal to the active electrode site.
4) The electrical stimulator should be preset at zero intensity and a timer set for fifteen or twenty minutes.
5) The electrical stimulator should be turned on and the intensity slowly increased until the patient feels a slight burning sensation. The current intensity should be slowly reduced if patient discomfort increases.
6) When the treatment time has elapsed, the electrical stimulation, if not terminated by a timer, should be reduced slowly back to zero and the electrical stimulator switched off.

Procedure utilized in application of solution (M07A2D)

1) The active electrode pad (positive for lidocaine) should be (a) saturated by a solution comprised of water or saline solution with lidocaine to a strength of at least 1% added and (b) firmly secured over the treatment site via taping, strapping or weighting. If a bath is utilized, the active electrode should be placed in a vessel adequately filled with enough solution of adequate chemical strength to cover the treatment site, and the body part to be stimulated submerged in it.
2) The water or saline-saturated nonactive electrode pad, negative in this case, should be placed over an area of skin distal to the active electrode. A foot or forearm bath may be used if the body part to be submerged is conveniently distant from the active electrode pad or bath.
3) The electrical stimulator should be preset at zero intensity and a timer set for fifteen or twenty minutes.
4) The electrical stimulator should be turned on and the intensity slowly increased until the patient feels a slight burning sensation. The current intensity should be slowly reduced if patient discomfort increases.

5) When the treatment time has elapsed, the electrical stimulation, if not terminated by a timer, should be reduced slowly back to zero and the electrical stimulator switched off.

Treatment termination   (M07A2E)

1) The electrodes should be removed from the treatment site or the body part removed from the bath, and the treatment site cleansed of all chemicals.

2) The treatment sites under both the positive and negative electrodes should be inspected for any signs of chemical or electrical burns, and the skin thoroughly towelled or blown dry.

3) The patient should be allowed to rest for a few minutes before leaving in warm, dry clothing.

4) The electrodes and any vessels used should be thoroughly cleansed and rinsed to get rid of any chemicals that may have accumulated in the sponges or on or near the metal component of the electrodes as a result of galvanism.

Partial List of Conditions Treated   (M07A3)

Treatable Causes   (M07A3A)

Bacterial infection   (C09)
Calcific deposit   (C08)
Interspinous ligamentous strain   (C03)
Joint ankylosis   (C42)
Joint approximation   (C47)
Ligamentous strain   (C25)
Restricted joint range of motion   (C26)
Soft tissue inflammation   (C05)
Tactile hypersensitivity   (C24)

Syndromes   (M07A3B)

ACHILLES TENDONITIS   (S52)
BACTERIAL INFECTION   (S63)
BURSITIS   (S26)
CALF PAIN   (S50)
CAPSULITIS   (S27)
CARPAL TUNNEL   (S35)
CERVICAL (NECK) PAIN   (S73)
CHEST PAIN   (S14)
ELBOW PAIN   (S47)
FACET   (S19)
FASCIITIS   (S20)
FOOT PAIN   (S51)
FOREARM PAIN   (S54)
HAND/FINGER PAIN   (S17)
HIGH THORACIC BACK PAIN   (S48)
JOINT SPRAIN   (S30)
LOWER ABDOMINAL PAIN   (S62)
LOW BACK PAIN   (S03)
MYOSITIS OSSIFICANS   (S67)
OSTEOARTHRITIS   (S21)
PIRIFORMIS   (S04)
POST IMMOBILIZATION   (S36)
REFERRED PAIN   (S01)
SHOULDER PAIN   (S08)
TENDONITIS   (S28)
TENNIS ELBOW (LATERAL EPICONDYLITIS)   (S32)
THIGH PAIN   (S49)
UNHEALED DERMAL LESION   (S39)
WRIST PAIN   (S16)

Precautions and Contraindications   (M07A4)

Galvanism   (M07A4A)

Direct current has effects not only on the free ions introduced into the tissues but has a physiological effect on the tissues themselves: an affect of "galvanism". Nondisassociated and charged colloid molecules from the adsorption of ions, such as droplets of fat, albumin, particles of starch, blood cells, bacteria and other single cells will be made to move toward the negative pole along with the ions (cataphoresis). Additionally, there are direct effects upon the chemicals which combine to make up and are contained within the various soft tissues, especially in the area of the electrodes. Essentially, the positive anode electrode will have the physiochemical effect of producing an acid reaction, repelling alkaloids and trace metals which have a positive charge, while producing the general physiological effects of "hardening" the soft tissues and decreasing nerve irritability. The negative cathode electrode will have the physiochemical effect of precipitating alkaline reactions while repelling acids and acidic radicals, "softening" the tissues and increasing nerve irritability. Both electrodes have the general effect of producing a mild heating of the tissues directly under them and stimulating vasomotor activity. These effects indicate the need for care when choosing electrode sites, and for consideration of the degree to which they may offset the beneficial effects of the iontophoresis process proposed.

Burn dangers   (M07A4B)

The chief danger inherent in iontophoresis is burns of the skin and excessive destruction of mucous membrane resulting from excessive current density. There is also danger of chemical burns resulting from acid and/or alkaloid reactions taking place below them, regardless of the chemical applied. Electrical burn danger may be reduced by taking care to avoid direct contact between the metal portion of the surface electrode and the patient's skin by covering all metal surfaces of the electrode with padding. Current strength must be regulated relative to the size of the electrodes, and current levels reduced according to electrode size.

Specific precautions (M07A4C)

In general, to qualify for iontophoresis, the patient's skin sensation should be normal. If it is not, tissue damage which ordinarily would be avoided by patient report of developing discomfort must be precluded by keeping current levels low.

Current should not be applied over denuded areas or over a recently developed scar.

Electrode pads should be evenly soaked with saline, water, or the ion-rich solution and must remain wet during the treatment session; a drying electrode may cause skin irritation or a burn.

The patient must be encouraged to report any uncomfortable burning sensation. Current levels should be adjusted to suit patient tolerance.

The patient should be allowed to leave the treatment area only with dry skin and clothing.

Bibliography (M07A5)

Bertolucci, pp. 103–108.
Delacerda, A Comparative Study of Three Methods of Treatment for Shoulder Girdle Myofascial Syndrome: pp. 51–54.
Delacerda, Iontophoresis for Treatment of Shinsplints: pp. 183–185.
Gordon, pp. 869–870.
Harris, pp. 109–112.
Kahn, TMJ Pain Control: pp. 14–15.
Shriber, pp. 124–135.
Stillwell, pp. 34–40.
Watkins, Pp. 114–127

IONTOPHORESIS INFLAMMATION AMELIORATION
(M07B)

Definition (M07B1)

Iontophoresis of either hydrocortisone or trolamine salicylate has been reported to be effective as a means of ameliorating soft tissue inflammation occurring in a superficial and localized area.

Hydrocortisone (M07B1A)

Hydrocortisone introduced through iontophoresis may diminish or stop a soft tissue inflammatory process by virtue of its stabilizing effect upon the cell membrane.

Trolamine salicylate (M07B1B)

Trolamine salicylate introduced through iontophoresis may inhibit prostaglandins production, a key chemical component of the inflammatory process (refer to Soft Tissue Inflammation [C05] for further discussion).

Procedures of Application (M07B2)

Concentrations (M07B2A)

Both the hydrocortisone (at a 0.5 to 1.0% concentration) and trolamine salicylate (usually at a strength of 10%) come in ointment forms.

Patient preparation (M07B2B)

1) The patient should be placed in a comfortable position which allows easy access to the treatment site.
2) The treatment site should be stripped of covering material and cleansed with a mild alcohol rub or saline solution.

Procedure utilized in application of ointment (M07B2C)

1) The ointment of choice should be spread evenly over the selected treatment site.
2) The active surface electrode pad (positive for the hydrocortisone and negative for the trolamine salicylate), generally made of a sponge-like material which has been saturated by water or a saline solution, should be placed directly over the ointment and firmly secured in place over the skin via taping, strapping or weighting.
3) The inactive surface electrode pad, also soaked with water or saline solution, should be placed over an area of skin distal to the active electrode site.
4) The electrical stimulator should be preset at zero intensity and a timer set for fifteen to twenty minutes.
5) The electrical stimulator should be turned on and the intensity slowly increased until the patient feels a slight burning sensation. The current intensity should be slowly reduced if patient discomfort increases.
6) When the treatment time has elapsed, and treatment has not been terminated by a timer, the electrical stimulation should be reduced slowly back to zero and the electrical stimulator switched off.

Treatment termination (M07B2D)

1) The electrodes should be removed from the treatment site and the treatment site cleansed of all chemicals.
2) The treatment sites under both the positive and negative electrodes should be inspected for any signs of chemical or electrical burns, and the skin towelled or blown dry.
3) The patient should be allowed to rest for a few minutes before leaving in warm, dry clothing.
4) The electrodes and any vessels used should be thoroughly cleansed and rinsed to get rid of the chemicals that may have accumulated in the sponges or on or near the metal component of the electrodes as a result of galvanism.

Partial List of Conditions Treated (M07B3)

Treatable Causes (M07B3A)

Hypermobile/instable joint (C10)
Interspinous ligamentous strain (C03)
Joint ankylosis (C42)
Joint approximation (C47)
Ligamentous strain (C25)
Nerve root compression (C04)
Restricted joint range of motion (C26)
Soft tissue inflammation (C05)

Tactile hypersensitivity (C24)
Trigger point formation (C01)

Syndromes (M07B3B)

ACHILLES TENDONITIS (S52)
BURSITIS (S26)
CALF PAIN (S50)
CAPSULITIS (S27)
CARPAL TUNNEL (S35)
CERVICAL (NECK) PAIN (S73)
CHEST PAIN (S14)
ELBOW PAIN (S47)
FACET (S19)
FASCIITIS (S20)
FOOT PAIN (S51)
FOREARM PAIN (S54)
HAND/FINGER PAIN (S17)
HIGH THORACIC BACK PAIN (S48)
KNEE PAIN (S46)
LOW BACK PAIN (S03)
MYOSITIS OSSIFICANS (S67)
OSTEOARTHRITIS (S21)
PIRIFORMIS (S04)
RADICULITIS (S29)
REFERRED PAIN (S01)
SHOULDER PAIN (S08)
TENDONITIS (S28)
TENNIS ELBOW (LATERAL
 EPICONDYLITIS) (S32)
THIGH PAIN (S49)
WRIST PAIN (S16)

Precautions and Contraindications (M07B4)

Galvanism (M07B4A)

Direct current has effects not only on the free ions introduced into the tissues but has a physiological effect on the tissues themselves: an affect of "galvanism". Nondisassociated and charged colloid molecules from the adsorption of ions, such as droplets of fat, albumin, particles of starch, blood cells, bacteria and other single cells will be made to move toward the negative pole along with the ions (cataphoresis). Additionally, there are direct effects upon the chemicals which combine to make up and are contained within the various soft tissues, especially in the area of the electrodes. Essentially, the positive anode electrode will have the physiochemical effect of producing an acid reaction, repelling alkaloids and trace metals which have a positive charge, while producing the general physiological effects of "hardening" the soft tissues and decreasing nerve irritability. The negative cathode electrode will have the physiochemical effect of precipitating alkaline reactions while repelling acids and acidic radicals, "softening" the tissues and increasing nerve irritability. Both electrodes have the general effect of producing a mild heating of the tissues directly under them and stimulating vasomotor activity. These effects indicate the need for care when choosing electrode sites, and for consideration of the degree to which they may offset the beneficial effects of the iontophoresis process proposed.

Burn dangers (M07B4B)

The chief danger inherent in iontophoresis is burns of the skin and excessive destruction of mucous membrane resulting from excessive current density. There is also danger of chemical burns resulting from acid and/or alkaloid reactions taking place below them, regardless of the chemical applied. Electrical burn danger may be reduced by taking care to avoid direct contact between the metal portion of the surface electrode and the patient's skin by covering all metal surfaces of the electrode with padding. Current strength must be regulated relative to the size of the electrodes, and current levels reduced according to electrode size.

Specific precautions (M07B4C)

In general, to qualify for iontophoresis, the patient's skin sensation should be normal. If it is not, tissue damage which ordinarily would be avoided by patient report of developing discomfort must be precluded by keeping current levels low.

Current should not be applied over denuded areas or over a recently developed scar.

Electrode pads should be evenly soaked with saline, water, or the ion-rich solution and must remain wet during the treatment session; a drying electrode may cause skin irritation or a burn.

The patient must be encouraged to report any uncomfortable burning sensation. Current levels should be adjusted to suit patient tolerance.

The patient should be allowed to leave the treatment area only with dry skin and clothing.

Bibliography (M07B5)

Bertolucci, pp. 103–108.
Delacerda, A Comparative Study of Three Methods of Treatment for Shoulder Girdle Myofascial Syndrome: pp. 51–54.
Delacerda, Iontophoresis for Treatment of Shinsplints: pp. 183–185.
Harris, pp. 109–112.

Kahn, Iontophoresis and Ultrasound for Postsurgical Temporomandibular Trismus and Paresthesia: pp. 307–308.
Kahn, TMJ Pain Control: pp. 14–15.
Kahn, Use of Iontophoresis in Peyronie's Disease: pp. 995–996.

Pasaki, Moss, Carroll, pp. 233–238.
Rothfield, pp. 974–975.
Shriber, pp. 124–135.
Stillwell, pp. 34–40.
Watkins, pp. 114–127.

IONTOPHORESIS EDEMA CONTROL (M07C)

Definition (M07C1)

Iontophoresis of either mecholyl or hyaluronidase has been reported to be effective as means of controlling or relieving edema.

Mecholyl (M07C1A)

Mecholyl introduced through iontophoresis may diminish or stop edema by fostering increased circulation (relieving ischemia).

Hyaluronidase (M07C1B)

Hyaluronidase so introduced may decrease edema through enzymatic action by causing hydrolyzation of hyaluronic acid (a viscous polysaccharide found in the interstices of the tissues), promoting diffusion and consequently absorption of fluids accumulated in the tissues.

Procedures of Application (M07C2)

Concentrations (M07C2A)

Hyaluronadase (Wydase) comes as a solution, generally at a concentration of 115 units/ml.

Mecholyl (in strengths ranging from 0.25 to 1%) comes in a lotion or ointment form.

Patient preparation (M07C2B)

1) The patient should be placed in a comfortable position which allows easy access to the treatment site.

2) The treatment site should be stripped of covering material and cleansed with a mild alcohol rub or saline solution.

Procedure utilized in application of ointment (M07C2C)

1) Mecholyl ointment of choice should be spread evenly over the selected treatment site.

2) The positive surface electrode pad, generally made of a sponge-like material, saturated by water or a saline solution, should be placed directly over the ointment and firmly secured in place via taping, strapping or weighting.

3) The negative surface electrode pad, also soaked with water or saline solution, should be placed over an area of skin distal to the active electrode site.

4) The electrical stimulator should be preset at zero intensity and a timer set for fifteen or twenty minutes.

5) The electrical stimulator should be turned on and the intensity slowly increased until the patient feels a slight burning sensation. The current intensity should be slowly reduced if patient discomfort increases.

6) When the treatment time has elapsed, the electrical stimulation, if not terminated by a timer, should be reduced slowly back to zero and the electrical stimulator switched off.

Procedure utilized in application of solution (M07C2D)

1) The positive pad-electrode should be (a) saturated by a solution containing the hyaluronadase or mecholyl, (b) placed over the treatment (edematous) site and (c) firmly secured in place via taping, strapping or weighting. If a bath is utilized, the positive electrode should be placed in a vessel adequately filled with enough solution to cover the treatment site.

2) The water or saline-saturated negative electrode pad, should be placed over an area of skin distal to the active electrode. A foot or forearm bath may be used if the body part to be submerged is conveniently distant from the active electrode pad or bath.

3) The electrical stimulator should be preset at zero intensity and a timer set for fifteen or twenty minutes.

4) The electrical stimulator should be turned on and the intensity slowly increased until the patient feels a slight burning sensation. The current intensity should be slowly reduced if patient discomfort increases.

5) When the treatment time has elapsed, the electrical stimulation, if not terminated by a timer, should be reduced slowly back to zero and the electrical stimulator switched off.

Treatment termination (M07C2E)

1) The electrodes should be removed from the treatment site or the body part removed from the bath, and the treatment site cleansed of all chemicals.

2) The treatment sites under both the positive and negative electrodes should be inspected for any signs of chemical or electrical burns, and the skin thoroughly towelled or blown dry.

3) The patient should be allowed to rest for a few minutes before leaving in warm, dry clothing.

4) The electrodes and any vessels used should be thoroughly cleansed and rinsed to get rid of the chemicals that may have accumulated in the sponges or on or near the metal component of the electrodes as a result of galvanism.

Partial List of Conditions Treated (M07C3)

Treatable Causes (M07C3A)

Habitual joint positioning (C48)
Joint ankylosis (C42)
Ligamentous strain (C25)
Nerve root compression (C04)
Peripheral nerve injury (C07)
Restricted joint range of motion (C26)
Soft tissue swelling C06)
Vascular insufficiency (C11)

Syndromes (M07C3B)

CARPAL TUNNEL (S35)
JOINT SPRAIN (S30)
MYOSITIS OSSIFICANS (S67)
OSTEOARTHRITIS (S21)
PITTING (LYMPH) EDEMA (S31)
POST IMMOBILIZATION (S36)
POST PERIPHERAL NERVE INJURY (S23)
RADICULITIS (S29)
SHOULDER/HAND (S57)
TENDONITIS (S28)
TENNIS ELBOW (LATERAL EPICONDYLITIS) (S32)

Precautions and Contraindications (M07C4)

Galvanism (M07C4A)

Direct current has effects not only on the free ions introduced into the tissues but has a physiological effect on the tissues themselves: an affect of "galvanism". Nondisassociated and charged colloid molecules from the adsorption of ions, such as droplets of fat, albumin, particles of starch, blood cells, bacteria and other single cells will be made to move toward the negative pole along with the ions (cataphoresis). Additionally, there are direct effects upon the chemicals which combine to make up and are contained within the various soft tissues, especially in the area of the electrodes. Essentially, the positive anode electrode will have the physiochemical effect of producing an acid reaction, repelling alkaloids and trace metals which have a positive charge, while producing the general physiological effects of "hardening" the soft tissues and decreasing nerve irritability. The negative cathode electrode will have the physiochemical effect of precipitating alkaline reactions while repelling acids and acidic radicals, "softening" the tissues and increasing nerve irritability. Both electrodes have the general effect of producing a mild heating of the tissues directly under them and stimulating vasomotor activity. These effects indicate the need for care when choosing electrode sites, and for consideration of the degree to which they may offset the beneficial effects of the iontophoresis process proposed.

Burn dangers (M07C4B)

The chief danger inherent in iontophoresis is burns of the skin and excessive destruction of mucous membrane resulting from excessive current density. There is also danger of chemical burns resulting from acid and/or alkaloid reactions taking place below them, regardless of the chemical applied. Electrical burn danger may be reduced by taking care to avoid direct contact between the metal portion of the surface electrode and the patient's skin by covering all metal surfaces of the electrode with padding. Current strength must be regulated relative to the size of the electrodes, and current levels reduced according to electrode size.

Specific precautions (M07C4C)

Excessive dosage or individual sensitivity may cause a systemic or autonomic reaction with mecholyl iontophoresis.

In general, to qualify for iontophoresis, the patient's skin sensation should be normal. If it is not, tissue damage which ordinarily would be avoided by patient report of developing discomfort must be precluded by keeping current levels low.

Current should not be applied over denuded areas or over a recently developed scar.

Electrode pads should be evenly soaked with saline, water, or the ion-rich solution and must remain wet during the treatment session; a drying electrode may cause skin irritation or a burn.

The patient must be encouraged to report any uncomfortable burning sensation. Current levels should be adjusted to suit patient tolerance.

Soon after beginning the process of Iontophoresis, mecholyl will (almost immediately) cause prickling of the skin followed by intense local hype remia and perspiration of the local area which will last for 6 to 8 hours after cessation of electrical stimulation. After 10 minutes of stimulation the mecholyl may penetrate the skin and cause more general effects.

The antidote for over exposure to mecholyl is the injection of 1cc (1/200 of a grain) of atropine.

The patient should be allowed to leave the treatment area only with dry skin and clothing.

Bibliography (M07C5)

Bertolucci, pp. 103–108.
Delacerda, Iontophoresis for Treatment of Shinsplints: pp. 183–185.
Harris, pp. 109–112.
Magistro, pp. 169–175.
Shriber, pp. 133–134.
Stillwell, pp. 34–40.

IONTOPHORESIS ISCHEMIA RELIEF (M07D)

Definition (M07D1)

Mecholyl (M07D1A)

Iontophoresis of mecholyl, a derivative of choline, has been reported to be effective in relieving ischemia, fostering increased circulation, by causing vasodilation through stimulation of the parasympathetic nerves and its antagonism to atropine.

Histamine (M07D1B)

Iontophoresis of histamine has also been reported to relieve ischemia by stimulating vasodilation of superficial blood vessels.

Procedures of Application (M07D2)

Concentrations (M07D2A)

Mecholyl comes in ointment forms at strengths of between .25 and 1.0%.

Histamine ions comes in an (imadyl) ointment form with a concentration of 1%.

Patient preparation (M07D2B)

1) The patient should be placed in a comfortable position which allows easy access to the treatment site.
2) The treatment site should be stripped of covering material and cleansed with a mild alcohol rub or saline solution.

Procedure utilized in application of ointment (M07D2C)

1) The ointment of choice should be spread evenly over the selected treatment site.
2) The positive surface electrode pad, generally made of a sponge-like material, should be a) saturated by water or saline solution, b) placed directly over the ointment and c) firmly secured in place via taping, strapping or weighting.
3) The negative surface electrode pad, also soaked with water or saline solution, should be placed over an area of skin distal to the active electrode site.
4) The electrical stimulator should be preset at zero intensity and a timer set for fifteen to twenty minutes. If histamine has been selected, stimulation should last for ten minutes or less to avoid general reaction to it.
5) The electrical stimulator should then be turned on and the intensity slowly increased until the patient feels a slight burning sensation. The current intensity should be slowly reduced if patient discomfort increases. If histamine is being applied, current levels should not be allowed to exceed one milliampere per square inch of active electrode.
6) When the treatment time has elapsed, and treatment has not been terminated by a timer, the electrical stimulation should be reduced slowly back to zero and the electrical stimulator switched off.

Treatment termination (M07D2D)

1) The electrodes should be removed from the treatment site or the body part removed from the bath, and the treatment site cleansed of all chemicals.
2) The treatment sites under both the positive and negative electrodes should be inspected for any signs of chemical or electrical burns, and the skin thoroughly towelled or blown dry.
3) The patient should be allowed to rest for a few minutes before leaving in warm, dry clothing.
4) The electrodes and any vessels used should be thoroughly cleansed and rinsed to get rid of the chemicals that may have accumulated in the sponges or on or near the metal component of the electrodes as a result of galvanism.

Partial List of Conditions Treated (M07D3)

Treatable Causes (M07D3A)

Bacterial infection (C09)
Extrafusal muscle spasm (C41)
Habitual joint positioning (C48)
Ligamentous strain (C25)
Peripheral nerve injury (C07)
Soft tissue inflammation (C05)
Soft tissue swelling (C06)
Unhealed dermal lesion (C12)
Vascular insufficiency (C11)

Syndromes (M07D3B)

ACHILLES TENDONITIS (S52)
BACTERIAL INFECTION (S63)

BUERGER'S DISEASE (S11)
DIABETES (S10)
HAND/FINGER PAIN (S17)
RAYNAUD'S DISEASE (S12)
SHOULDER/HAND (S57)
UNHEALED DERMAL LESION (S39)

Precautions and Contraindications (M07D4)

Galvanism (M07D4A)

Direct current has effects not only on the free ions introduced into the tissues but has a physiological effect on the tissues themselves: an affect of "galvanism". Nondisassociated and charged colloid molecules from the adsorption of ions, such as droplets of fat, albumin, particles of starch, blood cells, bacteria and other single cells will be made to move toward the negative pole along with the ions (cataphoresis). Additionally, there are direct effects upon the chemicals which combine to make up and are contained within the various soft tissues, especially in the area of the electrodes. Essentially, the positive anode electrode will have the physiochemical effect of producing an acid reaction, repelling alkaloids and trace metals which have a positive charge, while producing the general physiological effects of "hardening" the soft tissues and decreasing nerve irritability. The negative cathode electrode will have the physiochemical effect of precipitating alkaline reactions while repelling acids and acidic radicals, "softening" the tissues and increasing nerve irritability. Both electrodes have the general effect of producing a mild heating of the tissues directly under them and stimulating vasomotor activity. These effects indicate the need for care when choosing electrode sites, and for consideration of the degree to which they may offset the beneficial effects of the iontophoresis process proposed.

Burn dangers (M07D4B)

The chief danger inherent in iontophoresis is burns of the skin and excessive destruction of mucous membrane resulting from excessive current density. There is also danger of chemical burns resulting from acid and/or alkaloid reactions taking place below them, regardless of the chemical applied. Electrical burn danger may be reduced by taking care to avoid direct contact between the metal portion of the surface electrode and the patient's skin by covering all metal surfaces of the electrode with padding. Current strength must be regulated relative to the size of the electrodes, and current levels reduced according to electrode size.

Specific precautions (M07D4C)

Excessive dosage or individual sensitivity may cause a systemic or autonomic reaction with mecholyl iontophoresis.

In general, to qualify for iontophoresis, the patient's skin sensation should be normal. If it is not, tissue damage which ordinarily would be avoided by patient report of developing discomfort must be precluded by keeping current levels low.

Current should not be applied over denuded areas or over a recently developed scar.

Electrode pads should be evenly soaked with saline, water, or the ion-rich solution and must remain wet during the treatment session; a drying electrode may cause skin irritation or a burn.

The patient must be encouraged to report any uncomfortable burning sensation. Current levels should be adjusted to suit patient tolerance.

The patient should be allowed to leave the treatment area only with dry skin and clothing.

Soon after beginning the process of Iontophoresis, mecholyl will almost immediately cause prickling of the skin followed by intense local hyperemia and perspiration of the local area which will last for six to eight hours after cessation of electrical stimulation. After ten minutes of stimulation the mecholyl may penetrate the skin and cause more general effects. The antidote for mecholyl overtreatment is atropine injection as described above.

Over-treatment with histamine is marked by headache, flushing of the head area and chest, excessive salivation, increased peristalsis and constriction of the pupils; the antidote for overexposure to histamine is injection of 1cc (1/200 of a grain) of atropine. Histamine iontophoresis is contraindicated for patients prone to allergy.

Bibliography (M07D5)

Shriber, pp. 133–134.
Stillwell, pp. 34–40.
Watkins, pp. 114–127.

IONTOPHORESIS FACILITATION OF OPEN LESION HEALING (M07E)

Definition (M07E1)

Zinc (M07E1A)

Iontophoresis of zinc ions has been reported to be effective in facilitating the healing of open lesions by producing a grey coagulum in the wound which promotes granulation.

Procedures of Application (M07E2)

Concentration (M07E2A)

Zinc ions are available in solutions of zinc chloride or zinc sulfate at concentrations of 2%.

Patient preparation (M07E2B)

1) The patient should be placed in a comfortable position which allows easy access to the treatment site.
2) The treatment site should be stripped of covering material and cleansed with a mild alcohol rub or saline solution.

Procedure utilized in application of solution (M07E2C)

1) The positive pad-electrode should be (a) saturated by a solution containing the zinc solution, (b) placed over the edematous treatment site and (c) firmly secured in place via taping, strapping or weighting. If a bath is utilized, the positive electrode should be placed in a vessel adequately filled with enough solution to cover the treatment site.
2) The water or saline saturated negative electrode pad should be placed over an area of skin distal to the active electrode. A foot or forearm bath may be used if the body part to be submerged is conveniently distant from the active electrode pad or bath.
3) The electrical stimulator should be preset at zero intensity and a timer set for fifteen or twenty minutes.
4) The electrical stimulator should be turned on and the intensity slowly increased until the patient feels a slight burning sensation. The current intensity should be slowly reduced if patient discomfort increases.
5) When the treatment time has elapsed, the electrical stimulation, if not terminated by a timer, should be reduced slowly back to zero and the electrical stimulator switched off.

Treatment termination (M07E2D)

1) The electrodes should be removed from the treatment site or the body part removed from the bath, and the treatment site cleansed of all chemicals.
2) The treatment sites under both the positive and negative electrodes should be inspected for any signs of chemical or electrical burns, and the skin thoroughly towelled or blown dry.
3) The patient should be allowed to rest for a few minutes before leaving in warm, dry clothing.
4) The electrodes and any vessels used should be thoroughly cleansed and rinsed to get rid of the chemicals that may have accumulated in the sponges or on or near the metal component of the electrodes as a result of galvanism.

Partial List of Conditions Treated (M07E3)

Conditions Treated in the Field (M07E3A)

Decubitus ulcers

Treatable Causes (M07E3B)

Necrotic soft tissue (C18)
Unhealed dermal lesion (C12)

Syndromes (M07E3C)

ACNE (S45)
DIABETES (S10)
UNHEALED DERMAL LESION (S39)

Precautions and Contraindications (M07E4)

Galvanism (M07E4A)

Direct current has effects not only on the free ions introduced into the tissues but has a physiological effect on the tissues themselves: an affect of "galvanism". Nondisassociated and charged colloid molecules from the adsorption of ions, such as droplets of fat, albumin, particles of starch, blood cells, bacteria and other single cells will be made to move toward the negative pole along with the ions (cataphoresis). Additionally, there are direct effects upon the chemicals which combine to make up and are contained within the various soft tissues, especially in the area of the electrodes. Essentially, the positive anode electrode will have the physiochemical effect of producing an acid reaction, repelling alkaloids and trace metals which have a positive charge, while producing the general physiological effects of "hardening" the soft tissues and decreasing nerve irritability. The negative cathode electrode will have the physiochemical effect of precipitating alkaline reactions while repelling acids and acidic radicals, "softening" the tissues and increasing nerve irritability. Both electrodes have the general effect of producing a mild heating of the tissues directly under them and stimulating vasomotor activity. These effects indicate the need for care when choosing electrode sites, and for consideration of the degree to which they may offset the beneficial effects of the iontophoresis process proposed.

Burn dangers (M07E4B)

The chief danger inherent in iontophoresis is burns of the skin and excessive destruction of mucous membrane resulting from excessive current density. There is also danger of chemical burns resulting from acid and/or alkaloid reactions taking place below them, regardless of the chemical applied. Electrical burn danger may be reduced by taking care to avoid direct contact between the metal portion of the surface electrode and the patient's skin by covering all metal surfaces of the electrode with padding. Current strength must be regulated relative to the size of the electrodes, and current levels reduced according to electrode size.

Specific precautions (M07E4C)

In general, to qualify for iontophoresis, the patient's skin sensation should be normal. If it is not, tissue damage which ordinarily would be avoided by patient report of developing discomfort must be precluded by keeping current levels low.

Current should not be applied over denuded areas or over a recently developed scar.

Electrode pads should be evenly soaked with saline, water, or the ion-rich solution and must remain wet during the treatment session; a drying electrode may cause skin irritation or a burn.

The patient must be encouraged to report any uncomfortable burning sensation. Current levels should be adjusted to suit patient tolerance.

The patient should be allowed to leave the treatment area only with dry skin and clothing.

Bibliography (M07E5)

Cornwall, pp. 359–360.
Shriber, pp. 124–135.
Stillwell, pp. 34–40.
Watkins, pp. 114–127.

IONTOPHORESIS MUSCLE SPASM RELIEF (M07F)

Definition (M07F1)

Magnesium (M07F1A)

Iontophoresis of magnesium has been reported to be effective as means of relieving muscle spasm. No description of its action was found in the literature reviewed except for the note that it must play an important role in muscle metabolism since hypomagnesemia may precipitate muscle spasm, "cramps", tetany, choreiform and athetoid movements, and distal limb paresthesias and burning sensations.

Procedures of Application (M07F2)

Concentration (M07F2A)

Magnesium comes in a 2% magnesium sulfate solution or lotion form.

Patient preparation (M07F2B)

1) The patient should be placed in a comfortable position which allows easy access to the treatment site.
2) The treatment site should be stripped of covering material and cleansed with a mild alcohol rub or saline solution.

Procedure utilized in application of solution (M07F2C)

1) The positive surface electrode should be (a) saturated by a solution containing the magnesium, (b) placed over the muscle spasm or "cramp" and (c) firmly secured in place via taping, strapping or weighting.
2) The water or saline-saturated negative electrode pad should be placed over an area of skin distal to the active electrode. A foot or forearm bath may be used if the body part to be submerged is conveniently distant from the active electrode pad or bath.
3) The electrical stimulator should be preset at zero intensity and a timer set for five or ten minutes.
4) The electrical stimulator should be turned on and the intensity slowly increased until the patient feels a slight burning sensation. The current intensity should be slowly reduced if patient discomfort increases.
5) When the treatment time has elapsed, the electrical stimulation, if not terminated by a timer, should be reduced slowly back to zero and the electrical stimulator switched off.

Treatment termination (M07F2D)

1) The electrodes should be removed from the treatment site or the body part removed from the bath, and the treatment site cleansed of all chemicals.
2) The treatment sites under both the positive and negative electrodes should be inspected for any signs of chemical or electrical burns, and the skin thoroughly towelled or blown dry.
3) The patient should be allowed to rest for a few minutes before leaving in warm, dry clothing.
4) The electrodes and any vessels used should be thoroughly cleansed and rinsed to get rid of the chemicals that may have accumulated in the sponges or on or near the metal component of the electrodes as a result of galvanism.

Partial List of Conditions Treated (M07F3)

Treatable Causes (M07F3A)

Extrafusal muscle spasm (C41)
Restricted joint range of motion (C26)
Trigger point formation (C01)

Syndromes (M07F3B)

CALF PAIN (S50)
FOREARM PAIN (S54)
HIGH THORACIC BACK PAIN (S48)
KNEE PAIN (S46)
LOW BACK PAIN (S03)
REFERRED PAIN (S01)
SCIATICA (S05)
SHOULDER PAIN (S08)
THIGH PAIN (S49)

Precautions and Contraindications (M07F4)

Galvanism (M07F4A)

Direct current has effects not only on the free ions introduced into the tissues but has a physiological effect on the tissues themselves: an affect of "galvanism". Nondisassociated and charged colloid molecules from the adsorption of ions, such as droplets of fat, albumin, particles of starch, blood cells, bacteria and other single cells will be made to move toward the negative pole along with the ions (cataphoresis). Additionally, there are direct effects upon the chemicals which combine to make up and are contained within the various soft tissues, especially in the area of the electrodes. Essentially, the positive anode electrode will have the physiochemical effect of producing an acid reaction, repelling alkaloids and trace metals which have a positive charge, while producing the general physiological effects of "hardening" the soft tissues and decreasing nerve irritability. The negative cathode electrode will have the physiochemical effect of precipitating alkaline reactions while repelling acids and acidic radicals, "softening" the tissues and increasing nerve irritability. Both electrodes have the general effect of producing a mild heating of the tissues directly under them and stimulating vasomotor activity. These effects indicate the need for care when choosing electrode sites, and for consideration of the degree to which they may offset the beneficial effects of the iontophoresis process proposed.

Burn dangers (M07F4B)

The chief danger inherent in iontophoresis is burns of the skin and excessive destruction of mucous membrane resulting from excessive current density. There is also danger of chemical burns resulting from acid and/or alkaloid reactions taking place below them, regardless of the chemical applied. Electrical burn danger may be reduced by taking care to avoid direct contact between the metal portion of the surface electrode and the patient's skin by covering all metal surfaces of the electrode with padding. Current strength must be regulated relative to the size of the electrodes, and current levels reduced according to electrode size.

Specific precautions (M07F4C)

In general, to qualify for iontophoresis, the patient's skin sensation should be normal. If it is not, tissue damage which ordinarily would be avoided by patient report of developing discomfort must be precluded by keeping current levels low.

Current should not be applied over denuded areas or over a recently developed scar.

Electrode pads should be evenly soaked with saline, water, or the ion-rich solution and must remain wet during the treatment session; a drying electrode may cause skin irritation or a burn.

The patient must be encouraged to report any uncomfortable burning sensation. Current levels should be adjusted to suit patient tolerance.

The patient should be allowed to leave the treatment area only with dry skin and clothing.

Introduction of magnesium sulfate to patients with impaired renal function may produce cardiac and central nervous system depression and should be avoided unless a definite hypomagnesemia has been established by laboratory evaluation.

Bibliography (M07F5)

Shriber, pp. 124–135.
Stillwell, pp. 34–40.
Watkins, pp. 114–127.

IONTOPHORESIS HYPERTONICITY RELIEF (M07G)

Definition (MO7G1)

Calcium chloride (M07G1A)

Hypertonicity of skeletal musculature has been found to result from a fall in the level of ionized calcium, and it has been reported that iontophoresis of calcium may be effective as a remedy.

Procedures of Application (M07G2)

Concentration (M07G2A)

Calcium comes in a 2% calcium chloride solution or lotion form.

Patient preparation (M07G2B)

1) The patient should be placed in a comfortable position which allows easy access to the treatment site.
2) The treatment site should be stripped of covering material and cleansed with a mild alcohol rub or saline solution.

Procedure utilized in application of solution (M07G2C)

1) The positive surface electrode should be (a) saturated by a solution containing the calcium ions, (b) placed over the general area of muscular hypertonicity (if the condition is a fairly localized phenomena) and (c) firmly secured in place via taping, strapping or weighting.
2) The water or saline-saturated negative electrode pad should be placed over an area of skin distal to the active electrode. A foot or forearm bath may be used if the body part to be submerged is conveniently distant from the active electrode pad or bath.
3) The electrical stimulator should be preset at zero intensity and a timer set for fifteen to twenty minutes.
4) The electrical stimulator should be turned on and the intensity slowly increased until the patient feels a slight burning sensation. The current intensity should be slowly reduced if patient discomfort increases.
5) When the treatment time has elapsed, the electrical stimulation, if not terminated by a timer, should be reduced slowly back to zero and the electrical stimulator switched off.

Treatment termination (M07G2D)

1) The electrodes should be removed from the treatment site or the body part removed from the bath, and the treatment site cleansed of all chemicals.
2) The treatment sites under both the positive and negative electrodes should be inspected for any signs of chemical or electrical burns, and the skin thoroughly towelled or blown dry.
3) The patient should be allowed to rest for a few minutes before leaving in warm, dry clothing.
4) The electrodes and any vessels used should be thoroughly cleansed and rinsed to get rid of the chemicals that may have accumulated in the sponges or on or near the metal component of the electrodes as a result of galvanism.

Partial List of Conditions Treated (M07G3)

Treatable Causes (M07G3A)

Extrafusal muscle spasm (C41)
Pathological neuromuscular hypertonus (C28)
Restricted joint range of motion (C26)
Trigger point formation (C01)

Syndromes (M07G3B)

CALF PAIN (S50)
CHEST PAIN (S14)
ELBOW PAIN (S47)
FOOT PAIN (S51)
FOREARM PAIN (S54)
HAND/FINGER PAIN (S17)
KNEE PAIN (S46)
LOW BACK PAIN (S03)
MENSTRUAL CRAMPING (S60)
REFERRED PAIN (S01)
THIGH PAIN (S49)

Precautions and Contraindications (M07G4)

Galvanism (M07G4A)

Direct current has effects not only on the free ions introduced into the tissues but has a physiological effect on the tissues themselves: an affect of "galvanism". Nondisassociated and charged colloid molecules from the adsorption of ions, such as droplets of fat, albumin, particles of starch, blood cells, bacteria and other single cells will be made to move toward the negative pole along with the ions (cataphoresis). Additionally, there are direct effects upon the chemicals which combine to make up and are contained within the various soft tissues, especially in the area of the electrodes. Essentially, the positive anode electrode will have the physiochemical effect of producing an acid reaction, repelling alkaloids and trace metals which have a positive charge, while producing the general physiological effects of "hardening" the soft tissues and decreasing nerve irritability. The negative cathode electrode will have the physiochemical effect of precipitating alkaline reactions while repelling acids and acidic radicals, "softening" the tissues and increasing nerve irritability. Both electrodes have the general effect of producing a mild heating of the tissues directly under them and stimulating vasomotor activity. These effects indicate the need for care when choosing electrode sites, and for consideration of the degree to which they may offset the beneficial effects of the iontophoresis process proposed.

Burn dangers (M07G4B)

The chief danger inherent in iontophoresis is burns of the skin and excessive destruction of mucous membrane resulting from excessive current density. There is also danger of chemical burns resulting from acid and/or alkaloid reactions taking place below them, regardless of the chemical applied. Electrical burn danger may be reduced by taking care to avoid direct contact between the metal portion of the surface electrode and the patient's skin by covering all metal surfaces of the electrode with padding. Current strength must be regulated relative to the size of the electrodes, and current levels reduced according to electrode size.

Specific precautions (M07G4C)

In general, to qualify for iontophoresis, the patient's skin sensation should be normal. If it is not, tissue damage which ordinarily would be avoided by patient report of developing discomfort must be precluded by keeping current levels low.

Current should not be applied over denuded areas or over a recently developed scar.

Electrode pads should be evenly soaked with saline, water, or the ion-rich solution and must remain wet during the treatment session; a drying electrode may cause skin irritation or a burn.

The patient must be encouraged to report any uncomfortable burning sensation. Current levels should be adjusted to suit patient tolerance.

The patient should be allowed to leave the treatment area only with dry skin and clothing.

Iontophoresis of calcium ions is contraindicated for those patients suffering from hypercalcemia. Hypercalcemic symptoms include depression of nervous and neuromuscular function (emotional disturbances, confusion, skeletal muscle weakness and constipation), impairment of renal function (polyuria, polydipsia, and azotemia), and precipitation of calcium salts (calcific keratopathy, nephrocalcinosis and renal calculi).

Bibliography (M07G5)

Kahn, Calcium Iontophoresis in Suspected Myopathy: pp. 276–277.
Shriber, pp. 124–135.
Stillwell, pp. 34–40.
Watkins, pp. 114–127.

IONTOPHORESIS GOUTY TOPHI DISSOLUTION
(M07H)

Definition (M07H1)

Lithium chloride or carbonate (M07H1A)

It has been reported that gouty tophi may be dissolved through the iontophoresis of lithium. The ionic exchange between lithium and sodium is said to favorably affect the solubility of urates formed, and to have a sclerolytic effect upon tophi which may be present; i.e., the lithium may replace sodium ions in the sodium urate crystal and hasten absorption.

Procedures of Application (M07H2)

Concentration (M07H2A)

Lithium comes in a 2% lithium chloride or carbonate solution or lotion form.

Patient preparation (M07H2B)

1) The patient should be placed in a comfortable position which allows easy access to the treatment site.
2) The treatment site should be stripped of covering material and cleansed with a mild alcohol rub or saline solution.

Procedure utilized in application of solution (M07H2C)

1) The positive surface electrode should be (a) saturated by a solution containing the lithium ions, (b) placed over the site of suspected or confirmed gouty tophi and (c) firmly secured in place via taping, strapping or weighting.
2) The water or saline saturated negative electrode pad should be placed over an area of skin distal to the active electrode. A foot or forearm bath may be used if the body part to be submerged is conveniently distant from the active electrode pad or bath.
3) The electrical stimulator should be preset at zero intensity and a timer set for twenty minutes.
4) The electrical stimulator should be turned on and the intensity slowly increased until the patient feels a slight burning sensation. The current intensity should be slowly reduced if patient discomfort increases.
5) When the treatment time has elapsed, and treatment has not been terminated by a timer, the electrical stimulation should be reduced slowly back to zero and the electrical stimulator switched off.

Treatment termination (M07H2D)

1) The electrodes should be removed from the treatment site or the body part removed from the bath, and the treatment site cleansed of all chemicals.
2) The treatment sites under both the positive and negative electrodes should be inspected for any signs of chemical or electrical burns, and the skin thoroughly towelled or blown dry.
3) The patient should be allowed to rest for a few minutes before leaving in warm, dry clothing.
4) The electrodes and any vessels used should be thoroughly cleansed and rinsed to get rid of the chemicals that may have accumulated in the sponges or on or near the metal component of the electrodes as a result of galvanism.

Partial List of Conditions Treated (M07H3)

Treatable Causes (M07H3A)

Calcific deposit (C08)
Habitual joint positioning (C48)
Joint ankylosis (C42)
Nerve root compression (C04)
Restricted joint range of motion (C26)

Syndromes (M07H3B)

BURSITIS (S26)
ELBOW PAIN (S47)
FOOT PAIN (S51)
FROZEN SHOULDER (S64)
HAND/FINGER PAIN (S17)
HIGH THORACIC BACK PAIN (S48)
KNEE PAIN (S46)
MYOSITIS OSSIFICANS (S67)
OSTEOARTHRITIS (S21)
POST IMMOBILIZATION (S36)
REFERRED PAIN (S01)

SHOULDER/HAND (S57)
SHOULDER PAIN (S08)
TENDONITIS (S28)
TENNIS ELBOW (LATERAL
 EPICONDYLITIS) (S32)
WRIST PAIN (S16)

Precautions and Contraindications (M07H4)

Galvanism (M07H4A)

Direct current has effects not only on the free ions introduced into the tissues but has a physiological effect on the tissues themselves: an affect of "galvanism". Nondisassociated and charged colloid molecules from the adsorption of ions, such as droplets of fat, albumin, particles of starch, blood cells, bacteria and other single cells will be made to move toward the negative pole along with the ions (cataphoresis). Additionally, there are direct effects upon the chemicals which combine to make up and are contained within the various soft tissues, especially in the area of the electrodes. Essentially, the positive anode electrode will have the physiochemical effect of producing an acid reaction, repelling alkaloids and trace metals which have a positive charge, while producing the general physiological effects of "hardening" the soft tissues and decreasing nerve irritability. The negative cathode electrode will have the physiochemical effect of precipitating alkaline reactions while repelling acids and acidic radicals, "softening" the tissues and increasing nerve irritability. Both electrodes have the general effect of producing a mild heating of the tissues directly under them and stimulating vasomotor activity. These effects indicate the need for care when choosing electrode sites, and for consideration of the degree to which they may offset the beneficial effects of the iontophoresis process proposed.

Burn dangers (M07H4B)

The chief danger inherent in iontophoresis is burns of the skin and excessive destruction of mucous membrane resulting from excessive current density. There is also danger of chemical burns resulting from acid and/or alkaloid reactions taking place below them, regardless of the chemical applied. Electrical burn danger may be reduced by taking care to avoid direct contact between the metal portion of the surface electrode and the patient's skin by covering all metal surfaces of the electrode with padding. Current strength must be regulated relative to the size of the electrodes, and current levels reduced according to electrode size.

Specific precautions (M07H4C)

In general, to qualify for iontophoresis, the patient's skin sensation should be normal. If it is not, tissue damage which ordinarily would be avoided by patient report of developing discomfort must be precluded by keeping current levels low.

Current should not be applied over denuded areas or over a recently developed scar.

Electrode pads should be evenly soaked with saline, water, or the ion-rich solution and must remain wet during the treatment session; a drying electrode may cause skin irritation or a burn.

The patient must be encouraged to report any uncomfortable burning sensation. Current levels should be adjusted to suit patient tolerance.

The patient should be allowed to leave the treatment area only with dry skin and clothing.

Lithium should generally not be introduced into patients with significant renal or cardiovascular disease, severe debilitation or dehydration or sodium depletion, since there is a high risk of lithium toxicity developing in such patients. During iontophoresis of lithium, progressive or sudden changes in renal function (as measured by urine specific gravity or osmolality, serum creatinine or creatinine clearance), even within the normal ranges, indicate the need for reevaluation of treatment choice.

Bibliography (M07H5)

Kahn, A Case Report: Lithium Iontophoresis for Gouty Arthritis: pp. 113–114.
Shriber, pp. 124–135.
Stillwell, pp. 34–40.
Watkins, pp. 114–127.

IONTOPHORESIS SCLEROLYTIC (SCAR TISSUE) FORMATION MODERATION (M07I)

Definition (M07I1)

Chlorine (M07I1A)

Iontophoresis of chlorine has been reported to be effective as means of moderating (reducing and/or softening) sclerotic (scar tissue) formations, increasing extensibility and increasing the effectiveness of manual stretching.

Iodine (M07I1B)

Iontophoresis of iodine has been reported to be effective as means of moderating (reducing and/or softening) sclerotic (scar tissue) formations, increasing extensibility and increasing the effectiveness of manual stretching.

Methyl salicylate (M07I1C)

Iontophoresis of methyl salicylate has been reported to be effective as means of moderating (reducing and/or softening) sclerotic (scar tissue) formations, increasing extensibility and increasing the effectiveness of manual stretching.

Procedures of Application (M07I2)

Concentrations (M07I2A)

Iodine and methyl salicylate is generally applied together as an ointment (iodex) which is comprised of 4.7% iodine and 4.8% methyl salicylate.

Chlorine (sodium chloride) is generally available as a 2% solution or lotion form.

Patient preparation (M07I2B)

1) The patient should be placed in a comfortable position which allows easy access to the treatment site.
2) The treatment site should be stripped of covering material and cleansed with a mild alcohol rub or saline solution.

Procedure utilized in application of ointment (M07I2C)

1) The (iodex) ointment should be spread evenly over the treatment site and the negative surface electrode, which has been saturated with water or saline solution, should be placed directly over the ointment and firmly secured in place over the skin (taping, strapping or weighting the electrodes in place may be appropriate).
2) The inactive surface electrode pad (also soaked with water or saline solution) should be placed over an area of skin distal to the active electrode site.
3) The electrical stimulator should be preset at zero intensity and the timer (if available) set for 20 minutes.
4) The electrical stimulator should be turned on and the intensity slowly increased until the patient feels a slight burning sensation; the current intensity should be slowly reduced if patient discomfort increases.
5) When the treatment time has elapsed, and treatment has not been terminated by a timer, the electrical stimulation should be reduced slowly back to zero and the electrical stimulator switched off.

Procedure utilized in application of solution (M07I2D)

1) The negative surface electrode should be saturated by a solution containing the chlorine ions, placed over the treatment (edematous) site and firmly secured over the treatment site (taping, strapping or weighting the electrodes in place may be appropriate); if a bath is utilized, the active electrode should be placed in a vessel adequately filled with enough solution (of adequate chemical strength) to cover the treatment site.
2) The water (or saline solution) saturated nonactive electrode pad (positive in this case), should be placed over an area of skin distal to the active electrode; a foot or forearm bath may be used if the body part to be submerged is conveniently distant from the active electrode pad or bath.

3) The electrical stimulator should be preset at zero intensity and the timer (if available) set for 15 or 20 minutes.

4) The electrical stimulator should be turned on and the intensity slowly increased until the patient feels a slight burning sensation; the current intensity should be slowly reduced if patient discomfort increases.

5) When the treatment time has elapsed, and treatment has not been terminated by a timer, the electrical stimulation should be reduced slowly back to zero and the electrical stimulator switched off.

Treatment termination   (M0712E)

1) The electrodes should be removed from the treatment site or the body part removed from the bath, and the treatment site cleansed of all chemicals.

2) The treatment sites (under both the positive and negative electrodes) should be inspected for any signs of chemical or electrical burns, and the skin thoroughly dried (towelled or blow dried).

3) The patient should be allowed to rest for a few minutes before leaving in warm dry clothing.

4) The electrodes (and any vessels used) should be thoroughly cleansed and rinsed to get rid of the chemicals applied or other chemicals which may have accumulated in the sponges or on or near the metal component of the electrodes, as a result of galvanism.

Partial List of Conditions Treated   (M0713)

Treatable Causes   (M0713A)

Habitual joint positioning   (C48)
Joint ankylosis   (C42)
Nerve root compression   (C04)
Restricted joint range of motion   (C26)
Scar tissue formation   (C15)

Syndromes   (M0713B)

ACNE   (S45)
ADHESION   (S56)
CARPAL TUNNEL   (S35)
DUPYTREN'S CONTRACTURE   (S34)
FROZEN SHOULDER   (S64)
KELOID FORMATION   (S38)
POST IMMOBILIZATION   (S36)
SHOULDER/HAND   (S57)

Precautions and Contraindications   (M0714)

Direct current has effects not only on the free ions introduced to the affect the tissues but (as an affect of "galvanism") has an physiological effect on the tissues themselves. Nondisassociated and charged (from adsorption of ions) colloid molecules like droplets of fat, albumin, particles of starch, blood cells, bacteria, and other single cells move toward the negative pole (cataphoresis). Additionally, there are effects upon the chemicals contained within and combined to compose the various tissues, especially in the area of the electrodes. Essentially, the positive (anode) electrode has the physiochemical effect of producing an acid reaction, and repelling alkaloids and trace metals (which have a positive charge), while producing the general physiological effects "hardening" tissues and decreasing nerve irritability. The negative (cathode) electrode has the physiochemical effect of precipitating alkaline reactions while repelling acids and acid radicals, and producing the general physiological effects of "softening" tissues and increasing nerve irritability. Both electrodes have the general effect of producing a mild heating of the tissues under them and stimulating vasomotor activity. These effects point at the need of care when choosing electrode sites, and for consideration of whether the physiological of the "galvanism" may off-set the beneficial effects of the iontophoresis process proposed.

The chief danger is the possibility of skin burns or excessive destruction of mucous membrane; these are almost always due to an excessive density of current, but chemical burns may occur to tissues under the electrodes from the acid and alkaloid reactions taking place under them, regardless of the chemical subject to ion transfer. Electrical burn danger may be reduced by taking care to avoid direct contact between the metal electrode with the skin (all metal surfaces of the electrode must be covered with padding); also, current strength must be evaluated in relation to the size of the electrodes, with a reduction of current levels as the electrode size decreases.

The patient's skin sensation should be normal; if not, current levels should remain low relative to the size of the electrodes used.

Current should not be applied over denuded areas or over a recent scar.

The electrodes should be evenly soaked with saline, water, or the ion rich solution and must remain so during the treatment session (a drying electrode may cause skin irritation or burning).

No metal associated with the electrodes or electrode cables should be in direct contact with the patient's skin.

The patient must be encouraged to report any uncomfortable burning sensation and the current levels should be adjusted to suit the patient's tolerance.

The patient should not be allowed to leave a treatment session wearing clothing made wet from the treatment.

Some swelling may be expected to occur with the iontophoresis of chloride ions because of the tendency for fluids to collect in the area of the negative electrode.

Bibliography (M0715)

Langley, pp. 1395–1396.
Shriber, pp. 124–135.
Stillwell, pp. 34–40.
Tannenbaum, pp. 790–791.
Watkins, pp. 114–127.

IONTOPHORESIS CALCIUM DEPOSIT ABSORPTION
(M07J)

Definition (MO7J1)

Acetic acid (M07J1A)

It has been reported that calcium deposits may be reabsorbed through the iontophoresis of acetic acid; i.e., the acetate radical may replace the carbonate radical in the calcium carbonate molecule and hasten absorption.

Procedures of Application (M07J2)

Concentration (M07J2A)

Acetic acid comes in a 2% solution or lotion form.

Patient preparation (M07J2B)

1) The patient should be placed in a comfortable position which allows easy access to the treatment site.
2) The treatment site should be stripped of covering material and cleansed with a mild alcohol rub or saline solution.

Procedure utilized in application of solution (M07J2C)

1) The negative surface electrode should be saturated by a solution containing the acetic acid ions, placed over the site of the calcium deposit, and firmly secured in place (taping, strapping or weighting the electrodes in place may be appropriate).
2) The water (or saline solution) saturated positive electrode pad, should be placed over an area of skin distal to the active electrode; a foot or forearm bath may be used if the body part to be submerged is conveniently distant from the active electrode pad or bath.
3) The electrical stimulator should be preset at zero intensity and the timer (if available) set for 20 minutes.
4) The electrical stimulator should be turned on and the intensity slowly increased until the patient feels a slight burning sensation; the current intensity should be slowly reduced if patient discomfort increases.
5) When the treatment time has elapsed, and treatment has not been terminated by a timer, the electrical stimulation should be reduced slowly back to zero and the electrical stimulator switched off.

Treatment termination (M07J2D)

1) The electrodes should be removed from the treatment site or the body part removed from the bath, and the treatment site cleansed of all chemicals.
2) The treatment sites (under both the positive and negative electrodes) should be inspected for any signs of chemical or electrical burns, and the skin thoroughly dried off (rubbed dry with toweling or blow dried).
3) The patient should be allowed to rest for a few minutes before leaving in warm dry clothing.
4) The electrodes (and any vessels used) should be thoroughly cleansed and rinsed to get rid of the chemicals applied or other chemicals which may have accumulated in the sponges or on or near the metal component of the electrodes, as a result of galvanism.

Partial List of Conditions Treated (M07J3)

Treatable Causes (M07J3A)

Calcific deposit (C08)
Extrafusal muscle spasm (C41)
Habitual joint positioning (C48)
Joint ankylosis (C42)
Nerve root compression (C04)
Restricted joint range of motion (C26)

Syndromes (M07J3B)

ACHILLES TENDONITIS (S52)
BURSITIS (S26)
CALF PAIN (S50)
CAPSULITIS (S27)
CARPAL TUNNEL (S35)
ELBOW PAIN (S47)
FOOT PAIN (S51)
FOREARM PAIN (S54)
FROZEN SHOULDER (S64)
HAND/FINGER PAIN (S17)

HEADACHE PAIN (S02)
HIGH THORACIC BACK PAIN (S48)
KNEE PAIN (S46)
LOW BACK PAIN (S03)
MYOSITIS OSSIFICANS (S67)
OSTEOARTHRITIS (S21)
POST IMMOBILIZATION (S36)
POST PERIPHERAL NERVE INJURY (S23)
REFERRED PAIN (S01)
SCIATICA (S05)
SHOULDER/HAND (S57)
SHOULDER PAIN (S08)
TENDONITIS (S28)
TENNIS ELBOW (LATERAL EPICONDYLITIS) (S32)
THIGH PAIN (S49)
WRIST PAIN (S16)

Precautions and Contraindications (M07J4)

Direct current has effects not only on the free ions introduced to the affect the tissues but (as an affect of "galvanism") has an physiological effect on the tissues themselves. Nondisassociated and charged (from adsorption of ions) colloid molecules like droplets of fat, albumin, particles of starch, blood cells, bacteria, and other single cells move toward the negative pole (cataphoresis). Additionally, there are effects upon the chemicals contained within and combined to compose the various tissues, especially in the area of the electrodes. Essentially, the positive (anode) electrode has the physiochemical effect of producing an acid reaction, and repelling alkaloids and trace metals (which have a positive charge), while producing the general physiological effects "hardening" tissues and decreasing nerve irritability. The negative (cathode) electrode has the physiochemical effect of precipitating alkaline reactions while repelling acids and acid radicals, and producing the general physiological effects of "softening" tissues and increasing nerve irritability. Both electrodes have the general effect of producing a mild heating of the tissues under them and stimulating vasomotor activity. These effects point at the need of care when choosing electrode sites, and for consideration of whether the physiological of the "galvanism" may off-set the beneficial effects of the iontophoresis process proposed.

The chief danger is the possibility of skin burns or excessive destruction of mucous membrane; these are almost always due to an excessive density of current, but chemical burns may occur to tissues under the electrodes from the acid and alkaloid reactions taking place under them, regardless of the chemical subject to ion transfer. Electrical burn danger may be reduced by taking care to avoid direct contact between the metal electrode with the skin (all metal surfaces of the electrode must be covered with padding); also, current strength must be evaluated in relation to the size of the electrodes, with a reduction of current levels as the electrode size decreases.

The patient's skin sensation should be normal; if not, current levels should remain low relative to the size of the electrodes used.

Current should not be applied over denuded areas or over a recent scar.

The electrodes should be evenly soaked with saline, water, or the ion rich solution and must remain so during the treatment session (a drying electrode may cause skin irritation or burning).

No metal associated with the electrodes or electrode cables should be in direct contact with the patient's skin.

The patient must be encouraged to report any uncomfortable burning sensation and the current levels should be adjusted to suit the patient's tolerance.

The patient should not be allowed to leave a treatment session wearing clothing made wet from the treatment.

If skin irritation or hypersensitivity are apparent after treatment this particular treatment form should be discontinued for the sensitive patient.

Bibliography (M07J5)

Kahn, pp. 658–659.
Langley, pp. 1395–1396.
Psaki, pp. 84–87.
Shriber, pp. 124–135.
Stillwell, pp. 34–40.
Watkins, pp. 114–127.

IONTOPHORESIS HYPERHIDROSIS CONTROL (M07K)

Definition (M07K1)

Aluminum sulfate (M07K1A)

It has been reported that hyperhidrosis may be treated through the iontophoresis of aluminum (sulfate) ions; aluminum has been found to reduce sweat gland output (this facility has been exploited in many anti-perspirant products).

Procedures of Application (M07K2)

Concentration (M07K2A)

Aluminum comes in a (2% aluminum sulfate) solution or lotion form.

Patient preparation (M07K2B)

1) The patient should be placed in a comfortable position which allows easy access to the treatment site.
2) The treatment site should be stripped of covering material and cleansed with a mild alcohol rub or saline solution.

Procedure utilized in application of solution (M07K2C)

1) The positive electrode should be placed at the bottom of a bath containing the aluminum ion rich solution; the patient should place either both hands or both feet in the solution.
2) The negative surface electrode, which has been saturated by water or saline solution, should be placed over an area of skin distal to the active electrode. A saline solution bath containing the negative electrode may be used as the dispersive; the patient should put either both feet or both hands in the nonactive bath (those not placed in the active bath).
3) The electrical stimulator should be preset at zero intensity and the timer (if available) set for 20 minutes.
4) The electrical stimulator should be turned on and the intensity slowly increased until the patient feels a slight burning sensation; the current intensity should be slowly reduced if patient discomfort increases.
5) When the treatment time has elapsed, and treatment has not been terminated by a timer, the electrical stimulation should be reduced slowly back to zero and the electrical stimulator switched off.

Treatment termination (M07K2D)

1) The electrodes should be removed from the treatment site or the body part removed from the bath, and the treatment site cleansed of all chemicals.
2) The treatment sites (under both the positive and negative electrodes) should be inspected for any signs of chemical or electrical burns, and the skin thoroughly dried off (rubbed dry with toweling or blow dried).
3) The patient should be allowed to rest for a few minutes before leaving in warm dry clothing.
4) The electrodes (and any vessels used) should be thoroughly cleansed and rinsed to get rid of the chemicals applied or other chemicals which may have accumulated in the sponges or on or near the metal component of the electrodes, as a result of galvanism.

Partial List of Conditions Treated (M07K3)

Treatable Causes (M07K3A)

Hyperhidrosis (C37)
Neurogenic dermal disorders (psoriasis, hives) (C49)
Sympathetic hyperresponse (C38)

Syndromes (M07K3B)

CERVICAL DORSAL OUTLET (SCALENICUS ATTICUS/CERVICAL RIB) (S09)
PHOBIC REACTION (S44)
POST PERIPHERAL NERVE INJURY (S23)
SHOULDER/HAND (S57)

Precautions and Contraindications (M07K4)

Direct current has effects not only on the free ions introduced to the affect the tissues but (as an affect of "galvanism") has an physiological effect on the tissues themselves. Nondisassociated and charged (from adsorption of ions) colloid molecules like droplets of fat, albumin, particles of starch, blood cells, bacteria, and other single cells move toward the negative pole (cataphoresis). Additionally, there are effects upon the chemicals contained within and combined to compose the various tissues, especially in the area of the electrodes. Essentially, the positive (anode) electrode has the physiochemical effect of producing an acid reaction, and repelling alkaloids and trace metals (which have a positive charge), while producing the general physiological effects "hardening" tissues and decreasing nerve irritability. The negative (cathode) electrode has the physiochemical effect of precipitating alkaline reactions while repelling acids and acid radicals, and producing the general physiological effects of "softening" tissues and increasing nerve irritability. Both electrodes have the general effect of producing a mild heating of the tissues under them and stimulating vasomotor activity. These effects point at the need of care when choosing electrode sites, and for consideration of whether the physiological of the "galvanism" may off-set the beneficial effects of the iontophoresis process proposed.

The chief danger is the possibility of skin burns or excessive destruction of mucous membrane; these are almost always due to an excessive density of current, but chemical burns may occur to tissues under the electrodes from the acid and alkaloid reactions taking place under them, regardless of the chemical subject to ion transfer. Electrical burn danger may be reduced by taking care to avoid direct contact between the metal electrode with the skin (all metal surfaces of the electrode must be covered with padding); also, current strength must be evaluated in relation to the size of the electrodes, with a reduction of current levels as the electrode size decreases.

The patient's skin sensation should be normal; if not, current levels should remain low relative to the size of the electrodes used.

Current should not be applied over denuded areas or over a recent scar.

The electrodes should be evenly soaked with saline, water, or the ion rich solution and must remain so during the treatment session (a drying electrode may cause skin irritation or burning).

No metal associated with the electrodes or electrode cables should be in direct contact with the patient's skin.

The patient must be encouraged to report any uncomfortable burning sensation and the current levels should be adjusted to suit the patient's tolerance.

The patient should not be allowed to leave a treatment session wearing clothing made wet from the treatment.

Iontophoresis may be contraindicated for those suffering from or with a family history of Alzhiemer's disease; autopsy examination has demonstrated abnormally high concentrations of aluminum to be associated with certain central nervous system structures in patients who have expired while suffering from the disease.

Bibliography (M07K5)

Abell, pp. 87–91.
Hill, pp. 69–72.
Levit, pp. 505–507.
Morgan, p. 45.
Shriber, pp. 124–135.
Stillwell, pp. 34–40.
Watkins, pp. 114–127.

IONTOPHORESIS FUNGAL INFECTION ERADICATION
(M07L)

Definition (M07L1)

Copper sulfate (M07L1A)

It has been reported that fungal infections may be treated or eradicated by the iontophoresis of copper ions.

Procedures of Application (M07L2)

Concentration (M07L2A)

Copper ions come in a (2% copper sulfate) solution or lotion form.

Patient preparation (M07L2B)

1) The patient should be placed in a comfortable position which allows easy access to the treatment site.
2) The treatment site should be stripped of covering material and cleansed with a mild alcohol rub or saline solution.

Procedure utilized in application of solution (M07L2C)

1) The positive electrode should be placed at the bottom of a bath containing the copper ion rich solution; the patient should place the infected body part or parts in the solution.
2) The negative surface electrode, which has been saturated by water or saline solution, should be placed over an area of skin distal to the active electrode. A saline solution bath containing the negative electrode may be used as the dispersive; the patient should put body parts in the nonactive bath (those not placed in the active bath).
3) The electrical stimulator should be preset at zero intensity and the timer (if available) set for 20 minutes.
4) The electrical stimulator should be turned on and the intensity slowly increased until the patient feels a slight burning sensation; the current intensity should be slowly reduced if patient discomfort increases.
5) When the treatment time has elapsed, and treatment has not been terminated by a timer, the electrical stimulation should be reduced slowly back to zero and the electrical stimulator switched off.

Treatment termination (M07L2D)

1) The electrodes should be removed from the treatment site or the body part removed from the bath, and the treatment site cleansed of all chemicals.
2) The treatment sites (under both the positive and negative electrodes) should be inspected for any signs of chemical or electrical burns, and the skin thoroughly dried off (rubbed dry with toweling or blow dried).
3) The patient should be allowed to rest for a few minutes before leaving in warm dry clothing.
4) The electrodes (and any vessels used) should be thoroughly cleansed and rinsed to get rid of the chemicals applied or other chemicals which may have accumulated in the sponges or on or near the metal component of the electrodes, as a result of galvanism.

Partial List of Conditions Treated (M07L3)

Conditions Treated in the Field (M07L3A)

Fungal infections of the hands and/or feet

Precautions and Contraindications (M07L4)

Direct current has effects not only on the free ions introduced to the affect the tissues but (as an affect of "galvanism") has an physiological effect on the tissues themselves. Nondisassociated and charged (from adsorption of ions) colloid molecules like droplets of fat, albumin, particles of starch, blood cells, bacteria, and other single cells move toward the negative pole (cataphoresis). Additionally, there are effects upon the chemicals contained within and combined to compose the various tissues, especially in the area of the electrodes. Essentially, the positive (anode) electrode has the physiochemical effect of producing an acid reaction, and repelling alkaloids and trace metals (which have a positive charge), while producing the general physiological effects "hardening" tissues and decreasing nerve irritability. The negative (cathode) electrode has the physiochemical effect of precipitating alkaline reactions while repelling acids and acid radicals, and producing the general physiological effects of "softening" tissues and increasing nerve irritability. Both electrodes have the general effect of producing a mild heating of the tissues under them and stimulating vasomotor activity. These effects point at the need of care when choosing electrode sites, and for consideration of whether the physiological of the "galvanism" may off-set the beneficial effects of the iontophoresis process proposed.

The chief danger is the possibility of skin burns or excessive destruction of mucous membrane; these are almost always due to an excessive density of current, but chemical burns may occur to tissues under the electrodes from the acid and alkaloid reactions taking place under them, regardless of the chemical subject to ion transfer. Electrical burn danger may be reduced by taking care to avoid direct contact between the metal electrode with the skin (all metal surfaces of the electrode must be covered with padding); also, current strength must be evaluated in relation to the size of the electrodes, with a reduction of current levels as the electrode size decreases.

The patient's skin sensation should be normal; if not, current levels should remain low relative to the size of the electrodes used.

Current should not be applied over denuded areas or over a recent scar.

The electrodes should be evenly soaked with saline, water, or the ion rich solution and must remain so during the treatment session (a drying electrode may cause skin irritation or burning).

No metal associated with the electrodes or electrode cables should be in direct contact with the patient's skin.

The patient must be encouraged to report any uncomfortable burning sensation and the current levels should be adjusted to suit the patient's tolerance.

The patient should not be allowed to leave a treatment session wearing clothing made wet from the treatment.

Bibliography (M07L5)

Shriber, p. 133.
Stillwell, pp. 34–40.
Watkins, pp. 114–127.

IRREGULAR MENSES (C19)

Definition (C19A)

Irregular menses is defined as menstruation which occurs too frequently (polymenorrhea), or nonmenstrual (intermenstrual) bleeding from the uterus via the vagina (metrorrhagia). It is commonly termed "break through bleeding" and generally occurs in the middle of the menstrual cycle. It may also appear as an extended menstrual bleed, with bleeding continuing long after the normal menstrual period should be over.

Break through bleeding may be copius or appear in the form of minuscule amounts of blood. The latter is commonly referred to as "spotting".

Evaluation (C19B)

Evaluation of irregular menses must depend upon patient and/or physician report.

Etiology (C19C)

The causes of irregular menses (not including those associated with complications of pregnancy) include (1) malignant or benign tumors of the uterus and/or cervix, (2) polyps of the cervix or uterus (including fibroid tissue), (3) malposition or subinvolution of the uterus, (4) laceration of the cervix, (5) vascular congestion accompanying inflammation of the cervix, endometritis or other pelvic inflammatory disease, (6) endometriosis, (7) diseases affecting the blood, (8) hypertension within the uterus, (9) cardiac decompensation, (10) chemical or heavy metal poisoning (lead, phosphorus, or alcoholism), (11) cirrhosis of the liver, (12) hypoprothrombinemia, (13) vitamin B deficiencies, (14) vaginitis, (15) failure or rupture of the graafian follicle which results in an absence of corpus luteum formation and (16) endocrine hormonal imbalance caused by dysfunction of the thyroid, pituitary, or adrenal glands, or from abnormal secretory discharge from atrophic endometrium, or sudden discontinuance of estrogen supplements.

Treatment Notes (C19D)

Auricular electroacustimulation of the appropriate points has proven most effective as an ameliorating treatment for irregular menses caused by simple hormonal imbalance or benign tumor (fibroid tissue) formation in the uterus.

Bibliography (C19E)

Holvey, pp. 674–679.

TREATMENT OF IRREGULAR MENSES (C19F)

Technique Options (C19F1)

Evaluation techniques (C19F1A)

Auricular electroacustimulation	**M06**
Auricular acupoint evaluative survey	M06C
Determination of pathological symptom source	M06C5B
Peripheral electroacustimulation	**M05**
Peripheral acupoint survey	M05D

Treatment techniques (C19F1B)

Auricular electroacustimulation	**M06**
Pathology control	M06B
Laser stimulation	**M27**
Massage	**M23**
Acupressure	M23C
Peripheral electroacustimulation	**M05**
Pathology control	M05C
Physiological dysfunction	M05C5A
Transcutaneous nerve stimulation	**M04**
Prolonged stimulation of acupoints	M04B
Ultrahigh frequency sound (ultrasound)	**M01**
Acupoint stimulation	M01K

Techniques Suggested for the Acute Condition (C19F2)

Evaluation techniques (C19F2A)

Auricular electroacustimulation	**M06**
Auricular acupoint evaluative survey	M06C
Determination of pathological symptom source	M06C5B

Treatment techniques (C19F2B)

Auricular electroacustimulation	**M06**
Pathology control	M06B

Techniques Suggested for the Chronic Condition (C19F3)

Eqvaluation techniques (C19F3A)

Auricular electroacustimulation   *M06*
    Auricular acupoint evaluative survey   M06C

    Determination of pathological
        symptom source   M06C5B

Treatment techniques (C19F3B)

Auricular electroacustimulation   *M06*
    Pathology control   M06B

JOINT ANKYLOSIS (C42)

Definition (C42A)

Ankylosis is the restriction of the normal joint range of motion by changes in the tissues associated with the joint or tissues within the joint cavity. It is generally associated with joint disease or changes in the joint and associated tissues as the result of prolonged joint disuse (as seen in prolonged joint stabilization associated with splinting or casting).

Evaluation (C42B)

Evaluation of the extent and nature of joint ankylosis (without x-ray evidence) is dependant upon range of motion measurement and joint manipulation assessment.

Etiology (C42C)

Joint ankylosis may result from incomplete healing or restoration of normal joint structures damaged by arthritic processes, infection, or trauma (including burn damage). Ankylosis may result from tissue adhesions between the articular surfaces or associated soft tissues (capsule, bursae, ligaments, synovium, and/or skin) without actual bony union; this is generally called fibrous, partial or false ankylosis. Ankylosis may also occur from the formation of solid new bone spontaneously formed between the articular surfaces as part of an abnormal healing process; this is generally called bony, complete or true ankylosis.

Fibrous ankylosis may follow joint inflammatory disease, intra-articular fractures, chronic intra-articular hemorrhage (as seen in some cases of hemophilia), or soft tissue changes occurring as sequelae of infection, trauma or prolonged immobilization by disuse, splinting or casting.

Bony ankylosis of the joint usually results from severe prolonged infection of the joint or from advanced rheumatoid or rheumatoid-like arthritis which has stripped the bone ends of cartilaginous surfaces and permitted the bone ends to come together, triggering the growth of new bone (osteophytes) and the loss of joint space. Lipping and spur formation at the articular margins of the bone ends are early characteristics.

Treatment Notes (C42D)

Treatment of joint ankylosis is usually directed at preventing further reduction of joint range of motion and restoring range which has been lost. Fibrotic or adhesive ankylosis which has occurred without the formation of new bony formations is most treatable.

Neuromuscular reeducation utilizing electromyometric feedback may be used to help the patient reestablish control of neuromuscular balances and to speed the process of restoring joint ranges of motion lost through prolonged immobilization of the joint.

Bibliography (C42E)

Bickley, pp. 255–256.
Holvey, pp. 961–963.
Salter, pp. 192–218.
Shands, pp. 170–173.

Treatment of Joint Ankylosis (C42F)

Technique Options (C42F1)

Evaluation techniques (C42F1A)

Auxiliary evaluation techniques	**M26**
Auscultation	M26C
Joint and/or tendon sounds	M26C2C
Cryotherapy	**M10**
Evaluation of capillary response to cold	M10H
Determination of hyperactivity to cold	M10H2A
Differential skin resistance survey	**M22**
Electromyometric feedback	**M18**
Electromyometric evaluation	M18L
Determination of neuromuscular hypotonicity	M18L2
Exercise	**M12**
Functional testing	M12I
Evaluation of joint range of motion	M12I1
Evaluation of muscle strength	M12I2
Hydrotherapy	**M15**
Hydrotherapeutic Tests	M15F
Test for hypersensitivity to cold	M15F3
Massage	**M23**
Palpation evaluation	M23G
Determination of extrafusal muscle spasm or hypertonicity	M23G2
Determination of soft tissue tenderness	M23G3
Determination of fibroid, scar, or calcific formation	M23G4
Peripheral electroacustimulation	**M05**
Peripheral acupoint survey	M05D

joint ankylosis

Sensory stimulation	**M17**	Wound healing facilitation	M24C
Neurological evaluation	M17E	Scar tissue formations	M24C2B
Sensory evaluation	M17E2	*Sensory stimulation*	**M17**
		Provoking reflex physiological responses	M17C
Treatment techniques (C42F1B)		Prolonged heating for reflex physiological responses	M17C2A
Auricular electroacustimulation	**M06**	Prolonged cooling for reflex physiological responses	M17C2B
Pain relief	M06A	Brief cooling for reflex physiological responses	M17C2C
Cryotherapy	**M10**	*Superficial thermal (heat) therapy*	**M11**
Hypertonic neuromuscular management	M10B	Muscle relaxation (wet and dry sauna)	M11C
Extrafusal muscle spasm treatment	M10B2B	*Transcutaneous nerve stimulation*	**M04**
Analgesia/anesthesia	M10C	Prolonged stimulation of acupoints	M04B
Electrical facilitation of endorphin production (acusleep)	**M03**	*Ultrahigh frequency sound (ultrasound)*	**M01**
Increasing pain tolerance	M0305A	Phonophoresis	M01A
Electrical stimulation (ES)	**M02**	The phonophoresis of vitamin E in wound healing	M01A5B
Muscle lengthening	M02A	Scar tissue management	M01D
Muscle lengthening	M02A5A	Calcium deposit management	M01E
Increasing blood flow	M02C	Acupoint stimulation	M01K
Increasing capillary density	M02C2B	*Vibration*	**M08**
Electromyometric feedback	**M18**	Neuromuscular management	M08A
Post immobilization neuromuscular reeducation	M18I	Muscle lengthening	M08A2B
Muscle toning	M18K		
Exercise	**M12**	**Techniques Suggested for the Acute Condition** (C42F2)	
Muscle strengthening	M12A		
Muscle relengthening	M12C		
Tonically shortened musculature	M12C1A	**Evaluation techniques** (C42F2A)	
Muscle relengthening through reciprocal inhibition	M12C1C	*Cryotherapy*	**M10**
Muscle stretching	M12F	Evaluation of capillary response to cold	M10H
Guided patterning	M12H	Determination of hyperactivity to cold	M10H2A
Hydrotherapy	**M15**	*Differential skin resistance survey*	**M22**
Muscle tension amelioration	M15C	*Electromyometric feedback*	**M18**
Range of motion enhancement and physical reconditioning	M15D	Electromyometric evaluation	M18L
Iontophoresis	**M07**	Determination of neuromuscular hypotonicity	M18L2
Pain relief	M07A	*Exercise*	**M12**
Ischemia relief	M07D	Functional testing	M12I
Muscle spasm relief	M07F	Evaluation of joint range of motion	M12I1
Hypertonicity relief	M07G	Evaluation of muscle strength	M12I2
Calcific deposit absorption	M07J	*Massage*	**M23**
Joint mobilization	**M13**	Palpation evaluation	M23G
Laser stimulation	**M27**	Determination of extrafusal muscle spasm or hypertonicity	M23G2
Massage	**M23**		
Acupressure	M23C	Determination of soft tissue tenderness	M23G3
Muscle atrophy prevention	M23E		
Deep soft tissue mobilization	M23F	Determination of fibroid, scar, or calcific formation	M23G4
Muscles	M23F2A		
Ligaments	M23F2B	*Peripheral electroacustimulation*	**M05**
Tendon sheaths	M23F2C	Peripheral acupoint survey	M05D
Tendons	M23F2D	*Sensory stimulation*	**M17**
Peripheral electroacustimulation	**M05**	Neurological evaluation	M17E
Acupoint electrical stimulation within an inflamed zone	M05A	Sensory evaluation	M17E2
Analgesia	M05B		
Positive pressure	**M24**		

joint ankylosis

Treatment techniques (C42F2B)

Auricular electroacustimulation	**M06**
Pain relief	M06A
Cryotherapy	***M10***
Hypertonic neuromuscular management	M10B
Extrafusal muscle spasm treatment	M10B2B
Analgesia/anesthesia	M10C
Electrical facilitation of endorphin production (acusleep)	***M03***
Increasing pain tolerance	M0305A
Electrical stimulation (ES)	***M02***
Muscle lengthening	M02A
Muscle lengthening	M02A5A
Electromyometric feedback	***M18***
Post immobilization neuromuscular reeducation	M18I
Exercise	***M12***
Muscle relengthening	M12C
Tonically shortened musculature	M12C1A
Muscle relengthening through reciprocal inhibition	M12C1C
Guided patterning	M12H
Joint mobilization	***M13***
Peripheral electroacustimulation	***M05***
Acupoint electrical stimulation within an inflamed zone	M05A
Transcutaneous nerve stimulation	***M04***
Prolonged stimulation of acupoints	M04B
Ultrahigh frequency sound (ultrasound)	***M01***
Phonophoresis	M01A
The phonophoresis of vitamin E in wound healing	M01A5B
Calcium deposit management	M01E
Vibration	***M08***
Neuromuscular management	M08A
Muscle lengthening	M08A2B

Techniques Suggested for the Chronic Condition (C42F3)

Evaluation techniques (C42F3A)

Auxiliary evaluation techniques	**M26**
Auscultation	M26C
Joint and/or tendon sounds	M26C2C
Cryotherapy	***M10***
Evaluation of capillary response to cold	M10H
Determination of hyperactivity to cold	M10H2A
Electromyometric feedback	***M18***
Electromyometric evaluation	M18L
Determination of neuromuscular hypotonicity	M18L2
Exercise	***M12***
Functional testing	M12I
Evaluation of joint range of motion	M12I1
Evaluation of muscle strength	M12I2
Massage	***M23***
Palpation evaluation	M23G
Determination of extrafusal muscle spasm or hypertonicity	M23G2
Determination of soft tissue tenderness	M23G3
Determination of fibroid, scar, or calcific formation	M23G4
Peripheral electroacustimulation	***M05***
Peripheral acupoint survey	M05D
Sensory stimulation	***M17***
Neurological evaluation	M17E
Sensory evaluation	M17E2

Treatment techniques (C42F3B)

Auricular electroacustimulation	**M06**
Pain relief	M06A
Cryotherapy	***M10***
Analgesia/anesthesia	M10C
Electrical facilitation of endorphin production (acusleep)	***M03***
Increasing pain tolerance	M0305A
Electrical stimulation (ES)	***M02***
Increasing blood flow	M02C
Increasing capillary density	M02C2B
Electromyometric feedback	***M18***
Post immobilization neuromuscular reeducation	M18I
Muscle toning	M18K
Exercise	***M12***
Muscle strengthening	M12A
Muscle relengthening	M12C
Tonically shortened musculature	M12C1A
Muscle stretching	M12F
Joint mobilization	***M13***
Massage	***M23***
Deep soft tissue mobilization	M23F
Muscles	M23F2A
Ligaments	M23F2B
Tendon sheathes	M23F2C
Tendons	M23F2D
Peripheral electroacustimulation	***M05***
Analgesia	M05B
Transcutaneous nerve stimulation	***M04***
Prolonged stimulation of acupoints	M04B
Ultrahigh frequency sound (ultrasound)	***M01***
Scar tissue management	M01D
Calcium deposit management	M01E
Vibration	***M08***
Neuromuscular management	M08A
Muscle lengthening	M08A2B

JOINT APPROXIMATION (C47)

Definition (C47A)

Joint approximation is the forcing together of articular surfaces of bones sharing a common joint.

Joint approximation may be utilized therapeutically to facilitate neuromuscular tonus.

Joint approximation may occur traumatically from external force resulting in pain, reflexive hypertonus (muscle spasm), and soft tissue injury.

Evaluation (C47B)

Evaluation of traumatic joint approximation is dependent upon patient report, range of motion evaluation, assessment of associated muscle tonus, assessment of soft tissue tenderness over the site of suspected joint approximation and evaluation of the presence of soft tissue inflammation via differential skin resistance (DSR) survey, etc.

Etiology (C47C)

Joint approximation is a normal consequence to weight bearing joints such as the knees and the vertebral facets. In normal circumstances, joint approximation actually serves to facilitate reflex support by the muscles which cross and help support the involved joint. Joint approximation becomes a problem only if the articular surfaces come in contact with each other with sufficient force to damage them or causes impingement upon soft tissues associated with the joint (synovial tissue).

Traumatic joint approximation sometimes occurs when the joint is extended beyond normal limits (hyperextension) or forced into abnormal ranges (as in excessive lateral rotation of the thoracic vertebrae or lateral flexion of the knee). It also can occur if the articular surfaces are forced together with excessive force (as when jumping from a height and landing with the knees locked in extension), or if calcium deposits in the joint serve to indirectly allow articular surfaces to communicate (joint mice, bone spurs, etc.).

Treatment Notes (C47D)

Treatment of the sequelae of abnormal joint approximation is generally dependent upon the treatment of the resulting symptoms and the reduction of the approximation. Treatment is usually directed at decreasing any soft tissue inflammation and/or swelling, relieving of any overt pain (albeit temporarily), relieving any associated muscle spasm, encouraging relaxation of associated musculature and providing joint traction as the musculature relaxes.

Bibliography (C47E)

Salter, pp. 422–423.
Scott, pp. 218–219.

Treatment of Joint Approximation (C47F)

Technique Options (C47F1)

Evaluation techniques (C47F1A)

Auxiliary evaluation techniques	*M26*
Auscultation	M26C
Joint and/or tendon sounds	M26C2C
Cryotherapy	*M10*
Evaluation of capillary response to cold	M10H
Determination of hyperactivity to cold	M10H2A
Differential skin resistance survey	*M22*
Exercise	*M12*
Functional testing	M12I
Evaluation of joint range of motion	M12I1
Hydrotherapy	*M15*
Hydrotherapeutic Tests	M15F
Test for hypersensitivity to cold	M15F3
Massage	*M23*
Palpation evaluation	M23G
Determination of extrafusal muscle spasm or hypertonicity	M23G2
Determination of soft tissue tenderness	M23G3
Determination of fibroid, scar, or calcific formation	M23G4
Determination of soft tissue inflammation	M23G5
Determination of soft tissue swelling	M23G6
Peripheral electroacustimulation	*M05*
Peripheral acupoint survey	M05D
Sensory stimulation	*M17*
Neurological evaluation	M17E
Sensory evaluation	M17E2

Treatment techniques (C47F1B)

Auricular electroacustimulation	*M06*
Pain relief	M06A
Cryotherapy	*M10*
Hypertonic neuromuscular management	M10B

joint approximation

Extrafusal muscle spasm treatment	M10B2B
Analgesia/anesthesia	M10C
Electrical facilitation of endorphin production (acusleep)	***M03***
Increasing pain tolerance	M0305A
Electrical stimulation (ES)	***M02***
Muscle lenghtening	M02A
Muscle lengthening	M02A5A
Exercise	***M12***
Muscle relengthening	M12C
Tonically shortened musculature	M12C1A
Muscle relengthening through reciprocal inhibition	M12C1C
Muscle stretching	M12F
Hydrotherapy	***M15***
Muscle tension amelioration	M15C
Range of motion enhancement and physical reconditioning	M15D
Suppression of sympathetic hyperresponsiveness	M15E
Iontophoresis	***M07***
Pain relief	M07A
Inflammation amelioration	M07B
Muscle spasm relief	M07F
Hypertonicity relief	M07G
Calcific deposit absorption	M07J
Joint mobilization	***M13***
Laser stimulation	***M27***
Massage	***M23***
Acupressure	M23C
Deep soft tissue mobilization	M23F
Muscles	M23F2A
Ligaments	M23F2B
Tendon sheaths	M23F2C
Tendons	M23F2D
Peripheral electroacustimulation	***M05***
Acupoint electrical stimulation within an inflamed zone	M05A
Analgesia	M05B
Sensory stimulation	***M17***
Provoking reflex physiological responses	M17C
Prolonged heating for reflex physiological responses	M17C2A
Prolonged cooling for reflex physiological responses	M17C2B
Brief cooling for reflex physiological responses	M17C2C
Desensitization	M17D
Superficial thermal (heat) therapy	***M11***
Muscle relaxation (wet and dry sauna)	M11C
Excretory (sweat) facilitation	M11D
Taping	***M16***
Transcutaneous nerve stimulation	***M04***
Chronic pain management	M04A
Prolonged stimulation of acupoints	M04B
Ultrahigh frequency sound (ultrasound)	***M01***
Desensitization	M01C
Acupoint stimulation	M01K
Vertebral Traction	***M09***
Cervical traction	M09A (1)
Lumbar traction	M09B (2)
Vibration	***M08***
Neuromuscular management	M08A
Muscle lengthening	M08A2B
Cervical traction enhancement	M08A2C
Lumbar traction enhancement	M08A2D

(1) If cervical or upper thoracic vertebrae are involved

(2) If midthoracic to lumbar vertebrae are involved

Techniques Suggested for the Acute Condition (C47F2)

Evaluation techniques (C47F2A)

Cryotherapy	***M10***
Evaluation of capillary response to cold	M10H
Determination of hyperactivity to cold	M10H2A
Differential skin resistance survey	***M22***
Exercise	***M12***
Functional testing	M12I
Evaluation of joint range of motion	M12I1
Massage	***M23***
Palpation evaluation	M23G
Determination of extrafusal muscle spasm or hypertonicity	M23G2
Determination of soft tissue tenderness	M23G3
Determination of fibroid, scar, or calcific formation	M23G4
Determination of soft tissue inflammation	M23G5
Determination of soft tissue swelling	M23G6
Peripheral electroacustimulation	***M05***
Acupoint electrical stimulation within an inflamed zone	M05A
Peripheral acupoint survey	M05D
Sensory stimulation	***M17***
Neurological evaluation	M17E
Sensory evaluation	M17E2

Treatment techniques (C47F2B)

Auricular electroacustimulation	***M06***
Pain relief	M06A
Cryotherapy	***M10***
Hypertonic neuromuscular management	M10B
Extrafusal muscle spasm treatment	M10B2B
Electrical facilitation of endorphin production (acusleep)	***M03***
Increasing pain tolerance	M0305A
Electrical stimulation (ES)	***M02***
Muscle lengthening	M02A
Muscle lengthening	M02A5A

Exercise	**M12**
Muscle relengthening	M12C
Muscle relengthening through reciprocal inhibition	M12C1C
Peripheral electroacustimulation	**M05**
Acupoint electrical stimulation within an inflamed zone	M05A
Sensory stimulation	**M17**
Provoking reflex physiological responses	M17C
Prolonged heating for reflex physiological responses	M17C2A
Prolonged cooling for reflex physiological responses	M17C2B
Brief cooling for reflex physiological responses	M17C2C
Transcutaneous nerve stimulation	**M04**
Prolonged stimulation of acupoints	M04B
Ultrahigh frequency sound (ultrasound)	**M01**
Desensitization	M01C
Vertebral Traction	**M09**
Cervical traction	M09A (1)
Lumbar traction	M09B (2)
Vibration	**M08**
Neuromuscular management	M08A
Muscle lengthening	M08A2B
Cervical traction enhancement	M08A2C
Lumbar traction enhancement	M08A2D

(1) If cervical or upper thoracic vertebrae are involved

(2) If midthoracic to lumbar vertebrae are involved

Techniques Suggested for the Chronic Condition (C47F3)

Evaluation techniques (C47F3A)

Auxiliary evaluation techniques	**M26**
Auscultation	M26C
Joint and/or tendon sounds	M26C2C
Cryotherapy	**M10**
Evaluation of capillary response to cold	M10H
Determination of hyperactivity to cold	M10H2A
Differential skin resistance survey	**M22**
Exercise	M12
Functional testing	M12I
Evaluation of joint range of motion	M12I1
Massage	**M23**
Palpation evaluation	M23G
Determination of extrafusal muscle spasm or hypertonicity	M23G2
Determination of soft tissue tenderness	M23G3
Determination of fibroid, scar, or calcific formation	M23G4
Determination of soft tissue inflammation	M23G5
Determination of soft tissue swelling	M23G6
Peripheral electroacustimulation	**M05**
Peripheral acupoint survey	M05D
Sensory stimulation	**M17**
Neurological evaluation	M17E
Sensory evaluation	M17E2

Treatment techniques (C47F3B)

Cryotherapy	**M10**
Analgesia/anesthesia	M10C
Electrical facilitation of endorphin production (acusleep)	**M03**
Increasing pain tolerance	M0305A
Electrical stimulation (ES)	**M02**
Muscle lengthening	M02A
Muscle lengthening	M02A5A
Exercise	**M12**
Muscle relengthening	M12C
Tonically shortened musculature	M12C1A
Muscle stretching	M12F
Joint mobilization	**M13**
Massage	**M23**
Deep soft tissue mobilization	M23F
Muscles	M23F2A
Ligaments	M23F2B
Tendon sheaths	M23F2C
Tendons	M23F2D
Peripheral electroacustimulation	**M05**
Analgesia	M05B
Transcutaneous nerve stimulation	**M04**
Chronic pain management	M04A
Prolonged stimulation of acupoints	M04B
Ultrahigh frequency sound (ultrasound)	**M01**
Desensitization	M01C
Vertebral Traction	**M09**
Cervical traction	M09A (1)
Lumbar traction	M09B (2)
Vibration	**M08**
Neuromuscular management	M08A
Muscle lengthening	M08A2B
Cervical traction enhancement	M08A2C
Lumbar traction enhancement	M08A2D

(1) If cervical or upper thoracic vertebrae are involved

(2) If midthoracic to lumbar vertebrae are involved

JOINT MOBILIZATION (M13)

Definition (M1301)

Joint mobilization (sometimes referred to as joint manipulation) is a form of passive joint movement (i.e., without patient volitional interference or active participation), performed to promote joint play, component motion, and/or restoration of range of motion.

Joint range of motion is not only made up of the normally measured ranges (flexion, extension, hyperextension, and rotation) but also includes various accessory ranges. One of these accessory ranges includes joint play. Joint play is an involuntary motion (most often medial/lateral play) present in all synovial joints, and it is necessary for pain free, unrestricted voluntary motion, since it reflects innate joint space or articular surface distances.

Another accessory joint motion is component motion. Component motion is an active movement which a joint allows, but which is not generally recognized as being part of a joint's normal range of motion. An example of this type of range of motion may be seen when the radius slides distally on the ulna during elbow extension.

Massage, passive stretching, muscle relengthening techniques and passive range of motion exercise are used to increase the mobility of soft tissues in general. However, their effectiveness is limited by the dynamics of the involved contractile tissues and their ability to appreciate a lengthening process, as well as the patient's tolerance of pain. This is an important factor, since these techniques are most often employed at the end of range and may cause pain for the patient. Joint mobilization, on the other hand, may be more effective for the restoration of joint mobility, because its action is directed at the noncontractile structures of the joint (the capsule, ligaments, etc.) and is generally employed with the joint in middle ranges and is, for the most part, painless to the patient.

Joint mobilization may be very important when soft tissue injury results in inflammation and swelling of the joint. If left alone, such joints may become hypomobile and eventually, if the joint inflammation and swelling continue long enough, will begin to degenerate, raising the possibility of fibrous or bony ankylosis developing.

For best results, joint mobilization should be applied early in the development of an inflammatory process, and the technique of mobilization utilized should be chosen for its ability to maintain or promote range of motion without further irritating the involved soft tissues. Timely treatment may serve to help avoid or reduce the process of ankylosing, bony block (calcium deposit) development and/or collagen fiber adhesion development which may result as a consequence of joint sprain, immobilization, capsulitis, tendonitis, bursitis and vertebral facet syndromes. Joint mobilization may be used to help restore function lost by the development of these conditions in the absence of adequate treatment in early development stages, but its most efficient use would seem to be as a preventative measure.

Joint mobilization generally employs the techniques of articulation, stretch, manipulation and traction.

Procedures of Application (M1302)

General rules of joint mobilization (M1302A)

1) The patient and practitioner should be relaxed and in a relatively comfortable positions.

2) The procedure utilized should be relatively pain free. The technique utilized should be selected only following evaluation and after due consideration.

3) Of the two bones which generally comprise a joint, one should be stabilized while the other is mobilized.

4) The initial treatment session should be relatively mild, relative to duration of treatment and the force utilized, to test patient tolerance and the potential responsiveness to treatment.

5) The affected joint's ranges should be compared with those of its contralateral counterpart to establish the norms or potential for that joint. This comparison is only appropriate when counterpart joints are not simultaneously affected. If both are affected, their ranges should be compared against the optimum standard of range of motion (refer to Evaluation of Joint Range of Motion [M12I1]).

6) During evaluation, each component of joint motion should be observed as an independent function, separate from all other components of motion.

7) Joint mobility should not be evaluated when joint inflammation is present. This makes technique selection difficult and potentially hazardous, and points at the need for experienced advice or adequate field experience to provide a basis for technique selection.

8) To avoid undue and overwhelming shearing forces into the joint, lever arms should be kept as short as possible.

9) Joint mobilization should not be applied against an actively contracting muscle or muscle group.

General techniques of joint mobilization (M1302B)

Joint mobilization generally employs the techniques of articulation, stretch, manipulation and/or traction.

Articulation (M1302B01) *Articulation* is a graded oscillation or deviation of movement which causes one articular surface to move or shear away from the opposing articular surface. Articulation is generally employed in medial/lateral or forward/backward directions. It is generally performed to restore joint play and component motion and is usually applied to hypomobile joint.

Grade 1 articulation employs gentle deviations of motion, performed at the beginning of an available range of motion.

Grade 2 articulation employs gentle deviations of motion performed at midrange.

Grade 3 articulation employs moderate deviations of motion, performed through the entire available range of motion and subsequently into the restricted range.

Grade 4 articulation employs moderate deviations of motion, performed through the available joint range of motion, with oscillating deviations at the end of range of motion and into restricted range.

The first two grades are utilized in the early stages of developing joint stiffness to maintain joint mobility and prevent range loses, while the last two grades are employed to regain lost ranges (as in frozen shoulder or other contracted joint syndromes).

Stretch (M1302B02) *Stretch* utilizes many of the procedures employed in the various articulation grades, except that rotary pressure is applied to the joint at the end of available range of motion in place of the oscillating deviations.

Progressive stretch involves a successive series of short amplitude stretch joint motions imparted at progressive increments of the range. It is most often utilized to decrease mechanical and/or soft tissue restriction.

Continuous stretch is a sustained, gradually increasing pressure upon the joint and associated soft tissue structures, which is applied without interruption. It is most often utilized when immediate increases in range of motion are desired (as when adaptive shortening of soft tissue structures associated with the joint has occurred). Continuous stretch is most commonly held for periods ranging from five to twenty seconds.

Manipulation (M1302B03) *Manipulation* employs the same procedures used with the articulation grades and during stretches, but thrust is additionally performed at the end of joint range of motion. Thrust is a high velocity, low amplitude forced movement of the involved joint. It requires practitioner skill and judgement relative to the amplitude and direction of the force applied. Thrust is often used to snap adhesions, alter vertebral position, normalize segmental motion and to reduce pain.

Traction (M1302B04) *Traction* is applied to increase joint space by pulling one articular surface away from the other. Traction may be applied by manual or mechanical means. Traction is generally utilized to facilitate the *articulation* and *stretch* by reducing articular approximation and making freer joint play possible.

Traction or joint separation is resisted by muscle tonus (tonic and phasic stretch reflexes) and the tensile strength of other soft tissues surrounding the joint. The force of traction must overcome the combined forces of these components before distraction can take place. The degree of distraction will be limited by the patient's tolerance and response to stretch. Traction is most often applied intermittently, with five to ten second hold and relax periods between each application.

Partial List of Conditions Treated (M1303)

Treatable Causes (M1303A)

Calcific deposit (C08)
Extrafusal muscle spasm (C41)
Habitual joint positioning (C48)
Hypermobile/instable joint (C10)
Interspinous ligamentous strain (C03)
Joint ankylosis (C42)
Joint approximation (C47)
Ligamentous strain (C25)
Nerve root compression (C04)
Neuromuscular tonic imbalance (C30)
Pathological neuromuscular
  hypertonus (C28)
Peripheral nerve injury (C07)
Psychoneurogenic neuromuscular
  hypertonus (C29)
Restricted joint range of motion (C26)

Soft tissue inflammation (C05)
Soft tissue swelling (C06)
Trigger point formation (C01)

Syndromes (M1303B)

ACHILLES TENDONITIS (S52)
ADHESION (S56)
BURNS (SECOND AND/OR THIRD DEGREE) (S72)
BURSITIS (S26)
CALF PAIN (S50)
CAPSULITIS (S27)
CEREBRAL PALSY (S25)
CERVICAL DORSAL OUTLET (SCALENICUS ATTICUS/CERVICAL RIB) (S09)
CERVICAL (NECK) PAIN (S73)
CHEST PAIN (S14)
DUPYTREN'S CONTRACTURE (S34)
EARACHE (S40)
ELBOW PAIN (S47)
FACET (S19)
FASCIITIS (S20)
FOOT PAIN (S51)
FOREARM PAIN (S54)
FROZEN SHOULDER (S64)
HAND/FINGER PAIN (S17)
HEADACHE PAIN (S02)
HIGH THORACIC BACK PAIN (S48)
KELOID FORMATION (S38)
KNEE PAIN (S46)
LOWER ABDOMINAL PAIN (S62)
LOW BACK PAIN (S03)
MENSTRUAL CRAMPING (S60)
MYOSITIS OSSIFICANS (S67)
NEUROMUSCULAR PARALYSIS (S22)
OSTEOARTHRITIS (S21)
POST CEREBRAL VASCULAR ACCIDENT (CVA) (S07)
POST IMMOBILIZATION (S36)
POST PERIPHERAL NERVE INJURY (S23)
POST SPINAL CORD INJURY (S24)
POST WHIPLASH (S55)
RADICULITIS (S29)
REFERRED PAIN (S01)
SCIATICA (S05)
SHOULDER/HAND (S57)
SHOULDER PAIN (S08)
TEMPOROMANDIBULAR JOINT (TMJ) PAIN (S06)
TENDONITIS (S28)
TENNIS ELBOW (LATERAL EPICONDYLITIS) (S32)
THIGH PAIN (S49)
TOOTHACHE/JAW PAIN (S41)
TORTICOLLIS (WRY NECK) (S65)
WRIST PAIN (S16)

Precautions and Contraindications (M1304)

Mobilization (stretch) of the hip joint(s) of patients who have suffered cervical or upper thoracic spinal cord injuries may precipitate autonomic hyperreflexia. Symptoms include a rapid rise in blood pressure accompanied by bradycardia, profuse sweating and headache.

Joint mobilization should not be attempted against active muscle contraction.

An inflamed joint should not be evaluated or treated directly with mobilization techniques.

The articular position of a joint must be considered when performing joint mobilization. If the capsule and ligaments are on maximum tension and the articular surfaces of the joint cannot be separated by traction applied across the joint, it is said to be in the *maximum close packed position*. This is the joint position in which the concave articular surface of one bone is in complete congruence with the convex articular surface of the other bone. In general, rotation of the involved joint causes it to assume a close packed position, as will all the extremes of joint range of motion (extension is the maximum close packed position of the vertebrae). Joint mobilization should not be attempted upon a joint which is in a maximum close pack position, to avoid joint soft tissue strain and/or sprain. All examination and the preliminary treatment sessions should take place with the involved joint(s) in the maximum open (loose) packed position (the capsule at its most relaxed). The maximum open packed position usually occurs in the midrange of joint motion.

Notes Aside (M1305)

In order to successfully perform joint mobilization, joint approximation by the muscles crossing the involved joint should be minimized. This may be accomplished by making the patient comfortable, placing the joint in its most open packed position (in this position the articular surfaces are not totally congruent and the capsule is at its most lax), and by inhibiting muscle tone. The inhibition of muscle tone may be accomplished through cryotherapy (ice pack or ice massage), fast brushing of the skin overlying muscular antagonistic to the hypertonic muscle(s) and/or by vibration applied to the origins, insertions or tendons of hypertonic musculature and then its antagonistic musculature (each site vibrated for a single minute, at amplitudes of between 1.0 and 5.0 pounds per square inch, at 30 or 60 cycles/second).

Pain may make it impossible for mobilization to be performed in the direction of the most restricted range of motion. If such is the case, traction must be performed in other directions. This may help relieve pain and may lead eventually to increased mobility. Indeed, mobilization in a rotational direction is likely to increase ranges of motion in flexion, extension, abduction and adduction, *and* rotation.

Bibliography (M1306)

DiFabio, pp. 51–54.
McGarry, pp. 30–31.
Olson, pp. 351–356.

JOINT SPRAIN SYNDROME (S30)

Definition (S30A)

A joint *strain* is defined as an over-stretching of the ligamentous joint support without disruption of the integrity of ligamentous fibrils or the avulsion of the tendon away from its bony attachment. A joint *sprain* occurs when the stress upon the joint is more severe than that producing a strain, and causes tendon fibril tearing to occur (usually at its proximal or distal point of attachment), or the avulsion of a small fragment of bone at the site of tendon insertion or origin. The severity of the sprain generally depends upon how much joint subluxation occurs.

Joint sprain usually results in pain at the site of injury, joint instability and distension of the joint (swelling) by effusion and/or hemorrhage (which further strains the involved ligaments). Unrelieved effusion may promote the formation of adhesions, between soft tissue structures, which may delay healing.

Related Syndromes (S30B)

ADHESION (S56)
BURSITIS (S26)
CALF PAIN (S50)
CAPSULITIS (S27)
CARPAL TUNNEL (S35)
ELBOW PAIN (S47)
FASCIITIS (S20)
FOOT PAIN (S51)
FOREARM PAIN (S54)
FROZEN SHOULDER (S64)
HAND/FINGER PAIN (S17)
KNEE PAIN (S46)
MYOSITIS OSSIFICANS (S67)
OSTEOARTHRITIS (S21)
PITTING (LYMPH) EDEMA (S31)
POST IMMOBILIZATION (S36)
REFERRED PAIN (S01)
TENDONITIS (S28)
TENNIS ELBOW (LATERAL EPICONDYLITIS) (S32)
THIGH PAIN (S49)
WRIST PAIN (S16)

Treatable Causes Which May Contribute To The Syndrome (S30C)

Calcific deposit (C08)
Extrafusal muscle spasm (C41)
Habitual joint positioning (C48)
Hypermobile/instable joint (C10)
Joint ankylosis (C42)
Joint approximation (C47)
Ligamentous strain (C25)
Muscular weakness (C23)
Peripheral nerve injury (C07)
Restricted joint range of motion (C26)
Soft tissue inflammation (C05)
Soft tissue swelling (C06)
Tactile hypersensitivity (C24)
Trigger point formation (C01)
Vascular insufficiency (C11)

Treatment Notes (S30D)

Of initial concern in the treatment of the acute joint sprain is the control of swelling (effusion). Swelling can be minimized by (1) immediate application of a firm bandage which will not stretch (either linen straps or athletic tape applied over a covering sleeve), applied in a manner which will avoid distal strangulation, (2) application of ice packing to help inhibit any accompanying inflammatory process*, and (3) elevation of the involved extremity. The bandage should remain in place from seven to ten days.

Nonweight bearing range of motion exercise should begin within a few days and include all the normal ranges of motion. The patient should perform the exercise actively without help from the practitioner.

When the swelling has disappeared, the patient may attempt to weight bear or take resistance into the involved joint, but only after it has been taped for joint support. If the injury has resulted in damage to the ligament(s) or structure, a sufficiently destabilizing orthotic device may have to be utilized to provide ongoing stability when the joint is under stress.

Refer to Table of Trigger Point Formation Locations & Descriptions of Their Referred Pain Patterns [T005] for a detailed description of trigger point formations which may, singly or in combination, imitate or contribute to the pain accompanying a particular joint sprain syndrome.

* Clinical evidence suggests that ice packing does not decrease or prevent swelling, but less swelling is produced then if heat packing or contrast packing or baths are used.

Bibliography (S30E)

Salter, pp. 510–521.
Scott, pp. 121–124, 66–68, 185–186, 360–362.
Shands, p. 422.

KELOID FORMATION SYNDROME (S38)

Definition (S38A)

A keloid formation is a nodular, frequently lobulated, moveable, nonencapsulated, and generally linear mass of peculiar hyperplastic fibrous connective tissue. It is generally comprised of relatively large and parallel bands of densely collagenous material separated by irregular bands of fibrous tissue. Developmentally, keloids at first appearance feel rubbery but eventually become dense and hard.

Keloids occasionally form after burns, scalds, lacerations or surgical wounds, and may occur anywhere on the skin but are most frequently found (for unknown reasons) in the area of the sternum. Keloid formations occur unpredictably following injury of the skin, and often over the site of previous scars. Surgical excision of keloid formation has reportedly been unsatisfactory with almost an immediate redevelopment keloid formations at the surgical site.

The cause of keloid formation is unknown. All racial groups seem to suffer from keloid formations but they seem to occur with greater frequency among brown and yellow skinned populations.

Related Syndromes (S38B)

ACNE (S45)
ADHESION (S56)
BACTERIAL INFECTION (S63)
BURNS (SECOND AND/OR THIRD DEGREE) (S72)
CARPAL TUNNEL (S35)
POST PERIPHERAL NERVE INJURY (S23)
SHINGLES (HERPES ZOSTER) (S37)
TENDONITIS (S28)
UNHEALED DERMAL LESION (S39)

Treatable Causes Which May Contribute To The Syndrome (S38D)

Extrafusal muscle spasm (S41)
Habitual joint positioning (C48)
Joint ankylosis (C42)
Necrotic soft tissue (C18)
Nerve root compression (C04)
Neuroma formation (C14)
Peripheral nerve injury (C07)
Restricted joint range of motion (C26)
Scar tissue formation (C15)
Soft tissue inflammation (C05)
Soft tissue swelling (C06)
Tactile hypersensitivity (C24)
Unhealed dermal lesion (C12)

Treatment Notes (S38D)

Historically, keloid formations have been treated by surgical excision, irradiation, or local injection of cortisone. Excision has often been unsatisfactory, but irradiation and local cortisone injections have been reported to have been more satisfactory as treatment forms.

The phonophoresis of vitamin E has proven to be clinically effective in reducing the rigidity and height of the keloid formation. From surface observation, it seems apparent that the phonophoresis of vitamin E actually retards keloid formation, and in some cases seems to have reversed the keloid formation process, turning keloid tissue into more normal connective tissue.

Bibliography (S38E)

Bickley, pp. 52–53.

KINESIOLOGY OF AMBULATION (R004)

Bipedal Ambulation (R004A)

Bipedal ambulation (gait) depends upon the ability to (1) independently support the body in an upright position, (2) balance the body in that upright position and (3) the ability to execute the stepping motion required of the legs to move the body forward. These functions all require the interplay and coordination of the musculature of the legs, hips, trunk and to some extent the musculature of the arms, shoulders and neck. In normal circumstances most of this process is performed without conscious awareness except for the general perception that it is occurring. Normal bipedal ambulation is characterized by (1) a wide range of rapid and comfortable walking speeds, (2) smooth forward translation of the trunk and (3) a rhythmicity of step length and duration of the temporal components of the walking cycle.

Phases of Ambulation (R004B)

Normal forward ambulation depends upon a cycle of "walking" maneuvers performed by the two lower extremities, providing continuous locomotion in a linear direction. These maneuvers are variously labeled and include (1) swing phase, (2) heel strike (initial contact of the foot with the floor), which may occur almost simultaneous with (3) the double support phase, when both feet are simultaneously in contact with the floor, (4) stance phase (sometimes called the propulsion phase, when one of the feet is in full contact with the floor and the opposite leg is in swing phase, taking the next step), and (5) the push off phase, which initiates the swing phase.

For normal, smooth and theoretically ideal ambulation to occur, stance and swing phases occur alternately and continuously. One lower extremity is required to provide support and balance while the opposing lower extremity performs swing phase. Within each walking cycle two periods of single lower extremity support and two periods of double lower extremity support are supposed to occur, with one lower extremity beginning the stance phase while the other is ending the stance phase. This ideal pattern of ambulation phases varies a good deal when individual ambulation patterns are compared. Normal variations are dependent upon individual ambulation skills, which include the individual's neuromuscular coordination ability and acquired ambulation habits, some of which may reflect emotional states or even general psychological attitudes (an interesting study, but beyond the scope of this discussion).

Normal Variations in Ambulation (R004C)

One of the most common normal variations in the ambulation (gait) pattern is the missing heel strike phase. Heel strike may be missing on one or both sides, as a symptom of a casual or "flat footed" gait pattern. Push off may also be greatly reduced, almost to the point of being nonexistent. This gait pattern may seen in individuals who tend to ambulate with their knees kept habitually flexed.

The push off phase may also be missing in individuals who habitually walk on their heels. Such individuals tend to ambulate with a slightly back leaning posture.

Heel strike may be missing when an individual habitually walks on the balls of the feet or on the lateral edge foot or feet, which may result in failure to put full weight upon one or both of the heels before and/or during stance phase.

Variations also exist in the habitual duration of stance and/or swing phases, relative to each other and/or between each lower extremity which may produce shambling, lurching or even jerky gait patterns. These variations may occur without underlying pathology, and therefore must be classified as being normal (albeit not pretty).

Characteristics of Normal Ambulation (R004D)

As mentioned earlier, musculature of the abdomen, shoulders, arms and neck play a role in ambulation, especially for maintaining balance and stabilizing the trunk and head. For our purposes here, however, we will only discuss the role of the musculature of the hips and lower extremities in the functions of ambulation.

During normal ambulation in the forward direction, at the beginning of stance phase, coordinated synergistic muscular activity directs the lower extremity in swing phase obliquely forward, while providing restraint from excessive forward motion or knee extension, as the opposite lower extremity in stance phase is directed obliquely backward to provide propulsion in a forward direction. During this process the trunk is continuously translated forward over the alternating bases of support, usually and ideally at a constant linear horizontal velocity. The speed of translation from one base of support to the other depends upon the rapidity of steps and the stride length (the distance between successive contact points of the opposite feet).

Muscles Involved (R004E)

Normal ambulation depends upon the complex coordination of antagonistic musculature employed in dynamic interplay of synergistic muscle groups which together produce the required combination of proximal stabilization simultaneously with distal joint motion (refer to Table of Antagonistic Muscles of Common Interest [T006] and/or Exercise, Functional Testing, Pathological Neuromuscular Synergistic Pattern Evaluation, Synergistic Pattern Assessment [M12I7B] for listings of lower extremity antagonistic muscles and synergistic muscle groups).

Basically, the hip flexor muscle group brings the hip forward while the knee extensor (quadriceps femoris) group extends the knee as the hip extensors, including the gluteus maximus and hamstring group, limit the speed and range of hip flexion. The knee flexion (hamstring) group limits the speed and range of knee extension. The ankle dorsi flexors (including the long toe extensors) act to pull the ball of the foot up in preparation for heel strike. This occurs as the quadratus lumborum and the external and internal oblique muscles on the homolateral side act synergistically with the gluteus medius on the contralateral side (in stance phase) to keep the pelvis from rotating forward too far or falling.

Following heel strike, the ankle dorsi flexors serve to slow the descent of the ball of the foot to the stance phase position. The hip extensors actively extend and then hyperextend the hip as the knee extension (popliteus included) and flexion groups stabilize the knee through mid stance phase while simultaneously working with the hip flexors to limit the speed and range of hip extension and hyperextension. When the mid point of stance phase is reached the knee is allowed to flex and the plantar flexion group of the ankle (gastrocnemius and soleus muscles) begins the dynamic process of the push off phase. Near the end of the push off phase, the opposite extremity is beginning heel strike. When the push off phase ends the whole process begins again with flexion of the hip and knee. The knee is flexed with hip flexion as the ankle is dorsiflexed to allow the ball of the foot to clear the floor at mid swing phase.

As important as the active muscles described above are the muscles of stabilization that indirectly guide the direction of motion of the hip, knee, and ankle. In the hip, these muscles include the internal and external rotators of the hip, which respectively include the gluteus minimus and the hip adductor group (adductor magnus, adductor brevis, adductor longus, pectineus, gracilis and sartorius), obturator externus, obturator internus, quadratus femoris, piriformis, gemellus superior and gemellus inferior. Knee extension and flexion functions are stabilized and indirectly guided by the various components of the quadriceps femoris and hamstring groups. The peroneus longus, peroneus brevis, peroneus tertius, anterior and posterior tibialis, toe extensors and flexors stabilize and guide ankle and foot during the various phases.

Summary (R004F)

Ambulation is a complex neuromuscular function which requires the interplay and coordination of a plethora of central and peripheral neuromuscular mechanisms coordinated with other supraspinal mechanisms (balance, judgement, emotions, etc.). The end result is a function which is for the most part unconscious (automatic) and allows the body to move itself along bipedally.

KNEE PAIN SYNDROME (S46)

Definition (S46A)

Knee pain may be caused by a torn ligament, a displaced or torn knee cartilage, an effusion (swelling) into the knee, abscess, referred pain from other structures and diseases which cause pain or swelling in or around the knee joint.

Torn ligaments and displaced or torn knee cartilage result from trauma to the knee, either from external blunt force around or to the joint, or from a shearing twisting force into the knee joint. In the latter case, generally the femur is forcibly rotated on the tibial plateau while the tibia is fixed in place, usually occurring during the stance phase of ambulation.

Effusions into the knee are caused by bleeding into the knee joint or from excessive fluid accumulation in the joint, resulting from septic arthritis, abscess formation (accompanied by redness, fever and swelling of lymph nodes) and other diseases processes of the joint, including chondromalacia.

Pain may be referred into the knee joint from nerve root impingement or from trigger point formations.

Related Syndromes (S46B)

ACHILLES TENDONITIS (S52)
ADHESION (S56)
BACTERIAL INFECTION (S63)
BURSITIS (S26)
CALF PAIN (S50)
CAPSULITIS (S27)
CHONDROMALACIA (S15)
HYSTERIA/ANXIETY REACTION (S59)
JOINT SPRAIN (S30)
KELOID FORMATION (S38)
MYOSITIS OSSIFICANS (S67)
NEUROMUSCULAR PARALYSIS (S22)
OSTEOARTHRITIS (S21)
PHLEBITIS (S68)
PHOBIC REACTION (S44)
PITTING (LYMPH) EDEMA (S31)
POST CEREBRAL VASCULAR ACCIDENT (CVA) (S07)
POST IMMOBILIZATION (S36)
POST PERIPHERAL NERVE INJURY (S23)
POST SPINAL CORD INJURY (S24)
REFERRED PAIN (S01)
SCIATICA (S05)
SHINGLES (HERPES ZOSTER) (S37)
TENDONITIS (S28)
THIGH PAIN (S49)
UNHEALED DERMAL LESION (S39)

Treatable Causes Which May Contribute To The Syndrome (S46C)

Bacterial infection (C09)
Calcific deposit (C08)
Extrafusal muscle spasm (C41)
Habitual joint positioning (C48)
Hypermobile/instable joint (C10)
Interspinous ligamentous strain (C03)
Joint ankylosis (C42)
Joint approximation (C47)
Ligamentous strain (C25)
Muscular weakness (C23)
Nerve root compression (C04)
Neuromuscular tonic imbalance (C30)
Peripheral nerve injury (C07)
Psychoneurogenic neuromuscular hypertonus (C29)
Restricted joint range of motion (C26)
Soft tissue inflammation (C05)
Soft tissue swelling (C06)
Sympathetic hyperresponse (C38)
Tactile hypersensitivity (C24)
Trigger point formation (C01)
Unhealed dermal lesion (C12)
Vascular insufficiency (C11)

Treatment Notes (S46D)

Treatment of knee pain is dependant upon the cause.

The following is a list of trigger point formations which may, singly or in combination, refer pain into the area of the knee. It should be noted that it is possible for the given patient to experience pain throughout the entire stereotypical referred pain pattern or only in part or parts of it. Opposite to each listing is a description of the parts of the stereotypical referred pain pattern most commonly experienced by previous patients, relative to the given trigger point formation. Any descriptive reference which is underlined is a part of the pattern which has *not* been commonly experienced by previous patients. For complete description of each stereotypical referred pain pattern, refer to the Table of Referred Pain Patterns of Trigger Point Formation Origin [T005]. Please note the key to abbreviated words at the bottom of the list.

LOCATION (MUSCLE)	REFERRED PAIN ZONES (GENERAL)
Gluteus minimus	Gluteus maximus (excepting central area) & gluteus medius, p.l.leg, l.calf to p.of l. malleolus
Adductor longus	A. s.iliac crest into groin & down the a.med. thigh, central med. calf, med. malleolus & med. foot
Biceps femoris (hamstring)	P. d.central thigh, p.prox. calf
Vastus medialis	Vastus medialis, patella, knee joint
Gastrocnemius	P. med. calf, central thigh, behind med. malleolus, arch of the foot

Key:
maj.—major
prox.—proximal
med.—medial
para.—paraspinous
l.—lateral
a.—anterior
s.—superior
&—and
bord.—border
i.—inferior
m.—middle
d.—distal
p.—posterior
bil.—bilateral

LASER STIMULATION (M27)

Definition (M2701)

Laser is a term used to represent the process of *l*ight *a*mplification by *s*timulated *e*mission of *r*adiation. A laser beam is a ray of very dense light or other radiant emission created by bouncing radiation back and forth between two polished surfaces through a translucent medium, either crystalline or gaseous.

Basically, light laser beams result from by bouncing light wave radiation (commonly produced by the flash from a xenon lamp) back and forth between two parallel coaxial mirrors through crystal or gas until the emission particles or waves are in the same phase and proceeding in the same direction. When this occurs, a resultant, very intense, beam of monochromatic nonspreading visible light is produced.

In the clinical treatment setting two types of laser radiation (not including the laser types used for surgical procedures) are currently being utilized. They include (1) the infrared pulsed laser produced by passing light through a crystal synthesized from gallium arsenic or gallium aluminum, and (2) the helium-neon (He-Ne) laser produced by passing the light through a chamber containing helium and neon gas. Neither type has a thermal (heating) effect upon the skin and are consequently called "cold" lasers.

The effects of cold laser upon the physiology of exposed soft tissues must still be scientifically explored, though extensive claims in literature originating from European sources have been made for its use.

To date, the verifiable therapeutic effects upon human physiology by cold laser radiation are said to be limited to its ability to raise skin resistance (increase impedance) at sites of relatively, low (and considered abnormal) skin resistance. These changes have been correlated by some investigators with simultaneous decreases in pain originating from areas of treatment or therapeutic effects ascribed to the stimulation of acupoints.

Procedures of Application (M2702)

1) The patient should be placed in a sitting, semi-reclining, or lying position which will be comfortable for the session duration and which provides easy access to the treatment site for the practitioner.

2) An ohm meter set to register the impedance range of acupoints should be utilized to locate sites of abnormally low skin resistance (sites of high conductance); for details refer to the Peripheral Electroacustimulation, Peripheral Acupoint Survey [M05D].

3) Located sites of abnormally low skin resistance should be immediately exposed to cold laser radiation.

4) Each site should be stimulated for from fifteen to twenty seconds.

5) Treatment protocols may follow those traditionally recommended for peripheral acupoint stimulation, or the points found within zones of high skin resistance (above zones of deep tissue inflammation) may be stimulated (refer to Peripheral Electroacustimulation, Pathology Control [M05C] and/or Electrical Stimulation of Acupoints Within an Inflamed Zone [M05A] for further details).

6) Session length will vary according to how many points are stimulated.

7) Treatment sessions may occur as often as once daily.

Partial List of Conditions Treated (M2703)

Treatable Causes (M2703A)

Calcific deposit (C08)
Dental pain (C22)
Extrafusal muscle spasm (C41)
Habitual joint positioning (C48)
Hiccup (C44)
Hypermobile/instable joint (C10)
Hypertension (C39)
Insomnia (C46)
Interspinous ligamentous strain (C03)
Irregular menses (C19)
Joint ankylosis (C42)
Joint approximation (C47)
Ligamentous strain (C25)
Localized viral infection (C27)
Menorrhalgia (C20)
Migraine (vascular headache) (C34)
Nausea/vomiting (C36)
Nerve root compression (C04)
Neurogenic dermal disorders (psoriasis, hives) (C49)
Neuroma formation (C14)
Peripheral nerve injury (C07)
Phlebitis (C43)
Psychoneurogenic neuromuscular hypertonus (C29)
Restricted joint range of motion (C26)
Sinusitis (C16)
Soft tissue inflammation (C05)

Soft tissue swelling (C06)
Spastic sphincter (C40)
Sympathetic hyperresponse (C38)
Tactile hypersensitivity (C24)
Tinnitus (C32)
Trigger point formation (C01)
Unhealed dermal lesion (C12)
Visceral organ dysfunction (C02)

Syndromes (M2703B)

ACHILLES TENDONITIS (S52)
BREAK THROUGH BLEEDING (S61)
BURNS (SECOND AND/OR THIRD DEGREE) (S72)
BURSITIS (S26)
CALF PAIN (S50)
CAPSULITIS (S27)
CARPAL TUNNEL (S35)
CERVICAL DORSAL OUTLET (SCALENICUS ATTICUS/CERVICAL RIB) (S09)
CERVICAL (NECK) PAIN (S73)
CHEST PAIN (S14)
CHONDROMALACIA (S15)
CHRONIC BRONCHITIS/EMPHYSEMA (OBSTRUCTIVE PULMONARY DISEASE) (S13)
COLITIS (S42)
EARACHE (S40)
ELBOW PAIN (S47)
FACET (S19)
FASCIITIS (S20)
FOOT PAIN (S51)
FOREARM PAIN (S54)
FROZEN SHOULDER (S64)
HAND/FINGER PAIN (S17)
HEADACHE PAIN (S02)
HIGH THORACIC BACK PAIN (S48)
HYPERTENSION (S43)
HYSTERIA/ANXIETY REACTION (S59)
INSOMNIA (S71)
JOINT SPRAIN (S30)
KNEE PAIN (S46)
LOWER ABDOMINAL PAIN (S62)
LOW BACK PAIN (S03)
MENSTRUAL CRAMPING (S60)
MIGRAINE (VASCULAR) HEADACHE (S18)
MYOSITIS OSSIFICANS (S67)
OSTEOARTHRITIS (S21)
PHLEBITIS (S68)
PHOBIC REACTION (S44)
PIRIFORMIS (S04)
POST IMMOBILIZATION (S36)
POST PERIPHERAL NERVE INJURY (S23)
POST WHIPLASH (S55)
RADICULITIS (S29)
REFERRED PAIN (S01)
SCIATICA (S05)
SHINGLES (HERPES ZOSTER) (S37)
SHOULDER/HAND (S57)
SHOULDER PAIN (S08)
SINUS PAIN (S33)
TEMPOROMANDIBULAR JOINT (TMJ) PAIN (S06)
TENDONITIS (S28)
TENNIS ELBOW (LATERAL EPICONDYLITIS) (S32)
THIGH PAIN (S49)
TIC DOULOUREUX (S53)
TINNITUS (S70)
TOOTHACHE/JAW PAIN (S41)
UNHEALED DERMAL LESION (S39)
WRIST PAIN (S16)

Precautions and Contraindications (M2704)

When correctly practiced, no contraindication has been sited for cold laser stimulation of acupoints, except for application on pregnant women; acupoint stimulation, misapplied with other means has been reported to have caused miscarriage.

Patients with heart disease, anemia, diabetes and hypotension should be watched carefully for signs of a developing autonomic response. Symptoms may include sensations of fatigue or general muscular relaxation, drowsiness and rapid decreases in both systolic and diastolic blood pressures. Patients with diabetes should additionally be monitored for symptoms of developing diabetic coma, which may develop as metabolic insulin demand decreases as sympathetic slowing occurs. These symptoms may include dry and flushed skin, fever, dry mouth, intense thirst, absence of hunger, nausea and/or vomiting, abdominal pain (frequently seen), exaggerated air hunger, acetone breath odor, low blood pressure, weak and rapid pulse, lack of affect and glassy stare. Such "diabetic shock" symptomatology may be relieved by having the patient consume a small quantity of simple sugar, like a banana.

A thorough preliminary evaluation of both acute and chronic conditions before instituting treatment is very important. Acupoint stimulation may eliminate pain, thereby masking symptoms of a disease process or pathology for which cold laser acupoint stimulation may not be indicated.

Notes Aside (M2705)

European claims of cold laser treatment effectiveness for the treatment of traumatized muscle, tendon, nerves and bone, rheumatoid arthritis, dermo-epidermic diseases, couperosis, acne, flaccid tissue and wrinkles, scars, ulcers, burns and hydrolipodystrophy are tantalizing and deserve further investigation by reliable clinical practitioners. Although unlikely to be the panacea some might claim, valuable, nonsurgical, therapeutic techniques utilizing the cold laser may be gleaned from the apparent over-enthusiastic bombast present in some of the general literature available on the subject.

Bibliography (M2706)

Greathouse, pp. 1184–1187.
Lehmann, pp. 27–28.
Snyder-Mackler, pp. 1087–1090.
Townes, pp. 153–155.

LIGAMENTOUS STRAIN (C25)

Definition (C25A)

A ligamentous strain is caused by the overstretching of a ligament without causing it to tear. The result of a ligament strain is a sharp pain originating from the injured ligament, when the injury occurs, with an increasing sensation of stiffness and soreness several hours after injury. Referred pain, usually in a particular pattern, may also be produced by the hyperstress upon a ligament. Additionally, stability of an involved joint may be permanently compromised by ligament stretch which exceeds its elastic limit, since a ligament stretched beyond its ability to rebound will not spontaneously return (shrink back) to its original length, even over time.

Evaluation (C25B)

The predominant acute symptom, sharp pain occurring at the site of ligamentous strain, usually occurs without swelling. The predominant chronic symptom of ligamentous strain is a decrease in joint stability. Joint stability may be evaluated through the use of joint manipulative testing such as drawer sign, joint play against resistance, etc.

Etiology (C25C)

Improper posture or prolonged stress form awkward joint positions which may cause protracted ligamentous strain, its resulting in discomfort. Ligamentous strain may also occur when sudden pressure is applied to a joint with sufficient force to exceed the tensile strength or elastic limits of the supportive ligament too rapidly for the cosupportive muscle(s) to compensate for the insult. Ligamentous over stretching results, as when the patient "turns an ankle" without spraining it.

Treatment Notes (C25D)

Treatment of the acute ligamentous strain should be directed at protecting the joint from further injury with supportive splinting, taping and/or rest.

Treatment of the over stretched ligament, after the acute stage, should be directed at providing artificial compensatory joint support via bracing, strapping or taping when the joint is expected to be subjected to stress, and/or by compensatory muscle toning and shortening to provide some restoration of joint integrity.

Bibliography (C25E)

Salter, pp. 418–423.
Scott, pp. 156–157, 189–204, 348–362.

TREATMENT OF LIGAMENTOUS STRAIN (C25F)

Technique Options (C25F1)

Evaluation techniques (C25F1A)

Auricular electroacustimulation	*M06*
Auricular acupoint evaluative survey	M06C
Determination of referred pain source	M06C5A
Auxiliary evaluation techniques	*M26*
Auscultation	M26C
Joint and/or tendon sounds	M26C2C
Cryotherapy	*M10*
Evaluation of capillary response to cold	M10H
Determination of hyperreactivity to cold	M10H2A
Differential skin resistance survey	*M22*
Exercise	*M12*
Functional testing	M12I
Evaluation of joint range of motion	M12I1
Evaluation of muscle strength	M12I2
Massage	*M23*
Palpation evaluation	M23G
Determination of soft tissue tenderness	M23G3
Determination of soft tissue inflammation	M23G5
Peripheral electroacustimulation	*M05*
Peripheral acupoint survey	M05D

Treatment techniques (C25F1B)

Auricular electroacustimulation	*M06*
Pain relief	M06A
Cryotherapy	*M10*
Analgesia/anesthesia	M10C
Electrical facilitation of endorphin production (acusleep)	*M03*
Increasing pain tolerance	M0305A
Electrical stimulation (ES)	*M02*
Muscle reeducation/muscle toning	M02B
Inflammation control	M02G

ligamentous strain

Decreasing soft tissue inflammation	M02G5A
Exercise	**M12**
Muscle strengthening	M12A
Hydrotherapy	**M15**
Range of motion enhancement and physical reconditioning	M15D
Iontophoresis	**M07**
Pain relief	M07A
Inflammation amelioration	M07B
Joint mobilization	**M13**
Laser stimulation	**M27**
Massage	**M23**
Acupressure	M23C
Deep soft tissue mobilization	M23F
Muscles	M23F2A
Ligaments	M23F2B
Peripheral electroacustimulation	**M05**
Acupoint electrical stimulation within an inflamed zone	M05A
Analgesia	M05B
Taping	**M16**
Transcutaneous nerve stimulation	**M04**
Chronic pain management	M04A
Prolonged stimulation of acupoints	M04B
Ultrahigh frequency sound (ultrasound)	**M01**
Phonophoresis	M01A
The inflammatory process: control with phonophoresis	M01A5A
Desensitization	M01C
Acupoint stimulation	M01K
Vibration	**M08**
Neuromuscular management	M08A
Muscle lengthening	M08A2B
Desensitization of soft tissue	M08E

Techniques Suggested for the Acute Condition (C25F2)

Evaluation techniques (C25F2A)

Auricular electroacustimulation	**M06**
Auricular acupoint evaluative survey	M06C
Determination of referred pain source	M06C5A
Auxiliary evaluation techniques	**M26**
Auscultation	M26C
Joint and/or tendon sounds	M26C2C
Cryotherapy	**M10**
Evaluation of capillary response to cold	M10H
Determination of hyperreactivity to cold	M10H2A
Differential skin resistance survey	**M22**
Exercise	**M12**
Functional testing	M12I
Evaluation of joint range of motion	M12I1
Evaluation of muscle strength	M12I2
Massage	**M23**
Palpation evaluation	M23G
Determination of soft tissue tenderness	M23G3
Determination of soft tissue inflammation	M23G5
Peripheral electroacustimulation	**M05**
Peripheral acupoint survey	M05D

Treatment techniques (C25F2B)

Auricular electroacustimulation	**M06**
Pain relief	M06A
Cryotherapy	**M10**
Analgesia/anesthesia	M10C
Electrical facilitation of endorphin production (acusleep)	**M03**
Increasing pain tolerance	M0305A
Peripheral electroacustimulation	**M05**
Acupoint electrical stimulation within an inflamed zone	M05A
Taping	**M16**
Transcutaneous nerve stimulation	**M04**
Prolonged stimulation of acupoints	M04B
Ultrahigh frequency sound (ultrasound)	**M01**
Phonophoresis	M01A
The inflammatory process: control with phonophoresis	M01A5A
Vibration	**M08**
Neuromuscular management	M08A
Muscle lengthening	M08A2B

Techniques Suggested for the Chronic Condition (C25F3)

Evaluation techniques (C25F3A)

Auricular electroacustimulation	**M06**
Auricular acupoint evaluative survey	M06C
Determination of referred pain source	M06C5A
Auxiliary evaluation techniques	**M26**
Auscultation	M26C
Joint and/or tendon sounds	M26C2C
Cryotherapy	**M10**
Evaluation of capillary response to cold	M10H
Determination of hyperreactivity to cold	M10H2A
Differential skin resistance survey	**M22**
Exercise	**M12**
Functional testing	M12I
Evaluation of joint range of motion	M12I1
Evaluation of muscle strength	M12I2
Massage	**M23**
Palpation evaluation	M23G

Determination of soft tissue tenderness	M23G3	Joint mobilization	M13
		Massage	M23
Determination of soft tissue inflammation	M23G5	Deep soft tissue mobilization	M23F
		Muscles	M23F2A
Peripheral electroacustimulation	*M05*	Ligaments	M23F2B
Peripheral acupoint survey	M05D	*Peripheral electroacustimulation*	*M05*
		Acupoint electrical stimulation within an inflamed zone	M05A
Treatment techniques (C25F3B)		Analgesia	M05B
Cryotherapy	*M10*	*Transcutaneous nerve stimulation*	*M04*
Analgesia/anesthesia	M10C	Chronic pain management	M04A
Electrical facilitation of endorphin production (acusleep)	*M03*	Prolonged stimulation of acupoints	M04B
		Ultrahigh frequency sound (ultrasound)	*M01*
Increasing pain tolerance	M0305A	Desensitization	M01C
Electrical stimulation (ES)	*M02*	*Vibration*	*M08*
Muscle reeducation/muscle toning	M02B	Neuromuscular management	M08A
Exercise	*M12*	Muscle lengthening	M08A2B
Muscle strengthening	M12A		
Hydrotherapy	*M15*		
Range of motion enhancement and physical reconditioning	M15D		

LOCALIZED VIRAL INFECTION
(C27)

Definition (C27A)

Of interest to us here are those viral infections of superficial tissues which assume and maintain parasitic relations with susceptible host cells in a localized manner, specializing in the infestation cells of a particular type and/or in a particular area. Examples of such infection include herpes zoster (shingles) and plantar keratosis (plantar wart).

Infection and disease are not synonymous. Infection describes the relationship of the parasite to the host, whereas the term disease encompasses all the changes observed in the host as the result of infection.

Evaluation (C27B)

The sites of localized viral infections of the type of interest here are determined by the appearance of dermal lesions.

The primary site of peripheral nerve root infection by the herpes zoster (shingles) virus may be deduced from the pattern of eruptions which occur throughout the infected nerve's dermatome distribution. There will also be eruptions directly over the nerve root site of infection, just lateral to the vertebral spinous processes on the involved side.

Plantar keratosis (plantar wart) may be found by visual inspection of the sole of the foot. The "wart" is characterized by a thickening of the outer cornifying layer of the epithelium or a hyper-keratosis, forming a finger-like projection or cutaneous horn.

"Warts" are usually found beneath the metatarsal heads but can be found anywhere on the sole of the foot. The keratosis most commonly does not project above the surface of the skin and is generally surrounded by callus formation. Pressure over the "wart" usually produces pain.

Etiology (C27C)

To be successful a virus (or any infecting parasite) must arrive at a relationship with its host which allows the host to go on living while permitting adequate replication and spread of the virus. Successful viral infections usually take the form of a cytocidal (cytolytic or necrotizing) infection.

In the cytocidal infection, the virus, an obligate intracellular parasite, contrives to penetrate to the interior of the cell by interaction with a specific receptor protein on the cell's surface. The virus then, in some way, induces the cell to engulf and ingest it through endocytosis (inoculation). Once inside, the virus sheds its protein coat, where upon its viral nucleic acid escapes from the vacuole and assumes the direction of the cell's synthetic activity, causing it to produce new viral components (incubation). The newly synthesized virus is then assembled and the host cell bursts, releasing new virions for distribution to other susceptible cells. In acute illness this is the stage at which the patient exhibits all of the symptoms of the syndrome particular to the virus present. The newly produced virions attack new potential host cells, concentrating on the preferred tissue (trophism), the cells of which have the proper receptor proteins on their surfaces. In cases of herpes zoster, when this secondary viremia subsides, exanthem (skin eruptions) breaks out and a remission of acute symptoms occurs. The patient's autoimmune system may begin to exert itself at this point, by sensitizing lymphocytes (killer cells) which are released from peripheral lymphoid tissue to track down the virus in infected epithelial cells, destroying the virus and the host cells. If dermis cells are involved, dermal tissue may be destroyed and scarring may result.

Variations of the viral disease process often occur, depending upon the nature of the virus. Mumps and influenza, for example, may not produce a significant viremia or exanthem as an essential part of the disease, while the wart (verrucal vulgaris) is representative of a localized viral infection with a very slow development, mostly confining itself to refinement of the incubation stage and devoid of most manifestations of the post incubation stages seen in other types of viral infection.

Treatment Notes (C27D)

Ultrahigh frequency sound (ultrasound) has been demonstrated to be capable of depolymerizing macromolecules, altering the dispersion and solubility of serum proteins, inactivating certain enzymes (pepsin and hyaluronidase), and of flattening viruses and disrupting bacteriophages. Consequently, ultrasound has proven to be an effective tool for the eradication of various infecting viruses. If ultrasound can be made to play upon a virus it may destroy it, thereby reducing population levels, or slowing infection development (refer to Ultrahigh Frequency Sound, Viral Infection Eradication [M01I]).

For efficient treatment of herpes zoster nerve root infection, application of ultrasound should be made directly over the primary site of infection (over the posterior nerve root ganglion) and not over the sites of more peripheral lesions. Sounding peripheral sites is not contraindicated but it is not necessary.

When treating plantar keratosis (plantar wart) ultrasound should be applied directly over the lesion(s). An oil base coupling agent is preferable over aqueous gels because of the obstructive effects the oil has on the keratosis pores which are noted for being detrimental to the life of the keratosis.

Bibliography (C27E)

Adles, pp. 115–123.
Antich, pp. 99–102.
Bickley, pp. 114–118.
Holvey, pp. 1053–1055, 981.
Kent, pp. 15–18.
Ultrasonic Energy in Medicine, pp. 803–806.
Vaughn, pp. 396–397.

TREATMENT OF LOCALIZED VIRAL INFECTION (C27F)

Technique Options (C27F1)

Evaluation techniques (C27F1A)

Cryotherapy	*M10*
Evaluation of capillary response to cold	M10H
Determination of hyperreactivity to cold	M10H2A
Differential skin resistance survey	*M22*
Peripheral electroacustimulation	*M05*
Peripheral acupoint survey	M05D

Treatment techniques (C27F1B)

Auricular electroacustimulation	*M06*
Pain relief	M06A
Cryotherapy	*M10*
Analgesia/anesthesia	M10C
Electrical facilitation of endorphin production (acusleep)	*M03*
Increasing pain tolerance	M0305A
Laser stimulation	*M27*
Massage	*M23*
Acupressure	M23C
Peripheral electroacustimulation	*M05*
Analgesia	M05B
Transcutaneous nerve stimulation	*M04*
Chronic pain management	M04A
Prolonged stimulation of acupoints	M04B
Ultrahigh frequency sound (ultrasound)	*M01*
Viral infection eradication	M01I
Herpes zoster (Shingles)	M01I2A
Plantar warts	M01K
Acupoint stimulation	M01I2B

Techniques Suggested for the Acute Condition (C27F2)

Evaluation techniques (C27F2A)

Cryotherapy	*M10*
Evaluation of capillary response to cold	M10H
Determination of hyperreactivity to cold	M10H2A
Differential skin resistance survey	*M22*
Peripheral electroacustimulation	*M05*
Peripheral acupoint survey	M05D

Treatment techniques (C27F2B)

Auricular electroacustimulation	*M06*
Pain relief	M06A
Cryotherapy	*M10*
Analgesia/anesthesia	M10C
Electrical facilitation of endorphin production (acusleep)	*M03*
Increasing pain tolerance	M0305A
Peripheral electroacustimulation	*M05*
Analgesia	M05B
Transcutaneous nerve stimulation	*M04*
Prolonged stimulation of acupoints	M04B
Ultrahigh frequency sound (ultrasound)	*M01*
Viral infection eradication	M01I
Herpes zoster (Shingles)	M01I2A
Plantar warts	M01I2B

Techniques Suggested for the Chronic Condition (C27F3)

Evaluation techniques (C27F3A)

Peripheral electroacustimulation	*M05*
Peripheral acupoint survey	M05D

Treatment techniques (C27F3B)

Auricular electroacustimulation	*M06*
Pain relief	M06A
Electrical facilitation of endorphin production (acusleep)	*M03*
Increasing pain tolerance	M0305A

Peripheral electroacustimulation   **M05**
    Analgesia   M05B
Transcutaneous nerve stimulation   **M04**
    Chronic pain management   M04A
    Prolonged stimulation of acupoints   M04B

Ultrahigh frequency sound (ultrasound)   **M01**
    Viral infection eradication   M01I
       Herpes zoster (Shingles)   M01I2A
       Plantar warts   M01I2B

LOW BACK PAIN SYNDROME (S03)

Definition (S03A)

Pain occurring in the dorsal area of the human body, distal from the midthoracic vertebrae (T10) to the level of the first sacral vertebrae (S1) is defined as low back pain. Pain in the area of the gluteus maximus and gluteus medius muscles which is said to radiate across the sacral region is commonly referred to, though erroneously, as low back pain. The distribution of low back pain varies from patient to patient, ranging from diffuse pain throughout the entire region to an isolated unilateral "pin point" locus of pain.

Low back pain may be associated with fever, fatigue, disease and/or injury.

Diseases which are noted to cause low back pain include ankylosing spondylitis, osteoporosis, tumors of the spine and diseases of the lungs, kidneys and abdominal organs. Disease or injury to the intervertebral disks may also cause low back pain when resultant impingement upon nervous tissue as it exits the spinal column occurs. Pressure upon peripheral nerves in the area of the piriformis muscle may create low back pain similar to that associated with lumbar disc disease (small sciatica or piriformis syndrome). Localized viral infection of peripheral nerve roots (herpes zoster or shingles) in the lower thoracic or lumbosacral area may be responsible for low back pain in some cases.

Muscular fatigue and/or stress, including external trauma, of muscular or connective tissue may in many cases to be shown to be responsible for with low back pain. Muscular imbalance between the muscles supporting the lower spine, one side versus the other, may be found to be responsible in some cases of low back pain.

Trigger points and stretch or pressure upon the lower thoracic or lumbar interspinous ligaments may be responsible for low back pain, in distinctive referred pain patterns.

Related Syndromes (S03B)

ADHESION (S56)
BACTERIAL INFECTION (S63)
FACET (S19)
FASCIITIS (S20)
HYSTERIA/ANXIETY REACTION (S59)
LOWER ABDOMINAL PAIN (S62)
MENSTRUAL CRAMPING (S60)
OSTEOARTHRITIS (S21)
PHOBIC REACTION (S44)
PIRIFORMIS (S04)
POST PERIPHERAL NERVE INJURY (S23)
POST SPINAL CORD INJURY (S24)
REFERRED PAIN (S01)
SCIATICA (S05)
SHINGLES (HERPES ZOSTER) (S37)
UNHEALED DERMAL LESION (S39)

Treatable Causes Which May Contribute To The Syndrome (S03C)

Bacterial infection (C09)
Calcific deposit (C08)
Extrafusal muscle spasm (C41)
Habitual joint positioning (C48)
Hypermobile/instable joint (C10)
Interspinous ligamentous strain (C03)
Joint ankylosis (C42)
Joint approximation (C47)
Ligamentous strain (C25)
Localized viral infection (C27)
Menorrhalgia (C20)
Muscular weakness (C23)
Nerve root compression (C04)
Neuromuscular tonic imbalance (C30)
Peripheral nerve injury (C07)
Psychoneurogenic neuromuscular hypertonus (C29)
Restricted joint range of motion (C26)
Soft tissue inflammation (C05)
Soft tissue swelling (C06)
Sympathetic hyperresponse (C38)
Tactile hypersensitivity (C24)
Trigger point formation (C01)
Visceral organ dysfunction (C02)

Treatment Notes (S03D)

The most common source of low back pain is referred pain from trigger point formations. Other sources of low back pain include soft tissue inflammation of the paraspinous regions from T12 to S4 (usually occurring unilaterally), extrafusal muscle strain and spasm, osteoarthritic calcific deposit, the piriformis syndrome, tonic neuromuscular imbalance and discogenic disease.

The following is a list of trigger point formations which may, singly or in combination, refer pain into the low back area. It should be noted that it is possible for the give patient to experience pain throughout the entire stereotypical referred pain pattern or only in part or parts of it. Opposite to each listing is a description of the parts of the stereotypical referred pain pattern most commonly experienced by previous patients, relative to the given trigger point formation. Any descriptive reference which is underlined is a part of the pattern which has *not* been commonly experienced by previous patients. For complete description of each stereotypical referred pain pattern, refer to the Table of Referred Pain Patterns of Trigger Point Formation Origin [T005]. Please note the key to abbreviated words at the bottom of the list.

LOCATION (MUSCLE)	REFERRED PAIN ZONES (GENERAL)
Latissimus dorsi (abnormal)	A. delt., l. lower abdomen
Serratus posterior inferior	Para. T9–L2 & l.over p.lower ribs
Multifidus (S4)	Bil. para. S2–coccyx, & bil. med. prox. gluteus maximus
Longissimus thoracis (L1)	*Para. from L1* to & over p.iliac crest
Longissimus thoracis (T10–T11)	*Para. from T10,* gluteus maximus to gluteal fold
Multifidi (L2)	Bil. para. T12–L4, *to p.iliac crest a.upper quadrant of the abdomen*
Multifidi (S1)	Bil. para. L4–S5, med. gluteus maximus, coccyx, *prox. p.hamstrings, a.lower abdominal quadrant*
Iliocostalis lumborum (L1)	L. from L1 iliocostalis, *down over* central gluteus maximus & *l.to head* of trochanter
Iliocostalis thoracis (T11)	*Para. T11–S1,* l. bottom ribs, *l.bord. of the scapula,* a.lower abdominal quadrant
Caudal rectus abdominus	Bil.across the low back (L4–S4) from l. extreme to l. extreme
Gluteus minimus	Gluteus maximus (excepting central area) & gluteal medius, p.l.leg, l.calf to p.of l. malleolus
Gluteus medius	Para. L4–Coccyx, *gluteus medius, gluteus maximus-head of trochanter*

Key:
TMJ—temporomandibular joint
A/C—acromioclavicular
med.—medial
brach.—brachioradialis
para.—paraspinous
MP—metacarpal phalangeal
sterno.—sternocleidomastoideus
pect.—pectoralis
a.—anterior
s.—superior
&—and
maj.—major
bord.—border
f.—finger
l.—lateral
m.—middle
d.—distal
p.—posterior
prox.—proximal
bil.—bilateral
f.s—fingers
i.—inferior

LOWER ABDOMINAL PAIN SYNDROME (S62)

Definition (S62A)

Pain in the lower abdominal area may arise from structures on the abdominal wall or from within the abdominal cavity. Pain from the abdominal wall can emanate from muscles, boils, abscesses, herpes zoster (shingles) or referred from structures in other areas exterior to the abdominal cavity (ligamentous strain, nerve impingement or muscles). Pain from within the lower abdomen can arise from internal structures distressed by disorder, including appendicitis (pain usually referring to the right lower quadrant), salpingitis, ureteric colic, infection of the bladder, colic (from intestinal obstruction, including adhesions), diarrhea (colitis), menstrual cramps (dysmenorrhea), and mittelscherz (lower abdominal pain resulting from the escape of blood into the peritoneal cavity as a result of ovulation).

Related Syndromes (S62B)

ADHESION (S56)
BACTERIAL INFECTION (S63)
COLITIS (S42)
FACET (S19)
FASCIITIS (S20)
HYSTERIA/ANXIETY REACTION (S59)
LOW BACK PAIN (S03)
MENSTRUAL CRAMPING (S60)
OSTEOARTHRITIS (S21)
PHOBIC REACTION (S44)
PIRIFORMIS (S04)
POST PERIPHERAL NERVE INJURY (S23)
POST SPINAL CORD INJURY (S24)
REFERRED PAIN (S01)
SCIATICA (S05)
SHINGLES (HERPES ZOSTER) (S37)
UNHEALED DERMAL LESION (S39)

Treatable Causes Which May Contribute To The Syndrome (S62C)

Bacterial infection (C09)
Calcific deposit (C08)
Extrafusal muscle spasm (C41)
Habitual joint positioning (C48)
Hypermobile/instable joint (C10)
Interspinous ligamentous strain (C03)
Joint ankylosis (C42)
Joint approximation (C47)
Ligamentous strain (C25)
Localized viral infection (C27)
Menorrhalgia (C20)
Muscular weakness (C23)
Nausea/vomiting (C36)
Nerve root compression (C04)
Neuromuscular tonic imbalance (C30)
Peripheral nerve injury (C07)
Psychoneurogenic neuromuscular hypertonus (C29)
Restricted joint range of motion (C26)
Soft tissue inflammation (C05)
Soft tissue swelling (C06)
Sympathetic hyperresponse (C38)
Trigger point formation (C01)
Vascular insufficiency (C11)
Visceral organ dysfunction (C02)

Treatment Notes (S62D)

The following is a list of trigger point formations which may, singly or in combination, refer pain into the lower abdominal area. It should be noted that it is possible for the given patient to experience pain throughout the entire stereotypical referred pain pattern or only in part or parts of it. Opposite to each listing is a description of the parts of the stereotypical referred pain pattern most commonly experienced by previous patients, relative to the given trigger point formation. Any descriptive reference which is underlined is a part of the pattern which has *not* been commonly experienced by previous patients. For complete description of each stereotypical referred pain pattern, refer to the Table of Referred Pain Patterns of Trigger Point Formation Origin [T005]. Please note the key to abbreviated words at the bottom of the list.

LOCATION (MUSCLE)	REFERRED PAIN ZONES (GENERAL)
Multifidi (S1)	Bil. para. L4-S5, med. gluteus maximus, coccyx, *prox. p.hamstrings, a.lower abdominal quadrant*
Iliocostalis thoracis (T11)	Para. T11-S1, l. bottom ribs, *l. bord. of the scapula*, a.lower abdominal quadrant
External Oblique [B]	Bil. a.abdomen, groin, a.med. thigh

Pyramidalis   From umbilicus to groin   a.—anterior
McBurney's point   A. lower abdominal quadrant from midline to the l. extreme   m.—middle
s.—superior
d.—distal
Dysmenorrhea   Rectus abdominus just distal to the umbilicus   &—and
p.—posterior
l.—lateral
i.—inferior

Key:
prox.—proximal
para.—paraspinous
med.—medial
bord.—border
bil.—bilateral

Bibliography (S62E)

Gardner, pp. 151–152.

LUNG FLUID ACCUMULATION (C13)

Definition (C13A)

Lung fluid accumulation occurs when the lung tissues become irritated sufficiently to produce an exudate which the lungs are unable to eliminate rapidly enough to avoid accumulation.

Evaluation (C13B)

Evaluation of lung fluid accumulation depends upon observation of symptoms including the presence of shaking chill, sharp chest pain, fever and headache. Cough, varying from dry and hacking to productive of purulent sputum and painful paroxysms, may be present. Dyspnea may be present with rapid respiration, the patient may sweat profusely, and may be cyanotic: Fever may range from normal to 105° F., if infection is present, pulse rate may be rapid, ranging from normal to 130 beats per minute, rales and suppressed breathing sounds may be heard over the involved area (a pleural friction rub is often heard in the early stages of development), and the patient may complain of feeling acutely ill. Gastrointestinal symptoms of abdominal distention, jaundice, diarrhea, nausea and vomiting may be noted, and the right upper quadrant of the trunk may be tender and rigid if the middle or lower lobes of the lungs are involved. Herpes simplex sores of the lips and face may be present.

Should lung disease be present, the cough will produce sputum with color developing from pinkish or blood-flecked to rusty at the height of the disease finally to yellow and mucopurulent during the stage of resolution.

Etiology (C13C)

Lung tissues may become irritated and produce excessive amounts of fluid as the result of inhalation of irritating substances (pollens, industrial chemicals, vegetable parts, dusts, molds, etc.), or as the result of infection and the toxins produced by bacterial metabolism.

Treatment Notes (C13D)

The treatment of the lung fluid accumulation should be directed at facilitating expectoration of exudate to reduce bronchial obstruction and improve ventilation through postural drainage performed by the therapist (with vibration) and passive postural drainage performed by the patient, without outside help: The patient should be encouraged to expel all sputum produced through these means.

In the later stages of resolution, the patient should be taught to promote lung elasticity by creating positive back pressure through forced expiration against passage resistance (blow bottle or garden hose) and to reactivate the diaphragm through exercise and the utilization of manual compression of the abdomen during expiration.

Bibliography (C13E)

Basmajian, pp. 280–282.
Bickley, pp. 264–265.
Holvey, pp. 1294–1298.
Salter, p. 406.
Scott, pp. 413–414.

TREATMENT OF LUNG FLUID ACCUMULATION (C13F)

Technique Options (C13F1)

Evaluation techniques (C13F1A)

Auxiliary evaluation techniques	*M26*
Auscultation	M26C
Lung sounds	M26C2B

Treatment techniques (C13F1B)

Exercise	*M12*
Improving lung function	M12B
Postural drainage	*M14*
Vibration	*M08*
Postural drainage facilitation	M08D

Techniques Suggested for the Acute Condition (C13F2)

Evaluation techniques (C13F2A)

Auxiliary evaluation techniques	*M26*
Auscultation	M26C
Lung sounds	M26C2B

Treatment techniques (C13F2B)

Exercise	*M12*
Improving lung function	M12B

lung fluid accumulation

Postural drainage — *M14*
Vibration — *M08*
  Postural drainage facilitation — M08D

Techniques Suggested for the Chronic Condition (C13F3)

Evaluation techniques (C13F3A)

Auxiliary evaluation techniques — *M26*
  Auscultation — M26C
    Lung sounds — M26C2B

Treatment techniques (C13F3B)

Exercise — *M12*
  Improving lung function — M12B
Postural drainage — *M14*
Vibration — *M08*
  Postural drainage facilitation — M08D